Handbook of Green Analytical Chemistry

Handbook of Green Analytical Chemistry

Edited by

MIGUEL DE LA GUARDIA
Department of Analytical Chemistry, University of Valencia, Valencia, Spain

SALVADOR GARRIGUES
Department of Analytical Chemistry, University of Valencia, Valencia, Spain

WILEY

A John Wiley & Sons, Ltd., Publication

This edition first published 2012

Registered Office
John Wiley & Sons, Ltd, The Atrium, Southern Gate, Chichester, West Sussex, PO19 8SQ, United Kingdom

For details of our global editorial offices, for customer services and for information about how to apply for permission to reuse the copyright material in this book please see our website at www.wiley.com.

Library of Congress Cataloging-in-Publication Data

Handbook of green analytical chemistry / edited by Miguel de la Guardia, Salvador Garrigues.
p. cm.
Includes bibliographical references and index.
ISBN 978-0-470-97201-4 (cloth)
1. Environmental chemistry–Industrial applications–Handbooks, manuals, etc. 2. Environmental chemistry–Handbooks, manuals, etc.
I. Guardia, M. de la (Miguel de la) II. Garrigues, Salvador.
TD193.H35 2012
543–dc23

2011051666

A catalogue record for this book is available from the British Library.

Print ISBN: 9780470972014

Set in 10/12pt Times by SPi Publisher Services, Pondicherry, India
Printed and bound in Malaysia by Vivar Printing Sdn Bhd

1 2012

Contents

List of Contributors

Julio Carlos Afonso Departamento de Química Analítica, Universidade Federal do Rio de Janeiro, Cidade Universitária, Rio de Janeiro, Brazil

Elisa I. Vereda Alonso Department of Analytical Chemistry, University of Málaga, Málaga, Spain

José Luis Gómez Ariza Departamento de Química y Ciencia de los Materiales 'Profesor José Carlos Vílchez Martín', Universidad de Huelva, Huelva, Spain

Sergio Armenta Department of Analytical Chemistry, University of Valencia, Valencia, Spain

Tamara García Barrera Departamento de Química y Ciencia de los Materiales 'Profesor José Carlos Vílchez Martín', Universidad de Huelva, Huelva, Spain

Maria de Fátima Batista de Carvalho Centro de Pesquisa e Desenvolvimento, Cidade Universitária, Rio de Janeiro, Brazil

Miguel A. Belarra Department of Analytical Chemistry, University of Zaragoza, Zaragoza, Spain

Marcos Almeida Bezerra Departamento de Química e Exatas, Universidade Estadual do Sudoeste da Bahia, Jequié, Brazil

Soledad Cerutti Instituto de Química de San Luis, Universidad Nacional de San Luis-CONICET, San Luis, Argentina

Arabinda K. Das Department of Chemistry, University of Burdwan, Burdwan, West Bengal, India

Suparna Dutta Sonamukhi Girls' High School, Bankura, West Bengal, India

Alberto Escarpa Department of Analytical Chemistry and Chemical Engineering, University of Alcala, Madrid, Spain

Debora Fabbri Department of Analytical Chemistry, V. Pietro Giuria 5, Torino, Italy

Aline Soares Freire Departmento de Química Analítica, Universidade Federal do Rio de Janeiro, Rio de Janeiro, Brazil

Juan Francisco García-Reyes Analytical Chemistry Research Group, Department of Physical and Analytical Chemistry, University of Jaén, Jaén, Spain

Esperanza García-Ruiz Department of Analytical Chemistry, University of Zaragoza, Zaragoza, Spain

Amir Bagheri Garmarudi Chemistry Department, Faculty of Science, Imam Khomeini International University, Qazvin, Iran

Salvador Garrigues Department of Analytical Chemistry, University of Valencia, Valencia, Spain

Raúl Andrés Gil Instituto de Química de San Luis, Universidad Nacional de San Luis-CONICET, San Luis, Argentina

Miguel de la Guardia Department of Analytical Chemistry, University of Valencia, Valencia, Spain

Lucas Hernández Department of Analytical and Instrumental Analysis, Universidad Autónoma de Madrid, Madrid, Spain

Mihkel Kaljurand Institute of Chemistry, Faculty of Science, Tallinn University of Technology, Tallinn, Estonia

Hideko Kanazawa Faculty of Pharmacy, Keio University, Tokyo, Japan

Mohammadreza Khanmohammadi Chemistry Department, Faculty of Science, Imam Khomeini International University, Qazvin, Iran

Mihkel Koel Institute of Chemistry, Faculty of Science, Tallinn University of Technology, Tallinn, Estonia

Miguel Ángel López Department of Analytical Chemistry and Chemical Engineering, Faculty of Chemistry, University of Alcala, Madrid, Spain

Chi-Yu Lu Department of Biochemistry, Kaohsiung Medical University, Kaohsiung, Taiwan

María Dolores Luque de Castro Department of Analytical Chemistry, Campus of Rabanales, Córdoba, Spain

Luis Dante Martínez Instituto de Química de San Luis, Universidad Nacional de San Luis-CONICET, San Luis, Argentina

Agata Mechlińska Department of Analytical Chemistry, Chemical Faculty, Gdansk University of Technology (GUT), Gdansk, Poland

Mariana Medina-Sánchez Nanobioelectronics and Biosensors Group, Institut Català de Nanotecnologia: Universitat Autónoma de Barcelona, Bellaterra, Barcelona, Spain

Arben Merkoçi Nanobioelectronics and Biosensors Group, Institute Català de Nanotechnologia & ICREA, Barcelona, Spain

Miguel Alcaide Molina Department of Analytical Chemistry, University of Córdoba, Córdoba, Spain

Antonio Molina-Díaz Analytical Chemistry Research Group, Department of Physical and Analytical Chemistry, University of Jaén, Jaén, Spain

Jacek Namieśnik Department of Analytical Chemistry, Chemical Faculty, Gdansk University of Technology (GUT), Gdansk, Poland

Tadashi Nishio Faculty of Pharmacy, Keio University, Tokyo, Japan

Catalina Bosch Ojeda Department of Analytical Chemistry, University of Málaga, Málaga, Spain

Eliane Padua Oliveira Departamento de Geoquímica, Universidade Federal Fluminense, Niterói, Brazil

José Manuel Cano Pavón Department of Analytical Chemistry, University of Málaga, Málaga, Spain

José M. Pingarrón Department of Analytical Chemistry, Faculty of Chemistry, University Complutense of Madrid, Madrid, Spain

Edmondo Pramauro Department of Analytical Chemistry, V. Pietro Giuria 5, Torino, Italy

Alessandra Bianco Prevot Department of Analytical Chemistry, V. Pietro Giuria 5, Torino, Italy

Lourdes Ramos Department of Instrumental Analysis and Environmental Chemistry, Institute of Organic Chemistry, CSIC, Madrid, Spain

Natividad Ramos-Martos Analytical Chemistry Research Group, Department of Physical and Analytical Chemistry, University of Jaén, Jaén, Spain

Martín Resano Department of Analytical Chemistry, University of Zaragoza, Zaragoza, Spain

Fuensanta Sánchez Rojas Department of Analytical Chemistry, University of Málaga, Málaga, Spain

Ricardo Erthal Santelli Departamento de Química Analítica, Universidade Federal do Rio de Janeiro, Rio de Janeiro, Brazil

Marek Tobiszewski Department of Analytical Chemistry, Chemical Faculty, Gdansk University of Technology (GUT), Gdansk, Poland

Amparo García de Torres Department of Analytical Chemistry, University of Málaga, Málaga, Spain

Paloma Yáñez-Sedeño Department of Analytical Chemistry, Faculty of Chemistry, University Complutense of Madrid, Madrid, Spain

Preface

Now it is time to move from the general principles to the practice. The efforts made by the analytical chemistry and chemistry community opinion during the 2011 International Year of the Chemistry have been focused on demonstrating to the public that our discipline is not the reason for the environmental damage and the health problems that have emerged from our developed societies. On the contrary, chemistry is one of the main reasons to extend the human life and to improve its quality level and the best tool to solve the environmental problems created in the past by uncorrect use of the available technologies. So, it is a happy coincidence that in recent months the first books especially devoted to Green Analytical Chemistry have been published and also that important journals like *Trends in Analytical Chemistry* have devoted special issues to the topic of Green Analytical Chemistry.

The handbook, which the reader has in hand, is an attempt to advance the ethics and practical objectives of Green Analytical Chemistry. The book has been possible due to the invitation of Wiley-Blackwell editors but also because of the critical mass of research teams who have contributed to establish a series of methodological and technological tools to prevent and reduce the deleterious effects of our analytical activities.

As a main difference to previously published texts, the readers will find in this book a deep and complete perspective of the Green Analytical Chemistry as a matter of facts guided by the most fundamental principles and also a catalogue of tools for greening the work on chemical analysis.

The structure of the text covers a fundamental part, a series of proposals for greening the different steps of the analytical process and some final chapters focused on different fields of applications.

In the fundamental part, the main idea has been to move from historical and theoretical considerations to proposals for authors, editors, and users of the analytical laboratories to move from the old practices, which take into consideration only the method figures of merit, to a new frame in which the side environmental and operator risk effects could pay an important role. However, the most important part of the handbook is the series of detailed chapters, written by specialists in each field, which have made a literature survey on efforts to avoid reagent consumption and waste generation and can provide to the reader many practical tools to do environmentally friendly analytical tasks and to take advantage of the economical opportunities that are offered by Green Analytical Chemistry.

In the different application fields considered in this text, the reader will identify that Green Analytical Chemistry can operate in all contexts; from the industrial to the sanitary and not only in environmental applications, thus contributing once again, to move from the theory to the practice.

For the aforementioned reasons, editors and authors are convinced of the necessity of this book and the fact that a prestigious analytical journal like *Analytical and Bioanalytical Chemistry* is preparing a special issue on Green Analytical Methods for 2012 confirms that this is a good opportunity to incorporate to our everyday work the main ideas and tools of Green Analytical Chemistry and to do it, we hope that this handbook will be the reference book.

We would like to express our thanks to the personnel of Wiley-Blackwell who have offered all the time their support, specially Sarah Hall and Sarah Tilley for their help to make this book possible, and Lynette James for her diligent and careful work on editing the final version. Obviously, also the generosity, patience and good work of all the authors are acknowledged. Many of these authors are old friends with whom we have collaborated on many occasions in the past and who have influenced our research. On other occasions, like in the case of Mihkel Kaljurand, Mihkel Koel and Jacek Namieśnik, they are excellent specialists in the field but we do not have any previous relationship with them. However, their generous acceptance to participate in this project has been of great value to sum the efforts for greening our analytical work and has contributed to improve the handbook. On the other hand, we are totally convinced that this book is also the starting point for future cooperation in a new analytical chemistry built to improve both the fundamental and green parameters of the methods and to increase the amount of information obtained from samples with the minimum consumption of reagents and solvents, and the maximum safety for operators and the environment.

Miguel de la Guardia and Salvador Garrigues
Valencia, September 2011

Section I
Concepts

1

The Concept of Green Analytical Chemistry

Miguel de la Guardia and Salvador Garrigues

Department of Analytical Chemistry, University of Valencia, Valencia, Spain

1.1 Green Analytical Chemistry in the frame of Green Chemistry

Three years ago, when we published our review paper on Green Analytical Chemistry [1] it was clear that, at this time, Green Chemistry was a well established paradigm well supported by more than 50 published books, an increasing number of research teams who influenced the scientific literature and involved the editions of special journals like *Green Chemistry* or *Green Chemistry Letters and Reviews*. However, there was a big contrast between the situation of green catalyst development and the scarce use of the term Green Analytical Chemistry in the literature. In spite of the fact that many studies from 1995 [2–5] were focused on the objective of reducing the analytical wastes and making the methods environmentally friendly and sustainable there was little conscience in the analytical community about the use of green or sustainable terms to define their work.

Fortunately, the efforts of research teams like those of Jacek Namieśnick in Poland [6–9] and Mihkel Koel and Mihkel Kaljurand in Estonia [10–11] have contributed to establish the main principles and strategies which support the green practices in analytical chemistry and, because of that, the publication of the books of Koel and Kaljuran [12] in 2010, de la Guardia and Armenta [13] in 2011, and that of de la Guardia and Garrigues [14] in 2011 evidenced that nowadays Green Analytical Chemistry is becoming a movement which can modify our perspective and practices in the analytical field in future years.

A simple idea could be to consider Green Analytical Chemistry as a part of the whole green chemistry idea, in the same way that someone could consider that analytical chemistry is the part of chemistry devoted to development and analysis. However, it is evident that analytical chemistry itself is not a part, but all chemistry, observed from an analytical viewpoint which consists of searching for the differences between atoms, molecules and chemical structures. Ahead of considering the links between the elements of the periodic table or evaluating the molecules from the presence of a functional groups, analytical chemistry focuses on the differences between atoms and

Handbook of Green Analytical Chemistry, First Edition. Edited by Miguel de la Guardia and Salvador Garrigues.

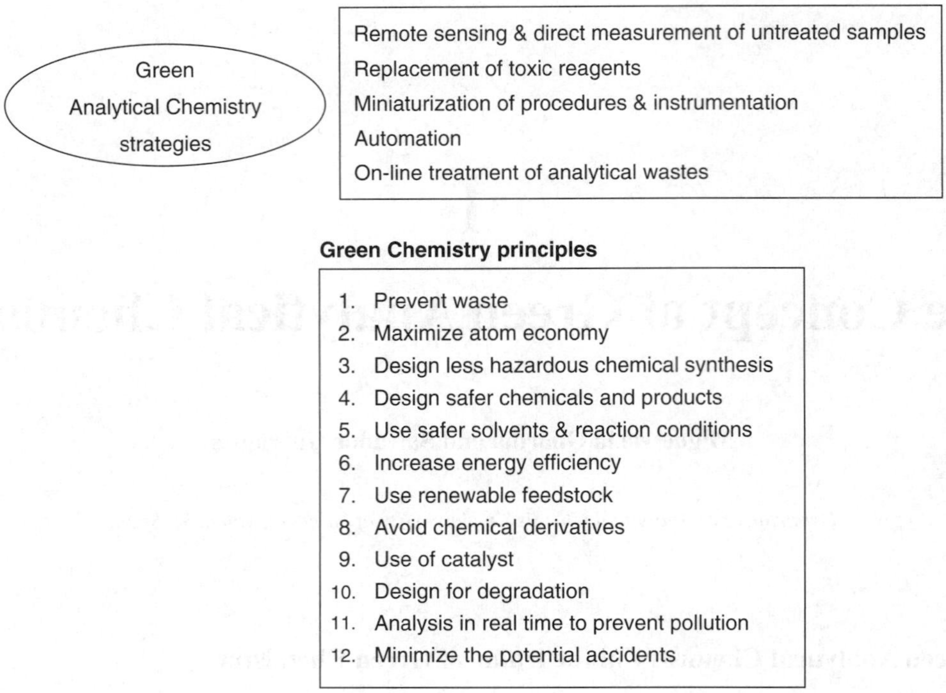

Figure 1.1 *The Green Analytical Chemistry strategies in the frame of the Green Chemistry principles.*

molecules which are apparently similar and thus there are many specificities of Green Analytical Chemistry which must be evaluated in order to be able to provide a clear orientation for greening the analytical tasks.

As Paul Anastas has established in his abundant literature on Green Chemistry [15–21], the idea to replace hazardous substances with less polluting ones or, if possible, innocuous products, and the prevention of waste products in origin together with the restricted use of the prime matters and energy can be summarized in 12 principles (see Figure 1.1). These principles focus on prevention more than on remediation of pollution effects of chemicals and provide guidelines for improving the synthesis methods through the use of renewable raw materials, the maximization of the final product in terms of total mass, the reduction of energy consumption and the search for the reduction of chemical toxicity of involved compounds, also improving the use of catalytic reagents instead of stoichiometric ones. In the aforementioned principles there is a direct reference to the analytical methodologies and the need that they must be improved to allow real time and in-process monitoring and control prior to the formation of hazardous substances.

However, the analytical work also involves the use of reagents and solvents, employs energy as well as data and results, and it generates waste. So, some of the Anasta's principles can be easily translated to the analytical field as those concerning the replacement of toxic reagents, energy saving, the reduction of reagents consumed and waste generation. However, there are several specific strategies of the analytical work which are of tremendous importance for greening our practices. As has been indicated in the scheme of Figure 1.1, remote sensing and direct measurements of untreated samples are the greenest methodologies which we can imagine and, because of that, the development of portable instruments and an instrumentation able to provide remote sample measurements without the use of reagents and solvents, will be a primary task in the future. Additionally, as is shown in Figure 1.2, all the developments in chemometrics will improve the multiparametric capabilities of the aforementioned instruments in order to provide as much information as possible with a reduced consumption of reagents and based on few measurements.

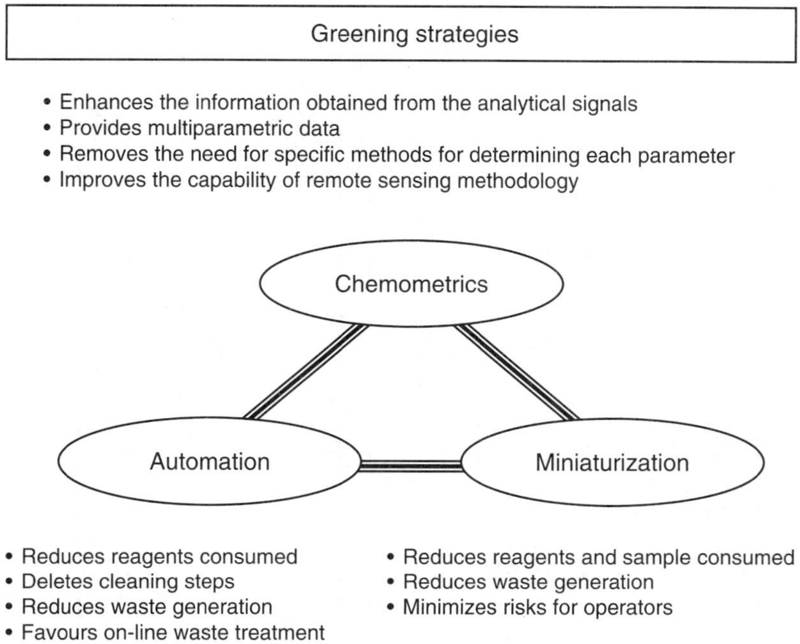

Figure 1.2 *The main tools for greening the analytical method.*

Miniaturization of processes and instruments will be also a key factor for the dramatic reduction of consumables and energy and many efforts have also been made in the literature to downsize the pretreatment and measurement steps, based on the development of microextraction technologies and micrototal analysis in order to move from gram and millilitre scales to micro- and nanoscales. So, it is clear that the strong reduction of reagents and solvents involved in miniaturization processes is welcome from the environmental point of view, but attention must be paid to the lack of representativity which can affect analytical results based on reduced amounts of bulk samples and thus, extra efforts must be made in order to avoid the potential drawbacks of using small amounts of samples.

Automation was a revolution in analytical chemistry in the mid1970s and the development of flow injection (FIA) [22], sequential injection analysis (SIA) [23] and multicommutation [24] provided essential tools for improving, at the same time, the main analytical figures of merit of the methods and their green parameters, based on scaling down the amount of reagents and sample employed and the use of pure solutions which are only mixed when necessary. That reduces drastically the reagents consumed and waste generated. An additional advantage offered by the automation in the analytical work is to avoid the cleaning of the glassware employed in former times in batch analysis, which also contributes to remove or minimize the use of solvents and detergents.

However, the fast, self-cleaning and reagent saving mechanized and automatized methods of analysis also produce waste, which in many cases are toxic residues containing small amounts of pollutant substances present in standards, employed reagents or injected samples. Because of that, the on-line treatment of analytical wastes has been emerged as an important contribution of Green Analytical Chemistry in order to move from the old practices, which do not take into account the deleterious environmental side effects of the analytical practices, to a new sustainable paradigm [5]. It is, from our point of view, a highly interesting contribution from the practical and also from the theoretical perspective, because it clearly shows that for deleting the pollution effects of chemicals an additional chemical effort

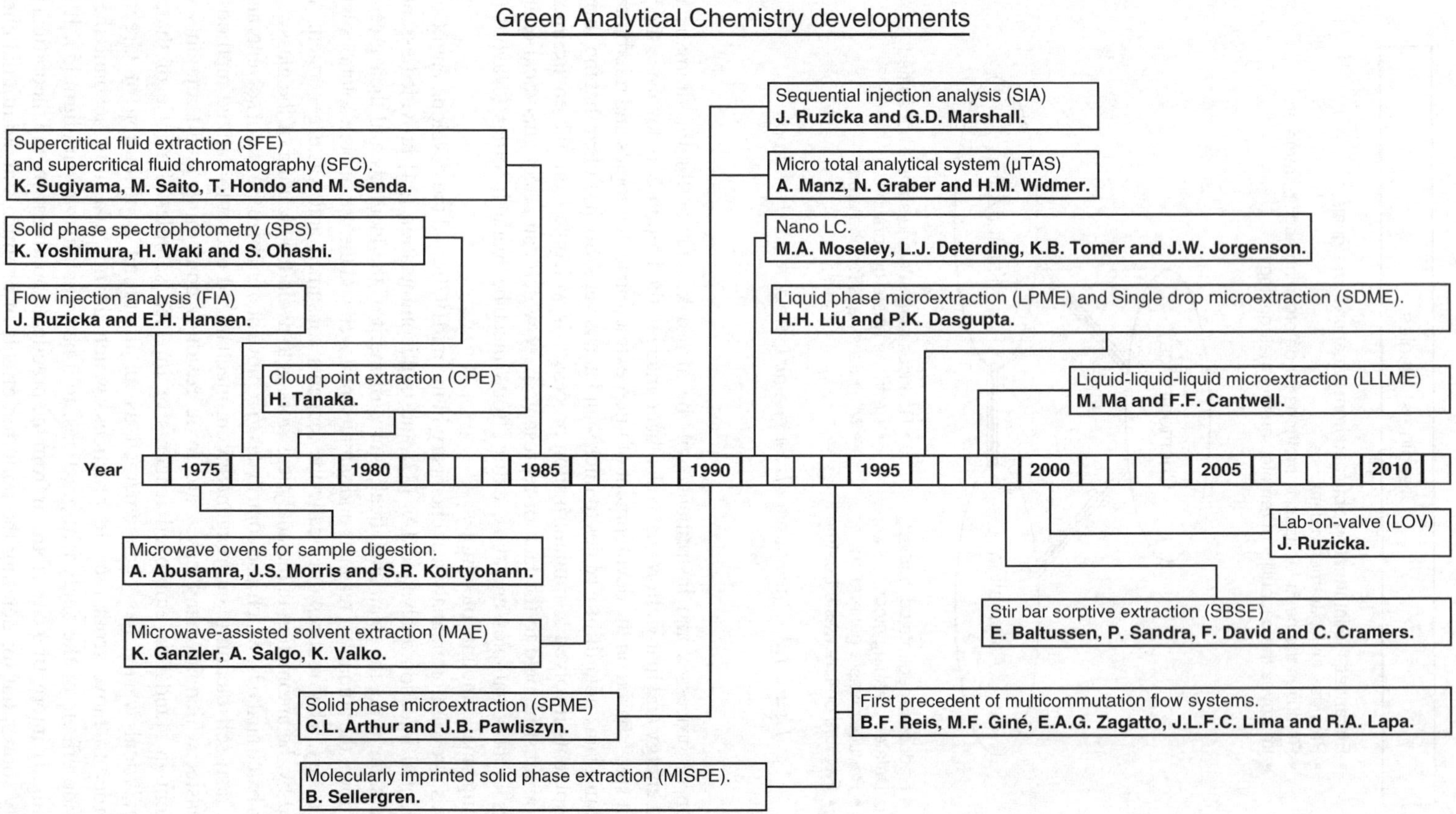

Figure 1.3 *Milestones in the development of Green Analytical Chemistry methods and strategies.*

is desirable. So it offers a clear example that chemistry is not only one of the reasons of the environmental pollution problems but also an important part of their solution.

The on-line reuse or recycling of solvents used in chromatography, flow or sequential analysis, the on-line decontamination of pollutant compounds through chemical oxidation, thermo or photodegradation, together with the use of biodegration systems and, in the case of pollutant mineral elements, their passivation and on-line removal, can be integrated in the whole analytical protocol. So, this strategy could provide clean methodologies which can improve the green parameters of a method without sacrificing any of its figures of merit.

In short, as is clearly shown in the scheme of Figure 1.2, the main tools available today for greening the analytical methods concern chemometrics, automation and miniaturization. From those, a drastic reduction of reagent consumption and waste generation can be made improving also the main analytical parameters.

On looking through the analytical work in the last 40 years (see Figure 1.3) it can be seen that the efforts made for greening the methods came from the objective to reduce the cost of analysis, to improve their speed and also to downsize the scale of work. We could mention, in addition to the development of FIA [22], SIA [23] and multicommutation [24], the use of microwave energy for sample digestion [25] and analyte extraction [26], developments in extraction techniques using solid phase and especially including a reduction of working scale in the case of solid phase microextraction (SPME) [27], the use of stir bar sorptive extraction (SBSE) [28], and measurements on solid phase spectrometry (SPS) [29]. Molecularly imprinted solid-phase extraction (MISPE) [30] has contributed to enhancing the selectivity of extraction techniques while reducing the amount of reagents employed.

From the initial contribution of cloud point techniques [31] liquid phase extraction also has been enhanced by reducing the volume of solvent required through the development of liquid phase microextraction (LPME) and single drop microextraction (SDME) [32,33], also including liquid-liquid-liquid microextraction (LLLME) [34,35]. The use of supercritical fluid extraction for both analytical and chromatographic separations was an important step in the development of new analytical applications [36], as well as the possibility of working at the nanoscale in liquid chromatography [37,38]. Finally, the proposal of miniaturized total chemical-analysis systems based sensors [39] or the development of lab-on-valve as a universal microflow analyser [40] are other examples of contributions to the development of today's analytical chemistry.

1.2 Green Analytical Chemistry versus Analytical Chemistry

We can understand that the environmental pollution is the matter of concern for all those who live and work on this planet but what value does Green Analytical Chemistry add to the essential importance of analytical chemistry? To answer this question we must think about the main aspects of the analytical methods and the challenges for the future.

On considering the essential aspects of the analytical work (see Figure 1.4), the analytical parameters emerge as the key factors to be considered. Accuracy, traceability, sensitivity, selectivity and precision are the essential and basic figures of merit which must be assured in order to provide to the industries, consumers and policy makers the appropriate tools to do their determinations. However, all the aforementioned parameters do not take into consideration the safety of operators or the environmental effects of the use of the analytical methods. Additional practical parameters, which must be also considered concern speed, cost and safety of the determinations which are called practical parameters but can affect also basic parameters such as precision, by increasing the number of replicate analyses based on their relative low cost and speed. So, at the end, an increase of practical parameters can reduce the standard deviation of determinations by increasing the number of analyses in the same sample and enhancing the analytical methodology in terms of precision.

Taking into consideration the objectives of Green Analytical Chemistry it could be enough to add to the aforementioned figures of merit the so called green parameters which involve the evaluation and quantification

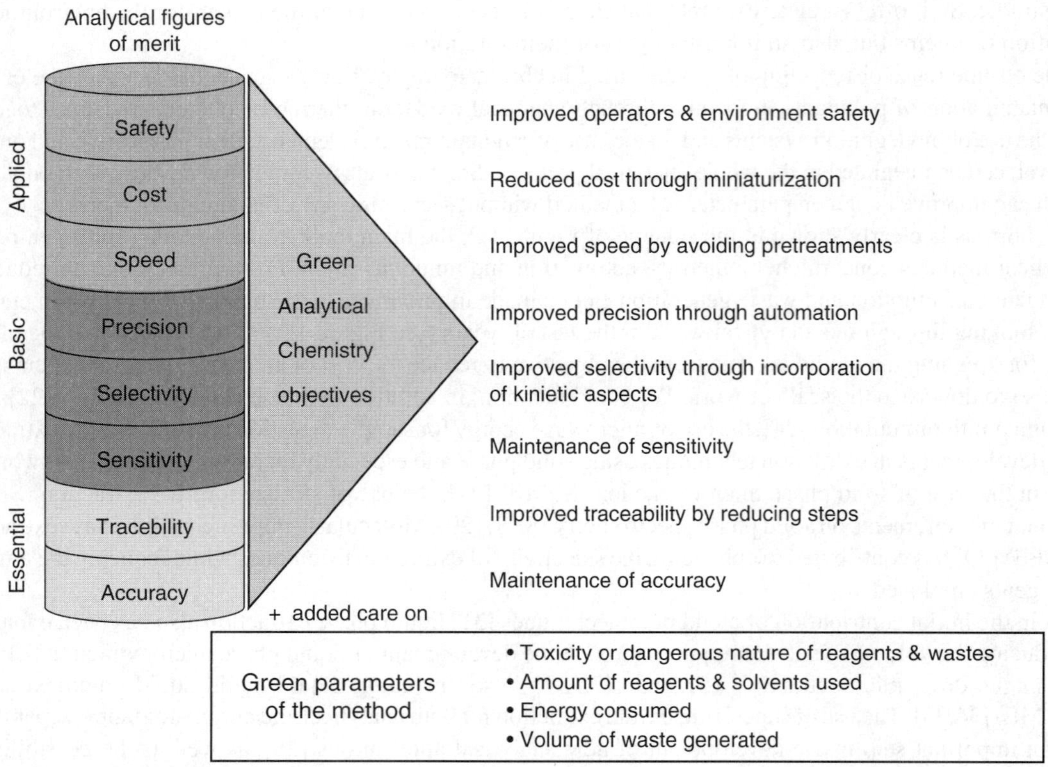

Figure 1.4 *Objectives of Green Analytical Chemistry in the frame of the analytical figures of merit.*

of: (1) the toxicity or dangerous nature of reagents and solvents employed, (2) the volume of reagents and solvents employed, (3) the energy consumed, and (4) the amount of waste generated.

In short, when we consider the Green Analytical Chemistry in the frame of Analytical Chemistry we must think that the basic idea is to preserve the main objectives and to try to improve the analytical figures of merit but at the same time, to add an extra effort to take into account the replacement of toxic reagents, to avoid or at least, to reduce the amount of reagents and solvents employed to do the analytical determinations, to evaluate and reduce the energy consumed and to avoid or minimize the volume of waste.

So, the Green Analytical Chemistry does not try to renounce to any one of the progress in method development but adds a compromise with the preservation of the environment, and, as it can be seen in the scheme of Figure 1.4, the main strategies involved in greening the analytical methods can also improve the traditional figures of merit. Because of that, there is no conflict between the work made in the past and that suggested for the future. Green Analytical Chemistry just adds an extra ethical value in front of environmental protection and thus, we can see the evolution of the analytical methodologies from the classical analytical chemistry to the green as a change of mentality and practices more drastic than modification of principles. In fact, Green Analytical Chemistry will continue to be an effort projected on the whole chemistry field to search for the best way to improve our knowledge on the composition and properties of all type of samples in order to provide a correct answer to any kind of problems in chemical terms.

When we look at the different steps of the so called analytical procedure and we consider sampling to sample preservation, sample transport and sample preparation to analyte preconcentration and analyte separation and determination, the translation from classical analytical chemistry to the green involves an

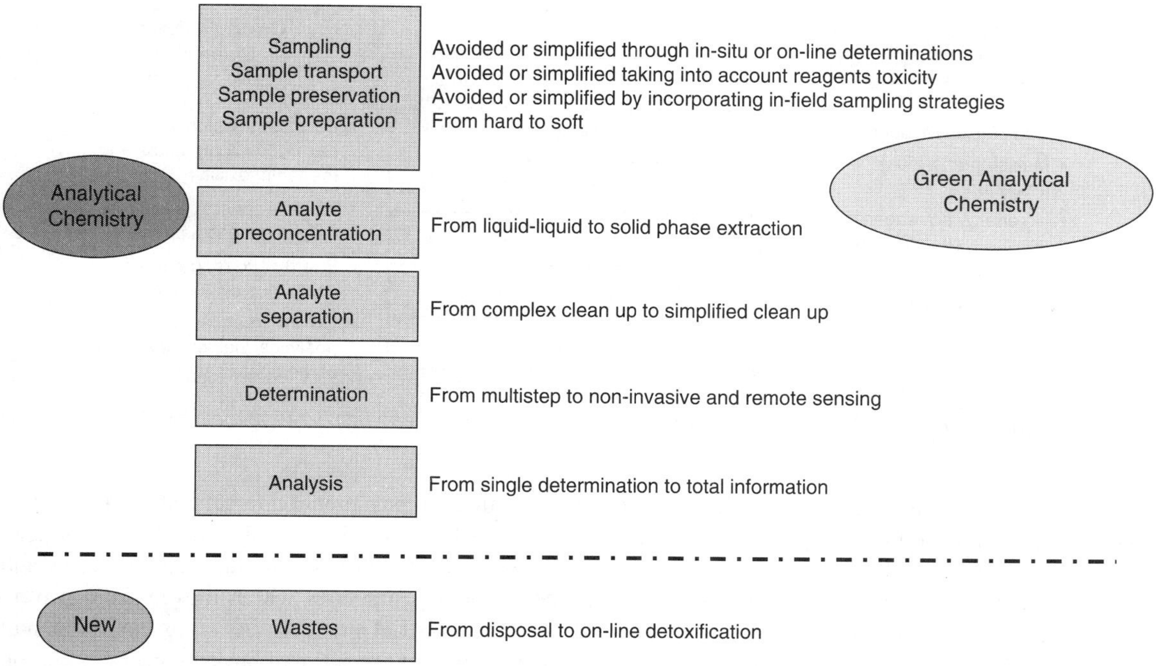

Figure 1.5 *The evaluation of methodologies from classical Analytical Chemistry to Green Analytical Chemistry.*

effort to avoid as many as possible steps, especially those concerning the movement of samples from their original environment to the laboratory, together with an evolution of our mentality from the hard methods of sample digestion or analyte extraction to the soft ones, involving a strong reduction of energy and reagents consumed. In many cases the aforementioned changes offer a simplification of matrix problems and opens exciting possibilities for the characterization of the specific chemical forms existing originally in the samples thus, also improving the main analytical parameters. As Figure 1.5 shows, additional efforts in greening the methods involve a transition from high reagent volume strategies like liquid-liquid extraction to microextraction ones and to solid phase extraction; and a general evolution from complex and multistep strategies to simplified alternatives and to non-invasive and remote sensing measurements. In short, the basic idea is to move from single determinations to methodologies providing total information from a reduced number of analytical measurements. Additionally, a new aspect to be included in our consideration of the analytical process is the waste generation and its treatment and, in this aspect, the change in mentality must move from disposal to on-line detoxification of residues generated though analytical measurements.

1.3 The ethical compromise of sustainability

Sustainability is a new concept emerged from the consideration of sustainable development [41] to describe an economy in equilibrium with basic ecological support systems [42]. So, this idea to recover the equilibrium between the man and the biosphere after many years of disordered technical development has not taken into consideration the environmental impact of human activities or all the risks involved of such activities in the long term, can explain new values established from the conscience about the limits of the development [43] and the need of the restoration of environmental equilibrium in order to assure the continuity of our life for the future

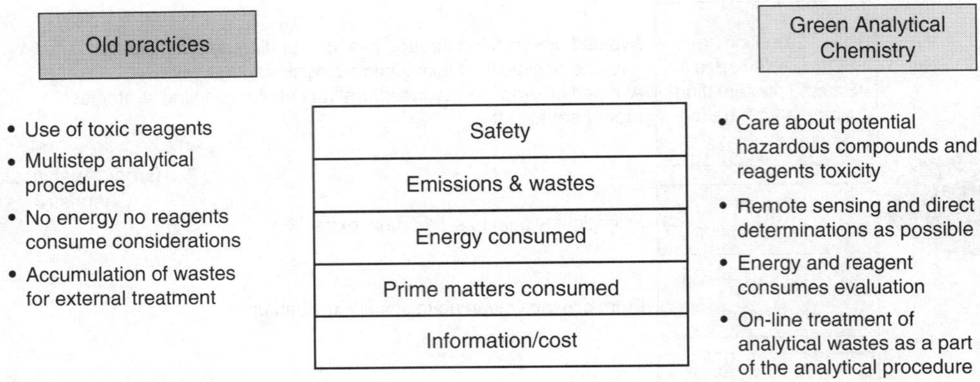

Figure 1.6 *The challenges of sustainability from the Green Analytical Chemistry viewpoint.*

generations [44]. Sustainability pushes the international community to pay attention to general problems such as ozone layer depletion and the generation of greenhouse gases which have dramatically affected the climatic.

In the aforementioned frame, the growth of the ecological mentality through the different countries and international forums has been tremendous, being influenced many practices which moved from private concern about water consumption or waste disposal [45] to industrial practices covering environmental aspects, in addition to quality improvement of their products [46] and the national laws regarding pollution control also affecting supranational norms like the Regulation for Registration, Evaluation, Authorisation and Restriction of CHemicals (REACH) [47] norm of the European Union which try to establish a safety frame for the control of chemical substances.

Important documents such as the Pimentel inform [48] and the *Silent Spring* by Rachel Carson [49] are in the foundation of decisions such as the foundation of the United States Environmental Protection Agency (US EPA) created by president Nixon in 1970, which controls the execution of environmental regulation in the US and the 1985 meeting of the Environmental Ministers of the Organization for Economic Co-operation and Development (OECD) which focused on three ideas: (2) the economic development and environment, (2) pollution prevention and control, and (3) environmental information and national reviews. These national and international actions provided a change in the environmental mentality from remediation to pollution prevention [50] thus improving good environmental practices in all sectors. So, the proposal of Green Analytical Chemistry can be of great importance inside the ecological paradigm of chemistry [51].

In short, pollution prevention is the key factor to be considered in the search for the sustainability of chemical activity and it is an important task because the tremendous development of the chemical industries and their impact on the environment have created the impression in the public eye and mass media that chemistry is the origin of environmental problems. Because of that, the chemistry itself is perceived as an intrinsically bad practice. So, it is our own responsibility to transmit to society the message that another chemistry is possible and that on considering chemistry problems from an environmental point of view, our practices could be very important for the pollution prevention and remediation.

On looking through the practices involved in the analytical methods Figure 1.6 shows that the environmental safety considerations, the worry about emissions and wastes, prime matters and energy consumption, can be compatible with the optimization of the information/cost relationship required for the selection of an analytical method. So, in the frame of Green Analytical Chemistry we can move our laboratories to avoid old practices like the use of toxic reagents and hazardous materials, the use of long and tedious multistep analytical procedures, to replace the excessive consumption of energy and reagents and to add to our methods a previous evaluation of the real needs, avoiding accumulation of waste. So, in this latter aspect we must move from the direct disposal of waste without an external treatment to the on-line process of residues. In front of the aforementioned practices, which remain part of the analytical work in many laboratories, greening the methods implies taking care of the potential hazards of reagents and solvents and considering their toxicity for the selection of a methodology. The tendency to use remote sensing and direct determinations if possible, in order to avoid sampling and sample transport, the use of reagents and the generation of waste must be also evaluated when considering also the energy and reagents consumed, and the possibility of incorporating on-line treatment of analytical wastes after analyte detection in order to save money and time derived from the waste accumulation and management.

The benefits of success on the aforementioned challenges is mainly from an ethical point of view and can transform the perception of our students and general society about the importance of chemistry and the beneficial effect of analytical practices, but also can provide economical opportunities.

1.4 The business opportunities of clean methods

Method development is a matter of science, but we cannot forget that method application is a matter of business and, because of that, any environmentally friendly proposal must be also quantified in economical terms of cost and benefits and not only based on ethical and scientific considerations. So, we stress in this section the business opportunities offered by greening the analytical methods.

Starting from the point that reduction of consumption of reagents, solvents and energy, is intrinsically a reduction of method costs, one can image that it could be of a great interest to move from the macro to microanalysis scale and that the use of remote sensing and direct analysis methodologies could be interesting alternatives to classical wet methods from the economical point of view. However, automation, another of the basic strategies for greening the analytical procedures, also offers good opportunities to save laboratory costs by reducing the needs of human intervention in method application. It is true that the aforementioned financial benefits are accompanied by increased costs in the acquisition of automation components, replacing macroanalysis systems with microanalysis ones and the cost of remote sensing and direct analysis instrumentation but, regarding the last aspect, there is no reason why the alternative set-ups must be more expensive than old macroanalysis tools. On the contrary, in some cases portable instruments and disposal systems are available in the market at a reasonable cost and the increase in the demand of such a system will lead to the reduction of their costs.

On the other hand, time is in many cases a matter of business and the need of a fast-as-possible analytical system for process monitoring and quality control is totally compatible with the reduction of analytical steps, the search for non-invasive direct and remote methods and thus, once again, it is clear that the objectives of greening the analytical work are compatible with economical opportunities. Because of that, the acquisition of new fast instruments must be considered as an investment in terms the benefits of a fast analytical response.

On concerning the search for multiparametric techniques the advantage of moving from an analytical instrumentation and a specific methodology focused to measure each required analyte, to the simultaneous determination of all parameters of interest from a single analytical response which can be processed mathematically in order to predict the values of the target analytes concentration and sample properties, becomes clear. In this sense, we are completely convinced that the chemometric treatment of non-invasive signals, like those obtained by infrared spectroscopy [52] offers the greenest technology and could replace many activities which are in current use in industrial and control laboratories. So, in some cases, fast multiparametric methods applied to untreated samples could replace the official methods and, in other occasions, the aforementioned methods could be of a great interest as screening tools.

On the other hand, remote or non-invasive methodologies have the additional advantage of their intrinsic flexibility to integrate additional parameters to those measured at present. So once again, there is a convergence between green and business objectives and we are absolutely convinced that the balance between cost and benefits of greening efforts in analytical chemistry is clearly favourable as indicated in Figure 1.7.

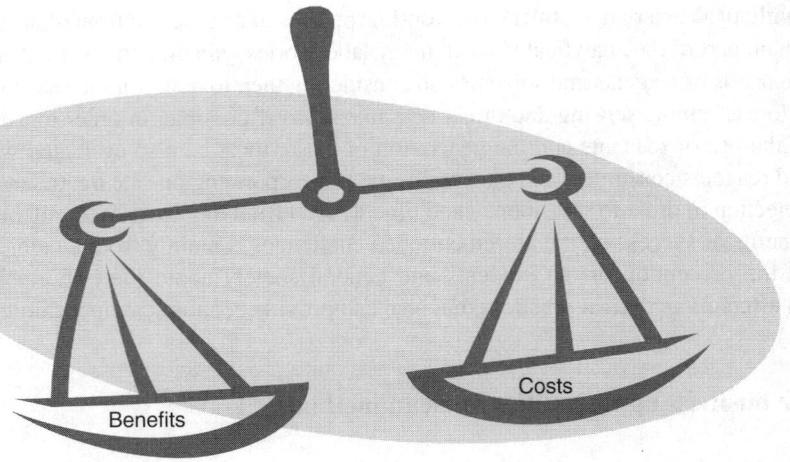

- Reduced consumption of reagents
- Reduced consumption of energy
- Reduced labour through automation
- No external waste treatment

- Acquisition of automation components
- Replacement of macroanalysis systems by microanalysis ones and portable remote sensing instruments

Figure 1.7 *The balance between cost and economical opportunities offered by Green Analytical Chemistry.*

On the other hand, the avoiding of waste generation or, at least, the minimization of analytical waste and the on-line treatment of those generated in the framework of the method, provide a drastic reduction of both risks and costs of the analytical determinations and offer new opportunities for on-line recycling of reagents. So, it is practically a no-expense effort which can reduce the costs of operation, especially when big series of samples of the same type must be treated every day using automatized procedures.

The economical consideration of the greening efforts in method development is, in our own opinion, the most attractive aspect of Green Analytical Chemistry and will be the reason for extended practice in the near future. However, to do it is our own responsibility and it will be possible if we can transmit the ethical, safety and economic benefits of the green alternatives proposed to the traditional practices in a clear way.

1.5 The attitudes of the scientific community

Tradition is a heavy heritage in all human practices and, in spite of the opportunities offered by a fast changing world, it is difficult to move from classical practices to new ones. In fact, in the past there was a big opposition to the instrumental methods of analysis from those who practiced the classical titrimetric and gravimetric analyses at the beginning of the twentieth century, based on well documented reactions and following stoichiometric proportions between analytes and reagents. However, nowadays nobody discusses that physicochemical methods of analysis are analytical methods, the most attractive and well adapted to the analytical needs. The same happened with the introduction of flow analysis methods, multivariate chemometric data processing, microwave-assisted sample treatments and kinetic analysis. However, the advantages offered by the emerging ideas and tools obliged the acceptance of these as valuable alternatives to previous ones and their incorporation to the regular practices. So, we think that the same will be do with the Green Analytical Chemistry if we are able to explain well the basic ideas that support it and to evaluate the benefits that operators and laboratories could obtain by greening their practices.

Attitudes	Reasons	Decisions
Ignorance	Lack of homogeneity between different approaches and efforts which reduces the visibility of ideas	The common use of homologated terms like "green"
Distrust	Lack of well evaluated practical applications	Incorporation of green parameters for the evaluation of new methodologies
Suspicion	The perception of environmental sustainability as a reason for additional cost	Correct evaluation of economical opportunities offered by Green Analytical Chemistry
Stubbornness	Refusal to see the green paradigm as a new approach	Correct explanation of principles and strategies
Agreement	Publication of special issues of journals and books success of Green Chemistry	Integration of Green Analytical Chemistry efforts in the frame of the Green Chemistry

Figure 1.8 *Attitudes of the scientific community towards Green Analytical Chemistry.*

Table 1.1 *Special issues of analytical journals devoted to Green Analytical principles and practices.*

Journal	Subject	Year
The Analyst	Environmental Friendly Analytical Chemistry	1995
Spectroscopy Letters	Green Spectroscopy	2010
TrAC – Trends in Analytical Chemistry	Green Analytical Chemistry	2010
Analytical and Bioanalytical Chemistry	Green Analytical Methods	2012

As we have summarized in Figure 1.8 the attitudes of the scientific community and the analytical method users regarding Green Analytical Chemistry can be identified in a big spectrum which covers everything from ignorance to distrust, suspiciousness or stubbornness and can move to the agreement. However, to do it we must be able to transmit the ideas and practices which support Green Analytical Chemistry in a clear way.

It is far from our objective to do any disqualification of the different attitudes that we can identify in the scientific community. On the contrary, we are absolutely convinced of the reasons for such attitudes and, as an example, the lack of homogeneity between the different approaches, efforts in avoiding the environmental side effects of the analytical procedures and the distrust in the capability of Green Analytical Chemistry, come from verbose excesses which forget to evaluate in a deep way the applications of the main strategies for greening the methods. Stubbornness is due to the lack of generalized evaluation of the common principles and general purposes of Green Analytical Chemistry and their relationship with the modern paradigm of analytical chemistry in order to clearly identify the rules and consequences of it.

In short, if we want to obtain the agreement of the scientific community and influence the practices of industrial and official laboratories, we must make a theoretical and practical effort to make both visible; principles and applications of Green Analytical Chemistry and to take advantages from the fact that nowadays Green Chemistry is considered a major topic in chemistry. To do it, the extended number of published books, papers and congress meetings which include reference to green ideas, and the increasing number of special issues of journals devoted to Green Analytical Methods in different fields (see Table 1.1) will influence the mentality and practices of the analytical chemistry and will reinforce the incorporation of the green parameters to the evaluation of alternative methodologies. The success of it is our own responsibility and starts from university teaching and analytical publication.

Acknowledgements

The authors acknowledge the financial support of the Generalitat Valenciana Project GV PROMETEO 2010-055 to write this book and to do the research in this field.

References

1. Armenta, S.; Garrigues, S. and de la Guardia, M. (2008) Green Analytical Chemistry, *TrAC - Trends Anal. Chem.*, **27**, 497–511.
2. de la Guardia, M. and Ruzicka, J. (1995) Towards environmentally conscientious Analytical Chemistry through miniaturization, containment and reagent replacement, *Analyst*, **120**, 17N–17N.
3. de la Guardia, M.; Khalaf, K.D.; Carbonell, V. and Morales-Rubio, A. (1995) Clean analytical method for the determination of propoxur, *Anal. Chim. Acta*, **308**, 462–468.
4. Pramauro, E.; Bianco-Prevot, A.; Augugliaro, V. and Palmisano, L. (1995) Photocatalytic treatment of laboratory wastes containing aromatic amines, *Analyst*, **120**, 237–242.
5. de la Guardia, M. (1999) An integrated approach of Analytical Chemistry, J. *Braz. Chem. Soc.*, **10**, 429–437.
6. Namiesnik, J. (2001) Green Analytical Chemistry - Some remarks, *J. Sep. Sci.*, **24**, 151–153.
7. Tobiszewski, M.; Mechlinska, A.; Zygmunt, B. and Namiesnik, J. (2009) Green Analytical Chemistry in sample preparation for determination of trace organic pollutants, *TrAC-Trends Anal. Chem.*, **28**, 943–951.
8. Rawa-Adkonis, M.; Wolska, L. and Namiesnik, J. (2003) Modern techniques of extraction of organic analytes from environmental matrices, *Crit. Rev. Anal. Chem.*, **33**, 199–248.
9. Tobiszewski, M.; Mechlinska, A. and Namiesnik, J. (2010) Green Analytical Chemistry-theory and practice, *Chem. Soc. Rev.*, **39**, 2869–2878.
10. Koel, M. and Kaljurand, M. (2006) Application of the principles of Green Chemistry in Analytical Chemistry, *Pure Appl. Chem.*, **78**, 1993–2002.
11. Kaljurand, M. and Koel, M. (2011) Recent Advancements on Greening Analytical Separation, *Crit. Rev. Anal. Chem.*, **41**, 2–20.
12. Koel, M. and Kaljurand, M. (2010) *Green Analytical Chemistry*, RSC Publishing, Cambridge.
13. de la Guardia, M. and Armenta, S. (2011) *Green Analytical Chemistry: Theory and Practice*, Elsevier, Amsterdam.
14. de la Guardia, M. and Garrigues, S. (eds) (2011) *Green Analytical Chemistry Challenges*, RSC Publishing, Cambridge.
15. Anastas, P.T. (1999) Green Chemistry and the role of analytical methodology development, *Crit. Rev. Anal. Chem.*, **29**, 167–175.
16. Poliakoff, M.; Fitzpatrick, J.M.; Farren, T.R. and Anastas, P.T. (2002) Green Chemistry: Science and politics of change, *Science*, **297**, 807–810.
17. Anastas, P.T. and Kirchhoff, M.M. (2002) Origins, current status, and future challenges of Green Chemistry, *Account. Chem. Res.*, **35**, 686–694.
18. Anastas, P.T. and Eghbali, N. (2010) Green Chemistry: Principles and Practice, *Chem. Soc. Rev.*, **39**, 301–312.
19. Anastas, P.T. and Williamson, T.C. (eds) (1996) *Green Chemistry: Designing Chemistry for the Environment*, American Chemical Society, Washington DC.
20. Anastas, P.T. and Williamson, T.C. (eds) (1998) *Green Chemistry: Frontiers in Benign Chemical Syntheses and Processes*, Oxford University Press, Oxford.
21. Anastas, P.T. and Warner, J.C. (1998) *Green Chemistry: Theory and Practice*, Oxford University Press, Oxford.
22. Ruzicka, J. and Hansen, E.H. (1975) Flow Injection Analyses. 1. New concept of fast continuous-flow analysis, *Anal. Chim. Acta*, **78**, 145–157.

23. Ruzicka, J. and Marshall, G.D. (1990) Sequential Injection - A new concept for chemical sensors, process analysis and laboratory assays, *Anal. Chim. Acta*, **237**, 329–343.
24. Reis, B.F.; Gine, M.F.; Zagatto, E.A.G.; Lima, J.L.F.C. and Lapa, R.A. (1994) Multicommutation in Flow-Analysis. 1. Binary sampling - Concepts, instrumentation and spectrophotometric determination of iron in plant digests, *Anal. Chim. Acta*, **293**, 129–138.
25. Abusamra, A.; Morris, J.S. and Koirtyohann, S.R. (1975) Wet ashing of some biological samples in a microwave oven, *Anal. Chem.*, **47**, 1475–1477.
26. Ganzler, K.; Salgo, A. and Valko, K. (1986) Microwave extraction - A novel sample preparation method for chromatography, *J. Chromatogr.*, **371**, 299–306.
27. Arthur, C.L. and Pawliszyn, J. (1990) Solid-phase microextraction with thermal-desorption using fused-silica optical fibers, *Anal. Chem.*, **62**, 2145–2148.
28. Baltussen, E.; Sandra, P.; David, F. and Cramers, C. (1999) Stir bar sorptive extraction (SBSE), a novel extraction technique for aqueous samples: Theory and principles, *J. Microcolumn Sep.*, **11**, 737–747.
29. Yoshimura, K.; Waki, H. and Ohashi, S. (1976) Ion-exchanger colorimetry. 1. Micro-determination of chromium, iron, copper and cobalt in water, *Talanta*, **23**, 449–454.
30. Sellergren, B. (1994) Direct drug determination by selective sample enrichment on an imprinted polymer, *Anal. Chem.*, **66**, 1578–1582.
31. Watanabe, H. and Tanaka, H. (1978) Nonionic surfactant as a new solvent for liquid-liquid - extraction of zinc (II) with 1-(2-pyridylazo)-2-naphthol, *Talanta*, **25**, 585–589.
32. Liu, H.H. and Dasgupta, P.K. (1996) Analytical Chemistry in a drop. Solvent extraction in a microdrop, *Anal. Chem.*, **68**, 1817–1821.
33. Jeannot, M.A. and Cantwell, F.F. (1996) Solvent microextraction into a single drop, *Anal. Chem.*, **68**, 2236–2240.
34. Ma, M.H. and Cantwell, F.F. (1998) Solvent microextraction with simultaneous back-extraction for sample cleanup and preconcentration: Quantitative extraction, *Anal. Chem.*, **70**, 3912–3919.
35. Ma, M.H. and Cantwell, F.F. (1999) Solvent microextraction with simultaneous back-extraction for sample cleanup and preconcentration: Preconcentration into a single microdrop, *Anal. Chem.*, **71**, 388–393.
36. Sugiyama, K.; Saito, M.; Hondo, T. and Senda, M. (1985) New double-stage separation analysis method - directly coupled laboratory-scale supercritical fluid extraction supercritical fluid chromatography, monitored with a multiwavelength ultraviolet detector, *J. Chromatogr.*, **332**, 107–116.
37. Moseley, M.A.; Deterding, L.J.; Tomer, K.B. and Jorgenson, J.W. (1991) Nanoscale packed-capillary liquid-chromatography coupled with mass-spectrometry using a coaxial continuous-flow fast-atom-bombardment interface, *Anal. Chem.*, **63**, 1467–1473.
38. Chervet, J.P.; Ursem, M. and Salzmann, J.B. (1996) Instrumental requirements for nanoscale liquid chromatography, *Anal. Chem.*, **68**, 1507–1512.
39. Manz, A.; Graber, N. and Widmer, H.M. (1990) Miniaturized total chemical-analysis systems - a novel concept for chemical sensing, *Sensor. Actuator. B-Chem.*, **1**, 244–248.
40. Ruzicka, J. (2000) Lab-on-valve: universal microflow analyzer based on sequential and bead injection, *Analyst*, **125**, 1053–1060.
41. United Nations (1987) *Report of the World Commission on Environment and Development*, General Assembly Resolution 42/187.
42. Stivers, R. (1976) *The Sustainable Society: Ethics and Economic Growth,* Westminster Press, Philadelphia.
43. Meadows, D.H.; Meadows, D.L.; Randers J. and Behrens III, W.W. (1972) *Limits to Growth*, Universe Books, New York.
44. Gore, A. (2006) *Earth in the Balance: Ecology and the Human Spirit*, Rodale books, New York.
45. Garrigues, S.; Armenta, S. and de la Guardia, M. (2010) Green strategies for decontamination of analytical wastes, *TrAC-Trends Anal. Chem.*, **29**, 592–601.
46. Clements, R.B. (1996) *Complete Guide to ISO 14000*, Prentice Hall, Upper Saddle River.
47. European Parliament, Council. Regulation (EC) (2006) No 1907/2006 of the European Parliament and of the Council of 18 December 2006 concerning the Registration, Evaluation, Authorisation and Restriction of Chemicals (REACH), establishing a European Chemicals Agency, amending Directive 1999/45/EC and repealing Council Regulation (EEC) No 793/93 and Commission Regulation (EC) No 1488/94 as well as Council Directive 76/769/EEC

and Commission Directives 91/155/EEC, 93/67/EEC, 93/105/EC and 2000/21/EC. *Official Journal of the European Union*, 396, 1–849.
48. Pimentel, G. (1995) *Opportunities in Chemistry*, National Academic Press, Washington DC.
49. Carson, R. (1962) *Silent Spring*, Houghton Mifflin Co., New York.
50. Stephan, D.G. and Atcheson, J. (1989) The EPAS Approach to Pollution Prevention, *Chem. Eng. Prog.*, **85**, 53–58.
51. Malissa, H. (1988) *Changes of paradigms in Analytical Chemistry*, in *Reviews on Analytical Chemistry, Euroanalysis VI* (ed. E. Roth), Les Editions de Physique, Paris.
52. Moros, J.; Garrigues, S. and de la Guardia, M. (2010) Vibrational spectroscopy provides a green tool for multi-component analysis, *TrAC-Trends Anal. Chem.*, **29**, 579–591.

2

Education in Green Analytical Chemistry

Miguel de la Guardia and Salvador Garrigues

Department of Analytical Chemistry, University of Valencia, Valencia, Spain

In Chapter 1, we justified the reasons not to consider analytical chemistry as a part of chemistry in the same sense so that we can focus on organic or inorganic compounds and in fact, there is an increasing difficulty in identifing the differences between biomolecules and organic ones; and to justify the totally different nature of organic and inorganic compounds with respect to organometallics. So, as evidenced in the scheme of Figure 2.1, in our opinion both physical chemistry and analytical chemistry can be considered as points of view on the chemical nature of the matter; in the specific case of analytical chemistry it can be considered as a look at the presence of atoms, molecules and their organization in all types of samples, which can justify their properties and behaviour and because of this, it is evident that the matter is the frame of analytical chemistry. Teaching analytical chemistry must be focussed on the analytical parameters and practices more than on sample composition. Because of this, in this chapter we will consider the main aspects of the Green Analytical Chemistry as a new paradigm and the integration of it with education at university level.

2.1 The structure of the Analytical Chemistry paradigm

As indicated in Figure 2.2 a paradigm is composed of a hard nucleus surrounded by theories and tools which create the core of the regular way to search and interpret the results obtained in a scientific discipline. Really, the term paradigm established by T.S. Kuhn in his book *The Structure of Scientific Revolutions* is a mixture of social perceptions and the effects of the researchers, scientific societies and journals which at the end, are the main actors in the scientific task [1]. In short, a paradigm can be interpreted in terms of the preconcepts that scientists applied to their search during periods of normal science between scientific revolutions. However, it is clear that the value of a scientific paradigm strongly depends on its possibilities to solve

Handbook of Green Analytical Chemistry, First Edition. Edited by Miguel de la Guardia and Salvador Garrigues.

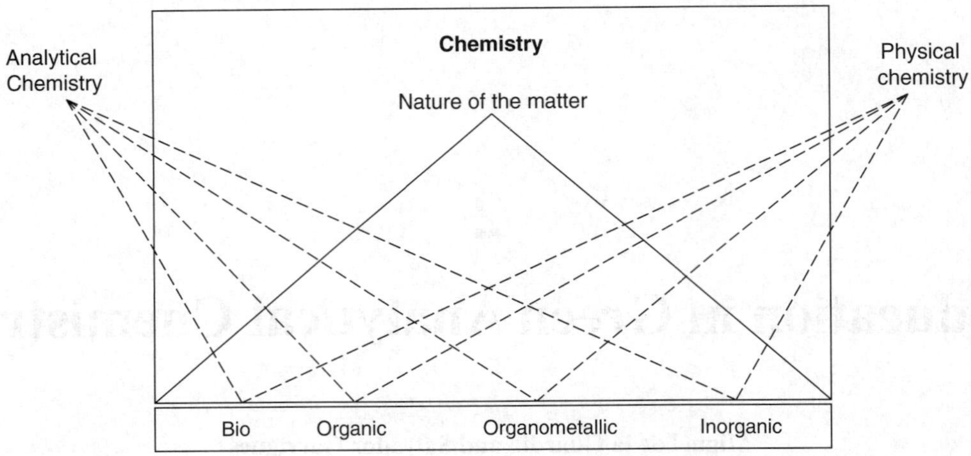

Figure 2.1 *Analytical Chemistry as a viewpoint on the nature of the matter.*

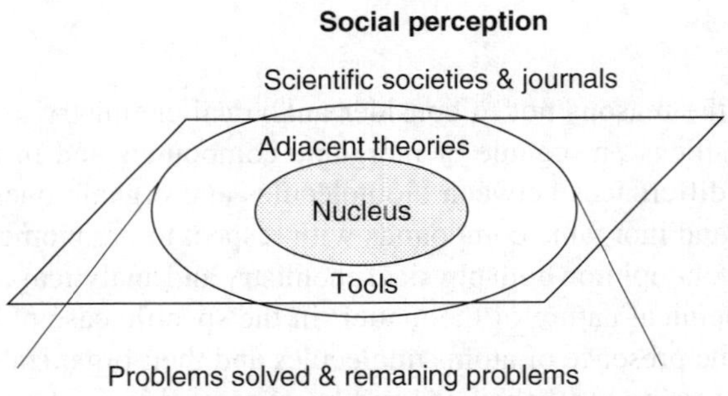

Figure 2.2 *A paradigm structure.*

problems and to predict new factors which also could be correctly interpreted using the established paradigm and that a crisis of a well-established paradigm can create a new paradigm, thus opening the door for a scientific revolution.

The science, in the aforementioned frame, progresses with a restricted number of fundamentals and during the periods of so called normal science everyone works comfortably, established in the tradition of their scientific discipline. A revolution takes place when the accepted paradigm is not appropriate enough to provide a correct answer to the new problems created by the advances in a discipline.

It is our opinion that nowadays the established paradigm in analytical chemistry has the same nucleus as the whole chemistry one. It would be based on the atomic and molecular theory which explains, together with the theory of the crystalline state, the relationship between sample composition and sample properties and thus, based on this core it would be astonishing for researchers to find that a property of a material could be based on a fraction of a molecule or that an atom could be totally destroyed during a reaction or can be exchanged in arbitrary proportions. In the case of analytical chemistry (see Figure 2.3) the analytical properties of methods, based on thermodynamic equilibrium and kinetic principles as well as the interaction between the matter and the electromagnetic radiation, and between matter and the electric field, can explain the basis of all the analytical procedures and form the basis of interpretation of all the problems

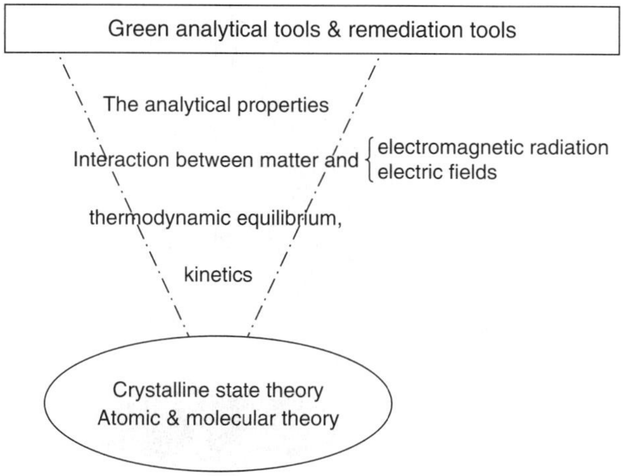

Figure 2.3 *The milestones of the today's research program in Analytical Chemistry.*

that we can solve today through the use of analytical chemistry; thus involving just surface changes of the paradigm which Lakatos terms a research program [2].

If we interpret the change of paradigm in terms of a revolution which creates doubts about the core theory of a discipline, it is clear that nowadays we continue to be in a normal period of the development of the analytical sciences. However, the introduction of green analytical tools and remediation tools come from a social demand about the present state-of-the-art analytical practices and because of that, we could agree with Malissa that the chemistry has been moved from a chemiological paradigm, based on the scientific principles established by Lavoisier, to a chemurgical paradigm and nowadays we must provide a social response in the frame of an ecological paradigm. In this last approach, the chemical practices must be considered to be in a close relationship with environmental equilibrium and the new social demands about health and safety [3]. So, the Green Analytical Chemistry paradigm is in fact an added value and an environmental responsibility imposed on the old practices without a drastic modification of the basic ideas exposed by Malissa on the primary paradigm of analytical chemistry which he defined from the equilibrium between rationalism and empiricism, explanation of a result through deductive analysis and extension of the aforementioned explanation through induction in a close interaction between axioms and facts, hypothesis and experiments in a way to search for the truth from a theory (see Figure 2.4). In the case of Green Analytical Chemistry, we will also consider environmental preservation.

So, we can conclude that the basic structure of the today's analytical chemistry is the same that at the end of the twentieth century, that the scientific method is the basis of the methodology employed to establish the correlation between the properties of the matter and its composition, that the interpretation of the analytical facts continues to be well supported by the atomic and molecular theory and by the crystalline state theory which both support the thermodynamic and kinetic principles of chemical reactions and the interaction between matter, electromagnetic radiation and electric fields [4]. So the simple aspect which has been drastically modified in the analytical chemistry paradigm has been the incorporation of the so called green parameters to the basic analytical properties. Accuracy, representativeness, traceability, sensitivity and selectivity in the renewed paradigm of Green Analytical Chemistry have been complemented and not changed by additional considerations on the safety of operators and the environment, the strong reduction of reagents, energy and solvents consumed, the search for as much as possible information about the samples from simple and direct measurements and the responsibility of the laboratories about the elimination, or at least the reduction and decontamination of analytical wastes.

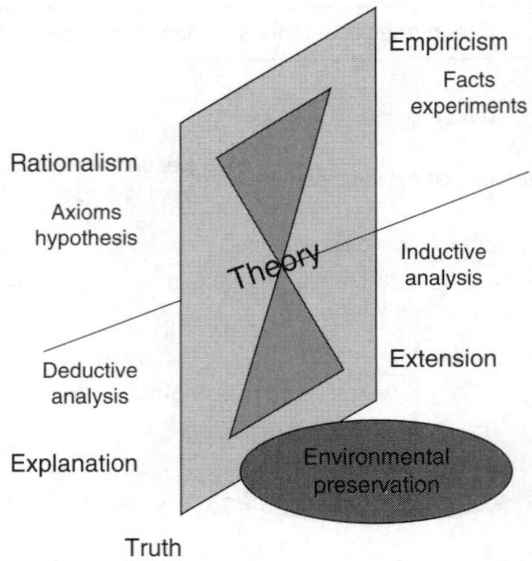

Figure 2.4 *The primary paradigm of Analytical Chemistry as an alteration between induction and deduction from the ideas of Malissa.*

2.2 The social perception of Analytical Chemistry

One of the problems which we have today as chemists is the mass media and bad image held by society about chemistry, the negative evaluation which everybody has about the benefits and drawbacks of the chemical activities. So, nowadays the images associated to the social perception of chemistry are those of polluted rivers, the black smoke of a chimney, the smog in the city and acid rain. Only those aspects which concern bio- or eco-chemistry escape from the aforementioned discredit of activities related to the synthesis and registration of chemicals, the chemical industries and all that is related to human efforts to create new molecules and to incorporate these new structures in our life. In such a frame analytical activities are considered just as an additional pollution focus. However, it is true also that there is a social perception of the need for analytical chemistry to evaluate the environmental side effects of basic chemical activities and we analytical chemists can take advantage of this fact.

At the middle of the 1980s George Pimentel, who was the president of the American Chemical Society (ACS) in 1986, presented a report to the National Academy of Sciences of the USA concerning *Opportunities in Chemistry* [5] which was a deep evaluation of the advantages offered by the chemical knowledge and the problems related to bad practices in this field. The aforementioned information can be considered as a starting point on the ecological mentality of the chemical community and, in this sense, the Pimentel's proposal to the Environmental Protection Agency (EPA), created in 1970 by the initiative of President Nixon [6] included a series of aspects that directly concerned analytical chemistry such as; the increase in the percentage of research and development funding devoted to exploratory research, the improvement of fundamental research on reaction pathways for substances of environmental interest, the detection of potentially undesirable environmental constituents at levels below their expected toxicity and the EPA support of analytical chemistry in a prominent way; thus clearly indicating that the analytical tools could be a key factor for pollution monitoring and to evaluate the deleterious side effect of the synthesis and fate of chemical compounds.

The Pimentel report also created the need for an increasing conscience of the chemical society about the side effects of all their activities. Based on this, many efforts can be identified in the literature which look for the reduction of prime matters and regents consumed, the deep control of chemical substances in all steps from the extraction of natural products to the different reactions involved in the synthesis of new products. It was also necessary to pay attention to the generation of by-products and the behaviour of chemical substances in the environment, taking into account the potential risks of hydrolysis and metabolite products. Also the risks involved by the analytical activities, as we have indicated before [7,8], were considered in order to avoid waste generation and to reduce the risks for operators through the search for miniaturization [9] and automatized methods of analysis [10] also looking for low energy sample treatment systems like the use of microwave-assisted methodologies [11]. All the aforementioned applied efforts were incorporated in the preliminary theoretical consideration about the so called environmentally friendly analytical methods [12] or sustainable analytical chemistry [13]. However, during the 1990s it was not so easy to find literature concerning the dispersed efforts for greening the analytical practices and it was recognized in a literature survey on green spectroscopy made in the frame of a special issue devoted to this topic by the journal *Spectroscopy Letters* [14]. Really, it was necessary to wait for the tremendous development of Green Chemistry made by the USA EPA and lead by Paul Anastas, who published a series of fundamental books from 1994 [15–18] trying to create a general conscience on the need for a Green Chemistry. In spite of that, until 2010 there has not been any specific published book on Green Analytical Chemistry [19].

The tremendous efforts made on greening both chemistry and analytical chemistry can be evaluated through the consideration of books and journals devoted to these aspects as it can be seen in Table 2.1. We think that theoretical and practical efforts are absolutely necessary to convince the members of the chemical societies about the need of such a revolution in our mentality and practices. On the other hand, it is also mandatory to be able to transmit a new message to society in terms that clearly show chemistry is a fundamental part of the solution of pollution problems and not just a part of the problem. The prize will be a new generation of chemists with a strong ethical compromise within society and the environment.

2.3 Teaching Analytical Chemistry

Analytical chemistry studies in the frame of chemistry degrees around the world have evolved in different ways as a function of the studies programs and national regulations. In Spain there is a great tradition in studying the existence of analytical chemistry departments as a specific area of knowledge in the frame of studies in chemistry, pharmacy, biology and other new studies like bromatology and toxicology, environmental sciences and chemical engineering.

Analytical chemistry teaching in the past in our country was closely related to inorganic analysis as it has been also the case in France and Italy. Because of that in former times, inorganic ion systematic identification approaches based on drop reactions, titrimetric and gravimetric methods of chemical analysis were the basis of analytical chemistry studies.

Theis discipline approach was removed in the last 30 years and replaced by the deep consideration of chemical equilibria. So, inorganic qualitative analysis and chemical methods of analysis based on stoichiometric reactions were extensively studied in the laboratory courses and the basic courses of analytical chemistry were focussed on the acid-base, complex formation, redox and precipitation equilibria developing many graphical and mathematical treatments in order to provide a complete picture on the ion reactions in aqueous media. So a change was produced from a descriptive approach to an essentially mathematical one that improved the level and complexity of the analytical studies.

However, the main part of present challenges in analysis remained absent from the content of the introductory courses, thus providing a false idea to the student on the objectives and the identity of analytical chemistry, which remained closely related to the inorganic analysis.

At present the main part of methods developed and applied focussed on organic molecules. So condensation and substitution reactions, which are of a main concern of organic analysis, were far from the simple scheme of the ion reactions considered in the analytical chemistry introductory courses. Fortunately efforts to create a specific personality of analytical chemistry in the frame of chemistry lead to the publication of totally new textbooks, like that of Professor Miguel Valcárcel, which focussed the basic studies of analytical chemistry in the analytical

Table 2.1 *Books and journals devoted to Green Chemistry and Green Analytical Chemistry.*

	Book Title	Year	Author/s – Editor/s
1	Green Technologies for the Removal of Emerging Pollutants from Water and Wastewater	2011	G. Lofrano
2	Green Chemistry for Environmental Remediation	2011	R. Sanghi and V. Singh
3	Green Chemistry	2011	R. Luque
4	Challenges in Green Analytical Chemistry	2011	M. de la Guardia and S. Garrigues (eds)
5	Green Organic Chemistry in Lecture and Laboratory (Sustainability)	2011	A. Dicks
6	Biocatalysis for Green Chemistry and Chemical Process Development	2011	J. Tao and R.J. Kazlauskas
7	Cellulose Fibers: Bio- and Nano-Polymer Composites: Green Chemistry and Technology	2011	S. Kalia, B.S. Kaith and I. Kaur
8	Green Polymerization Methods: Renewable Starting Materials, Catalysis and Waste Reduction	2011	R.T. Mathers and M.A.R. Meier
9	Monograph on Green Chemistry	2011	A. Parashar
10	Green Chemistry: Greener Alternatives to Synthetic Organic Transformations	2011	V.K. Ahluwalis
11	Green Chemistry and Engineering: A Practical Design Approach	2011	C. Jiménez-González and D.J.C. Constable
12	Green Polymer Chemistry: Biocatalysis and Biomaterials (ACS Symposium Series)	2011	H. Cheng and R. Gross
13	Green Chemistry and the Ten Commandments of Sustainability	2011	S. Manahan
14	Green Analytical Chemistry: Theory and Practice	2011	M. de la Guardia and S. Armenta
15	Environmental Chemistry: Green Chemistry and Pollutants in Ecosystems	2010	E. Lichtfouse, J. Schwarzbauer and D. Robert
16	Green Chemistry and Sustainable Science and Technology: A Brief Introduction	2010	S. Manahan
17	Green Chemistry for Environmental Sustainability	2010	S.K. Sharma and A. Mudhoo
18	Handbook of Green Chemistry - Green Solvents	2010	W. Leitner, P.G. Jessop, C.J. Li and P. Wasserscheid
19	Green Analytical Chemistry	2010	M. Koel and M. Kaljurand
20	Green Chemistry in the Pharmaceutical Industry	2010	P. Dunn, A. Wells and M.T. Williams
21	Experiments in Green and Sustainable Chemistry	2009	H.W. Roesky, D. Kennepohl and J.M. Lehn
22	Green Chemistry Education: Changing the Course of Chemistry (ACS Symposium Series)	2009	P.T. Anastas, I.J. Levy and K.E. Parent
23	Green Chemistry Research Trends	2009	J.T. Pearlman
24	Handbook of Green Chemistry - Green Catalysis: Volume 1 - Homogeneous Catalysis	2009	R.H. Crabtree and P.T. Anastas
25	Handbook of Green Chemistry - Green Catalysis: Volume 2 - Heterogeneous Catalysis	2009	R.H. Crabtree and P.T. Anastas
26	Handbook of Green Chemistry - Green Catalysis: Volume 3 - Biocatalysis	2009	R.H. Crabtree and P.T. Anastas
27	Alternative Solvents for Green Chemistry (RSC Green Chemistry)	2009	F.M. Kerton, J.H. Clark and G.A. Kraus
28	Green Chemistry: Environmentally Benign Reactions	2009	V. K. Ahluwalia
29	Green Chemistry Metrics: Measuring and Monitoring Sustainable Processes	2008	A. Lapkin and D. Constable
30	Methods and Reagents for Green Chemistry: An Introduction	2007	A. Perosa, F. Zecchini and P. Tundo

31	Green Chemistry and Engineering	2007	M. Doble and A. Kumar
32	Green Chemistry and Catalysis	2007	R.A. Sheldon, I. Arends and U. Hanefeld
33	Exploring Opportunities in Green Chemistry and Engineering Education: A Workshop Summary to the Chemical Sciences Roundtable	2007	P.T. Anastas, F. Wood-Black, T. Masciangioli, E. McGowan and L. Ruth
34	Green Chemistry in Undergraduate Practical Courses	2006	J. Hardy and J.H. Clark
35	Agricultural Applications in Green Chemistry (ACS Symposium)	2004	W.M. Nelson
36	New Trends in Green Chemistry	2004	V.K. Ahluwalia and M. Kidwai
37	Green Chemistry Using Liquid and Supercritical Carbon Dioxide (Green Chemistry Series)	2003	J.M. DeSimone and W. Tumas
38	Green Chemistry: Environment Friendly Alternatives	2003	R. Sanghi and M. M. Srivastava
39	Green Organic Chemistry: Strategies, Tools, and Laboratory Experiments	2003	K. Doxsee and J.E. Hutchison
40	Introduction to Green Chemistry. Instructional activities for introductory chemistry	2002	M.A. Ryan and M. Tinnesand (eds)
41	Green Chemistry: and introductory text	2002	M. Lancaster
42	Handbook of Green Chemistry and Technology	2002	J.H. Clark and D. Macquarrie
43	Ionic Liquids: Industrial Applications for Green Chemistry (ACS Symposium)	2002	R.D. Rogers and Kenneth R. Seddon
44	Green Chemistry: A Teaching Resource	2001	D. Warren
45	Introduction to Green Chemistry	2001	A.S. Matlack
46	Real-world cases in Green Chemistry	2000	M.C. Cann and M.E. Connelly
47	Green Chemistry: Theory and Practice	1998	P.T. Anastas and J.C. Warner
48	Green Chemistry: Frontiers in Benign Chemical Syntheses and Processes	1998	P.T. Anastas and T.C. Williamson
49	Green Chemistry: Designing Chemistry for the Environment (ACS Symposium Series)	1996	P.T. Anastas and T.C. Williamson
50	Benign by Design: Alternative synthetic design for pollution prevention	1994	P.T. Anastas and C.A. Farris
51	Green chemistry concepts and experiments	1991	W. Min, S.Z. Guo and W.X. Li
52	Ionic liquids and Green Chemistry	1991	Z. Suo and J.B. Zhu
53	Learning from the textbook: Green Chemistry Experimental	1991	R.Y. Jie
	Journal Title (Publisher)	**Year**	**Web site**
1	Green Chemistry (Royal Society of Chemistry)	1999	http://pubs.rsc.org/en/journals/journalissues/gc
2	Green Chemistry Letters and Reviews (Taylor & Francis)	2007	http://www.tandf.co.uk/journals/titles/17518253.asp

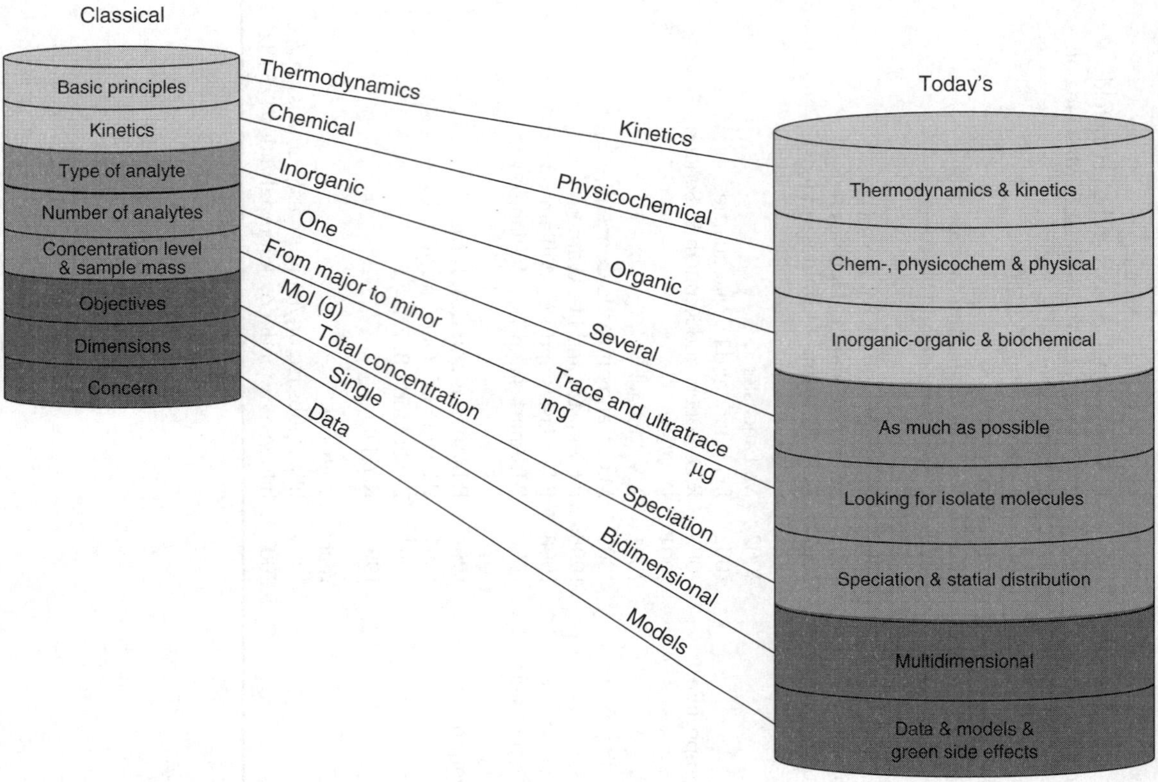

Figure 2.5 *Evolution of Analytical Chemistry from classical analysis to the current real work.*

properties of the methods and related topics, like traceability, screening and process monitoring, which are really the aspects which differentiate the analytical practice as a metrological discipline devoted to problem solving [20].

In our opinion, Valcárcel's book of together with his activity in the Federation of European Chemical Societies (FECS) Working Party on Analytical Chemistry (WPAC) played an important role in the strong modification of analytical chemistry studies in Spain [21] and also regarding the European consideration of our discipline [22,23].

Between the recent revolution of the content of analytical chemistry at university level, the change can be identified in the basic principles from the thermodynamic ones to the close integration between thermodynamic and kinetic aspects, considering both physical and physicochemical kinetics. On considering the type of analytes, it is clear that our activity have moved from the inorganic field to the organic one also considering biochemical analysis. The same extension, not replacement, has been done for a number of considered analytes; which has moved from one to several (as many as possible elements and/or compounds per sample) and also for concentration levels of target analytes; which has moved from major and minor components to trace and ultratrace analysis with an increasing demand on analyses at micro and submicro sample scales. On the other hand new challenges in analytical chemistry correspond to the need to move from total concentration determinations to speciation analysis, from average concentration determinations to layer by layer complete characterization of samples and from simple to bidimensional and multidimensional analyses.

In such a changing context (see Figure 2.5), nowadays we must include the change of analyst conscience from a simple interest in data analysis to interest in models and the strong consideration of the environmental side effects of our practice (as a consequence of the high demand of analytical information).

2.4 Teaching Green Analytical Chemistry

Teaching analytical chemistry today is in our opinion, maintaining the advances of the past in order to improve the main analytical figures of merit of the available approaches and also to improve them. At the same time it is necessary to answer adequately the problems related to our social compromise with the safety of operators and the environment. Especially in Europe, the present situation created by the REACH norm concerning the Register Evaluation Authorization and Restriction of Chemicals, imposes new responsibilities on us in order to educate well our future chemists in basic principles and methodologies of chemical and instrumental analysis as well as in trace analysis, chemometrics, automation and sensors. In this framework it will become especially important to consider the deleterious effects of chemicals and chemical reactions in order to ensure that our university students could have ethical behaviour and evaluate the immediate changes to make in the classical approaches in order to take advantage of the new economic opportunities offered.

So, teaching analytical chemistry will include thinking about the analytical problems and their solutions in terms of sustainability, considering both the classical figures of merit reported in the past, but also evaluating the persistent, bioaccumulative and toxic characteristics (PBC) of some reagents, the use of hazardous or corrosive reagents or solvents and the generation of analytical wastes. These latter aspects are important in order to make the appropriate reagent replacement, to move from hard to soft analytical practices and to evaluate energy consumption additionally than to minimize waste generation and to incorporate their on-line treatment in the body of the whole analytical process. So, the challenge today is to fix new objectives without sacrificing the former ones and it must transform all our teaching practices from the content of the theoretical lessons to the laboratory practices.

In fact teaching Green Analytical Chemistry cannot mean to add lessons on the side effects of our methods. On the contrary, all the contents must be modified by introducing the environmental ethical compromise, from beginning to end of the analytical process. It could be a nonsense to speak about safety and pollution risks of the old practices without providing alternatives. So, from our point of view, greening our teaching practices must involve a strong theoretical effort together with a change in our practices, starting from the seminars and practical work suggested to students and incorporating decontamination steps in our laboratory experiments [24].

In recent years many efforts have been made in order to incorporate Green Chemistry principles to education contained in the short text from 2002 by the American Chemical Society *Introduction to Green Chemistry: Instructional Activities for Introductory Chemistry* [25]. Unfortunately Green Chemistry in many cases has remained the matter of study in specific master's programs remaining and optional matters, in spite of efforts made in prestigious journals like the *Journal of Chemical Education* devoted to education in chemistry [26,27]. Because of that, we continue to be far from the desire expressed by Daryle Busch, when he was president of the American Chemical Society, that Green Chemistry could represent one of the milestones which could contribute to a sustainable future by 2000 and because of that it was absolutely necessary to teach the value of the Green Chemistry to the chemists of tomorrow [28]. Unfortunately, to our knowledge there is no prospect of a textbook in analytical chemistry written from the perspective of Green Chemistry and we are far from the integration of the research efforts for greening the analytical procedures in our everyday teaching activity. For this reason (see Figure 2.6) we are convinced that a pedagogical effort must be made in all different aspects from the integration of theoretical principles of Green Analytical Chemistry in the university text books to the generalized use of material safety data sheets (MSDS) on both the laboratory notebook documents and as complementary data from analytical method protocols. Additionally, the incorporation of green pictograms and green parameters in the seminars devoted to evaluate different alternative methods to solve real problems could be really helpful in creating a new mentality in our young students.

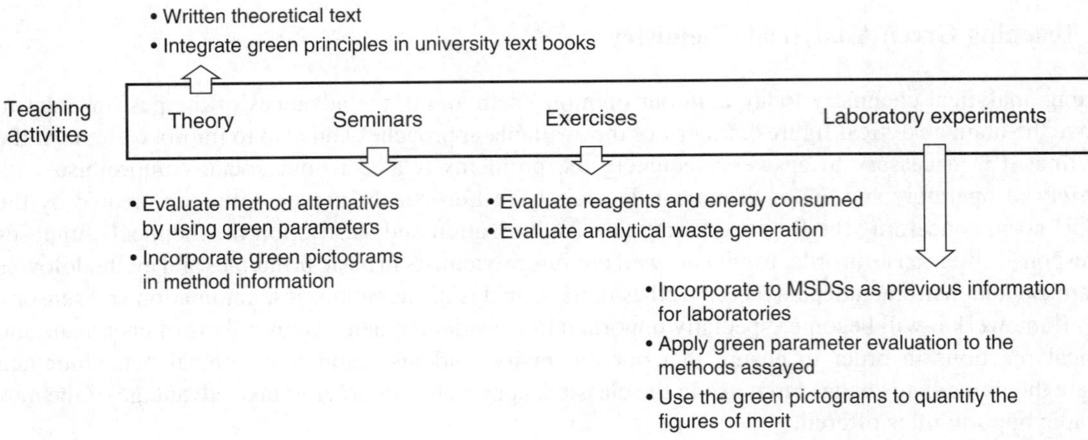

Figure 2.6 *Aspects to be considered for greening Analytical Chemistry teaching practices.*

2.5 From the bench to the real world

Nowadays there is a big effort in scientific literature to search for new tools in analytical chemistry which can provide a reduction of reagents and energy consumed and could avoid or minimize toxic wastes [29,30] additional to efforts to look for the replacement of toxic reagents by innocuous ones [31]. However, it is time to cross the line between the academic and the real world and to look for practical applications of the green developed methods. To do it, the new green tools must be tested in depth to evidence their advantages in terms of both environmental safety and economy and in this sense it is a priority task to correctly evaluate the green alternatives in these two ways. Parameters, like the amount of reagents consumed for each determination and the volume of waste generated by 100 determinations must be included between the figures of merit of the green approaches and compared with those of previously available methodologies. It seems not enough to include general sentences indicating that the use of ethanol in liquid chromatography is greener than the use of acetonitrile or to generalize that flow analysis methods offer a reduction of reagent consumption in front of in-batch methods. If we could move from our desks and university laboratories to the implementation of green methods in the industrial world, we must quantify the green alternatives in order to provide solid arguments for a change of mentality and practices.

On the other hand, it could be of a great interest to make a deep evaluation of the green alternative methods, looking on the present advantages and drawbacks but also thinking about the future and to do it, the strengths, weaknesses, opportunities and threats (SWOT) methodology offers a clear picture for the comparison between two alternative procedures [32].

The SWOT methodology consists of a deep evaluation of the strengths and weaknesses of a proposal, a new procedure or a change of conditions in our case. So, the reader can evaluate from this situation the opportunities offered by the evaluated proposal and its threats. The aforementioned methodology thus provides a clear comparison between the alternative suggested and the previous situation, and also a look into the future.

Someone could think that SWOT methodology is an economic or social tool more than a scientific one but do not hesitate; if we would convince the industry and laboratory managers to move their old practices, we must use the same arguments employed in the economic field and Figure 2.7 shows, as an example, the evaluation of incorporating a photoassisted waste decontamination step on-line in the determination of

S	W
• Well documented method in batch • Clear advantage of using FIA to save reagents and time • Complete mineralization of toxic wastes • On-line recovery of the catalyst • The figures of the method without decontamination step were not modified	• Needs the incorporation of an UV lamp • Adds a new FIA line • Requires the continuous filling of catalyst • Increases the energy consumption
O	**T**
• Avoids waste treatment outside • The system could be improved moving from continuous filling to discrete injection of same reagents • Easy adaptability of the decontamination unit to other methods	• The photo-assisted decontamination does not affect the presence of toxic elements additionally than organic molecules

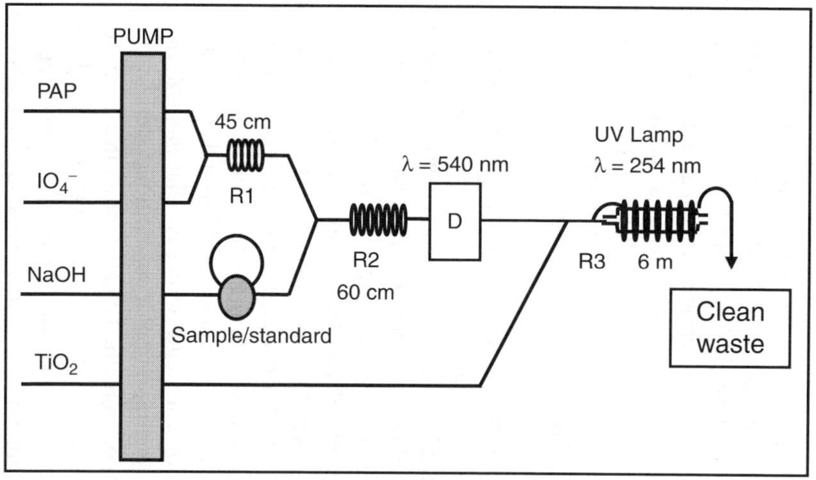

Figure 2.7 *Analysis of the advantages of incorporating the on-line photo-assisted waste decontamination step in the determination of resorcinol with p-aminophenol (PAP). Inset: the schematic diagram of the FIA set-up including the waste treatment step following the method proposed by de la Guardia et al.* [33].

resorcinol with p-aminophenol (PAP). As can be seen in the figure, the proposed procedure [33] maintains all the benefits of the fast and complete reaction of PAP with the target analyte when it is due in-batch, but incorporating this well known reaction to the advantages of the in-line generation of the active form of PAP created by its reaction with IO_4^-. The use of the classical flow injection approach also improves selectivity of this reaction by adding the physical and physicochemical kinetic reasons to avoid interferences of other phenolic compounds. However, the main advantage comes from the fact that the system is filled with pure solutions, which are merged only when required, thus avoiding the generation of waste coming from unemployed mixtures of reagents, and that the FIA methodology avoids the use of glassware and simplifies the cleaning of the system also reducing waste.

On considering the addition of the in-line decontamination of the FIA waste, this step does not modify at all the analytical figures of merit of the method, because it is incorporated at the end of the analytical process after detection. So, this added chemistry to the remaining PAP and the dye formed between PAP and resorcinol permits their complete mineralization based on the combined action of UV light and TiO_2. The process is fast and the catalyst can be recovered on-line from the clean waste just by a simple filtration or flocculation. Regarding the photo-assisted unit, it is well adapted to the FIA system because the Teflon tubes, commonly used to transport the solutions, are transparent to UV radiation and, because of that, the photodegradation reactor is just a Teflon tube rolled on the surface of an UV lamp. Concerning the weaknesses of the process, it is clear that we need to incorporate the UV lamp and a new FIA line to the process, in addition to the use of a new reagent (TiO_2). However, all the aforementioned elements are very cheap and common in a chemistry laboratory and could be employed in other applications; a single weakness that has not discussed is the increment of the energy consumption involved by the incorporation of an additional step.

On looking into the future, the application of the reported method clearly shows that chemistry, can be the best solution to solve pollution trouble like the generation of a toxic waste. On the other hand, the in-line detoxification of the waste (which was reduced by the incorporation of the FIA methodology as compared with the previous in bath process) avoids the external treatment of it, thus reducing the costs of analysis and avoiding the troubleswhich can be created by the accumulation of toxic residues in our laboratories. Additionally the method can be improved again by moving from classical FIA to sequential injection analysis (SIA) or multicommutation strategies of mechanization methods [34], being the detoxification unit easily adapted to treat the effluents of these method modifications and also being useful for the treatment of other organic molecules which could be mineralized through photo-assisted treatment.

The threats of the aforementioned strategy only come from the complex nature of some samples which could contain mineral toxic elements. In this case the detoxification of the waste generated from samples cannot be completed as it is necessary to add a precipitation step after photodegradation of organics to reduce the toxicity of the analytical wastes: here, a minimum amount of a solid residue could be obtained. The objective of the aforementioned discussion is just to evidence the kind of work to be done to move from academic studies to actual green alternatives for the industrial field.

Now, we think that it is time to move from the development of new green alternatives of methods of analysis to a deep evaluation of their advantages in order to improve the incorporation of green methods to solve real problems.

2.6 Making sustainable professionals for the future

In all the universities around the world there is a strong discussion about the preference on the three fundamental tasks of a professor such as; education, research and innovation and there are many professionals who prefer the knowledge transfer actions, based on the economical benefits this kind of work offers the economy of both the laboratory and the professor himself. On the other hand, many university professors like to publish so much because in the top scientific publications the prestige of a scientist is measured and thus, the number of publications, the H index of an author, the impact index, or citations of their work are the main figures of merit for many university professors. However, the aforementioned criteria are totally correct for evaluating the quality of our work, we cannot forget that the original task of our professional activity is to teach and because of that, the teaching practices of a university professor must be considered in his evaluation.

From our point of view and our own experience, it is evident that one of the most satisfactory facts of our professional activity is verifying the influence of our courses on the mentality of our young students. Do not forget that the philosophical transcendence is rooted in our deepest desires related to the success of professional activity, apart from the economical benefits and in this sense we would like to share with the reader an argument related to the possibilities in order to touch on a bit of immortality in our job. If you think about the impact of your research you can measure, compared to others, the time elapsed after publication of a paper during which our study remained to be cited and, in such a case, we can confirm that 10–15 years is a reasonable period of survival of a scientific publication (please do not think about exceptions, like that of the structure of the DNA double helix structure) if we look at the year of publication of the references employed in our last works. However, if we think that at the end of our teaching activity our age could be more than three times that of our young students, we can conclude that our words could remain for longer in the memory of our students than on the pages of the journals. So, it is important to think about it and, if you are convinced that Green Analytical Chemistry is a challenge for the future of analytical chemistry, it is of great importance to make an extra effort to transmit these ideas to the students in order that they could incorporate the principles and strategies of Green Analytical Chemistry to their future activities. We are convinced that it could be a great benefit for the environment but, additionally, it could offer exciting opportunities for their own professional future.

So, in conclusion, we consider that in order to prepare sustainable professionals in analytical chemistry there is a need for the success of the Green Analytical Chemistry project and a guarantee of convergence between the interest of both society and the scientific community.

Acknowledgements

The authors acknowledge the financial support of the Generalitat Valenciana Project GV PROMETEO 2010-055 to write this book and to research in this field.

References

1. Kuhn, T.S. (1962) *The Structure of Scientific Revolutions.* University of Chicago Press, Chicago.
2. Lakatos, I. (1978) *The Methodology of Scientific Research Programmes: Philosophical Papers,* Vol. 1, (eds J. Worrall and G. Currie), Cambridge University Press, Cambridge.
3. Malissa, H. (1988) Changes of paradigms in Analytical Chemistry, in Reviews on Analytical Chemistry, in *Euroanalysis VI*, (ed. E. Roth), Les Editions de Physique, Paris.
4. de la Guardia, M. (1991) *Proyecto Docente de Química Analítica*, Valencia.
5. Pimentel, G. (1995) *Opportunities in Chemistry*, National Academic Press, Washington D.C.
6. Origins of the EPA (1992) *The Guardian*, Spring issue. Available at http://www.epa.gov/aboutepa/history/publications/print/origins.html (accessed 3 November, 2011).
7. Armenta, S.; Garrigues, S. and de la Guardia, M. (2008) Green Analytical Chemistry, *TrAC - Trends Anal. Chem.*, **27**, 497–511.
8. de la Guardia, M. and Armenta, S. (2011) *Green Analytical Chemistry: Theory and Practice*, Elsevier, Amsterdam.
9. Manz, A.; Graber, N. and Widmer, H.M. (1990) Miniaturized total chemical-analysis systems - a novel concept for chemical sensing, *Sensor. Actuator. B-Chem.*, **1**, 244–248.

10. Ruzicka, J. and Hansen, E.H. (1975) Flow Injection Analyses. 1. New concept of fast continuous-flow analysis, *Anal. Chim. Acta*, **78**, 145–157.
11. Abusamra, A.; Morris, J.S. and Koirtyohann, S.R. (1975) Wet ashing of some biological samples in a microwave oven, *Anal. Chem.*, **47**, 1475–1477.
12. de la Guardia, M. and Ruzicka, J. (1995) Towards environmentally conscientious Analytical Chemistry through miniaturization, containment and reagent replacement, *Analyst*, **120**, 17N–17N.
13. de la Guardia, M. (1999) An integrated approach of Analytical Chemistry, J. *Braz. Chem. Soc.*, **10**, 429–437.
14. Armenta, S. and de la Guardia, M. (2009) Green Spectroscopy: A scientometric picture, *Spectrosc. Lett.*, **42**, 277–283.
15. Anastas, P.T. and Farris, C.A. (1994) *Benign by Design, Alternative Synthetic Design for Pollution prevention*, American Chemical Society, Washington D.C.
16. Anastas, P.T. and Williamson, T.C. (eds) 1996) *Green Chemistry: Designing Chemistry for the Environment*, American Chemical Society, Washington D.C.
17. Anastas, P.T. and Warner, J.C. (1998) *Green Chemistry: Theory and Practice*, Oxford University Press, Oxford.
18. Anastas, P.T. and Williamson, T.C. (eds) (1998) *Green Chemistry: Frontiers in Benign Chemical Syntheses and Processes*, Oxford University Press, Oxford.
19. Koel, M. and Kaljurand, M. (2010) *Green Analytical Chemistry*, RSC Publishing, Cambridge.
20. Valcárcel, M. (2000) *Principles of Analytical Chemistry: a Textbook*, Springer-Verlag, Heidelberg.
21. Agencia Nacional de Evaluación de la Calidad y Acreditación (2005) *Libro Blanco: Título de Grado en Química*, ANECA, Madrid.
22. Salzer, R.; Mitchell, T.; Mimero, P.; Karayannis, M.; Efstathiou, C.; Smith, A. and Valcárcel, M. (2005) Analytical Chemistry in the European higher education area, *Anal. Bioanal. Chem.*, **381**, 33–40.
23. Kellner, R.l Mermet, J.M.; Otto, M.; Valcárcel, M. and Widmer, H.M. (eds) (2004) *Analytical Chemistry: A Modern Approach to Analytical Science*, Wiley-VCH, Weinheim.
24. Armenta, S. and de la Guardia, M. (2011) Determination of mercury in milk by cold vapor atomic fluorescence: a Green Analytical Chemistry laboratory experiment, *J. Chem. Educ.*, **88**, 488–491.
25. Ryan, M.A. and Tinnesand, M. (eds) (2002) *Introduction to Green Chemistry: Instructional Activities for Introductory Chemistry*, American Chemical Society, Washington, D.C.
26. Cann, M.C. (1999) Bringing state-of-the-art, applied, novel, Green Chemistry to the classroom by employing the Presidential Green Chemistry Challenge Awards, *J. Chem. Educ.*, **76**, 1639–1641.
27. Hjeresen, D.L.; Schutt, D.L. and Boese, J.M. (2000) Green Chemistry and education. J. *Chem. Educ.*, **77**, 1543–1547.
28. Henry, C.M. (2000) Color me green - Five environmentally friendly technologies are recognized by the Presidential Green Chemistry Challenge Awards, *Chem. Eng. News*, **78**, 49–55.
29. Namiesnik, J. (2001) Green Analytical Chemistry - Some remarks, J. *Sep. Sci.*, **24**, 151–153.
30. Tobiszewski, M.; Mechlinska, A. and Namiesnik, J. (2010) Green Analytical Chemistry-theory and practice, *Chem. Soc. Rev.*, **39**, 2869–2878.
31. Keith, L.H.; Gron, L.U. and Young, J.L. (2007) Green analytical methodologies, *Chem. Rev.*, **107**, 2695–2708.
32. Deetlefs, M. and Seddon, K.R. (2010) Assessing the greenness of some typical laboratory ionic liquid preparations, *Green Chem.*, **12**, 17–30.
33. de la Guardia, M.; Khalaf, K.D.; Hasan, B.A.; Morales-Rubio, A. and Carbonell, V. (1995) In-line, titanium dioxide-catalyzed, ultraviolet mineralization of toxic aromatic-compounds in the waste stream from a flow injection-based resorcinol analyzer, *Analyst*, **120**, 231–235.
34. Reis, B.F.; Morales-Rubio, A. and de la Guardia, M. (1999) Environmentally friendly Analytical Chemistry through automation: comparative study of strategies for carbaryl determination with p-aminophenol, *Anal. Chim. Acta*, **392**, 265–272.

3

Green Analytical Laboratory Experiments

Suparna Dutta[1] and Arabinda K. Das[2]

[1] *Sonamukhi Girls' High School, Bankura, West Bengal, India*
[2] *Department of Chemistry, University of Burdwan, Burdwan, West Bengal, India*

In keeping with the current trend to avoid hazardous chemicals and to use environmentally clean methods, chemists all over the world are engaged in designing and developing green schemes for carrying out experiments in laboratory. As development of clean methods is cheaper than the cleaning of polluted environments, Green Analytical Chemistry could be more attractive from both aesthetic and economical viewpoints. The green analytical laboratory experiments depicted here avoid the use of excess reagents and energy, focus on prevention and reuse of wastes. In fact, analytical chemistry is an area in which efforts have been conducted from early times to make the newly used methods sustainable, safe and environmentally friendly. This chapter compiles 15 such green analytical laboratory experiments which will help students and researchers alike.

3.1 Greening the university laboratories

In the frame of an ecological paradigm, the concept of green chemistry, sustainable chemistry or environmentally friendly chemistry appeared towards the end of the twentieth century. Generally speaking, green chemistry seeks to reduce hazardous chemicals and chemical processes from the environment. The most effective green method is to avoid the use or creation of toxic materials rather than relegating toxins to post-processing cleanup. Green chemistry goes to the root of the problem and aims to eliminate the hazard itself. It is the only science that focuses on the intrinsic hazard of chemicals or chemical processes. It seeks to minimize or eliminate that hazard so that we do not have to worry about exposure.

According to Anastas, one of the pioneers of the green chemistry concept, 'It is the use of chemistry techniques and methodologies that reduce or eliminate the use or generation of feed stocks, products,

Handbook of Green Analytical Chemistry, First Edition. Edited by Miguel de la Guardia and Salvador Garrigues.

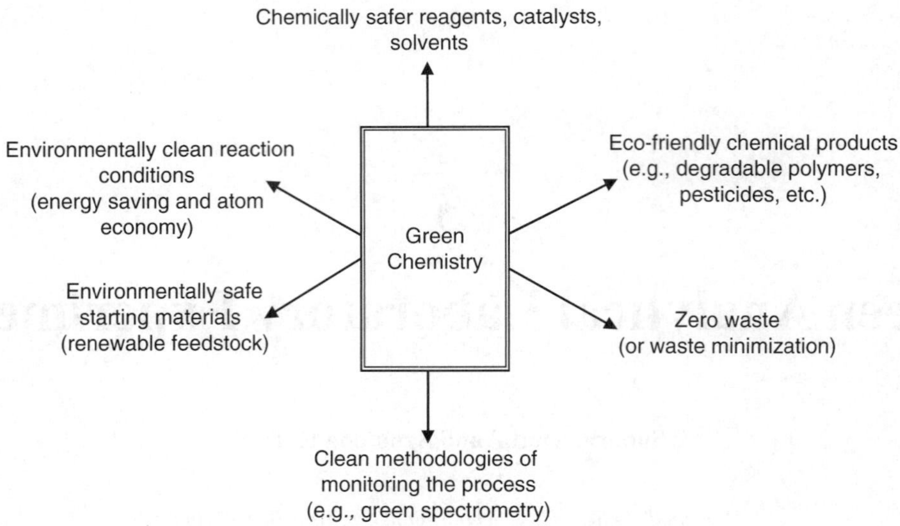

Figure 3.1 *Components of green chemistry.*

by-products, solvents, reagents, etc. that are hazardous to human health or the environment' [1]. Keeping this goal in mind, the components of green chemistry may be depicted in Figure 3.1, from which it is evident that green chemistry aims to design the new chemical products to replace the existing hazardous chemical products provided the new chemicals are having the same desirable properties of the existing one. For example, development of a new pesticide which is toxic to the target species only and at the same time its residue biodegrades easily into harmless products.

Green materials have been applied in the design of benign synthetic processes to eliminate or minimize pollution at the source, as well as in the remediation of a variety of existing environmental pollution problems. In fact, green chemistry aims to develop the greener way of obtaining products by using renewable resources (e.g., biomass) rather than nonrenewable resources (e.g., crude oil). Use of biotechnology involving biological feedstock, biocatalysts and biosynthesis is the most important aspect in chemical industries. Green chemistry aims establish greener reaction conditions so that energy consumption can be minimized rather than using drastic reaction conditions (e.g., high temperature).

In the aforementioned frame analytical chemists have concentrated on developing alternative analytical methodologies not requiring solvents or reagents, in reducing the amount of solvents required in sample pretreatment and also reducing the amount as well as the toxicity of solvents and regents employed in the measurement step, especially by automation and miniaturization. Naturally, green analytical techniques are being practiced currently which encompass screening methodologies, replacement of toxic reagents, minimization of wastes, recovery of reagents, on-line decontamination of wastes and reagent free methodologies.

On the basis of the literature data and observation of current trends in chemical analysis and monitoring, it is evidenced that rapid progress is being made in the development of those analytical methodologies that assure observation of the principles of Green Analytical Chemistry [2]. As development of clean methods is cheaper than the cleaning of polluted environments, Green Analytical Chemistry could be more attractive from both aesthetic and economical points of view.

Recent efforts to develop green chemistry applications to all fields of chemistry have started to receive attention from an academic perspective [3]. The main objectives of the green laboratory include: teaching the need for green chemistry concepts in chemistry and chemical analysis; illustrating the principles of the environmentally clean processes; and exposing students to the principles and protocols followed for handling environmental samples.

No doubt expansion of the practice and application of Green Analytical Chemistry will require educating students to green principles while advancing the state-of-the-art. This chapter deals with 15 Green Analytical Chemistry experiments which can be practiced by the graduate students of chemistry.

3.2 Green laboratory experiments

There are numerous studies that provide direct analytical methodologies for reduction of reagent consumption or waste minimization, the recycling of used solvents or the replacement of toxic components by non-toxic or at least less toxic ones. In this section we have included 15 green analytical laboratory experiments which may be described as solvent free, pressurized-solvent extractions, microwave-assisted treatments, ultrasound-assisted leaching, solid phase extractions, green spectrometry and so on. These experiments are chosen because when the students complete their course work they can understand that sustainability involves not only avoiding pollution, and decontaminating wastes, but also reducing the consumption of regents and energy. Thus the students become familiar with basic principles of analytical, environmental and green chemistry through a real world application. We feel laboratory experiments demonstrating these efforts could be included in academic curricula.

3.2.1 Green methods for sample pretreatment

Sample preparation is a bottleneck with traditional methods using high reagent consumption, large-scale and labor-intensive requirements based on environmentally burden procedures. Nowadays, aspects involving sample preparation and little or no sample dissolution have been introduced as principles of Green Analytical Chemistry and the sample pretreatment step has received special attention to make the process greener [4]. Because of a more efficient contact between sample and extractant, dilute acids and/or non-toxic reagents can be used. The main mechanisms involve the use of ultrasound, microwave irradiation and pressurization.

3.2.1.1 Ultrasound-assisted leaching of metals from biological materials

Ultrasounds are used to extract analytes with a drastic reduction of solvents and energy consumed or without use of organic solvents. The application of ultrasound by means of ultrasonic waterbaths or ultrasonic probes for assisting or accelerating extraction of both organic and inorganic compounds from solid materials is a current practice [5]. This can be coupled to spectrometric detection like atomic absorption spectrometry (AAS). Under these conditions, analyte transport from solid particles to the liquid phase is more efficient and quantitative extractions in short times can be expected. Additionally, due to effective interaction between the solid sample and the extracting solution, chemical bonds between the analytes and the matrix components can be disrupted by the action of diluted acids. Ultrasonic extractions are relatively simple and inexpensive techniques for green extractions.

The use of ultrasound whilst using a waterbath device at room temperature and 0.5 M HNO_3 as a leaching agent seems to be very convenient and the method is capable of extracting analytes like cadmium, copper, manganese and lead from biological materials within 5 min and the metals can be analyzed by AAS.

Table 3.1 Typical analytical results and analytical figures of merit.

	Cd	Cu	Mn	Pb
NIST SRM 1572 Citrus Leaves	0.02 ± 0.01	15.6 ± 0.9	19 ± 1	9.6 ± 0.1
Found (Certified value) $\mu g\ g^{-1}$	(0.03 ± 0.01)	(16.5 ± 1.0)	(23 ± 2)	(13.3 ± 2.4)
Sensitivity/ng (absorbance 0.0044)	0.001	0.061	0.005	0.017
Precision/% (RSD of 5 determinations)	3.4	3.1	3.1	3.9

***Green laboratory proposal* [6]** For the preparation of sample solution by using the solid-liquid extraction technique, 1–100 mg of powdered biological sample was mixed in a clean plastic beaker (10 ml) with 5 ml of 1 M nitric acid by using the Bransonic 3200 J-4 (USA) ultrasonic cleaner for 5 min at 40°C, and then the solid phase in the sample solution was separated from the liquid phase by centrifuging for 10 min at 4000 rev/min. The supernatant of sample solution (10 µl) was measured by graphite furnace AAS using graphite tube cuvette. Typical analytical results, sensitivities, and precisions for Cd, Cu, Mn and Pb are presented in Table 3.1.

Comments The analytical results for different metals obtained by means of the environmentally clean pretreatment method of powdered biological samples are in satisfactory agreement with the certified values. This technique can overcome the problems associated with slurry sampling, such as mixing of slurry during measurement, use of stabilizing agents, influence of particle size, and so on. Moreover, the present procedure accounts for minimization of leaching agent. Above all, there is no acid fume emission in the laboratory, thus providing an eco-friendly leaching methodology.

3.2.1.2 Microwave-assisted dissolution of complicated solid matrices

Most of the wet digestion techniques for complicated solid matrices involve constant supervision and prolonged time for complete dissolution. There are also possibilities of loss of trace elements, contamination and pollution during these steps. In search of an alternative wet digestion technique, microwave-assisted treatment was found to be an important tool. The heating of a solid sample in microwave oven increases against power supplied and the irradiation time. Better contact between the particles and acids is the key to rapid solubility. Heated dielectric liquids (water/acids) in contact with the dielectric particles will generate heat. This may create large thermal convection currents, which can agitate and sweep away the stagnant surface layers of dissolved solution and thus expose fresh surface to solution. Many researchers have reported the use of microwave–assisted techniques for digestion of various samples. It is chosen not only for sample pretreatment but also for accelerating chemical reactions [7].

***Green laboratory proposal* [8]** About 20 mg of a sample of fly ash or soil or igneous rock were accurately weighed in an hermetically sealed polytetrafluoroethylene (PTFE) reactor as shown in Figure 3.2 (115 ml internal volume, 1 cm wall thickness) and 4 ml aqua regia were added. After 2 min irradiation at 550W in a domestic Moulinex Super Crousty (Paris, France) microwave oven with 2450 MHz frequency, the reactor was cooled. Then 1.5 ml HF were added while performing the second step of digestion for 4 min at a power of 440W. The third step of digestion was carried out by treating the resulting mixture with 3 ml of 33% (*v*/*v*) H_2O_2 for 4 min at a power of 330 W. An additional digestion step was performed at a power of 550W for 2 min to complete the dissolution of the remaining portion, if any. The final step of treatment with 4 ml saturated H_3BO_3 solution for 10 min on a boiling waterbath helps to destroy the excess HF in order to avoid the damage

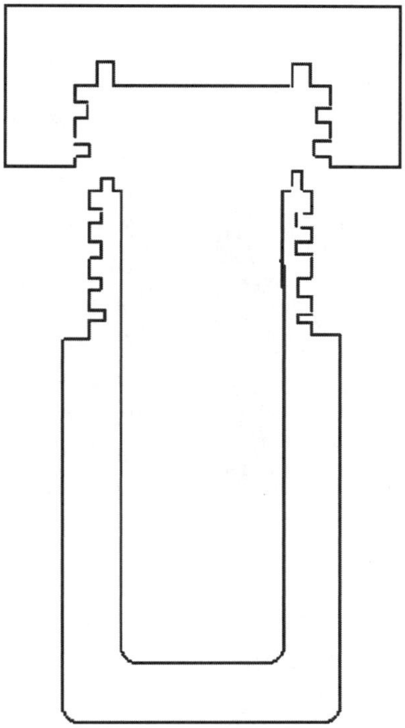

Figure 3.2 *Diagram of PTFE reactor. Reproduced from [7] with permission from Elsevier, © 1993.*

of glassware and so on. The digested samples were diluted to a definite volume in a volumetric flask for further analysis by using a suitable detection method like atomic spectrometry.

The aforementioned procedure, developed for dissolution of elemental analysis was validated by different ways. For example, results obtained by microwave assisted digestion/ICPMS (inductively coupled plasma mass spectrometry) have been comparable with those found by NAA (neutron activation analysis) for some selected elements in the same fly ash sample (Figure 3.3).

Comments Dissolution in a PTFE reactor with the help of microwave power is extremely safe for both the operator and laboratory environment avoiding acid fumes. The methodology not only shortens the sample dissolution time but also requires less acid consumption than conventional wet digestion method. As many as 58 trace elements could be analyzed in a fly ash matrix using ICPMS.

3.2.1.3 Pressurized liquid extraction of metals

Pressurized liquid extraction (PLE) technique has emerged in the field of green analytical technique in drying and inert solvent. It is based on using solvents at high pressure and/or high temperature without reaching the critical point. When using water as a solvent, the technique is commonly called pressurized hot water extraction (PHWE) [9]. In general, the system comprises a stainless steel extraction container, in which the sample, previously dispersed in an inert support, is placed and for which parameters such as temperature, pressure, extraction time, extraction cycles and so on are programmed electronically. The use of water or aqueous solution makes the technique environmentally friendly.

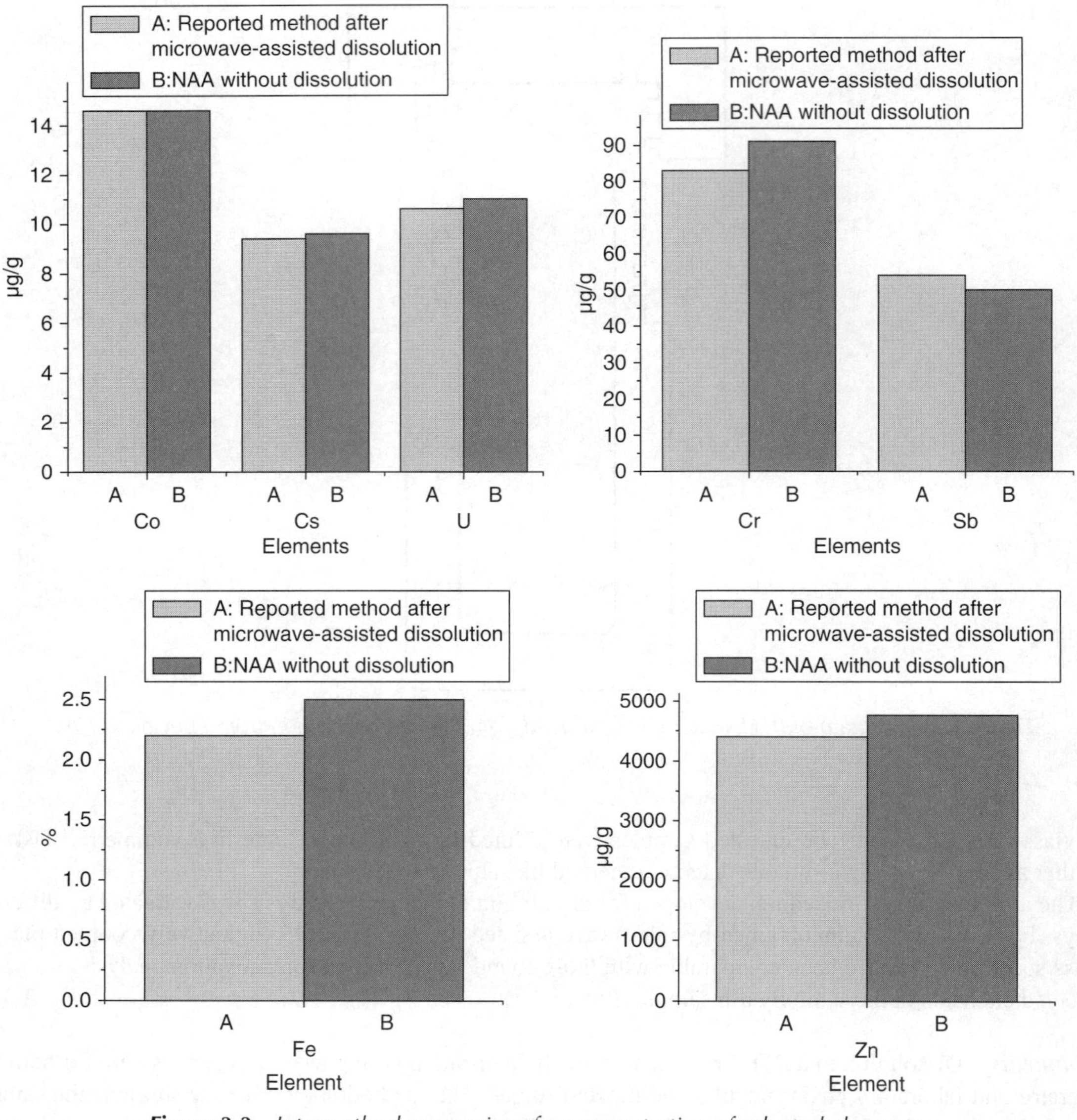

Figure 3.3 *Intermethod comparison for concentration of selected elements.*

Green laboratory proposal **[10]** Superheated liquid-liquid extraction was carried out using an extractor [9], shown in Figure 3.4 consisted of the following components: (1) an electrically heated oven (HP 5720A, Hewlett-Packard) for keeping the extraction coil at the desired temperature (the coil is a stainless steel tube of 6 m × 0.4 mm i.d.); (2) a high pressure pump (Shimadzu) for propelling the extractant; (3) a dosifier piston pump (Euromotori) for propelling the oil; (4) a T-junction as confluence point for allowing both phases to be merged; (5) a cooling system (consisting of a coil located in a water bath at 25°C) for cooling the effluent from the oven. All connections were made with stainless steel tubes/coil (i.d. 6 mm).

For extraction of copper, iron, nickel, vanadium, zinc from industrial oils, acidified water containing 20% (*v/v*) HNO_3 + 1 M KCl at a temperature of 175°C has been found to be effective. The contaminated oil and

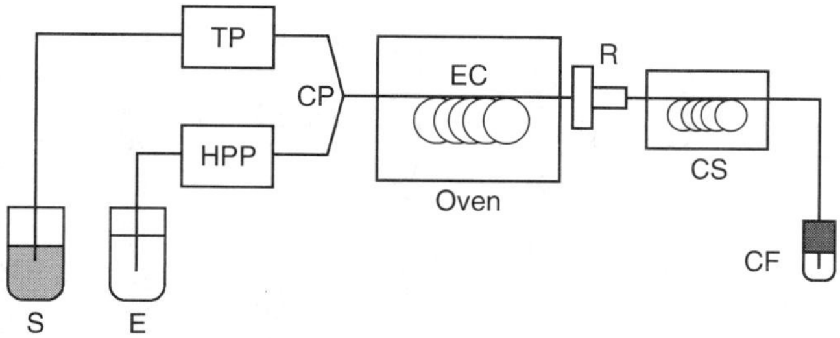

Figure 3.4 *Schematic diagram of the extraction system (S - oil sample; E - extractant; TP - triphasic pump; HPP - high pressure pump; CP - confluence point; EC - extraction coil; CS - cooling system; R - restrictor; CF - collection flask). Reproduced from [10] with permission from Elsevier, © 2004.*

the extractant were continuously pumped at the pressure needed to maintain the liquid state at preset temperature by the dosifer piston pump and the high pressure pump, respectively. The oil and extractant streams reached at the confluence point and then, the segments of the immiscible phases led to the extraction coil in the oven at the working temperature. The segments of oil and extractant circulated through the extraction coil at high temperature and pressure allowed the analytes to be transferred from the oil into the aqueous phase. After extraction, the oil-water segments were cooled at the end of the oven by passage through another coil located in a waterbath at 25°C. Both phases were spontaneously separated after collection in a vessel. The analytes were determined in treated and untreated oil by flame atomic absorption spectrometry.

Comments This experiment is highly significant with respect to solution of a solid waste disposal problem. The uncontrolled dumping of used tires is of great environmental concern. The solution to this problem comes from an innovative technology based on the recycle of this solid waste. A liquid-liquid extraction method using modified superheated steam as an extractant for removal of various metals from oil resulting from recycled tires has been effective. The study provides useful data for the cheapest working conditions to be used in order to obtain a given demetalization level.

3.2.2 Green separation using liquid-liquid, solid-phase and solventless extractions

The most widespread techniques for separation and preconcentration are liquid-liquid and solid-liquid extractions. Typically, preconcentration is achieved by addition of a suitable reagent to the sample solution and extraction of the resulting product to an organic phase or a hydrophobic solid phase. Various greener alternatives have been developed by minimizing both the use of organic solvents and the extraction times in comparison to traditional liquid-liquid extractions [11]. Solid-phase extraction offers a number of important benefits when compared with liquid-liquid extraction, such as elimination of solvents, higher enrichment factors, high recoveries, rapid phase separation, lower cost and the facility for coupling to different detection techniques.

3.2.2.1 Environmentally clean extraction of mercury species

Aqueous biphasic extraction system (ABS) offers a clean approach for extracting metal ions in an eco-friendly way. ABS may be thought of as relatively green separation medium since they are made of water and rather harmless polymer such as polyethylene glycol (PEG). The phase separation in an ABS occurs due to water structuring and the partial dehydration of the polymer chain within the two phases. Applications of the polymer-salt ABS in the separation of organic molecules, metal ions, radiochemicals, and the recovery of nanoparticles and minerals have been demonstrated. Thus the use of ABS instead of traditional solvent

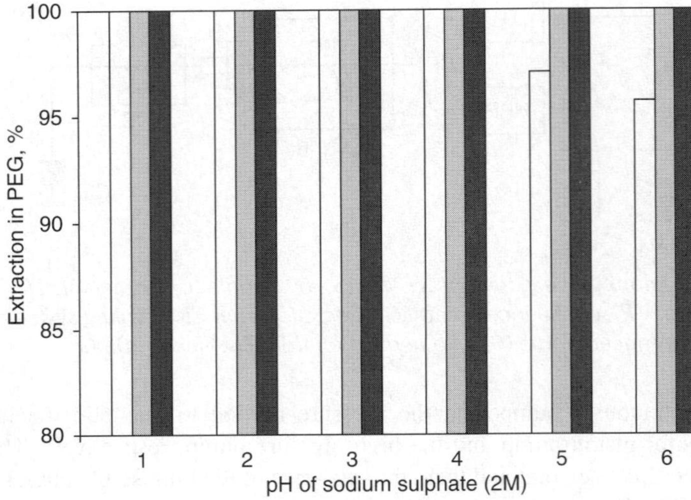

Figure 3.5 *Extraction profile of different salts of mercury in PEG rich phase of an ABS against different pH of Na_2SO_4 salt rich phase (in each pH, left column is for CH_3HgCl, middle for $HgCl_2$ and right one for Hg_2Cl_2). Reproduced from [13] with permission from Elsevier, © 2009.*

extraction systems has attracted considerable attention in the field of green chemistry [12]. ABS is a powerful system for differentiating among various species of the same element. Extractability of total mercury irrespective of its chemical form using ABS is demonstrated below. The method may find an extensive application for the total removal and decontamination of mercury from any system.

***Green laboratory proposal* [13]** The aqueous biphasic system was prepared by mixing 3 ml of 50% (*w/w*) PEG-4000 (MW = 3500–4000 g mol^{-1}) and 3 ml of 2 M Na_2SO_4. The irradiated salts of mercury (^{203}Hg was obtained by neutron activation of $HgCl_2$ and Hg_2Cl_2 while $CH_3{}^{203}HgCl$ was synthesized from $^{203}HgCl_2$ using a Grignard reagent) were dissolved in diluted hydrochloric acid and triple distilled water was added to make the stock solution. The concentrations of $^{203}HgCl_2$, $^{203}Hg_2Cl_2$, $CH_3{}^{203}HgCl$ in the stock solutions were 0.036, 0.021 and 0.040 M respectively. From each species 100 μl aliquot of a solution was added to the ABS system and shaken mechanically for 10 min. The system was centrifuged for 5 min and 1 ml from the PEG rich phase was carefully taken out for γ spectrometric studies and ^{203}Hg showed γ energy peak at 279.188 keV (81.5%). The results of the biphasic extractions of different mercury species in salt rich phase indicate that all the three forms of mercury can be extracted appreciably from high ionic strength of Na_2SO_4. Figure 3.5 demonstrates that partitioning of mercury species to PEG phase is possible at lower pH range (1–4).

Comments The present method offers green way of preconcentration of mercury species in place of classical solvent extraction method. ABS offers this unique condition when extraction is done with PEG (50%) solution and Na_2SO_4 (2 M) rich phase. No other condition could be found at which all of these species of mercury are extractable using conventional systems. Thus ABS can be considered as a clean technology to remove total mercury.

3.2.2.2 *Biosorption of uranium*

Biosorption has recently received a great deal of attention due to low costs of materials used and for their environmentally friendly nature. Chitin is the second most abundant natural amino polysaccharide found in the outer skeleton of insects, crabs, shrimps and other marine animals. Chitin is structurally similar to

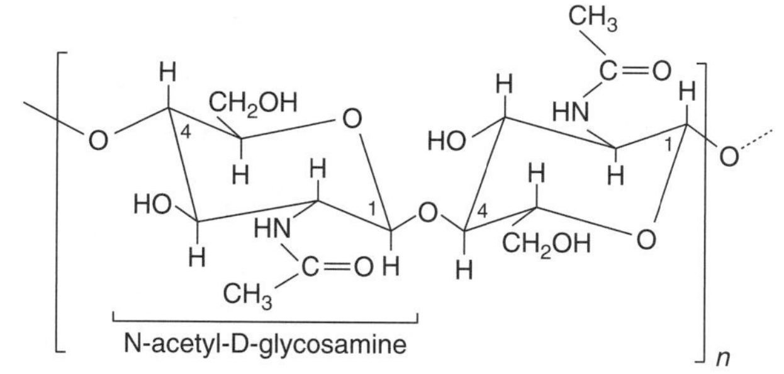

Chitin *i.e.* poly[β–(1→4)–2–acetamido–2–deoxy–D–glucopyranose]

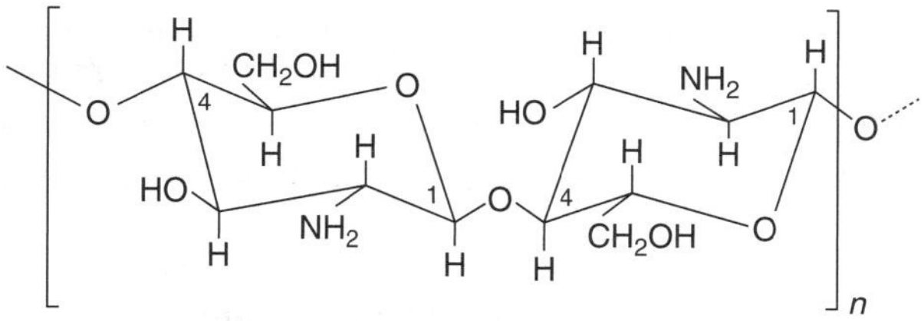

Chitosan *i.e.* poly[β–(1→4)–2–amino–2–deoxy–D–glucopyranose]

Scheme 3.1 *Structures of chitin and chitosan.*

cellulose and chitosan, a heteropolymer constituted by glucosamine and a fraction of acetylglucosamine residues (Scheme 3.1), is a well-known biopolymer characterized by its high sorption capacity. Amine groups of chitosan are responsible for the high sorption of metal ions through the chelation mechanism or by ion exchange depending on the nature of metal ions and the pH conditions. Chitosan soaked in carbon tetrachloride when irradiated with γ rays leads to a cross-linked product with $>CCl_2$ group as a spacer and the resulting polymer (CRC) and its hydrolyzed product (CRCH) are found to have good chemical stability [14]. Chitosan and its derivatives have wide range of applications for example, biosensors for removal of heavy metal ions including uranium.

***Green laboratory proposal* [15]** Commercially available medium viscosity grade chitosan was used for preparing CRC and CRCH and the irradiation was carried out in a Co-60 γ chamber of radiation dose of 5 kGy/h. 1.0571 g CRCH or 1.05 g CRC were immersed in Millipore water and allowed to swell. The glass column (120 mm × 8 mm) was than packed with the swollen polymer. After the column was conditioned at pH 6 with nitric acid, 100 ppm uranium (VI) solution at pH 6 spiked with U-233 tracer was allowed to pass through the column at a flow rate of 1 ml min^{-1}. The effluent was collected in fractions and the uranium concentration was determined either by radiometry using a scintillation detector (measuring α activity of U-233) or by spectrophotometry where Arsenazo III was used as the chromogenic agent. More than 97% elution of uranium was possible from the CRCH column using Na_2CO_3 (1M) eluent which is known to form strong carbonato complex. The distribution coefficient (Kd) values of uranium with CRCH and CRC were found to be 384

$[Kd = \frac{(Co - C)/W}{C/V}$ where C_0 = initial concentration of metal ion, C = total concentration of metal ion after equilibration, W = weight of biopolymer, V = volume of the aqueous phase in ml] and 125, respectively. On the other hand, the exchange capacities for uranium with those polymers were found to be respectively 0.050 and 0.034 mmol g^{-1}.

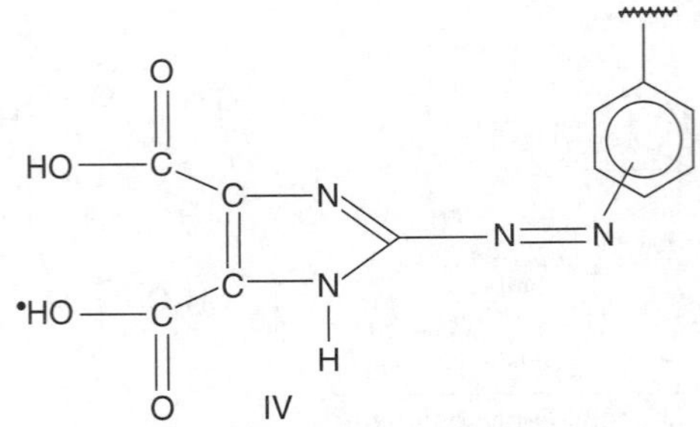

Scheme 3.2 *Anchoring of imidazole 4,5-dicarboxylic acid moiety onto DVB polystyrene matrix.*

Comments The work shows a comparative study between the two radiation grafted chitosan derivatives – cross-linked chitosan (CRC) and cross-linked chitosan after hydrolysis (CRCH). The polymers can be effective for the preconcentration and separation of uranium from different matrices. Evidently, the method of biosorption is simple, eco-friendly and could be effective for the determination of trace amounts of uranium.

3.2.2.3 Green speciation of vanadium using a solid phase exchanger

Metal speciation is currently very important from the perspective of environmental, biological, geological and medical applications. Vanadium is a real challenge in the field of speciation. This is because of its complex chemistry and also this element has been proved to be essential after the discovery of its insulin like properties [16]. Different forms of vanadium like V_2O_5, vanadates as vanadium (V) species and vanadyl as vanadium (IV) are of different physiological action and toxicity. Hence differentiation and quantification of vanadium in these two oxidation states are important. Co-precipitation followed by instrumental analysis is the common method for speciation of vanadium but it is tedious, time consuming and applicable to higher concentrations. Solvent extraction has also been used for the preconcentration and separation of vanadium but due to excess solvent requirement, the use of such technique is restricted. On the other hand, solid phase extraction could be a method of choice for preconcentration of vanadium species at very low level.

***Green laboratory proposal* [17]** The use of chelating resin has been popular with the introduction of chelating group like imidazole 4,5-dicarboxylic acid anchored by –N = N- into the polystyrene –DVB matrix (Scheme 3.2) and its application to the preconcentration and separation of vanadium species from natural water. A glass column of 130 mm × 10 mm was used for the present work. Air dried resin (1 g) was immersed in deionized water for 24 h and allowed to swell. The column was packed with swollen beads of the resin. The bed volume of the column was 2 ml. The sorption and recovery characteristics for V (IV) and V (V) were thoroughly examined under the flow rate of 0.5 ml min^{-1}. The sorbed V (IV) and V (V) were eluted by 0.1 mol l^{-1} malonic acid solution and 0.15 mol l^{-1} NaOH solution respectively. The samples of natural water were filtered using 0.45 µm Millipore membrane filter. Taking 500 ml of each sample, the pH was adjusted to 3 and passed through the column at a flow rate of 0.5 ml min^{-1}. After elution using the proposed method (10 ml of eluent was used in each portion) the concentration of vanadium was measured in the final solution. The results of vanadium speciation in different samples are shown in Figure 3.6.

Comments The speciation study of vanadium in natural water can be effectively carried out using solid phase extractor containing imidazole 4,5-dicarboxylic acid as the ligating site. Both V (IV) and V (V) are sorbed at pH 3 and thereby eluted with 0.1M malonic acid for V (IV) and 0.15 M NaOH for V (V). Therefore,

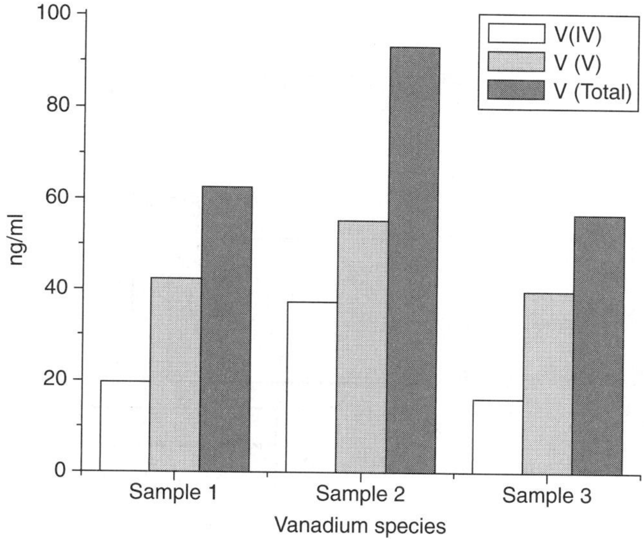

Figure 3.6 *Vanadium speciation in natural water.*

functionalized resin can be used as a good sorbent for preconcentration and determination of vanadium species at trace levels. The method is simple, low cost, easy to handle and environmentally benign.

3.2.2.4 Solventfree extraction of essential oils

During the last two and a half decades, solventfree extraction methods have been introduced or rediscovered, forced not only by the concern and/or cost associated with extensive solvent use but also by the new opportunities offered by green devices. Microwave-assisted extraction has been proved to be effectively applicable for extractions from difficult sample matrices, previously treated by longer time and solvent intensive Soxhlet extractions and hydrodistillations. Solventfree microwave extraction (SFME) apparatus is a combination of microwave heating and dry distillation at atmospheric pressure. SFME has been successfully introduced for laboratory scale applications in the extraction of essential oils from different kinds of aromatic plants. The method for extracting natural products without using any solvent and using microwave energy has been reported in 2003 [18].

***Green laboratory proposal* [19]** SFME was performed in a Milestone DryDist microwave laboratory oven working at a frequency of 2455 MHz with a maximum power of 1000 W. The dimensions of the PTFE-coated cavity are 35 cm × 35 cm × 35 cm. During such experiments, time, temperature, pressure and power can be controlled.

In a typical SFME procedure (Figure 3.7) performed at atmospheric pressure, 250 g of fresh plant materials were heated using a fixed power 500 W for 30 min without adding any solvent or water. A cooling system outside the microwave cavity condensed the distillate continuously. Condensed water was refluxed to the extraction vessel in order to provide uniform conditions of temperature and humidity for extraction. The extraction was continued at 100°C until no more essential oil was obtained. The essential oil was collected, dried under anhydrous Na_2SO_4 and stored at 0°C until further use.

Comments SFME has been compared with the conventional hydrodistillation technique for the extraction of essential oil from three herbs; basil (*Ocimum basilicum L.*), garden mint (*Mentha crispa L.*) and thyme

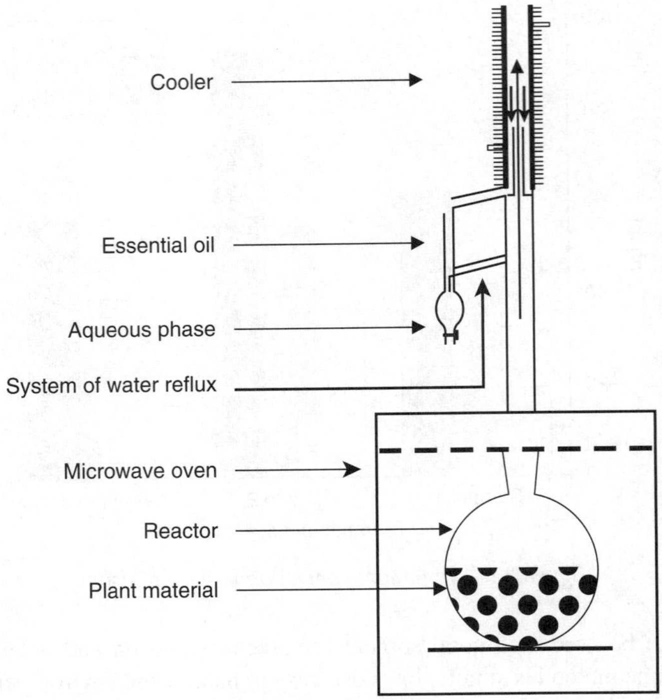

Figure 3.7 *Apparatus for solvent free microwave assisted extraction. Reproduced from [19] with permission from Elsevier, © 2004.*

(*Thymus vulgaris L.*). The essential oils extracted by SFME for 30 min were similar in yield and in their aromatic profiles to those obtained by conventional hydrodistillation for 4.5 h. The SFME method yields an essential oil with higher amounts of more valuable oxygenated compounds, and saves cost, time, energy and plant material. SFME is a green technology and appears as a better alternative for the extraction of essential oils from aromatic plants.

3.2.3 Green alternatives for chemical reactions

Several strategies can be adopted aiming greener analytical procedures, involving replacement of toxic reagents, minimization of reagent consumption, recycling and waste treatment. However, the development should be carried out without critically affecting key analytical features such as sensitively, accuracy and precision. This challenge becomes more serious for analytes in low concentration, especially in samples with complex matrices [20]. For this reason, laboratory experiments with an environmentally friendly orientation have emerged covering a wide variety of issues concerning chemical reactions. Following discussion is focused on an overview of some well known greener chemical reactions for analytical laboratory.

3.2.3.1 The green chemical approach for nitration of aromatic compounds

In several industrially important processes like nitration of aromatic compounds, a large excess of sulfuric acid is required because the by-product water slows down the reaction by diluting the acid [21]. At the end of these processes, a large amount of 'spent acid' is obtained which, in batch reactions, is usually neutralized and disposed off, while in continuous processes, it has to be cycled by complex methodologies. Also, the separation

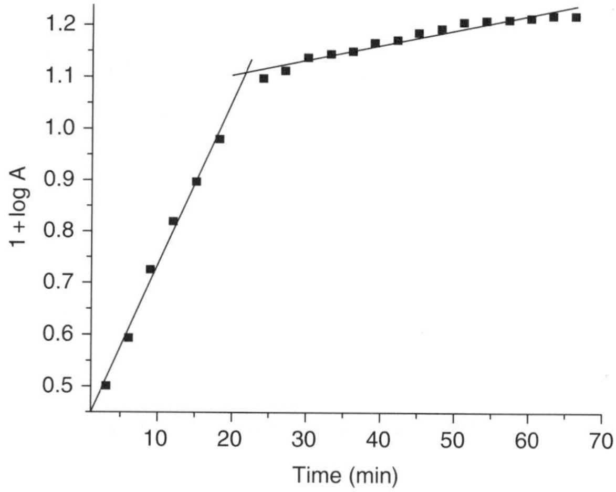

Figure 3.8 *Plot of (1 + log A) versus t (min) for anthranilic acid.*

of the products from the acid is often a difficult task and energy consuming process that habitually implies a basic aqueous work up. Additionally, sulfuric acid is corrosive in nature and is difficult to transport and handle. Naturally, it has been proposed to develop an alternative green chemical pathway to obtain nitro compounds under ambient conditions. Thus photochemical aromatic nitration of anthranilic acid has been carried out in the presence of ultraviolet radiations and the formation of the product could be studied spectrophotometrically.

***Green laboratory proposal* [22]** 0.1370 g of anthranilic acid was dissolved in 100 ml of doubly distilled water. 0.690 g of sodium nitrite and 0.0680 g of sodium formate were dissolved separately in 100 ml of doubly distilled water. All these three solutions were used as stock solutions; of which 6 ml of anthranilic acid solution, 8 ml of sodium nitrite solution and 0.7 ml of sodium formate solution were mixed together, so that the concentration of each of them in the resulting solution was 2.4×10^{-3} M, 3.2×10^{-3} M and 2.8×10^{-4} M respectively. The reaction mixture was exposed to ultraviolet lamp (254 nm). The optical density of this solution at different time intervals was determined at the wavelength of 380 nm for anthranilic acid with the help of a spectrophotometer. Some control experiments were also carried out and it was confirmed that light is necessary for photochemical nitration. The major product could be identified by gas chromatography combined with mass spectrometry as 2-amino-5-hydroxy-3-nitrobenzoic acid. The results of a typical run for photochemical nitration show absorbance (A) values increase with time of exposure which indicates the formation of nitro compounds. A plot of (1 + logA) versus exposure time was found to be linear (Figure 3.8) and hence this reaction follows pseudo-first order kinetics (in two stages: the first stage being faster than second stage). The rate constant was determined by the expression, k = 2.303 × slope and the values were $k_1 = 1.228 \times 10^{-3}\ s^{-1}$, $k_2 = 1.150 \times 10^{-4}\ s^{-1}$.

Comments In the reaction shown in Scheme 3.3 formate ions were used as •OH radical scavengers in order to increase the rate of nitration, but since •OH radicals are very reactive and present in higher concentrations than $\cdot NO_2$ radicals, so –OH group is also introduced in the ring even in the presence of formate ion. Nevertheless, ordinary nitration of aromatic system is accompanied by use of corrosive nitric and sulfuric acids and liberation of hazardous brown nitrous fumes which add to environmental pollution. The present work reports a facile route for nitration. Although the yield of the product is 42% but the photochemical nitration of aromatic compounds by alkali nitrite provides an eco-friendly route.

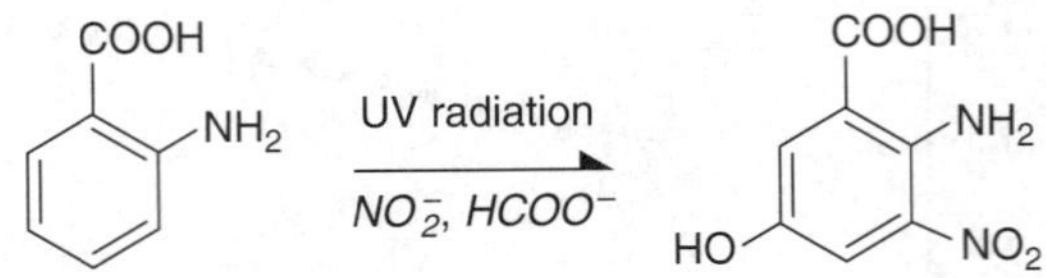

Scheme 3.3 *Conversion of anthranilic acid to 2-amino-5-hydroxy-3-nitrobenzoic (yield: 42%).*

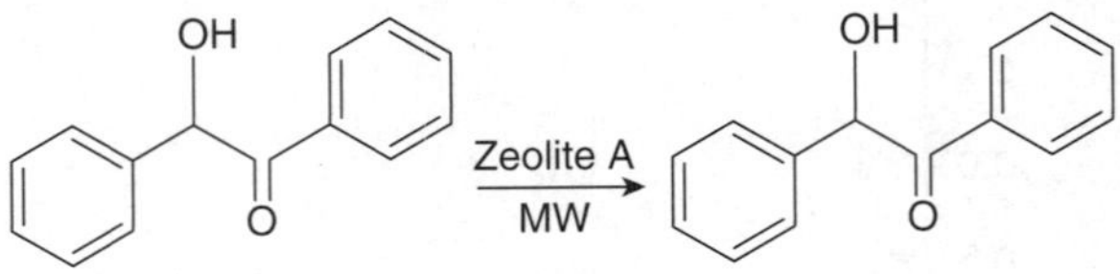

Scheme 3.4 *Oxidation of benzoins.*

3.2.3.2 Clean oxidation of benzoins on zeolite A

Benzils have received a great deal of attention because of their practical applications, that is, as photosensitive agents and synthetic agents in organic and pharmaceutical chemistry. Oxidation of benzoin derivatives with an oxidizing agent is a common method for the preparation of benzils. There are many reagents for this oxidation but long reaction periods, use of toxic metallic compounds and corrosive acids are some disadvantages of these methods. Recently, the combination of supported reagents and microwave irradiation are used to carry out a wide range of reactions in short times, high conversion and selectivity, without using any solvent. Zeolite A has been used in many organic reactions [23].

Green laboratory proposal **[24]** Oxidation of benzoins were efficiently performed with zeolite A catalyst without any oxidizing agent under solvent-free conditions and microwave (MW) irradiation (Scheme 3.4):

The solvent-free condition was achieved by mixing benzoins intimately with zeolite A followed by irradiation in a Moulinex (Paris, France) 2735A domestic microwave oven, at 2450 MHz (100% power, 850 W). Benzoin (2 mmol) and zeolite A (2 g) were mixed thoroughly in a mortar, transferred to a beaker, kept in a silica jacket and irradiated for 6 min. The whole procedure was monitored by thin layer chromatography (TLC). The final mixture was extracted in methylene chloride and the solvent was removed under rotary evaporator to get pure benzil. This procedure gives 80% yield of the product. In this reaction, both the symmetrical and unsymmetrical benzoins are rapidly oxidized to the corresponding benzils in the absence of organic solvent or oxidizing agents. The catalyst can be reused (zeolite after washing twice with methanol was activated again at 500°C before reuse). The catalytic efficiency is essentially retained; however, the yields decrease significantly after using the zeolite three or four times.

Comments The aforementioned solid phase synthesis of benzils with zeolite A in a solvent-free condition under microwave irradiation is facile, clean and environmentally friendly. The method has several advantages such as higher yields, shorter reaction times, easy set-up and easy work-up, reusable catalyst and elimination of organic solvent and oxidizing agent. This is a good example of one of the essential principles of green chemistry, that is, to minimize the components using only substrate and catalyst and no waste.

3.2.3.3 Green synthesis of a solid phase exchanger

Anchoring the active site to a solid support in a polymeric network provides an immobilized active surface capable of forming coordinate bond through donor atoms, thus separating specific metals from

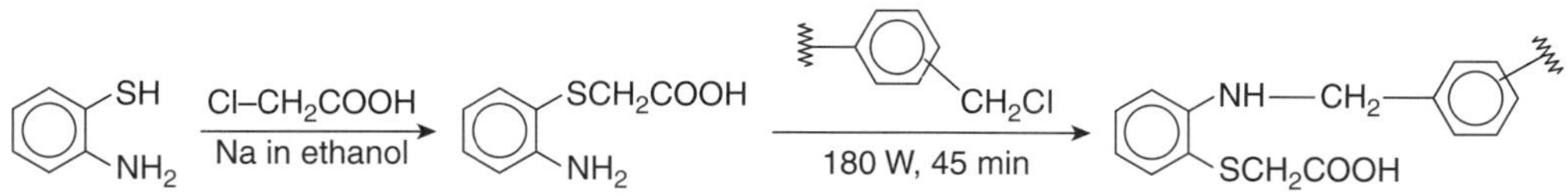

Scheme 3.5 *Synthesis of functionalized chelating resin containing 2-aminothiophenyl S-acetic acid.*

complex mixtures [25]. The use of microwave-assisted treatment in organic synthesis is an attractive area of research nowadays, because most of the reactions are completed within a few minutes instead of classical prolonged refluxing. The use of this eco-friendly technique for functionalization of polymer matrix is scant.

***Green laboratory proposal* [26]** Functionalized resins are used to extract analytes in situ and store for analysis at a secondary location; thus providing a greener way of analyte preconcentration. Air-dried chloromethylated polystyrene copolymer containing 3% divinylbenzene (DVB) was used as the starting material. The copolymer bead (50 mg, 100–150 mesh) was treated with 2-amiothiophenyl S-acetic acid (35 mg) in the presence of Na_2CO_3 (40 mg) in N, N-dimethyl formamide (15 ml) taken in a PTFE reactor (Figure 3.2, 115 ml internal volume, 1 cm wall thickness, hermetic screw cap). The reactor was then placed inside a microwave oven (180 W) for 45 min. The functionalized chelating resin containing 2-aminothiophenyl S-acetic acid thus obtained is stable in 1–12 M HCl/NaOH for at least 7 d and was characterized by IR spectral study (Scheme 3.5).

Comments The special feature of this method of functionalized resin synthesis is the shortening of time to 45 min against 20 h using traditional method of refluxing. The exchangers obtained by two methods are identical in nature that obtained by traditional method. It is noticed that the resin containing 2-aminothiophenyl S-acetic acid moiety is selective for soft metal ions. This high selectivity may be due to the soft basic sulfur atom which plays the key role in binding and may be followed by chelation via nitrogen atoms.

3.2.4 Green spectroscopy

Spectroscopy is probably the field of analytical chemistry which comparably involves the maximum opportunities for green technologies. The wide electromagnetic spectrum offers opportunities for the determination of trace elements, trace organic compounds and other components in various types of samples and the versatility of the spectroscopic techniques can be applied to obtain maximum information about sample composition with minimum sample treatments. The state-of-the-art of green spectroscopy, as a sustainable and eco-friendly alternative to the classical spectrometric methods of analysis, has been established from the evaluation of the scientific literature published about this topic [27].

3.2.4.1 Green UV/V is spectrometric determination of bismuth by use of an internal standard

Among different quantification methods, internal standard (IS) is specially well suited if the amount of sample reaching analytical cell or the instrument response varies slightly from run to run for various reasons that are difficult to control. As general rule, IS is a chemical species other than analyte, which is added in a constant and known amount to all samples, calibration and blank solutions. Analytical signals for analyte ($S_{analyte}$) and IS (S_{IS}) are then registered and the ratios of analyte response versus internal standard response ($S_{analyte}/S_{IS}$) are used for calibration and quantification. If the amount of IS is kept constant in all solutions, $S_{analyte}/S_{is}$ value remains directly proportional to analyte concentration. The IS

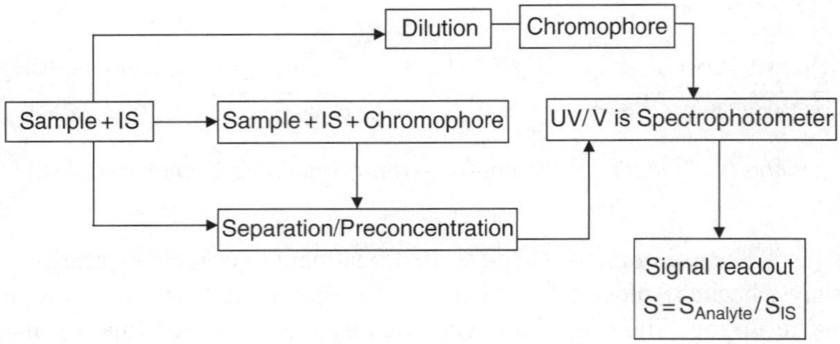

Figure 3.9 *General scheme of UV/V is spectrophotometric procedure with the application of IS.*

Table 3.2 *Comparison of the characteristics of classical batch procedure versus micro-scale procedure for extractive-spectrophotometric determination of bismuth.*

Procedure	Phase	Final volume	Detection Limit	% Extraction
Batch Procedure	Aqueous Phase: sample + H_2SO_4 + 5% KI + 5% ascorbic acid + 1% BTAB	25 ml		
	Organic Phase: chloroform	10 ml	0.5 μg ml^{-1}	99.5 ± 0.6%
Micro-scale Procedure	Aqueous Phase: sample + H_2SO_4 + 5% KI + 5% ascorbic acid + 1% BTAB + IS	800 μl		
	Organic Phase: chloroform	400 μl (chloroform only)	0.6 μg ml^{-1}	99.8 ± 0.5%

should match closely the physicochemical characteristics of analyte(s) which would assure similar behavior of these species during sample preparation and signal registration. The general scheme of microscale UV/Vis spectrophotometric procedure with the application of IS method is shown schematically in Figure 3.9 [28].

***Green laboratory proposal* [29]** In the pharmaceutical industry quantification of the active compound is an issue. For example, different batch spectrophotometric procedures are available for pharmaceuticals containing bismuth compounds. Among them, the procedure based on the formation of tetraiodobismuthate (III) in aqueous sulfuric acid medium, ion-pair formation with benzyltributylammonium cation (BTAB) and extraction to chloroform has been selected and downscaled. Nile blue was selected as the IS. The spectral bands of bismuth complex (λ_{max} = 491 nm) and IS (λ_{max} = 623 nm) did not present any overlapping, so the analytical signal was measured from zero-order spectra ($S = A_{491nm}/A_{623nm}$). Another important selection criterion was similar extraction yield of analyte and IS to chloroform. The effect of reagents and ion-pairing agent concentrations on the IS and bismuth signal was systematically studied, which enable extraction yields 99.5±0.6% and 99.8±0.5% respectively for the two species. In Table 3.2, the experimental parameters and some analytical figures of merit evaluated for classical batch procedure and micro-assay by the present method are compared.

Comments According to green chemistry principles, the micro procedure offers reduced consumption of reagents and consequently lower waste generation. One special feature of the method is that there is no need

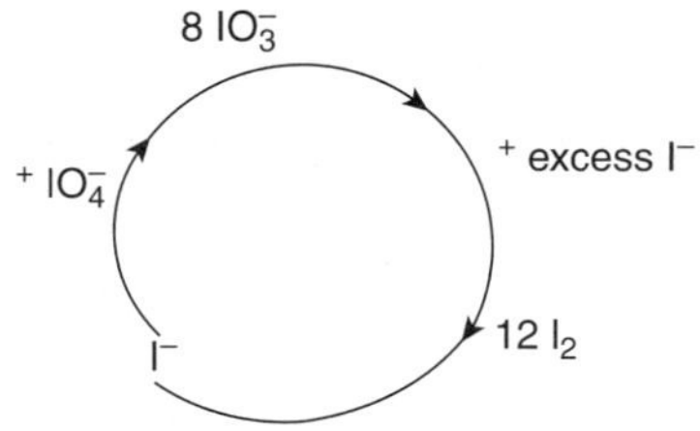

$$2\,I^- + IO_4^- + 2H^+ \rightarrow I_2 + IO_3^- + H_2O$$

$$I_2 + 5\,IO_4^- + H_2O \rightarrow 7\,IO_3^- + 2H^+$$

Figure 3.10 *Amplification reaction of iodide.*

Table 3.3 *Analytical characteristics for the determination of iodide. Reproduced by permission of The Royal Society of Chemistry © 1994.*

Analytical Characteristics	Microwave-assisted oxidation	Classical oxidation
R	0.9997	0.9900
DR (μg ml^{-1})	0.25–10.0	0.75–8.5
LOD (μg ml^{-1})	0.072	0.077
RSD (%)	1.9	4.3

R = regression coefficient; DR = dynamic range; LOD = limit of detection for a 99.86% probability level (k =3); RSD = relative standard deviation for 6 determinations of a 3 μg ml^{-1} iodide.

to use volumetric vessels; small volumes of samples and reagents can be handled, and the unique precise operation required is the measurement of sample and IS volumes. The most important feature of spectrophotometric micro-assays with the use of IS is their low environmental impact, operational simplicity and low cost, without sacrificing the accuracy and precision of the analytical results.

3.2.4.2 Amplification of trace iodide before spectrophotometric determination

Conversion of the iodine to iodate gives a six-fold increase in sensitively, although in the determination of iodide, after a previous oxidation to iodate by periodate and reaction of the iodate formed with an excess of iodide, an amplification of 24-fold can be obtained. The strategy has been shown in Figure 3.10. Oxidation of iodide by periodate to produce iodate is slow and quantitative oxidation of iodide is achieved only after heating in a boiling water-bath for 2 h in presence of 300 fold molar excess of periodate [30]. Also the consequent very strong solution of molybdate (25% *m/v)* presents serious difficulties relating to laboratory wastes, high blank value and poor precision in the determination of iodine.

***Green laboratory proposal* [31]** The use of microwave energy for conducting organic and inorganic reactions has been widely accepted. A simple amplification experiment can be described here for green laboratory. 1.5 ml of 0.013 M periodate solution were added to 1 ml iodide solution (75 μg ml^{-1}) taken in a 115 ml internal volume PTFE reactor with 1 cm wall thickness and hermetic screw cap (Figure 3.2). The mixture was allowed to stand at room temperature for 5 min then the reactor was introduced inside a microwave oven with 2450 MHz frequency in which it was radiated for 4 min at 520 W. To the reaction mixture taken in a volumetric flask, the following reagent solutions were added in the given order: 0.5 ml of 2.5% (*m/v*) ammonium molybdate, 2 ml acetate buffer (pH 2.2) and 1 ml same iodide solution. The resulting solution was mixed thoroughly and allowed to stand for 2 min before extraction with 18 ml (in three fractions) of chloroform. The combined organic phase was than made up to 25 ml in a volumetric flask and the absorbance measurements were carried out at 510 nm against a reagent blank. Table 3.3 shows the results of the calibrations obtained for the spectrophotometric measurement of standard iodide solutions, amplified after microwave-assisted oxidation (calibration line: 0.0040+0.0666 $\mu g\,ml^{-1}$) and classical oxidation (0.0297+0.0527 $\mu g\,ml^{-1}$) by using periodate along with the corresponding analytical characteristics found following both procedures. It can be seen that the microwave-assisted procedure provides the better precision and low blank value.

Comments The microwave-assisted procedure described here is a fast and reproducible method for the quantitative oxidation of iodide, which can be applied as a previous step requiring periodate oxidation. As compared to the classical method of periodate oxidation, the use of microwave energy for activation of amplification reaction has made the reaction not only shortening of time (4 min instead of 2 h) but also consumption of less energy, closed reactor for amplification (24-fold) was effective for avoiding any sample loss.

3.2.4.3 *Use of naturally occurring reagents for extraction - spectrophotometric determination of iron*

Today flow injection analysis (FIA) has become a very powerful analytical tool. By introducing FIA, not only several-fold increase in the sensitivity could be achieved by coupling on-line between preconcentration and detection techniques, but also eco-friendly and low-cost procedures can be improved based on the on-line separation followed by analyte determination. The potential of flow systems to develop greener analytical procedures, by replacement of toxic reagents, waste minimization, and in-line treatment and recycling has been emphasized [32]. Gaining popularity, the Green Analytical Chemistry aims to explore the use of alternative reagents or alternative synthetic methods that minimize the use of toxic chemicals. Thus a natural guava (*Psidium guajava L.*, Myrtaceae family) extract has been investigated as an alternative natural indicator for iron quantification by the flow injection (FI) technique.

***Green laboratory proposal* [33]** Fresh guava leaves (10 g) were treated in 150 ml water (pH 7), diluted hydrochloric acid (pH 4.8) or sodium acetate buffer (pH 4.8) in a blending machine for 5 min. Then the suspension was filtered through filter paper (Whatman no. 1) and the filtrate was kept at room temperature for further use. It was found that guava leaf extract is stable for at least 4 h, which is sufficient for a continuous experiment using FI. Thus the extract was prepared daily before use.

Both Fe (II), Fe (III) solutions (10 and 100 ppm each) were prepared in water, HCl solution and acetate buffer. Guava leaf extracts in the three media were mixed with iron solutions prepared in the matching medium at the ratio of 1:1 (*v/v*) for example, 10 ml extract:10 ml iron solution. Guava leaf extract without iron solution was used as the blank solution. Mixtures of the guava leaf extract and iron solution were scanned for absorption spectra, measured against each medium in the visible region using a UV/Visible spectrophotometer. Finally, it was set at 570 nm and connected to a computer to record the FIA program. The FI manifold with two lines, where the reagent was injected into the stream of buffer that was later mixed with the stream of sample solution, is depicted is Figure 3.11.

The typical analytical characteristics of the reverse FI method are as follows. It has much better sensitivity (1 ppm) and low blank value in comparison to normal FI system. It has the linear calibration with $R^2 = 0.9997$ in the working range 1–10 ppm iron. Percent recoveries in tap water spiked with Fe(III) was 95–97%. Sample throughput was 20 injections/h.

Comments The striking aspect of this green method is that no sample treatment is needed for iron concentrations at or below 30 ppm since Fe (III) and Fe (II) absorbances are similar, probably due to complete reduction of Fe (III) by some chemicals like tannin naturally present in the guava leaves. Higher concentrations, however, might be measured by adding a reducing agent such as hydroxylamine to convert Fe (III) to Fe (II).

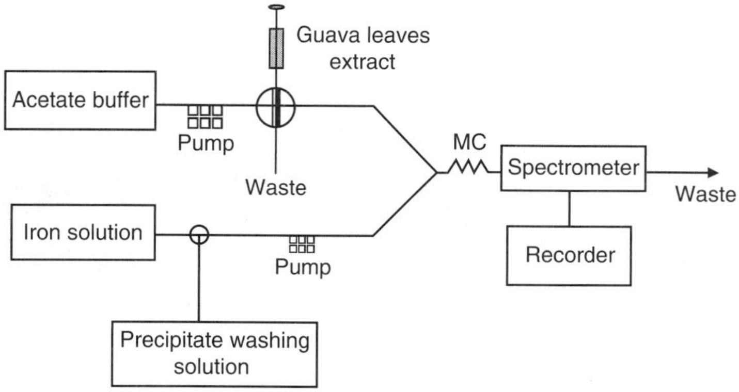

Figure 3.11 *Flow injection manifold used for iron determination. Reproduced from [33] with permission from Elsevier, © 2005.*

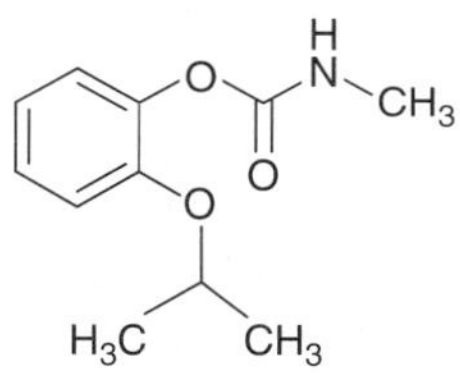

Scheme 3.6 *Propoxur.*

3.2.4.4 A clean spectrometric method for the determination of propoxur

One of the most important carbamates which is widely used in agriculture is propoxur [2-(1-methylethoxy) phenol methylcarbamate] (Scheme 3.6), but in addition to the insecticidal properties exhibited by this compound, other biological effects have been reported.

Hence analytical methods have been developed for the determination of propoxur and its metabolites in different matrices of the environment. Spectrophotometric method has been developed for the determination of propoxur based on alkaline hydrolysis to yield 2-isopropoxyphenol followed by coupling with different chemical reagents such as p-aminophenol (PAP). But one of the problems of such spectrophotometric methods is that not only the compound to be determined is toxic but also the reagents employed have also toxic effects. So photo-assisted catalytic procedures using TiO_2 illuminated with UV radiation appeared to be very promising, efficient and economically viable that can be used to detoxify organic molecules. Flow injection is widely used in analytical determinations because it provides better flexibility for carrying out enormous chemical processes including on-line hydrolysis, derivatization, photochemical reactions, and detoxification of organic molecules [34].

***Green laboratory proposal* [35]** In actual practice 100 µl of alkaline solutions of propoxur standards in 0.33 M NaOH or samples previously made alkaline, were directly injected into a carrier solution of 0.33 M NaOH, at the same time a flow of a solution of 545 µg ml^{-1} PAP and a 0.004 M KIO_4 solution were continuously fed in. By using the 5-channel manifold shown in Figure 3.12, propoxur was determined spectrophotometrically at 600 nm. After the measurement step, the waste was diluted with water and irradiated at 254 nm in order to

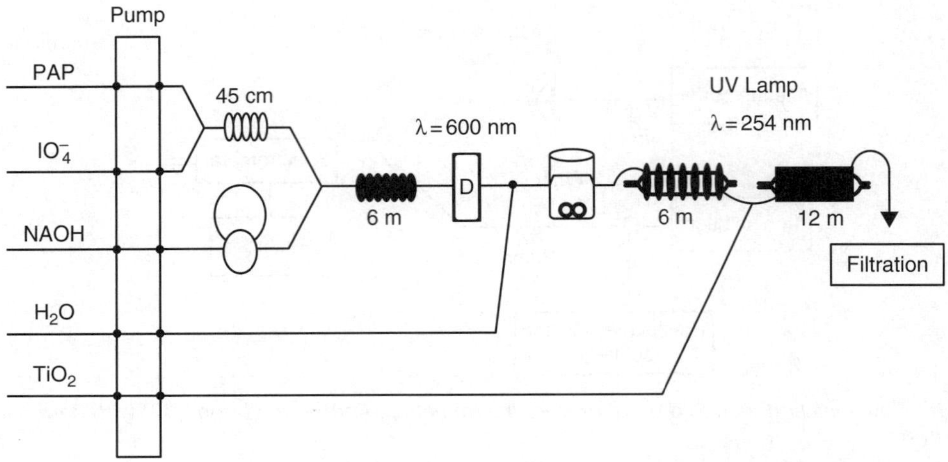

Figure 3.12 *Manifolds employed for both spectrophotometric determination of propoxur and degradation of the waste. Reproduced from [35] with permission from Elsevier, © 1995.*

photodegrade the propoxur. After this treatment, the waste was merged with a TiO_2 slurry flow (0.5 g l^{-1} prepared in 2.4 M HCl) and irradiated in a second irradiation coil at 254 nm. At the end, the waste was filtered in order to recover TiO_2, which could be used several times. Complete degradation of propoxur can be achieved in less than 100 min.

Comments The clean flow injection-spectrophotometric procedure for the determination of propoxur in water includes an on-line methodology as well as on-line detoxification of flowing solution after the analytical measurement has been over. Complete mineralization of solutions of PAP, propoxur, and the reaction product between propoxur and PAP was achieved without affecting the low limit of detection (0.1 μg ml^{-1}) and good injection frequency (80 h^{-1}). Thus the method demonstrates the development of greener concepts in FI methodology which will minimize the effect of analytical measurements on the environment.

3.2.4.5 Green strategies for the spectrophotometric determination of phenol

The development of flow injection (FI) now commonly applied in analytical chemistry, offers immense possibilities for automation of environmental analysis. Large quantities of phenolic compounds are produced as raw materials, resulting environmental contamination. Phenols can be present in aquatic environments and the monitoring can be done by the operational determination of total phenols. The method of the American Public Health Association (APHA), requires a tedious and time-consuming extraction of phenols in chloroform from water sample [36]. So eco-friendly procedures with improved sensitivity have been reported.

Green laboratory proposal **[37]** A greener and sensitive flow-based procedure has been presented for fast determination of phenols in waters by oxidative coupling to 4-aminoantipyrine (4-AAP) in the presence of potassium hexacyanoferrate (III) (shown in Scheme 3.7).

This method utilized a 100 cm long pathlength optical cell that, together with the multicommutation analysis, increased the method sensitivity enough to obviate the need for preconcentration in chloroform.

The flow diagram of the system is shown in Figure 3.13. In the initial status, all valves are switched off and the carrier flows through the analytical path while other solutions recycle to their vessels. Sample and reagent

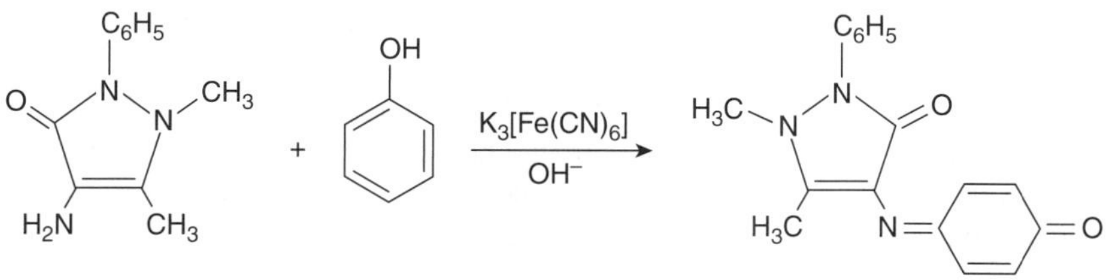

Scheme 3.7 *Oxidative coupling of phenol with 4-AAP in alkaline medium.*

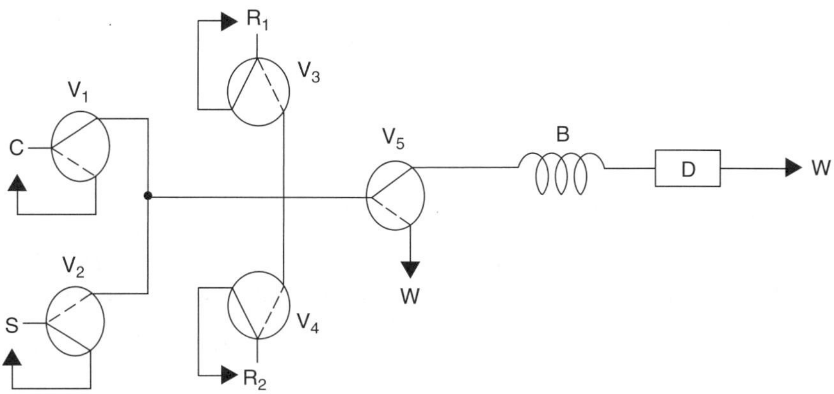

Figure 3.13 *Flow diagram of the system for determination of phenols [V_i - three-way solenoid valves; B - reaction coil (80 cm); D- long pathlength flow cell (100 cm optical path); C - water carrier (5.4 ml min^{-1}); S- sample (6.0 ml min^{-1}), R_1- 0.1% (m/v) 4-AAP buffered at pH 10.0 (0.8 ml min^{-1}); R_2 -0.2% (m/v) $K_3[Fe(CN)_6]$ (0.6 ml min^{-1}); W - waste. Dashed lines represent the flow paths when the values are switched on]. Reproduced from [37] with permission from Elsevier, © 2004.*

solutions are introduced by simultaneously switching the valves V_1-V_4 on for 7 s. Then all values are switched off and data acquisition is started. For sample replacing, valves V_2 and V_5 are simultaneously switched on for 10 s and the path between these valves is washed with the sample. Spectrophotometric measurements are carried out at 500 nm. Sensitivity was 80-fold higher than that achieved with a 1 cm flow cell, making feasible the determination of phenols in the range 10–100 μg l^{-1} with a detection limit of 1 μg l^{-1}.

The sampling rate and the coefficient of variation were estimated to be 90 determinations per hour and 0.6% (n = 10) respectively. The multicommunication approach allowed a 200-fold reduction of the reagent consumption in comparison to reference batch method. Recoveries within 93.3 and 106% were achieved for determination of phenol in natural and wastewater sample.

Comments The improved flow system yielded unique analytical features for phenols determination, obtaining the sensitivity which can be achievable by the procedures that exploit analyte preconcentration. Reagent consumption and toxic waste production are significantly reduced in comparison to both flow-based and batch methods. Also, the use of toxic organic solvents is avoided. Hence the procedure is inherently greener than the earlier reported spectrophotometric methods and can produce highly beneficial effects by reducing both costs and waste generation.

3.3 The place of Green Analytical Chemistry in the future of our laboratories

Making our analytical methodologies greener is something of interest not only for the future but also for the present [38]. The experiments presented above are green schemes for already existing conventional materials. The green analytical experiments are concerned with the design, development and implementation of products and chemical processes that include elimination of toxic organic solvents/reagents, solventless synthesis, less waste generation, reuse, facile methods of preparation at low temperature range, shorter reaction time, high yields and so on. Use of on-line determination helps the students to learn various instrumentation methods of analysis for example, UV/Vis and IR spectroscopy, AAS to name a few. Green chemistry experiments are introduced not to drastically replace the conventional ones; rather they are considered complementary to the existing protocols. This not only provides a wider view of various techniques but also imbibes inquest in innovative minds for future development and growth of the subject in general with due emphasis to green chemistry context.

References

1. Anastas, P.T. (1999) Green Chemistry and the role of analytical methodology development. *Crit. Rev. Anal. Chem.*, **29**, 167–175.
2. Lopez-Avila, V. (1999) Sample preparation for environmental analysis. *Crit. Rev. Anal. Chem.*, **29**, 195–230.
3. Reed, S.M. and Hutchison, J.E. (2000) Green Chemistry in the organic teaching laboratory: An environmentally benign synthesis of adipic acid, *J. Chun. Ed.*, **77**, 1627–1628.
4. He, Y.; Tang, L.; Wu, X.; Hou, X.D. and Lee, Y.I. (2007) Spectroscopy: The best way toward Green Chemistry?, *Appl. Spectrosc. Rev.*, **42**, 119–138.
5. Santos Junior, D.; Krug, F.J.; de Godoy Percira, M. and Korn, M. (2006) Currents on ultrasound-assisted extraction for sample preparation and spectroscopic analytes determination. *Appl. Spetrosc. Rev.*, **41**, 305–321.
6. Minami, H. ;Hanjyo, T. and Atsuya, I. (1996) A new solid-liquid extraction sampling technique for direct determination of trace elements in biological materials by graphite furnace atomic absorption spectrometry, *Spectrochim. Acta, Part B*, **51**, 211–220.
7. Khalaf, K.D.; Morales-Rubio, A.and de la Guardia, M. (1993) Rapid microwave assisted hydrolysis of formetanate, *Anal. Chim. Acta*, **281**, 249–257.
8. Das, A.K.; Chakraborty, R.; de la Guardia, M.; Cervera, M.L. and Goswami, D. (2001) ICP-MS multielement determination in fly ash after microwave-assisted digestion of samples. *Talanta*, **54**, 975–981.
9. Luque de Castro, M.D.; Jimenez-Carmona, M.M. and Fernandez-Perez, V. (1998) *Spanish Patent*, No. P9802299.
10. Morales-Munoz, S.; Luque-Garcia, J.L. and Luque de Castro, M.D. (2004) Demetalization of oils resulting from recycled tires by liquid-liquid extraction using modified superheated water – An environmentally friendly approach, *Talanta*, **63**, 391–396.
11. Keith, L.H.; Gron, L.U. and Young, J.L. (2007) Green analytical methodologies, *Chem. Rev.*, **107**, 2695–2708.
12. Chen, J.; Spear, S.K.; Huddleston, J.G.; Holbrey, J.D. and Rogers, R.D. (2004) Application of polyethylene glycol-based aqueous biphasic reactive extraction to the catalytic oxidation of cycle olefins, *J. Chromatogr. B*, **807**, 145–149.
13. Roy, K. and Lahiri, S. (2009) Extraction of Hg (I), Hg (II) and methylmercury using polyethylene glycol based aqueous biphasic system, *Appl. Radiation Isotopes*, **67**, 1781–1784.
14. Peter, M.G. (2002) *Chitin and Chitosan from Animal Sources*, in *Biopolymers (Polysaccharides II)*, Vol. 2 (eds S. De Bates, P.J. Vandamme and A. Steinbuchel), Wiley-VCH Verlag, Weinheim, Ch.15.
15. Dutta, S.; Mohapatra, P.K.; Ramnani, S.P.; Sabharwal, S.; Das, A.K. and Manchanda, V.K. (2008) Use of chitosan derivatives as solid phase extractors for metal ions, *Desalination*, **232**, 234–242.
16. Das, A.K.; Banerjee, D.; Cervera, M.L. and de la Guardia, M. (2006) Recent developments in vanadium speciation, *Current Topics Anal. Chem.*, **6**, 41–52.
17. Banerjee, D.; Mondal, B.C.; Das, D. and Das, A.K. (2003) Use of imidazole 4,5-dicarboxylic acid resin in vanadium speciation, *Microchim. Acta*, **141**, 107–113.

18. Chemat, F.; Smadja, J. and Lucchesi, M.E. (2003) *European Patent Demand*, 03001183.7.
19. Lucchesi, M.E.; Chemat, F. and Smadja, J. (2004) Solvent-free microwave extraction of essential oil from aromatic herbs: comparison with conventional hydro-distillation, *J. Chromatgr. A*, **1043**, 323–327.
20. Rocha, F.R.P.; Teixeira, L.S.G. and Nobrega, J.A. (2009) Green strategies in trace analysis: A glimpse of simple alternatives for sample pretreatment and analyte determination, *Spectrosc. Lett.*, **42**, 418–429.
21. Furniss, B.S.; Hannaford, A.J.; Rogers, V.; Smith, P.W.G. and Tatchell A.R. (Revisors), (1978) *Vogel's Textbook of Practical Organic Chemistry Including Qualitative Organic Analysis*, 4th edn, English Language Book Society, London, p. 621.
22. Dugar, A. and Ameta, S.C. (2010) A green chemical approach for nitration of aromatic compounds, *J. Indian Chem. Soc.*, **87**, 507–510.
23. Smith, K. (1992) *Solid Supports and Catalysts in Organic Chemistry*, Ellis Horwood, Chichester, pp. 130–170.
24. Balalaie, S.; Golizeh, M. and Hashtroudi, M.S. (2000) Clean oxidation of benzions on zeolite A using microwave irradiation under solvent-free conditions, *Green Chem.*, **2**, 277–278.
25. Mondal, B.C.; Das, D. and Das, A.K. (2002) Synthesis and characterization of a new resin functionalized with 2-naphthol-3,6-disulfonic acid and its application for the speciation of chromium in natural water, *Talanta*, **56**, 145–152.
26. Mondal, B.C. and Das, A.K. (2002) Microwave-assisted synthesis of a new chelating resin containing 2-aminothiophenyl S-acetic acid and its application to the determination of lead, *Reactive Functional Polym.*, **53**, 45–52.
27. Armenta, S. and de la Guardia, M. (2009) Green spectroscopy: A Scientometric picture, *Spectrosc. Lett.*, **42**, 277–283.
28. Wrobel, K. and Wrobel, K. (2009) Micro-scale UV/Vis spectrometric batch procedures by use of an internal standard –A Green Chemistry approach, *Spectrosc. Lett.*, **42**, 327–333.
29. Wrobel, K.; Wrobel, K. and Urbina, E.M.C. (2000) Application of internal standard for micro-extraction spectrophotometric determination of bismuth in pharmaceutical formulations, *Microchim. Acta*, **135**, 87–90.
30. Belcher, R.; Hamya, J.W. and Townshend, A. (1970) Titrimetric determination of iodide after twenty four-fold amplification, *Anal. Chim. Acta*, **49**, 570–572.
31. Das, A.K.; Cervera, M.L. and de la Guardia, M. (1994) Rapid microwave-assisted amplification for the spectrophotometric determination of iodine, *Anal. Proc. Including Anal. Comm.*, **31**, 345–347.
32. Rocha, F.R.P.; Nobrega, J.A. and Fatibello-Filho, O. (2001) Flow analysis strategies to greener analytical chemistry. An overview, *Green Chem.*, **3**, 216–220.
33. Settleeworrarit, T.; Hartwell, S.K.; Lapanatnoppakhun, S.; Jakmunee, J.; Christian, G.D. and Grudpan, K. (2005) Exploiting guava leaf extract as an alternative natural reagent for flow injection determination of iron, *Talanta*, **68**, 262–267.
34. Pramauro, E. (1990) *Los Pesticidas y el Medioambiente*, University of Valencia, Valencia.
35. de la Guardia, M.; Khalaf, K.D.; Carbonell, V. and Morales-Rubio, A. (1995) Clean analytical method for the determination of propoxur, *Anal. Chim. Acta*, **308**, 462–468.
36. Clesceri, L.S.; Greenberg, A.E. and Eaton, A.D. (1998) *Standard Methods for the Examination of Water and Wastewater*, 20th edn, American Public Health Association, Washington DC, pp. 5–42.
37. Lupetti, K.O.; Rocha, F.R.P. and Fatibello-Filho, O. (2004) An improved flow system for phenols determination exploiting multicommutation and long pathlength spectrophotometry, *Talanta*, **62**, 463–467.
38. Das, A.K. and de la Guardia, M. (2009) Greener the spectroscopy, *Spectrosc. Lett.*, **42**, 275–276.

4

Publishing in Green Analytical Chemistry

Salvador Garrigues and Miguel de la Guardia

Department of Analytical Chemistry, University of Valencia, Valencia, Spain

Scientific literature continues to be the main diffusion channel for new developments and knowledge in analytical chemistry and thus, the interest of the community on a subject can be evaluated from the number of published papers, related patents and communications in international congresses which year after year, include the characteristic keywords of the topic. So, if the analytical chemistry community is interested in Green Analytical Chemistry it can be easily verified by the evolution of the scientific literature and thus, one of the objectives of those who have been involved in this task must be the diffusion of the ideas and developments on greening the analytical methods. This task then must be done together with the efforts indicated in previous chapters about the integration of the Green Analytical Chemistry ideas and strategies in the teaching activities.

On considering the classical diffusion channels of the new developments in analytical chemistry (see Figure 4.1) it must be accepted that scientific publication in international journals is the most common method of diffusion and in this sense, original papers and review articles are the most commonly cited sources of knowledge. However, other less frequently used communication methods, like books, presentations in international meetings and patents must be also considered in order to have a complete description of the state of a topic.

Regarding the time elapsed from the development of new ideas and results and their availability for the scientific community, presentation in international congresses, MSc and PhD memories, together with national and international patents, are the faster communication methods. However, it must be considered that patents involve a series of filters and costs and these facts limit the use of this method for the communication of new results as well as providing a delay in being easily available to the community. On the other hand congress presentations are limited by the dates and geographical characteristics of the meetings and the availability of these data is terribly conditioned as it is for academic documents. So publication in international journals followed by writing books is mandatory if we would like to spread our results to the whole scientific

Handbook of Green Analytical Chemistry, First Edition. Edited by Miguel de la Guardia and Salvador Garrigues.

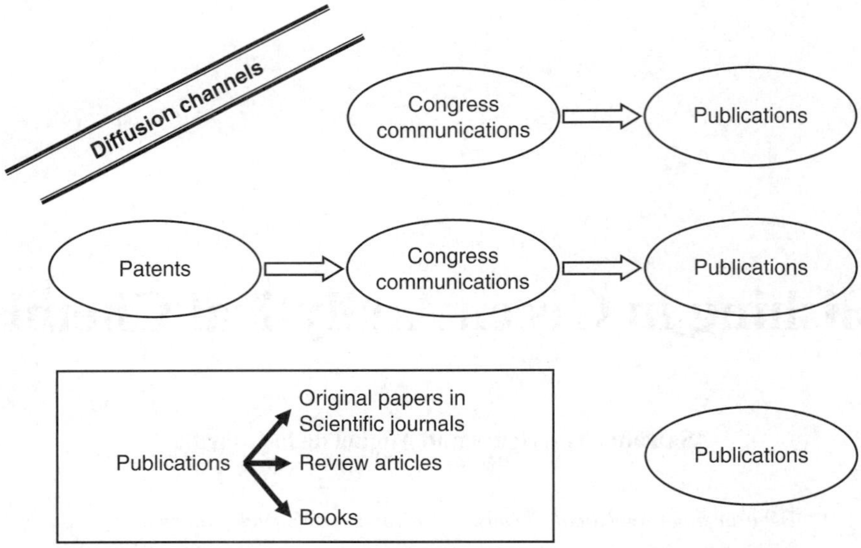

Figure 4.1 Diffusion channels of the new developments in Analytical Chemistry.

community and in this task the available scientific databases are the main tools to access to such knowledge. However, nowadays we must consider also internet channels as a fast and easily accessible way which has an increased importance in the diffusion of scientific results.

4.1 A bibliometric study of the literature in Green Analytical Chemistry

We have evaluated the use of green analytical terms in the scientific literature using the databases included in the Thomson Reuters Web of Knowledge (previously named ISI Web of Knowledge).

As in Figure 4.2, in the period from 1970 until May 2011 we have found 87 published papers which include the main terms *Green Analytical Chemistry*, *Green Analytical Methods* or, more generically, *green methods*. The aforementioned terms were introduced in 1999, 2004 and 2007, respectively.

On the other hand, the inclusion of terms, such as *environmentally friendly methods*, *clean methods*, *sustainable methods* and the associated terms linked to analytical chemistry increases the number of papers related to greening analytical procedures to 149, thus evidencing that, especially in former times from 1995 till 2003, there was a general agreement on the terminology to be used for describing efforts to reduce the deleterious environmental side effects of the development and application of analytical procedures. So, we could conclude that it is absolutely vital to homogenize the terminology to be used in titles, abstracts, and keywords, if we would like the efforts of the analytical community in this field to be visible.

Considering the journals in which the green analytical papers were published (see Table 4.1) it can be appreciated that they correspond to journals devoted to general analytical chemistry (*Analytica Chimica Acta*, *Talanta*, *TrAC-Trends in Analytical Chemistry* and *Analytical and Bioanalytical Chemistry*), chromatography (*Journal of Chromatography A* and *Chromatographia*), spectroscopy (*Spectroscopy Letters*) in addition to electronalytical journals, national journals, education, food and environmental journals.

The most productive authors in the field have been listed in Table 4.2, indicating the number of published papers in the considered period of time which include the selected extended keywords and as it can be seen, Spanish and Brazilian together with Polish authors have been in the most involved teams up until this moment.

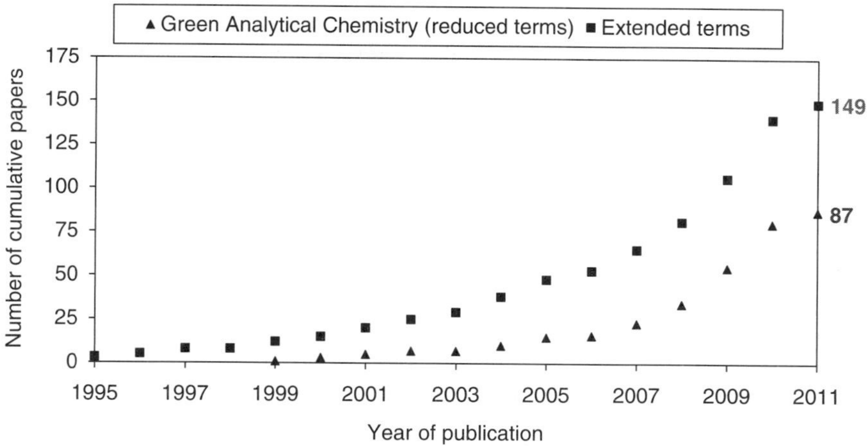

Notes:

Reduced terms: 'green analytical chemistry', 'green analytical method', 'green methods'
Extended terms: 'green analytical chemistry', 'green analytical method', 'green methods', 'environmentally friendly method', 'clean method*', 'clean analytical method*', 'sustainable method*', 'sustainable analytical method*', 'environmentally friendly method*', 'environmentally friendly analytical chemistry'

Figure 4.2 *Evolution of the scientific literature concerning Green Analytical Chemistry (for details about database and keywords employed see text).*

So, from data reported in Tables 4.1 and 4.2 it can be concluded that editors and many authors are seriously involved in the integration of environmental considerations in their publications and it has created the foundation for the acceptance of Green Analytical Chemistry as a hot topic of current research and we hope that the size of the literature in this field will grow exponentially in the next decade.

On evaluating the subject area of Green Analytical Chemistry papers published to current times (see Figure 4.3) it is clear that general analytical chemistry, biochemical research methods, followed by spectroscopy are the main areas concerned, spreading the applications to different fields and that clearly shows that there is common interest in greening the analytical procedures from both research and application fields.

4.2 Milestones of the literature on Green Analytical Chemistry

As has been explained in Chapter 1 the greenest hits in the field of analytical chemistry come from: (1) the reduction of the size of sample and measurement units through miniaturization, (2) the automation of procedures, (3) the reduction of energy consumption obtained by moving from hard procedures involving the use of high temperatures during long periods of time, to soft ones by incorporating low or focussed energy; in this case the use of sonication at room temperature using probes or the general use of microwave ovens which involve short heating periods, are now common green practices, and (4) the development of direct analysis procedures without any sample treatment based on both, remote sensing and non-invasive methodologies. The developments in portable instruments and in chemometrics have been the key factors for the evolution of the last considered aspect.

We have summarized the milestones previously (see Figure 1.4, Chapter 1) which, in our opinion, have created the tools for greening the analytical methods but in many of these cases the origin of

Table 4.1 *Analytical journals which have published Green Analytical works.*

Journal Title	Papers N°
Analytica Chimica Acta	18
Talanta	16
Journal of Chromatography A	13
TrAC-Trends in Analytical Chemistry	10
Analytical and Bioanalytical Chemistry	9
Spectroscopy Letters	9
Chromatographia	6
Journal of the Brazilian Chemical Society	5
Analyst	3
Polish Journal of Environmental Studies	3
Quimica Nova	3
American Laboratory	2
Analytical Letters	2
Analytical Methods	2
Chemosphere	2
Critical Reviews in Analytical Chemistry	2
Electroanalysis	2
International Journal of Electrochemical Science	2
Journal of Analytical Atomic Spectrometry	2
Journal of Chemical Education	2
Journal of the Food Hygienic Society of Japan	2
Progress in Environmental Science and Technology, Vol II, Pts A and B	2
Spectrochimica Acta Part B-Atomic Spectroscopy	2
Abstracts of Papers of the American Chemical Society	1
Accounts of Chemical Research	1
ACS Symposium Series	1
Analusis	1
Analytical Chemistry	1
Analytical Sciences	1
Annali di Chimica	1
Applied Spectroscopy Reviews	1
Chemical Society Reviews	1
Chinese Journal of Analytical Chemistry	1
Clean Solvents	1
Ecological Chemistry and Engineering S-Chemia i Inzynieria Ekologiczna S	1
Electrochemistry Communications	1
Forest Pathology	1
Fresenius Journal of Analytical Chemistry	1
Green Chemistry	1
Journal of Analytical and Applied Pyrolysis	1
Journal of AOAC International	1
Journal of Capillary Electrophoresis	1
Journal of Electroanalytical Chemistry	1
Journal of Environmental Monitoring	1
Journal of Environmental Science and Health Part A-Toxic/Hazardous Substances & Environmental Engineering	1

(continued)

Table 4.1 (*continued*)

Journal Title	Papers N°
Journal of Near Infrared Spectroscopy	1
Journal of Pharmacy and Pharmacology	1
Journal of Separation Science	1
Microchemical Journal	1
Pure and Applied Chemistry	1
Restaurator-International Journal for the Preservation of Library and Archival Material	1
Rivista Italiana delle Sostanze Grasse	1
Soil Biology & Biochemistry	1
Spectrochimica Acta Part A-Molecular and Biomolecular Spectroscopy	1

Table 4.2 *Most productive authors in the field of Green Analytical Chemistry (three or more papers).*

Author	Paper N°	Percentage
M. de la Guardia	21	14.1%
S. Garrigues	10	6.7%
F.R.P. Rocha	10	6.7%
J. Namiesnik	9	6.0%
S. Armenta	8	5.4%
A. Morales-Rubio	6	4.0%
M.D. Luque de Castro	5	3.4%
H.R. Pezza	4	2.7%
L. Pezza	4	2.7%
G. Quintas	4	2.7%
C. Bendicho	3	2.0%
D. Fabbri	3	2.0%
O. Fatibello	3	2.0%
X.D. Hou	3	2.0%
I. Lavilla	3	2.0%
J.L.F. Costa-Lima	3	2.0%
J.L. Luque-Garcia	3	2.0%
W.R. Melchert	3	2.0%
D. Nematollahi	3	2.0%
J.A. Nobrega	3	2.0%
M.L.M.F.S. Saraiva	3	2.0%
J. Wang	3	2.0%
X. Wang	3	2.0%
W. Wardencki	3	2.0%
P.L. Weinert	3	2.0%
L.L. Zhang	3	2.0%
R.S. Zhao	3	2.0%

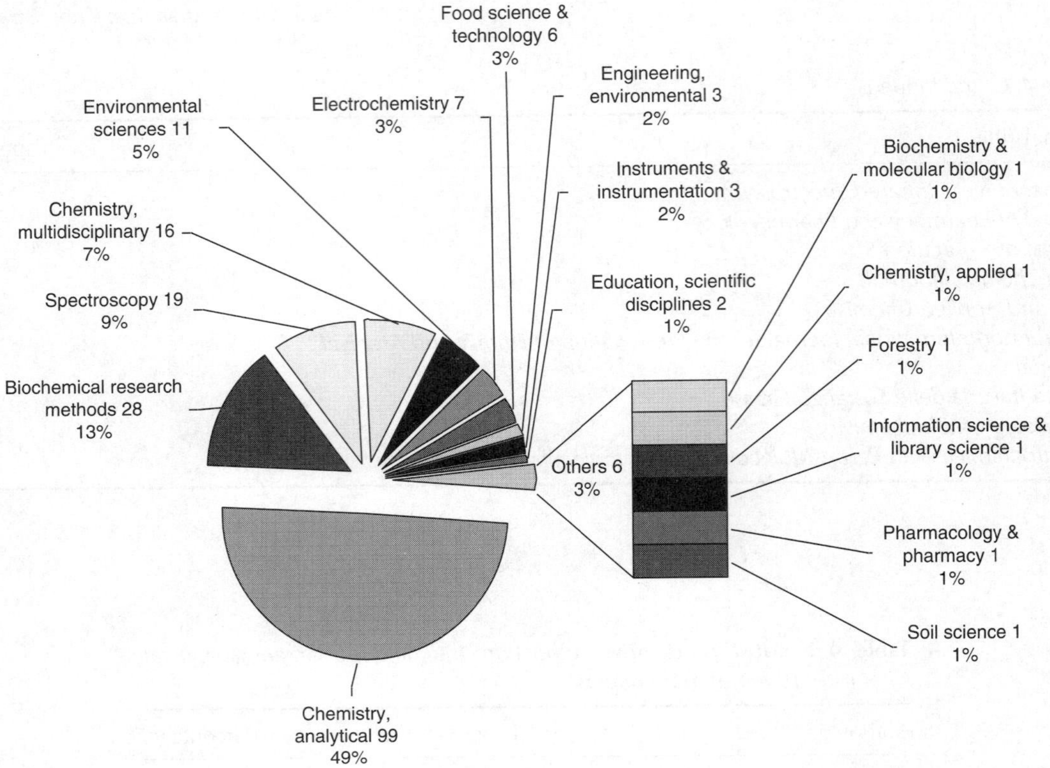

Figure 4.3 Distribution of publications on Green Analytical Chemistry as a function of field of research and application.

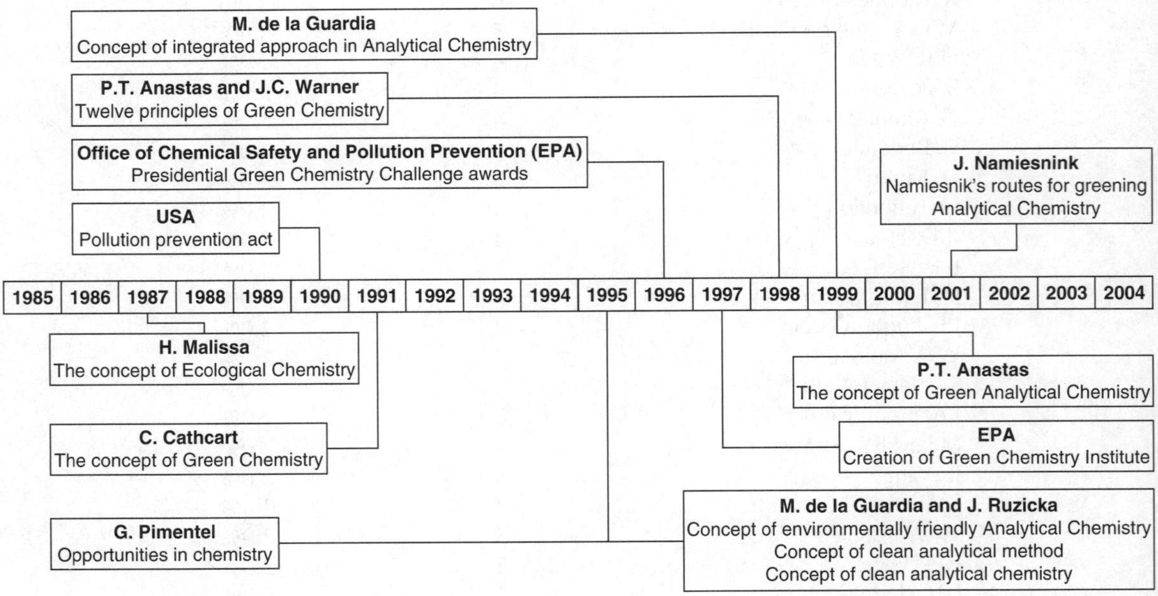

Figure 4.4 Milestones in Green Analytical Chemistry: theoretical contributions.

such a contribution was the search for new methodologies or just to look for improved methodology regarding laboratory production and costs.

On the other hand, we have summarized in Figure 4.4 the theoretical contributions which have been of a great importance to fix the objectives of Green Analytical Chemistry and to focus the aforementioned tools by way of avoiding environmental damage as a collateral effect of the analytical practices. So, we can conclude that in fact the tools for greening the analytical methods were developed years before starting the movement of Green Analytical Chemistry, and previous to the development of Green Chemistry.

The concepts of Ecological Chemistry [1], Green Chemistry [2], and Green Analytical Chemistry [3], developed in 1987, 1991 and 1999, respectively, have contributed to put the automation, miniaturization and direct measurement analytical tools in the frame of an environmentally friendly activity.

If we consider the public actions which have created the institutional support to the Green Analytical Chemistry, we are convinced that the Pollution Prevention Act of the USA in 1990 [4], the Pimentel report on *Opportunities in Chemistry* of 1995 [5], together with the creation of the Office of Chemical Safety and Pollution Prevention (EPA) in 1996 [6], and the creation of the EPA Green Chemistry Institute in 1997 [7] are the cornerstones of the developments in both Green Chemistry and Green Analytical Chemistry.

In the specific case of analytical chemistry, the concepts of clean analytical method and environmentally friendly methods in 1995 [8], integrated approaches in analytical chemistry [9], and the contribution of Namiesnik [10] and Kaljurand and Koel [11] have been of a great importance to involve the analytical community on greening the analytical practices.

4.3 The need for powerful keywords

As a conclusion of the previous sections we must accept that, in spite of the recent publication of three books and several special issues of reference journals in the field of analytical chemistry and spectroscopy devoted to Green Analytical Chemistry, we are far from the general use of environmental friendly keywords in the analytical papers. So, in contrast with the growing interest on reducing the amount of reagents and energy consumption and the generation of waste there are only a few papers in which authors evaluate the advantages of greening the analytical methods together with the deep validation of classical analytical parameters (accuracy, precision, sensitivity and selectivity) or with the evaluation of the increased speed and limited cost which offers new methods. So, it is important to insist on the common aspects of our waste and to not dilute the impact of our efforts, and to do this it is necessary to insist on the evaluation of the environmental side effects of both proposed and previously available methodologies and to make visible in literature the green approaches to solve analytical problems. To do it, the appropriate selection of powerful keywords which could be useful to recover these works from scientific literature is mandatory.

Figure 4.5 summarizes the environmental and practical keywords employed up until today to describe the efforts made to avoid the environmental deleterious effects of the analytical methods and the need for a general use of a green terminology because, as Linthorst has clearly shown regarding Green Chemistry in his review *An Overview: Origins and Developments of Green Chemistry* [12] the green terms become the most popular alternative regarding Green Chemistry, being found in less than 10 references per year which focus on being sustainable or clean chemistry and less than 50 which call their efforts environmental contributions, in spite of the fact that the number of articles per year which include the term Green Chemistry has been growth to most of 350–400 in the last two years.

In such a way, the analytical chemistry must also move to the general use of the term Green Analytical Chemistry in order to be clearly identified with other environmental conservation movements and it would beneficial for the whole community that efforts for greening our practices could be easily identified in the scientific literature.

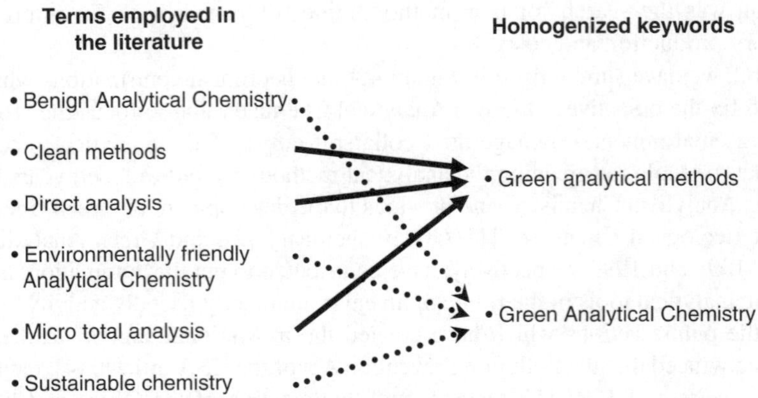

Figure 4.5 *From environmental keywords to Green Analytical Chemistry. The need for homogenized keywords in the field.*

On the other hand, there is no problem in continuing to use the sustainable or environmentally friendly terms in the body of the text, because they describe better the real world than the green one. However, as keyword it is totally necessary to establish a clear identification of our efforts and we are convinced that the use of green terms in titles, abstracts and as additional keywords of our papers is mandatory to do it.

4.4 A new attitude of authors faced with green parameters

As indicated in Chapter 2 concerning *Education in Green Analytical Chemistry*, the single thing which could guarantee the success of the Green Analytical Chemistry movement is the deep change in the mentality of people working in the development and application of analytical methods and to do it, additionally than to spread well the green alternatives in the scientific literature, it will be necessary that authors include in their evaluation of methods a clear identification of the side effects of their use, thus paying attention to the amounts of reagents and energy involved in all the analytical steps, to the volumes of waste obtained in each case and to the potential toxicity of both, employed reagents and generated wastes. We are convinced that the inclusion of the aforementioned parameters in the list of data to be evaluated, will contribute to the replacement of toxic reagents by innocuous ones, to the limitation of amounts of reagents employed and to the introduction of detoxification steps after the measurement of standards and samples. All those parameters will clearly show the economical opportunities offered by the green methods as compared with the classical ones and also contribute to encouraging these applications and the industrial laboratories to greening their practices.

In some cases, this change in mentality could created just by adding a section in the 'Results and Discussion' part regarding the evaluation of green parameters and in other cases, in which a new methodology is proposed, it could be enough to just evaluate the reagents and energy consumption and the waste generated by the different approaches in a comparative table. These simple modifications in the basic structure of the analytical publications will contribute to create the objective conditions to a silent revolution in our practices because data, more than arguments, are terribly convincing when you evaluate the different methods of analysis.

In such a framework, it will be clear that the direct methodologies, which are able to provide multiparametric information about the composition and properties of untreated samples without using reagents or creating dangerous wastes, will be imposed as the methods of choice when they are suitable. Additionally, the old in-batch and macroanalysis practices will be identified as costly and dangerous in preference to automatized

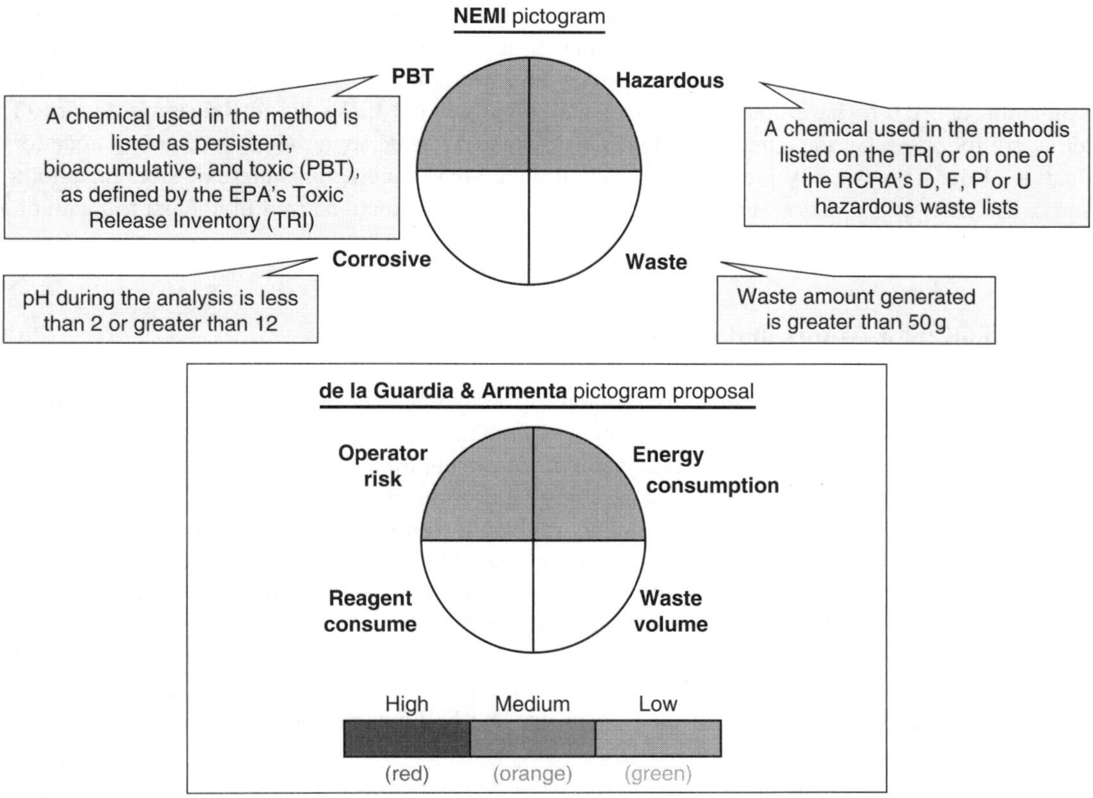

Figure 4.6 *Green pictograms proposed in the literature.*

and miniaturized ones, just from the identification of their cost in terms of time and money of our practices. However, another important aspect to be considered is the need to think about the methods after data acquisition and thus, to incorporate inside the protocol the system for on-line treatment of waste including recovery or decontamination steps when they could be possible, or at least, passivation methodologies. These treatments could reduce the cost of analysis and limit the risk of residue accumulation whilst providing a good feeling for the method's users and improving their environmental conscience. So, such a method which incorporates additional detoxification steps or no use of reagents or the generation of waste could be identified as a totally green one.

In short, we are convinced that green terms must be reflected as keywords in method development and method application and green parameters must be evaluated in the body of the text as a part of the results and discussion section. Really, the idea is not to replace the typical figures of merit of the analytical methods but to add new aspects such as, potential toxicity or risks of reagents employed, amounts of solvents required, reagents and energy necessaries for sample analysis, and volume of wastes generated. On the other hand, the use of green pictograms as proposed by the NEMI [13] and by us [14] could provide a fast and adequate environmental evaluation of the method, suitable for incorporation in the figures of the paper as a complementary and easy-to-understand information. In this sense, Figure 4.6 provides a description of the meaning of the green pictograms.

The NEMI proposal includes the consideration of green or not green character of a method based on four parameters: Persistent/Bioaccumulative/Toxic (PBT), Hazardous, Corrosive and Waste aspects which must be evaluated to consider the methodology as green or not. In this sense it must be understood that the corrosive aspect concerns the use of a pH lower than 2 or greater than 12, and that the waste requirements include that the amount of generated waste must be lower than 50 per determination. To accept the classification of PBT or hazardous of the reagents employed, the EPA's Toxic Release Inventory (TRI) criterion are followed. On the other hand, de la Guardia and Armenta suggested a three colour code for the evaluation of the side effects of methods and each of these kind of pictograms makes visible the efforts for greening the analytical practices in order to move to dangerous situations (red in colour) to medium and *green* safe alternatives.

4.5 A proposal for editors and reviewers

It is clear that Green Analytical Chemistry is a hot topic in today's analytical work and many journals organize special issues on it. However, we think that now it is time to move from the theoretical considerations to practical efforts to integrate the environmental point of view in our everyday work. To do it, we are convinced that it is necessary to introduce evaluation of the side effects of the use of the analytical methods as a formal requirement of papers to be published, in the same way that the adequacy of statistical evaluation of results obtained in each manuscript is a basic question to be answered by both authors and reviewers.

So, it could be convenient to include in publication norms the need to provide information about the potential toxicity of all reagents and solvents required for method application, together with an evaluation of energy and reagent consumption per sample to be analysed and wastes generated for a working time or a fixed number of determinations. This information could be of a great value to encourage authors themselves to think about the potential side effects of their chemistry and be fundamental for the evaluation of method greenness.

Additionally, in the instruction for reviewers, it could be convenient to ask about their opinion on the green aspects of the publication proposal thus creating the appropriate conditions for a rigorous evaluation of the methodology advantages provided by the new procedures and their environmental cost, the matter of interest being to improve not simply the main figures of merit but, also, to evaluate the risks for both, operators and the environment.

The aforementioned recommendations will contribute to create an increased environmental conscience about method proposals and also will impact on the readers who will be obliged to think about the side effects of the different alternative methods. All that will be highly positive in order to create new needs like those of toxic reagent replacement, reagent and energy consume reduction and waste minimization, thus offering tremendous opportunities for method development and method improvement which in part, will contribute to reinforce the position of analytical chemistry in the scientific literature and to solve real problems in an efficient way. An approach like that suggested will involve taking into consideration not only social and economical considerations but also environmental ones.

So, as indicated in the scheme of Figure 4.7, these suggestions will increase the environmental conscience of both analytical method developers and users, and contribute to provide complete information to the analytical chemistry readers about the side effects of methods available. We think that these suggestions offer new business and science opportunities and they are so easy to implement that there is no reason to delay the appropriate decisions of the journal editors in this sense.

On the other hand, we are sure that any effort in the way of increasing information about the risks of the analytical methods will be welcome by the readers. At the end, they are the potential users of the authors and journal proposals. We are living in a world with an increased environmental conscience, thus it is important potential method users can find relevant details in published papers on the amount of energy and reagents to be used and their potential toxicity, and because of that, a small extra effort must be made to evaluate

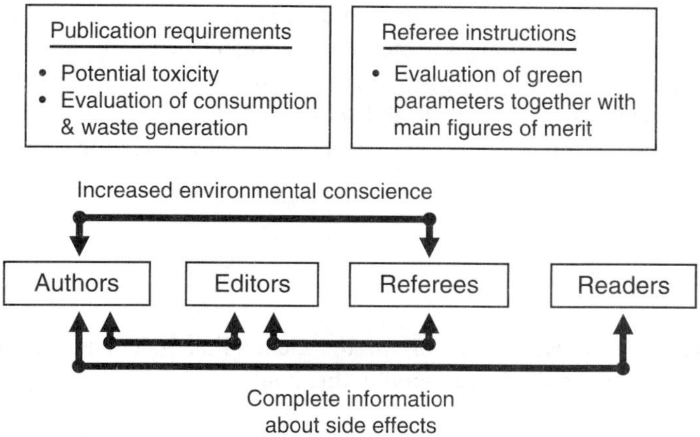

Figure 4.7 *Suggestions for the improvement of Green Analytical Chemistry publications.*

consumption and waste in method publications. It could open up tremendous opportunities for the readers to select the best procedure from methodological, environmental and economical point of view. Thus once again, to be environmentally friendly does not limit at all the possibilities but enhances and expands the capabilities of the analytical methods. In this sense, green information about the methods (1) offers new lines of method improvement, (2) increases the multiparametric information obtained from direct or quasi-direct measurements, (3) reduces cost, time of analysis, costs and risks, (4) reduces physical and chemical treatments and replaces those by improved mathematical ones, and (5) incorporates solutions to the analytical troubles, like on-line waste treatment.

In short, by greening analytical publications we will contribute to the development of analytical chemistry and enhance the safety and applied figures of merit of methods.

4.6 The future starts now

Up until this point we have spoken about the origin of the green conscience at the end of the twentieth century and the tremendous development made in the first decade of the twenty-first century concerning the environmental conscience applied to the analytical methods.

The year 2011 was the International Year of Chemistry and contributed to increase the relationship between chemistry and society, and in such a way we have been obliged to transmit to the public the idea that chemistry is the key to solve the pollution problem and not the reason for environmental damage. So, the aforementioned proposals for greening the publications on analytical chemistry, together with the efforts for greening our teaching activities, can move the public opinion about our discipline.

Now it is time to move from the special issues devoted to Green Analytical Chemistry to new practices, totally integrated in our everyday activity. To think about the collateral environmental and personal damage that can result from our work and to prevent those risks which can arise from method application cannot be fashion or a temporally interesting proposal. The objective is to preserve our planet. In this sense, the contribution of each measurement made in a sustainable way can seem irrelevant but all those practices together can guarantee a better future for our children. This is the challenge and now it is time to take the appropriate measures to avoid pollution problems and to increase the safety of the analytical laboratories.

International agencies have shown the way for careful control of chemicals, policy makers have in their hands a series of environmental protection laws and the governments in the different countries are evaluating different strategies and specific decisions to reduce the pollution. However, all these decisions will be nothing if the scientists and technicians do not participate and do not re-evaluate their own activities on thinking about the environmental problems. So, on greening analytical chemistry publications we will contribute to make our world safer.

Acknowledgements

The authors acknowledge the financial support of the Generalitat Valenciana Project GV PROMETEO 2010–055 to write this book and to do the research in this field.

References

1. Malissa, H. (1988) Changes of paradigms in Analytical Chemistry, in *Reviews on Analytical Chemistry, Euroanalysis VI* (ed. E. Roth), Les Editions de Physique, Paris.
2. Cathcart, C. (1990) Green Chemistry in the Emerald Isle, *Chem. Ind-London*, **21**, 684–687.
3. Anastas, P.T. (1999) Green chemistry and the role of analytical methodology development, *Crit. Rev. Anal. Chem.*, **29**, 167–175.
4. Pollution Prevention Act of 1990 (1990) Public Law, 101–508, 13101. (Public Law 101–508, Nov. 5, 1990; 104 Stat. 1388, 42 U.S.C. 13101-13109).
5. Pimentel, G. (1995) *Opportunities in Chemistry*, National Academic Press, Washington D.C.
6. US Environmental Protection Agency, (n.d.) *Presidential Green Chemistry Challenge*. Available at http://www.epa.gov/gcc/pubs/pgcc/presgcc.html (accessed 3 November, 2011).
7. Raber, L. (2000) Green Chemistry Institute joins ACS, *Chem. Eng. News*, **78**, 13–14.
8. de la Guardia, M. and Ruzicka, J. (1995) Towards environmentally conscientious Analytical Chemistry through miniaturization, containment and reagent replacement, *Analyst*, **120**, 17N-17N.
9. de la Guardia, M. (1999) An integrated approach of Analytical Chemistry, *J. Braz. Chem. Soc.*, **10**, 429–437.
10. Namiesnik, J. (2001) Green Analytical Chemistry – Some remarks, *J. Sep. Sci.*, **24**, 151–153.
11. Koel. M. and Kaljurand, M. (2006) Application of the principles of Green Chemistry in Analytical Chemistry, *Pure Appl. Chem.*, **78**, 1993–2002.
12. Linthorst, J.A. (2010) An overview: origins and development of Green Chemistry, *Found. Chem.*, **12**, 55–68.
13. Keith, L.H.; Gron, L.U. and Young, J.L. (2007) Green analytical methodologies, *Chem. Rev.*, **107**, 2695–2708.
14. de la Guardia, M. and Armenta, S. (2011) *Green Analytical Chemistry: Theory and Practice*. Elsevier, Amsterdam.

Section II

The Analytical Process

5

Greening Sampling Techniques

José Luis Gómez Ariza and Tamara García Barrera

Departamento De Química Y Ciencia De Los Materiales 'Profesor José Carlos Vílchez Martín', Universidad De Huelva, Huelva, Spain

Sampling is a very critical operation in analysis which can dramatically affect final results. Preservation and stabilization reagents such as acids, solvents and buffers, as well as filters, containers and refrigeration/cryogenic conditions, represent important environmentally detrimental toxic inputs. Since billions of samples are annually collected for environmental, health or food quality monitoring, the use of environmental friendly sampling practices can have important consequences. This paper reviews the more significant proposal for this purpose under the concept of Green Analytical Chemistry. The application of direct techniques of analysis that avoid sample collection, transport, pretreatment or preparation, such as ion-selective electrodes, portable X-ray fluorescence spectrometers or remote analysis techniques represent some examples of new trends in Green Analytical Chemistry. Other approaches are centered on solventless techniques for sample extraction, especially supercritical fluid extraction (SFE), membrane extraction systems, and solid phase extraction (SPE). The reduction of reagents and solvent volumes together with the increase of rate samples in flow-through solid phase spectroscopic, hollow-fiber GC/HPLC/CE and nanoparticle-based analytical approaches also represent relevant contributions in this field. Further options in green sampling is related the use of sensitive probes that work in streams, at in vivo samples and in-field analysis. Finally, automation of analysis contributes to labor and energy consumption and can be combined with the above approaches for green analysis and friendly environmental sampling, although many further studies are still required in this area.

Handbook of Green Analytical Chemistry, First Edition. Edited by Miguel de la Guardia and Salvador Garrigues.

5.1 Greening analytical chemistry solutions for sampling

The concept of green chemistry is closely related to the principles of sustainable chemistry and the clear trend towards their implementation in chemical plants and laboratories. From these premises guidelines have to be established for chemists and production engineers to make their activities less harmful towards the environment. This new way of action is supported by rules and principles [1,2]. A great number of analytical methods have been proposed to monitor environmental friendly activities, but these analytical activities have also been performed in a friendly way from the environmental point of view [3]. Therefore, any development towards green chemistry is not possible without the existence of Green Analytical Chemistry (GAC).

The features related to GAC consider the reduction of reagents, particularly organic reagents, from analytical procedures; the reduction in emissions of gases and vapors, liquid and solid waste produced in the analytical laboratories; the elimination and substitution of toxic reagents; and the reduction of energy necessary for analytical procedure development. Other important features in Green Analytical Chemistry are related to sampling. In routine analytical procedures, sample collection can be avoided by applying in-line systems that allow researchers to perform measurements directly in the sample, in-field analysis (e.g., analysis in a pipeline or a river). In this way, the problems related to sample preparation, transportation to the laboratory and storage can be avoided. The application of direct techniques of analysis that avoid sample transport, pretreatment or preparation, such as ion-selective electrodes for water analysis, portable X-ray fluorescence spectrometers for soil metal contamination assessment or remote analysis techniques (LIDAR or SODAR) to measure air pollution, represent some examples of new trends in Green Analytical Chemistry. Other approaches are centered on solventless techniques, especially for sample extraction. Several possibilities can be mentioned in relation to this stage of the analysis: supercritical fluid extraction, membrane extraction systems coupled directly (on-line) with measuring devices and solid phase extraction.

5.2 New green approaches to reduce problems related to sample losses, sample contamination, transport and storage

Sample interaction with wall sampling containers or transfer lines in on-line analytical arrangements can provoke significant losses or contamination of the analytes. In addition, problems related to sample transport and storage in batch analysis also affect significantly the precision of the results and require the use of considerable amounts of reagents and waste against GAC features. Direct analytical systems or analytical arrangements intending to reduce the minimum contact of analytical systems with the sample, and moreover to reduce or even eliminate sample transport and storage, have been proposed. However, ideal approaches based in direct analysis are difficult to achieve and alternative procedures based on a combination of simplified sample treatment methods, environmentally friendly, and high-rate sample devices based on the use of flow-through solid phase spectroscopy, hollow-fiber GC/HPLC/CE and nanoparticles have been proposed under the focus of Green Analytical Chemistry.

5.2.1 Methods based on flow-through solid phase spectroscopy

The immobilization of the species of interest (either the analyte or any derivative product) on an appropriate solid support (usually microbeads from either a polymeric or a non-polar material) and the direct measurement of the analytical property (typically absorbance or luminescence) on the solid is the base of Flow-Through Solid-Phase Spectroscopy (FSPS), in which retention (sampling), preconcentration, and detection processes are all performed simultaneously in the flow cell. These systems allow increasing of both sensitivity and selectivity to be reached, together with a drastic decrease of reagent consumption. This instrumental implementation is called optosensor, flow-through optosensor or, simply, sensor [4]. An interesting alternative has been the use of sequential injection analysis (SIA) instead of flow injection analysis (FIA) [5] providing strong reduction of both reagents and waste. Therefore, the combination of multicommutation principles with SPS detection [6] has become a significant contribution to the GAC concept, since each reagent is operated independently in multicommutated optosensors and only the required volume for each sample is introduced in the analytical system which represents a drastic minimization of both reagents and waste. Therefore, FSPS can be envisaged as a variant of analysis based on the use of immobilized reagents.

Tables 5.1 and 5.2 summarize several typical applications of solid phase spectroscopy in GAC, as well as sample treatments proposed in these approaches which intend to reduce or avoid the amount of waste and the use of toxic reagents.

Simpler optosensors use monochannel devices in which the carrier solution also acts as an eluent and use intrinsic properties of the analytes for sensing (Table 5.1). These systems provide increased sensitivity in relation to analogous approaches in homogeneous solution, possibly due to the higher concentration of the analyte in the solid phase, this support can be reused which represents a considerable reduction of solid waste. In addition, the amount of sample is drastically reduced and the only reagent necessary is the carrier solution, all these characteristics represent valuable contributions in relation to GAC. Similar benefits can be mentioned in multiparameter sensors.

Sequential injection analysis in combination with solid phase spectrometry (SPS) (Table 5.2) is another interesting alternative in GAC based in optosensors, since a remarkable reduction in reagents and waste is achieved as well as a higher automation level of analysis.

Multicommutation devices allow us to handle micro-volumes of samples and reagents in simple and complex instrumental manifolds, minimizing both reagents and waste production, therefore the combination of multicommutation with SPS has become a good implementation in the optosensor field [17].

An important point to be considered is that samples in typical case-studies that use flow-through solid phase spectroscopic systems (Tables 5.1 and 5.2) are mainly liquids (waters, serum, urine, syrup) which require very little preparation. However, in pharmaceuticals excipients, such as tablets or suppositories and especially condiments such as pepper, sample handling is more complex with many steps that can contribute to analyte losses and lower accuracy of results. In addition, these treatments involve the use of reagents and solvents reducing the greening character of these procedures. Therefore, further studies on solid sample treatment that complement the green possibilities of flow-through solid phase spectroscopic sensors are necessary.

5.2.2 Methods based on hollow-fiber GC/HPLC/CE

The use of porous hydrophobic membranes has been proposed for air, headspace and aqueous phase analysis. These systems allow a direct analysis of volatile compounds from air or water (headspace extraction) and quasi-direct analysis of polar compounds in liquid samples, which reduce the problems associated to sampling and sample storage and treatment. Due to its low cost, the hollow-fiber extraction device can be disposed after a single extraction which eliminates the possibility of carry over effects. In addition, because a small volume of organic solvent is required and little waste generated, these approaches are environmentally friendly and compatible with the 'Green Chemistry' concept.

A poly(dimethylsiloxane) (PDMS) hollow-fiber membrane has proved useful to perform surface-water monitoring using an especially designed cap [18]. The extraction cap was placed on the aqueous surface in a glass water bath container which contained aqueous sample to a 1 cm depth. The pressure of the headspace was ambient. For underwater monitoring, the extraction cap was positioned in an aqueous sample at a depth of 25 cm. The pressure of the headspace was higher than the room pressure because of the height of the water. To allow a 1 cm depth of water inside the cap, the headspace pressure was adjusted by opening and closing

Table 5.1 *Applications of solid phase spectroscopy in GAC (FIA-SPS sensors).*

Methodology	Spectrometric basis	Interaction Solid phase - analyte	Type of sample or matrix	Analyte	Sampling rate (h^{-1})	Effluent (ml/ determination)	Sample treatment	References
Monoparameter sensor	Photochemically induced fluorescence (PIF)	Photoproduct $\lambda_{emission}$ 443 nm (thiamine UV irradiation) Retained in cells filled with C18 silica gel	Human serum and urine	Thiamine	50	2.9	*Urine*: dilution with water (1:1000 or 1:2000, v/v) *Serum* was diluted with water (1:200, v/v)	[7]
	Photochemically induced fluorescence (PIF)	Derivative $\lambda_{emission}$ 443 nm (imidacloprid UV irradiation) Retained in cells filled with C18 silica gel	Pepper and waters	Imidacloprid	–	–	*Water*: pH at 11.5 (0.01 M Na_2HPO_4/NaOH). Filtration, cellulose acetate filter (0.45 μm) *Pepper*: – extraction (acetone) – shaking (dichloro methane–light petroleum (1:1) – Evaporation to dryness – Reconstitution: water–acetonitrile (3:1) – Clean-up: C_{18} cartridge	[8]
	Spectrophotometric absorption (λ_{abs} 443 nm)	Complex of Fe(II) with 1-(2-thiazolylazo)-2-naphthol immobilized on C18-bonded silica	Waters	Fe(II)	24	3.6	Water samples collection – polyethylene vessels – filtration (0.45 μm)	[9]
Multiparameter sensor	Spectrophotometric absorption $\lambda_{267 nm}$ (AA) $\lambda_{262 nm}$ (PCT)	Dextran type anion exchanger resin AA: pH = 5.6 PCT: pH = 12.5 One sensing zone	Pharmaceuticals	Ascorbic acid (AA) Paracetamol (PCT)	AA: 20 PCT: 25	AA: 3 ml PCT: 2.5 ml	*Tablets (AA):* powdered + water + sonication *Liquid preparation/PCT*: water dilution *Suppository*/PCT: CH_3Cl + water (pH 8-9)	[10]
	Spectrophotometric absorption $\lambda_{250 nm}$ (TH, AA)	SEPHADEX SP C-25 (TH) SEPHADEX QAE C.25 (AA) Two sensing zones	Pharmaceuticals	Thiamine (TH) Ascorbic acid (AA)	–	–	*Tablet (TH)*: powdered, 0.1 M HCl 0.1, sonication, filtration *Tablet (AA)*: powdered + water + sonication	[11]

Table 5.2 *Applications of solid phase spectroscopy in GAC (SIA-SPS and Multicommutation-SPS sensors).*

Methodology	Spectrometric basis	Interaction Solid phase – analyte	Type of sample or matrix	Analyte	Sampling rate (h^{-1})	Effluent (ml/determination)	Sample treatment	References
SIA-SPS Monoparameter sensor	Renewable Surface Solid Phase Spectrofluorimetry (λ_{exc} = 385 nm; λ_{em} = 443 nm)	Thiochrome from Thiamine (Thiamine oxidation by hexacyanoferrate (III)) Sorption of thiochrome on C18-SP(DB) Solid-phase fluorescence emission (λ_{em} = 443 nm)	Pharmaceuticals (tablet, syrup)	Thiamine	30	2.0	*Tablet*: powdered, 0.1 M HCl 0.1, sonication, filtration *Syrup*: 0.1 M HCl, dilution	[12]
Multiparameter sensor	Solid-phase fluorimetry Native fluorescence (B6, B2)	C18 silica gel (vit. B6, B2) Anionic QAE-A25 (vit. C)	Pharmaceuticals	Vitamins (B_6, B_2 and C)	12	Vit B_6, B_2: 3.5 Vit C: 2.5	Deionized water, sonication + 5×10^{-3} M H_2SO_4	[13]
	Chemiluminiscence ($KMnO_4 + H_2SO_4$) (vit. C)	Two sensing zones						
	Hypoxantine: Solid-phase spectrophotometry (λ_{abs} = 290 nm) *Potassium*: Potenciometry with selective electrodes	*Hypoxantine (Hy)* Aminoalkylated glass beads incubated in 2.5ml of 2.5% glutaraldehyde for 1 h, 4 u of xanthine oxidase. (Oxid. of Hy to acid uric) *Potassium (K)* Tubular K ion-selective electrode	Vitreous humor	Hypoxanthine Potassium	30	2.7	Sample dilution, filtration 0.20 μm filters	[14]
Multicommutation-SPS Monoparameter sensor	LED-based photometer – long pathlength flow cell (100 mm)	Molybdenum blue method	River water	PO_4^{3-}	20	2.4	Filtration (cellulose acetate 0.45 μm)	[15]
Multiparameter sensor	Solid Phase Fluorimetry: – fuberidazole (FBZ) – o-phenylphenol (OPP) (native fluorescence) FBZ: λ_{exc} = 314; λ_{em} = 356 nm OPP: λ_{exc} = 250; λ_{em} = 345 nm	C_{18} silica gel solid support (separation + determination)	Well and river water	*Pesticides*: fuberidazole o-phenylphenol	12	5.0	Filtration (cellulose acetate 0.45 μm)	[16]

a pressure release switch. During the sampling/extraction, the position of the extraction cap was fixed using a clamp to hold the neck of the cap, and the cap was tightly in contact with the bottom of the bath container in each extraction. This contact ensures the proper sealing of the aqueous sample inside the extraction cap and prevents the sample inside from exchanging with the solution outside of the cap. The monitoring of VOCs (benzene, toluene, ethylbenzene, *o*-xylene, and trichloroethylene) was performed using 4 min of trapping and 2 s of thermal desorption for further analysis by gas chromatography with flame ionization detection (FID). Compared with direct water extraction, headspace extraction has many advantages. Membrane headspace extraction allows the membrane to be isolated from sample matrix, which results in a longer membrane life without performance change. VOCs have much larger diffusion coefficients in air than in water, which results in fast mass transfer and a smaller boundary layer (membrane/air) effect in headspace extraction, hence an improved response. In addition, most VOCs have larger partition coefficients for membrane/air than membrane/water, which leads to a higher driving force the diffusion into the membrane in headspace extraction and, hence, a good sensitivity. Therefore, compared with direct water extraction, headspace extraction could be the choice and reduce significantly sample handle, transport and storage and if the probe is combined with a portable micro gas chromatograph can be potentially useful for field monitoring because of its simplicity, enhanced sensitivity, convenience of deployment and automation, and reliability [18]. In addition, the solventless character of the approach and unnecessary use of reagents as well as absence of waste makes this method very promising in the field of GAC.

Ionizable and charged species such as drugs and contaminants (nitrophenols, pesticides, etc.) can be extracted from water, human urine and plasma, and milk for their analysis by capillary electrophoresis (CE) and HPLC, using supported liquid membranes (SLM). In this technique a thin film of organic non-polar organic solvent immobilized in the pores of a porous polypropylene hollow fiber allows the extraction of acid or basic compounds from the exterior of the hollow fiber, donor aqueous phase, at pH suitable to get the target compounds in neutral form; in the internal channel of the hollow fiber is the acceptor aqueous phase, the pH of which is adjusted to ionize again the target compounds. Under stirring the neutral compounds in the donor phase are extracted into the organic film on the fiber and then back extracted into the acceptor phase inside the fiber (liquid-liquid-liquid microextraction – LLLME). Due to the high ratio of donor phase volume to acceptor phase, the preconcentration factor is usually very high and sample cleanup occurrs simultaneously because macromolecules, ionic compounds with acid-basic character differ to the extraction system used, and neutral components are extracted into the acceptor solution. In addition, very little sample handle is necessary since water, urine and plasma only require sample acid or basic pH adjustment depending on the analysis of basic or acid analytes. Therefore, sampling is very simple and non-specialist samples requirements or handling are necessary, in addition, the extraction unit is disposed of after a single extraction which avoids cross-contamination problems from sample to sample [19]. The low cost of each extraction unit allows a large number of different samples to be processed providing a high sample capacity that reduces the need for sample storage. Finally, low amount of reagent and solvents are required [19,20].

Hollow-fiber combined with liquid chromatography or capillary electrophoresis has been proposed for GAC of methamphetamine in human urine, and human plasma utilizing 10^{-1} M HCl as the acceptor phase and 10^{-1} M NaOH in the donor solution. The preconcentration factor was 75, the number of samples extracted in parallel were 20 and the detection limit of methamphetamine utilizing LLLME/CE was 5 ng ml^{-1} in both human urine and plasma [19]. Other approach proposes a simple, inexpensive and disposable device for liquid phase microextraction (LPME) that can be combined with GC, HPLC and CE for the analysis of pharmaceuticals in human urine and plasma [20]. Two benzodiazepines, diazepam and prazepam were selected as model compounds for the compatibility study of LPME-GC. The acceptor solvent was *n*-octanol in both the pores and the internal volume of the hollow fiber. Diazepam and prazepam were extracted efficiently from the aqueous sample solutions, such as plasma, by adjusting the pH at 5.5. Methamphetamine was used as model basic substance for the LPME-CE system in the extraction and analysis from human plasma, and naproxen was selected as the acidic model compound to be extracted from human urine. A three phase system was used in this case with *n*-octanol as an effective barrier between the two aqueous phases, donor (plasma pH about 13) and acceptor (water pH 1), methamphetamine was extracted from the plasma with a recovery of 75% and a preconcentration factor of 75. The acidic drug naproxen present in a human urine sample (4 ml acidified with HCl until pH 1) was extracted with 25 µl of 0.02 M NaOH as the acceptor solution, in this case resulted in a 125-time preconcentration (78% extraction efficiency). The antidepressant citalopram and its metabolite *N*-desmethylcitalopram are basic drugs present in the plasma that can be analyzed by LPME-HPLC, using of 0.01 M phosphate (pH 3) as an acceptor solution and the hollow fiber impregnated with *n*-octanol. Samples containing the drug were adjusted to pH 13 with NaOH. For both citalopram and the metabolite, a 30 time preconcentration was obtained corresponding to a 75% extraction efficiency.

Finally, the use of LLLME and HPLC has been proposed for the analysis and preconcentration of phenoxy herbicides from bovine milk, using NaOH (0.1 M) as an internal acceptor solution into the fiber and HCl (0.5 M) in the donor phase (the sample) [21]. The type of solvent immobilized in the fiber is also very critical to achieve an efficient analyte extraction and preconcentration for the five different phenoxy herbicides considered: 2,4-Dichlorobenzoic acid (2,4-DCBA), 2,4-dichlorophenoxyacetic acid (2,4-D), 2-(2,4,5-trichlorophenoxy) propionic acid (fenoprop), 2-(4-chloro-phenoxy)-2-methylpropionic acid (mecoprop) and 3,5-dichlorobenzoic acid (3,5-DCBA). 1-octanol could extract all five herbicides with a suitable enrichment factor (about 900), which is probably due to its relatively higher polarity and its greater affinity for the acidic herbicides.

The preparation of a milk sample is very simple and only requires acidification with HCl to make the final concentration of 0.5 M and centrifugation for 15 min at 2000 rpm.

5.2.3 Methods based on the use of nanoparticles

The possibility of integrating functionalized nano-materials to improve the performance of the analysis is an alternative approach to Green Analytical Chemistry due to the possibility of reducing sample handling and manipulation, as well as to get a considerable decrease in the amount of reagents. Nanoparticles can be used for direct analyte extraction and preconcentration from liquid samples such as water [22], seawater [23] and urine [23], with practically untreated samples increasing analytical probes' or sensors' sensitivity or selectivity, especially in the case of electrochemical sensors [24].

A more detailed description about the use of nanoparticles in green liquid sample analysis is developed in the following case-studies:

1. An on-line configuration of microdialysis (MD), Au/TiO2 nanoparticle preconcentration, and high performance liquid chromatography-ultraviolet (HPLC-UV) detection method was developed for the simultaneous measurement of cobalt and nickel concentrations in water. The sample matrix was first cleaned with an MD system using a MD probe. A continuously flowing dialysate stream was introduced into tubing coated with Au/TiO2 nanoparticles to adsorb metals, followed by elution by an acidic eluent. The enriched samples were derivatized on-line using 8-hydroxyquinoline. The separation of Co and Ni were achieved by using a LC-C18 column. The three aforementioned system components were connected on-line using a valve control. The UV detection was performed at 319 nm. The proposed method offers a simple and reliable procedure to determine the levels of Co and Ni in environmental water samples. Organic solvents were absent in the MD sampling and extraction processes [22].
2. A novel system for trace analysis of multiple heavy metals was developed integrating a magnetic nano-adsorbent with a lab-on-valve system and coupling this to an inductively coupled plasma mass spectrometer [23]. For this purpose, a magnetic nano-adsorbent was prepared by surface modification of iron-based magnetic nanoparticles with polyacrylic acid (MNPs-PAA), this sorbent exhibited

superparamagnetic characteristics which allowed the immobilization of nanoparticles in a PTFE tubing using an external magnetic force. The high surface density of PAA (3.30×10^{11} molecules cm^{-2}) on MNPs-PAA significantly reduced the amounts of adsorbent needed for adsorption of heavy metals (Mn^{2+}, Co^{2+}, Cu^{2+}, Zn^{2+}, and Pb^{2+}). Good analytical performances could be obtained by direct injection of an aliquot of 20 µL of the sample which is analyzed in 5 min. The system is useful for trace analysis of multiple heavy metals in environmental and biological aqueous samples. The potential for integrating functionalized nanomaterials to improve the performance of existing analytical systems is an alternative approach to Green Analytical Chemistry.

3. Carbon-based screen-printed electrodes (SPCEs) modified with carbon nanotubes (CNTs), using Osteryoung square-wave voltammetry (OSWV), has been proposed for determination of enzymes activities as bioindicators of soil quality [24]. The sensor was applied to determine the activities of arylsulphatase (AS), acid phosphatase (ACP) and alkaline phosphatase (ALP) in agricultural soils. The activities of these enzymes were detected using a non-electroactive substrate, which produces an electroactive product. Therefore, p-aminophenyl phosphate (pAPP) was used as a substrate that is converted to p-aminophenol (pAP) after enzymatic dephosphorylation, and 4-nitrocatechol sulfate (4-NCS) was used as a substrate for AS activity based on its catalytic effect on the hydrolysis of 4-NCS into 4-nitrocatechol (4-NC). The products of both enzymatic reactions were then quantified with (SPCEs) modified with carbon nanotubes (CNTs). This assay allows the direct analysis of AS, ACP and ALP in soils with very little sample preparation: sieving ($\leq$ 2 mm) after removing the plant material and roots and storage in plastic bags at 4°C for a few days to stabilize the microbiological activity disturbed during soil sampling. The analyses were achieved within two weeks after the sample collection. The assay reduce drastically the amount of solvents required in more conventional approaches based on spectrophotometric, fluorescence or chemiluminescence detection and also diminishes the generation of waste.

5.3 Greening analytical in-line systems

On-line analysis is perhaps the more frequent approach used in routine monitoring analysis requiring both real time and reliable information about some analytical problem or process. In on-line analysis an automated sampling system is used to collect and preparing the sample before the introduction into the on-line analytical system just in an intermittent or continuous mode. Many on-line analytical procedures have been proposed under the GAC principles; however, the more important drawback in on-line analysis, consisting of the need for an separate analytical line to drive the samples to the analytical instrument, is overcome by in-line analysis that perform the chemical analysis in situ, directly on the process line or the analytical issue. In-line analysis generally uses a sensitive probe that works in a reliable manner in a process stream or reactor, at in vivo monitoring, at in-field environmental problems, and other issues, such as the monitoring of water- and wastewater-quality parameters [25]. However, strict in-line analysis does not cover all the demands of in-field analysis due to the usual use of combined on-line/in-line analysis that does not meet the requirements of in-field analysis, especially easy portability.

Two examples can be cited to illustrate the use of in-line analysis, the determination of nitrate and nitrite in waters. These parameters are especially important in relation to the eutrophication of natural waters and aquifer salification, in addition some health problems caused by cyanosis, and in-vivo production of carcinogenic, mutagenic and teratogenic nitrosamines and nitrosamides [26] are also related. For these reasons, water nitrite and nitrate monitoring is frequent to assure environmental quality, and in-line methods could provide fast and direct procedures that reduces considerably sample treatment and manipulations.

Nitrate analysis in water is usually performed with cadmium filings for nitrate reduction and carcinogenic reagents for the diazocoupling reaction with nitrite. Some procedures have intended to avoid the use of toxic reagents, with this purpose, photochemical or enzymatic reduction of nitrate by nitrate reductase have been used instead of cadmium. However, some toxic reagents are still used for the final spectrophotometric determination of nitrite such as sulfanilamide (SAN) and N-1-naphthyl ethylenediamine dihydrochloride (NED) or luminol for chemiluminiscent detection. To avoid the use of toxic reagents in nitrate measurement, direct spectrophotometric analysis at low wavelength, around 200 nm, can be used which reduces considerably sample preparation. However, the procedure has important drawbacks because other substances represent serious interferences, such as Cl^-, NO_2^-, and Fe^{3+}, as well as organic matter. For this reason interferences elimination by in-line separation of nitrate using an anion-exchange resin allows selective nitrate determination without employing toxic reagents [27].

An example of green approach for nitrite determination in natural water was proposed by Melchert *et al.* [26] based on flow-injection spectrophotometry. The flow system was designed with solenoid micro-pumps in order to minimize reagent consumption and waste generation. The approach is based on the Griess diazocoupling reaction with sulfanilamide and N-(1-naphthyl)ethylenediamine (NED) yielding an azo dye, followed by photodegradation of the low amount of waste generated based on the photo-Fenton reaction. The sampling rate was estimated at 108 determinations per hour. The consumption of the most toxic reagent (NED) in the conventional Griess reaction is reduced 55-fold and 20-fold in comparison to batch method and flow injection with continuous reagent addition, respectively. A colorless residue was obtained by in-line photodegradation with reduction of 87% of the total organic carbon content. The results obtained for natural water samples were in agreement with those achieved by the reference method at the 95% confidence level.

An alternative procedure on the basis of the reaction of nitrite and iodide to yield iodine has been also proposed. In this case the sampling rate was estimated at 44 determinations per hour. Despite avoiding the use of toxic reagents, the nitrite–iodide method presented the worst performance in terms of selectivity and sensitivity.

5.4 In-field sampling

Several in-field sampling techniques environmentally friendly have been developed until now. In this context, the use of sorbents is very important since they offer good analytical features while they are solvent-free techniques or consume little volume of reagents. Solid phase extraction (SPE) and solid phase microextraction (SPME) allow in situ sample extraction and store samples for analysis. New field samplers have been developed for SPME [28] and SPME-based approach for sampling and analysis has been used for the determination of transformation products of highly reactive and toxic unsymmetrical dimethylhydrazine (UDMH) [29], for propofol exhalation in pigs, humans and sheep during anesthesia [30] and even in vivo sampling, as applications of SPME for analysis of volatile emissions and drugs in biological fluids [31]. A rolling stir bar sampling procedure has been developed for sampling volatile compounds from agricultural products, plant materials and human skin for thermal desorption/GC-MS [32].

As mentioned, SPE presents important advantages in relation with the environmental sampling as it requires less volume of solvent. Furthermore, the cartridges or disks can be used for storage of the species and SPE provides high enhancement factors if a large volume of water (1–21) is passed through the sorbent without breakthrough [33. SPE has been used as 'in field-sampling' technique for the simultaneous determination of selected endocrine disrupters in water [34] and for extraction and storage of tributyltins (TBT, DBT and MBT) in sea waters for tin speciation using C_{18} cartridges. In this latter example, the sampling operation is very simple and only requires passing a suitable volume of water through the cartridge with the help of a syringe, therefore it is not necessary the use of reagents and allows in-field sampling. Stability of tributyltin in C_{18} cartridges was checked for almost two years, although phenyltin species were only stable for

60 days. The cartridges loaded with the organotin compounds, after drying under nitrogen stream, can be stored at room temperature in the dark, which reduce the space needed for storage [35].

Other examples of on-line in-field sampling and monitoring is the use of ion-pair solid-phase extraction for determination of benzene- and naphthalenesulfonates in waste water [36] and a novel solid phase-extraction device integrated onto a centrifugal microfluidic platform has also been developed for organic pollutants in aqueous samples [37].

Another alternative for in-field sampling is SFE (supercritical fluid extraction). The main advantages of SFE are high preconcentration rates, cleanliness and safety, quantitative capability, possibility for in-field sampling, on-line coupling with hyphenated techniques and automation [33].

The solvents used in the analytical techniques for sampling and analysis should be eco-friendly such as supercritical fluids or ionic liquids. However, supercritical fluids are not in widespread use due to the lack of standard procedures and the scarcity of studies about the handling of liquid samples. In the case of ionic liquids, there are few data on toxicity (bioaccumulation and biodegradability), few data on physical properties and not clear understanding of solvent/material properties with regards to the structural features of ionic liquids. Also there are not official analytical methods where ionic liquids are involved; they are expensive and not commercially available.

Nevertheless, the ideal green analysis would run in situ integrating sampling and analysis, with real time measurement techniques that do not use reagents for preservatives or extractions or that use small amount of them. The 'Triad approach' is promoted by the US EPA for decision making that proactively exploits new characterization and treatment tools [38]. The goal of the Triad approach is to assure that project decisions about contaminant presence, location, fate, exposure, and risk reduction choices and design, are made correctly and cost-effectively. 'Correct' decisions are defined as the decisions that would be made if fully completely accurate knowledge of contamination nature and extent and receptor exposure were available to decision-makers. The foundation for site-related decisions that are both correct and optimized (from a cost-benefit standpoint) is the conceptual site model (CSM). A CSM uses all available historical and current information to estimate [39]:

1. where contamination is (or might be) located,
2. how much is (or might be) there,
3. how variable concentrations may be and how much spatial patterning may be present,
4. what is happening to contaminants as far as fate and migration,
5. who might be exposed to contaminants or harmful degradation products, and
6. what might be done to manage risk by mitigating exposure.

The triad refers to three primary components; systematic planning, dynamic work strategies, and real time measurement systems. The most important component for Green Analytical Chemistry is real time measurement systems such as X-ray fluorescence spectrometry (XRF) [40], attenuated total reflectance infrared spectroscopy, lab-on-a-chip technology, field sensors or microfluidic devices [38]. Also in the literature are some examples of capillary electrophoresis microchips [41], single-drop biosensors [42], microfluidic SPE for sample preconcentration [43] and membrane implanted analysis for cholesterol [44]. In field determination of metals can be also performed by a spectrophotometric analysis system [45], by electrochemical stripping 46] or by a miniaturized stripping-based carbon modified sensor [47]. In-field determination of arsenic in natural waters can be performed by a bacteria based bioassay with biosensor-reporter strains as elements for the detection [48].

The liquid-phase microextraction techniques can also be used for in-field sampling in the framework of Green Analytical Chemistry using less toxic extractant phases and new materials (nanomaterials, monolithic silica, etc.) [49].

5.5 Environmentally friendly sample stabilization

Sample stabilization is required when analyte losses or degradations occur immediately after sampling. If the stabilization of the samples implies the use of chemical preservatives the time between the sampling and the analysis would be reduced until the period of time in which the analytes are stable. The analysis in situ is a good option to overcome this problem; however it is not possible in all cases. Also, the reduction of the size of the analyzed samples (reduction of scale of determination) makes possible the elimination of chemical preservatives to stabilize the sample during their transportation and storage [3], in this sense the lab-on-a-chip analysis and immunoassays are very useful.

An important aspect in the analysis of biological samples is gathering and collection of samples and their preservation until the analysis. Today, billions of biological specimens and samples collected by researchers in academia, research institutes, hospitals and commercial organizations are often stored in cold environments (refrigeration at –4°C, low- and ultralow- temperature freezers at –50°C to –80°C, and liquid nitrogen at –170°C) but this practice has shortcomings since cold refrigeration/freezers produce hydrofluorocarbons, which are some of the most potent greenhouse gas pollutants. In addition, the typical ultra-low-temperature freezer consumes a lot of energy releasing carbon dioxide. In recent years, new technologies have been developed for the stabilization and storage of biological samples at room temperature. These technologies can be based on extremophile biology that allows some organisms to survive in a dry state (anhydrobiosis) for more than 100 years. Anhydrobiotic organisms can protect their DNA, RNA, proteins, membranes and cells for long term survival in a dried state and later be revived by simple rehydration [50,51].

On the other hand, polyethylene glycol (PEG) is found to be an inexpensive, non-toxic and an environmentally friendly stabilization medium for the samples. Moreover PEG can be recovered and reused. PEG has been added to hydroxyapatite nano-particles medium system for sample stabilization [52].

Another example is that insoluble Fe(III) particulates can be formed in samples that could hinder subsequent extraction and analysis. To overcome this problem, the samples are usually stored in an acidic medium but acids are not environmentally friendly; they are also toxic and dangerous. The use of ascorbic acid to reduce Fe(III) colloids and/or to prevent further oxidation of Fe(II) by dissolved oxygen during sampling was investigated and compared to sample stabilization with hydrochloric acid resulting with the first one being the more efficient [53].

Finally, SPE cartridges can also be used to stabilize and store the samples. Some cartridges (Hysphere-GP, Hysphere-SH, PRP-1 and PLRP-S) have been investigated in order to evaluate the stability of different analytes [54].

5.6 Sampling for automatization

The reduction of labor and energy consumption can be achieved by automation and robotization of analysis, the parallel determination of many analytes in a single analytical cycle and by the utilization of hyphenated analytical techniques such as solid phase extraction-gas chromatography, solid phase extraction-liquid chromatography (SPE-GC/SPE-HPLC), and supercritical fluid extraction-supercritical fluid chromatography (SFE-SFC). In this way we can reduce energy consumption per analysis or per analyte. The automatization also reduces sample size as well as solvent and reagent consumption. Some examples of analytical techniques designed for this purpose are flow injection analysis (FIA), sequential injection analysis (SIA) and multicommutation. Liquid phase microextraction procedures (LPME) [49] and SPME (solid phase microextraction) can also be automatized.

Generally, these techniques allow successive sampling and automatized analysis, especially SPME and LPME. As above mentioned, SPME integrates sampling, extraction, concentration and sample introduction into a single solvent-free step. SPME reduces the time necessary for sample preparation, decreases purchase and disposal costs of solvents and can improve detection limits. There are several examples in the literature such as the sampling and sample preparation by SPME for analysis of indoor and ambient air [55]. In this work the authors demonstrated that SPME has great potential to be used for sampling VOCs.

The automation of LPME improves reproducibility compared to manual operation. Most efforts made in automating LPME have focused on dynamic LPME, mainly due to the difficulty in performing repeated movements of a microsyringe manually. Also, a semi-automated method, based on sequential injection analysis, for combining direct SDME and ETAAS has been reported [56].

5.7 Future possibilities in green sampling

Green sampling is perhaps one of the more important and critical aspects of Green Analytical Chemistry. The reduction or elimination of toxic reagents used for sample preservation and stabilization, as well as waste produced in sampling, such as filters and contaminated containers represent a serious challenge for the analytical chemist. Additional contaminants agents, in particular energy consumption to produce cryogenic conditions for sample storage and greenhouse gases emission, increase the problems related to sampling and sample preparation. Due to the great amount of samples produced for environmental, health and food quality monitoring, green sampling is a valuable issue. However, most of contributions in this area are focused on sample preparation and green analytical techniques, but more attention is required on the sampling itself.

The more relevant issues in greening sampling are related to techniques for in situ sample extraction and storage such us the solid phase extraction and solid phase microextraction. SPME integrates sampling, extraction, concentration and sample introduction into a single solvent-free step. Another alternative for in-field sampling is supercritical fluid extraction that allows high preconcentration rates, cleanliness and safety and quantitative capability. The in-field analysis also overcome some problems, but requires the development of reliable and selective probes, solventless techniques for sample collection and analytes extraction and the use of non-conventional methods of analysis involving flow-through solid-phase spectroscopy, hollow-fiber combined with chromatography and nanoparticle-based analytical methods that allow reagents and solvent reduction, and sample rate increasing. The real time measurements systems like X-ray fluorescence spectrometry, attenuated total reflectance infrared spectroscopy, lab-on-a-chip technology, field sensors or microfluidic devices seems to be good alternatives for Green Analytical Chemistry. Of course, the use of capillary electrophoresis microchips, single-drop biosensors, microfluidic SPE for sample preconcentration and membrane implanted analysis represents very advisable tools in this field.

Finally, automation of analysis contributes to labor and energy consumption and can be combined with the above approaches for green analysis and friendly environmental sampling, although many further studies are still required in this area. In this sense, techniques for successively sampling and automatized analysis such us SPME and LPME can be reliable alternatives in Green Analytical Chemistry.

References

1. Anastas, P.T. (1998) *Green Chemistry. Theory and Practice*, Oxford University Press, New York, p. 30
2. ACS Green Chemistry Institute, (n.d.) *Green Chemistry*. Available at www.acs.org/education/greenchem/principles.html (accessed November 1, 2011).
3. Namiesnik, J. (2001) Green Analytical Chemistry - Some remarks, *J. Sep. Sci.*, **24**, 151–153.
4. Ruzicka, J. and Hansen, E. (1985) Optosensing at active surfaces—A new detection principle in flow injection analysis. *Anal. Chim. Acta*, **173**, 3–21.

5. Llorent-Martinez, E.J.; Ortega-Barrales, P. and Molina-Diaz, A. (2008) Sequential injection multi-optosensor based on a dual-luminescence system using two sensing zones: Application to multivitamin determination, *Microchim. Acta*, **162**, 199–204.
6. Llorent-Martinez, E.J.; Dominguez-Vidal, A.; Ortega-Barrales, P.; de la Guardia, M. and Molina-Diaz, A. (2005) Implementation of multicommutation principle with flow-through multioptosensors, *Anal. Chim. Acta*, **545**, 113–118.
7. López-Flores, J.; Fernández-de Córdova, M.L. and Molina-Díaz, A. (2005) Implementation of flow-through solid phase spectroscopic transduction with photochemically induced fluorescence: Determination of thiamine, *Anal. Chim. Acta*, **53**, 161–168.
8. López-Flores, J.; Fernández-De Córdova, M.L. and Molina-Díaz, A. (2007) Development of a photochemically induced fluorescence-based optosensor for the determination of imidacloprid in peppers and environmental waters, *Talanta*, **72**, 991–997.
9. Teixeira, S.G.L. and Rocha, F.R.P. (2007) A green analytical procedure for sensitive and selective determination of iron in water samples by flow-injection solid-phase spectrophotometry, *Talanta*, **71**, 1507–1511.
10. Ruiz-Medina, A.; Fernández-de Córdova, M.L.; Ayora-Cañada, M.J.; Pascual-Reguera, M.I. and Molina-Díaz, A. (2000) A flow-through solid phase UV spectrophotometric biparameter sensor for the sequential determination of ascorbic acid and paracetamol, *Anal. Chim. Acta*, **404**, 131–139.
11. Ruiz-Medina, A.; Ortega Barrales, P.; Fernández-de Córdova, M.L. and Molina-Díaz, A. (2002) A continuous flow solid phase spectroscopic sensor using two sensing zones: Determination of thiamine and ascorbic acid, *J. AOAC Intern.*, **85**, 369–374.
12. Zhu, H.; Chen, H. and Zhou, Y. (2003) Determination of thiamine in pharmaceutical preparations by sequential injection renewable surface solid-phase spectrofluorometry, *Anal. Sci.*, **19**, 289–294.
13. Llorent-Martinez, E.J.; Ortega-Barrales, P. and Molina-Diaz, A. (2008) Sequential injection multi-optosensor based on a dual-luminescence system using two sensing zones: Application to multivitamin determination, *Microchim. Acta*, **162**, 199–204.
14. Passos, M.L.C.; Santos, A.M.; Pereira, A.I.; Santos, J.R.; Santos, A.J.C.; Saraiva, M.L.M.F.S. and Lima, J.L.F.C. (2009) Estimation of postmortem interval by hypoxanthine and potassium evaluation in vitreous humor with a sequential injection system, *Talanta*, **79**, 1094–1099.
15. Silva, M.B.; Borges, S.S.; Perdigo, S.R.W. and Reis, B.F. (2009) Green Chemistry-Sensitive Analytical Procedure for Photometric Determination of Orthophosphate in River and Tap Water by Use of a Simple LED-Based Photometer, *Spectros. Lett.*, **42**, 356–362.
16. Llorent-Martinez, E.J.; Ortega-Barrales, P. and Molina-Diaz, A. (2006) Multi-commutated flow-through multi-optosensing: A tool for environmental analysis, *Spectrosc. Lett.*, **39**, 619–629.
17. García-Reyes, J.F.; Gilbert-López, B. and Molina-Díaz, A. (2009) Flow-Through Solid-Phase Spectroscopy: A Contribution to Green Analytical Chemistry, *Spectrosc. Lett.*, **42**, 383–393.
18. Luo, Y.Z. and Pawliszyn, J. (2000) Membrane Extraction with a Sorbent Interface for Headspace Monitoring of Aqueous Samples Using a Cap Sampling Device, *Anal. Chem.*, **72**, 1058–1063.
19. Pedersen-Bjergaard, S. and Rasmussen, K.E. (1999) Liquid-Liquid-Liquid Microextraction for Sample Preparation of Biological Fluids Prior to Capillary Electrophoresis, *Anal. Chem.*, **71**, 2650–2656.
20. Rasmussen, K.E.; Pedersen-Bjergaard, S.; Krogh, M.; Ugland, H.G. and Grønhaug, T. (2000) Development of a simple in-vial liquid-phase microextraction device for drug analysis compatible with capillary gas chromatography, capillary electrophoresis and high-performance liquid chromatography, *J. Chromatogr. A*, **873**, 3–11.
21. Zhu, L.; Ee, K.H.; Zhao, L. and Lee, H.K. (2002) Analysis of phenoxy herbicides in bovine milk by means of liquid–liquid–liquid microextraction with a hollow-fiber membrane, *J. Chromatogr. A*, **963**, 335–343.
22. Cheng, G.W.; Lee, Ch.F.; Hsu, K.Ch.; Wu, H.L and Huang, Y.L. (2008) On-line microdialysis–nano-Au/TiO2–high-performance liquid chromatography system for the simultaneous determination of cobalt and nickel in water, *J. Chromatogr. A*, **1201**, 202–207.
23. Lee, P.L.; Sun, Y.Ch. and Ling, Y.Ch. (2009) Magnetic nano-adsorbent integrated with lab-on-valve system for trace analysis of multiple heavy metals, *J. Anal At. Spectrom.*, **24**, 320–327.
24. Stege, P.W.; Messina, G.; Bianchi, G.; Olsina, R.A. and Raba, J. (2009) Determination of arylsulphatase and phosphatase enzyme activities in soil using screen-printed electrodes modified with multi-walled carbon nanotubes, *Soil Biol. Biochem.*, **41**, 2444–2452.

25. Bonastre, A.; Ors, R.; Capella, J.V.; Fabra, M.J. and Peris, M. (2005) In-line chemical analysis of wastewater: present and future trends, *Trends Anal. Chem.*, **24**, 128–137.
26. Melchert, W.R.; Infante, C.M.C. and Rocha, F.R.P. (2007) Development and critical comparison of greener flow procedures for nitrite determination in natural waters, *Microchem. J.*, **85**, 209–213.
27. Melchert, W.R. and Rocha, F.R.P. (2005) A green analytical procedure for flow-injection determination of nitrate in natural waters, *Talanta*, **65**, 461–465.
28. Chen, Y. and Pawliszyn, J. (2004) Kinetics and the on-site application of standards in a solid-phase microextraction fiber, *J. Anal. Chem.*, **76**, 5807–5815.
29. Kenessov, B.N.; Koziel, J.A.; Grotenhuis, T. and Carlsen, L. (2010) Screening of transformation products in soils contaminated with unsymmetrical dimethylhydrazine using headspace SPME and GC-MS, *Anal. Chim. Acta*, **674**, 32–39.
30. Grossherr, M.; Varadarajan, B.; Dibbelt, L.; Schmucker, P.; Sedemund-Adib, B.; Gehring, H. and Hengstenberg, A. (2009) Measurement of propofol exhalation in pigs, humans and sheep during anesthesia - The impact of reference methods, *IFMBE Proceedings*, **25**, 678–680.
31. Musteata, F.M. and Pawliszyn, J. (2007) In vivo sampling with solid phase microextraction, *J. Biochem. Biophys. Methods*, **70**, 181–193.
32. Soini, H.A.; Bruce, K.E.; Klouckova, I.; Brereton, R.G.; Penn, D.J. and Novotny, M. (2006) In situ surface sampling of biological objects and preconcentration of their volatiles for chromatographic analysis, *Anal. Chem.*, **78**, 7161–7168.
33. Wrobel, K.; Kannamkumarath, S.; Wrobel, K. and Caruso, J.A. (2003) Environmentally friendly sample treatment for speciation analysis by hyphenated techniques, *Green Chem.*, **5**, 250–259.
34. López-Roldán, P.; López De Alda, M.J. and Barceló, D. (2004) Simultaneous determination of selected endocrine disrupters (pesticides, phenols and phthalates) in water by in-field solid-phase extraction (SPE) using the prototype PROFEXS followed by on-line SPE (PROSPEKT) and analysis by liquid chromatography-atmospheric pressure chemical ionisation-mass spectrometry, *Anal. Bioanal. Chem.*, **378**, 599–609.
35. Gómez-Ariza, J.L.; Giraldez, I.; Morales, E.; Ariese, F.; Cofino, W. and Quevauviller, Ph. (1999) Stability and storage problems in organotin speciation in environmental samples, *J. Environ. Monitor.*, **1**, 197–202.
36. Alonso, M.C. and Barceló, D. (2002) Stability study and determination of benzene- and naphthalenesulfonates following an on-line solid-phase extraction method using the new programmable field extraction system, *Analyst*, **127**, 472–479.
37. Lafleur, J.P.; Rackov, A.A.; McAuley, S. and Salin, E.D. (2010) Miniaturised centrifugal solid phase extraction platforms for in-field sampling, pre-concentration and spectrometric detection of organic pollutants in aqueous samples, *Talanta*, **81**, 722–726.
38. Keith, L.H.; Gron, L.U. and Young, J.L. (2007) Green Analytical Methodologies, *Chem. Rev.*, **107**, 2695–2708.
39. Crumbling, D.M. (2004) *Summary of the Triad Approach, US Environmental Protection Agency*. Available at: Triad Resource Centre, *http://www.triadcentral.org/ref/doc/triadsummary.pdf.* (accessed November 1, 2011).
40. Bamford, S.A.; Wegrzynek, D.; Chinea-Cano, E. and Markowicz, A. (2004) Application of X-ray fluorescence techniques for the determination of hazardous and essential trace elements in environmental and biological materials, *Nukleonika*, **49**, 87–95.
41. Wang, J.; Chatrathi, M.P. and Tian, B. (2000) Micromachined separation chips with a precolumn reactor and end-column electrochemical detector, *Anal. Chim. Acta*, **416**, 9–14.
42. Moore, E.J.; Kreuzer, M.P.; Pravda, M. and Guilbault, G.G. (2004) Development of a rapid single-drop analysis biosensor for screening of phenanthrene in water samples, *Electroanalysis*, **16**, 1653–1659.
43. Yu, C.; Davey, M.H.; Svec, F. and Frechet, J.M. (2001) Monolithic porous polymer for on-chip solid-phase extraction and preconcentration prepared by photoinitiated in situ polymerization within a microfluidic device, *Anal. Chem.*, **73**, 5088–5096.
44. Kim, J.E.; Cho, J.H. and Paek, S.H. (2005) Functional Membrane-Implanted Lab-on-a-Chip for Analysis of Percent HDL Cholesterol, *Anal. Chem.*, **77**, 7901–7907.
45. Callahan, M.R.; Kaltenbacher, E.A. and Byrne, R.H. (2004) In-situ measurements of Cu in an estuarine environment using a portable spectrophotometric analysis system, *Environ. Sci. Technol.*, **38**, 587–593.

46. Wang, J.; Larson, D.; Foster, N.; Armalis, S.; Lu, J. and Rongrong, X. (1995) Remote electrochemical sensor for trace metal contaminants, *Anal. Chem.*, **67**, 1481–1485.
47. Palchetti, I.; Laschi, S. and Mascini, M. (2005) Miniaturised stripping-based carbon modified sensor for in field analysis of heavy metals, *Anal. Chim. Acta*, **530**, 61–67.
48. Diesel, E.; Schreiber, M. and Van Der Meer, J.R. (2009) Development of bacteria-based bioassays for arsenic detection in natural waters, *Anal. Bioanal. Chem.*, **394**, 687–693.
49. Peña-Pereira, F.; Lavilla, I. and Bendicho, C. (2010) Liquid-phase microextraction techniques within the framework of Green Chemistry, *Trends. Anal. Chem.*, **29**, 617–628.
50. Clement, O. (2009) *Biological Sample Storage and Management.* Lab Manager, 4, 26–29. Available at: http://www.labmanager.com/?articles.view/articleNo/3455/) (accessed November 1, 2011). Address for free magazine issue: http://photos.labmanager.com/magazinePDFs/2009/LMM-Oct-2009.pdf.
51. Frippiat, C.; Zorbo, S.; Leonard, D.; Marcotte, A.; Chaput, M.; Aelbrecht, C. and Noel, F. (2011) Evaluation of novel forensic DNA storage methodologies, *Forensic Sci. Int.: Genetics*, **5**, 386–392.
52. Tian, A.; Wang, C.; Xue, X.X.; Wu, A.H.; Guan, G.F.; Wang, L. and Qiu, B. (2010) Inhibitory effect of nano-HA particles on human U87 glioblastoma cells viability, *J. Inorg. Mat.*, **25**, 101–106.
53. Doran, G.; Helliwell, S. and Eberbach, P. (2005) Extraction of fipronil and thiobencarb from anaerobic water samples using solid-phase extraction, *J. AOAC Int.*, **88**, 854–859.
54. Patsias, J. and Papadopoulou-Mourkidou, E. (2000) Development of an automated on-line solid-phase extraction-high-performance liquid chromatographic method for the analysis of aniline, phenol, caffeine and various selected substituted aniline and phenol compounds in aqueous matrices, *J. Chromatogr. A*, **904**, 171–188.
55. Koziel, J.A. and Novak, I. (2002) Sampling and sample-preparation strategies based on solid-phase microextraction for analysis of indoor air, *Trends Anal. Chem.*, **21**, 840–850.
56. Peña, F.; Lavilla, I. and Bendicho, C. (2008) Immersed single-drop microextraction interfaced with sequential injection analysis for determination of Cr(VI) in natural waters by electrothermal-atomic absorption spectrometry, *Spectrochim. Acta, Part B*, **63**, 498–503.

6

Direct Analysis of Samples

Sergio Armenta and Miguel de la Guardia

Department of Analytical Chemistry, University of Valencia, Valencia, Spain

Direct analysis of samples is probably the greenest way to make an analytical determination because, on directly making the analyte measurements on untreated samples, the main part of the slow and expensive steps (in terms of time and handling) can be avoided as indicated in Figure 6.1; thus providing direct measurements suitable to obtain the analytical composition of samples without creating any trouble like the generation of waste and avoiding the use of reagents or solvents. So, in this chapter we will focus on the different strategies available to perform the analytical determinations and to obtain as information much as possible from the sample without any contact or after a simple physical treatment whilst also considering those techniques which can provide information about samples with a reduced sample damage.

6.1 Remote environmental sensing

Field-portable and low cost methods which avoid sample pretreatment for monitoring contaminants and toxins in the environment have received increased attention during recent decades due to the new green conscience of scientists. The ability to perform fast and on-site measurements provides significant advantages in environmental and industrial areas. Rapid decision-making offered by on-site monitoring saves costs and provides additional safety for workers and for the environment ahead of overexposure to toxic substances. Stand-off techniques which avoid the operator making contact with the sample, avoiding any sample treatment and those which involve just a physical interaction with the sample will be described, presenting some applications of this key strategy for greening the analytical methods. In this section we will present the main tools available in the scientific literature for remote sensing of chemicals in different media.

Handbook of Green Analytical Chemistry, First Edition. Edited by Miguel de la Guardia and Salvador Garrigues.

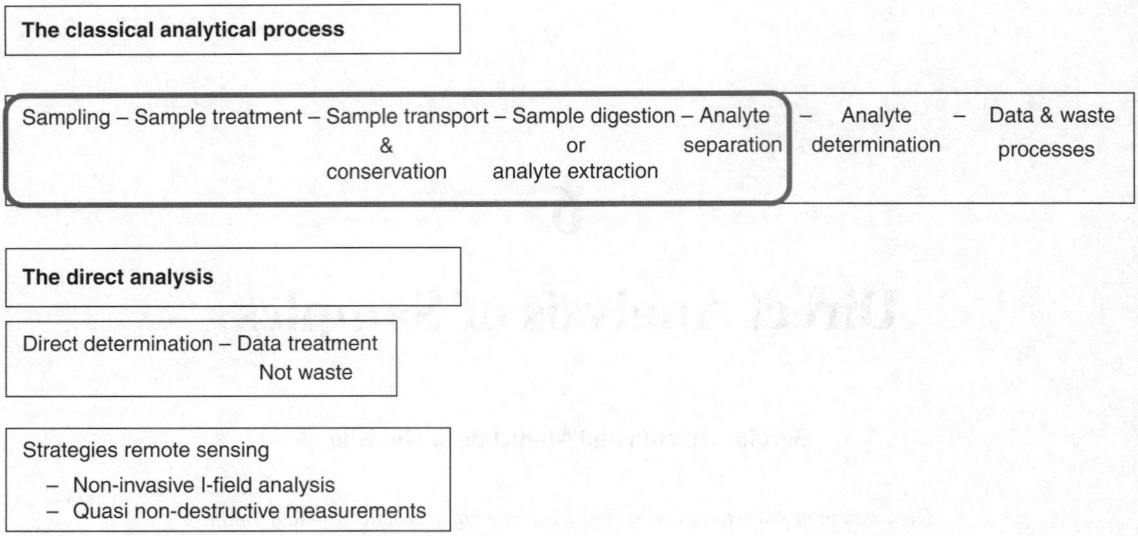

Figure 6.1 Evolution from classical multi-step analytical procedures to direct analysis, also indicating the main strategies available.

6.1.1 Synthetic Aperture Radar (SAR) images (satellite sensors)

Synthetic Aperture Radar (SAR) is a remote sensing system which captures two-dimensional images based on the reflection of the microwave backscattering properties of the surface. SAR deployed on satellites is an important tool today in remote sensing of chemicals and mapping of their distribution due to wide area coverage and day and night all-weather capabilities. SAR has been successfully employed for oil spill monitoring [1], surface soil moisture [2] and for monitoring vegetation recovery [3] among others.

6.1.2 Open-path spectroscopy

Open-Path Technologies (OP) measure the concentration of chemicals in an open path of air. In these techniques a beam of electromagnetic energy is emitted into the air and its interactions with the air components are measured. While some techniques are capable of measuring only a few compounds, others can measure a large number of analytes simultaneously.

6.1.2.1 FTIR (non-coherent and coherent light sources)

Open-path FTIR spectrometry has been under investigation since the beginning of the 1970s [4], but in recent decades the growing concern for the environment has moved this methodology to the forefront of environmental analytical techniques [5]. In Figure 6.2 different open-path FTIR configurations can be seen demonstrating the versatility of sampling geometries that can be achieved. The source is placed in front of the detection systems or the reflector optics, requiring an unobstructed line between them for VOCs monitoring. Three types of open-path FT-IR configurations are commonly used in air monitoring known as monostatic, bistatic and passive. In the monostatic configuration the source, optics and detector are in one end of the infrared path, while at the other end a reflector optic is placed. In the bistatic mode the source and source optics are separated from the collection optics and detector. In the passive configuration, the infrared source is generally

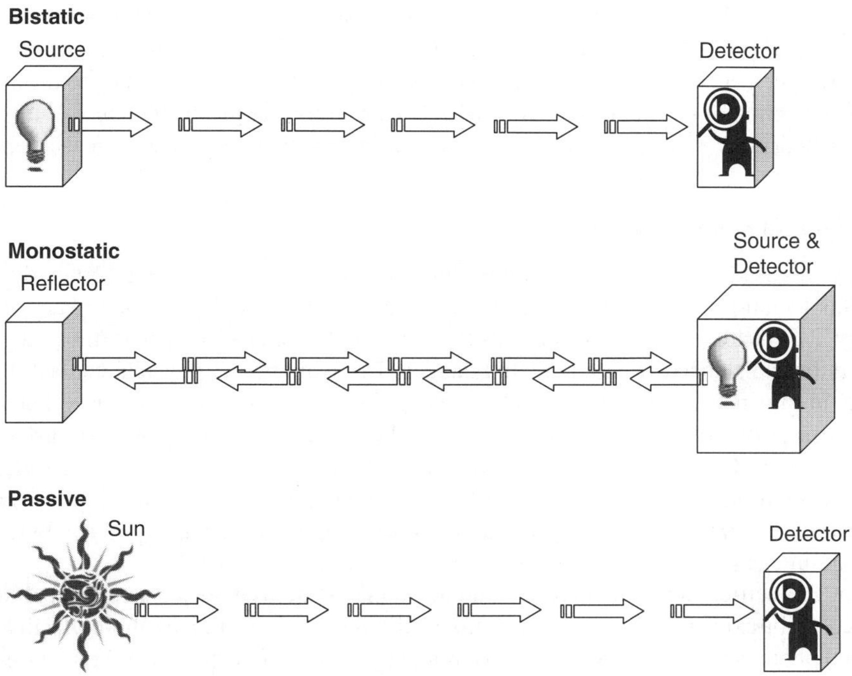

Figure 6.2 *Open-path IR spectroscopy geometries available.*

the Sun, a preheated plume, hot buildings or process equipment and the OP-FTIR consists of receiving optics and a detector. The sensitivity of the passive configuration is generally lower than that of the active method, but the monitored distances are longer (up to several kilometers).

The open path FTIR is a long-path technique with a relatively high sampling volume which means that a single open-path FTIR system, rather than numerous point samplers, can be used to locate and define a toxic gas plume. Another advantage is that the open-path FTIR method can do measurements in situ, thus avoiding sample collection and transport to the laboratory for analysis. Moreover, the open-path FTIR method measures in near real time, enabling the technology to be used in process control or to detect toxic species which represent an imminent hazard, thus providing fast information for making decisions.

Obviously, meteorological conditions can affect the monitoring system [6];for example, fog water droplets scatter infrared radiation precluding data collection. On the other hand, rain degrades the quality of data because of temperature fluctuations affecting the instrumentation and the large amount of water absorption observed in the spectra, but the data will be still usable. Variations of temperature also degrade quality of obtained results. The distance that can be monitored depends on the site geometry and the optical components used to generate the infrared beam. Although optical systems can be assembled for more than one kilometer path lengths, the practical limit of monitored distances for most commercially available active systems is approximately 500 m. Another drawback of the technique is the limit of detection (LOD), which for most of the molecules under study is between 1 and 100 parts per billion by volume (ppbv), whereas parts-per-trillion sensitivity is required in some cases [7].

Routine applications include monitoring of toxic chemicals in industrial environments [8] and the environmental determination of the most abundant and important greenhouse gases, such as H_2O, CO_2, CH_4, N_2O and CO [9].

Traditionally the OP-FTIR technique employs non-coherent light sources based on a resistive arrangement heated from 1000 to 1800°C. On the other hand, mid-IR coherent light sources such as lead-salt tunable diode lasers (TDL) have been successful used for the detection of NO, NO_2, CO and CO_2 for distances of several meters [10] and detection of CO at distances of 610 m [11]. Moreover, open path ozone, water vapor and CO_2 detection over distances of 5800 m, employing two quantum cascade lasers have been also reported [12].

6.1.2.2 Stand-off Raman spectroscopy

Raman spectroscopy is also a useful technique for interrogating the sample from a certain distance by using a telescope. This technique is usually called remote or stand-off Raman spectroscopy and it was explored theoretically by T. Hirschfeld [13]. As can be easily assumed, the main advantage of those stand-off systems is that the operator is distanced from a potential danger [14] avoiding the physical contact with the sample and enhancing in situ measurements possibilities [15]. Stand-off measurements have been employed to investigate the mineralogy of meteorites [16] and mineral characterization on planetary surfaces [17].

Raman-based stand-off systems have the potential to characterize materials through glass and plastic containers, allowing the analysis of the content without opening the container. This property has been used to discriminate between sodas, juices and liquors and hazardous liquids such as liquid explosives or chemical warfare agent simulants from a distance of 7 m approximately [18,19].

Two different geometries can be adopted in stand off Raman systems named coaxial and oblique which can also be referred as bi-axial (see Figure 6.3). Coaxial illumination maximizes the collection efficiency of Raman scattered photons and allows Raman measurements at different distances without system realignment [20]. In the oblique mode the laser is focused directly to the target sample, ensuring that all the laser power reaches the sample, and the telescope collects the scattered radiation at an oblique angle, determined by the distance from the sample to the Raman instrument. This geometry has provided increased sensitivity as compared to the coaxial mode [21].

The distances that separate sample and operator in stand-off Raman spectroscopy using the oblique geometry vary from 10 to 50 m. On the other hand, coaxial systems have provided useful Raman spectra of minerals organic and inorganic material at 9 to 120 m distance [22].

6.1.2.3 UV Differential Optical Absorption Spectroscopy (DOAS)

The DOAS technique is based on the fact that all chemical compounds absorb light at specific wavelengths. The spectroscopic absorption from various molecular species over a path length is proportional to the absorptivity and concentration of a compound and to the length of the light path. The technique has many advantages, including real-time measurements and the ability to simultaneously measure several species [23]. DOAS can be also considered as a clean technique, involving operation in the open air.

The DOAS equipment consists of an emitter, a receiver and an analyzer. The distance from the emitter to the spectrometer can vary from several meters to kilometers. DOAS can be grouped in active, in both monostatic and bistatic modes (similar to OP-FTIR), and passive mode, depending on the emitter used. In passive DOAS sunlight is employed as light source and it has been successfully used to measure stratospheric and tropospheric trace gases [24] and halogen compounds at various sites [25]. In the active DOAS, the ultraviolet light source can be a xenon arc lamp, a tungsten halogen or a deuterium lamp. Technological progress during recent decade has made LEDs as an interesting alternative to conventional broadband light sources [26].

A recently developed imaging DOAS (IDOAS) technique [27] has been used to obtain trace gas information resolved in two dimensions, allowing for a better understanding of atmospheric emissions, their transport, and relevant chemical reactions.

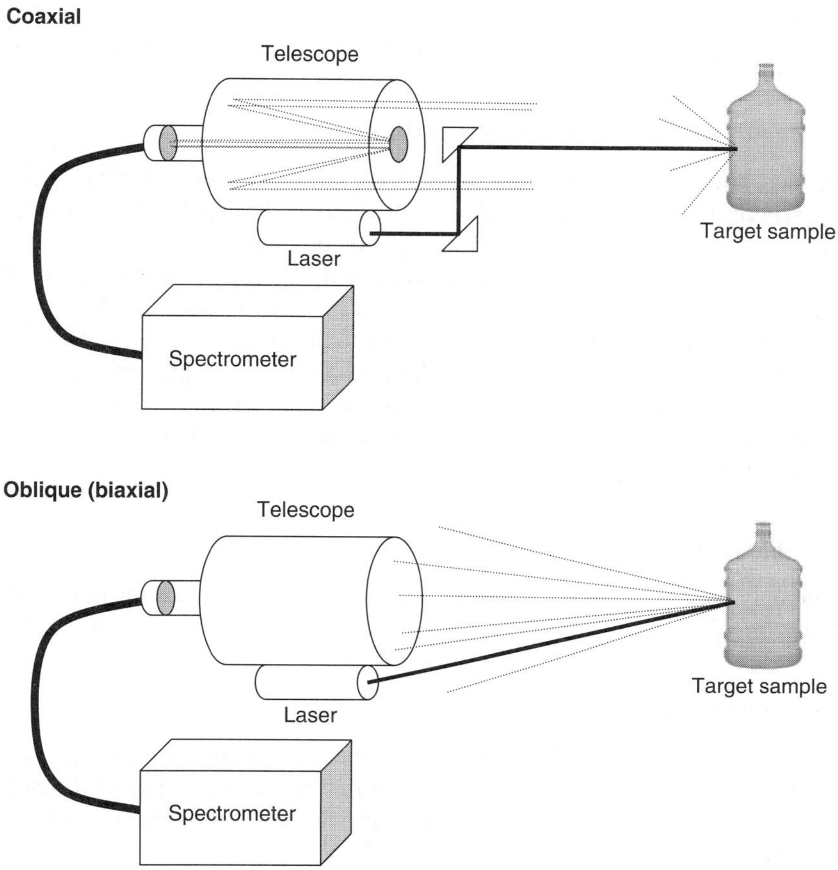

Figure 6.3 *Stand-off Raman spectroscopy geometries available.*

6.1.2.4 Light Detection and Ranging (LIDAR) and Differential Absorption LIDAR (DIAL)

LIDAR is based on similar principles than RADAR but it uses light to collect useful information [28]. The simplest LIDAR approach is used to determine the distance to a solid or hard target and, because of that, it has been used for coastal mapping [29] and to measure vegetation heights [30] among others.

On the other hand, differential absorption LIDAR (DIAL) can measure chemical concentrations in the atmosphere. The DIAL system remotely assesses concentrations of an atmospheric pollutant from the ratio between the backscattered light on and off a spectroscopic absorption line of the pollutant. The range resolution is obtained by emitting short pulses of light, combined with time-resolved acquisition of the elastically backscattered light. DIAL system consists of a dual laser system, a telescopic receiver and a detector system. It can be deployed in monostatic or bistatic configurations, depending on the position of the laser and the detector system. The monostatic configuration is usually employed for contaminant profiling and it can be performed in either a coaxial configuration, where the laser beam is transmitted within the receiver front, or in a biaxial mode, in which the laser is adjacent to the receiving unit. Tunable laser systems can be used as light source allowing for a fast change in the wavelengths to measure a wider range of contaminants.

Lidar techniques are useful for three-dimensional mapping of pollutants [31], remote monitoring of ambient air [32], industrial emissions [33] and natural emissions due to various geographical phenomena [34].

6.1.2.5 Laser Induced Breakdown Spectroscopy (LIBS)

In the LIBS technique, powerful laser pulses are focused on the target sample to form a light spark or plasma, resulting from the vaporization and atomization of a small amount of the target material. The spark contains the emission spectra of the elements within the plasma. Collection of the plasma light, followed by spectroscopic detection, allows identification of the elements due to their specific spectral signatures. LIBS technique has been shown to be useful at distances of up to 45m [35]. The main advantages of LIBS are: (1) fast analyses, (2) small analysis area of ≤1 mm diameter, (3) simultaneous quantitative and qualitative multi-element analysis capability, (4) relatively low detection limits and (5) and the possibility to provide depth profiles into the sample [36].

Stand-off LIBS has been used mainly for the analysis of solid samples with ranges up to 100 m for applications in industrial process control [37], nuclear power stations [38], environment survey [39], planetary exploration [40] and cultural heritage [41].

6.1.3 Field-portable analyzers

While many commonly used portable analyzers are based on electrochemical measures [42] some rely on spectroscopic means [43].

6.1.3.1 Electrochemical based analyzers

As has been mentioned, electrochemical analyzers meets many of the requirements for on-site and in situ measurements due to their intrinsic characteristics of portability and low power demand [44], offering tremendous potential for obtaining analytical information in a faster, simpler, greener and cheaper mode than laboratory-based assays.

Electrochemical devices have been successfully used for blood-glucose monitoring obtaining the desired clinical information in a fast and simple manner [45]. The pocket-sized analyzer is based on a ferricyanide or ferrocene-derivative mediator and its use is very simple: the diabetic patient pricks the finger, places the small blood droplet on the sensor strip, and obtains the blood-glucose concentration in few seconds [46]. Another example is the i-STAT Portable Clinical Analyzer which is able to perform eight common clinical tests simultaneously in about 90 s [47]. These tests are based on various enzyme electrodes, gas sensors, or ionselective electrodes, connected to both, amperometric and potentiometric, detection schemes. Moreover, portable (hand-held), battery-powered, easy-to-use stripping analyzers have been developed for measuring trace metals in clinical samples [48].

6.1.3.2 Ion Mobility Spectrometry (IMS) based analyzers

The interest of ion mobility spectrometry (IMS) as fast vapor sensors relies mainly on its high sensitivity, low cost and suitability for real-time monitoring purposes. IMS is based on the fact that ions created at ambient pressure from radioactive materials can be separated and characterized rapidly under weak electric fields [49]. The core of a mobility spectrometer is the drift tube, where ion formation and separation take place. The basic elements of the drift tubes are the radioactive ion source, ion shutter, drift rings and Faraday plate detector (see Figure 6.4).

A drawback of IMS is its reduced linear ranges, which has relegated the method to threshold-detector status. This was partially solved through the addition of a mechanical servo inlet for dilution of sample flow [50], expanding the linear range by a factor of 100 to 1000.

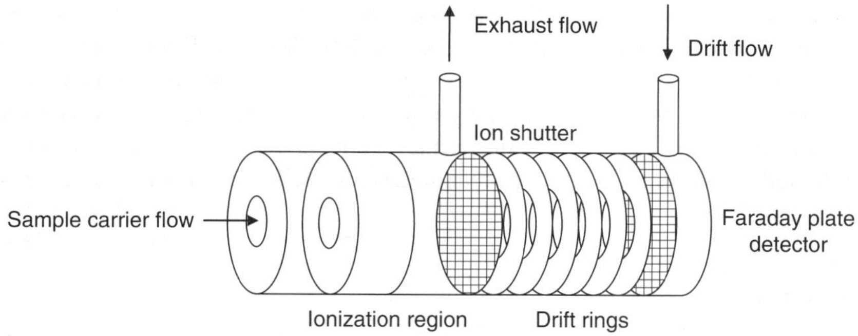

Figure 6.4 *Basic scheme of an ion mobility spectrometer.*

Several applications of IMS can be highlighted such as fast screening of explosives in security checkpoints [51], meat freshness through the analysis of biogenic amines [52] and air quality monitoring in industrial or laboratory areas [53]. Although IMS has been conceived for monitoring volatile analytes, the on-line coupling of efficient sample introduction systems allows the fast monitoring of solid and liquid samples [54].

6.1.3.3 Other spectroscopy based portable analyzers

Colorimetric and UV/Vis spectrophotometric monitors for gases and vapors are widely used in the industry. Colorimetric analyzers use a reaction between a selective reagent and the target analyte in order to form a colored complex which is measured in the visible range. Those analyzers provide limits of detection of the order of parts per million (ppm), offering a continuous monitoring capability. They are available as portable, battery-powered devices for personal monitoring and equipped with external power sources for static monitoring of a predetermined area. There are also chemiluminescent detectors for measuring ozone and nitrogen oxides, based on the excitation of O_3 molecules via chemical reaction and the subsequent detection of photon emission from the excited state species. Luminol and ethylene have been used as reactants in the determination of NO_2 and ozone. Photometric analyzers for the determination of sulfur or phosphorus compounds are commonly used for real time monitoring with detection limits in the parts per billion range. Moreover, fluorescence analyzers are commercially available for monitoring CO and SO_2 providing detection limits in the part per billion (ppb) range.

It should be also mentioned the usefulness of IR hand held analyzers. Their lightweight and compact size is perfect for applications that require repetitive analyses on large surfaces or in locations that are physically restricted. An example of this type of IR analyzers is the Flexscan and Exoscan from Agilent Technologies [55]. At the same line, Axsun has developed the AnavoTM Analyzer which is a hand held portable NIR analyzer designed specifically for industrial applications [56].

6.2 Process monitoring: in-line, on-line and at-line measurements

Measurement of safety and quality in industrial processes requires an on-line detection technique with the following properties: (1) it should be compact and robust to be mounted in or close to the production line and take place under realistic environment, (2) provide real time, (3) non-destructive measurements, (4) relatively low cost and (5) permanent monitoring. Raman and FTIR, in the mid (MIR) and near (NIR) range, spectroscopies are considered as privileged methods for in- or on-line process monitoring applications due to their real time analysis capability, molecular specific nature and well-resolved bands which offer the possibility of qualitative and quantitative analysis. Moreover, NIR, MIR and Raman spectroscopic techniques also provide the possibility to use the spectrometers in a remote manner through the use of fiber-optic waveguides. In the case of in situ ATR FTIR monitoring, the use of remote probes with long distances is limited by the absorption of light by fiber optics, which is not the case for NIR or Raman spectroscopic equipments. Additionally, both Raman and NIR techniques are well suited for measurements through glass or plastic media.

6.2.1 NIR spectroscopy

NIR spectra can be measured by different sampling modes fitting different applications. In practice, the common modes are transmittance, interactance, transflectance, diffuse transmittance, and diffuse reflectance, with the last two being most frequently used. Diffuse transmittance measurements are usually performed from 800 to 1100 nm using sample thickness of 1–2 cm. On the other hand, diffuse reflectance measurements are performed in the range from 1100 to 2500 nm with sample thickness of 1 cm.

At present, NIR technique is widely accepted as one of the most promising on/in-line process control techniques due to its nondestructive nature, reliability and accuracy for monitoring chemical and physical parameters during industrial processing [57]. Filter instruments, which fulfill the criteria of compactness, robustness and reduced cost, working in the NIR range can be considered a rapid technique. There are three types of NIR on-line analyzers: remote sensors, bypass analyzers and fiber optic probes.

The first on-line NIR sensor was based on a sensing head remote approximately 200 mm at an angle of 60° from the horizontal of the flow sample. Their advantages rely on the low cost of instrumentation and simplicity of installation. However, it is susceptible to interference from ambient light variations, dust build-up on the optical surfaces and atmospheric humidity variations. Applications of this type of sensor in the food industry include moisture determination on both continuous samples such as powders and discontinuous samples such as bread [58], control of protein in flour [59] and meat composition analysis [60].

NIR by-pass samplers were first developed to measure the protein content of flour and this is still the most popular application of a NIR feedback control system. In this method, the sample must be held against the optics window during the measurement cycle. These conditions were met by taking direct samples from the flour stream, either from gravity-fed spouting or a positive pressure blow line, into a mechanism designed to reproduce the action of a human operator in packing flour inside a sample cup [61].

Finally, fiber-optic probes have the widest range of applications in on-line analysis. Laboratory NIR analyzer platforms have been modified, enclosed in process hardened housing, and used extensively for material identifications in the pharmaceutical [62] and the nutraceutical industries [63].

6.2.2 Raman spectroscopy

Raman spectroscopy is a non-destructive technique which provides rapid analysis of untreated samples. Raman spectra are relatively less complicated than NIR ones presenting well-resolved peaks. Thus, the interpretation of Raman data is easier and does not necessarily require a chemometric analysis. Moreover, water is a weak Raman scatterer, which makes the technique attractive for many industrial applications involving aqueous media. The main drawbacks of Raman spectroscopy are its poor sensitivity, which limits the application range and the possible fluorescence of samples when high energy lasers are used.

Different laser sources can be used in Raman spectroscopy, from Ar^+- based, He–Ne or Kr^+ lasers with lines in the range of 470–650 nm to diode (780 nm) and Nd:YAG lasers that provide the 1064 nm line.

Increasing the laser power increases the analytical sensitivity but at the same time increases the possibility of sample fluorescence. There are two possible configurations in Raman spectroscopy depending on the way the light is delivered and collected from the sample, 180° and 90° configurations. Moreover, optical fibers can be used to excite the sample and collect the Raman scattering.

Related to quantitative analysis of samples by Raman spectroscopy it should be highlighted a possible problem of sampling representivity. With homogeneous liquid samples there is no problem in acquiring spectra which accurately reflect the composition of the sample. However, for solid samples, such as tablets or pills, this could be a problem. Thus, the accuracy of the measurements will depend on the heterogeneity of the sample and the spot diameter of the laser beam. Fourier transform instruments provide a 1 mm spot diameter and more than 1 mm penetration depth. Even the relatively large diameter spot sizes used in FT spectrometers are typically not enough to eliminate sampling errors. A way to resolve sampling problems in Raman spectroscopy is to rotate the sample under the beam so that the laser probes a circular track in the sample. Another alternative is the use of grid sampling using motorized and automated x,y stages.

For on-line and in-line applications of Raman spectroscopy, the spectrometer is interfaced to the sampling location using conventional optical fiber cables [64], enabling remote sampling at tens or even hundreds of meters from the spectrometer. The main advantage of Raman spectroscopy is the possibility to qualitatively and quantitatively analyze materials inside glass containers, polymeric bottles and blisters. Raman spectroscopy has been used to detect liquid explosives within bottles and other plastic packaging [65], illegal drugs dissolved in beverages [66], active ingredients in pharmaceuticals inside plastic bottles [67], amber vials [68] or capsules [69], and ethanol content of beverages inside glass bottles [70] among others.

6.2.3 MIR spectroscopy

The advent of the attenuated total reflectance (ATR) sampling technique [71] has liberated the users of mid-IR spectroscopy from laborious sample preparation techniques for obtaining high-quality spectra suitable for performing quantitative analysis. An infrared beam is directed to an optically dense crystal with a high refractive index, called an internal reflection element, at a certain angle, being created an evanescent wave that extends beyond the surface of the crystal into the sample held in contact with the crystal some micrometers [72].

ATR spectroscopy can perform at-line or on-line measurements, being the most commonly used the rapid at-line measurements where samples are taken from the production line and analyzed in the laboratory within a matter of minutes [73]. It is clear that such at-line measurements provide an improvement over traditional off-line methods. However, a most appropriate approach is the on-line monitoring using fiber optic ATR probes [74], enabling accurate determination of analytes in real time. It is possible to construct chalcogenide glass fibers, but they have poor physical and transmission characteristics compared to those available for the NIR and UV-visible regions [reference]. Therefore, these fibers are restricted in length, from 1 to 5 m, and have to be heavily armored and protected from physical strain.

6.2.4 Imaging technology and image analysis

Chemical imaging systems are capable to provide spectral and spatial information simultaneously. The spatial locations of the spectra identify chemical species inside the samples and map their distributions. It is possible to obtain a chemical image using three techniques: (1) Point mapping, (2) line imaging and (3) focal plane array (FPA) detectors. The first technique is based on the combination of a classic spectrometer and a moving stage [75], obtaining the image point by point. In the in-line imaging, the detector acquires the spatial and spectral dimensions simultaneously, similarly to point by point but line by line [76].

FPA enables thousands of spectra to be acquired simultaneously [77]. Development of FPA detectors has increased the popularity of chemical imaging [78]. Acquisition of high quality images is possible in the NIR and MIR ranges and also with Raman spectroscopy in a few minutes. Applications of chemical imaging have since increased in various fields, from waste sorting [79] to biological tissue [80] and food quality [81], but especially in the pharmaceutical sector [82].

6.3 At-line non-destructive or quasi non-destructive measurements

6.3.1 Photoacoustic Spectroscopy (PAS)

Photoacoustic spectroscopy (PAS) is a versatile sampling technique which does not require that the sample be transmitting of the incident light and can provide depth profile analysis from several micrometers to more than 100 μm. PAS indirectly measures IR absorption of sample by sensing absorption-induced heating of the sample that is transferred to the surrounding gas and produces a thermal-expansion-driven pressurization, which is detected by a microphone (see Figure 6.5) [83]. The capability of varying sampling depth is a key feature of PAS.

Some limitations can be found to analyze heterogeneous samples due to the fact that commonly used sample holder cups are 10 mm in diameter by 6 mm thick. Thus, to assure that the obtained data are representative of the whole sample a fine or coarse powdering of the sample may be necessary followed by splitting out and analyzing triplicate samples or it may be necessary to average the spectra of a number of randomly selected samples [84]. PAS analysis of liquid samples involves situations where the sample is unsuitable for transmission measurements due to its high opacity, high viscosity, suspended components and/or risk of cross contamination. PAS sampling of liquids usually uses disposable cups, which eliminates cross contamination and clean-up problems.

Applications of PAS include determination of drug content in semi-solid formulations [85], pesticide determination in agrochemicals [86] and trace gas monitoring [87] among others.

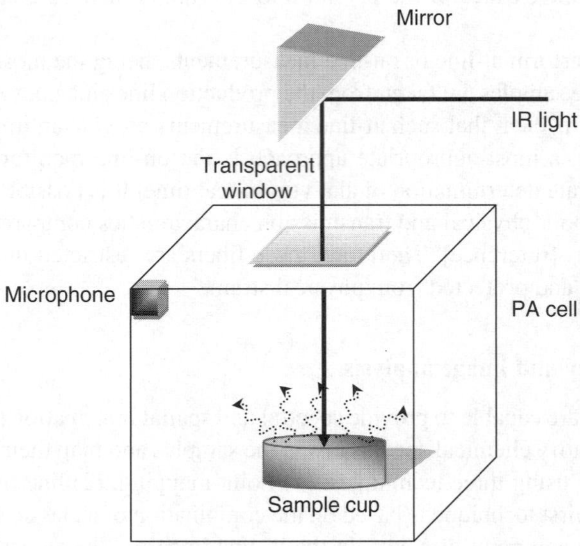

Figure 6.5 *The main principle of photoacoustic spectroscopy.*

6.3.2 Ambient Mass Spectrometry (MS)

Traditional ionization sources such as electrospray ionization (ESI), atmospheric pressure chemical ionization (APCI) and atmospheric pressure photo ionization (APPI) provided an efficient way to separate analytes from the solution-phase matrix. So, to transfer free ions from solution at atmospheric pressure into the high-vacuum environment required for analysis by MS. However, those ionization methods still require extensive sample-preparation steps before the sample can be dissolved in an appropriate solvent which is commonly required by the analytical system. The main advantage of ambient desorption ionization MS is that it provides the direct analysis of samples in the open atmosphere with little or no sample preparation, which offers a much simplified work flow and ease of use to MS analyses.

The development in the past decade of new ambient desorption ionization sources such as desorption electrospray ionization (DESI), direct analysis in real time (DART) and flowing atmospheric-pressure afterglow (FAPA), became possible to analyze samples directly in their native form, avoiding sample preparation steps in the analytical procedure. The development of DESI in 2004 by Takats *et al.* [88] created an awareness of the potential of open ambient environment analysis and sparked a new sub-field in MS and the utility of this concept can be demonstrated by the rapid introduction of a lot of different new ambient desorption ionization methods in the last decade [89]. In this excellent review article, the different developed ambient ionization techniques (more than 15) can be grouped according to the traditional technique that plays the central role in the ionization process (ESI, APCI or APPI).

Since the introduction in 2004, ambient desorption ionization MS has been used in numerous areas of application, including forensics, imaging, metabolomics, pharmaceuticals, characterization of natural products, bacteria, polymers, proteins, and explosives detection, providing detection limits in the femtomol range. The introduction in recent years of the use of internal standardization has been improved the quantitative capabilities of this technique [90].

6.3.3 Solid sampling plasma sources

Solid sampling devices have been used to vaporize liquid and solid samples into flame and plasma for many years [91]. Different sampling devices have been successfully developed for the vaporization of the analytes directly from solid samples. Among them, arc and spark, glow discharge and laser ablation should be highlighted. The combination of these solid sampling systems with inductively coupled plasma (ICP) optical emission spectrometry (OES) or mass spectrometry (MS) have become a powerful tool for a fast and simple quantitative multi-elemental determination. In principle, any sample introduction method which is appropriate to be used in ICP-OES is also suitable for ICP-MS.

An arc is a continuous electrical discharge of high energy between two electrodes. In the direct current (DC) arc, the sample is usually packed into an anode electrode, or the sample itself becomes the anode [92]. One advantage of this methodology is the low detection limits achieved due to the relatively large amounts of excited sample. However, the unstable nature of the discharge leads to poor precision and the intensity of emission is highly dependent on the matrix. The use of an alternate current (AC) spark, which provides an intermittent discharge, greatly improves the precision of measurements but the sensitivity is poorer than in the DC arc due to the relatively small amount of vaporized sample. Modern electrical discharges combine the characteristics of both arc and spark to obtain optimal detectability and precision for quantitative work [93]. These instruments are referred to as direct reading spectrometers and, in conjunction with arc/spark sources, are widely used in the steel industry because they can be used for the rapid analysis of solid samples.

Glow discharge (GD), coupled to optical emission (GDOES) or mass spectrometry (GDMS), is an established technique nowadays for direct analysis of both, solid conducting and non-conducting samples [94]. A glow discharge is formed between two electrodes in an inert gas atmosphere at low pressure. The sample is usually the cathode and the wall of the discharge chamber is the anode. With DC glow discharges, non-conducting samples must be mixed with a conducting material and pressed into a pellet, while radiofrequency glow discharges allow the direct analysis of non-conducting samples. Glow discharges are used as versatile tools in analytical chemistry for the direct analysis of homogeneous solids, conducting and non-conducting materials, although the difficulties of standardization and the lack of certified reference materials often does not allow precise quantification for non-conductors [95].

The use of a laser beam for evaporating, atomizing and ionizing a sample is known as laser ablation (LA) [96]. The principle is similar to that used in LIBS, the sample is evaporated and partially atomized and ionized by the laser, but in LA-ICP-OES and LA-ICP-MS the vapor is transferred to the plasma via a transfer tube. Obviously, any type of solid sample can be ablated by LA and there are no sample-size requirements and no sample preparation procedures involved. Moreover, a focused laser beam allows spatial characterization of heterogeneity in solid samples, with micron resolution in terms of lateral and depth conditions. Applications of LA-ICP-OES/MS span a great range of industrial fields that include environmental [97], geological [98], archaeological [99] and forensic samples [100].

6.3.4 Nuclear Magnetic Resonance (NMR)

NMR spectroscopy probes atomic environments based on the different resonance frequencies exhibited by different nuclei in a strong magnetic field. Many nuclei are observable by NMR but those of hydrogen and carbon atoms are the most frequently studied. NMR spectroscopy of solutions is commonly used for structure elucidation and quantitative analysis. However, solid-state NMR measurements provide a tremendous capability to characterize solid samples with no sample treatment. The main advantages of solid-state NMR are its sensitivity and minimal sample preparation. However, the main drawback of the technique is the long data acquisition time and the relatively high price and maintenance cost of the instruments. That is why the technique is often not considered for routine analysis of samples [101].

Magnetic resonance imaging (MRI) is a non-invasive technique which can provide cross-sectional images from inside solid materials and living organisms. MRI is formed from the nuclear magnetic resonance (NMR) signal, which is generated by certain nuclei as 1H, 19F, 31P, and 13C, when subjected to a strong magnetic field and irradiated with radio waves [102].

6.3.5 X-ray spectroscopy

X-ray based spectroscopy methodologies involve the excitation of an atom by the removal of an electron from an inner energy level. Different methodologies can be identified based on the interaction of the X-ray and the samples such as X-ray fluorescence, X-ray absorption and Auger emission spectroscopy among others [103].

The excitation of the atoms by irradiation of the sample with X-rays of shorter wavelength than analyte elements and later emission of characteristic secondary fluorescent X-ray spectra is the principle of X-ray fluorescence analysis. After energy absorption, the atoms are in a highly excited states and an electron jumps from a high to a low shell thus emitting an X-ray photon. These secondary X-ray emissions are characteristics of the sample elements as their energy or wavelength is different for each element and each transition [104]. X-ray fluorescence is a valuable tool for the analysis of major constituents. Their main advantages are the capability to perform direct analysis of the elemental composition of solid substances, and the possibility of quantitative determinations with errors of the order of 5 to 10% without the use of a set of standards. However, the main limitation of the technique is its low sensitivity, the limit of detection being of the order of 10 ng.

It is possible to perform a point-by-point spatial distribution (100 μm^3) of the elements across the surface of the sample, by using a high-energy electron beam, as in an electron microscope and an X-ray fluorescence spectrometer. This technique is known as electron probe microanalysis or microbeam X-ray fluorescence spectrometry (μ-XRF) [105]. Excitation is restricted to thin surface layers because the electron beam penetrates to a depth of 1 or 2 mm into the specimen.

In X-ray photoelectron spectroscopy (XPS), the probe is an X-ray photon and the detected particle is the photoelectron emitted by it [106].

The excitation of the atoms by direct bombardment of the sample with electrons is the basis of Auger emission spectroscopy [107]. Once an atom is ionized, it relaxes by emitting an electron with an energy characteristic of the atom. The basic advantages of this technique are its high sensitivity for chemical analysis in the 5 to 20 Å region near to the surface, a rapid data acquisition speed, its ability to detect all elements above helium and its capability of high-spatial resolution making Auger emission spectroscopy a really important tool for solving problems related to surface analysis [108].

Although complex experimental equipment is required, measurement in X-ray absorption methods is straightforward. In X-ray absorption the intensity of an X-ray beam is diminished as it passes through the sample material.

6.3.6 Other surface analysis techniques

Other commonly used surface spectroscopy techniques for analyzing the composition and chemistry of solid surfaces are secondary-ion mass spectrometry (SIMS) and ion scattering spectroscopy (ISS).

In SIMS, secondary ion emission results from the collision of high energy particles with the sample. After the impact, most chemical bonds are broken and atoms or combinations of atoms are released from a few atomic layers of the sample surface [109]. In practice, the penetration depth is around a few tens of nm. However, the major part of the emitted ions arises from the atoms present in the first 2 nm. The use of mass analyzers; such as double focusing magnetic sector, quadrupole mass analyzer and time of flight spectrometer, allows the recording of sample images with an appropriate spatial resolution. Two types of SIMS can be identified: 'static'' SIMS, introduced by Benninghoven [110] and 'dynamic' SIMS [111]. In dynamic SIMS, elemental information can be extracted from the sample. On the other hand, static SIMS measurements provide molecular information.

Low-energy ion scattering (LEIS), also called ISS, is an unique tool in surface analysis, since it provides the atomic composition of the outer atomic layer. In LEIS a sample is bombarded with noble gas ions (He^+, Ne^+ or Ar^+) at an angle smaller than 60° respect to the surface normal and only projectiles that are backscattered into a certain angle, typically 140°, are analyzed [112]. For quantitative surface analysis, only the signal of scattered ions, S+, is analyzed. One of the main advantages of LEIS is that it offers the possibility to analyze a great diversity of materials, from metals, oxides and polymers to even liquid surfaces.

6.4 New challenges in direct analysis

In the previous sections a literature survey of the main techniques available for direct analysis of samples without any sample damage or with a reduced physical erosion of the sample surface has been made. In fact, all the mentioned techniques offer fast, easy to perform and environmentally friendly alternatives to the classical methods based on sample digestion and/or analyte extraction and, from our point of view, the main aspects of all discussed alternatives comes from the fact that no reagents and no sample treatments are required; thus avoiding the use of toxic reagents and the generation of dangerous wastes. However, additional efforts are required in order to improve standardization of measurements and also chemometric efforts will be welcome in order to obtain as much as possible information from the sample signals. So, in the next few years we could perform many new applications of remote sensing and non-invasive methods based on the improvement of data analysis and it will also be time for ablation and surface analysis techniques which can enhance our knowledge about new materials and their behavior. On considering vibrational techniques, the surface enhanced Raman and FTIR spectroscopies will also offer tremendous capabilities, especially based on the use of nanomaterials.

References

1. Brekke, C. and Solberg, A.H.S. (2005) Oil spill detection by satellite remote sensing, *Remote Sens. Environ.*, **95**, 1–13.
2. Giacomelli, A.; Bacchiega, U.; Troch, P.A. and Mancini, M. (1995) Evaluation of surface soil-moisture distribution by means of SAR remote-sensing techniques and conceptual hydrological modeling, *J. Hydrol.*, **166**, 445–459.
3. Minchella, A.; Del Frate, F.; Capogna, F.; Anselmi, S. and Manes, F. (2009) Use of multitemporal SAR data for monitoring vegetation recovery of Mediterranean burned areas, *Remote Sens. Environ.*, **113**, 588–597.
4. Hanst, P.L. (1971) Spectroscopic methods for air pollution measurement, *Adv. Environ. Sci. Technol.*, **2**, 91–213.
5. Hammaker, R.M.; Fateley, W.G.; Chaffin, C.T.; Marshall, T.L.; Tucker, M.D.; Makepeace, V.D. and Poholarz, J.M. (1993) FT-IR remote sensing of industrial atmospheres for spatial characterization, *Appl. Spectosc.*, **47**, 1471–1475.
6. Shao, L.; Pollard, M.J.; Griffiths, P.R.; Westermann, D.T. and Bjorneberg, D.L. (2007) Rejection criteria for open-path Fourier transform infrared spectrometry during continuous atmospheric monitoring, *Vibr. Spectrosc.*, **43**, 78–85.
7. Griffiths, P.R.; Shao, L. and Leytem, A.B. (2009) Completely automated open-path FT-IR spectrometry, *Anal. Bioanal. Chem.*, **393**, 45–50.
8. Visser, T. (2000) Infrared Spectroscopy in Environmental Analysis, in *Encyclopedia of Analytical Chemistry*, Vol. 1 (ed. R.A. Meyers), John Wiley & Sons, Ltd, Chichester.
9. Griffith, D.W.T. and Jamie, I.M. (2000) Fourier-Transform Infrared Spectrometry in Atmospheric and Trace Gas Analysis, in *Encyclopedia of Analytical Chemistry*, Vol. 3 (ed. R.A. Meyers), John Wiley & Sons, Ltd, Chichester.
10. Nelson, D.D.; Zahniser, M.S.; McManus, J.B.; Shorter, J.H.; Wormhoudt, J.C. and Kolb, C.E. (1996) Recent improvements in atmospheric trace gas monitoring using mid-infrared tunable diode lasers, in *Application of Tunable Diode and Other Infrared Sources for Atmospheric Studies and Industrial Process Monitoring* (ed. A. Fried), Proceedings of the Society of Photo-Optical Instrumentation Engineers (SPIE), vol. 2834, pp. 148–159.
11. Ku, R.T.; Hinkley, E.D. and Sample, J.O. (1975) Long-path monitoring of atmospheric carbon-monoxide with a tunable diode laser system, *Appl. Opt.*, **14**, 854–861.
12. Taslakov, M.; Simeonov, V. and van den Bergh, H. (2006) Open path atmospheric spectroscopy using room temperature operated pulsed quantum cascade laser, *Spectrochim. Acta A*, **63**, 1002–1008.
13. Hirschfeld, T. (1974) Range independence of signal in variable focus remote Raman spectrometry, *Appl. Optics*, **13**, 1435–1437.
14. Angel, S.M.; Kulp, T.J. and Vess, T.M. (1992) Remote Raman spectroscopy at intermediate ranges using low-power CW lasers, *Appl. Spectrosc.*, **46**, 1085–1091.
15. Vandenabeele, P.; Castro, K.; Hargreaves, M.; Moens, L.; Madariaga, J.M. and Edwards, H.G.M. (2007) Comparative study of mobile Raman instrumentation for art analysis, *Anal. Chim. Acta*, **588**, 108–116.
16. Sharma, S.K.; Lucey, P.G.; Ghosh, M.; Hubble, H.W. and Horton, K.A. (2003) Stand-off Raman spectroscopic detection of minerals on planetary surfaces, *Spectrochim. Acta A*, **59**, 2391–2407.
17. European Space Agency, The ExoMars Instruments (last updated: February 1, 2008). Available at http://www.esa.int/SPECIALS/ExoMars/SEMSZIAMS7F_0.html (accessed November 3, 2011).
18. Ramirez-Cedeno, M.L.; Ortiz-Rivera, W.; Pacheco-Londono, L.C. and Hernandez-Rivera, S.P. (2010) Remote detection of hazardous liquids concealed in glass and plastic containers, *IEEE Sensors J.*, **10**, 693–698.
19. Pacheco-Londono, L.C.; Ortiz-Rivera, W.; Primera-Pedrozo, O.M. and Hernandez-Rivera, S.P. (2009) Vibrational spectroscopy standoff detection of explosives, *Anal. Bioanal. Chem.*, **395**, 323–335.
20. Carter, J.C.; Scaffidi, J.; Burnett, S.; Vasser, B.; Sharma, S.K. and Angel, S.M. (2005) Stand-off Raman detection using dispersive and tunable filter based systems, *Spectrochim. Acta A*, **61**, 2288–2298.

21. Misra, A.K.; Sharma, S.K.; Chio, C.H.; Lucey, P.G. and Lienert, B. (2005) Pulsed remote Raman system for daytime measurements of mineral spectra, *Spectrochim. Acta A*, **61**, 2281–2287.
22. Sharma, S.K.; Misra, A.K.; Lucey, P.G.; Angel, S.M. and McKay, C.P. (2006) Remote pulsed Raman spectroscopy of inorganic and organic materials to a radial distance of 100 meters, *Appl. Spectrosc.*, **60**, 871–876.
23. Platt, U. and Stutz, J. (2008) *Differential Optical Absorption Spectroscopy (DOAS)—Principles and Applications*, Springer, Berlin.
24. Richter, A.; Burrows, J.P.; Nuss, H.; Granier, C. and Niemeier, U. (2005) Increase in tropospheric nitrogen dioxide over China observed from space, *Nature*, **437**, 129–132.
25. Arpag, K.H.; Johnston, P.V.; Miller, H.L.; Sanders, R.W. and Solomon, S. (1994) Observations of the stratospheric BRO column over Colorado, 40-degrees-N, *J. Geophys. Res.*, **99**, 8175–8181.
26. Kim, K.C.; Schmidt, M.C.; Sato, H.; Wu, F.; Fellows, N.; Saito, M.; Fujito, K.; Speck, J.S.; Nakamura, S. and DenBaars, S.P. (2007) Improved electroluminescence on nonpolar m-plane InGaN/GaN quantum wells LEDs, *Phys. Status Solidi-R.*, **1**, 125–127.
27. Lohberger, F.; Honninger, G. and Platt, U. (2004) Ground-based imaging differential optical absorption spectroscopy of atmospheric gases, *Appl. Optics*, **43**, 4711–4717.
28. Measures, R.M. (1984) Laser Remote Sensing: Fundamentals and Applications, John Wiley & Sons, Inc., New York.
29. Elaksher, A.F. (2008) Fusion of hyperspectral images and lidar-based DEMs for coastal mapping, *Opt. Laser. Eng.*, **46**, 493–498.
30. Streutker, D.R. and Glenn, N.F. (2006) LiDAR measurement of sagebrush steppe vegetation heights, *Remote Sens. Environ.*, **102**, 135–145.
31. Zanzottera, E. (1990) Differential absorption Lidar techniques in the determination of trace pollutants and physical parameters of the atmosphere, in *Critical Reviews in Analytical Chemistry, 21* (ed. W. Zielinski), CRC Press, N.W., Boca Raton,.
32. Hardesty, R.M. (1984) Coherent DIAL measurement of range-resolved water-vapor concentration, *Appl. Optics*, **23**, 2545–2553.
33. Edner, H.; Rangnarson, P.; Spannare, S. and Svanberg, S. (1993) Differential optical-absorption spectroscopy (DOAS) system for urban atmospheric-pollution monitoring, *Appl. Optics*, **32**, 327–333.
34. Edner, H.; Faris, G.W.; Sunesson, A.; Svanberg, S.; Bjarnason, J.O.; Kristmansdottir, H. and Sigurdsson, K.H. (1991) LIDAR search for atmospheric atomic mercury in Icelandic geothermal fields, *J. Geophys. Res.*, **96**, 2977–2986.
35. Palanco, S.; Baena, J.M. and Laserna, J.J. (2002) Open-path laser-induced plasma spectrometry for remote analytical measurements on solid surfaces, *Spectrochim. Acta B*, **57**, 591–599.
36. Knight, A.K.; Scherbarth, N.L.; Cremers, D.A. and Ferris, M.J. (2000) Characterization of laser-induced breakdown spectroscopy (LIBS) for application to space exploration, *Appl. Spectrosc.*, **54**, 331–340.
37. Cremers, D.A. (1987) The analysis of metals at a distance using laser-induced breakdown spectroscopy, *Appl. Spectrosc.*, **41**, 572–578.
38. Botheroyd, I.M.; Young, J.; Whitehouse, A.I. and Duckworth, A. (1998) Remote analysis of steels and other solid materials using Laser-Induced Breakdown Spectroscopy (LIBS), *CLEO/Eur. Conference Proceedings, Glasgow*, p. 198, abstr. CWF79.
39. López-Moreno, C.; Palanco, S. and Laserna, J.J. (2004) Remote laser-induced plasma spectrometry for elemental analysis of samples of environmental interest, *J. Anal. At. Spectrom.*, **19**, 1479–1484.
40. Blacic, J.D.; Pettit, D.R. and Cremers, D.A. (1992) Laser-induced breakdown spectroscopy for remote elemental analysis of planetary surfaces, *Proc. of the International Symposium on Spectral Sensing Research*, Springer Netherlands, Maui, HI.
41. Grönlund, R.; Lundqvist, M. and Svanberg, S. (2006) Remote imaging Laser-Induced Breakdown Spectroscopy and Laser-Induced Fluorescence Spectroscopy using nanosecond pulses from a mobile lidar system, *Appl. Spectrosc.*, **60**, 853–859.
42. Ashley, K. (1994) Electroanalytical applications in occupational and environmental-health, *Electroanalysis*, **6**, 805–820.
43. Draper, W.M.; Ashley, K.; Glowacki, C.R. and Michael, P.R. (1999) Industrial hygiene chemistry: Keeping pace with rapid change in the workplace, *Anal. Chem.*, **71**, 33R–60R.

44. Simion, M.; Iorgulescu, E.E.; Angelescu, A.; Kleps, I.; Miu, M. and Bragaru, A. (2002) Micro and nanoelectrode voltammetric measurements, *International Semiconductor Conference CAS 2002 Proceedings*, 123–126.
45. Heller, A. and Feldman, B. (2008) Electrochemical glucose sensors and their applications in diabetes management, *Chem. Rev.*, **108**, 2482–2505.
46. Green, M. and Hilditch, P. (1991) Disposable Single-Use Sensors, *Anal. Proc.*, **28**, 374–376.
47. Erickson, K. and Wilding, P. (1993) Evaluation of a novel point-of-care system, the I-STAT portable clinical analyzer, *Clin. Chem.*, **39**, 283–287.
48. Wang, J.; Tian, B.; Lu, J.; Olsen, C.; Yarnitsky, C.; Olsen, K.; Hammerstrom, D. and Bennett, W. (1999) Stripping analysis into the 21st century: faster, smaller, cheaper, simpler and better, *Anal. Chim. Acta*, **385**, 429–435.
49. Eiceman, G.A. and Karpas, Z. (1994) Ion Mobility Spectrometry, CRC Press, Boca Raton.
50. Young, D.; Thomas, C.L.P.; Breach, J.; Brittain, A.H. and Eiceman, G.A. (1999) Extending the concentration and linear dynamic range of ion mobility spectrometry with a sheath flow inlet, *Anal. Chim. Acta*, **381**, 69–83.
51. Ewing, R.G.; Atkinson, D.A.; Eiceman, G.A. and Ewing, G.J. (2001) A critical review of ion mobility spectrometry for the detection of explosives and explosive related compounds, *Talanta*, **54**, 515–529.
52. Karpas, Z.; Tilman, B.; Gdalevsky, R. and Lorber, A. (2002) Determination of volatile biogenic amines in muscle food products by ion mobility spectrometry, *Anal. Chim. Acta*, **463**, 155–163.
53. Eiceman, G.A.; Nazarov, E.G.; Tadjikov, B. and Miller, R.A. (2000) Monitoring volatile organic compounds in ambient air inside and outside buildings with the use of a radio-frequency-based ion-mobility analyzer with a micromachined drift tube, *Field Anal. Chem. Technol.*, **4**, 297–308.
54. Arce, L.; Menendez, M.; Garrido-Delgado, R. and Valcarcel, M. (2008) Sample-introduction systems coupled to ion-mobility spectrometry equipment for determining compounds present in gaseous, liquid and solid samples, *TrAC-Trend Anal. Chem.*, **27**, 139–150.
55. Agilent Technologies (2011) Compact, portable, and handheld FTIR solutions for solid and liquid analysis, publication number: 5990-8674EN. Available at http://www.chem.agilent.com/Library/flyers/Public/5990-8674EN_Flyer_4100_4200_4500_5500.pdf (accessed November 3, 2011).
56. Axsun Technologies, Inc. (2007) AnavoTM Handheld NIR material Analyzer, data sheet. Available at http://www.axsun.com/PDF/Anavo_Handheld_NIR.pdf (accessed November 3, 2011).
57. Kemeny, G.J. (2008) Process Analysis *in Handbook of Near Infrared Analysis*, 3rd edn (eds. D.A. Burns and E.W. Ciurczak), CRC Press, Boca Raton.
58. Wesley, I.J.; Larsen, N.; Osborne, B.G. and Skerritt, J.H. (1998) Non-invasive monitoring of dough mixing by near-infrared spectroscopy, *J. Cereal Sci.*, **27**, 61–69.
59. Determination of Protein by Near Infrared Reflectance (NIR) Spectroscopy (1995) Standard Method of the International Association for Cereal Science and Technology. Available at http://www.icc.or.at/standard_methods/159 (accessed 3 November 2011).
60. Hildrum, K.I.; Nilsen, B.N.; Westad, F. and Wahlgren, N.M. (2004) In-line analysis of ground beef using a diode array near infrared instrument on a conveyor belt, *J. Near Infrared Spectrosc.*, **12**, 367–376.
61. Fearn, T. and Maris, P.I. (1991) An application of box–jenkins methodology to the control of gluten addition in a flour mill, *Appl. Statist.*, **40**, 477–484.
62. Blanco, M. and Romero, M.A. (2001) Near-infrared libraries in the pharmaceutical industry, *Analyst*, **126**, 2212–2217.
63. Woo, Y.A.; Kim, H.J. and Cho, J. (1999) Identification of herbal medicines using pattern recognition techniques with near-infrared reflectance spectra, *Microchem. J.*, **63**, 61–70.
64. Lewis, I.R. and Rosenblum, S.S. (2002) General introduction to fiber-optics, in *Handbook of Vibrational Spectroscopy* (eds J. Chalmers and P.R. Griffiths), John Wiley & Sons, Ltd, Chichester.
65. Eliasson, C.; Macleod, N.A. and Matousek, P. (2007) Noninvasive detection of concealed liquid explosives using Raman spectroscopy, *Anal. Chem.*, **79**, 8185–8189.
66. Eliasson, C.; Macleod, N.A. and Matousek, P. (2008) Non-invasive detection of cocaine dissolved in beverages using displaced Raman spectroscopy, *Anal. Chim. Acta*, **607**, 50–53.
67. Kim, M.; Chung, H.; Woo, Y. and Kemper, M.S. (2007) A new non-invasive, quantitative Raman technique for the determination of an active ingredient in pharmaceutical liquids by direct measurement through a plastic bottle, *Anal. Chim. Acta*, **587**, 200–207.

68. McCreery, R.L.; Horn, A.J.; Spencer, J. and Jefferson, E. (1998) Noninvasive identification of materials inside USP vials with Raman spectroscopy and a Raman spectral library, *J. Pharm. Sci.*, **87**, 1–8.
69. Eliasson, C.; Macleod, N.A.; Jayes, L.C.; Clarke, F.C.; Hammond, S.V.; Smith, M.R. and Matousek, P. (2008) Non-invasive quantitative assessment of the content of pharmaceutical capsules using transmission Raman spectroscopy, *J. Pharm. Biomed. Anal.*, **47**, 221–229.
70. Nordon, A.; Mills, A.; Burn, R.T.; Cusick, F.M. and Littlejohn, D. (2005) Comparison of non-invasive NIR and Raman spectrometries for determination of alcohol content of spirits, *Anal. Chim. Acta*, **548**, 148–158.
71. Smith, B.C. (1996) *Fundamentals of Fourier Transform Infrared Spectroscopy*, CRC Press, Boca Raton.
72. Andrews, J. and Dallin, P. (2004) Frequency Modulation Spectroscopy, *Spectrosc. Europe*, **15**, 24–26.
73. Roychoudhury, P.; Harvey, L.M. and McNeil, B. (2006) At-line monitoring of ammonium, glucose, methyl oleate and biomass in a complex antibiotic fermentation process using attenuated total reflectance-mid-infrared (ATR-MIR) spectroscopy, *Anal. Chim. Acta*, **561**, 218–224.
74. Ruckebusch, C.; Duponchel, L. and Huvenne, J.P. (2002) Interpretation and improvement of an artificial neural network MIR calibration, *Chemometr. Intell. Lab.*, **62**, 189–198.
75. Harthcock, M.A. and Atkin, S.C. (1988) Imaging with functional-group maps using infrared microspectroscopy, *Appl. Spectrosc.*, **42**, 449–455.
76. Sasic, S. and Clark, D.A. (2006) Defining a strategy for chemical imaging of industrial pharmaceutical samples on Raman line-mapping and global illumination instruments, *Appl. Spectrosc.*, **60**, 494–502.
77. Treado, P.J. and Nelson, M.P. (2002) Raman Imaging, in *Handbook of Vibrational Spectroscopy*, Vol. 2 (eds. J.M. Chalmers and P.R. Griffiths), John Wiley & Sons, Ltd, Chichester.
78. Treado, P.; Levin, I.W. and Lewis, E.N. (1994) Indium-Antimonide (InSb) focal-plane array (FPA) detection for Near-Infrared imaging microscopy, *Appl. Spectrosc.*, **48**, 607–615.
79. van den Broek, W.H.A.M.; Derks, E.P.P.A.; van de Ven, E.W.; Wienke, D.; Geladi, P. and Buydens, L.M.C. (1996) Plastic identification by remote sensing spectroscopic NIR imaging using kernel partial least squares (KPLS), *Chemometr. Intell. Lab.*, **35**, 187–197.
80. Mansfield, J.R.; McIntosh, L.M.; Crowson, A.N.; Mantsch, H.H. and Jackson, M. (1999) Search engines for infrared microscopic Maps: linear discriminant analysis using an external training set, *Appl. Spectrosc.*, **53**, 1323–1330.
81. Gowen, A.A.; O'Donnell, C.P.; Cullen, P.J.; Downey, G. and Frias, J.M. (2007) Hyperspectral imaging - an emerging process analytical tool for food quality and safety control, *Trends Food Sci. Tech.*, **18**, 590–598.
82. Amigo, J.M.; Cruz, J.; Bautista, M.; Maspoch, S.; Coello, J. and Blanco, M. (2008) Study of pharmaceutical samples by NIR chemical-image and multivariate analysis, *TrAC-Trend Anal. Chem.*, **27**, 696–713.
83. Michaelian, K.H. (2003) *Photoacoustic Infrared Spectroscopy*, John Wiley & Sons, Ltd, Hoboken.
84. McClelland, J.F.; Jones, R.W. and Bajic, S.J. (2002) FT-IR Photoacoustic Spectroscopy, in *Handbook of Vibrational Spectroscopy*, Vol. 2 (eds. J.M. Chalmers and P.R. Griffiths), John Wiley & Sons, Ltd, Chichester.
85. Neubert, R.; Collin, B. and Wartewig, S. (1997) Quantitative analysis of drug content in semisolid formulations using step-scan FT-IR photoacoustic spectroscopy, *Vibr. Spectrosc.*, **13**, 241–244.
86. Armenta, S.; Moros, J.; Garrigues, S. and de la Guardia, M. (2006) Direct determination of Mancozeb by photoacoustic spectrometry, *Anal. Chim. Acta*, **567**, 255–261.
87. Koskinen, V.; Fonsen, J.; Kauppinen, J. and Kauppinen, I. (2006) Extremely sensitive trace gas analysis with modern photoacoustic spectroscopy, *Vibr. Spectrosc.*, **42**, 239–242.
88. Takats, Z.; Wiseman, J.M.; Gologan, B. and Cooks, R.G. (2004) Mass spectrometry sampling under ambient conditions with desorption electrospray ionization, *Science*, **306**, 471–473.
89. Venter, A.; Nefliu, M. and Cooks, R.G. (2008) Ambient desorption ionization mass spectrometry, *TrAC-Trend Anal. Chem.*, **27**, 284–290.
90. Gonzalvez, A.; Shelley, J.; Wiseman, J.M.; Armenta, S.; de la Guardia, M. and Hieftje, G.M. (2011) Direct quantification of veterinary residues in foodstuff through desorption electrospray ionization-mass spectrometry, *Analyst*, in press.
91. Baumann, H. (1992) Solid sampling with inductively coupled plasma-mass spectrometry, *Fresenius J. Anal. Chem.*, **342**, 907–916.
92. Gunther, D.; Jackson, S.E. and Longerich, H.P. (1999) Laser ablation and arc/spark solid sample introduction into inductively coupled plasma mass spectrometry, *Spectrochim. Acta B*, **54**, 381–409.

93. Potts, P.J. (1987) *Handbook of Silicate Rock Analysis*, Blackie, New York.
94. Payling, R.; Jones, D.G. and Bengtson, A. (1997) *Glow Discharge Optical Emission Spectrometry*, John Wiley & Sons, Ltd, Chichester.
95. Belenguer, Ph.; Ganciu, M.; Guillot, Ph. and Nelis, Th. (2009) Pulsed glow discharges for analytical applications, *Spectrochim. Acta Part B*, **64**, 623–641.
96. Russo, R.E.; Mao, X.; Liu, H.; Gonzalez, J. and Mao, S.S. (2002) Laser ablation in analytical chemistry - a review, *Talanta*, **57**, 425–451.
97. Narewski, U.; Wemer, G.; Schulz, H. and Vogt, C. (2000) Application of laser ablation inductively coupled mass spectrometry (LA-ICP-MS) for the determination of major, minor, and trace elements in bark samples, *Fresenius J. Anal. Chem.*, **366**, 167–170.
98. Horn, I.; Rudnick, R.L. and McDonough, W.F. (2000) Precise elemental and isotope ratio determination by simultaneous solution nebulization and laser ablation-ICP-MS: application to U-Pb geochronology (Vol 164, p. 283), *Chem. Geol.*, **167**, 405.
99. Devos, W.; Moor, C. and Lienemann, P. (1999) Determination of impurities in antique silver objects for authentication by laser ablation inductively coupled plasma mass spectrometry (LA-ICP-MS), *J. Anal. At. Spectrom.*, **14**, 621–626.
100. Walling, R.J.; Lynch, B.F. and Herring, D. (1997) Use of laser ablation inductively coupled plasma mass spectrometry for fingerprinting scene of crime evidence, *J. Anal. At. Spectrom.*, **12**, 195–203.
101. Sanders, J.K.M. and Hunter, B.K. (1987) *Modern NMR Spectroscopy*, Oxford University Press, Oxford.
102. Richardson, J.C.; Bowtell, R.W.; Mader, K. and Melia, C.D. (2005) Pharmaceutical applications of magnetic resonance imaging (MRI), *Adv. Drug Delivery Rev.*, **57**, 1191–1209.
103. Fifield, F.W. and Kealey, D. (2000) *Principles and Practice of Analytical Chemistry*, Blackwell Science Ltd, Oxford.
104. Szoboszlai, N.; Polgári, Z.; Mihucz, V.G. and Zaray, G. (2009) Recent trends in total reflection X-ray fluorescence spectrometry for biological applications, *Anal. Chim. Acta*, **633**, 1–18.
105. Lankosz, M.; Szczerbowska-Boruchowska, M.; Chwiej, J.; Ostachowicz, J.; Simionovici, A. and Bohic, S. (2004) Research in quantitative microscopic X-ray fluorescence analysis, *Spectrochim. Acta B*, **59**, 1517–1521.
106. Barr, T.L. (1994) *Modern XPS*, CRC Press, Boca Raton.
107. Auger, M.P. (1925) Sur les rayons secondaires produit dans un gal par des rayons, *Compt. Rend.*, **180**, 65.
108. Chourasia, A.R. and Chopra, D.R. (1997) Auger Electron Spectroscopy in *Handbook of Instrumental Techniques for Analytical Chemistry* (ed. F. Settle), Prentice-Hall Inc., New Jersey.
109. Pacholski, M.L. and Winograd, N. (1999) Imaging with mass spectrometry, *Chem. Rev.*, **99**, 2977–3005.
110. Benninghoven, A. (1973) Surface investigation of solids by the statical method of secondary ion mass spectroscopy (SIMS), *Surf. Sci.*, **35**, 427–437.
111. Guerquin-Kern, J.L.; Wu, T.D.; Quintana, C. and Croisy, A. (2005) Progress in analytical imaging of the cell by dynamic secondary ion mass spectrometry (SIMS microscopy), *BBA-Gen Subjects*, **1724**, 228–238.
112. Brongersma, H.H.; Draxler, M.; de Ridder, M. and Bauer, P. (2007) Surface composition analysis by low-energy ion scattering, *Surf. Sci. Reports*, **62**, 63–109.

7

Green Analytical Chemistry Approaches in Sample Preparation

Marek Tobiszewski, Agata Mechlińska and Jacek Namieśnik

Department of Analytical Chemistry, Chemical Faculty, Gdansk University of Technology (GUT), Gdansk, Poland

This chapter presents environmentally friendly solutions for sample preparation. Attention is focused on extraction techniques and the minimization of organic solvent consumption, as the large quantities used in analytical laboratories are a pressing environmental problem. Solventless sample preparation techniques, alternative solvents, mediated extractions are discussed. The principle of each technique is briefly described, together with its main areas of application and some of its analytical parameters.

7.1 About sample preparation

Rapid advances in technology have led to the design of highly sensitive analytical instruments. Unfortunately however, most of them are unable to directly handle samples with a complex matrix composition. Nevertheless, modern chemical analysis requires the determination of certain target compounds at trace levels. To meet these demands, a sample preparation step is performed. On the one hand, the sample matrix is converted to one that is easily handled by analytical apparatus; on the other, analytes are concentrated to improve the method's sensitivity.

Nowadays, most determinations of trace organic compounds involve chromatographic techniques. Since chromatographic columns are sensitive to impurities often present in the matrix, this should be converted to one of simple composition. Most pre-chromatographic sample preparation techniques involve the extraction of analytes. Liquid-liquid extraction (LLE) is one of the oldest of these techniques, but it is time-consuming,

Handbook of Green Analytical Chemistry, First Edition. Edited by Miguel de la Guardia and Salvador Garrigues.

laborious, difficult to automate and requires large amounts of organic solvents. Organic solvents are toxic and volatile; they are also expensive because they are required to be of high purity. Despite these drawbacks and the existence of other, more environmentally friendly sample preparation methods, LLE is still used in analytical practice. For instance, the method for determining PAHs in drinking water recommended by EPA (method 550) involves LLE with 180 ml methylene chloride. Another LLE method recommended by EPA (method 506) for determining phthalate and adipate esters requires 180 ml methylene chloride and 40 ml hexane. When such methods are routine, hundreds of such extractions may be performed daily, which seriously endangers the environment. Ironically, analytical chemistry serves as a tool for assessing the state of the environment, yet at the same time it may pollute that very environment, mainly with organic solvents. A very important aspect of the problem is the location of the analytical instrument vis-à-vis the sample. Four modes can be distinguished [1]:

1. *off-line* – the sample is collected and transported to laboratory for analysis.
2. *at-line* – the sample is collected manually and analysed at the sampling site. The analytical device is usually portable and is periodically carried to the medium to be analysed.
3. *on-line* – the analytical device is located at the sampling site. Periodically, samples are collected automatically and analysed by the device.
4. *in-line* – the analytical sensor is placed in the investigated medium.

The *off-line* mode is the least advantageous, because transportation and storage themselves require material and energy inputs. Moreover, this approach usually involves a sample preparation step, whereas the other modes often have no need for this step.

Direct analytical techniques do not require a sample preparation step, even in *off-line* mode. The sample is collected and analysed without any prior preparation. Such techniques include graphite furnace atomic absorption spectroscopy and neutron activation analysis. Direct chromatographic techniques include direct aqueous injection, when a water sample is injected without preparation by the on-column injector. However, the technique is applicable only to samples with clean matrices.

Most analytical techniques are indirect, so they require a sample preparation step. But at the same time, there is an urgent need to reduce the negative impact on the environment caused by analysts during sample preparation.

7.2 Miniaturized extraction techniques

The increasing demand for faster, more cost-effective and more environmentally friendly analytical methods is a major incentive to improve the classical procedures used for sample treatment in chemical analysis. In most classical procedures, the use of rapid and powerful instrumental techniques for the final separation and detection of analytes contrasts with the time-consuming and usually manual methods used for sample preparation, which slows down the total analytical process. The efforts made in this field in the past 10 years have led to the adaptation of existing methods and the development of new techniques to save time and chemicals and to improve overall performance [2].

7.2.1 Solid-phase extraction (SPE)

Disposable cartridges for solid-phase extraction have been used for more than 20 years (the first cartridges were introduced in 1978, syringe format types in 1979, and precolumns for on-line coupling with liquid chromatography (LC) in the early 1980s) [3]. Today, in the case of liquid samples the most common extraction technique is probably solid-phase extraction – in all of its modes and formats – but especially when more polar analytes are involved [2].

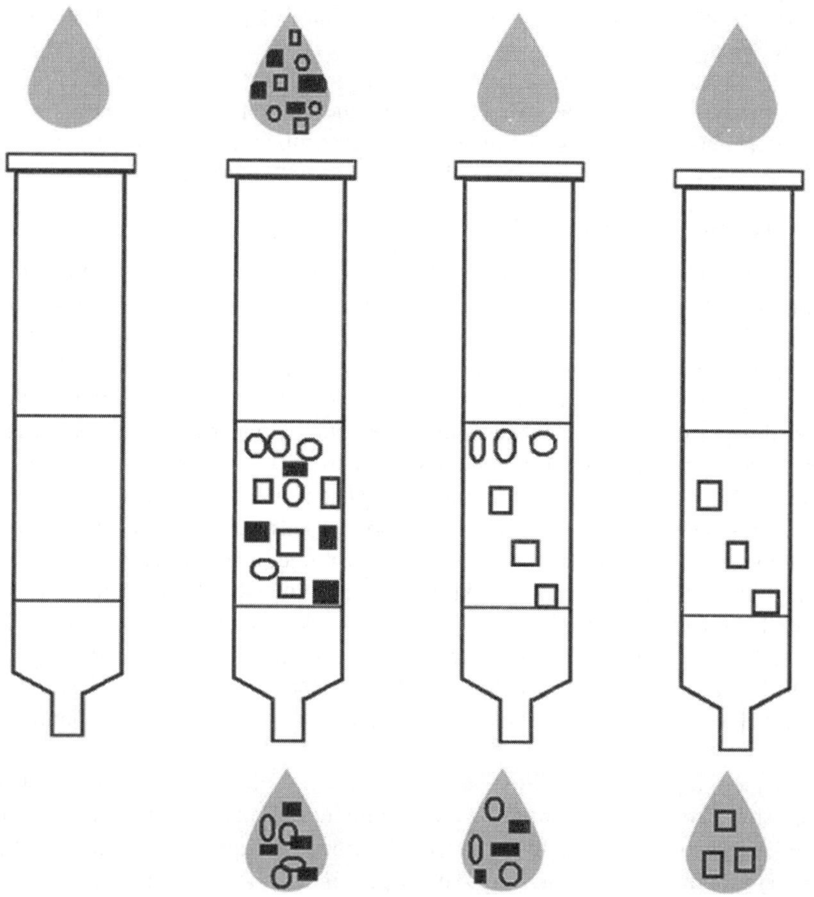

Figure 7.1 *Steps of extraction cycle with use of SPE.*

This technique consists of the partitioning of organic compounds dissolved in a water sample between a solid sorbent and the liquid phase, as a result of which analytes are selectively trapped by the sorbent. Trapping is based on the adsorption of the analytes dissolved in water on the sorbent surface (solid phase) and/or their partitioning between the liquid sorbent (stationary phase) coated on the support and the aqueous sample. In SPE, extraction columns (sorbent, e.g. modified silica gel packed into a column), Empore discs (sorbent with a porous Teflon® membrane) and speed discs are used. The trapped analytes can be released by solvent extraction or thermal desorption [4]. Organic solvents are not removed completely from the procedure – they are used for column conditioning and analyte release after extraction (see Figure 7.1).

7.2.2 Solid-phase microextraction (SPME)

Solid-phase microextraction was introduced in 1989 by Arthur and Pawliszyn as a new sampling and sample-preparation method. It is a simple and effective adsorption/absorption and desorption technique that eliminates the need for solvents and combines sampling, isolation, and enrichment in one step. SPME has been widely applied to the sampling and analysis of environmental, food, aroma, forensic and pharmaceutical samples [5].

SPME involves exposing a small amount of extracting phase dispersed on a solid support to the sample matrix in two basic steps:

1. partitioning of analytes between the extraction phase and the sample matrix, and
2. desorption of concentrated extracts into an analytical instrument [6,7].

7.2.3 Stir-bar sorptive extraction (SBSE)

Stir-bar sorptive extraction was introduced in 1999 as a solventless sample preparation method for the isolation and enrichment of organic compounds from aqueous matrices. The method is based on sorptive extraction, whereby the solutes are extracted into a polymer coating on a magnetic stirring rod. The extraction is controlled by the partition coefficient of the solutes between the polymer coating and the sample matrix and by the phase volume ratio of the polymer coating and the sample [8]. For a polydimethylsiloxane coating and aqueous samples, this partition coefficient resembles the octanol–water partition coefficient ($K_{o/w}$). SBSE is mostly used for extracting compounds from aqueous matrices; it is also applied in headspace mode for liquid and solid samples and in passive air sampling mode [9]. The most widely used sorptive extraction phase is polydimethylsiloxane (PDMS). This compound is commonly used as a stationary phase in gas chromatography (GC), is thermostable, can be used over a broad temperature range (it acts as a liquid between −20 and 320°C) and has interesting diffusion properties. After extraction, the solutes can be introduced quantitatively into the analytical system by thermal desorption (TD) or after back-extraction with an organic solvent (liquid desorption, LD). SBSE is a standard method in environmental analysis for the extraction of volatile organic compounds (VOCs) and semi-volatile compounds [9], polycyclic aromatic hydrocarbons (PAHs) [10–21], pesticides and polychlorinated biphenyls (PCBs) [22–29] and malodorous compounds in drinking water [30–33].

Although the extraction phase in SBSE is the same as that in SPME, the extract obtained is 50–250 times larger in volume. After extraction, the solutes are thermally desorbed and analysed by GC in much the same way as in SPME. Alternatively, the analytes can be desorbed by liquid extraction. The basic principles of SPME and SBSE are therefore identical. Sorptive extraction is, by nature, an equilibrium technique and for water samples, the extraction of solute from the aqueous phase into the extraction phase is controlled by the partition coefficient of the solute between the silicone phase and the aqueous phase [9]. Figure 7.2 gives a theoretical comparison between SPME and SBSE, showing the recovery in function of the octanol-water distribution coefficient. While high recoveries (>50%) are only obtained for solutes with $K_{(o/w)}$ >10 000 using SPME, the recovery obtained by stir-bar sorptive extraction is already higher than 50% for solutes with $K_{(o/w)}$ >100.

7.2.4 Liquid-liquid microextraction

Miniaturization of the liquid-liquid extraction step led to the development of single drop microextraction (SDME) [34]. The idea is to extract a fraction of the analytes present in the sample into a small amount of organic solvent (usually one or a few microlitres) placed on the top of a microsyringe. After extraction, the solvent drop is sucked into the microsyringe (compatible with a GC injection port) and the extract injected into a chromatograph [35]. SDME is a convenient, one-step sample preparation approach [36] and the equipment required is cheap, simple, available in all analytical laboratories and consumes negligible amounts of solvents [37].

The two basic modes of operation are direct immersion DI-SDME, when the acceptor organic phase is in direct contact with the aqueous sample; and headspace HS-SDME, when the acceptor phase is placed in the gaseous phase above the liquid sample (see Figure 7.3). HS-SDME is used for the extraction of volatile and

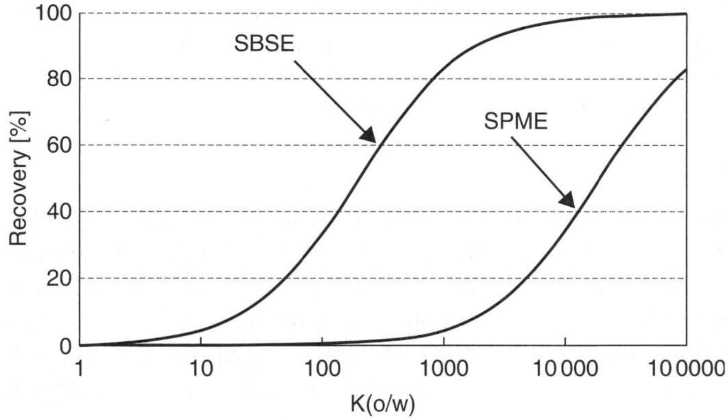

Figure 7.2 *Theoretical recoveries of SBSE and SPME. Theoretical recoveries for solutes as a function of the octanol–water partition coefficient $K_{(o/w)}$ for SBSE (10 ml sample, 50 μl PDMS phase) and SPME (10 ml sample, 0.5 μl PDMS phase).*

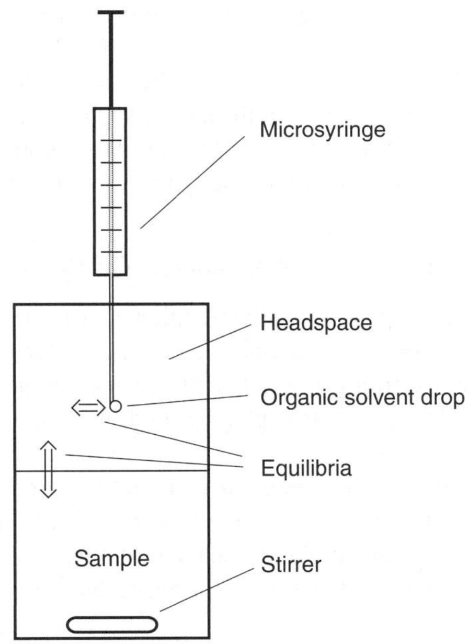

Figure 7.3 *SDME setup in headspace mode.*

semi-volatile organic compounds. The advantage of HS-SDME over DI-SDME is the better stability of the solvent drop and that the aqueous sample can be stirred more quickly. Further improvements involve addressing the problem of solvent evaporation by the use of ionic liquids as the receptor phase [38]. The factors influencing the analytical parameters of the method are the extraction time, solvent properties, solvent drop and sample volumes, stirring rate, temperature and ionic strength of sample solution [39].

The main disadvantage of SDME is the instability of the solvent drop during extraction. If the sample is stirred too vigorously or the syringe is operated by an inexperienced analyst, the drop may be dislodged. In the case of HS-SDME there is the problem of proper solvent selection. If the solvent is too volatile, evaporation during extraction may affect the method's analytical parameters. If the solvent is not volatile enough, it may be incompatible with the GC device. Also, samples with a complex matrix composition may cause the stability parameters of the solvent drop to deteriorate.

7.2.5 Membrane extraction

A great variety of membranes are used in sample preparation. Membrane extraction may be based on polymeric (symmetric or asymmetric) or liquid membranes. Polymeric membranes are usually made of low-density polyethylene (LDPE), polypropylene (PP) or polydimethyl siloxane (PDMS), whereas liquid membranes consist of solvent trapped in the pores of PP or polytetrafluoroethylene (PTFE). The membranes can be in the form of a hollow fibre (HF) or a flat sheet (FS). Membrane extraction techniques have great potential for automation and miniaturization, which reduce the consumption of reagents and energy. Most of these techniques fall into one of the following categories: supported liquid membrane extraction, microporous membrane liquid-liquid extraction, membrane extraction with sorbent interface, or membrane-assisted solvent extraction [40].

7.2.5.1 Supported liquid membrane (SLM)

Aqueous phases are present on both sides of an organic liquid membrane (in FS or HS form). The liquid present in the membrane pores – often kerosene or undecane – is insoluble in water. Analyte transfer across the membrane is driven by the difference in concentration. To obtain such a difference, the analyte in the donor phase is present in nonionic form, whereas in the acceptor phase the analyte is transferred to its ionic form [41]. This can be achieved by adjusting the pH. SLM is used to extract ionic compounds, and the device can work in *off-line* or *on-line* mode in relation to the analytical instrument.

7.2.5.2 Microporous membrane liquid-liquid extraction (MMLLE)

The concentration gradient between two phases and the partition coefficient between the aqueous and organic phase are what drives the analyte across the membrane. An organic, water-immiscible solvent is the acceptor phase on one side of the membrane, filling the pores of the membrane. The receptor organic phase is compatible with gas chromatography and normal-phase liquid chromatography. MMLLE is used for the extraction of organic compounds.

Liquid-phase microextraction (LPME) involves the use of a hollow fibre filled with organic solvent [42]. The hollow fibre is supported on a microsyringe, which is also used for collecting the receptor phase and injecting it into the GC column [43]. LPME combines the ideas of membrane extraction and SDME; the main advantage of LPME over SDME is that it reduces acceptor phase dissolution in the aqueous sample.

7.2.5.3 Membrane extraction with sorbent interface (MESI)

Non-porous polymeric membranes are applied in this technique in both FS and HF modes. The aqueous or gaseous sample is on one side of the membrane barrier, the gaseous receptor phase is on the other [44]. The analytes are collected in a trap connected to the GC column. MESI is used to extract volatile and semi-volatile analytes.

7.2.5.4 Membrane assisted solvent extraction (MASE)

A three-phase system is formed in MASE: aqueous + polymeric nonporous membrane (usually FS) + organic phase. Solvents like heptanes, hexane or cyclohexane are usually used as the organic receptor phase. The consumption of solvent is about 1 ml per analysis. Significantly, before the membrane is first used, it has to

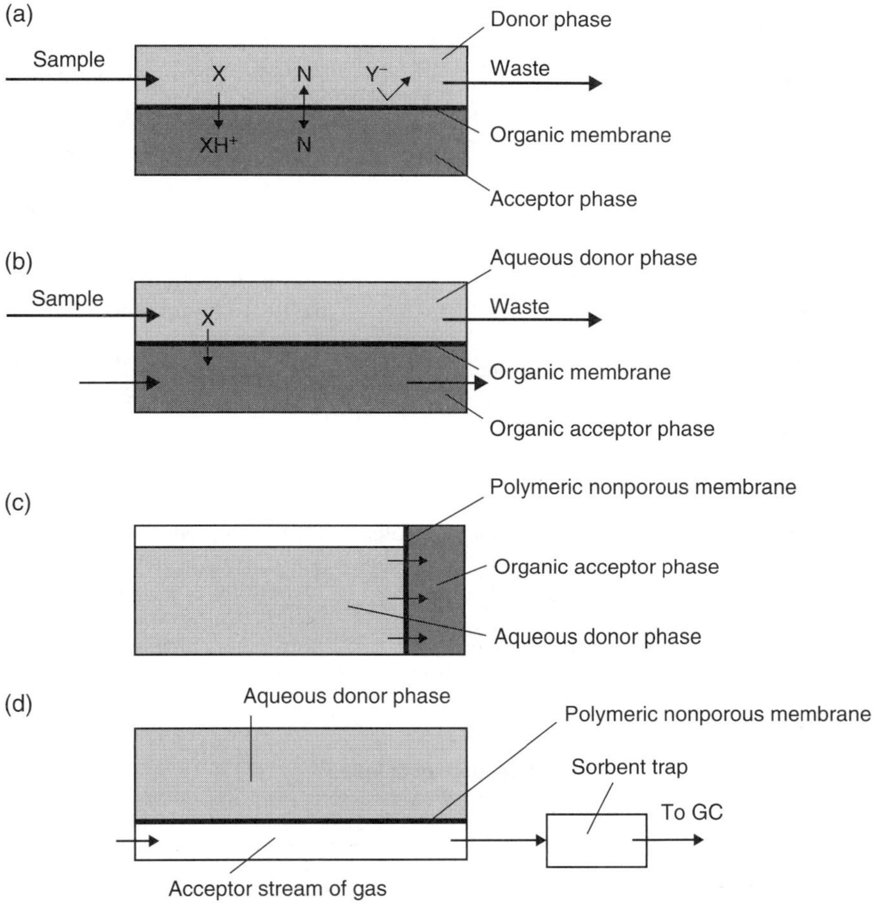

Figure 7.4 *Schemes of different solutions based on membrane techniques (a) SLM; (b) MMLLE; (c) MASE; (d) MESI.*

be conditioned, which consumes more organic solvent [45]. After conditioning, however, the membrane is reusable for extraction purposes. With MASE hydrophobic compounds are extracted from aqueous samples.

For comparison of different solutions see Figure 7.4.

7.2.5.5 Applications of membrane extraction

Applicable in many areas of analytical sample preparation (see Table 7.1), membrane techniques significantly reduce the consumption of reagents.

7.2.6 Gas extraction

One of the most commonly used techniques for the isolation of VOCs from water samples is extraction into the gaseous phase. Headspace analysis techniques are based on the partition of analytes between the liquid and the gaseous phase. The gaseous phase (headspace) is analysed together with the volatile compounds

Table 7.1 Membranes in sample preparation.

Extraction mode	Analyte	Matrix	Acceptor phase consumption per analysis	Ref.
SLM	Chlorophenols	Water	100 μl water pH 11; diethyl ether used for unit regeneration	[46]
	Cadmium	Sea water	23.5 ml acidic solution	[47]
MMLLE	17 pesticides	Red wines	100 μl toluene	[48, 49]
	Herbicides	Water	960 μl chloroform	[50]
	Vinclozolin	Water	100 μl 1-chlorobutane	[51]
	2-Ethylhexyl-4-(dimethylamino) benzoate	Urine	150 μl decane	[52]
MESI	BTEX	Tap water	Thermal trap	[44]
	Aliphatic and aromatic organics	Water	Sorbent trap	[53]
MASE	PCBs	Environmental water	800 μl cyclohexane	[54]
	PAHs	Water and beverages	400 μl ethyl acetate, but the membrane was conditioned with 50 ml cyclohexane prior to use	[55]
	Phenols	Ground water	800 μl ethyl acetate	[56]

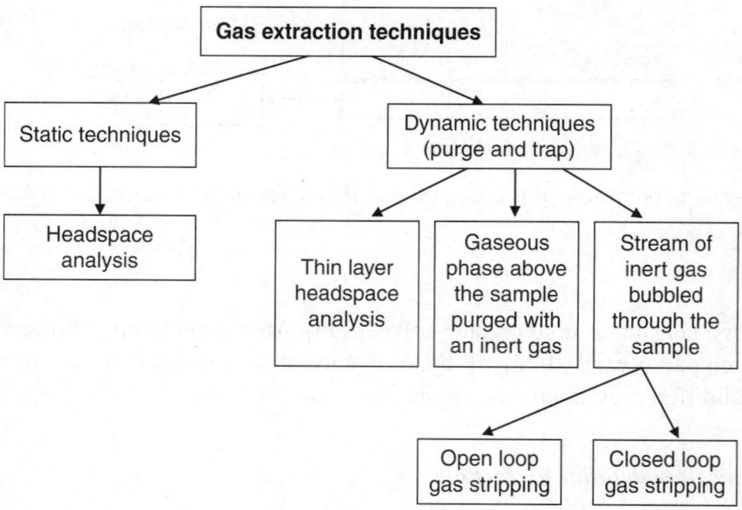

Figure 7.5 The classification of gas extraction techniques.

liberated from the liquid sample [4, 57, 58, 59]. Effective release from the liquid phase (especially aqueous) is possible for volatile, semi-volatile, non-polar or weakly polar compounds. This method is fast, eliminating the negative influence of compounds contained in the matrix, which may impede the analysis and contaminate the chromatographic system. Figure 7.5 presents a schematic diagram of the classification of gas extraction techniques.

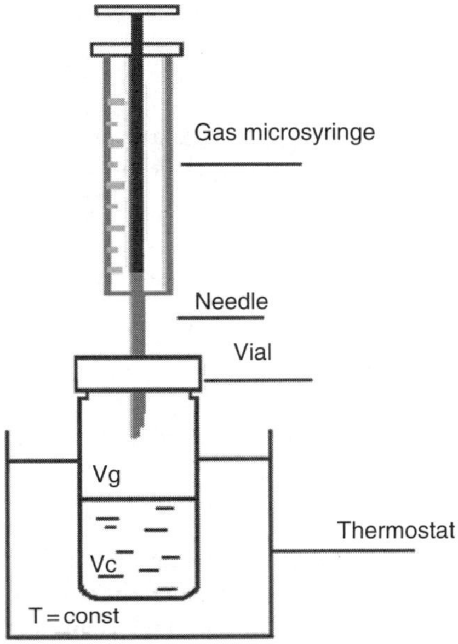

Figure 7.6 *Headspace analysis in a static mode.*

7.2.6.1 *Static headspace sampling*

Static headspace sampling is employed with GC in numerous fields and with a variety of applications. It has been a primary tool for analysis of volatile organic compounds in:

- environmental analysis,
- flavour and fragrance analysis,
- pharmaceuticals,
- clinical and biological analysis.

Because of its limited sensitivity, static headspace sampling is mostly employed for applications in the high-ppb to percent concentration ranges [57].

In static headspace sampling the liquid phase (sample) and the gaseous phase (receiving matrix) both remain static (see Figure 7.6). The analytical process can be divided into two stages:

1. a state of equilibrium is reached between the analysed water sample (placed in an appropriate container) and the gaseous phase,
2. a sample of the gas from above the surface of the liquid is collected either manually (with a syringe) or automatically, which is subsequently sent for chromatographic analysis [59].

The detection limit of headspace analysis in a static system is influenced by:

- detector sensitivity,
- analyte boiling point,
- the partition coefficient of the analyte between the liquid (aqueous) and gaseous phase.

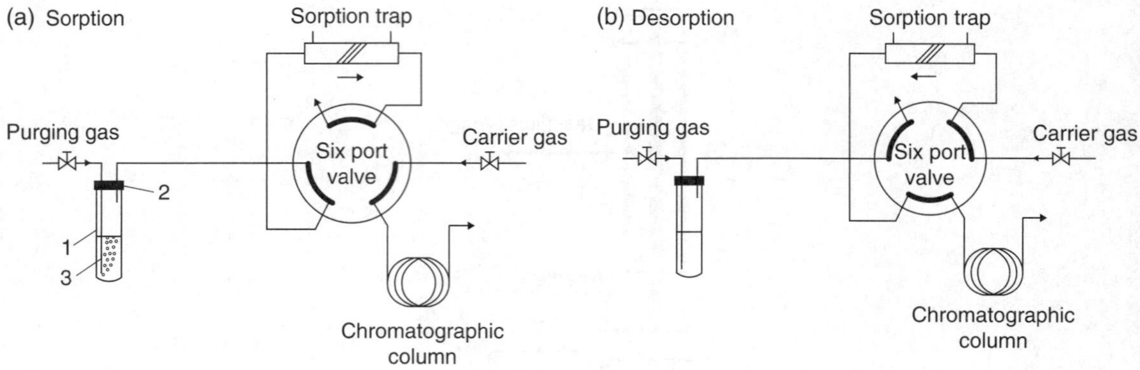

Figure 7.7 *The principle of the Purge and Trap technique operation.*

For most compounds the value of the partition coefficient may change by as much as two orders of magnitude if the temperature and/or pH are altered, and also if the sample is salted out (usually with NaCl) [59].

A more common extraction technique, owing to the significant shortening of the extraction time and elimination of matrix effects is headspace SPME (HS-SPME), which has been successfully applied to determine volatile and semi-volatile compounds [60].

7.2.6.2 Headspace analysis in a dynamic system

The techniques of headspace analysis in a dynamic system can be divided into two groups. The techniques based on purging the gaseous space above the sample with a stream of an inert gas belong to the first group. The second one embraces those techniques where a stream of an inert gas is bubbled through the analysed water sample. The following gases are employed for purging analytes: purified or synthetic air, nitrogen, helium and argon [59].

In order to eliminate the broadening of the band of liberated compounds (caused by the slow rate of the purging process) and to enrich the extracted analytes, the following methods are employed:

Cryotrapping – the purged analytes are frozen in a capillary tube cooled with liquid nitrogen; after the trapping stage the cryogenic trap is rapidly heated (ballistic heating) and the liberated analytes are introduced onto the head of the chromatographic column [61];

Purging with Whole Column Cryotrapping (P/WCC) – in this case the purged analytes in the stream of purging gas are introduced directly into the chromatographic column cooled with liquid nitrogen to a temperature of 80°C; after the purge and trap process is finished, the chromatographic column is rapidly heated to the initial temperature of the temperature programme employed [62];

Open Loop Stripping (OLS) – another name for this technique is *Purge and Trap* (PT) – developed by Bellar and Lichtenberg in 1974; the analytes purged with a stream of a neutral gas are directed to a trap filled with a layer of solid sorbent, and the gas stream, free of the analytes, is removed from the system (see Figure 7.7). The following sorbents are most commonly used as sorption trap fillers: Tenax, Carbopack, Carbosieve, silica gel, or active carbon. After the trapping stage, the analytes are released with a suitable solvent or by thermal desorption [57, 58, 63].

Closed Loop Stripping (CLS) – in this technique the gaseous phase, after running through the sample and the trap, is directed back to circulate in a closed loop system and carry a new portion of analytes onto the sorbent bed [4, 64].

7.2.6.3 Thin-layer headspace extraction

In this technique, the analysed water sample is continually fed by a pump into a special TLHS column. Inside the column, the sample runs down under gravity inside a spiral thermostatted glass tube, as a countercurrent to the stream of clean purging gas. Volatile compounds contained in the aqueous phase are released and enter the gaseous phase. Next, the water vapour and the volatile analytes are transferred to a second column, where condensation of the water vapour together with the analytes takes place and the condensed water vapour becomes a stream of liquid sorbent. Samples of this liquid effluent, which is an aqueous solution of the organic analytes isolated from the water sample, are analysed further by DAI-GC. In this technique, the analytes should not be too volatile or immiscible with water, otherwise they will either not condense, or they will form a two-phase mixture that cannot be injected directly into the GC [4, 59].

7.3 Alternative solvents

Solvents are widely used in analytical laboratories and in industry. Despite all the precautions taken, they inevitably contaminate the air, land and water because they are difficult to contain and recycle. Researchers have therefore focused on reducing solvent use through the development of solvent-free processes and more efficient recycling protocols. But these approaches have their limitations, thus necessitating a pollution prevention approach and the search for environmentally benign alternative solvents. Over the last 15 years new, unconventional solvent alternatives have attracted much attention from academia and industry. They include ionic liquids, carbon dioxide, supercritical water and fluorous phases.

7.3.1 Analytical applications of ionic liquids

As a result of studies of new solvents that do not emit harmful vapours, a new class of chemical compounds has been discovered. These are the so-called ionic liquids, which fulfil the requirements of 'green solvents' [65].

Ionic liquids (ILs) are a group of organic salts with melting points ranging from −81°C to 125°C [66]. The physicochemical properties of ILs depend on the nature and size of both their cationic and anionic constituents [67]. Their application in analytical chemistry, especially in separating analytes, is merited because ILs have a number of unique properties, such as negligible vapour pressure, good thermal stability, tuneable viscosity and miscibility with water and organic solvents, as well as good extractability for various organic compounds and metal ions. ILs are used as promising alternatives to the traditional organic solvents employed in sample preparation. They have been used as extraction solvents for a wide range of analytes, from environmental contaminants to biomacromolecules and nanomaterials, and as solvents for various detection techniques [68, 69]. ILs have also been successfully applied in various areas of analytical chemistry, especially in the separation of analytes and other applications are being developed. The use of ILs in supported liquid membrane extraction, which has been studied for industrial separation, is reasonably expected to be applied in sample preparation. Chiral ILs may well come to be applied in the chromatographic/CE separation of chiral mixtures. Moreover, ILs appear to have significant advantages as matrices for MALDI-MS, so more research into this subject may broaden the application of MALDI-MS in the analysis of both low and high molecular weight substances. As ILs have wide electrochemical windows and high conductivities, they have great potential for use as sensors in electrochemical analysis [68].

Table 7.2 Critical properties of solvents used in SFE.

Solvent	Critical property	
	Temperature [°C]	Pressure [atm]
Ethene	10	51
Water	374	221
Carbon dioxide	32	72
Ethane	32	48
Nitrous oxide	37	72
Sulfur hexafluoride	46	38
Pentane	–77	33
Fluoroform	26	48
Ammonia	133	113

7.3.2 Supercritical fluid extraction

Supercritical fluid extraction was introduced into laboratory practice by K. Zosel in 1978, since when there has been a steady growth of interest in this analytical technique. Environmentally-benign and efficient, SFE has been extensively with respect to the extraction of organic compounds from various matrices (e.g. medicinal plants [70], food [71], seeds, fruits, leaves, flowers, rhizomes [72], soil [73], drugs [74], sediments [75], honey [76] and water [77]). Unfortunately, some organic pollutants are polar, so they cannot be extracted quantitatively from environmental solids with $SFCO_2$ [78]. But by modifying $SFCO_2$ with methanol, the solvent becomes more polar and is then suitable for the extraction of more polar target compounds.

SFE has been adopted by the EPA as a reference method for extracting Petroleum Hydrocarbons (Method 3560, in 1996), PAHs (Method 3561, in 1996) and PCBs (Method 3562, in 2007) from solid environmental matrices. SFE has recently been used for extracting POPs from different plant materials and several analytical applications dealing with the extraction of POPs from different animal tissues have also been reported [79].

A wide variety of solvents is available for use as SFs, including carbon dioxide, nitrous oxide, ethane, propane, n-pentane, ammonia, fluoroform, sulfur hexafluoride and water. Table 7.2 lists the most common compounds that can be used as supercritical fluids. Carbon dioxide is widely used in SFE because it is safe (low critical parameters: $Tc = 32°C$, $Pc = 72$ atm) (see Figure 7.8), readily available, inexpensive and non-toxic [78]. Carbon dioxide is a gas at room temperature, so once the extraction is completed, and the system decompressed, substantial elimination of CO_2 is achieved without residues, yielding a solvent-free extract [80].

The only serious drawback of SFE is that the investment costs are higher compared to those of the traditional process (extraction plus separation), which is relatively cheap and very easily scaled up for industrial purposes [72, 74]. Because of their properties, supercritical fluids are uniquely versatile for controlling extraction and chromatographic selectivity. Furthermore, the solvating power of supercritical fluids can be adjusted by altering pressure and temperature.

As an extraction medium, supercritical fluids have many advantages, the most important of which are:

- the minimization or complete exclusion of organic solvents;
- the possibility of selective extraction if minor changes are made to the pressure and temperature in the extraction system;
- a much higher degree of extracting agent purity than is achievable with organic solvents, because carbon dioxide and nitrous oxide, commonly used as supercritical fluids, are gases under ambient conditions;

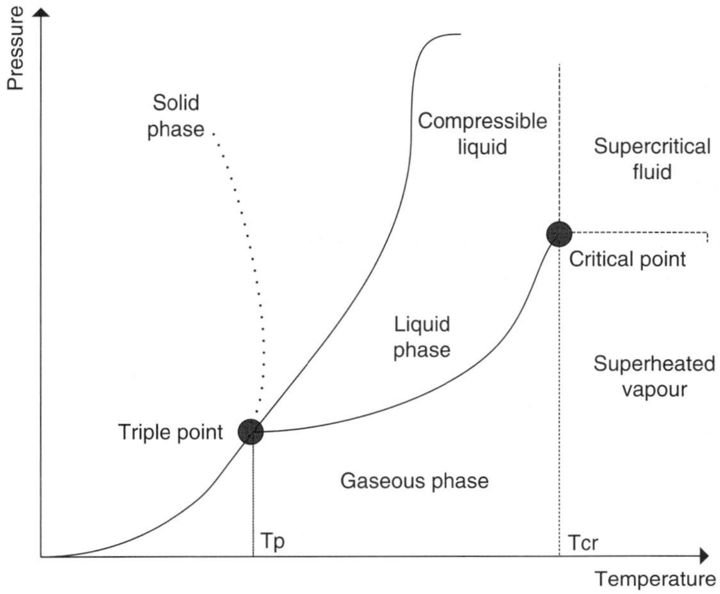

Figure 7.8 *Phase diagram.*

- the simplicity of concentrating and removing the extracting agent from the extract;
- the possibility of analysing the whole extract, not only a small part of it [77].

7.3.3 Subcritical water extraction

Water is probably the most common chemical compound in human experience and also the most necessary. Water is a very special solvent because of its unique structure, which includes vast numbers of hydrogen bonds. At room temperature it has a very high boiling point for its mass, a high dielectric constant and a high polarity. But when water is heated, its properties change markedly: the hydrogen-bonded lattice is disrupted with the increase in thermal motion. As the temperature rises, permittivity drops markedly and systematically, the diffusion rate increases, and viscosity and surface tension decrease. At elevated temperatures and moderate pressures the polarity of water falls significantly, so that it acts like a typical organic solvent, for example methanol, ethanol or acetone; water can therefore replace organic solvents for the extraction of semi- or poorly volatile compounds [81].

Superheated water is a general term for denoting the region of the condensed phase between 100°C and the critical point. This phase is often described as subcritical water, but by analogy with supercritical fluid extraction, the term should preferably be employed for the region close to the critical point. Superheated water extraction is also called 'subcritical water extraction' (SWE), 'hot water extraction' (HWE) or 'pressurized water extraction' (PWE).

First reported by Hawthorne in 1994 as an environmentally friendly solvent, superheated water can be effective for extracting a wide variety of organic compounds from diverse matrices but not for many other organic compounds, which because of their polarity are very poorly soluble in water [82]. It has been used to extract pesticides and polycyclic aromatic hydrocarbons, isoflavones [83], lignans [84], saponins [85] anthocyanins [86], dioxins, PCBs and pesticides [87]. As the temperature of liquid water is raised under pressure, the polarity decreases and it can be used as an extraction solvent for a wide range of compounds [88].

7.3.4 Fluorous phases

Fluorous chemistry has been studied intensively from the perspective of 'Green Chemistry', as the products can be readily separated and the solvents are reusable. The term 'fluorous' was introduced as an analogue to the term 'aqueous', meaning 'dissolved in fluorocarbon solvents'. Although highly fluorinated compounds do not dissolve either in common organic solvents or in water, they dissolve well in fluorous solvents such as perfluoroalkane. Among the solvents commonly used to separate substances from mixtures, to isolate dissolved compounds from solutions and to remove soluble impurities from mixtures, there are toxic ones like diethyl ether, methylene chloride, chloroform and acetone. The process exploits the immiscibility of two liquid phases, one usually aqueous and the other organic. Highly fluorinated (fluorous) compounds are referred to as the 'third liquid phase', because organics, water and fluorous compounds are mutually immiscible (Figure 7.9). Often a strong function of temperature, this property presents an important new opportunity for separations and for solution-phase organic synthesis [8].

Fluorous solvents have the following characteristics:

- they form homogeneous solutions with some organic solvents at elevated temperatures;
- their boiling points are almost equal to those of the corresponding hydrocarbons regardless of their molecular mass;
- many gases are very soluble in them.

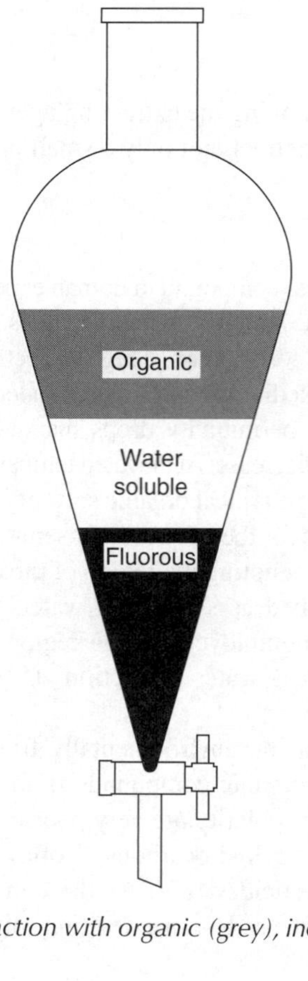

Figure 7.9 Three-phase liquid-liquid extraction with organic (grey), inorganic (white) and fluorous (black) phases.

Because they can be used in an integrated way that allows for product isolation, analogies can be drawn between fluorous-phase synthesis and solid-phase synthesis on supports like polymer resins. Extensively adopted for automated synthesis and combinatorial chemistry, solid-phase synthesis has revolutionized chemical synthesis while reducing the need for large amounts of solvents, especially for the separation of compounds. Fluorous phases may well be adopted on a similarly widespread scale, especially in view of the burgeoning commercial interest from such companies as Fluorous Technologies Inc. [89].

7.4 Assisted extractions

Liquid-liquid or solid-liquid extraction can be performed in the presence of external agents like microwaves, ultrasound or elevated pressures and temperatures. In this way the analytical parameters of extraction are improved and the negative environmental impact is reduced.

7.4.1 Microwave-assisted extraction

Microwaves offer an alternative way of delivering energy to the sample: a sample can be heated, digested or homogenized, water can be removed, and extraction enhanced. In contrast to conventional heating, microwaves are delivered to the whole volume of the sample (if it contains dipole molecules), thus heating it uniformly.

Microwaves are much used in analytical chemistry at the sample preparation step [90] to improve LLE extraction efficiency. Solvent volumes for extracting analytes are reduced, and the extraction time is usually shortened from about 24 h to 30 min compared with Soxhlet extraction, thus saving time and energy. The extraction of drugs from serum takes just one minute, mainly because microwave energy can release these substances from protein binding sites [91]. With microwave-assisted extraction, the sample size can be reduced, so organic solvent consumption is likewise diminished [92].

The digestion of environmental, biological, food and industrial samples is another field of application of microwave energy [93]: in this approach the process time is reduced to just a few seconds.

Exposure to microwaves removes water and volatile compounds from a sample, so its water content can be determined, or it can be dried before analysis. With microwave drying, local hot spot formation is less likely, and energy can be saved in comparison with conventional heating, especially when the process is performed under reduced pressure [94]. The drying time is often 10 times shorter than that of traditional drying methods [95]. Another application of microwaves is sample clean-up; here, the precipitation of proteins from blood samples is enhanced [93].

7.4.2 Ultrasound-assisted extraction

When a solution is exposed to ultrasonic energy, acoustic cavitation takes place. Bubbles of gas are formed which then implode, resulting in local increases of temperature and pressure. There are two basic types of ultrasound device; ultrasonic probes and ultrasonic baths. The most common application of ultrasound in analytical laboratories is ultrasound extraction (USE). This is used mainly for the extraction of analytes from solid samples (e.g. plant material [96]), but ultrasound-assisted LLE is also carried out. Ultrasonic energy acts in several ways to enhance extraction. High temperatures increase the solubility and diffusivity of compounds, and high pressures increase the penetration power of the solvent into the solid or second liquid phase. Both effects result in better recoveries of analytes and shortened extraction times [97]. With USE, extraction conditions are milder – for example, temperature and pressure are lower [98].

Ultrasound energy can also stimulate sample digestion, that is, when a substance is dissolved in the presence of moisture and heat. This process is often used to convert the matrix into one that is more conveniently handled by analytical apparatus and ultrasound substantially shortens the digestion time. In addition, combining ultrasound with microwave energy gives better results than if these agents are used separately to digest food and mineral samples [99].

In ultrasound-assisted methods of cleaning glassware, the ultrasonic energy causes trace contaminants to be desorbed from container or sampler walls. In another application of this kind, contaminants are removed from the surfaces of the membranes used for separating components from mixtures or for purifying analytical reagents [100].

Ultrasound significantly increases the rate of sample dissolution: mass transfer coefficients are greater in its presence than in its absence [101]. Dissolution of solid samples is necessary mainly in food analysis (i.e. powdered milk and other powdered food products [97]).

Ultrasound can also be used to homogenize or emulsify a sample (if it consists of two immiscible liquid phases): it can alter its physical properties without altering its chemical composition. In this way its chemical structure is made uniform, even though two phases are still present (one phase dispersed in the other). The relevant physical processes are mixing and stirring, which are caused by the propagation of ultrasonic waves through the liquid medium [102]. Many samples of food containing fats require homogenization before chemical analysis can begin.

Ultrasound is frequently used to speed up chemical reactions (the term sonochemistry has been introduced), for example, the derivatization of inorganic and organic analytes in sample preparation. But this application is rare; ultrasound is, however, commonly applied to reactions like depolymerization, esterification, alkylation, redox reactions and complex formation, to facilitate the production of target derivatives [102]. The ultrasound-assisted combination of extraction and derivatization, for example, in the extraction of haloacetic acids from vegetables and their simultaneous derivatization to methyl esters, enables both processes to be performed in less than 15 min [103].

An important aspect in process analysis is on-line sampling, which makes frequent use of membranes. These have the disadvantage of fouling up, which impairs their quality and shortens their operational lifespan [104].

Some liquid samples need to be degassed before analysis: with ultrasound this is done quickly and efficiently.

7.4.3 Pressurized liquid extraction

Pressurized liquid extraction (PLE) (or pressurized fluid extraction or enhanced solvent extraction) is considered to be the next generation of extraction procedures after Soxhlet extraction. The main advantages of PLE over Soxhlet extraction are the smaller volumes of solvent required (reduced from up to 200 ml to 1–20 ml) and the shorter extraction time (reduced from 24 h to 30 min). PLE uses a variety of environmentally friendly solvents like water, ethanol and methanol, but also traditional organic solvents [105]. The high pressure applied during extraction enables work at temperatures well above the boiling point at standard pressure. Elevated temperatures increase analyte solubility and diffusion rates, but decrease viscosity and surface tension of the solvent. Elevated pressure improves the penetrability of the extractant into the pores of the target material [106].

PLE is a common method for extracting organic analytes from environmental [107], food [108] and biological [109] samples; it has also been used to extract trace elements from food samples [110].

Finally, it should be mentioned that the term 'accelerated solvent extraction (ASE®)' is often misused as a synonym for PLE. Although its principle of operation is similar to that of PLE, ASE is the registered name of a method and devices designed by a certain company.

7.5 Final remarks

Though considered to be one of the most polluting steps in chemical analysis, sample preparation can be performed in an environmentally friendly manner. The main environmental issue in this context is the high consumption of organic solvents. This can be reduced by the application of 'green' analytical methods, which not only reduce the negative impact of analyses, but are often economically viable. The methods eliminating or reducing solvent volumes are often based on extraction on a smaller scale. Apart from the environmental benefits, the analytical parameters of alternative extraction techniques are usually better.

References

1. Tobiszewski, M.; Mechlińska, A.; Zygmunt, B. and Namieśnik, J. (2009) Green Analytical Chemistry in sample preparation for determination of trace organic pollutants, *TrAC-Trend. Anal. Chem.*, **28**, 943–951.
2. Koel, M. and Kaljurand, M. (2006) Application of the principles of Green Chemistry in analytical chemistry, *Pure Appl. Chem.*, **78**, 1993–2002.
3. Hennion, M.C. (1999) Solid-phase extraction: method development, sorbents, and coupling with liquid chromatography, *J. Chromatogr. A*, **856**, 3–54.
4. Jakubowska, N.; Zygmunt, B.; Polkowska, Ż.; Zabiegała, B. and Namieśnik, J. (2009) Sample preparation for gas chromatographic determination of halogenated volatile organic compounds in environmental and biological samples, *J. Chromatogr. A*, **1216**, 422–441.
5. Ouyang, G. and Pawliszyn, J. (2006) SPME in environmental analysis, *Anal. Bioanal. Chem.*, **386**, 1059–1073.
6. Risticevic, S.; Niri, V.H.; Vuckovic, D. and Pawliszyn, J. (2009) Recent developments in solid-phase microextraction, *Anal. Bioanal. Chem.*, **393**, 781–795.
7. Ouyang, G. and Pawliszyn, J. (2006) Recent developments in SPME for on-site analysis and monitoring, *TrAC-Trend. Anal. Chem.*, **25**, 692–703.
8. Kawaguchi, M.; Ito, R.; Saito, K. and Nakazawa, H. (2006) Novel stir bar sorptive extraction methods for environmental and biomedical analysis, *J. Pharm. Biomed. Anal.*, **40**, 500–508.
9. David, F. and Sandra, P. (2007) Stir bar sorptive extraction for trace analysis, J. *Chromatogr. A*, **1152**, 54–69.
10. Baltussen, E.; Sandra, P.; David, F. and Cramers, C.A. (1999) Stir bar sorptive extraction (SBSE), a novel extraction technique for aqueous samples: Theory and principles, *J. Microcol. Sep.*, **11**, 737–747.
11. Popp, P.; Bauer, C. and Weinrich, L. (2001) Application of stir bar sorptive extraction in combination with column liquid chromatography for the determination of polycyclic aromatic hydrocarbons in water samples, *Anal. Chim. Acta*, **436**, 1–9.
12. Popp, P.; Bauer, C.; Hauser, B.; Keil, P. and Wennrich, L. (2003) Extraction of polycyclic aromatic hydrocarbons and organochlorine compounds from water: A comparison between solid-phase microextraction and stir bar sorptive extraction, *J. Sep. Sci.*, **26**, 961–967.
13. Liu, W.M.; Wang, H.M. and Guan, Y.F. (2004) Preparation of stir bars for sorptive extraction using sol-gel technology, *J. Chromatogr. A*, **1045**, 15–22.
14. Popp, P.; Bauer, C.; Paschke, A. and Montero, L. (2004) Application of a polysiloxane-based extraction method combined with column liquid chromatography to determine polycyclic aromatic hydrocarbons in environmental samples, *Anal. Chim. Acta*, **504**, 307–312.
15. Kolahgar, B.; Hoffmann, A. and Heiden, A.C. (2002) Application of stir bar sorptive extraction to the determination of polycyclic aromatic hydrocarbons in aqueous samples, *J. Chromatogr. A*, **963**, 225–230.
16. Leon, V.M.; Alvarez, B.; Cobollo, M.A.; Munoz, S. and Valor, I. (2003) Analysis of 35 priority semivolatile compounds in water by stir bar sorptive extraction-thermal desorption-gas chromatography-mass spectrometry I. Method optimisation, *J. Chromatogr. A*, **999**, 91–101.
17. Serodio, P. and Nogueira, J.M.F. (2004) Multi-residue screening of endocrine disrupters chemicals in water samples by stir bar sorptive extraction-liquid desorption-capillary gas chromatography-mass spectrometry detection, *Anal. Chim. Acta*, **517**, 21–32.

18. Garcia-Falcon, M.S.; Perez-Lamela, C. and Simal-Gandara, J. (2004) Strategies for the extraction of free and bound polycyclic aromatic hydrocarbons in run-off waters rich in organic matter, *Anal. Chim. Acta*, **508**, 177–183.
19. Garcia-Falcon, M.S.; Cancho-Grande, B. and Simal-Gandara, J. (2004) Stirring bar sorptive extraction in the determination of PAHs in drinking waters, *Water Res.*, **38**, 1679–1684.
20. Roy, G.; Vuillemin, R. and Guyomarch, J. (2005) On-site determination of polynuclear aromatic hydrocarbons in seawater by stir bar sorptive extraction (SBSE) and thermal desorption GC-MS, *Talanta*, **66**, 540–546.
21. Leon, V.M.; Llorca-Porcel, J.; Alvarez, B.; Cobollo, M.A.; Munoz, S. and Valor, I. (2006) Analysis of 35 priority semivolatile compounds in water by stir bar sorptive extraction-thermal desorption-gas chromatography-mass spectrometry Part II: Method validation, *Anal. Chim. Acta*, **558**, 261–266.
22. Bicchi, C.; Cordero, C.; Rubiolo, P. and Sandra, P. (2003) Impact of water/PDMS phase ratio, volume of PDMS, and sampling time on Stir Bar Sorptive Extraction (SBSE) recovery of some pesticides with different K-O/W, *J. Sep. Sci.*, **26**, 1650–1656.
23. Montero, L.; Popp, P.; Paschke, A. and Pawliszyn, J. (2004) Polydimethylsiloxane rod extraction, a novel technique for the determination of organic micropollutants in water samples by thermal desorption-capillary gas chromatography-mass spectrometry, *J. Chromatogr. A*, **1025**, 17–26.
24. Mothes, S.; Popp, P. and Wennrich, R. (2003) Analysis of organophosphorus insecticides in natural waters by use of stir-bar-sorptive extraction then gas chromatography with atomic emission detection, *Chromatographia*, **57**, S249–S252.
25. Standler, A.; Schatzl, A. and Klampfl, C.W. (2004) Determination of the insect repellent Bayrepel (R) in pool and lake water by gas chromatography after preconcentration with solid-phase extraction and stir-bar-sorptive extraction, *Microchim. Acta*, **148**, 151–156.
26. Popp, P.; Keil, P. Montero, L. and Rückert, M. (2005) Optimized method for the determination of 25 polychlorinated biphenyls in water samples using stir bar sorptive extraction followed by thermodesorption gas chromatography mass spectrometry, *J. Chromatogr. A*, **1071**, 155–162.
27. Brossa, L.; Marce, R.M.; Borrull, F. and Pocurull, E. (2005) Determination of endocrine disruptors in environmental water samples by stir bar sorptive extraction-liquid desorption – Large volume injection-gas chromatography, *Chromatographia*, **61**, 61–65.
28. Nakamura, S. and Daishima, S. (2005) Simultaneous determination of 64 pesticides in river water by stir bar sorptive extraction and thermal desorption-gas chromatography-mass spectrometry, *Anal. Bioanal. Chem.*, **382**, 99–107.
29. Serodio, P. and Nogueira, J.M.F. (2005) Development of a stir-bar-sorptive extraction-liquid desorption-large-volume injection capillary gas chromatographic-mass spectrometric method for pyrethroid pesticides in water samples, *Anal. Bioanal. Chem.*, **382**, 1141–1151.
30. Benanou, D.; Acobas, F.; de Roubin, M.R.; David, F. and Sandra, P.D.P. (2003) Analysis of off-flavors in the aquatic environment by stir bar sorptive extraction-thermal desorption-capillary GC/MS/olfactometry, *Anal. Bioanal. Chem.*, **376**, 69–77.
31. Nakamura, S.; Nakamura, N. and Ito, S. (2001) Determination of 2-methylisoborneol and geosmin in water by gas chromatography-mass spectrometry using stir bar sorptive extraction, *J. Sep. Sci.*, **24**, 674–677.
32. Ochiai, N.; Sasamoto, K.; Takino, M.; Yamashita, S.; Daishima, S.; Heiden, A.C. and Hoffmann, A. (2002) Simultaneous determination of preservatives in beverages, vinegar, aqueous sauces, and quasi-drug drinks by stir-bar sorptive extraction (SBSE) and thermal desorption GC-MS, *Anal. Bioanal. Chem.*, **373**, 56–63.
33. Benanou, D.; Acobas, F. and de Roubin, M.R. (2004) Optimization of stir bar sorptive extraction applied to the determination of odorous compounds in drinking water, *Water Sci. Technol.*, **49**, 161–170.
34. Zhao, E.C.; Han, L.J.; Jiang, S.R.; Wang, Q.X. and Zhou, Z.Q. (2006) Application of a single-drop microextraction for the analysis of organophosphorus pesticides in juice, *J. Chromatogr. A*, **1114**, 269–273.
35. Lambropoulou, D.A.; Psillakis, E.; Albanis, T.A. and Kalogerakis, N. (2004) Single-drop microextraction for the analysis of organophosphorous insecticides in water, *Anal. Chim. Acta*, **516**, 205–211.
36. Xu, L.; Basheer, C. and Lee, H.K. (2007) Developments in single-drop microextraction, *J. Chromatogr. A*, **1152**, 184–192.
37. Zhao, E.C.; Shan, W.L.; Jiang, S.R.; Liu, Y. and Zhou, Z.Q. (2006) Determination of the chloroacetanilide herbicides in waters using single-drop microextraction and gas chromatography, *Microchem. J.*, **83**, 105–110.
38. Vidal, L.; Psillakis, E.; Domini, C.E.; Grane, N.; Marken, F. and Canals, A. (2007) An ionic liquid as a solvent for headspace single drop microextraction of chlorobenzenes from water samples, *Anal. Chim. Acta*, **584**, 189–195.

39. Wardencki, W.; Curyło, J. and Namieśnik, J. (2007) Trends in solventless sample preparation techniques for environmental analysis, *J. Biochem. Biophys. Methods*, **70**, 275–288.
40. Jakubowska, N.; Polkowska, Ż.; Namieśnik, J. and Przyjazny, A. (2005) Analytical applications of membrane extraction for biomedical and environmental liquid sample preparation, *Crit. Rev. Anal. Chem.*, **35**, 217–235.
41. Jonsson, J.A. and Mathiasson, L. (1999) Liquid membrane extraction in analytical sample preparation I. Principles, *TrAC-Trend. Anal. Chem.*, **18**, 318–325.
42. Psillakis, E. and Kalogerakis, N. (2003) Developments in liquid-phase microextraction, *TrAC-Trend. Anal. Chem.*, **22**, 565–574.
43. Rasmussen, K.E. and Petersen-Bjergaard, S. (2004) Developments in hollow fibre-based, liquid-phase microextraction, *TrAC-Trend. Anal. Chem.*, **23**, 1–10.
44. Segal, A.; Górecki, T.; Mussche, P.; Lips, J. and Pawliszyn, J. (2000) Development of membrane extraction with a sorbent interface-micro gas chromatography system for field analysis, *J. Chromatogr. A*, **873**, 13–27.
45. Barri, T. and Jonsson, J.A. (2008) Advances and developments in membrane extraction for gas chromatography - Techniques and applications, *J. Chromatogr. A*, **1186**, 16–38.
46. Almeda, S.; Nozal, L.; Arce, L. and Valcarcel, M. (2007) Direct determination of chlorophenols present in liquid samples by using a supported liquid membrane coupled in-line with capillary electrophoresis equipment, *Anal. Chim. Acta*, **587**, 97–103.
47. Irigoyen, L.; Moreno, C.; Mendiguchia, C. and Garcia-Vargas, M. (2006) Application of liquid membranes to sample preconcentration for the spectrometric determination of cadmium in seawater, *J. Membr. Sci.*, **274**, 169–172.
48. Luthje, K.; Hyotylainen, T. and Riekkola, M.L. (2004) On-line coupling of microporous membrane liquid-liquid extraction and gas chromatography in the analysis of organic pollutants in water, *Anal. Bioanal. Chem.*, **378**, 1991–1998.
49. Hyotylainen, T.; Luthje, K.; Rautiainen-Rama, M. and Riekkola, M.L. (2004) Determination of pesticides in red wines with on-line coupled microporous membrane liquid-liquid extraction-gas chromatography, *J. Chromatogr. A*, **1056**, 267–271.
50. Zhou, Q.X.; Liu, J.F.; Cai, Y.Q.; Liu, G.G. and Jiang, G.B. (2003) Micro-porous membrane liquid-liquid extraction as an enrichment step prior to nonaqueous capillary electrophoresis determination of sulfonylurea herbicides, *Microchem. J.*, **74**, 157–163.
51. Sandahl, M.; Ulfsson, E. and Mathiasson, L. (2000) Automated determination of Vinclozolin at the ppb level in aqueous samples by a combination of microporous membrane liquid-liquid extraction and adsorption chromatography, *Anal. Chim. Acta*, **424**, 1–5.
52. March, J.G.; Palou, J.; Chisvert, A. and Salvador, A. (2009) A simple novel configuration for in-vial microporous membrane liquid-liquid extraction, *J. Chromatogr. A*, **1216**, 5160–5163.
53. Matz, G.; Kibelka, G.; Dahl J. and Lannemann, F. (1999) Experimental study on solvent-less sample preparation methods - Membrane extraction with a sorbent interface, thermal membrane desorption application and purge-and-trap, *J. Chromatogr. A*, **830**, 365–376.
54. Schellin, M. and Popp, P. (2003) Membrane-assisted solvent extraction of polychlorinated biphenyls in river water and other matrices combined with large volume injection-gas chromatography-mass spectrometric detection, *J. Chromatogr. A*, **1020**, 153–160.
55. Rodil, R.; Schellin, M. and Popp, P. (2007) Analysis of polycyclic-aromatic hydrocarbons in water and beverages using membrane-assisted solvent extraction in combination with large volume injection-gas chromatography-mass spectrometric detection, *J. Chromatogr. A*, **1163**, 288–297.
56. Schellin, M. and Popp, P. (2005) Membrane-assisted solvent extraction of seven phenols combined with large volume injection-gas chromatography-mass spectrometric detection, *J. Chromatogr. A*, **1072**, 37–43.
57. Snow, N.H. and Slack, G.C. (2002) Head-space analysis in modern gas chromatography, *TrAC-Trend. Anal. Chem.*, **21**, 608–617.
58. Voice, T.C. and Kolb, B. (1993) Static and dynamic headspace analysis of volatile organic-compounds in soils, *Environ. Sci. Technol.*, **27**, 709–713.
59. Marczak, M.; Wolska, L.; Chrzanowski, W. and Namieśnik, J. (2006) Microanalysis of volatile organic compounds (VOCs) in water samples - methods and instruments, *Microchim. Acta*, **155**, 331–348.
60. Lambropoulou, D.A.; Konstantinou, I.K. and Albanis, T.A. (2007) Recent developments in headspace microextraction techniques for the analysis of environmental contaminants in different matrices, *J. Chromatogr. A*, **1152**, 70–96.

61. Kolb, B. and Ettre, L.S. (2006) *Static Headspace – Gas Chromatography: Theory and Practice,* 2nd edn, John Wiley & Sons, Inc., Hoboken.
62. Pankow, J.F. and Rosen, M.E. (1984) The analysis of volatile compounds by purge and trap with whole column cryotrapping (WCC) on a fused-silica capillary column, HRC, *J. High Res. Chromatogr.*, **7**, 504–508.
63. Heikes, D.L.; Jensen, S.R. and Fleming-Jones, M.E. (1995) Purge-and-trap extraction with GC-MS determination of volatile organic-compounds in table-ready foods, *J. Agric. Food Chem.*, **43**, 2869–2875.
64. Eggers, N.; Kenefick, S.; Richardson, S.; Wigglesworth, T. and Girard, B. (2003) Evaluation of closed-loop stripping for the isolation of wine aroma compounds from aqueous solution, *Am. J. Enol. Vitic.*, **54**, 92–98.
65. Baker, G.A.; Baker, S.N.; Pandey, S. and Bright, F.V. (2005) An analytical view of ionic liquids, *Analyst*, **130**, 800–808.
66. Huddleston, J.G.; Visser, A.E.; Reichert, W.M.; Willauer, H.D.; Broker, G.A. and Rogers, R.D. (2001) Characterization and comparison of hydrophilic and hydrophobic room temperature ionic liquids incorporating the imidazolium cation, *Green Chem.*, **3**, 156–164.
67. Liu, J.F.; Jönsson, J.A. and Jiang, G.B. (2005) Application of ionic liquids in analytical chemistry, *TrAC-Trend. Anal. Chem.*, **24**, 20–27.
68. Liu, R.; Liu, J.F.; Yin, Y.G.; Hu, X.L. and Jiang, G.B. (2009) Ionic liquids in sample preparation, *Anal. Bioanal. Chem.*, **393**, 871–883.
69. Kuhne, E.; Santarossac, S.; Perez, E.; Witkamp, G.J. and Peters, C.J. (2008) New approach in the design of reactions and separations using an ionic liquid and carbon dioxide as solvents: Phase equilibria in two selected ternary systems, *J. Supercrit. Fluids*, **46**, 93–98.
70. Tang, F.; Zhang, Q.L.; Nie, Z.; Chen, B. and Yao, S.Z. (2009) Sample preparation for analyzing traditional Chinese medicines, *TrAC-Trend. Anal. Chem.*, **28**, 1253–1262.
71. Askin, R.; Goto, M. and Sasaki, M. (2008) Supercritical fluid extraction in food analysis, in *Handbook of Food Analysis Instruments*, CRC Press, p. 25.
72. Reverchon, E. and De Marco, I. (2006) Supercritical fluid extraction and fractionation of natural matter, *J. Supercrit. Fluids*, **38**, 146–166.
73. Sporring, S.; Bøwadt, S.; Svensmark, B. and Bjorklund, E. (2005) Comprehensive comparison of classic Soxhlet extraction with Soxtec extraction, ultrasonication extraction, supercritical fluid extraction, microwave assisted extraction and accelerated solvent extraction for the determination of polychlorinated biphenyls in soil, *J. Chromatogr. A*, **1090**, 1–9.
74. Davies, O.R.; Lewis, A.L.; Whitaker, M.J.; Tai, H.; Shakesheff, K.M. and Howdle, S.M. (2008) Applications of supercritical CO(2) in the fabrication of polymer systems for drug delivery and tissue engineering, *Advan. Drug Delivery Rev.*, **60**, 373–387.
75. Rezaee, M.; Yamini, Y.; Moradi, M.; Saleh, A.; Faraji, M. and Naeeni, M.H. (2010) Supercritical fluid extraction combined with dispersive liquid-liquid microextraction as a sensitive and efficient sample preparation method for determination of organic compounds in solid samples, *J. Supercrit. Fluid.*, **55**, 161–168.
76. Rissato, S.R.; Galhiane, M.S.; Knoll, F.R.N. and Apon, B.M. (2004) Supercritical fluid extraction for pesticide multiresidue analysis in honey: determination by gas chromatography with electron-capture and mass spectrometry detection, *J. Chromatogr. A*, **1048**, 153–159.
77. Glazkov, I.N.; Revelsky, I.A.; Efimov, I.P. and Zolotov, Y.A. (1999) Supercritical fluid extraction of water samples containing ultratrace amounts of organic micropollutants, *J. Microcolumn Sep.*, **11**, 729–736.
78. Armenta, S.; Garrigues, S. and de la Guardia, M. (2008) Green Analytical Chemistry, *TrAC-Trend. Anal. Chem.*, **27**, 497–511.
79. Onofrejová, L.; Vašíčková, J.; Klejdus, B.; Stratil, P.; Mišurcová, L.; Kráčmar, S.; Kopecký, J. and Vacek, J. (2010) Bioactive phenols in algae: The application of pressurized-liquid and solid-phase extraction techniques, *J. Pharm. Biomed. Anal.*, **51**, 464–470.
80. Herrero, M.; Cifuentes, A. and Ibañez, E. (2006) Sub- and supercritical fluid extraction of functional ingredients from different natural sources: Plants, food-by-products, algae and microalgae - A review, *Food Chem.*, **98**, 136–148.
81. Smith, R.M. (2002) Extractions with superheated water, *J. Chromatogr. A*, **975**, 31–46.
82. Tollback, J.; Bigata, M.B.; Crescenzi, C. and Strom, J. (2008) Feasibility of analyzing fine particulate matter in air using solid-phase extraction membranes and dynamic subcritical water extraction, *Anal. Chem.*, **80**, 3159–3167.

83. Hsun, C.L.; Chuan, C.Y. and Ming, C.C. (2004) Extracting and purifying isoflavones from defatted soybean flakes using superheated water at elevated pressures, *Food Chem.*, **84**, 279–285.
84. Cacace, J.E. and Mazza, G. (2006) Pressurized low polarity water extraction of lignans from whole flaxseed, *J. Food Eng.*, **77**, 1087–1095.
85. Güçlü-Üstündağ, Ö.; Balsevich, J. and Mazza, G. (2007) Pressurized low polarity water extraction of saponins from cow cockle seed, *J. Food Eng.*, **80**, 619–630.
86. Ju, Z.Y. and Howard, L.R. (2003) Subcritical water and sulfured water extraction of anthocyanins and other phenolics from dried red grape skin, *J. Food Sci.*, **70**, S270–S276.
87. Anitescu, G. and Tavlarides, L.L. (2006) Supercritical extraction of contaminants from soils and sediments, *J. Supercrit. Fluids*, **38**, 167–180.
88. Saim, N.; Osman, R.; Yasin, W.A.H.M. and Hamid, R.D. (2008) Subcritical water extraction of essential oil fromcoriander (*Coriandrum sativum L.*) seeds, *Malaysian J. Anal. Sci.*, **12**, 22–24.
89. DeSimone, J.M. (2002) Practical approaches to green solvents, *Science*, **297**, 799–803.
90. Smith, F.E. and Arsenault, E.A. (1996) Microwave-assisted sample preparation in analytical chemistry, *Talanta*, **43**, 1207–1268.
91. Franke, M.; Winek, C.L. and Kingston, H.M. (1996) Extraction of selected drugs from serum using microwave irradiation, *Forensic Sci. Int.*, **81**, 51–59.
92. Purcaro, G.; Moret, S. and Conte, L.S. (2009) Optimisation of microwave assisted extraction (MAE) for polycyclic aromatic hydrocarbon (PAH) determination in smoked meat, *Meat Sci.*, **81**, 275–280.
93. de Nascimento, P.C.; Bohrer, D.; Garcia, S. and Ritzel, A.F. (1997) Liquid chromatography with ultraviolet absorbance detection of ethylenethiourea in blood serum after microwave irradiation as an auxiliary cleanup step, *Analyst*, **122**, 733–735.
94. Jin, Q.H.; Liang, F.; Zhang, H.Q.; Zhao, L.W.; Huan, Y.F and Song, D.Q. (1999) Application of microwave techniques in analytical chemistry, *TrAC-Trend. Anal. Chem.*, **18**, 479–484.
95. Veillet, S.; Tomaoa, V.; Visinonib, F. and Chemat, F. (2009) New and rapid analytical procedure for water content determination: Microwave accelerated Dean-Stark, *Anal. Chim. Acta*, **632**, 203–207.
96. Vinatoru, M. (2001) An overview of the ultrasonically assisted extraction of bioactive principles from herbs, *Ultrason. Sonochem.*, **8**, 303–313.
97. Priego-Capote, F. and Luque de Castro, M.D. (2004) Analytical uses of ultrasound – I. Sample preparation, *TrAC-Trend. Anal. Chem.*, **23**, 644–653.
98. Romdhane. M. and Gourdon, C. (2002) Investigation in solid-liquid extraction: influence of ultrasound, *Chem. Eng. J.*, **87**, 11–19.
99. Chemat, S.; Lagha, A.; Amar, H.A. and Chemat, F. (2004) Ultrasound assisted microwave digestion, *Ultrason. Sonochem.*, **11**, 5–8.
100. Chai, X.; Kobayashi, T. and Fujii, N. (1999) Ultrasound-associated cleaning of polymeric membranes for water treatment, *Sep. Purif. Technol.*, **15**, 139–146.
101. Kannan, A. and Pathan, S.K.H. (2004) Enhancement of solid dissolution process, *Chem. Eng. J.*, **102**, 45–49.
102. Luque de Castro, M.D. and Priego-Capote, F. (2007) Ultrasound-assisted preparation of liquid samples, *Talanta*, **72**, 321–334.
103. Alvarez-Sanchez, B.; Priego-Capote, F. and Luque de Castro, M.D. (2008) Ultrasonic enhancement of leaching and in situ derivatization of haloacetic acids in vegetable foods prior to gas chromatography-electron capture detection, *J. Chromatogr. A*, **1201**, 21–26.
104. Kobayashi, T.; Chai, X. and Fujii, N. (1999) Ultrasound enhanced cross-flow membrane filtration, *Sep. Purif. Technol.*, **17**, 31–40.
105. Mendiola, J.A.; Herrero, M.; Cifuentes, A. and Ibanez, E. (2007) Use of compressed fluids for sample preparation: Food applications, *J. Chromatogr. A*, **1152**, 234–246.
106. Conte, E.; Milani, R.; Morali, G. and Abballe, F. (1997) Comparison between accelerated solvent extraction and traditional extraction methods for the analysis of the herbicide diflufenican in soil, *J. Chromatogr. A*, **765**, 121–125.
107. Li, D.H.; Oh, J.R. and Park, J. (2003) Direct extraction of alkylphenols, chlorophenols and bisphenol A from acid-digested sediment suspension for simultaneous gas chromatographic-mass spectrometric analysis, *J. Chromatogr. A*, **1012**, 207–214.

108. Boselli, E.; Velazco, V.; Caboni, M.F. and Lercker, G. (2001) Pressurized liquid extraction of lipids for the determination of oxysterols in egg-containing food, *J. Chromatogr. A*, **917**, 239–244.
109. Schafer, K. (1998) Accelerated solvent extraction of lipids for determining the fatty acid composition of biological material, *Anal. Chim. Acta*, **358**, 69–77.
110. Moreda-Pineiro, J.; Alonso-Rodrıguez, E.; Lopez-Mahıa, P.; Muniategui-Lorenzo, S.; Fernandez-Fernandez, E.; Prada-Rodrıguez, D.; Moreda-Pineiro, A.; Bermejo-Barrera, A. and Bermejo-Barrera, P. (2006) Pressurized liquid extraction as a novel sample pre-treatment for trace element leaching from biological material, *Anal. Chim. Acta*, **572**, 172–179.

8

Green Sample Preparation with Non-Chromatographic Separation Techniques

María Dolores Luque de Castro and Miguel Alcaide Molina

Department of Analytical Chemistry, Campus of Rabanales, University of Córdoba, Spain

8.1 Sample preparation in the frame of the analytical process

The analytical process involves preliminary operations that can be dealt with as a whole in relation to the concept "sample preparation" (SP). This term is widely used at present, but is occasionally confused with 'sample pretreatment" because the boundary between the two (i.e. where sample pretreatment ends or what precedes and follows sample preparation), if any, is rather ill-defined. In the authors' opinion, sample preparation includes every step required to make a sample, or rather, the target analytes it contains, ready for insertion into a measuring instrument and may involve more than one step; this is consistent with IUPAC's statement that 'sample preparation is intended to transfer or transform the analytes into measurable forms' [1]. On the other hand, sample pretreatment can be envisaged as the first step in a sample preparation process or as one preceding some specific action (e.g. a pretreatment for either subsequent liquid–liquid extraction or insertion of a solution into a chromatograph). From this point of view, it is unclear when such a step finishes and detection starts (e.g. ionization in methods using MS detection or vaporization and atomization treatments in atomic detectors).

Sample preparation is most often a necessity since even the simplest samples are frequently unsuitable for direct analysis. In most cases, the need arises from excessive dilution or concentration of the target analytes in the sample or from incompatibility with the standard operational procedures for some instruments. A large number of SP approaches have been devised to enable the detection of analytes that range from classical operations - many of which including calcination, wet digestion or Soxhlet leaching have been in use for centuries - to modern operations developed less than two decades ago (e.g. supercritical fluid leaching). In any case, both classical and modern SP operations are being continuously improved.

Handbook of Green Analytical Chemistry, First Edition. Edited by Miguel de la Guardia and Salvador Garrigues.

Sample preparation is the most crucial step, and the bottleneck, of most analytical methodologies; also, it is the primary source of error in the analytical process and remains one of its most time-consuming steps. In fact, modern uncertainty evaluation methods consider the uncertainty associated with chemical preparation of the sample to be a part of the uncertainty budget of the final result [2]. As a result, improving SP can greatly help the analytical process. This foregoing has led to the increasing recognition of SP as a legitimate area of specialization in analytical chemistry in response to the growing advances and reliance on instrumental analysis. In fact, SP equipment is evolving in much the same was as chromatography and spectroscopy did in the past. However, SP is still consolidating, albeit at an incredibly fast pace.

Efforts at introducing novel sample preparation modes are frequently not accompanied by their use in the development of official analytical methods. For example, most methods of analysis endorsed by official bodies rely on traditional SP techniques. This can be ascribed to the fact that traditional methods are developed in routine analytical laboratories and analysts know quite well how they operate. However, traditional SP methods pose well-known problems such as those derived from the use of toxic organics and environmentally hazardous, multi-step procedures that often result in the loss of analytes and preclude integration with the other steps of the analytical process. Also, recycling pure solvents usually causes more problems than it solves and is generally unfeasible in spite of the increasing availability of integrated solvent management programmes.

Fortunately, the situation is changing in many respects and new, better SP techniques are being introduced at a steady pace. Automation and miniaturization are two currently widespread trends in sample preparation; both meet the requirements of analytical chemistry and are implicitly included among the research priorities for the twenty-first century. This is a consequence of the development of miniaturized separation systems, which has been recognized as the most important endeavour in this context since such systems fulfil many analytical requirements and enable fast analyses at low operational costs and with little environmental pollution. The success of micro-scale separation techniques has encouraged many scientists to seek ways to miniaturize SP processes especially suited to micro-scale work. In fact, one of the most recent trends in SP development is miniaturization, with special emphasis on the process and medium to be used for the highly efficient extraction and preconcentration of sample components. The benefits are obvious: shorter analysis times, reduced solvent and sample consumption, and the ability to process large numbers of samples simultaneously. This trend to miniaturization is not always properly understood by scientists, who occasionally downscale analytical systems unnecessarily and are confronted with the typical shortcomings of small-scale work as a result [3].

The greatest advantage of miniaturized techniques as regards 'greenness' is that they use very small amounts - if any - of environmentally hazardous solvents and sample; an added advantage with polluting samples such as pesticides, polycyclic aromatic hydrocarbons, volatile organic compounds and various other substances as the resulting waste solutions are less polluting. This is consistent with existing policies aimed at reducing environmental pollution in response to the increased awareness of the polluting, hazardous nature of the organic solvents and inorganic acids used by many SP methods at present; and has promoted international initiatives to suppress them. This phasing out of solvents is poised to induce a major change in analytical methodologies and also to provide an opportunity to formulate practical alternatives to existing SP methods. Such is the case with the recent approaches based on the use of water as their only solvent [4, 5].

The present trend as regards SP approaches is to develop new methods from scratch or improve existing, classical SP methods in order to circumvent their typical shortcomings. Classical SP methods can be improved by using auxiliary forms of energy to assist some separation steps or developing new, more effective materials for solid-phase extraction. The ideal sample preparation technique would be one using little or no solvent and applicable to a wide range of matrices in order to cater for an also wide variety of potential problems. Also, it would afford simultaneous separation and concentration of the target analytes, and be amenable to on-site use.

Although chromatographies are the most widely used techniques in analytical laboratories, the undeniable importance of non-chromatographic techniques as more economical alternatives to their chromatographic counterparts or effective choices for implementing auxiliary steps, warrants their consideration towards the development of environmentally friendly SP approaches. This chapter is intended to expose readers to the latest improvements in green sample preparation methods based on non-chromatographic techniques including pervaporation, microdialysis and solid-phase microextraction, some of which are directly applicable to solid samples. Consequently, the techniques discussed here are organized according to the particular phases involved.

8.2 Separation techniques involving a gas–liquid interface

The green chemistry approach to separation techniques involving a gas–liquid interface relies mainly on a sparing use of solvents, which is often accomplished by using a downscaled methodology or manifold. Automation plays a major role here as it dispenses with need for human intervention and avoids potential over-wasting as a result. Flow injection (FI) and sequential injection (SI) manifolds - which have lately evolved into so-called 'laboratory-in-a-valve', 'lab-on-valve' or 'LOV' systems - have been widely used for this purpose and provided substantial improvements in recent years [6–8].

8.2.1 Gas diffusion

The gas-diffusion technique relies on the permeability of microporous membranes to gases. The well-known mechanism of transport across a gas diffusion membrane (viz., sorption of volatile components at the membrane surface, diffusion of sorbed components through the polymer matrix and evaporation from the polymer into the vapour phase on the permeate side of the membrane) governs the efficiency of the process, via the intrinsic properties of the polymers used to prepare the membranes mainly.

Conventional gas-diffusion modules of the sandwich and tubular types (Figure 8.1) [9] have been miniaturized for reduced consumption of both sample and auxiliary solutions. Although sandwich-type modules use lower liquid volumes, their efficiency is heavily dependent on the way channels on the donor and acceptor sides are designed. Similarly to the earliest conventional modules, most miniaturized systems are designed and constructed by researchers or operators themselves as the need arises to reduce costs for a specific application. These self-devised modules are typically designed for easy connection to FI manifolds of low-diameter tubing (usually 0.3 mm i.d.) and, preferably, SI systems [10]. No LOV–gas diffusion approach has so far been reported, however; when a gas phase of analyte is formed in an LOV, it is removed from the sample matrix by means of a gas–liquid separator [11].

The most frequent field of application of continuous (FI and SI) systems in this context is gas diffusion in environmental analysis, where they have been used to develop new methods for the determination of chlorine dioxide based on the fluorescence quenching of chromotropic acid [12]; that of ethyl xanthate by gas diffusion and UV detection as CS_2 in the process monitoring of sulfide ore flotation [13]; and that of urea in alcoholic beverages using microgas diffusion to separate CO_2 from urease (the catalyst) and photometric detection with an acid–base indicator [10]. Further reduction of costs and waste is achieved by using the enzyme in immobilized form to exploit the typical reusability of biocatalysts.

8.2.2 Pervaporation

Analytical pervaporation is intrinsically a near-green membrane-based separation technique inasmuch as it is always implemented in a low-size format and almost invariably uses non-polluting solvents. This technique combines continuous evaporation and gas diffusion through a gas-permeable membrane, both processes

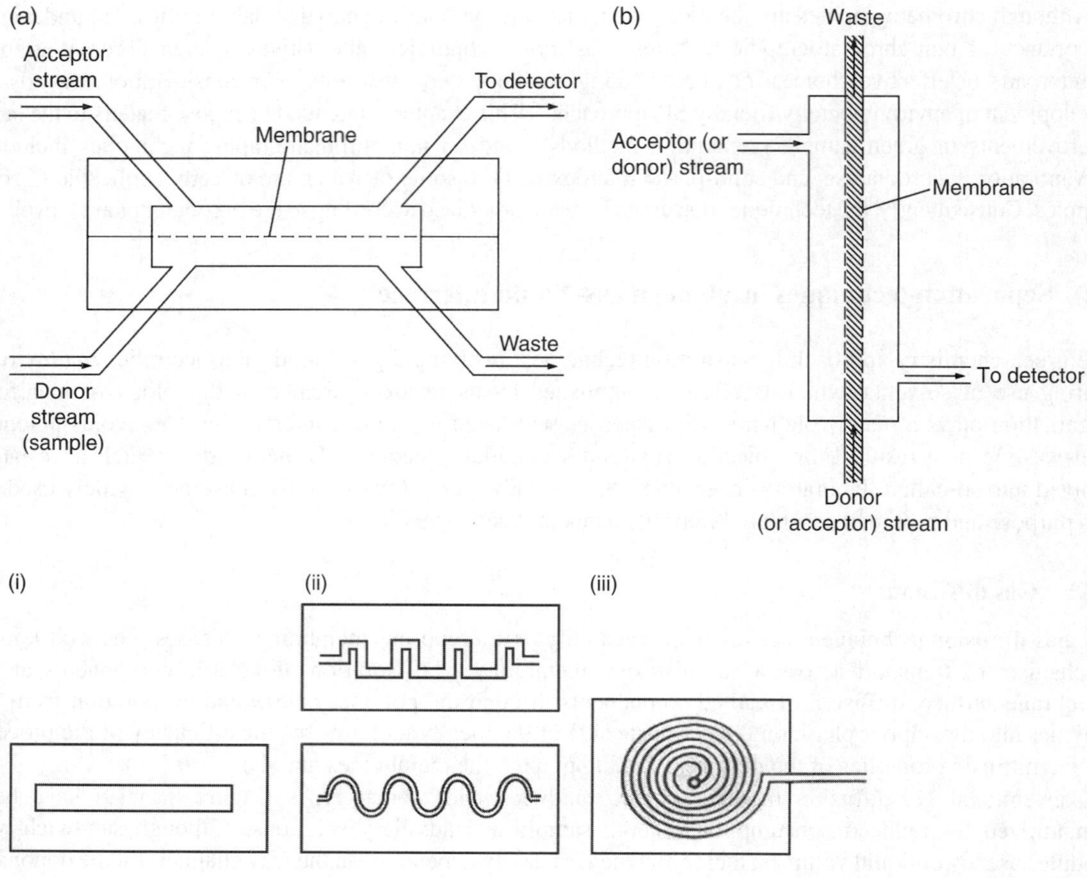

Figure 8.1 Types of membrane-based units. (a) Sandwich membrane in various chamber designs: (i) parallelepipedal, (ii) winding and (iii) spiral. (b) Tubular (hollow-fibre) type membrane.

taking place in a single step in the same unit. The volatile analyte - or its volatile reaction product - present in a heated donor phase, evaporates through a porous membrane and is collected in an acceptor stream for appropriate detection [14]. An air gap is left between the sample in the donor chamber and the membrane (see Figure 8.2) in order to avoid clogging of membrane pores when processing dirty samples and also to permit the presence in the donor chamber of species such as high-molecular-weight components, acids, bases, organic solvents, and so on, which could damage the membrane on contact with it. This is the most salient advantage of analytical pervaporation over its industrial counterpart and other membrane-based non-chromatographic techniques such as dialysis or gas diffusion. As with gas-diffusion modules, the pervaporator can be designed and constructed by the operators themselves as the need arises.

A dramatic reduction in acceptor volume - and a consequent increase in concentration - can be obtained by using a spiral acceptor chamber with its outlet pointing to the bottom instead of a flat chamber with its inlet and outlet facing each other (Figure 8.2 b). This chamber design is especially useful with a view to implementing pervaporation prior to capillary electrophoresis (CE) separation since the small amount of solution used, enriched with volatile analytes, suffices for insertion into the capillary. This combined

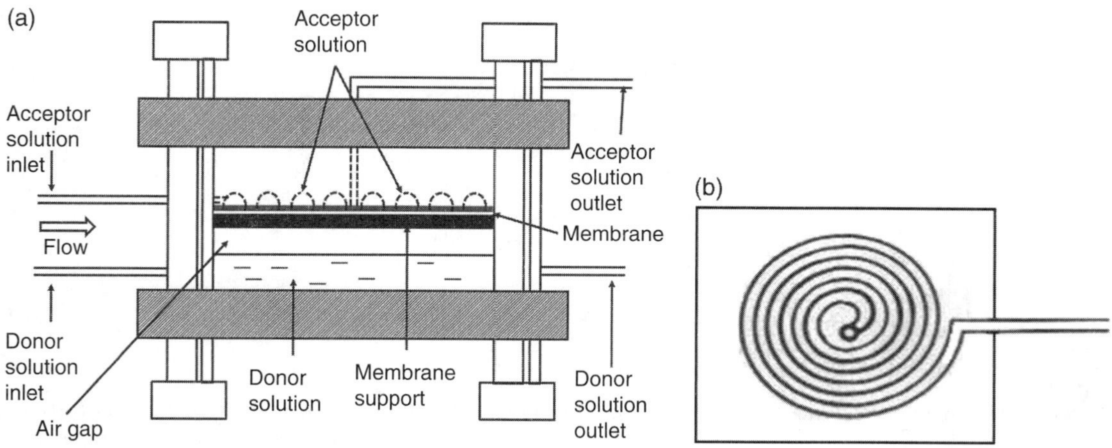

Figure 8.2 (a) Front view of a pervaporator. (b) Schematic of the spiral acceptor chamber.

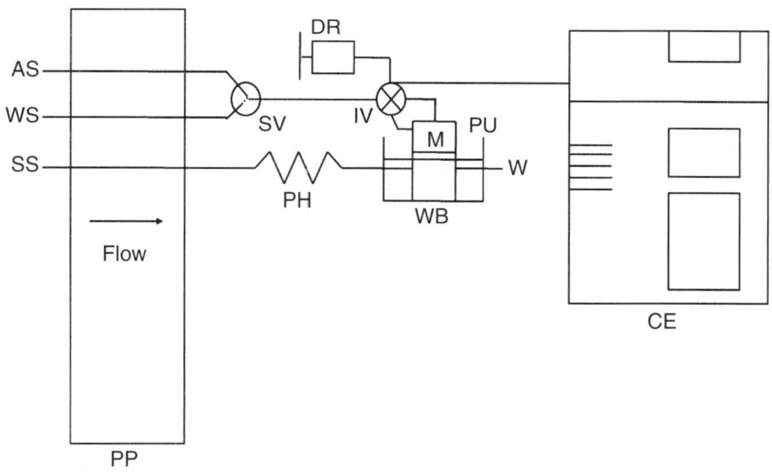

Figure 8.3 *Experimental set-up for the determination of aldehydes and acetone by pervaporation–capillary electrophoresis. AS, air stream; CE, capillary electrophoresis instrument; DR, derivatization reagent; IV, injection valve; M, membrane; PH, pre-heater; PP, peristaltic pump; PU, pervaporation unit; SS, sample stream; SV, switching valve (—, position 1; - - -, position 2); W, waste; WB, water bath; WS, water stream.*

approach has been successfully used to analyse volatile compounds - and also non-volatile ones that can be easily converted into volatile derivatives - in samples in various physical states. Its applications include the determination of various pollutants such as acetone, acetaldehyde, formaldehyde, hexenal and 2-*trans*-hexenal - which are used as indicators of quality degradation in foodstuff [15] - in yogurt and juice. The samples were in slurry form, which allowed the analytical system to be automated (see Figure 8.3). With solid samples - which are also amenable to analytical pervaporation - the pervaporator requires cleaning by hand and must therefore be disassembled between consecutive analyses. Although this increases the operational time, it still affords practical application in, for example, the on-line capillary electrophoresis determination of amines in solid food samples [16]. Finally, a properly optimized pervaporation system can be highly efficient and environmentally friendly; in fact, some of the factors to be optimized for this

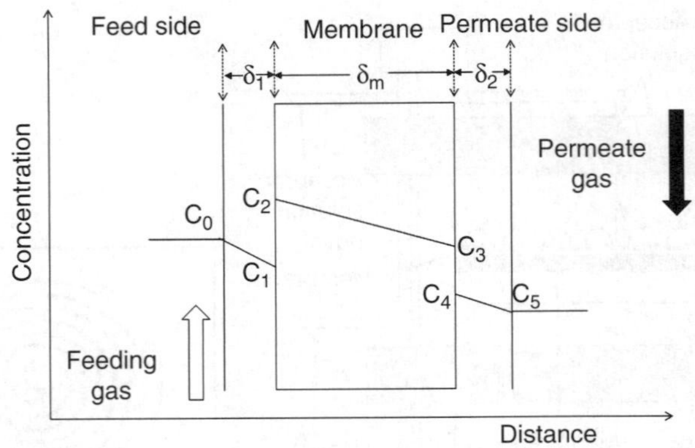

Figure 8.4 *Process and concentration profile during semi-volatile analyte permeation through a non-porous membrane.* C_0, *analyte concentration in the bulk feed gas;* C_1, *analyte concentration in the feed side boundary layer at the interface between the membrane and the feed side boundary layer;* C_2, *analyte concentration in the membrane at the interface between the membrane and the feed side boundary layer;* C_3, *analyte concentration in the membrane at the interface between the membrane and the permeate side boundary layer;* C_4, *analyte concentration in the permeate side boundary layer at the interface between the membrane and the permeate side boundary layer; and* C_5, *analyte concentration in the bulk permeate gas.*

purpose are closely related with green chemistry (e.g. the solvents used as donor or acceptor solutions, or the carrier driving the separated analytes to the detector).

8.2.3 Membrane extraction with a sorbent interface

Membrane extraction using a sorbent interface (MESI) was developed to address the need for continuous monitoring of volatile and semi-volatile organic compounds, with on-site analysis in mind.

MESI provides a simple, flexible, sensitive method for the extraction and preconcentration of relatively non-polar, low-molecular weight compounds from gaseous, solid and liquid matrices. Also, it has proved a useful on-line sample insertion interface for chromatographic and electrophoretic separations [17, 18]. Analyte permeation through a non-porous polymer membrane usually involves several consecutive steps [19, 20], namely: (1) mass transfer on the feed side boundary layer; (2) sorption–desorption at the feed side interface; (3) mass transfer across the membrane; (4) sorption–desorption at the permeate side interface; and (5) mass transfer across the permeate side boundary layer (see Figure 8.4).

The high selectivity of MESI, which facilitates the isolation of analytes from complex matrices, is accompanied by other advantages such as compatibility with any type of gas chromatograph (GC) and applicability to continuous monitoring and automation of industrial processes [21]. In any case, the greatest advantage of MESI, particularly as regards the subject matter of this chapter, is the need for no extraction solvent by virtue of its using a stripping gas stream to carry the analytes to the detector following passage through the membrane. In fact, the membrane acts as a selective barrier allowing passage of some analytes and retaining others [21]. Also, because the components of an MESI system are connected on-line, samples can be analysed in real time. The need for no analyte transport or additional preparation steps saves time and reduces the potential for analyte losses, which can be very helpful towards developing green methods for pollutants or other, environmentally hazardous compounds. MESI can be used for a variety of purposes including the monitoring of compounds present in air, water, plants and breath [19, 22–24].

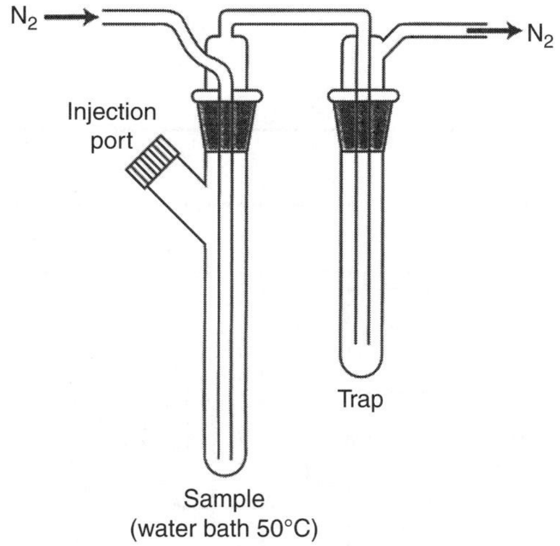

Figure 8.5 *Microdistillation apparatus.*

8.2.4 Distillation and microdistillation

Distillation, which is based on the compositional difference between a liquid and the vapour it forms, is an old separation technique the polluting effects of which can be minimized by using an appropriate sample volume and vapour-collection medium. The efficiency of this technique depends on the differences in volatility (vapour pressure) between the species present in the liquid. Similarly to other gas–liquid techniques discussed in this chapter, the greenest distillation approaches are those involving miniaturization and dynamic manifolds in which the vapour formed is retained in the tubing system in order to avoid losses of environmental pollutants. To this end, the vapour is trapped - usually by a liquid - without user intervention and partial or total automation introduced as a result.

Miniaturization of distillation has led to microdistillation. A continuous analyser with a built-in microdistillation systems usually comprises; (1) a propulsion module with the two-fold purpose of continuously feeding the microdistiller with liquid samples - optionally preconditioned with either an additional buffer or a reagent stream - and removing the outgoing streams from all other units, also in a continuous fashion; (2) a suitable microdistillation module for gas–liquid separation; (3) a condenser and/or absorption module to transfer the gaseous analyte to an outgoing liquid stream; and (4) a continuous detection module.

Microdistillation has been successfully applied to the analysis of biological samples such as blood and gut contents for the determination of hydrogen sulfide without the typical problems posed by the volatility of this analyte and viscosity of the samples in other techniques [25]. Figure 8.5 shows the microdistiller used, where detection was done by ion chromatography. Finally, Figure 8.6 shows the experimental set-up used for the determination of fluoride ion in water samples, based on a flow system combining microdistillation and spectrophotometric detection [26].

8.2.5 Head-space separation

Head-space separation is probably one of the greenest separation techniques, particularly when the target analyte is volatile enough to be separated from the sample matrix simply by heating (i.e. without a reagent). Current head-space devices, which are usually coupled to a gas-chromatograph, consist of an oven, a sampling device and a transfer line to the separation–detection instrument. The system is internally equipped with a

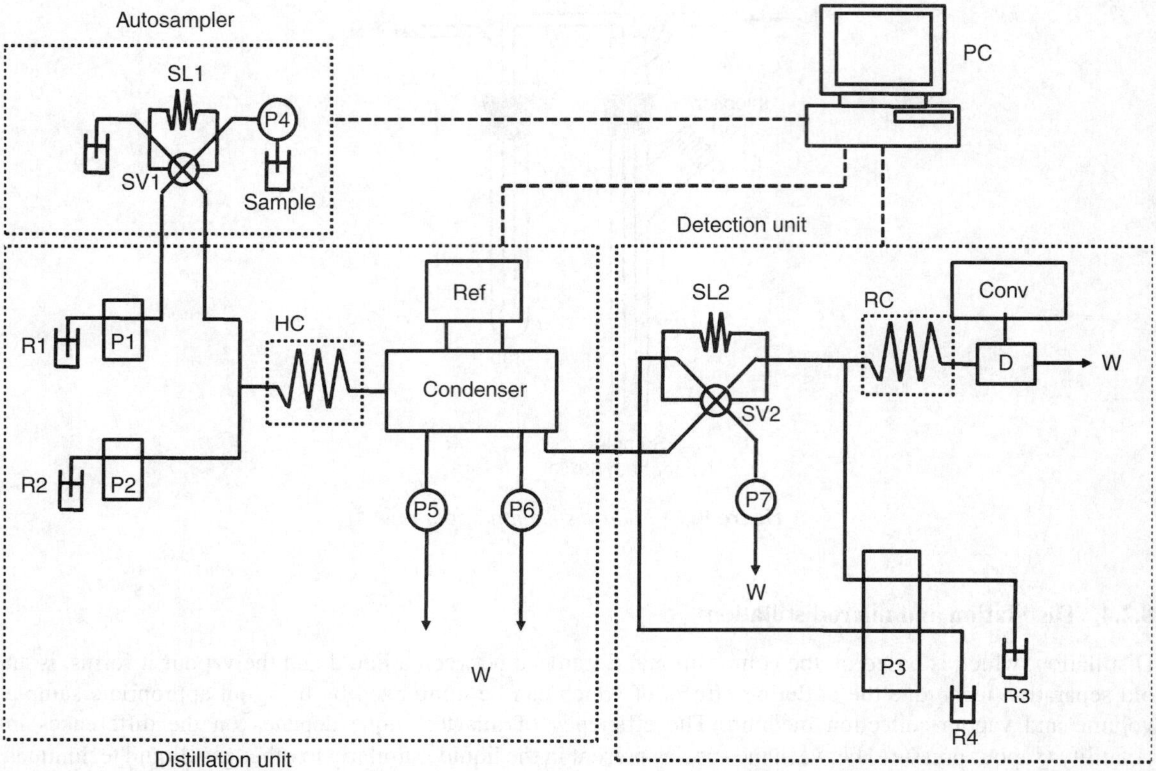

Figure 8.6 *Schematic diagram of an automated microdistillation/flow-injection system. Conv, converter; D, detector; HC, heating coil; P1–P7, pump; RC, reaction coil; Ref, condenser; R1–R4, reservoir; SL1, SL2, sample loop for distillation and detection, respectively; SV1, SV2, six-way valves; T, thermostated bath; W, waste.*

double circuit: one for the carrier gas (generally helium) and the other for the auxiliary gas, which is used to pressurize the vial. The two circuits are connected by a six-way valve which operates automatically as required in each step of the head-space procedure (basically, conditioning, pressurization, sample-loop filling, injection and purging). Automating head-space procedures increases their reproducibility, and reduces operator involvement and risks. Modern head-space units use relatively small volumes: typically less than 5 ml for samples and 10 ml or less for vials.

Pawliszyn *et al.* [27, 28] combined head-space separation and hollow-fibre membrane extraction in a single module to develop the simplest, least problematic choice by virtue of its avoiding the need for a sampling pump and a flow-meter. In addition, head-space methodology avoids plugging of the extraction system by placing the membrane above the sample without physical contact. The head-space membrane module can be made rugged and suitable for long-term field monitoring applications. MESI systems can be operated in a batch or multiplex mode. Unlike conventional single-injection or batch GC, a multiplex approach introduces a random series of sample concentration pulses through the inlet of the GC column, which responds to each concentration pulse independently. Component peaks from different injection pulses overlap to form a multiplex chromatogram at the GC detector. The multiplex technique adds the response obtained from a multiplicity of small samples to give a response reflective of a single large sample, thus increasing the sensitivity of the analysis. However, the most salient feature of the multiplex technique as regards environmental friendliness is the sorbent interface, which can be much simpler than in batch MESI.

8.2.6 Hydride generation and cold-mercury vapour formation

Hydride generation and cold-mercury vapour formation are two traditional ways of converting in a selective and specific manner respectively, and separating non-volatile ions in solution into volatile species through a chemical reaction. These methods require very acid media (usually HCl) and strong reductants (usually sodium borohydride, but also stannous salts for mercury) in dynamic manifolds where the resulting molecular or atomic vapour is driven to an atomic spectrometer for detection.

Continuous advances in atomic detectors have enabled the miniaturization of manifolds and reduced the consumption of polluting reagents as a result. For example, Pawliszyn [29] developed a rapid, sensitive method for the simultaneous multi-channel hydride generation atomic fluorescence spectrometry (HG-AFS) determination of total arsenic, total bismuth, total tellurium and total selenium in tea leaves. Coupling this separation approach to atomic detection via an FI system has enabled the detection of heavy metals such as lead in geosamples [30].

Also, coupling cold-mercury vapour formation to atomic detection has enabled the determination of mercury in edible fish meat; using atomic absorption spectrometry [31] allowed checking whether the mercury levels ingested by eating the target species were above or below the recommendations of the Food and Agriculture Organization (FAO).

8.3 Techniques involving a liquid–liquid interface

8.3.1 Dialysis and microdialysis

Dialysis is an old technique especially well-known in its passive mode, where solutes are separated by effect of a concentration gradient of the analytes between two miscible liquid phases separated by a membrane.

Green approaches to dialysis rely on miniaturization to reduce reagent consumption. Unsurprisingly, the rise in popularity of miniaturized dialysis coincided in time with that of FI, which has boosted its development with the aid of dynamic manifolds. Implementing dialysis in FI systems has in fact facilitated automation of the separation process with high precision and low reagent consumption [9] in a dramatically shortened time thanks to the high mass-transfer efficiency that results from FI operating under non-equilibrium conditions. The efficiency of the dialysis process is thus increased at the expense of a reduction in throughput derived from the need to stop the acceptor phase over a preset interval while the donor phase (sample) is kept still or flowing.

Dialysers usually differ from gas-diffusers only in the nature of the membrane they use (see Figure 8.1). As in previous applications, it is increasingly common for practitioners themselves to construct their own, custom modules to meet their specific needs. This has led to the development of a number of only slightly different units.

Both FI and SI manifolds have been coupled to dialysers for similar purposes (viz., analyte dilution and removal of high-molecular-weight sample components), albeit with substantially reduced sample and reagent consumption in the latter. This is not always the case, however; for example, both FI and SI provided excellent precision with similar sample and reagent volumes (142 and 750 μl, respectively, per analysis) in the separation of chloride from milk prior to the conductimetric determination of the analyte in the dialysate [33].

No LOV devices have so far been coupled to dialysis even though their joint use would afford a dramatic decrease in sample and reagent consumption, and hence greener chemistry, probably without detracting from analytical performance.

Microdialysis (MD) warrants separate discussion here inasmuch as it can be regarded as the ultimate miniaturization stage of dialysis, where the dialysate constitutes the sample itself. Effectively reducing the sample size for dialysis has enabled the development of the microdialyser or microdialysis probe for direct in

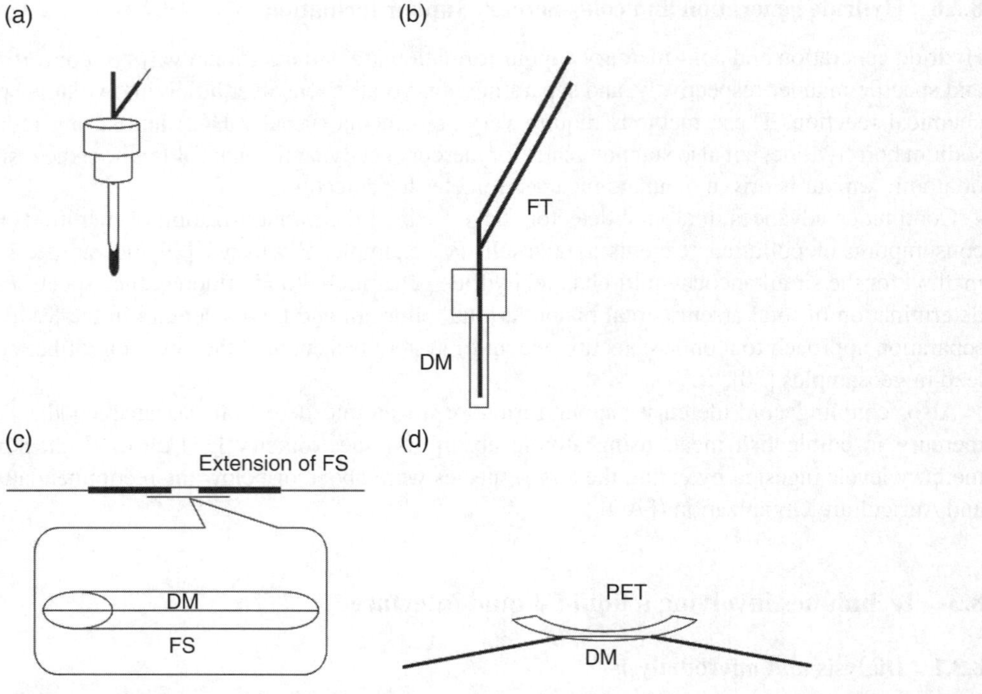

Figure 8.7 Types of microdialysis probes. (a) Concentric cannula probe. (b) Flexible probe. (c) Linear probe. (d) Flow-through probe. DM, dialysis membrane; FS, fibre skeleton; FT, flexible tubing; PET, polyethylene tubing.

vivo and/or in vitro measurements. Because the sampled system can in fact be a living system or a portion thereof, a substantial reduction in operational time and sample preparation costs is thus accomplished. After the probe is inserted in the zone to be sampled, its inner fibre is slowly filled with - or circulated by in dynamic microdialysis - the sampling solution. As with conventional diffusion, the perfusate solution must be similar in pH and ionic composition to the sampled system. The membrane, whether porous or non-porous, allows passage of small molecules of the sample into the acceptor solution for subsequent collection or transport with a view to on-line separation and/or measurement of the analytes (e.g. by capillary electrophoresis, which has been successfully coupled to microdialysis for various purposes [34–36]). Figure 8.7 shows the most common types of MD probes. One added advantage of microdialysis is that flow rates can be up to 10 times lower than in conventional dialysis in FI or SI systems. A special technique known as 'superslow microdialysis' circumvents some of the typical shortcomings of microdialysis such as the need for time-consuming calculations in order to offset depletion and incomplete recovery of some analytes.

8.3.2 Liquid–liquid extraction

Conventional liquid–liquid extraction (LLE), which involves mass transfer between two immiscible liquids, is frequently used in SP procedures to separate an analyte from interfering substances present in the sample matrix, preconcentrate the analyte for increased sensitivity or improve limits of detection. This highly useful technique has been widely used in pharmaceutical, environmental, agricultural and industrial analyses. The main shortcoming of conventional LLE is the need for large volumes of organic, volatile solvents, which make it a far-from-green technique. In addition, LLE is labour-intensive, time-consuming and usually implemented in open spaces - which can pose health hazards associated to the high volatility of some organic solvents. This is especially worrisome in preparing large batches of samples [37]. Similarly to other techniques discussed above, coupling an FI manifold to a conventional unit such as a solvent segmenter, extraction coil or phase separator for continuous LLE can be considered the first step towards the development of a green LLE approach [9]. A number of improvements in continuous LLE have allowed some or even all units involved in the process to be excluded (e.g. in liquid–liquid extraction without phase separation [38, 39]).

Liquid–liquid extraction can be performed in a continuous manner (CLLE) or with the aid of membranes (MLLE). As with the above-described techniques, coupling LLE to a continuous system can make it suitable for Green Analytical Chemistry to an extent dependent on the volumes of sample and solvent needed, among other factors.

Conventional continuous liquid–liquid extraction (CCLLE) has allowed chemists to overcome traditional problems of batch LLE and to develop green methods in some cases. In fact, CCLLE methods use reagents and expensive organic solvents more sparingly and are less time-consuming than conventional LLE. In any case, CCLLE shares the sequence of steps of traditional LLE, namely: dispensing appropriate volumes of the immiscible phases, bringing the phases into contact to facilitate mass transfer and separating the phases prior to determining the analytes. Concerning the subsequent determination, CCLLE is most often used to extract metals from aqueous samples, so detection is usually atomic or molecular spectrometric [40, 41]. Anions such as perchlorate, nitrate and nitrite, and organic compounds, have also been extracted with CCLLE, which is thus suitable for a wide range of applications [39, 42–44]. Moreover, CCLLE methods are accurate and repeatable, but require skilled operators [45].

Other, non-conventional CLLE approaches dispensing with one or more of their typical units have also been proposed to develop green methods. Thus, CCLLE has been implemented without phase separation [46]; by using an extraction coil only and changing the flow direction iteratively [47]; or even without the three modules, with a cell containing an organic phase heavier than the sample, the latter being bubbled at the bottom of the cell to effect liquid–liquid separation while bubbles travel through the extractant layer to the upper section of the cell [48].

Worth a special note here are the attempts at enhancing CCLLE performance by using an auxiliary form of energy to shorten extraction times and reduces or avoids operator exposure to organic solvents in the waste and their release into the atmosphere. Figures 8.8 and 8.9 show selected examples of ultrasound-assisted

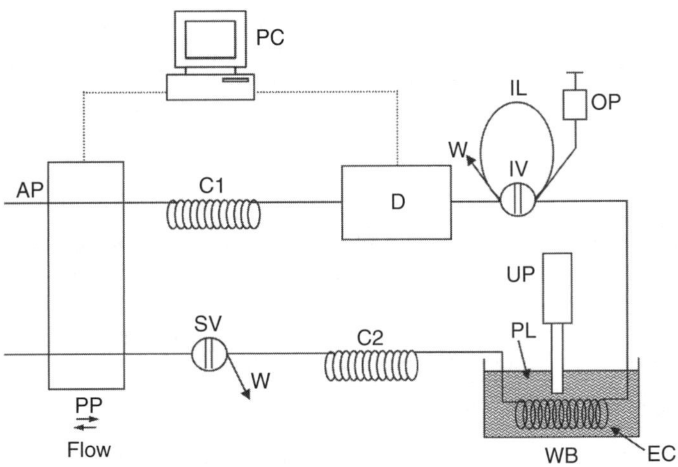

Figure 8.8 *Flow injection manifold for continuous liquid–liquid extraction without phase separation and subsequent hydrolysis in an alkaline medium, both assisted by ultrasound, without phase separation. AP, aqueous phase; C, coil; DAD, diode array detector; ERC, extraction–reaction coil; IV, injection valve; IL, injection loop; OP, organic phase; PC, personal computer; PP, peristaltic pump; SV, switching valve; UP, ultrasonic probe; WB, water bath; W1 and W2, wastes.*

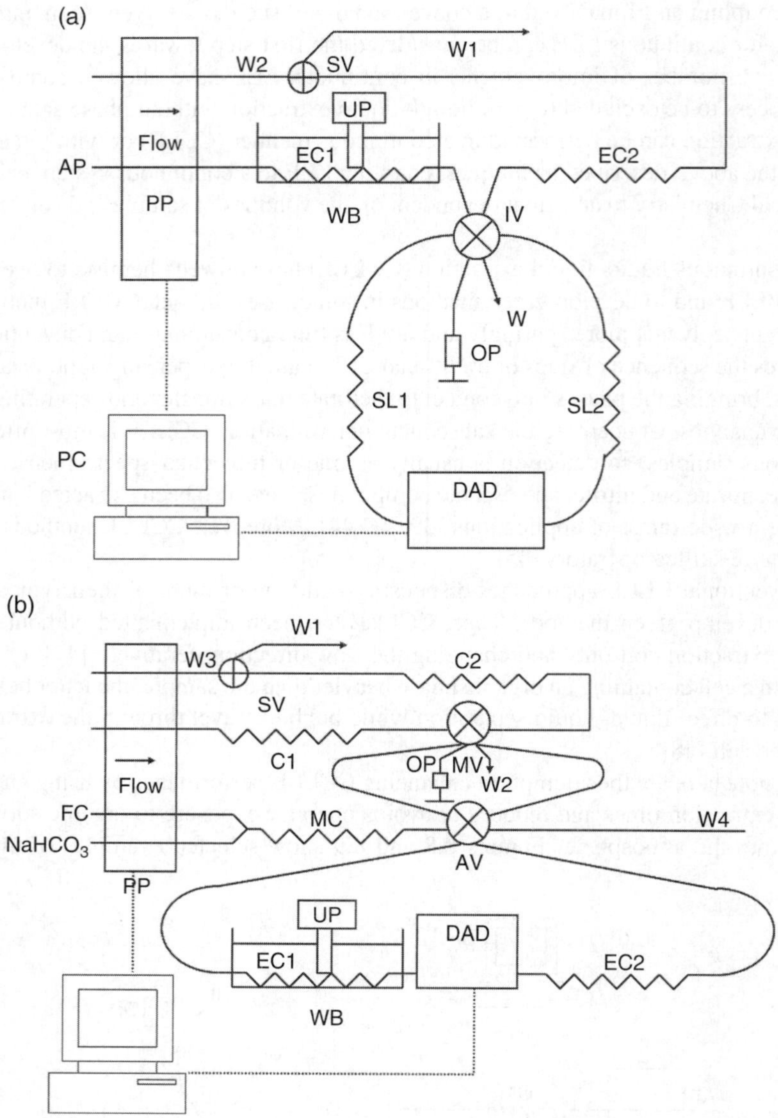

Figure 8.9 *Flow injection manifolds for continuous ultrasound-assisted liquid–liquid extraction without phase separation under ultrasound irradiation, and alternative monitoring of the irradiated and non-irradiated zones of the extractant plug. AP, aqueous phase; DAD, diode array detector; EC, extraction coil; FC, Folin–Ciocalteu reagent; IV, injection valve; MC, mixing coil; OP, organic phase; PC, personal computer; PP, peristaltic pump; SL, sub-loop; SV, switching valve; UP, ultrasonic probe; W, waste; WB, water bath.*

liquid–liquid extraction (USALLE). Figure 8.8 depicts the manifold used to extract paracetamol from suppositories without phase separation; using ultrasound energy facilitated extraction and hydrolysis of the target analyte, and enabling continuous monitoring of the overall process with a diode array (DAD) instrument [49]. The manifolds in Figure 8.9 are somewhat more complex. Figure 8.9(a) shows the dynamic system used to monitor the irradiated and non-irradiated zone of organic phase enriched with analyte from the aqueous

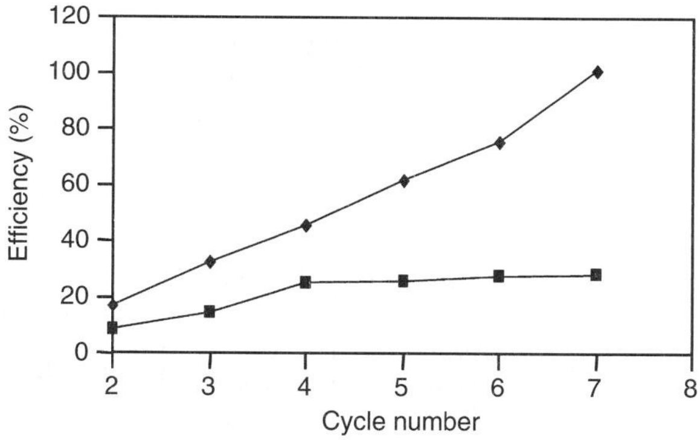

Figure 8.10 *Comparison of extraction and subsequent hydrolysis in an alkaline medium in the presence and absence of ultrasound.*

phase; ultrasound-accelerated mass transfer between the two immiscible phases. A more complex manifold (Figure 8.9(b)) was also developed for ultrasound-assisted extraction, albeit from an organic phase to an aqueous phase, with monitoring of the two mass-transfer zones (one in the presence of ultrasound and the other in its absence). Avoiding passage of the organic phase through the peristaltic pump dispensed with the need for special, expensive tubing and its frequent replacement [50].

Overall, ultrasound energy increases separation efficiency, thereby shortening extraction times and reducing reagent consumption. This is clearly apparent from Figure 8.10, which compares the results obtained in the extraction of paracetamol with and without ultrasound. USALLE experimental set-ups require the ultrasound application variables (particularly probe position, radiation amplitude and pulse duration) to be optimized since extraction times are typically short and, with low reagent and samples volumes, it is crucial to produce as little environmentally hazardous waste as possible.

Attempts at reducing solvent and sample volumes have led to the development of LLE in microfluidic chips consisting of two microfabricated glass plates with a microporous membrane sandwiched in between. Gravity was used to drive the aqueous and organic flows through channels separated by the membrane. The native fluorescence of the analyte (butyl rhodamine B) extracted into the organic phase was monitored in situ with a laser-induced fluorescence detector. The operational complexity of this system makes it impractical for routine use, however [51].

8.3.3 Single-drop microextraction

By virtue of its simplicity and good analytical results, single-drop microextraction (SDME) has become a very popular non-chromatographic separation technique since the mid-1990s, when Dasgupta used a very small amount of solvent holding from a tip of a syringe to separate analytes from their matrix.

There are seven different modes of solvent microextraction falling under the SDME category; all use two or three phases (see Figure 8.11) [52]. The most common modes are probably head-space SDME (41% of all reported procedures) and direct immersion SDME (38%). Both are quite simple and use inexpensive equipment. Also, both use a very small drop, in the microlitre range, of an appropriate organic solvent held by the tip of a syringe to effect extraction. This dramatically reduces the consumption of organic phase relative to alternative liquid–liquid extraction approaches.

Because they use similar volumes of organic solvent and sample, direct immersion and head-space SDME are also similar in environmental friendliness. However, the two differ operationally; thus, unlike other

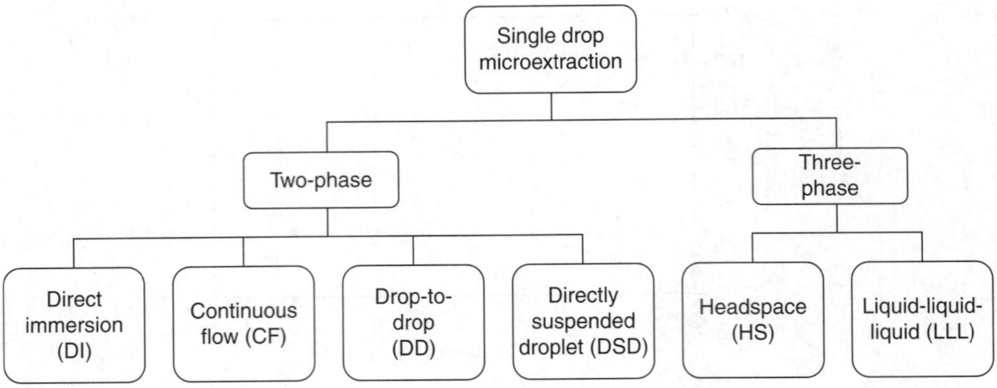

Figure 8.11 *Single-drop microextraction (SDME) modes.*

SDME modes, head-space single-drop microextraction requires equilibrium conditions, which considerably increases analysis times. Also, because head-space SDME is a three-phase technique, it requires both the aqueous sample and the organic solvent to reach equilibrium with the head space. Ultimately, this is a hindrance to the development of green methods owing to the long stirring times needed and energy consumption needed. This drawback, however, can be minimized by using temperature control to accelerate both equilibria. The problem here is that aqueous sample/head-space distribution increases with increasing temperature, whereas the head-space/organic solvent equilibrium requires a low temperature to be reached. Using dual temperature control complicates the experimental set-up and increases the energy requirements; as a result, this approach is only practical with analytes at very low concentrations.

Using direct immersion with dirty matrices potentially containing suspended solids can be a problem because the drop, held by the syringe, can easily be detached and separation fails. Using head-space SDME instead is recommended for this purpose in order to avoid the need for filtration or some other pretreatment and hence the use of additional energy or the production of more waste in the form of disposable materials such as filters or other cleaning devices.

Direct immersion and head-space SDME differ somewhat in scope, even though they can be applied indifferently to some groups of analytes. Because the microdrop of extracting solvent used in direct immersion SDME is brought into direct contact with the aqueous sample, the solvent must obviously immiscible with water (i.e. a non-polar or very slightly polar solvent unless an ionic liquid is used as extractant). Consequently, direct immersion SDME is best suited to the separation and enrichment of non-polar analytes from relatively clean matrices such as tap or ground water. Since volatile compounds are best preconcentrated by head-space SDME, the preferred use of direct immersion SDME is for semi-volatile or low-volatile compounds.

Head-space SDME is usually followed by gas chromatographic separation coupled with any type of detection (75% of all reported applications). This is an advantage towards the development of green methods as it avoids the need for chromatographic solvents. Gas chromatography is also the main choice for direct immersion SDME, which, however, is also often used in combination with liquid chromatography [52]. As a result, direct immersion SDME typically uses greater amounts of organic solvents, not only for chromatographic separation, but also for solvent replacement in liquid chromatography because the organic solvents typically used in SDME are incompatible with HPLC. This shortcoming can be circumvented by using an ionic liquid (IL) as acceptor phase in SDME, as proposed for the determination of UV filters by ILs–SDME followed by LC with UV detection [53]. Using an aqueous solvent as acceptor phase for the analytes can facilitate direct implementation of HPLC without the need for solvent replacement. Since samples are most often aqueous, the three-phase mode (liquid–liquid–liquid microextraction) is usable here; however, because it requires attainment of two equilibria, it increases extraction times. The organic solvent must be completely immiscible with water and usually employed in large volumes, which increases the risk of atmospheric pollution and restricts the use of this mode to those cases where other, greener techniques are unsuitable for the intended purpose.

Not always is SDME followed by a chromatographic–detection step, however. In fact, direct immersion ILs–SDME has also been used as a pretreatment to electrothermal atomic absorption spectrometry in the determination of lead [54], as has head-space SDME for the separation of chlorine and its determination by fibre optics-based cuvetteless microspectrophotometry after oxidation with permanganate [55].

Direct immersion and head-space SDME affords efficient separation with very low volumes amounts of organic solvents and typical volumes of sample. Reducing the volume of sample has enabled miniaturization of the SDME technique in the form of drop-to-drop solvent microextraction [52], where both the sample and organic solvent volumes are in the microlitre region. Drop-to-drop microextraction is especially effective when the sample is available in small amounts (e.g. blood, serum or even less abundant biological fluids). In addition to the green advantage of using very small amounts of sample and solvents, this technique features fast equilibration of the two without stirring, which saves energy and simplifies the experimental set-up.

One less common SDME mode uses a continuous flow of sample and a small drop of organic solvent as acceptor. This entails circulating large volumes of sample through an extraction chamber accommodating the syringe, with a microdrop at the tip of the needle. As shown by several studies, this mode provides no better detection limits than direct immersion SDME [56–60].

Using SDME in combination with gas chromatography usually requires derivatizing the analytes into more volatile compounds prior to injection into the chromatograph. Derivatization should be accomplished with as little reagent and acceptor organic solvent as possible since most derivatizing reagents are also volatile pollutants. The most effective way of meeting this requirement is by performing derivatization in the syringe itself following SDME but preceding injection. This approach, called 'in-syringe derivatization' is 'doubly green' since it minimizes the amounts of hazardous reagents and solvents needed for extraction and derivatization. A recent method of this type has enabled the determination of volatile organic acids in tobacco by head-space SDME with in-syringe methyl ester formation prior to insertion into GC–MS equipment [61].

8.4 Techniques involving a liquid–solid interface

8.4.1 Solid-phase extraction

Liquid–liquid extraction (LLE) has traditionally been the most widely used non-chromatographic technique for liquid samples. As noted earlier, however, LLE requires large volumes of organic solvents and sample and provides limited automation capabilities. Solid-phase extraction (SPE) has emerged as an effective solution to these shortcomings. In fact, SPE is suitable for aqueous samples. Analytes are concentrated on a solid sorbent and then desorbed with an organic solvent which, however, is used in lower volumes than in LLE. Thus, replacing LLE with SPE - which is possible provided no sample conditioning is required - can be one step towards green chemistry.

Some authors include SPE among solventless separation techniques on the grounds that it uses a solid sorbent when, the analytes must nearly always be eluted and transferred to the detection system by a liquid phase, and some organic solvent, however little, is required as a result. Moreover, SPE sorbents need some conditioning and this is usually done with an organic or acid solution that produces potentially hazardous waste requiring proper disposal.

The previous shortcomings have been addressed by downsizing SPE devices and automating their operation in order to reduce reagent and solvent consumption, and operator risks, as far as possible. The earliest SPE cartridges were used in FI systems. Flow injection still provides a simple, easy means for automating SPE in accordance with the particular requirements of each application. In fact, FI significantly increases preconcentration rates and expedites sample treatment while substantially reducing sample and organic solvent volumes in relation to conventional SPE. Also, it allows the use of smaller SPE columns and hence of reduced amounts of sorbent in packed microcolumns.

One further step towards miniaturizing and automating SPE was taken by coupling it on-line to SI. The first ensuing approach was reported in the early 2000s and has been followed by many other successful applications [62–68]. The most salient advantages of SI-SPE manifolds are their high simplicity, automatability and throughput, which, together with their need for little sample and solvent, provide an effective means for developing green methods.

The 'laboratory on a valve' (LOV) is a suitable platform for further miniaturization and automation in SPE. The design flexibility of LOV facilitates the implementation of this technique in a variety of situations with the sole limit of the user's imagination and the availability of appropriate hardware and software. A wide variety of analytes can be separated and preconcentrated according to affinity by using several cartridges loaded with different sorbents and arranged serially or in parallel for their subsequent elution and on-line determination. One problem with LOV systems is the difficulty of handling large volumes of sample owing to the limited capacity of the syringe pumps used as propulsion systems. In fact, sample volumes exceeding the syringe capacity require repeated flushing and increase analysis times as a result.

Most SPE sorbents are re-conditioned after elution. This requires the use of small volumes of organic solvents which can add to the polluting load of the procedure. The recent wide expansion of carbon nanomaterials and their growing number of uses have led to the proposal of carbon nanotubes (CNTs) and nanohorns as efficient sorbents for SPE [69–71].

Carbon nanotubes are large molecules containing thousands of carbon atoms in an electronically delocalized aromatic system that are rolled up into a cylinder and closed by two caps. Depending on the number of carbon atom layers forming their walls, nanotubes can be single-walled CNTs or multiwalled CNTs. CNTs have high surface areas, mechanical strength and chemical stability. Specially enthusiastic researchers have named these materials 'green sorbents' on the grounds of their assumed low polluting power. The vast – though not necessarily exhaustive – range of applications of nanomaterials devised so far includes the administration of drugs, which has aroused great hopes and attracted much effort. Nevertheless, the controversial belief that nanotubes may be toxic to animals and humans has considerably reduced their use in the pharmaceutical field and so have the potential hazards for operators and the environment in analytical chemistry.

Recent experiments on mice have shown that animals exposed to carbon nanotubes by inhalation or direct injection in the abdominal region are unable to break down that material and develop lung problems – and occasionally, cancer – as a result. Some authors have compared the biopersistence of nanomaterials with that of asbestos. Research into nanomaterials in this context has focused on ways to suppress their toxicity. A study of the effects of myeloperoxydase (MPO), contained in white-blood corpuscles, over carbon nanotubes in vitro and in mice revealed that the enzyme can decompose CNTs into carbon and water, but only when present in amounts within the capacity of the neutrophile enzymatic system [72].

Other compounds which have recently came into widespread use and the environmental friendliness of which remains to be assessed are ionic liquids (ILs), which are inorganic or organic salts with melting points below 100°C. These substances were discovered in the early twentieth century but have only recently emerged as highly useful solvents in chemistry. Their analytical applications rely on their unique characteristics, which include a near-zero vapour pressure, good thermal stability, tunable viscosity, miscibility with water and organic solvents, and good extractability for various organic compounds and metal ions. These properties have attracted much interest in ILs as solvents for a wide range of applications in separation techniques such as chromatography, capillary electrophoresis, electrochromatography and liquid–liquid extraction, among others.

Ionic liquids have been used as both sorbents and eluents for SPE. For use as sorbents, ILs are immobilized onto an appropriate support that is then used as packing material. Sorbents consisting of ILs immobilized onto silica have recently been used to isolate tanshinones [73], enrich polyunsaturated fatty acid methyl esters from complex samples [74] and isolate active drugs from liquorice [75]. Also, a polymeric phase was recently evaluated as a support for CF_3COO^- in packed cartridges furnished with two frits; the supported IL proved effective in retaining pharmaceuticals from wastewater [76]. This new sorbent for reversed-phase, weak anion-exchange, strong anion-exchange and strong cation-exchange SPE was found to selectively extract a group of acid compounds quantitatively from aqueous samples while washing basic analytes also present in the sample.

The intrinsic non-volatility and non-flammability of ILs make them potentially good substitutes for conventional organic solvents; however, their unknown fate is a strong restriction for wider use of these theoretically ideal green solvents.

8.4.2 Solid-phase microextraction

Solid-phase microextraction (SPME) is a solvent-free, miniaturized extraction technique providing a convenient alternative to most conventional extraction techniques. SPME was originally developed by Arthur and Pawliszyn [77] and has since then been increasingly used to preconcentrate a vast number of compounds in a wide variety of matrices, especially for the extraction of volatile and semi-volatile organic compounds, and usually in combination with GC. Typical targets for SPME include phenols, which have been successfully preconcentrated from environmental water [78–84], soil [85] and landfill leachates [86], among others include; UV filters [87], chemical preservatives [88] and nicotine [89] to name a few.

This separation technique has grown enormously in use since various types of SPME fibres were made commercially available in 1993 [90]. SMPE fibres can be obtained from a variety of materials including 50 m carbowax/templated resin (CW-TPR), 60 m polydimethylsiloxane/divinylbenzene (PDMS-DVB) and 85 m polyacrylate (PA) [78–80, 84–86]. Also, SPME fibres have been custom-made from 25,27-dihydroxy-26,28-oxy(2,7-dioxo-3,6-diazaoctyl) oxy-*p-tert*-butylcalix [91] arene/oil (amide-bridged C [91] /OH-TSO) coating [81], electropolymerized aniline-based fibre [82], polyethylene glycol-coated solid-phase microextraction fibre [92], and poly(acrylamide–vinylpyridin–*N,N*-methylene*bis*acrylamide) monolithic capillary for in-tube-solid-phase microextraction [93]. Such a wide spectrum of fibres provides ample choice for effective retention of a variety of analytes without interference.

Conventional SPME is performed on a fused-silica capillary coated by one of the many available fibre types. Such a simple scheme (Figure 8.12) is the basis for a number of commercial solid-phase microextractors. Its simplicity reduces manipulation hazards, improves reproducibility and facilitates complete automation of the experimental design, especially when combined with thermal desorption and GC. Analytes can be desorbed with a thermal desorption system (TD) or an organic solvent. Although this makes SPME compatible with both GC and LC, TD is a greener choice as it avoids atmospheric pollution by volatile solvents. In any case, the amount of organic solvents required is very small (typically in the microlitre range), so SPME can still be considered a green technique.

Like SDME, SPME can be used in the direct immersion and head-space modes. Head-space–SPME affords the separation of analytes from dirty samples without the need for filtering or any other steps generating waste from disposable materials and requiring additional consumption of energy. Both modes can be expedited with magnetic stirring.

SPME can be performed in a GC or micro-LC capillary column [94–96]. This approach, called 'in-tube SPME', is generally coupled on-line to a liquid chromatograph through an autosampler. This affords extraction, desorption and injection of the analytes in a fully automated way. In-tube SPME has also been used in combination with GC and CE [97, 98].

Because SPME is usually coupled to GC and this chromatographic technique often requires derivatization of the target analytes, the derivatizing step should involve green chemistry. For example, DI-SPME followed

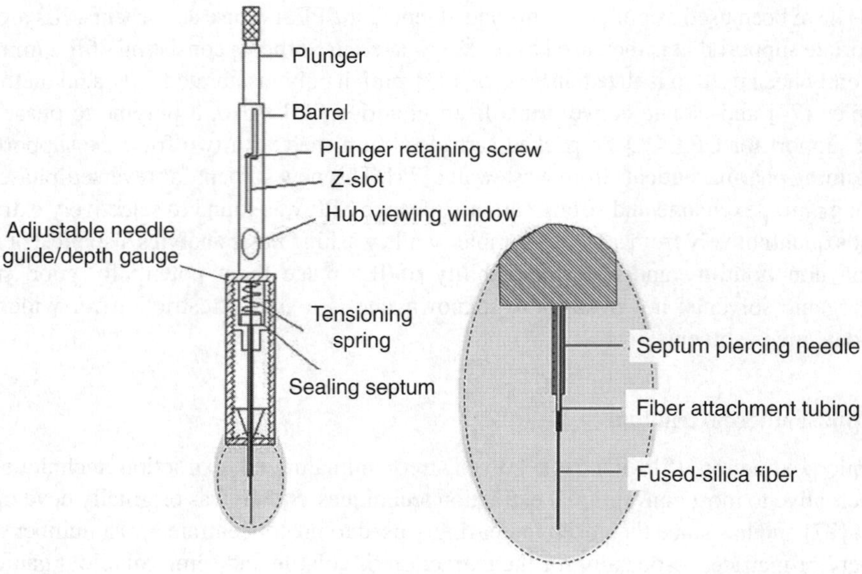

Figure 8.12 *Commercial solid-phase microextraction device.*

by head-space derivatization with BSTFA requires preventing contact of the derivatizing reagent with the atmosphere [99].

In addition to the previous fibre-coating materials, CNTs [100] have been successfully used in SPME to remove many types of pollutants such as dioxins from air [101]; and lead [102], cadmium [103], zinc [104], fluoride [105], 1,2-dichlorobenzene [106] and trihalomethanes [107] from water, to name a few. Also, multiwalled CNTs have a great potential as SPE sorbents [108–110] and SPME fibre coatings [111] for the analysis of organic compounds. Commercially available CNTs can be used as such or after appropriate treatment for adaptation to the target application. Treating CNTs with acid mixtures such as concentrated H_2SO_4 (98%) or HNO_3 (70%) alters their physico–chemical properties by effect of the partial formation of carboxyl groups at their sidewalls. The oxygenated functional groups thus formed act as sites for the binding of a variety of chemical molecules onto nanotube surfaces [112–114] with differential adsorption selectivity relative to raw CNTs.

SPME has inevitably been compared with other separation techniques in terms of performance. Thus, SPME was used for the extraction of volatile components in Angelica species and compared with supercritical-CO_2 extraction (SFE); both techniques being combined with GC–MS. Although the two methods were similar in environmental friendliness, SPME provided easier, on-line connection to the chromatograph, used simpler equipment, was more expeditious and incurred lower costs; on the other hand, SFE required less sample preparation work prior to extraction [115].

8.4.3 Stir-bar sorptive extraction

Solventless sample preparation techniques based on sorptive extraction with gum base have aroused interest for the development of green methods. Specifically, stir-bar sorptive extraction (SBSE) has proved an environmentally friendly alternative to liquid extraction.

SBSE was intended to circumvent the shortcomings occasionally associated with the use of other solvent-free techniques (e.g. SPME). Such problems include a low extraction efficiency for solutes that are only partially soluble in water [116]; also, very polar compounds can lead to competition between the aqueous phase, SPME fibres, glass walls of the extraction vessel and the surface of polytetrafluoroethylene stir-bar [117, 118]. Coating stir-bars with an amount of polydimethylsiloxane (PDMS) 50–250 times greater than that typically used in SPME substantially increases the amount of extractant to be handled. Typically, the stir-bar in SBSE is a magnetic cylinder covered with a glass layer that is turn coated with organic material (usually PDMS). In recent years, however, sol–gel technology has been successfully used as an alternative choice for stir-bar coating.

Similarly to SPME, SBSE dispenses with thermal desorption and the need for polluting organic solvents. However, SBSE is slower than SPME owing to the larger amount of sorbent used; also, it requires cold trapping to avoid analyte losses.

Although thermal desorption is the most environmentally friendly way to desorb the target compounds, liquid desorption with a very low volume (100–200 μl) of an organic solvent such as methanol or acetonitrile is also usually effective. Extraction can be improved by applying ultrasound energy and the extract is most often ready for injection into an HPLC or GC instrument.

One salient feature of SBSE is its availability for in situ derivatization, if required. This is especially interesting when extracting relatively high-polar compounds, the extraction efficiency for which is diminished by the non-polar nature of PDMS. Examples of in situ derivatization in SBSE include the extraction of hydroxyl groups in phenols with acetic acid anhydride [119–121], carboxyl groups with ethyl chloroformate [119–121] and carbonyl compounds with *O*-(2,3,4,5,6-pentafluorobenzyl)[122]. In situ derivatization also improves sensitivity when used prior to GC. Simultaneous extraction and derivatization shortens analysis times and increases throughput as a result. In addition, SBSE affords de-conjugation of biological samples simultaneously with extraction (i.e. in situ de-conjugation). The advantages of this procedure are similar to those of in situ derivatization (shorter sample preparation times, mainly).

8.4.4 Continuous filtration

Filtration is a very simple and old separation technique outdated by modern analytical procedures. However, it has regained some importance after its joint use with continuous flow systems. The ensuing dynamic approaches have allowed downsizing equipment and reducing sample and reagent volumes as a result. In addition, dynamic systems facilitate automatic filtration, thus suppressing or reducing operational and handling hazards, and making analytical procedures safer for humans and environment. Filtration in continuous systems can be done with or without a filtering device. In fact, filtering can be accomplished by using an auxiliary form of energy such as ultrasound or a filterless device such as a knotted reactor (see Figure 8.13(a)) [45] to separate phases. Continuous filtration is commonly used to remove solid particles from a liquid sample or from the leachate of a solid sample–leachant system. It is also used after precipitation, either to remove interferents from the sample or to isolate the analyte as a solid for indirect measurement or subsequent dissolution prior to detection [45].

The filters used in continuous filtration systems are usually in the form of stainless-steel cylinders or plates (Figures 8.13(b) and 8.13(c)), disposable membranes or packed beds. Stainless-steel filters are the most widely used and have the advantage that they can be reused over long periods simply by alternating filtration and cleaning. Filters can be cleaned by hand, flushing an appropriate solvent or applying ultrasound.

Disposable membrane filters are usually made of nylon or cellulose. Unlike steel filters, their construction requires too much energy and material for so short a useful life; so much so that filters must be replaced rather frequently. This is also the case with packed-bed filters owing to their frequent need for refilling (usually with polystyrene granules, cotton or filter paper pulp) and poor reproducibility; a result of the difficulty of packing the material to the same degree of tightness every time.

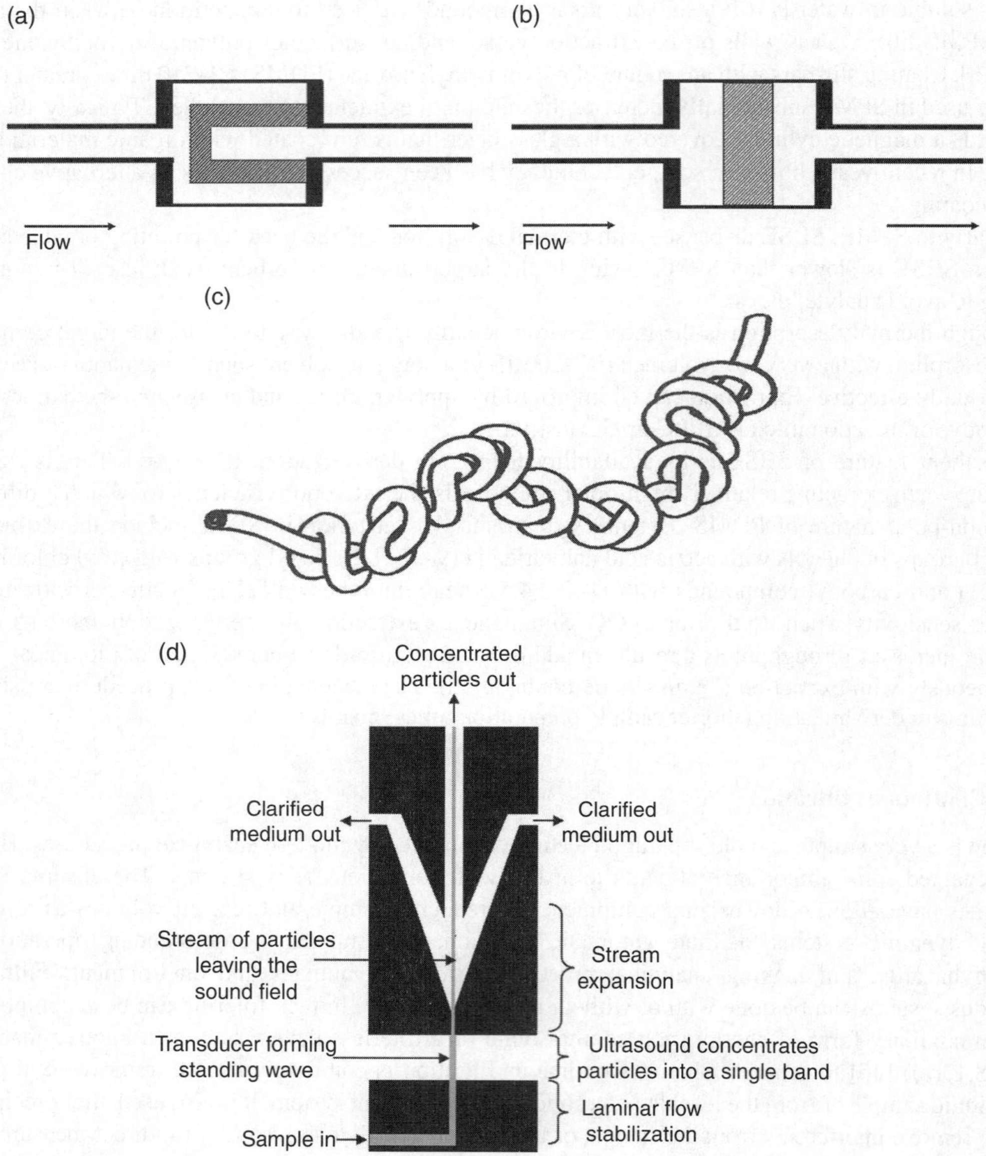

Figure 8.13 *Filter and filterless devices for filtration. Stainless-steel filters: (a) cylindrical and (b) planar (disk-shaped). (c) Knotted reactor. (d) Filterless ultrasound-assisted filtration chamber.*

Knotted reactors are very often the preferred choice among filterless continuous filtration systems on account of the low dispersion and efficient mixing they provide; a result of radial dispersion being favoured over axial dispersion inside the device [123]. Also, knotted reactors are typically made of PTFE, a highly inert, environmentally friendly material, are easy and inexpensive to construct, and have a long useful life, all of which reduce production and material costs.

Filterless filtration is also enabled by the use of ultrasound as auxiliary energy. The US filtration chamber of Figure 8.13(d) works by promoting collisions between the solid particles contained in the liquid phase, which tend to stick together. This is accomplished by creating points where the particles are held in the fluid solely by ultrasonic forces and by agglomerating with other particles [124]. Although the use of ultrasound avoids the production of disposable materials (waste), it should be minimized by using the optimum US frequency, intensity, power and propagation direction in each case. Downscaling ultrasound-assisted filterless filtration avoids the need to handle the separation cell and allows separation to be integrated in microfluidic systems. Ultrasound-assisted filtration has proved an effective continuous filtration approach [125].

Filtration can be preceded by a precipitation step. In this case, precipitation and filtration can be both implemented in a continuous flow system to exploit the analytical and environmental advantages of continuous approaches. Continuous precipitation–filtration can be followed by dissolution if needed. This introduces an additional reagent (solvent) in the process. The precipitation–filtration–dissolution sequence has been used in immunochemical methods involving the formation of high-molecular weight immunocomplexes such as that for the determination of anti-canine immunoglobulin G [126]. Again, filtration can rely on a filterless device (a knotted reactor, preferably). In fact, a knotted reactor was recently used to immobilize the ligand on its inner walls for the simultaneous separation–preconcentration of traces of various metals before detection by ICP-TOF/MS [127]. The typical purposes of using precipitation–filtration without precipitate dissolution include interference removal and indirect determinations of analytes based on the signal decrease caused by analyte precipitation, which will be proportional to the analyte concentration in the sample [40]. Inorganic anions such as halogenides [128] and sulfate [129] have been determined in this way, using Ag(I) and Pb(II), respectively, as precipitants, and so have organic compounds such as sulfonamides [130] and local anaesthetics [131] with Cu(II) and Co(II), respectively. The main operational problem of this methodology is the need for frequent cleaning of the filter, which can be overcome by placing it in the loop of an injection valve and cleaning in the opposite direction to filtration after each precipitation step.

8.5 A Green future for sample preparation

As shown in this chapter, non-chromatographic separation techniques provide a wide range of choices for sample preparation. Sustained research in this field and today's growing environmental awareness are bound to boost the development of increasingly greener analytical methods. Analytical chemists can play a prominent role in the search for environmentally benign non-chromatographic separation methods, and so can other professionals involved in the design of analytical equipment and procedures.

Acknowledgement

The Spanish Ministerio de Ciencia e Innovación (MICINN) and FEDER program are thanked for financial support through project CTQ2009-07430.

References

1. Paschke, A. (2003) Consideration of the physicochemical properties of sample matrices – an important step in sampling and sample preparation, *Trends Anal. Chem.*, **22**, 78–89.
2. Bièvere, P.D. (2002) The uncertainty of the chemical sample preparation is part of the uncertainty of the measurement result, *Accred. Qual. Assur.*, **7**, 181.
3. Luque de Castro, M.D. and Priego-Capote, F. (2008) Miniaturization of analytical steps: necessity and snobbism. *Anal. Bioanal. Chem.* **390**, 67–69.
4. Morales-Muñoz, S.; Luque-García, J.L. and Luque de Castro, M.D. (2003) Approaches for accelerating sample preparation in environmental analysis, *Crit. Rev. Environ. Sci.*, **33**, 391–421.

5. Morales-Muñoz, S.; Luque-García, J.L. and Luque de Castro, M.D. (2006) Pure and modified water assisted by auxiliary energies: An environmental friendly extractant for sample preparation, *Anal. Chim. Acta*, **557**, 278–286.
6. Luque de Castro, M.D.; Ruiz-Jiménez, J. and Perez-Serradilla, J.A. (2008) Lab-on-valve: a useful tool in biochemical analysis, *Trends Anal. Chem.*, **27**, 118–126.
7. Almendral-Parra, M.J.; Alonso-Mateos, A. and Fuentes-Prieto, M.S. (2010) A gas diffusion technique coupled with flow injection systems. Optimization of the process in its application to the fluorimetric determination of ammonium in water samples, *J. Fluoresc.*, **20**, 55–65.
8. Oliveira, S.M.; Lopes, T.I.M.S.; Toth I.V. and Rangel, A.O.S.S. (2009) Development of a gas diffusion multicommuted flow injection system for the determination of sulfur dioxide in wines, comparing malachite green and pararosaniline chemistries, *J. Agr. Food Chem.*, **57**, 3415–3422.
9. Luque de Castro, M.D. (2008) Membrane-based separation techniques: dialysis, das diffusion and pervaporation, *in Advances in Flow Injection Analysis and Related Techniques*, (eds S.D. Kolev and I.D. McKelvie), Comprehensive Analytical Chemistry 54, Wilson and Wilson's, Victoria,.
10. Ilida, Y.; Suganuma, Y.; Matsumoto, K. and Satoh, I. (2006) Novel determination system for urea in alcoholic beverages by using an FIA system with an acid urease column, *Anal. Sci.*, **22**, 173–176.
11. Erxleben, H. and Ruzicka, J. (2005) Atomic absorption spectroscopy for mercury, automated by sequential injection and miniaturized in Lab-on-Valve system, *Anal. Chem.*, **77**, 5124–5128.
12. Themelis, D.G. and Kika, F.S. (2006) Gas-diffusion flow injection assay for the selective determination of chlorine dioxide based on the fluorescence quenching of chromotropic acid, *Microchem. J.*, **82**, 108–112.
13. Cordeiro, T.G.; Hidalgo, P.; Gutz, I.G.R. and Pedrotti, J.J. (2010) Flow injection analysis of ethyl xanthate by gas diffusion and UV detection as CS_2 for process monitoring of sulfide ore flotation, *Talanta*, **82**, 790–795.
14. Luque de Castro, M.D. and Luque-García, J.L. (2002) *Acceleration and Automation of Solid Sample Treatment*, Elsevier, Amsterdam.
15. Ruiz-Jiménez, J. and Luque de Castro, M.D. (2006) On-line pervaporation–capillary electrophoresis for the determination of volatile analytes in food slurries, *J. Chromatogr. A*, **1128**, 251–258.
16. Ruiz-Jiménez, J. and Luque de Castro, M.D. (2006) Pervaporation as interface between solid samples and capillary electrophoresis. Determination of biogenic amines in food, *J. Chromatogr. A*, **1110**, 245–253.
17. Jönsson, J.Å. and Mathiasson, L. (1999) Liquid membrane extraction in analytical sample preparation. I. Principles, *Trends Anal. Chem.*, **18**, 318–325.
18. Cordero, B.M.; Perez Pavón, J.L.; García Pinto, C.; Fernández Laespada, M.E.; Carabias Martínez, R. and Rodríguez Gonzalo, E. (2000) Analytical applications of membrane extraction in chromatography and electrophoresis, *J. Chromatogr. A*, **902**, 195–204.
19. Liu, X. and Pawliszyn, J. (2006) Determination of membrane permeability without calibration using solid-phase microextraction (SPME), *J. Membrane Sci.*, **268**, 65–73.
20. Pawliszyn, J. (2002) Sampling and sample preparation for field and laboratory, *in Comprehensive Analytical Chemistry* (eds D. Barceló and J. Pawliszyn), Elsevier, Amsterdam.
21. Kaykhaii, M.; Sarafraz-Yazdi, A.; Chamsaz, M. and Pawliszyn, J. (2002) Membrane extraction with sorbent interface–gas chromatography as an effective and fast means for continuous monitoring of thermal degradation products of polyacrylonitrile, *Analyst*, **127**, 912–916.
22. Liu, X.; Pawliszyn, R.; Wang, L. and Pawliszyn, J. (2004) On-site monitoring of biogenic emissions from Eucalyptus dunnii leaves using membrane extraction with sorbent interface combined with a portable gas chromatograph system, *Analyst*, **129**, 55–62.
23. Segal, A.; Gorecki, T.; Mussche, P.; Lips, J. and Pawliszyn, J. (2000) Development of membrane extraction with a sorbent interface–micro gas chromatography system for field analysis, *J. Chromatogr. A*, **873**, 13–27.
24. Ma, W.; Liu, X. and Pawliszyn, J. (2006) Analysis of human breath with microextraction techniques and continuous monitoring of carbon dioxide concentration, *Anal. Bioanal. Chem.*, **385**, 1398–1408.
25. Richardson, C.J.; Magee, E.A.M. and Cummings, J.H. (2000) A new method for the determination of sulphide in gastrointestinal contents and whole blood by microdistillation and ion chromatography, *Clin. Chim. Acta*, **293**, 115–125.

26. Shimada, K.; Shimoda, T.; Kokusen, H. and Nakano, S. (2005) Automatic microdistillation flow-injection system for the spectrophotometric determination of fluoride, *Talanta*, **66**, 80–85.
27. Luo, Z. and Pawliszyn, J. (2000) Membrane extraction with a sorbent interface for headspace monitoring of aqueous samples using a cap sampling device, *Anal. Chem.*, **72**, 1058–1063.
28. Yang, M.J. and Pawliszyn, J. (1996) Headspace membrane extraction combined with multiplex gas chromatography and mass selective detector for monitoring of volatile organic compounds, *J. Microcolumn Separations*, **8**, 89–98.
29. Zhang, N.; Fub,N.; Fang, Z.; Feng, Y. and Ke, L. (2011) Simultaneous multi-channel hydride generation atomic fluorescence spectrometry determination of arsenic, bismuth, tellurium and selenium in tea leaves, *Food Chem.*, **124**, 1185–1188.
30. Chun-Hua, T. and Xu-Guang, H. (2009) Trace lead measurement and online removal of matrix interference in geosamples by ion-exchange coupled with flow injection and hydride generation atomic fluorescence spectrometry, *J. Autom. Method. Manag.*, **1**, 1–7.
31. Saei-Dehkordi, S.S.; Fallah, A.A. and Nematollahi, A. (2010) Arsenic and mercury in commercially valuable fish species from the Persian Gulf: Influence of season and habitat, *Food Chem. Toxicol.*, **48**, 2945–2950.
33. Silva, F.V.; Souza, G.B.; Ferraz, L.F.M. and Nogueira, A.R.A. (1999) Determination of chloride in milk using sequential injection automated conductometry, *Food Chem.*, **67**, 317–332.
34. Arnett, S.D.; Osbourn, D.M.; Moore, K.D.; Vandaveer, S.S. and Lunete, C.E. (2005) Determination of 8-oxoguanine and 8-hydroxy-2′-deoxyguanosine in the rat cerebral cortex using microdialysis sampling and capillary electrophoresis with electrochemical detection, *J. Chromatrogr. B*, **827**, 16–25.
35. Benturquia, N.; Couderc, F.; Sauvinet, V.; Orset, C.; Parrot, S.; Bayle, C.; Renaud, B. and Denoroy, L. (2005) Analysis of serotonin in brain microdialysates using capillary electrophoresis and native laser-induced fluorescence detection, *Electrophoresis*, **26**, 1071–1079.
36. Mayer, B.X.; Petsch, M.; Tschernko, E.M. and Müller, M. (2003) Strategies for the determination of cefazolin in plasma and microdialysis samples by short-end capillary zone electrophoresis, *Electrophoresis*, **24**, 1215–1222.
37. Motomizu, S. and Sakai, T. (2008) On-line sample pretreatment: extraction and preconcentration, in *Advances in Flow Injection Analysis and Related Techniques* (eds. S.D. Kolev and I.D. McKelvie), Comprehensive Analytical Chemistry 54, Wilson and Wilson's, Victoria,.
38. Cañete, F.; Ríos, A.; Luque de Castro, M.D. and Valcárcel, M. (1988) Liquid–liquid extraction in continuous flow systems without phase separation, *Anal. Chem.*, **60**, 2354–2357.
39. Karlberg, B. and Thelander, S. (1978) Extraction based on the flow-injection principle. Part I. Description of the extraction system, *Anal. Chim. Acta*, **98**, 1–7.
40. Valcárcel, M. and Luque de Castro, M.D. (1991) *Non-Chromatographic Continuous Separation Techniques,* The Royal Society of Chemistry, Cambridge.
41. Fang, Z. (1993) *Flow Injection Separation and Preconcentration*, VCH, Weinheim.
42. Ahlmann, N.; Niehoff, A.; Rinas, U.; Scheper, T. and Schügerl, K. (1986) Continuous on-line monitoring of intracellular enzyme activity, *Anal. Chim. Acta*, **190**, 221–228.
43. Nyberg, L. (1970) Selective determinations of amines and quaternary ammonium compounds as ion pairs with methyl orange by an automated method applicable to single dose analysis, *J. Pharm. Pharmacol.*, **22**, 500–506.
44. Rutter, E.R. (1972) Automated method for screening urine for amphetamine and some related primary amines, *Clin. Chem.*, **18**, 616–620.
45. Luque de Castro, M.D. and Álvarez-Sánchez, B. (2008) Membrane-based separation techniques: liquid–liquid extraction and filtration, in *Advances in Flow Injection Analysis and Related Techniques*, (eds. S.D. Kolev and I.D. McKelvie), Comprehensive Analytical Chemistry 54, Wilson and Wilson's, Victoria.
46. Ortiz-Boyer, F.; García-Mesa, J.A. and Luque de Castro, M.D. (1994) Phase discrimination in continuous liquid–liquid extraction with diode array photometric detection, *Anal. Chem.*, **66**, 2794–2798.
47. García-Mesa, J.A.; Linares, P.; Luque de Castro, M.D. and Valcárcel, M. (1990) Direct automatic determination of polyphenols in olive oils in the aqueous phase of a flow-injection liquid–liquid extraction system without phase separation, *Anal. Chim. Acta*, **235**, 441–450.
48. López-Fernández, J.M.; Ríos, A. and Valcárcel, M. (1996) Automatic determination of total aliphatic amines by online photometric liquid–liquid microextraction, *Fresenius J. Anal. Chem.*, **356**, 49–55.

49. Priego-Capote, F. and Luque de Castro, M.D. (2003) Ultrasound-assisted continuous liquid–liquid extraction without phase separation and hydrolysis of paracetamol in suppositories, *Anal. Chim. Acta*, **489**, 223–231.
50. Ruiz-Jiménez, J. and Luque de Castro, M.D. (2003)Flow injection manifolds for liquid–liquid extraction without phase separation assisted by ultrasound, *Anal. Chim. Acta*, **489**, 1–10.
51. Cai, Z.X.; Fang, Q.; Chen, H.W. and Fang, Z.L. (2006) A microfluidic chip based liquid–liquid extraction system with microporous membrane, *Anal. Chim. Acta*, **556**, 151–156.
52. Jeannot, M.A.; Przyjazny, A. and Kokosa, J.M. (2010) Single drop microextraction—Developments, applications and future trends, *J. Chromatrogr. A*, **1217**, 2326–2336.
53. Vidal, L.; Chisvert, A.; Canals, A. and Salvador, A. (2010) Ionic liquid-based single-drop microextraction followed by liquid chromatography-ultraviolet spectrophotometry detection to determine typical UV filters in surface water samples, *Talanta*, **81**, 549–555.
54. Martinis, E.M.; Bertón, P.; Altamirano, J.C.; Hakala, U. and Wuilloud, R.G. (2010) Tetradecyl(trihexyl)phosphonium chloride ionic liquid single-drop microextraction for electrothermal atomic absorption spectrometric determination of lead in water samples, *Talanta*, **80**, 2034–2040.
55. Pillai, A.K.K.V.; Jain, A. and Verma, K.K. (2010) Headspace single-drop microextraction and fibre optics-based cuvetteless micro-spectrometry for the determination of chloride involving oxidation with permanganate, *Talanta*, **80**, 1816–1822.
56. Liu, X.; Chen, X.; Yang, S. and Wang, X. (2007) Comparison of continuous-flow microextraction and static liquid-phase microextraction for the determination of *p*-toluidine in Chlamydomonas reinhardtii, *J. Sep. Sci.*, **30**, 2506–2512.
57. Liu, X.; Wang, X.; Chen, X. and Yang, S. (2007) LC determination of 4-Bromoaniline in green algae Chlamydomonas reinhardtii after continuous-flow microextraction, *Chromatographia*, **65**, 447–451.
58. Liu, X.J.; Chen, X.W.; Yang, S. and Wang, X.D. (2007) Continuous-flow microextration coupled with HPLC for the determination of 4-chloroaniline in Chlamydomonas reinhardtii, *Bull. Environ. Contam. Toxicol.*, **78**, 368–372.
59. Liu, Y.; Hashi, Y. and Lin, J.-M. (2007) Continuous-flow microextraction and gas chromatographic–mass spectrometric determination of polycyclic aromatic hydrocarbon compounds in water, *Anal. Chim. Acta*, **585**, 294–299.
60. Xiao, Q.; Hu, B.; Yu, C.; Xia, L. and Jiang, Z. (2006) Optimization of a single-drop microextraction procedure for the determination of organophosphorus pesticides in water and fruit juice with gas chromatography-flame photometric detection, *Talanta*, **69**, 848–855.
61. Sha, Y.; Meng, J.; Zhang, Y.; Deng, C. and Wu, D. (2010) Determination of volatile organic compounds in tobacco by single-drop microextraction with in-syringe derivatization followed by GC–MS, *J. Sep. Sci.*, **33**, 212–217.
62. Erustes, J.A.; Andrade-Eiroa, A.; Cladera, A.; Forteza, R. and Cerda, V. (2001) Fast sequential injection determination of benzo[a]pyrene using variable angle fluorescence with on-line solid-phase extraction, *Analyst*, **126**, 451–456.
63. Miró, M.; Cladera, A.; Estela, J.M. and Cerda, V. (2000) Sequential injection spectrophotometric analysis of nitrite in natural waters using an on-line solid-phase extraction and preconcentration method, *Analyst*, **125**, 943–948.
64. Šatínský, D.; Karlíček, R. and Svoboda, A. (2002) Using on-line solid-phase extraction for flow-injection spectrophotometric determination of salbutamol, *Anal. Chim. Acta*, **455**, 103–109.
65. Legnerová, Z.; Šatínský, D. and Solich, P. (2003) Using on-line solid phase extraction for simultaneous determination of ascorbic acid and rutin trihydrate by sequential injection analysis, *Anal. Chim. Acta*, **497**, 165–174.
66. Idris, A.M. (2007) On-line coupling of solid-phase extraction, derivatization reaction and spectrophotometry by sequential injection analysis: application to trifluoperazine assay in human urine, *J. Pharmacol. Toxicol.*, **56**, 330–335.
67. Ma, J.; Yuan, D. and Liang, Y. (2008) Sequential injection analysis of nanomolar soluble reactive phosphorus in seawater with HLB solid phase extraction, *Mar. Chem.*, **111**, 151–159.
68. León, Z.; Chisvert, A.; Balaguer, A. and Salvador, A. (2010) Development of a fully automated sequential injection solid-phase extraction procedure coupled to liquid chromatography to determine free 2-hydroxy-4-methoxybenzophenone and 2-hydroxy-4-methoxybenzophenone-5-sulphonic acid in human urine. *Anal. Chim. Acta*, **664**, 178–84.
69. Katsumata, H.; Kojima, H.; Kaneco, S.; Suzuki, T. and Ohta, K. (2010) Preconcentration of atrazine and simazine with multiwalled carbon nanotubes as solid-phase extraction disk, *Microchem. J.*, **96**, 348–351.

70. Fang, G.; Wang, X. and Wang, S. (2010) Multiwalled carbon nanotubes as SPE adsorbents for simultaneous determination of seven sulfonylurea herbicides in environmental water by LC–MS-MS, *Chromatographia*, **72**, 403–409.
71. Zhang, Z.; Zhang, H.; Hu, Y.; Yang, X. and Yao, S. (2010) Novel surface molecularly imprinted material modified multi-walled carbon nanotubes as solid-phase extraction sorbent for selective extraction gallium ion from fly ash, *Talanta*, **82**, 304–311.
72. Kagan, V.E. *et al.* (2010) Carbon nanotubes degraded by neutrophil myeloperoxidase induce less pulmonary inflammation, *Nat. Nanotechnol.*, **5**, 354–359.
73. Tian. M.; Yan, H. and Row, K.H. (2009) Solid-phase extraction of tanshinones from Salvia Miltiorrhiza Bunge using ionic liquid-modified silica sorbents, *J. Chromatogr. B*, **877**, 738–742.
74. Li, M.; Pham, P.J.; Wang, T.; Pittman Jr., C.U. and Li, T. (2009) Selective extraction and enrichment of polyunsaturated fatty acid methyl esters from fish oil by novel π-complexing sorbents, *J. Sep. Purif.*, **66**, 1–8.
75. Tian, M.; Bi, W. and Row, K.H. (2009) Solid-phase extraction of liquiritin and glycyrrhizic acid from licore using ionic liquid-based silica sorbent, *J. Sep. Sci.*, **32**, 4033–4039.
76. Fontanals, N.; Ronka, S.; Borrull, F.; Trochimczuk, A.W. and Marcé, R.M. (2009) Supported imidazolium ionic liquid phases: a new material for solid-phase extraction, *Talanta*, **80**, 250–256.
77. Arthur, C.L. and Pawliszyn, J. (1990) Solid phase microextraction with thermal desorption using fused silica optical fibers, *Anal. Chem.*, **62**, 2145–2148.
78. Bartfák, P. and Čáp, L. (1997) Determination of phenols by solid-phase microextraction, *J. Chromatogr. A*, **767**, 171–175.
79. González-Toledo, E.; Prat, M.D. and Alpendurada, M.F. (2001) Solid-phase microextraction coupled to liquid chromatography for the analysis of phenolic compounds in water, *J. Chromatogr. A*, **923**, 45–52.
80. Peñalver, A.; Pocurull, E.; Borrull, F. and Marcé, R.M. (2002) Solid-phase microextraction coupled to high-performance liquid chromatography to determine phenolic compounds in water samples, *J. Chromatogr. A*, **953**, 79–87.
81. Zhou, F.R.; Li, X.J. and Zeng, Z.R. (2005) Determination of phenolic compounds in wastewater samples using a novel fiber by solid-phase microextraction coupled to gas chromatography, *Anal. Chim. Acta*, **538**, 63–70.
82. Bagheri, H.; Mir, A. and Babanezhad, E. (2005) An electropolymerized aniline-based fiber coating for solid phase microextraction of phenols from water, *Anal. Chim. Acta*, **532**, 89–95.
83. Olejniczak, J. and Staniewski, J. (2007) Enrichment of phenols from water with in-situ derivatization by in-tube solid phase microextraction-solvent desorption prior to off-line gas chromatographic determination with large-volume injection, *Anal. Chim. Acta*, **588**, 64–72.
84. Simões, N.G.; Cardoso, V.V.; Ferreira, E.; Benoliel, M.J. and Almeida, C.M.M. (2007) Experimental and statistical validation of SPME–GC–MS analysis of phenol and chlorophenols in raw and treated water, *Chemosphere*, **68**, 501–510.
85. Baciocchi, R.; Attina, M.; Lombardi, G. and Boni, M.R. (2001) Fast determination of phenols in contaminated soils, *J. Chromatogr. A*, **911**, 135–141.
86. Ribeiro, A.; Neves, M.H.; Almeida, M.F.; Alves, A. and Santos, L. (2002) Direct determination of chlorophenols in landfill leachates by solid-phase micro-extraction–gas chromatography–mass spectrometry, *J. Chromatogr. A*, **975**, 267–274.
87. Liu, H.; Liu, L.; Xiong, Y.; Yang, X. and Luan, T. (2010) Simultaneous determination of UV filters and polycyclic musks in aqueous samples by solid-phase microextraction and gas chromatography–mass spectrometry, *J. Chromatogr. A*, **1217**, 6747–6753.
88. Fernández-Álvarez, M.; Lamas, J.P.; Sánchez-Prado, L.; Llompart, M.; García-Jares, C. and Lores, M. (2010) Development of a solid-phase microextraction gas chromatography with microelectron-capture detection method for the determination of 5-bromo-5-nitro-1,3-dioxane in rinse-off cosmetics, *J. Chromatogr. A*, **1217**, 6634–6639.
89. Pandey, S.K. and Kim, K.H. (2010) Determination of hazardous VOCs and nicotine released from mainstream smoke by the combination of the SPME and GC–MS methods, *TheScientificWorldJOURNAL*, **10**, 1318–1329.
90. Nerín, C.; Salafranca, J.; Aznar, M. and Batlle, R. (2009) Critical review on recent developments in solventless techniques for extraction of analytes, *Anal. Bioanal. Chem.*, **389**, 809–833.

91. Lee, H.B.; Weng, L.D. and Chau, A.S.Y. (1984) Chemical derivatization analysis of pesticide residues. VIII. Analysis of 15 chlorophenols in natural water by in situ acetylation, *J. Assoc. Off. Anal. Chem.*, **67**, 789–794.
92. Wang, Z.Y.; Xiao, C.H.; Wu, C.Y. and Han, H.M. (2000) High-performance polyethylene glycol-coated solid-phase microextraction fibers using sol-gel technology, *J. Chromatogr. A*, **893**, 157–168.
93. Fan, Y.; Zhang, M. and Feng, Y.Q. (2005) Poly(acrylamide-vinylpyridine-N,N'-methylene bisacrylamide) monolithic capillary for in-tube solid-phase microextraction coupled to high performance liquid chromatography, *J. Chromatogr. A*, **1099**, 84–89.
94. Tomkins, B.A. and Barnard, A.R. (2002) Determination of organochlorine pesticides in ground water using solid-phase microextraction followed by dual-column gas chromatography with electron-capture detection, *J. Chromatogr. A*, **964**, 21–33.
95. Achten, C. and Püttmann, W. (2000) Determination of methyl tert-butyl ether in surface water by use of solid-phase microextraction, *Environ. Sci. Technol.*, **34**, 1359–1364.
96. Hernández, F.; Beltrán, J.; López, F.J. and Gaspar, J.V. (2000) Use of solid-phase microextraction for the quantitative determination of herbicides in soil and water samples, *Anal. Chem.*, **72**, 2313–2322.
97. Gonçalves, C. and Alpendurada, M.F. (2002) Comparison of three different poly(dimethylsiloxane)-divinylbenzene fibres for the analysis of pesticide multiresidues in water samples: structure and efficiency, *J. Chromatogr. A*, **963** 19–26.
98. Vercauteren, J.; De Meester, A.; De Smaele, T.; Vanhaecke, F.; Moens, L.; Dams, R. and Sandra, P. (2000) Headspace solid-phase microextraction-capillary gas chromatography–ICP mass spectrometry for the determination of the organotin pesticide fentin in environmental samples, *J. Anal. At. Spectrom.*, **15**, 651–656.
99. Viñas, P.; Campillo, N.; Martínez-Castillo, N. and Hernández-Córdoba, M. (2010) Comparison of two derivatization-based methods for solid-phase microextraction-gas chromatography-mass spectrometric determination of bisphenol A, bisphenol S and biphenol migrated from food cans, *Anal. Bioanal. Chem.*, **397**, 115–125.
100. Saridara, C.; Brukh, R.; Iqbal, Z. and Mitra, S. (2005) Preconcentration of volatile organics on self-assembled, carbon nanotubes in a microtrap, *Anal. Chem.*, **77**, 1183–1187.
101. Long, Q.R. and Yang, R.T. (2001) Carbon nanotubes as superior sorbent for dioxin removal, *J. Am. Chem. Soc.*, **123**, 2058–2059.
102. Li, Y.H.; Zhu, Y.Q. and Zhao, Y.M. (2006) Different morphologies of carbon nanotubes effect on the lead removal from aqueous solution, *Diamond. Relat. Mater.*, **15**, 90–94.
103. Li, Y.H.; Wang, S.; Luan, Z.; Ding, J.; Xu, C. and Wu, D. (2003) Adsorption of cadmium(II) from aqueous solution by surface oxidized carbon nanotubes, *Carbon*, **41**, 1057–1062.
104. Lu, C.X.; Chiu, H.S. and Liu, C.T. (2006) Removal of Zinc(II) from aqueous solution by purified carbon nanotubes: kinetics and equilibrium studies, *Ind. Eng. Chem. Res.*, **45**, 2850–2855.
105. Li, Y.H.; Wang, S.; Wei, J. and Zhang, X. (2003) Adsorption of fluoride from water by aligned carbon nanotubes, *Mater. Res. Bull.*, **38**, 469–476.
106. Peng, X.; Li, Y.H.; Luan, Z.; Di, Z. and Wang, H. (2003) Adsorption of 1,2-dichlorobenzene from water to carbon nanotubes, *Chem. Phys. Lett.*, **376**, 154–158.
107. Lu, C.; Chung, Y.L. and Chang, K.F. (2005) Adsorption of trihalomethanes from water with carbon nanotubes, *Water Res.*, **39**, 1183–1189.
108. Cai, Y.Q.; Jiang, G.B.; Liu, J.F. and Zhou, Q.X. (2003) Multiwalled carbon nanotubes as a solid-phase extraction adsorbent for the determination of bisphenol A, 4-n-nonylphenol, and 4-tert-octylphenol, *Anal. Chem.*, **75**, 2517–2521.
109. Liu, G.H.; Wang, J.L.; Zhu, Y.F. and Zhang, X.R. (2004) Application of multiwalled carbon nanotubes as a solid-phase extraction sorbent for chlorobenzenes, *Anal. Lett.*, **37**, 3085–3104.
110. Cai, Y.Q.; Jiang, G.B.; Liu, J.F. and Zhou, Q.X. (2003) Multi-walled carbon nanotubes packed cartridge for the solid-phase extraction of several phthalate esters from water samples and their determination by high performance liquid chromatography, *Anal. Chim. Acta.*, **494**, 149–156.
111. Wang, J.X.; Jiang, D.Q.; Gu, Z.Y. and Yan, X.P. (2006) Multiwalled carbon nanotubes coated fibers for solid-phase microextraction of polybrominated diphenyl ethers in water and milk samples before gas chromatography with electron-capture detection, *J. Chromatogr. A*, **1137**, 8–12.
112. Banerjee, S.; Kahn, M.G.C. and Wong, S.S. (2003) Rational chemical strategies for carbon nanotube functionalization, *Chem. Eur. J.*, **9**, 1898–1908.

113. Niyogi, S.; Hamon, M.A.; Hu, H.; Zhao, B.; Bhowmik, P. and Sen, R. (2002) Chemistry of single-walled carbon nanotubes, *Acc. Chem. Res.*, **35**, 1105–1113.
114. Banerjee, S. and Wong, S.S. (2003) In situ quantum dot growth on multiwalled carbon nanotubes, *J. Am. Chem. Soc.*, **125**, 10342–10350.
115. Kim, M.R.; Abd El-Aty, A.M.; Choi, J.H.; Lee, K.B. and Shim, J.H. (2006) Identification of volatile components in Angelica species using supercritical-CO_2 fluid extraction and solid phase microextraction coupled to gas chromatography–mass spectrometry, *Biomed. Chromatogr.*, **20**, 1267–1273.
116. Arthur, C.L.; Pratt, K.; Motlagh, S.; Pawliszyn, J. and Belardi, R.L. (1992) Environmental analysis of organic compounds in water using solid phase micro extraction, *J. High Resolut. Chromatogr.*, **15**, 741–744.
117. Yang, Y.; Hawthorne, S.B.; Miller, D.J.; Liu, Y. and Lee, M.L. (1998) Adsorption versus absorption of polychlorinated biphenyls onto solid-phase microextraction coatings, *Anal. Chem.*, **70**, 1866–1869.
118. Baltussen, E.; Sandra, P.; David, F.; Janssen, H.G. and Cramers, C. (1999) Study into the equilibrium mechanism between water and poly(dimethylsiloxane) for very apolar solutes: adsorption or sorption?, *Anal. Chem.*, **71**, 5213–5216.
119. Tienpont, B.; David, F.; Desmet, K. and Sandra, P. (2002) Stir bar sorptive extraction-thermal desorption-capillary GC-MS applied to biological fluids, *Anal. Bioanal. Chem.*, **373**, 46–55.
120. Tienpont, B.; David, F.; Benijts, T. and Sandra, P. Stir bar sorptive extraction-thermal desorption-capillary GC-MS for profiling and target component analysis of pharmaceutical drugs in urine, *J. Pharm. Biomed. Anal.*, **32**, 569–579 (2003).
121. Tienpont, B.; David, F.; Stopforth, A. and Sandra, P. (2003) Comprehensive profiling of drugs of abuse in biological fluids by stir-bar sorptive extraction-thermal desorption–capillary gas chromatography–mass spectrometry, *LC-GC Eur.*, **16**, 5–13.
122. Ochiai, N.; Sasamoto, K.; Daishima, S.; Heiden, A.C. and Hoffmann, A. (2003) Determination of stale-flavor carbonyl compounds in beer by stir bar sorptive extraction with in-situ derivatization and thermal desorption-gas chromatography–mass spectrometry, *J. Chromatogr. A*, **986**, 101–110.
123. Fang, Z.; Welz, B. and Sperling, M. (1991) Contribution of system components to dispersion in the analysis of microvolume samples by flow injection flame atomic absorption spectrometry, *J. Anal. At. Spectrom.*, **6**, 179–185.
124. Luque de Castro, M.D. and Priego-Capote, F. (2006) *Analytical Applications of Ultrasound,* Elsevier, Amsterdam.
125. Wang, R.Y.; Jarrat, J.A.; Keay, P.J.; Hawkes, J.J. and Coakley, W.T. (2000) Development of an automated on-line analysis system using flow injection, ultrasound filtration and CCD detection, *Talanta*, **52**, 129–139.
126. Delgado-Reyes, F.; Arce, C.; Moreno, A.; Fernández Romero, J.M. and Luque de Castro, M.D. (2001) Determination of anti-canine IgG using a continuous filtration/dissolution system based on the formation of a high-molecular size immunocomplex, *Talanta*, **55**, 821–829.
127. Dimitrova-Koleva, B.; Benkhedda, K.; Ivanova, E. and Adams, F. (2007) Determination of trace elements in natural waters by inductively coupled plasma time of flight mass spectrometry after flow injection preconcentration in a knotted reactor, *Talanta*, **71**, 44–50.
128. Martínez, P.; Gallego, M. and Valcárcel, M. (1987) Indirect atomic absorption determination of chloride by continuous precipitation of silver chloride in a flow injection system, *J. Anal. At. Spectrom.*, **2**, 211–217.
129. Zorro, J.; Gallego, M. and Valcárcel, M. (1988) Automated determination of sulfate by continuous precipitation–filtration of lead sulfate, *Microchem. J.*, **39**, 71–78.
130. Montero, R.; Gallego, M. and Valcárcel, M. (1988) Indirect atomic absorption spectrometric determination of sulfonamides in pharmaceutical preparations and urine by continuous precipitation, *J. Anal. At. Spectrom.*, **3**, 725–731.
131. Montero, R.; Gallego, M. and Valcárcel, M. (1988) Indirect atomic absorption spectrometric determination of local anesthetics in pharmaceutical preparations with a flow-injection precipitation technique, *Anal. Chim. Acta*, **215**, 241–247.

9
Capillary Electrophoresis

Mihkel Kaljurand

Faculty of Science, Tallinn University of Technology, Tallinn, Estonia

9.1 The capillary electrophoresis separation techniques

Capillary electrophoresis (CE) is comprised of a group of related methods that use a narrow-bore fused-silica capillary as a separation column to separate a complex array of large and small molecules. The separation of molecules is driven by the strength of the electric field. CE has been fully described in many publications [1,2]. This article provides a description of CE for the reader who is unfamiliar with the concept. Separation is based on the difference in the speed of charged particles (ions, biopolymers, molecular clusters, cells, etc.) that migrate through the capillary. To state it in the simplest terms, the velocity of migration depends on the charge and size of the species, as follows:

$$v = \frac{zE}{6\pi\eta r}, \tag{9.1}$$

where v is the velocity, z is the charge and r is the radius of the species. E is the strength of the applied electric field and η is the viscosity of the medium in the capillary. The sample is introduced by immersing the end of the capillary into a vial of the sample and applying pressure, vacuum or voltage. In most applications, the migration of analytes is accompanied by the desired or undesired movement of the (aqueous) separation media, which is known as electroosmosis and is due to the drag of the hydrogen ions on the separation buffer. The speed of electroosmosis depends on the pH of the media. CE equipment commonly includes an on-column detector (UV, conductivity or fluorescence).

Depending on the types of capillaries and electrolytes used, CE can be divided into several separation techniques:

Handbook of Green Analytical Chemistry, First Edition. Edited by Miguel de la Guardia and Salvador Garrigues.

1. *Capillary Zone Electrophoresis* (CZE) is the simplest form of CE. The separation occurs in a bare capillary and the separation mechanism is based on differences in the charge-to-mass ratio of the analytes, as shown in the formula above. CZE uses a homogeneous buffer solution and constant field strength throughout the length of the capillary. Simple inorganic ions can be separated by CZE. Controlling the pH of the media, dissociating the acidic groups in the solute or protonating the basic functions of the solute can also be used for separation.
2. *Capillary Gel Electrophoresis* (CGE) uses polymers in solution to create a molecular sieve (the adaptation of traditional gel electrophoresis to the capillary). This allows analytes with similar charge-to-mass ratios to be resolved by size. This technique is employed for the analysis of proteins and DNA sequencing.
3. *Capillary Isoelectric Focusing* (CIEF) uses a pH gradient generated between the cathode and anode. A solute (amphoteric molecules, such as proteins) will migrate to a point where its net charge is zero, migration stops, and the sample is focused into a tight zone (at the isoelectric point, pI). CIEF is commonly employed in protein characterization.
4. *Isotachophoresis* (ITP) is a separation technique based on the migration of the sample components between high mobility (leading) and low mobility (terminating) electrolytes located at the inlet and outlet of the capillary. Solutes with mobility values between those of the leading and terminating electrolytes stack into sharp, focused zones.
5. *Electrokinetic Chromatography* (EKC) combines electroosmosis, electrophoresis and chromatographic phenomena in a single capillary. Chromatography is introduced by adding an extra component to the separation buffer, which can interact with analytes. An example is cyclodextrin-mediated EKC. In this process, the enantiomers interact with the cyclodextrins, which separates chiral compounds.
6. *Micellar Electrokinetic Capillary Chromatography* (MEKC) is a mode of EKC in which surfactants are added to the buffer solution at concentrations that form micelles. The separation principle of MEKC is based on the partition differential between the micelle and the solvent. In addition to charged species, MEKC can separate neutral solutes and use stationary or mobile micelles.
7. *Micro Emulsion Electrokinetic Chromatography* (MEEKC) is a mode of EKC in which solutes partition between moving oil droplets and a buffer. The oil droplets are usually formed by sonicating immiscible heptane or octane with water. MEEK separates both aqueous and water-insoluble compounds.
8. *Non-Aqueous Capillary Electrophoresis* (NACE) separates analytes in a medium composed of organic solvents (methanol, acetonitrile).
9. *Capillary Electrochromatography* (CEC) uses an electric field instead of hydraulic pressure to force the mobile phase through a packed bed in a capillary. It is a hybrid separation method that combines the high separation efficiency of CZE with HPLC.

This long list of different modes of CE demonstrates the flexibility of the method. When it was invented in the middle of the last century, [3–6] capillary electrophoresis (CE) functioned so well that it was expected to quickly replace high performance liquid chromatography (HPLC). This did not happen, and HPLC is still the main analytical separation method used by chemists. The main probable reason is that HPLC has better reproducibility than CE. This restricts the use of CE in industrial laboratories where large numbers of samples must be analysed with the high precision needed for product quality control. The migration time of a compound also depends on the sample matrix. Careful stabilization of the capillary temperature in modern instruments can improve the reproducibility of migration times significantly. However, the migration speed of a compound also depends on the strength of the electric field in the capillary and this may vary between samples in complicated matrices, such as biofluids or plant extracts. This makes it difficult to identify compounds in matrices with which the use of CE might have the greatest impact. Mass spectrometric (MS) detection could be a great help in such cases; however, HPLC-MS is also much easier to implement than CE-MS. The potential benefits of CE-MS have been investigated for the last 20 years. Reliable solutions have emerged only recently [7]. However, the development of CE-MS technology does not contribute significantly to the discussion of CE as a green analytical method and thus remains outside of this review.

Nevertheless, CE is popular in academia (approximately 2000 research papers are published yearly) because it provides interesting material for studies (there are endless possibilities to alter the composition of the separation buffer, and the technique demonstrates superior efficiency compared to other separation methods). Still, the development of CE has stagnated in recent years. The disadvantage of CE with regard to its inferior reproducibility as compared to HPLC is such an obstacle that the industry is still clinging to the convenience of HPLC. Relatively few new publications on CE development have been published in the last year as compared to the number of publications related to its application. This indicates that CE is maturing rapidly and that its field of application could be quite narrow.

Green chemistry might affect such dismissive and pessimistic attitudes towards the future of CE. If the demand for green analytical methods continues, it could create new interest in CE. The further advancement of CE as a green analytical technique could open new research directions in this field – especially those aimed at improving its reproducibility. CE is one of a number of techniques that meet the definition of a green analytical method. It is perhaps the only liquid phase separation method that is sophisticated, selective and sensitive enough to solve a large number of analytical problems and can be made green. Surprisingly, the green chemistry community has only recently recognized this fact [8–11]. CE consumes little power and it can be miniaturized and made portable, due to the fact that it requires very little solvent (μl). The latter attribute is a major advantage of CE in light of the well-known acetonitrile (ACN) crisis in the year 2009 [12]. ACN is a popular HPLC solvent that the typical analytical laboratory consumes by the litre every day. Up until the recent downturn of the global economy, which led to decreased automobile parts production, of which process ACN is a byproduct [12], no one thought seriously about replacing ACN as an HPLC eluent. Now, faced with the ACN crisis, which resulted in a shortage and enormous price increases, companies are looking at green chromatographic solutions in earnest. However, green chromatography that uses solvents other than ACN and/or elevated temperatures for separations might not provide the required solutions and this might lead to CE becoming a viable alternative [13].

The commercially available CE instruments (e.g. from Agilent and Beckman) with their robotic autosamplers are bulky and hardly fit the definition of a green instrument. The green features of CE have been incorporated and demonstrated in several types of custom-made portable CE devices. In this chapter, we will discuss the characteristics of CE that make it a promising candidate as a genuinely green analytical technique compared to other separation methods. Special attention is given to comparing CE with different modes of HPLC and microfluidics-based lab-on-a-chip (LOC) devices. Microscale HPLC and LOC are new, cutting edge research directions in analytical separations. We will attempt to prove that, contrary to widespread belief, the greening of these methods faces more serious obstacles than classical CE. The possible ways of overcoming the disadvantages of CE will also be discussed.

9.2 Capillary electrophoresis among other liquid phase separation methods

9.2.1 Basic instrumentation for liquid phase separations

A typical analytical separation instrument has the following functional elements: a device for creating eluent flow, separation column, detector and data recording system. In order to actuate the eluent in an HPLC instrument, pressure is applied. Highly technical mechanical pumps generate the pressure. The electrolyte flow in CE is established by electroosmosis, which is generated by a high voltage power supply. The separation mechanism in HPLC is the partition of a sample between mobile and stationary phases. In contrast, CE separates sample compounds by differences in their migration in an electric field; however, the complete separation process could be rather elaborate and include participation as well. Capillary electrophoresis (CE)

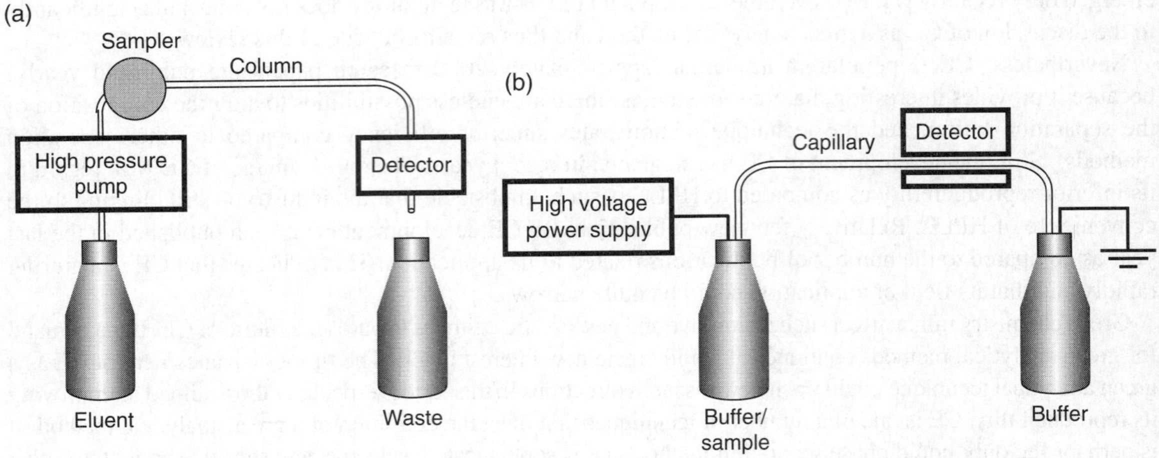

Figure 9.1 *(a) Schematic diagram of an HPLC instrument. (b) Schematic diagram of CE equipment. For sampling, the buffer reservoir is replaced by a sample reservoir.*

typically utilizes high electric fields for separation of an analyte, which might be based on its charge, size, or hydrophobicity. The same detector could be used for both HPLC and CE, but the column, which is the essence of the separation method, differs significantly. CE uses capillaries 50 cm in length with an internal diameter of 10–100 µm. In contrast, HPLC uses standard columns of 15 cm length with a 4.7 mm diameter. Whereas most HPLC separations are performed using an H_2O/ACN gradient, which elutes samples in a reverse phase column, the choice of CE buffer systems is much more flexible. In CE, the sample is introduced by immersing the end of the capillary into a sample vial and applying pressure – vacuum or voltage. HPLC uses mechanical six-port two-position rotary valves for sample introduction. The basic schematics of typical CE and HPLC instruments are shown in Figure 9.1. Note the functional similarity of both separation methods. However, there are differences: the driving force of the mobile phase in CE is fundamentally different from that of HPLC and detection in CE must be performed on the column in order to not compromise the efficiency of the separation. HPLC needs a special device for sample introduction, whereas CE sampling, as mentioned above, can be accomplished by exchanging sample/buffer vials.

9.2.2 CE versus HPLC from the point of view of Green Analytical Chemistry

CE is a powerful separation technique that uses minimal amounts of solvent. As already mentioned, a typical CE column is a narrow capillary made of fused silica with an inner diameter (i.d.) of 50 µm and a length of approximately half a meter. It has a volume of about 5 µl. Because electroosmosis – the force that propels the eluent in CE – has a flow rate of sub-µl min^{-1}, given a typical runtime of 10 min, it follows that eluent consumption during a CE run is almost nonexistent. ACN and other harmful solvents are rarely used as eluents in CE because most separations are achieved in non-toxic buffer systems such as phosphate and borate.

Despite the striking functional similarity between the two techniques (as demonstrated in Figure 9.1), the capillary electropherograph has a much simpler construction because of the lack of moving mechanical parts. Instead of high precision, elevated pressure pumps and rotary valves, only a high voltage power supply is required for CE.

In Green Analytical Chemistry, reduction of the usage of harmful eluents (like ACN) seems to be the main criterion for a method to be considered green. How does CE compare with different modes of HPLC in this respect? The typical eluent in CE is a non-toxic salt solution, and the amount of this buffer is determined not by the inner diameter of the capillary but by the total volume of the microvials that a human operator can manipulate. This is commonly 100 μl for typical commercial or research instruments, although smaller volumes can be easily employed as well. For purposes of comparison, Table 9.1 (adapted from Noga *et al.* [14]). contains the inner diameters of columns, flow rates and volumes of eluent consumed during a 20 min run for various modes of LC.

Table 9.1 reveals that only capillary HPLC (also known as nano-LC) is comparable with CE for eluent economy. The other modes of HPLC consume much more eluent per run than does CE. This highlights a problem associated with this technique. It is not easy to generate the flow rates necessary for capillary LC with a typical HPLC pump, which can deliver no less than 100 μl min^{-1}. The solution is to split the eluent just after the pump. The splitter can be a simple 'T' connection or an active device that compensates for viscosity changes during gradient elution. The major part of the flow is expelled as waste, and a small percentage is introduced into a capillary column. Therefore, more than 99% of an ultrapure, expensive eluent is wasted, so nano-LC cannot be called a green method when splitters are used. A few manufacturers have recently introduced splitless nano-LC instruments, which eliminate almost all the disadvantages of splitters. Such solutions have been provided by Agilent Technologies (Santa Clara, California) and the Waters Corporation (Milford, Massachusetts) [15]. Eksigent Technologies, Inc. (Livermore, California) [16] recently introduced its commercial cHPipLC Nanoflex system. The glass chip contains etched, microfabricated channels with C18 reversed-phase packing, which accommodates a direct coupling with a nano-HPLC pump and delivers the mobile phase to the chip.

The eluent flow in a splitless nano-LC instrument is controlled as follows: flow meters in each mobile phase path continuously monitor the flow rate and feed a signal back to a microprocessor. Pressure is generated by connecting laboratory air or nitrogen to a pneumatic amplifier that increases the pressure in the system. A microfluidic controller regulates this pressure to produce the required flow rate. For example, 100 psi of incoming air pressure from the laboratory air system can be used to generate a range of hydraulic pressure up to 10 000 psi [16]. However, apart from the cost of nano-LC, the technology involved in nanoflow pumping for HPLC mobile phases would sacrifice the portability and simplicity required of a truly green analytical system.

Other innovations are being sought, in addition to using pressure to actuate the liquids. Much research and development has been driven by the need to effectively manipulate small volumes of chemical and biological liquids at micro and/or nano rates of flow. The principles of these pumping techniques are based on several well-known phenomena. Electrokinetically-driven continuous flow pumps (such as electrophoretic and electroosmotic pumps), surface chemistry-based continuous flow micropumps (such as the opto-electrowetting-based pump),

Table 9.1 *Typical internal diameters, flow rates and amounts of eluent consumed for various LC columns.*

Column ID	Common name	Typical flow rate	Amount of eluent consumed (per 20 min run)
75 μm	Capillary	100–300 nl min^{-1}	2–6 μl
150, 250 μm	Capillary	300–500 nl min^{-1}	6–10 μl
500 μm	Capillary/narrowbore	0.5–10 μl min^{-1}	10–200 μl
0.75, 0.8 mm	Narrowbore/ microbore	10–100 μl min^{-1}	0.2–2 ml
1.1, 2.1 mm	Microanalytical	100–500 μl min^{-1}	2–10 ml
4.6 mm	Analytical	1–2.5 ml min^{-1}	20–50 ml
8 – 20 mm	Semi-preparative	>2 ml min^{-1}	>40 mls

optically-driven, electrochemical and constant gravity-driven pumps, and a combination of techniques such as hydrodynamic flow and electrokinetic/gravity/magnetophoretic pumping have been proposed [17]. Among the plethora of solutions, electroosmotic pumps are receiving increasing attention [18]. These have several advantages over other micropumps: bi-directionality, constant and pulse-free flows well suited to microanalytical systems and an absence of moving parts. Therefore, the difference between CE and HPLC blurs when HPLC is conducted in nanoscale.

The fact that CE is inherently green increases the pressure to make HPLC greener as well. Several research groups are pursuing this objective (see recent reviews [19–21] and references therein) but the route to green HPLC is not easy. Some methods of making HPLC greener are as follows:

- Solvent reduction:
  - fast chromatography using a short column packed with smaller particles, and conventional HPLC with elevated pressures;
  - fast separation using ultra-high performance liquid chromatography (UHPLC) systems. Recent years have witnessed remarkable interest in a new mode of chromatographic analysis, based upon increased use of smaller particle (sub-2 μm) stationary phases operating at higher back-pressures;
  - conventional HPLC systems with minor modifications for greener separations. Significant solvent savings can be obtained simply by reducing column dimensions;
  - newer HPLC column technologies for greener separations, and
  - microflow and capillary HPLC. This mode has already been discussed at length above. Recent interest in this method is driven largely by the need to couple HPLC with MS and has little to do with waste reduction. However, the green-chemistry advantages of microflow chromatography are evident.
- Solvent replacement:
  - replacement of acetonitrile with methanol or ethanol in RP-HPLC. Using the greener solvent may be an option for those analytical laboratories where its poorer elutropic strength and UV cut-off (compared to ACN) would not be a problem;
  - superheated water chromatography. Water at elevated temperatures becomes less polar and can be used for the RP elution of many compounds including natural products, pharmaceuticals, and environmental samples. At elevated temperatures (>100°C), water is incompatible with most conventional silica-based stationary phases. More temperature-resistant packing materials are usually required (e.g. polystyrene-divinylbenzene or zirconia-based materials, and
  - carbon dioxide-based chromatography.

One can see that the research follows the general lines of green chemistry. The possible solutions can be expressed by the 'three R-s': Reduce, Replace, Recycle (the latter approach is not yet very popular). The solutions are mainly self-evident, and are compromises between the quest for greenness and decreased analysis performance (reduction of efficiency, increase of detection limits, increase of instrument complexity and cost).

An overlooked aspect of elevated temperature HPLC (with water as the eluent) and supercritical fluid chromatography is the fact that the greenness of the methods is achieved by using large amounts of energy (electricity), which might lead to an overall increase in the size of a method's environmental footprint. Only one study to date has been performed along these lines with regard to the preparative mode of chromatography [22]. The results of the study are not encouraging: the green advantages of using supercritical CO_2 as an eluent instead of ACN are evident only from the point of view of a single laboratory, but not when the whole processes of producing ACN and CO_2 are considered. This may be true for other modes of green HPLC as well. Or, to speak metaphorically, green HPLC may simply sweep its waste under the rug.

What are the expectations of Green Analytical Chemistry for CE? The method was very actively studied at the beginning of the 1990s, but its development has lost momentum in recent years. More than 2000 papers on fundamental studies in CE have been published each year of the last decade. Nevertheless, the industry has been slow to accept the method. CE is struggling to replace HPLC for the analysis of conventional small molecule pharmaceuticals. A certain amount of avoidance can be explained by the lack of familiarity with CE among analysts in general, which can lead to the development of methods that are not robust, perform poorly and give CE a negative reputation. Analytical chemists have learned that CE is highly sensitive to parameter changes and that it is not a very reproducible technique, but modern instrumentation has eliminated most of the early variability issues. As robust interfaces with mass spectrometers are developed, the obstacle of the incompatibility of CE with MS will soon be overcome [23].

As we already pointed out, CE has received little recognition as a genuinely green separation method despite the fact that the solvents it employs are usually harmless aqueous buffers[1]. As the ACN shortage continues to influence the field of analytical separations, CE is gathering increased attention, as confirmed in the report from one panel discussion [13]. Suppliers of CE equipment have responded to the need for improved robustness and reliability. They have made improvements in newer equipment. There is also a trend towards well-monitored and validated chemistry/capillary kits to improve performance, for example, for inorganic anion and metal ion determinations. The emergence of more powerful, automated instruments makes this technique more accessible than ever. There are some signs that the future of CE is bright. It is probable [13] that the use of CE for small inorganic anion and metal ion analyses will replace ion-exchange chromatography. Chiral analysis by CE is well established. The method is also used for screening and characterizing compounds (determination of *pKa*, solubility, etc.). Dedicated equipment and related kits/capillaries will be available for specific protein characterization/assays. CE is far superior to either SDS-PAGE or IEF gels. However, CE should not completely replace HPLC. It is a technique that is complementary and orthogonal to HPLC. Proponents of CE believe that outdated thinking on the part of separation scientists is the biggest obstacle to acceptance of the technique. As highlighted by the forum cited above, [13] the method still needs the support of instrument manufacturers to build better and more robust instrumentation, of vendors of consumables to continue to devise kits and reagents, and of scientists themselves to develop novel methodologies and applications. Without this type of investment, the technique cannot expand, but without expansion, no one wants to invest. Wider acceptance of green chemistry by analysts may help to break this impasse.

9.2.3 CE as a method of choice for portable instruments

In addition to its small consumption of chemicals, another overlooked quality of CE is its potential for portability, via miniaturization of all its components. Portability is believed to be an important and obvious feature of green analytical instruments. Portable instruments are considered to be more economical than their stationary counterparts. They consume fewer resources (either power or chemical) and generate less waste. Moreover, they are designed to be taken to the site of the analysis, that is, a point-of-care (POC) such as a hospital, home or crime scene. The portability of an instrument is generally understood to be its ability to

---

[1] Although arguments in favour of greenness for the benefit of mankind may seem too philosophical for a chromatographer who wrestles with everyday problems, the political-economic situation in a particular country may persuade scientists to accept green solutions. The authors have personal experience in this regard. The rise of CE at the beginning of the 1990s coincided with political changes in Eastern Europe that caused a dramatic reduction in funding for fundamental research. In this situation, it was almost impossible to use HPLC due to the lack of supplies and its requirement for large amounts of solvents and spare parts. On the other hand, it was relatively easy to assemble CE instruments from old colour television sets (which contained a high voltage power supply) and to modify the cells of discarded optical HPLC detectors an thereby carrying on research at a reasonably competitive level.

work where a sample can be collected and analysed *in-situ*. In fact, many field analyses could be accomplished without traditional sampling.

Field Analytical Chemistry (FAC) is a growing discipline that promises to liberate the analyst from tedious and inconvenient sample manipulations. Analytical methods typically involve sample preparation, and a time- and labour-consuming collection process usually needs to take place before samples can be transported to a laboratory for analysis. Sample collection can often be problematic. For instance, samples of polluted soil have to be collected from many locations according to a laborious sampling plan to avoid missing the 'hot spot'. Samples must usually be treated with specific reagents and stored in containers under certain conditions to maintain their integrity prior to analysis. Furthermore, in many cases sampling is fraught with difficulties, such as at hazardous polluted sites or with precious cultural relics and archaeological objects. FAC can not only eliminate the need for sample transportation but also greatly shorten the analysis time or even provide real-time results, thereby providing rapid warning and accurate feedback. Thus, the logic and nature of FAC are such that analysis is done with little or no sample collection and preparation. Transportation of the sample to a laboratory is eliminated.

He *et al.* summarize the green characteristics of FAC instruments [24]. In order to perform field analyses, the ideal analytical instrument should meet several requirements: it should (1) have a fast response time to be able to acquire the necessary information on a real-time or near real-time basis, (2) be capable of in situ/at-site rather than just on-site[2] analysis and need little or no sample preparation, (3) be portable for field use with a minimum requirement for energy (battery-powered is desirable), consumables (gases /solvents), or clean space for handling samples, and (4) perform a cost-effective analysis. Turl and Wood are more specific with regard to the characteristics of an FAC instrument to be useful at a point-of-care [25]. A step out of the lab and into the field requires a giant leap from technology to capability. The instrument, in addition to consuming very little power and being of a size and weight that enables portability, must:

- perform adequately;
- work in a harsh environment;
- have a minimal number of failures and be easy to maintain;
- ensure the security of classified data, and
- be easy to operate.

Furthermore,

- an appropriate sampling method must be developed;
- health and safety risks must be minimized, and
- operator training must be uncomplicated.

Considering the technical sophistication of nano-LC, it is difficult to imagine that a portable gas or liquid chromatograph could meet all these conditions; although it might be technically possible (portable gas chromatographs are used successfully in space applications). There are even reports of field-portable gas [26,27] and liquid chromatographs [28]. Nevertheless, chromatography is not well suited to portability because eluent is required for the analysis. On the other hand, CE is a promising technology for field instruments.

---

[2] He *et al.* [9] provide a definition of the terms 'in situ' and 'on site'. In this chapter, 'on-site analysis' is understood to be a common analysis procedure that involves sample collection/preparation using a field-portable instrument. 'In situ' analysis leaves the sample site virtually undisturbed. 'In situ' analysis could be done with an X-ray spectrometer but it is difficult to imagine how a chromatographic analysis of art or soil samples would be possible.

CE can easily be made portable because the electrophoresis process in a capillary consumes little power, so small-sized high voltage power supplies can be used. Optical detection is not currently well suited to portable instruments; however, the emergence of small-sized light-emitting diodes will soon change that situation. Da Silva [29] and Zemann [30] have developed a useful detection device: a contactless conductivity detector (CCD) that measures the conductance of a small gap between tubular electrodes laid on the separation capillary. The device is inherently small, and most analytes can be detected with a CCD. Several groups have developed designs for new portable CE instruments. Hauser's group [31] has developed (and optimized) a portable CE instrument with CCD for the sensitive field measurement of ionic compounds in environmental samples. It is battery-powered, and the high voltage modules are capable of delivering up to 15 kV at either polarity for more than one day. Inorganic cations and anions, including ions of heavy metals and arsenates, can be determined with detection limits of approximately 0.2 to 1 mM. The instrument was field-tested in a remote region of Tasmania. Nitrite and ammonium were determined on-site at concentrations as low as 10 ppb, in the presence of other common inorganic ions at concentrations two to three orders of magnitude higher. In another publication, Haddad's group demonstrated the use of CE for the detection of explosives in the environment [32]. Instead of CCD, Haddad's team used indirect photometric detection. They proved that it is possible to analyse blast residues at a crime scene, where they can be sampled simply by wiping hard surfaces with a wet cloth, rather than by transporting the residues back to the laboratory. They found that they could separate and detect the 12 cations at concentrations as low as 0.11 mg l^{-1} and separate and detect the 15 anions at concentrations as low as 0.24 mg l^{-1}. In both cases, the analyses took less than 10 minutes. However, they found that CCD performed better than indirect photometric detection [33].

Seiman *et al.* [34] attempted to develop a robust sampling procedure for on-site analysis. In this project, the CE analyser consisted of two pieces of capillary that were separated by a narrow gap (30 μm). To introduce the sample, a plastic syringe was inserted into a socket connected to the gap, and the background electrolyte in the cross-section of the sampler was flushed out by the sample stream injected by the syringe. The sample between the capillaries was then carried into the separation channel by electroosmosis flow (EOF), and the background electrolyte filled the junction between the two capillaries when high voltage was applied. This technique reduces the manipulation of buffer vials. The method developed for this instrument has an LOD of 4–8 μM for phosphonic acids and 0.3–0.5 μM for cations, and an RSD (internal standard) of 8%.

Sampling complexity can be reduced even further by means of a simple manual buffer/sample vial exchange at the capillary inlet. A possible design for what is conceivably the simplest portable instrument is shown in Figure 9.2. An analysis compartment with two buffer vials and a CCD sensor is located at the front of the instrument and a touch-screen computer on top. A typical electropherogram of the phosphonic acids recorded with this instrument is shown in Figure 9.3. This figure is discussed further in Section 9.3.

The design of a sampling instrument can vary according to the analyst's requirements. For example, the robustness of a portable sampling instrument meant for field use must take into account that the operator might have to work in a protective suit and gloves. In this case, split sampling or manipulating the sample with a syringe rather than operating with vials would be more convenient (see Figure 9.4).

Electrophoresis is a key technology for further micronizing analytical separation methods by making use of an advanced concept based on lab-on-a-chip platforms. It is believed that this technology will open the way to many inexpensive point-of-care medical diagnostic devices. Many reports on portable CE analysers based on microfluidics or lab-on-a-chip (LOC) platforms have been published in recent years (see the following publications for examples [35,36]). CE-based microfluidic devices will be described fully in Chapter 17 on *Miniaturization*. We will now discuss one overlooked feature that impedes the potential greenness of LOC devices.

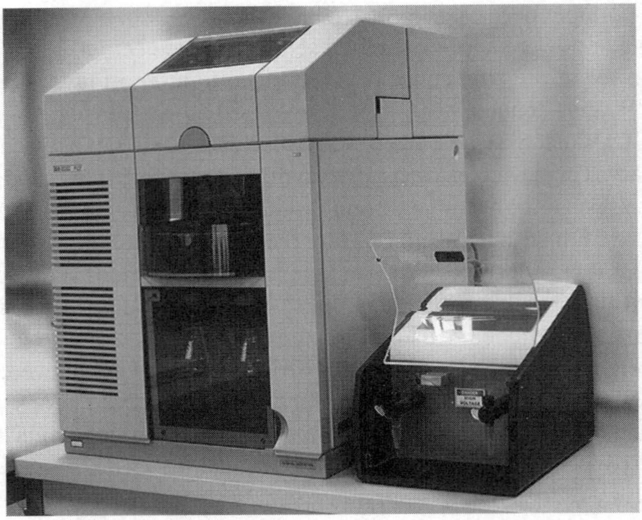

Figure 9.2 On the right: portable CE instrument controlled via a touch-screen computer (the screen can be seen on top of the instrument). The electronics and high voltage power supply are located below the computer. The CE compartment with two sample/buffer vials and a CCD block is at the front. High voltage is applied to the vials via electrodes located inside the two plastic insulators (black rods). On the left for comparison: a commercial CE instrument (photograph of the working instruments from the author's laboratory).

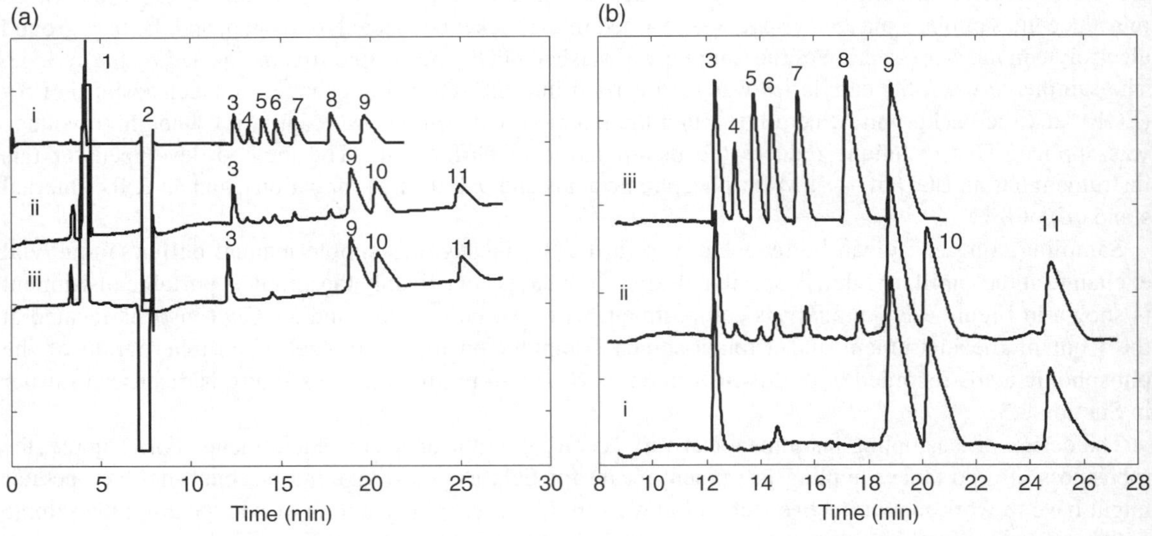

Figure 9.3 An example of on-site analysis of phosphonic acids using a portable CE instrument. Procedure: an aqueous solution of 75 μM phosphonic acids and internal standards was sprayed over a 10 cm^2 area of sand. After 24 h the 5 mg of the polluted sand was collected and extracted with MilliQ water; (a) Raw elecrtopherograms; (b) Corrected elelctropherograms. (i) blank, (ii) electropherogram of the extract, (iii) electropherogram of the standards. Peaks: 1 – cations, 2 – EOF, 3 – sodium decanoate (internal standard), 4 – pinacolylmethylphosphonic acid, 5 – 1-butylphosphonic acid, 6 – propylphosphonic acid, 7 – ethylphosphonic acid, 8 – methylphosphonic acid, 9 – lactic acid (internal standard), 10 – salicylic acid, 11- unknown. (From the authors' laboratory).

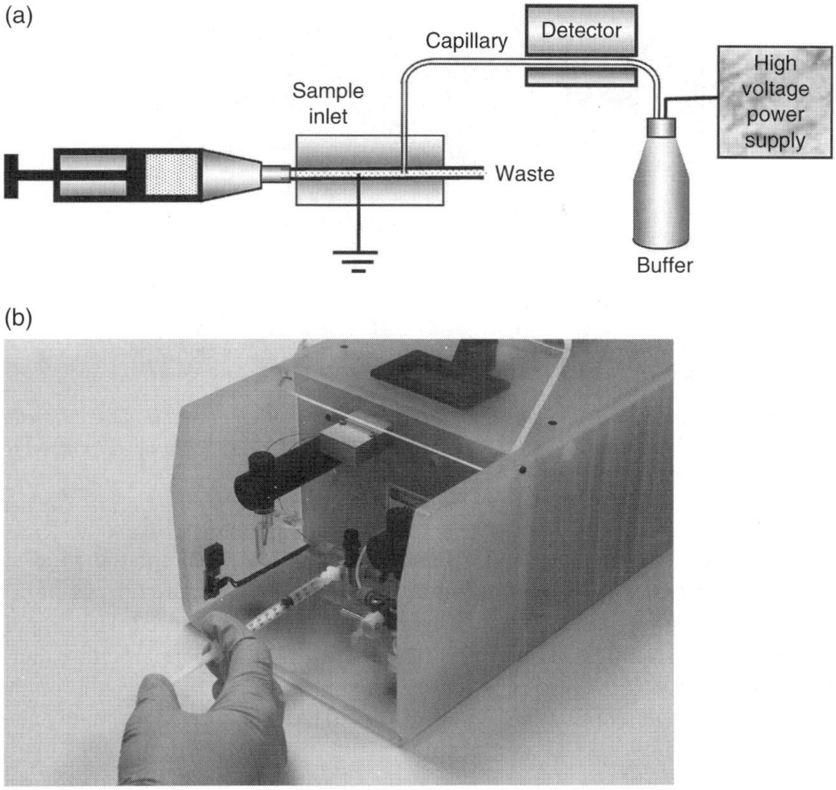

Figure 9.4 *Portable CE field sampling method using a syringe. (a) Schematic; (b) Practical implementation. (Photo from the authors' laboratory).*

9.2.4 World-to-chip interfacing and the quest for a 'killer' application for LOC devices

Continuous-flow microfluidics deals with the precise control and manipulation of fluids that are geometrically constrained to minute (typically sub-millimetre) channels etched into a glass or some polymeric substrate. 'Micro' in this context signifies small volumes (nl, pl, fl), small size, low energy consumption, and other effects of a micro scale on fluids. It appears from this description that LOC devices satisfactorily fulfil the criteria for a green analytical method, and their miniature size and consumption of very small amounts of solvents enables portability and field operation. However, the portability of LOC devices is not that easy to achieve.

The connection between the components of microfluidic devices and the macro-environment is called the macro-to-micro interface [37], interconnect [38–41], or world-to-chip interface [42–46]. It is this need to connect with the macro-environment that jeopardizes the greenness of microfluidic devices. The difficulty stems from the fact that samples and reagents are typically transferred in quantities ranging from microlitres to millilitres (or even litres), but microfluidic devices can accommodate only nanolitres or picolitres of samples/reagents due to the size of their reaction chambers and channels, the dimensions of which are commonly on the order of microns. This challenge has been overlooked in research environments such as academic laboratories, but it erodes a founding pillar of green chemistry: minimal use of organic solvents. The world-to-chip problem cannot be ignored in routine applications. A good illustration is provided by the photograph in Figure 9.5 selected at random by a Google search on the keyword 'microfluidics': a tiny

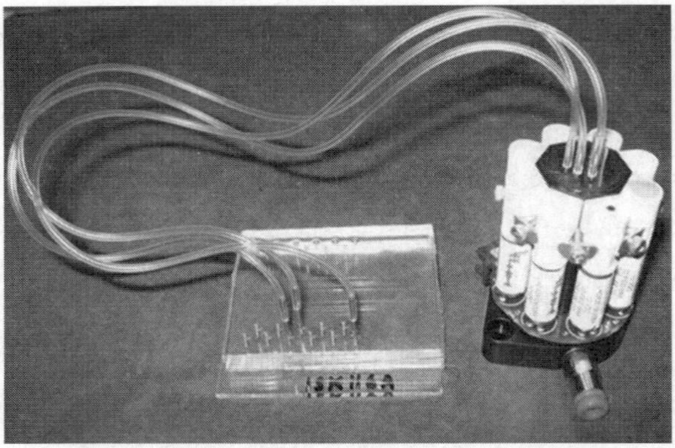

Figure 9.5 *Connecting microfluidic devices to the world: an image of an actual device. Microfluidic device for combinatorial chemistry integrated to an eight-valve manifold through tygon tubing. Reprinted from [47] with kind permission from Springer Science + Business Media. © 2009.*

microchip is surrounded by bulky valves and a bundle of tubing needed for the fluid supply (pumps), actuation (by gas) and detection [47]. One can find examples where there are dozens of connecting tubes, indicating the number of actuating (perhaps syringe-type) pumps that are required. The supporting equipment is usually not shown in illustrations of microfluidic devices. Although microchips can perform many sophisticated operations, the analytical process could only be considered truly microminiaturized if the instrumentation supporting the unit operations in the chip could be miniaturized by many orders of magnitude as well.

The problem of the so-called 'world-to-chip' interface has plagued microfluidics since its inception. Microfluidics must integrate all components of the system on the same chip to ensure the portability and minimum energy consumption required by the principles of Green Analytical Chemistry. Pumps, valves, mixers, and so on must be miniaturized in order to achieve an integrated system, which forces a choice between active methods – efficient, but requiring energy and difficult to miniaturize – and the passive methods provided by non-instrumental microfluidics – easier to integrate, but less efficient. This is a huge challenge, especially in biotechnology, where the volume of the targets of study located in the macroscopic environment must be reduced to accommodate the microscopic environment of the microfluidic device. Finally, the huge surface/volume ratio of miniaturized systems could modify the physical behaviour of the system and create new problems, such as target molecules adhering to the solid walls, or the fluid being prevented from entering the microchannels due to the effect of capillary forces. According to some researchers, the dilemma of the world-to-chip interface is one of the bottle-necks in the development of μTAS [48]. Solving this problem is critical for high-throughput applications where manual manipulation is not economical and a macro-to-micro interface must be developed.

Nevertheless, the main thrust of portable CE systems development is towards chip-based miniaturized CE. However, capillary-based (non-chip) portable CE systems have certain unmatched advantages as recently described by Ryvolová *et al.* [49]. These include the relatively simple cylindrical geometry of the CE capillary, maximum volume-to-surface ratio, no requirement to design and fabricate a chip, the low costs of capillary compared to chip, and better performance with some detection techniques. On the other hand, microfluidic chips – the key components of LOC devices – have frequently been designed for very specific applications. They are relatively expensive and unique. Compared to portable field analysers based on 'classical' CE, in which a capillary can easily be discarded if a problem occurs, custom-made microfluidic chips are definitely not yet disposal products. This condition restricts their use in field applications.

Academic institutions and corporations have developed a plethora of lab-on-a-chip devices for different applications; however, the overwhelming need is for a universal application that could trigger widespread use of microchips in biomedicine. There is a general consensus that miniaturization should have advanced more quickly than it has. A few years ago, many analytical chemists were enthusiastic about chip-based analyses, but one can now sense disappointment and disillusionment among the members of that community. If there is no widely used application, the development of microfluidic devices will not be economically profitable. This phenomenon is known as the absence of a 'killer' application. The term

> *... is commonly used to describe a product which has such highly desirable properties that it generates very large revenues with attractive margins in a comparatively short amount of time. In addition to this purely economic description it also helps to promote the underlying technology, thus helping typically 'disruptive technologies' (a technology that enables products which dramatically change markets due to their (often unexpected) performance and which are not achievable by simple linear extrapolation of existing products or technologies'* [50]

Examples of killer applications are digital photography and flat panel television screens. Let us hope that a killer application will soon be found, that the 'world-to-chip' problem will be solved as quickly as possible, and that obstacles to developing the ultimate method of green analysis will be surmounted. Interestingly, the development of LOC systems can be described by a 'hype cycle' model introduced in 1995. According to Mukhopadhyay [51], the model has five stages: a technology trigger, a peak of inflated expectations, a trough of disillusionment, a slope of enlightenment, and finally, a plateau of productivity. He believes that microfluidics is now on the slope of enlightenment. Many microfluidics proofs-of-concepts have been advanced, but the gap between academic proofs-of-concept and the need of industry is wide. Only 1% of proofs-of-concept become commercialized [51] and most innovative technologies take longer than anticipated to develop large markets.

9.2.5 Gradient elution moving boundary electrophoresis and electrophoretic exclusion

Despite the stagnation of the development of CE instrumentation, some exciting proposals are still coming forward. A mode of CE called gradient elution moving boundary electrophoresis (GEMBE) is a further step in the direction of making CE technology portable and minimizing sample preparation. GEMBE was developed by Ross *et al.* of the US NIST Biochemical Science Division [52–54]. GEMBE operates by pumping a buffer solution under controlled pressure into the sample vessel in the opposite direction of the rapidly moving analytes. The opposing buffer flow acts as a fluid gate between the sample reservoir and the capillary. The pressure is gradually reduced during the analysis runtime and a specific analyte is detected when the flow becomes sufficiently weak that the electrophoretic migration of the analyte pushes it against the pressure flow into the capillary. GEMBE can analyse samples in complex matrices, such as whole milk, blood serum and dirt in solution. Usually, samples in complex matrices are difficult to analyse because the heavy components (e.g. fat globules in milk, proteins in blood or particulates in environmental samples) can foul the separation capillary. In the case of GEMBE, this does not happen because the pressure flow keeps out the unwanted material during analysis and the various analytes enter the channel at different times based on their particular electrophoretic motion.

The analytes are detected by contactless conductivity. The detector signal is in the form of steps, because the conductivity of the capillary at the detection point changes when a new analyte band enters the detector window. Those steps can be differentiated to create peaks on an electropherogram. However, detection can be simplified even further. Because the separation channel is so short, only one moving boundary step is present in the channel at any given time, and the current through the channel can be measured and used to send a signal comparable to that which is normally generated by more complicated means of detection (see Figure 9.6).

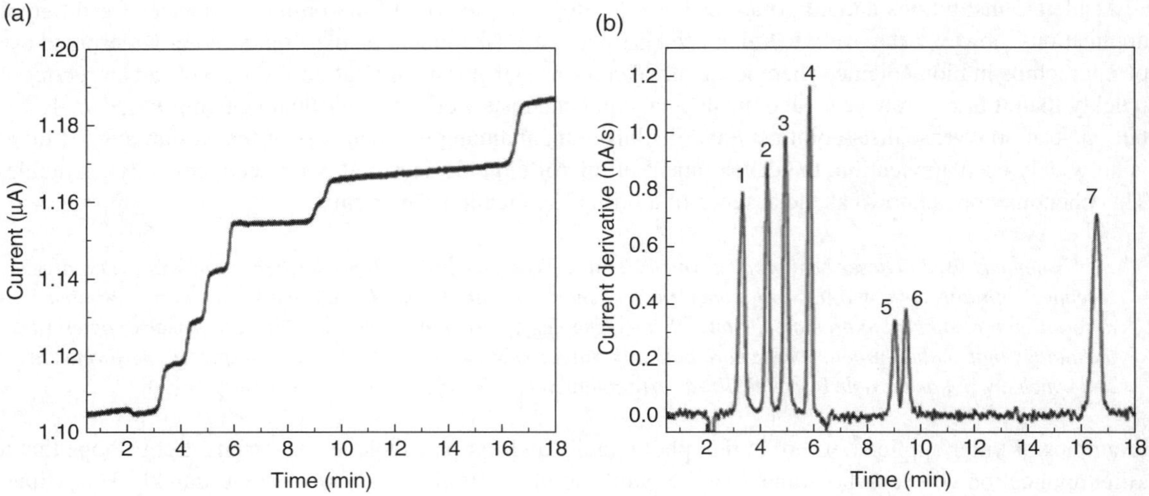

Figure 9.6 Example of GEMBE separation with channel current detection. Sample: 200 mmol l^{-1} each of tartaric acid (1), malic acid (2), citric acid (3), succinic acid (4), acetic acid (5), lactic acid (6), and ascorbic acid (7). Separation conditions: 5 mm i.d., 2.5 mm length capillary, 8000 V cm^{-1}; run buffer: 20 mmol l^{-1} His, 100 mmol l^{-1} MES, pH 5.5. (a) Plot of electrophoresis current over time. Each step corresponds to one of the analytes in the sample. (b) Plot of current derivative over time. Reprinted with permission from [53]. © 2009 American Chemical Society.

GEMBE is arguably the simplest method of electrophoretic separation and detection. Its detection limits are typically in the range of a few micromoles per litre, which is comparable to the detection limits for CE with UV absorbance, or conductivity detection in capillary or microchip electrophoresis. It is adequate for a large variety of chemical and biochemical analyses.

The GEMBE technique has been validated by testing it with solutions of whole milk, dirt, estuarine sediment, coal fly ash, pulverized leaves and blood serum. In all cases, without any sample preparation, it performed a reproducible separation and quantification of specific analytes (including potassium, calcium, sodium, magnesium, lithium and melamine). Researchers are optimistic about the prospects of the technique: 'GEMBE shows promise for many applications (e.g. monitoring contaminants in food or water supplies, determining nutrient levels in soil, detecting biochemical warfare agents, and diagnosing medical conditions)' [55]. They are working to miniaturize the equipment and integrate all the components to develop a 'lab-on-a-chip' field analyser. However, a requirement for the controlled and precise regulation of pressure complicates an otherwise very promising technique.

One way to overcome this disadvantage is simply by using gravity instead of pressure to generate the hydrodynamic flow as described by Meighan *et al.* [56]. The authors call this method 'electrophoretic exclusion'. In this technique, hydrodynamic flow counters electrophoretic velocity not by pressure but by the differences between the input and output vials of the separation capillary. Hydrodynamic flow prevents a species from entering the capillary. The method accumulates analytes in the bulk solution of the sample vial rather than within the capillary. When the electrophoretic velocity of a species is greater than the hydrodynamic flow into the capillary, the species is excluded from the channel. It is arrested at the entrance of the capillary, whereas the lighter species is relatively unaffected by the applied force. Once the potential is removed, the collected species are released and detected (in this research, by absorbance). The separation can be controlled by changing the flow rate or applied electric field in order to selectively exclude certain species while allowing others to pass through the capillary. In this experiment, the vials and capillaries were mounted on a rotatable board so that the flow rate could be controlled. The distance between the menisci in the sample and buffer vials was 1.5 cm, and the calculated flow rate was 0.88 nl s^{-1}. The concept was proven using the cationic dye, Methyl Violet and the neutral dye, Martius Yellow.

The potential of separation methods that counter hydrodynamic flow with electrophoretic migration seems very promising for field analysis because it greatly simplifies the instrumentation and increases its overall robustness. However, the type of samples that can be determined by this technique might be limited.

9.3 Possible ways of surmounting the disadvantages of CE

It has been mentioned above that some reluctance on the part of analysts to use CE is due to poor reproducibility of peak migration times and areas. This is illustrated in Figure 9.3. In this example, an electropherogram of a standard of phosphonic acids solution (Figure 9.3, a, i) is provided together with an electropherogram of an extract of the same solution applied to sand (Figure 9.3, a, ii). It is evident that there is some resemblance between these two electropherograms but the peaks of the standards can hardly be recognized among the matrix peaks. The identification and quantification of pollutants is almost impossible. The reason is hardly surprizing: the electropherograms were recorded by different analysts, using a portable custom-made instrument without column temperature stabilization. Sampling was performed manually by means of a syringe that forced a fixed volume of a sample to flow past the separation capillary inlet (see Figure 9.4). Obviously, the results are not encouraging. Simplifying construction of the POC instrument to make it more robust (by sacrificing a thermostat and making the sampling procedure more suitable for field operations) decreases performance. Ways to improve it must be sought.

One way to improve the performance of portable CE is to look for more sophisticated and systematic approaches to data processing that take advantage of the algorithms that modern software can provide. This might appear to be making up data, but in fact it is not. By knowing the model of the signal, reliable corrections can be made to the raw data using mathematical calculations. For example, base line correction is included in most chromatographic software. Much more elaborate procedures can be developed. Moreover, using sophisticated software to improve an analytical method without changing the instrumentation is obviously a thoroughly green approach to analysis. This idea has existed since the first attempts to mathematically separate chromatographic peaks were made in the 1970s. But it has not achieved the type of success needed to impress a chromatographer who prefers to develop better columns. The appearance and widespread use of diode array and mass-spectrometric detectors can significantly improve the prospects of using software to improve analytical signals. Whether it is too optimistic to speak of 'mathematical' chromatography [57] is difficult to judge at the moment – a column is still necessary – but it does appear that increasing the level of sophistication of data processing enables the complexity of an instrument to be reduced, in clear conformity with the goals of Green Analytical Chemistry.

Using software to improve the reproducibility of migration times has been of recent interest to several researchers [58] who have studied the use of CE in metabolomics. The nonreproducibility of migration times threatens the application of chemometrics to the sets (e.g. of samples of body fluids). Various algorithms have been proposed [59]. The characteristic feature of these algorithms is their capability to process a large number of electropherograms without human intervention. These elaborate algorithms require a representative reference electropherogram and the rest of the set of recorded chromatograms must be very similar to the reference electropherogram. This approach is not promising for field analyses where the composition of the samples varies. One approach in this case is to mix the sample solution with known components of which the migration time is outside the window of those of the analytes of interest and can therefore serve as anchors. Assuming that the axis of deformation of the migration time is linear during the analysis process, the axis can be corrected by software through simple linear interpolation. Contrary to metabolomic studies, the number of samples in field analysis is relatively small. Here, the analyst can take advantage of human superiority over the computer to recognize patterns, and can directly intervene in the analysis process by identifying common points on both the reference and running electropherograms. This approach is illustrated in Figure 9.3(b). Adding compounds #3 (sodium decanoate) and #9 (lactic acid) to the standards and sand extract solutions makes it possible to correct the migration time axis of the raw electropherogram and facilitate identification. This example uses an important class of analytes – phosphonic acids – which are of great interest as degradation products of certain chemical weapons. In this case, the useful anchor compounds could be found relatively easily. Considerable research and experimental design might be needed to develop proper methods of field analysis for other compounds of interest (e.g. toxic industrial chemicals).

Correct and reproducible quantization is an even more complicated issue for field analysis using portable CE instruments. Sampling by wiping or by extracting affects reproducibility. Reproducibility is also influenced by the sampling process, because different operators cannot manipulate the syringes and vials in exactly the same way. The author's experience indicates that peak area reproducibility can vary about 20%, which is definitely not acceptable. The anchor compounds described above can be used to improve reproducibility. The author's have found that using anchor compounds as internal standards can reduce the standard deviation of the peak area measurements approximately five times.

The shortcomings of CE that result from the robustness needed for field analysis can therefore be partially overcome by the systematic sophistication of the analysis process. This approach does require sacrificing one feature of a portable instrument: operational simplicity. However, kits can be prepared for use beforehand, so the process of analysis in the field still remains straightforward for the technician who is not a chemist.

CE has much higher detection and quantification limits than ordinary HPLC due to the smaller sample capacity of the capillary columns. This adds one more element of sophistication to the development of field CE instruments. If regulations require that the concentration of some pollutant in the environment be below the detection limit of the portable CE instrument, then various methods of sample concentration (perhaps solid phase extraction) would be necessary. Fortunately, solid phase extraction is a well-developed method of sample preparation. The author does not foresee any major obstacles to its application in field analysis using portable CE instruments.

9.4 Sample preparation in CE

A microscale separation method will be greener than its larger-scale counterpart, but appropriate sample preparation procedures must still be considered. How green is an analytical method if sample preparation eliminates the desired outcome? Green approaches to this issue have been largely ignored until recently. Xie and He researched a group of green sample-pretreatment techniques known as liquid-phase microextraction (LPME), which is used as a method of sample preparation mainly in connection with gas and liquid chromatography, but also for CE [9]. Single-drop microextraction (SDME) [60,61] is a popular LPME technique. A droplet of several μL is suspended at the tip of a microsyringe, which serves as both the solvent holder and sample injector for performing the extraction and extract injection procedures. The extract can then be analysed off-line by CE. Various possibilities for microextraction in CE have been described [62].

Most of the methods of green sample preparation based on LPME and used in CE were developed for gas and liquid chromatography and do not take advantage of the specific features of a CE apparatus. Those methods are effectively described in a special issue of *Analytica Chimica Acta* [63] and in other chapters of this book, and we refer the reader to those descriptions. However, SDME, an LPME technique adapted for CE, deserves a more thorough description. It takes full advantage of the CE column format: the droplet is suspended on the top of the capillary instead of at the tip of the syringe needle. This ingenious approach developed by Choi *et al.* [64,65]. According to that approach, SDME consists of four steps (Figure 9.7):

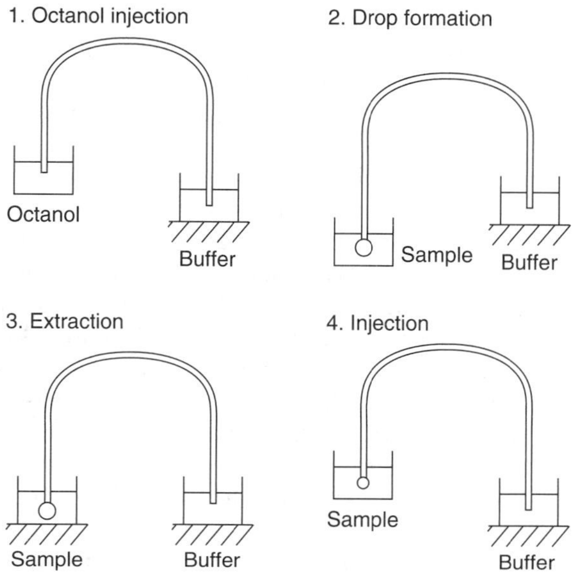

Figure 9.7 *Procedures for 3-phase SDME-CE. (1) Injection of the organic phase, (2) two-phase drop formation by ejection of the organic phase, (3) extraction of the sample from the donor phase into the acceptor phase, and (4) injection of the enriched acceptor phase into the capillary. Reprinted with permission from [64]. © 2004 American Chemical Society.*

(1) pressure injection of the organic phase (octanol) into the capillary,
(2) formation of a two-layer drop in the donor phase using backpressure in the outlet vial of the capillary to eject the organic phase; the organic phase forms a thin layer over the CE buffer droplet,
(3) extraction of the sample in the donor phase into the acceptor phase, and
(4) injection of the enriched acceptor phase into the capillary.

In the final step, a portion of the enriched drop is injected hydrodynamically. The remainder of the drop is automatically detached when the capillary is removed from the donor phase vial and placed in a run buffer vial.

If the pH of the donor phase is low, the analytes are mainly protonated and penetrate the aqueous phase of the acceptor through the organic layer. If the pH is high, the analytes deprotonate and remain trapped in the droplet. Enrichment factors of thousands can be achieved by this process. By combining this method of sampling with stacking, the enrichment can be even higher.

Suspending the droplet at the capillary inlet is somewhat tricky. In order to stabilize the position of the droplet, the capillary inlet must be hydrophobized by proper chemical treatment, or, in simple cases, Teflon sleeves must be used.

9.5 Is capillary electrophoresis a green alternative?

We have attempted to prove in this chapter that CE meets all the requirements of a green analytical technique and should be seriously considered as an alternative to the more conventional HPLC.

As we saw above, both of classical CE's competitors in the field of liquid phase separations – HPLC and LOC devices – have serious obstacles to miniaturization and portability, frequently due to the simple fact that pressure is required for solvent actuation. This could also be a concern with GEMBE. However, the differences between the three techniques diminish if actuation is effected by an electric field. It appears from this discussion that CE is not immediately about to become the main method of separation. The leap from academia to national and industrial laboratories is a challenge for analysts and provides fertile ground for research[3].

In principle, a completely 'green' world would be free from any human influence on the environment – an absurd concept. Moreover, a lesson from thermodynamics is that systems that produce more entropy and, consequently, more environmental pollution, are likelier to survive [66]. This makes protection of the environment objectively difficult. On the other hand, the prospect of an ecological catastrophe has elevated concerns that current values must be replaced with radically new ones that are not merely based on the Darwinian struggle for survival. Aiming for sustainability is arguably a better strategy for human behaviour and for Green Analytical Chemistry as well. The author of this chapter understands sustainability to be the ability to survive in a competitive environment with limited resources. Of course, it is strange to contemplate that analytical chemistry could be the bottleneck to the ultimate survival of the human species. However, the choice of a green method plays an essential role in every analytical laboratory. The purpose of analytical chemistry is clear: to produce specific information about the identity and amount of a particular substance. As a branch of information science, analytical chemistry does not need to consume many resources at all. The most sustainable analytical chemistry is one in which the required results are obtained less expensively, while simultaneously complying with the regulations and laws.

There is an unfortunate issue that Green Analytical Chemistry research must confront. Corporate policy that is only directed towards the self-interested exploitation of the consumer market and environmental resources is not overly supportive of the green worldview. Companies that produce analytical instrumentation are no exceptions. A willingness to respond to new discoveries in analytical science with new instruments must be acknowledged; however, a trend towards lowering the level of service for existing analytical instruments, or even a reluctance to perform such service, has become noticeable in recent years. Competition motivates instrument manufacturers to sell more and more new instruments, and providing service is not profitable. This trend can be widely observed in everyday life. The author's position might be criticized as naïve with regard to the requirements of economic development, but it is clear that the advancement of Green Analytical Chemistry will remain an obscure curiosity on the part of academics, and the effects of their research will have negligible impact, if the attitude of suppliers of analytical instrumentation does not change.

References

1. Schmitt-Kopplin, P. (ed.) (2008) *Capillary Electrophoresis: Methods and Protocols*, Humana Press, Totowa, NJ.
2. Landers, J.P. (ed.) (2008) *Handbook of Capillary and Microchip Electrophoresis and Associated Microtechniques*, 3rd edn, CRC Press, Boca Raton, FL.
3. Hjertén, S. (1967) Free zone electrophoresis, *Chromatogr. Rev.*, **9**, 122–219.
4. Virtanen, R. (1974) Zone electrophoresis in a narrow-bore tube employing potentiometric detection, *Acta Polytech. Scand. Chem.*, **123**, 1–67.
5. Everaerts, F.M. and Hoving-Keulemans, W.M.L. (1970) Zone electrophoresis in capillary tubes, *Sci. Tools.*, **17**, 25–28.
6. Jorgenson, J.W. and Lukacs, K.D. (1981) Zone electrophoresis in open-tubular glass capillaries, *Anal. Chem.*, **53**, 1298–1302.

[3] Ironically, such research need not be green at all: due to the small amount of academic research in separation science as compared to the rest of the analytical chemical world, even non-green and expensive methods could be used to prove the greenness of CE.

7. Mischak, H.; Coon, J.J.; Novak, J.; Weissinger, E.M.; Schanstra, J.P. and Dominiczak, A.F. (2009) Capillary electrophoresis–mass spectrometry as a powerful tool in biomarker discovery and clinical diagnosis: An update of recent developments, *Mass Spectrom. Rev.*, **28**, 703–724.
8. Koel, M. and Kaljurand, M. (2006) Application of the principles of Green Chemistry in Analytical Chemistry, *Pure Appl. Chem.*, **78**, 1993–2002.
9. Xie, H.Y. and He, Y.Z. (2010) Green analytical methodologies combining liquid-phase microextraction with capillary electrophoresis, *TrAC-Trend. Anal. Chem.*, **29**, 629–635.
10. Stege, P.W.; Messina, G.A.; Bianchi, G. and Olsina, R.A. (2010) Determination of the β-Glucosidase Activity in Different Soils by Pre Capillary Enzyme Assay Using Capillary Electrophoresis with Laser-Induced Fluorescence Detection, *J. Fluorescence*, **20**, 517–523.
11. Pena-Pereira, F.; Lavilla, I. and Bendicho, C. (2010) Liquid-phase microextraction techniques within the framework of Green Chemistry, *TrAC-Trend. Anal. Chem.*, **29**, 617–628.
12. Tullo, A. (2008) A Solvent Dries Up, *Chem. Eng. News*, **86**, 27.
13. Altria, K.D.; Ciccone, W. and Nunnally, B. (2009) *Capillary Electrophoresis, LC-GCTechnology Forum*. Available at http://chromatographyonline.findanalytichem.com/lcgc/Technology-Forum-Capillary-Electrophoresis/ArticleStandard/Article/detail/618413 (accessed 7 November, 2011).
14. Noga, M.; Sucharski, F.; Suder, P. and Silberring, J. (2007) A practical guide to nano-LC troubleshooting, *J. Sep. Sci.*, **30**, 2179–2189.
15. Henion, J. (2009) The reality of lab-on-a-chip technology for the mass spectrometry laboratory, *LC GC N. Am.*, **27**, 1 Oct.
16. Eksigent Technologies (2005) *Microfluidic Flow Control, Technical Note*. Available at http://www.eksigent.com/hplc/tech/mfc.php (accessed 7 November, 2011).
17. Chen, L.; Lee, S.; Choo, J. and Lee, E.K. (2008) Continuous dynamic flow micropumps for microfluid manipulation, *J. Micromech. Microeng.*, **18**, Article Number 013001.
18. Wang, X.; Wang, S.; Gendhar, B. and Cheng, C. (2009) Electroosmotic pumps for microflow analysis, *TrAC-Trend. Anal. Chem.*, **28**, 64–74.
19. Hartonen, K. and Riekkola, M. (2008) Liquid chromatography at elevated temperatures with pure water as the mobile phase, *TrAC-Trend. Anal. Chem.*, **27**, 1–14.
20. Koel, M. and Kaljurand, M. (2010) *Green Analytical Chemistry*, Royal Society of Chemistry, Cambridge.
21. Welch, C.J.; Wu, N.; Biba, M.; Hartman, R.; Brkovic, T.; Gong, X.; Helmy, R.; Schafer, W.; Cuff, J.; Pirzada, Z. and Zhou, L. (2010) Greening analytical chromatography, *TrAC-Trend. Anal. Chem.*, **29**, 667–680.
22. Van der Vorst, G.; Van Langenhove, H.; DePape, F.; Aelterman, W.; Dingenen, J. and Dewulf, J. (2009) Exergetic life cycle analysis for the selection of chromatographic separation processes in the pharmaceutical industry: preparative HPLC versus preparative SFC, *Green Chem.*, **11**, 1007–1012.
23. Maxwell, E.J. and Chen, D.D.Y. (2008) Twenty years of interface development for capillary electrophoresis–electrospray ionization–mass spectrometry, *Anal. Chim. Acta*, **627** 25–33.
24. He, Y.; Tang, L.; Wu, X. and Hou, X. (2007) Spectroscopy: the best way toward Green Analytical Chemistry?, *Appl. Spectrosc. Rev.*, **42**, 119–138.
25. Turl, D.E.P. and Wood, D.R.W. (2008) A step out of the lab and into the field: a giant leap from technology to capability, *Analyst*, **133**, 558–562.
26. Contreras, J.A.; Murray, J.A.; Tolley, S.E.; Oliphant, J.L.; Tolley, H.D.; Lammert, S.A.; Lee, E.D.; Later, D.W. and Lee, M.L. (2008) Hand-portable gas chromatograph-toroidal ion trap mass spectrometer (GC-TMS) for detection of hazardous compounds, *J. American Soc. Mass Spect.*, **19**, 1425–1434.
27. Lin, H.Q.; Ye, Q.; Deng, C.H. and Zhang, X.M. (2008) Field analysis of acetaldehyde in mainstream tobacco smoke using solid-phase microextraction and a portable gas chromatograph, *J. Chromatogr. A*, **1198**, 34–37.
28. Nelson, M.A.; Gates, A.; Dodlinger, M. and Hage, D.S. (2004) Development of a portable immunoextraction-reversed-phase liquid chromatography system for field studies of herbicide residues, *Anal. Chem.*, **76**, 805–813.
29. da Silva, J.A.F. and do Lago, C.L. (1998) An oscillometric detector for capillary electrophoresis, *Anal. Chem.*, **70**, 4339–4343.
30. Zemann, A.J;. Schnell, E.; Volgger, D. and Bonn, G.K. (1998) Contactless conductivity detection for capillary electrophoresis, *Anal. Chem.*, **70**, 563–567.

31. Kuban, P.; Nguyen, H.T.A.; Macka, M.; Haddad, P.R. and Hauser, P.C. (2007) New fully portable instrument for the versatile determination of cations and anions by capillary electrophoresis with contactless conductivity detection, *Electroanalysis*, **19**, 2059–2065.
32. Hutchinson, J.P.; Evenhuis, C.J.; Johns, C.; Kazarian, A.A.; Breadmore, M.C.; Macka, M.; Hilder, E.F.; Guijt, R.M.; Dicinoski, G.W. and Haddad, P.R. (2007) Identification of inorganic improvised explosive devices by analysis of postblast residues using portable capillary electrophoresis instrumentation and indirect photometric detection with a light-emitting diode, *Anal. Chem.*, **79**, 7005–7013.
33. Hutchinson, J.P.; Johns, C.; Breadmore, M.C.; Hilder, E.F.; Guijt, R.M.; Lennard, C.; Dicinoski, G. and Haddad, P.R. (2008) Identification of inorganic ions in post-blast explosive residues using portable CE instrumentation and capacitively coupled contactless conductivity detection, *Electrophoresis*, **29**, 4593–4602.
34. Seiman, A.; Jaanus, M.; Vaher, M. and Kaljurand, M. (2009) A portable capillary electropherograph equipped with a cross-sampler and a contactless-conductivity detector for the detection of the degradation products of chemical warfare agents in soil extracts, *Electrophoresis*, **30**, 507–514.
35. Kaigala, G.V.; Behnam, M.; Bliss, C.; Khorasani, M.; Ho, S.; McMullin, J.N.; Elliott, D.G. and Backhouse, C.J. (2009) Inexpensive, universal serial bus-powered and fully portable lab-on-a-chip-based capillary electrophoresis instrument, *IET Nanobiotechnol.*, **3**, 1–7.
36. Kaigala, G.V.; Hoang, V.N.; Stickel, A.; Lauzon, J.; Manage, D.; Pilarski, L.M. and Backhouse, C.J. (2008) An inexpensive and portable microchip-based platform for integrated RT-PCR and capillary electrophoresis, *Analyst*, **133**, 331–338.
37. Jesson, G.; Kylberg, G. and Andersson, P. (2003) A versatile micro-to-macro dispensing system, in *Micro Total Analysis Systems* (eds. M.A. Nothrup, K.F. Jensen, D.J. Harrison), *Transduders Research Foundation*, San Diego, CA, USA, pp. 155–158.
38. Nittis, V.; Fortt, R.; Legge, C.H. and de Mello, A.J. (2001) A high-pressure interconnect for chemical microsystem applications, *Lab. Chip*, **1**, 148–152.
39. Puntambekar, A. and Ahn, C.H. (2002) Self-aligning microfluidic interconnects for glass- and plastic-based microfluidic systems, *J. Micromech. Microeng*., **12**, Article number 35.
40. Gonzalez, C.; Collins, S.D. and Smith, R.L. (1998) Fluidic interconnects for modular assembly of chemical microsystems, *Sens. Actuators, B*, **49**, 40–45.
41. Chen, H.; Acharya, D.; Gajraj, A. and Melners, J.C. (2003) Robust interconnects and packaging for microfluidic elastomeric chips, *Anal. Chem.*, **75**, 5287–5291.
42. Ramsey, J.M. (1999) The burgeoning power of the shrinking laboratory, *Nat. Biotechnol.*, **17**, 1061–1062.
43. Attiya, S.; Jemere, A.; Tang, T.; Fitzpatrick, G.; Seiler, K.; Chiem, N. (2001) and Harrison, D.J. Design of an interface to allow microfluidic electrophoresis chips to drink from the fire hose of the external environment, *Electrophoresis*, **22**, 318–327.
44. Bings, N.H.; Wang, C.; Skinner, C.D.; Colyer, C.L.; Thibault, P. and Harrison, D.J. (1999) Microfluidic devices connected to fused-silica capillaries with minimal dead volume, *Anal. Chem.*, **71**, 3292–3296.
45. Yang, Z. and Maeda, R. (2002) A world-to-chip socket for microfluidic prototype development, *Electrophoresis*, **23**, 3474–3478.
46. Liu, J.; Hansen, C. and Quake, S.R. (2003) Solving the 'world-to-chip' interface problem with a microfluidic matrix, *Anal. Chem.*, **75**, 4718–4723.
47. Kondapalli, S. and Kirby, B.J. (2009) Refolding of β-galactosidase: microfluidic device for reagent metering and mixing and quantification of refolding yield, *Microfluid. Nanofluid.*, **7**, 275–281.
48. Zhong, F.; Chen, Y.; Marcus, J.S.; Scherer, A.L.; Quake, S.R.; Taylor, C.R. and Weiner, L.P. (2008) A microfluidic processor for gene expression profiling of single human embryonic stem cells, *Lab. Chip*, **8**, 68–74.
49. Ryvolova, M.; Preisler, J.; Brabazon, D. and Macka, M. (2010) Portable capillary-based (non-chip) capillary electrophoresis, *TrAC-Trend. Anal. Chem.*, **29**, 339–353.
50. Becker, H. (2009) Hype, hope and hubris: the quest for the killer application in microfluidics, *Lab. Chip*, **9**, 2119–2122.
51. Mukhopadhyay, R. (2009) Microfluidics: On the Slope of Enlightenment, *Anal. Chem.*, **81**, 4169–4173.
52. Shackman, J.G.; Munson, M.S. and Ross, D. (2007) Gradient elution moving boundary electrophoresis for high-throughput multiplexed microfluidic devices, *Anal. Chem.*, **79**, 565–571.

53. Ross, D. and Romantseva, E.F. (2009) Gradient elution moving boundary electrophoresis with channel current detection, *Anal. Chem.*, **81**, 7326–7335.
54. Strychalski, E.A.; Henry, A.C. and Ross, D. (2009) Microfluidic analysis of complex samples with minimal sample preparation using gradient elution moving boundary electrophoresis, *Anal. Chem.*, **81**, 10201–10207.
55. Anonymous (2010) GEMBE gets results from dirty samples, *TrAC-Trend. Anal. Chem.*, **29**, IV-V.
56. Meighan, M.M.; Keebaugh, M.W.; Quihuis, A.M.; Kenyon, S.M. and Hayes, M.A. (2009) Electrophoretic exclusion for the selective transport of small molecules, *Electrophoresis*, **30**, 3786–3792.
57. Bro, R.; Viereck, N.; Toft, M.; Toft, H.; Hansen, P.I. and Engelsen, S.B. (2010) Mathematical chromatography solves the cocktail party effect in mixtures using 2D spectra and PARAFAC, *TrAC-Trend. Anal. Chem.*, **29**, 281–284.
58. Tomasi, G.; van den Berg, F. and Andersson, C. (2004) Correlation optimized warping and dynamic time warping as preprocessing methods for chromatographic data, *J. Chemometrics*, **18**, 231–241.
59. Nielsen, N.P.V.; Carstensen, J.M. and Smedsgaard, J. (1998) Aligning of single and multiple wavelength chromatographic profiles for chemometric data analysis using correlation optimised warping, *J. Chromatogr. A*, **805**, 17–35.
60. Jeannot, M.A. and Cantwell, F.F. (1996) Solvent microextraction into a single drop, *Anal. Chem.*, **68**, 2236–2240.
61. He, Y. and Lee, H.K. (1997) Liquid phase microextraction in a single drop of organic solvent by using a conventional microsyringe, *Anal. Chem.*, **69**, 4634–4640.
62. Nozal, L.; Arce, L.; Simonet, B.M.; Rios, A. and Valcarcel, M. (2007) In-line liquid-phase microextraction for selective enrichment and direct electrophoretic analysis of acidic drugs, *Electrophoresis*, **28**, 3284–3289.
63. de la Guardia, M. (ed.) (2010) Green Analytical Chemistry (special issue), *TrAC-Trend. Anal. Chem.*, **29**, 577–780.
64. Choi, K.; Kim, Y. and Chung, D.S. (2004) Liquid-phase microextraction as an on-line preconcentration method in capillary electrophoresis, *Anal. Chem.*, **76**, 855–858.
65. Choi, K.; Kim, S.J.; Jin, Y.G.; Jang, Y.O.; Kim, J.S. and Chung, D.S. (2009) Single drop microextraction using commercial capillary electrophoresis instruments, *Anal. Chem.*, **81**, 225–230.
66. Rebane, K.K. (1995) Energy, entropy, environment - why is protection of the environment objectively difficult, *Ecol. Econ.*, **13**, 89–92.

10

Green Chromatography

Chi-Yu Lu

Department of Biochemistry, Kaohsiung Medical University, Kaohsiung, Taiwan

Chromatography has been widely used in several fields for a long time. One kind of technique, liquid chromatography, is a powerful tool for sample separation and purification but organic waste is dramatically produced at the same time. Green chemistry consists of chemicals and chemical processes designed to minimize the use and generation of hazardous substances [1]. It is important to respect Green Analytical Chemistry when green chemistry is developed. Green liquid chromatography is one part of Green Analytical Chemistry. The major goal of green liquid chromatography is to reduce the production of organic waste in order to decrease these harmful chemicals and protect our environment.

10.1 Greening liquid chromatography

Greening liquid chromatography (LC) has two main purposes: to reduce the use of organic solvents and minimize the production of organic waste. How can the production of organic waste under reasonable analytical time be minimized? Table 10.1 shows four strategies to attain this goal.

Compared with normal-phase LC that uses a lot of organic solvents, reversed-phase LC could attain this aim. For reversed-phase LC, the composition of the mobile phase is water and water miscible solvents. Reversed-phase C18, made up of aliphatic alkyl chains, is the most common stationary phase used in several fields. Except for very hydrophobic compounds, this kind of strategy could reduce the use of organic solvents because one of the eluents is water.

For very aliphatic analytes, C8 or C4 may be a suitable alternative for the stationary phase. Using more hydrophobic stationary phase to separate aliphatic compounds will increase the usage of organic solvents in order to elute the analytes in a suitable time. Under an acceptable range of separation and selectivity, the change of stationary phase from C18 to C8 or C4 could shorten retention time and then reduce the use of

Handbook of Green Analytical Chemistry, First Edition. Edited by Miguel de la Guardia and Salvador Garrigues.

Table 10.1 *How to minimize the production of organic waste.*

	Strategy
I	select reversed-phase LC
II	change stationary phase from C18 to C8,C4 or C1
III	shorten column length
IV	use isocratic mode to replace gradient mode

organic solvents. If the compounds are easy to separate, C4, diol, CN or amino columns may be suitable for sample separation.

Short column replacement with a long column is a good strategy to minimize the production of organic waste. The shorter the analytical column, the smaller the amount of organic solvents used for sample separation. Traditional lengths for analytical columns are 25, 15 and 10 cm. Using shorter columns (such as 5 cm) is a good idea when the resolution, selectivity and retention time are reasonable to analyze the desired compounds. An ideal LC column and mobile phase for separation science are guard column and water, respectively.

In order to obtain good separation, gradient mode is very popular in LC for sample analysis. The demerit of gradient mode is the production of organic waste because the mobile phase is needed to balance to the initial situation before the next injection. Using isocratic mode to replace gradient mode is workable when baseline separation is easily attainable under practical analytical time. It is assured that the isocratic mode is the best choice when LC conditions are effortless in establishment and the separation results are acceptable.

10.2 Green solvents

For green LC, the reduction of the waste of organic solvent is an important issue for protecting our environment. Consumption of organic solvents is necessary when the pumps of the LC system are running for analysis or separation purposes. Organic solvents are expensive, but the cost for recycling and dealing with organic waste is also very expensive.

10.2.1 Hydrophilic solvents

Compared with hydrophobic solvents, water miscible solvents have low volatility and toxicity. Hydrophilic solvents are commonly used in reversed-phase LC, such as tetrahydrofuran, acetone, acetonitrile, methanol, ethanol and isopropanol (or propanol). Due to the high UV cutoff [2], acetone is unsuitable when used as the mobile phase. Hence, acetone has high background absorbance under 330 nm and the baseline of the chromatogram is not easily controlled at the constant situation. Because of the expense and an unpleasant odor, tetrahydrofuran is seldom used as the mobile phase. Ethanol and isopropanol (or propanol) are rarely used as the mobile phase because of the high cost and viscosity, respectively. Besides the high cost, ethanol is the less toxic solvent used as the mobile phase. Methanol and acetonitrile are the most used solvents for sample analysis, but acetonitrile is more expensive than methanol. Methanol is the cheapest hydrophilic solvent and this compound exists ubiquitously in the natural surroundings, so methanol is the best choice of solvent for the LC mobile phase under acceptable analytical conditions. Ethanol and isopropanol (or propanol) could be used as modifiers to adjust the retention time.

After sample analysis, these alcohols containing eluents could be reused for different purposes, such as for fuel to solve the problem of organic waste. Recently alcohol fuel is an alternative (supplementary) fuel for gasoline. If all the alcoholic eluent could be recycled and reused as alternative energy, the mobile phase after the LC analysis is a large source of alcohol fuel.

10.2.2 Ionic liquids

An ionic liquid, also called liquid electrolyte or ionic fluid, is a salt that could be used as a solvent. Because of the property of low vapor pressure, it reduces the release of substances in the environment and can be used as a green solvent [3, 4]. Ionic liquids also could be used as clean media for green sample extraction [5, 6]. They contain many merits in analytical chemistry, such as negligible vapor pressure, good thermal stability, tunable viscosity and miscibility with water and organic solvents [7]. Recently, ionic liquids have been used as additives in LC to improve the peak shape by pairing ion mechanism [8–12]. For green solvent selection, an ionic liquid could be a potential candidate for LC [13]. The physicochemical and solvation properties of ionic liquids are dependent on the combination of cations and anions; many such combinations of cations and anions of ionic liquids have been reported and their properties are well documented [14–19]. General ionic liquids used in separations are listed here and shown in Figure 10.1 [18]:

(a) alkylimidazolium
(b) dimethyldinonylammonium bromide
(c) (R)- N,N,N-trimethyl-2-aminobutanol-bis(trifluoromethanesulfone)imidate
(d) phenylcholine-bis(trifluoro-methanesulfone)imidate
(e) ethylcholine-bis(trifluoromethanesulfone)imidate
(f) undecenoxycarbonyl-L-leucinol bromide (L-UCLB)
(g) undecenoxycarbonyl-L-pyrrolidinol bromide (L-UCPB)
(h) N-alkyl-N-methylpyrrolidinium bromide
(i) tetraalkyl-ammonium tetrafluoroborate

Ionic liquids are widely used in many analytical areas, such as water sample extraction [20], separation of ephedrines [21], extraction of anabolic androgenic steroids [22], determination of aromatic analytes [23, 24], determination of active compounds from plants [25–27], quantification of phenoxyl acid herbicides and phenols [28], and analysis of pharmaceutical components [29–31].

10.2.3 Supercritical Fluid Chromatography (SFC)

In the past, scientists tried to find green solvents for separation, extraction and purification purposes. Water is the best choice for green LC, but using water for LC as the mobile phase is an ideal condition. For practical experimentation, it is difficult to use pure water as the eluent. Hence, scientists are attempting to discover a new mobile phase in LC. Gas chromatography (GC) is a kind of chromatography that uses inert gas as the carrier gas. Could gas be used as a mobile phase for LC? This is an amazing concept for LC. Supercritical fluid has special properties between gas and liquid, so widely used for desired analytes extraction [32–36]. Supercritical fluid chromatography (SFC) is a special technique that combines the advantages of LC and GC. SFC usually uses carbon dioxide as the mobile phase and this technique is nearly a solvent-free method with the exception of adding methanol, ethanol or isopropyl alcohol as the modifiers. Because of the physical and chemical properties, such as low viscosity and high diffusivity, SFC coupled with switching valve is suitably used in chiral separation [37–42]. Different additives can give the best resolution to chiral compounds and these compounds are affected obviously by charged additives [42]. Switching valve is a smart design to use as the modifier or column selector. Choosing different modifiers coupled with chiral columns to separate different chiral compounds can obtain good separation efficiency (see Figure 10.2).

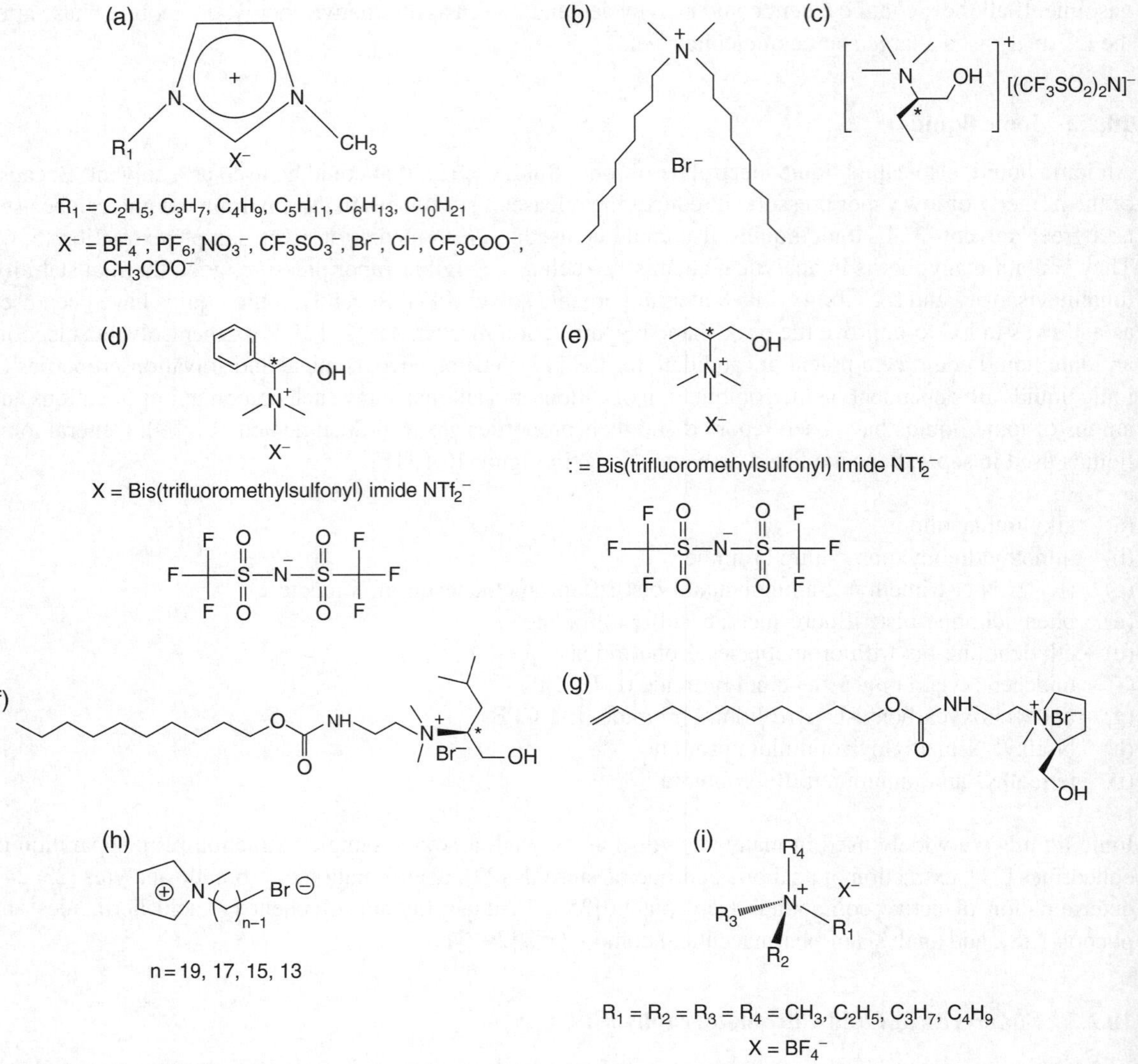

Figure 10.1 The structures of primary classes of ionic liquids [18]. *Reprinted from [18] with permission from Wiley-VCH Verlag GmbH & Co. KGaA © 2007.*

Recently, application of SFC in drug and biosample analysis has been well-documented [43–60]. For a green approach, SFC is a potential strategy for reduction of the organic waste. Unfortunately, SFC is not as popular as LC in separation and purification science use until now.

10.3 Green instruments

The aim of the green instrument is to reduce the production of organic waste and maintain suitable separation efficiency and sensitivity at the same time. For traditional liquid chromatography (LC), the flow rate of this analytical-scale pump is usually over 1 ml min^{-1} (the preparative-scale pump is ignored because the flow rate of this kind of pump is faster than the analytical-scale pump). Some designs slow down the flow rate of the

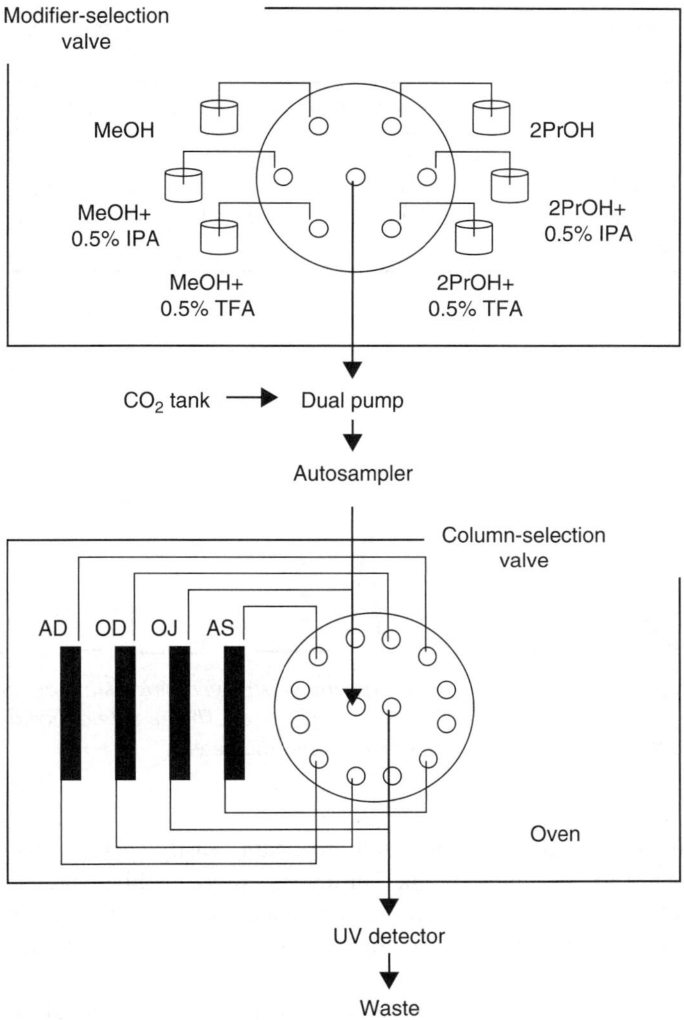

Figure 10.2 *Chiral column- and modifier-switching system used in the SFC screen* [42]. *IPA, PrOH and TFA are isopropylamine, isopropanol and trifluoroacetic acid. AD, OD, OJ and AS are four chiral columns.*

analytical-scale pump: microbore, capillary and nano LC. These instruments could help us to reduce the production of organic waste. Their approximate column inner diameter and pump flow rate are shown in Table 10.2.

10.3.1 Microbore Liquid Chromatography (microbore LC)

Microbore LC is a simple alternative approach for traditional LC. Compared with traditional LC, the inner diameter of these columns used in microbore equipment is about 1 mm (or <1 mm). The pump flow rate of microbore LC is about 100 ~ 500 µl min^{-1}. All the tubing system in microbore LC is also smaller than the traditional LC system. Microbore LC uses low flow rate and narrow tubing; this design could reduce the usage of organic solvents and increase the sensitivity simultaneously. The same concentration of sample in

Table 10.2 *Column inner diameter and pump flow rate of different types of LC.*

LC type	Column inner diameter	Pump flow rate
traditional LC	<10 mm; >1 mm	>1 ml min^{-1}
microbore LC	1 mm ~ 500 μm	100 ~ 500 μl min^{-1}
capillary LC	100 ~ 500 μm	1 ~ 25 μl min^{-1}
nano LC	<100 μm	<1 μl min^{-1}

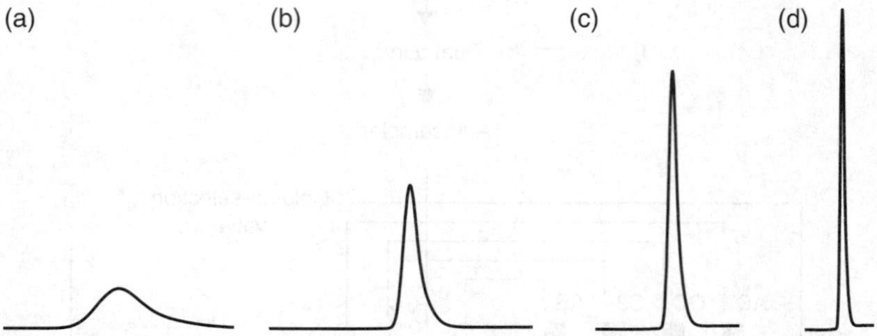

Figure 10.3 *The cartoon diagram of the band broadening versus LC tubing diameter and column at the same stationary phase: tubing diameter and column, (a > (b) > (c) > (d). The analyte concentration is the same and peak dispersion easily occurs in the tubing with a broad inner diameter.*

traditional LC is diluted by mobile phase and band broadening easily occurs. This phenomenon could be avoided when the tubing of LC system is shrunk. For low rate of the mobile phase and narrow tubing of the LC system, peak dispersion (or peak spreading) in this condition is small. Microbore LC is suitable for developing the low dispersion LC (LDLC). The relationship between the band broadening and the diameter of LC tubing is shown in Figure 10.3.

Theoretically, the peak of sample in microbore LC is sharper than traditional LC and the sensitivity is increased obviously at the same concentration. Microbore LC is the fast and simple way to reduce organic waste from traditional LC. There are many areas using microbore LC as the analytical system, such as drug (or drug-like compound) analysis [61–69], protein or peptide analysis [70–78], inorganic analysis [79–88], hormone analysis [89–100], neurotransmitter analysis [101–112] and toxin analysis [113–116].

10.3.2 Capillary Liquid Chromatography (capillary LC)

Capillary LC is a good strategy to reduce organic waste effectively. Capillary LC is derived from microbore LC, but the tubing and flow rate of capillary LC are all narrowed down. Sometimes it is difficult to distinguish between microbore LC and capillary LC. The pump flow rate of capillary LC generally is about 1 ~ 25 μl min^{-1}. Capillary LC has two major designs: capillary flow pump and traditional LC pump coupled with a splitter. The later is a traditional LC system; its flow rate at microliter level is generated by a flow rate controller or a splitter or both. A capillary flow pump could generate the exact flow rate at microliter level; this kind of pump is expensive. Because of the low detected volume, the commercial detector of capillary LC is UV, photo diode array (PDA) and mass spectrometry (see Figure 10.4).

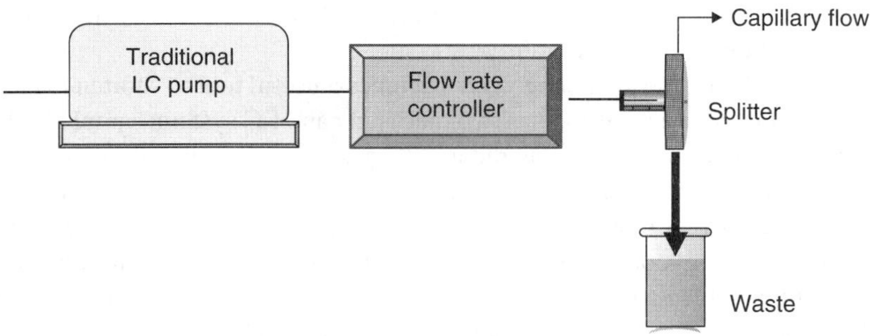

Figure 10.4 Capillary flow is generated by a flow rate controller or a splitter or both. Many cheap Cap LC systems are manufactured by this design. However, the flow rate of this design is not very constant. A high quality Cap LC system could generate stable flow rate, and the cost of this instrument is very high.

Capillary LC is not capillary electrophoresis (CE), but the fused-silica capillary in CE could be used as the tubing system of capillary LC. The capillary inner diameters commonly used in capillary LC as the tubing are 150, 100, 75 and 50 µm. The capillary is very flexible and easily bent because it is coated with a kind of protective polymer called polyimide. Intact polyimide film could help to shield the capillary tubing, but capillary tubing becomes fragile when the polyimide layer is destroyed or scraped. An alternative polymer plastic, polyether ether ketone (PEEK), is used to replace fused-silica capillary. PEEK has chemical resistance, robust and flexible properties, and is a good material to use as the LC tubing. The PEEK inner diameters commonly used in capillary LC as the tubing are 150, 125, 100, 75, 65 and 50 µm. Recently, a new tubing, fused-silica capillary sheathed in firm PEEK, called PEEKsil has been developed. This fused-silica-lined polymer tubing is an excellent replacement for PEEK or capillary tubing. The PEEKsil inner diameters commonly used in capillary LC as the tubing are 150, 100, 75 and 50 µm.

Using capillary LC as the analytical system has not been a widespread technique in separation science up to the present. Some applications of capillary LC are documented, such as capillary column preparation protein [117–130], peptide analysis [131–145], drug analysis [146–156], metabolite analysis [157–161], neurotransmitter analysis [162–172], flavonoid analysis [173–179], pesticide analysis [180–187] and environmental analysis [185, 188–194] and nucleotide analysis [195–200].

10.3.3 Nano Liquid Chromatography (nano LC)

Nano LC is a special and novel design for the LC pump in recent years. Compared with modern microfluidics using capillary and electrokinetic forces to drive liquid, the mobile phase of nano LC is pushed by a delicate and exquisite pump. The pump flow rate of nano LC is <1 µl min^{-1} and this instrument has the least flow rate of LC pump until now. The generally used pump flow rate of this system is below 200 nl min^{-1}. Like capillary LC, nano LC also has two designs: nano flow pump and traditional LC pump coupled with a flow rate controller or splitter. Compared with traditional LC, nano LC is nearly a solvent-free system for analytical and separation science. Nano LC coupled with mass spectrometry is widely used in proteomics research. Exception for protein or peptide analysis, nano LC is rarely used as a regular instrument for analytical and separation technology. Because of the low flow rate and tiny detected volume, the suitable detector of nano LC is mass spectrometry coupled with nano spray source. For nano LC, the connect tubing of this nano system is very important because of the dead volume and band broadening issues. The tubing inner diameters used in nano LC are below 75 µm, so 65, 50, 25 and 20 µm tubings are usually used. Fused-silica capillary, PEEK and PEEKsil tubing are suitable used in nano LC system. Because of the low flow rate, the dead volume question of the tubing system is significant. For instant, 1 μl dead volume will increase 5 min retention time of analyte when the flow rate is 200 nl min^{-1}.

In order to solve this problem, the switching valve becomes a useful tool to shorten the analytical time and avoid band broadening. It is inexpensive and has two merits in nano LC system: sample on-line concentration and reduction of the analytical time (it means reduce the production of organic waste simultaneously). In order to decrease the sample loading time, high flow rate (usually 10–30 μl min^{-1}) is utilized. In this step, a low percentage of organic solvent (sometimes water is used for sample loading) is used to concentrate the sample in the trapping (desalting) column. In the next step, the switching valve will rotate to another position to elute the sample by high percentage of organic solvent from the trapping column to the analytical column. For the analytical step, the low flow rate of the mobile phase is <200 nl min^{-1} and the interested analytes will be eluted to the detector directly. This design not only can avoid peak dispersion by the mobile phase, but can also shorten the retention time of the analytes. Furthermore, decrease of the usage of organic solvents could be expected and an increase the sensitivity of the analytical method could be attainable simultaneously.

Using nano LC as the analytical system has been rare until now because of the high cost of this instrument. Compared with nano LC and traditional LC, nano LC is more difficult than traditional LC to handle and operate. It is arduous to discover leakage problems of tubing and connector, maintain the stability of the flow rate, and avoid the obstruction of this analytical system. The last problem easily occurs because the inner diameter of the tubing is very thin. Any small particle and high concentration of the analytes would cause this problem. After sample injection, any small particle may also affect the flow rate and pressure of this sophisticated system. Systemic obstruction may cause the leak problem, and this issue is hard to solve because the leak mobile phase is too tiny (nano scale). Proteomics research or peptide analysis is the most studied area using this system for protein identification and quantitation [201–215]. Other applications, such as enantiomeric separation [216–224] and biosample analysis [225–231] also use nano LC as the analytical instrument (see Figure 10.5).

10.3.4 How to transfer the LC condition from traditional LC to microbore LC, capillary LC or nano LC

Much experience shows that it is difficult to narrow down the LC scale dramatically, such as changing the LC conditions from traditional LC to nano LC directly. Transferring the LC conditions step-by-step is a good method – this is a simple way to approximately calculate the desired flow rate or column diameter at the similar retention time. For example, a compound separated by a 15 cm column and the original flow rate and column diameter are 1 ml min^{-1} and 4.6 mm respectively. If your expected column diameter is 920 μm and column length is 15 cm, your desired flow rate is approximately 40 μl min^{-1} at the similar retention time and the same particle size of the column. The answer could be calculated by the following equation:

$V_R = F \times t_R$

V_R = retention volume of the analyte

F = flow rate

t_R = retention time of the analyte

$t_R = V_R/F$

column volume = radius × radius × 3.14 × column length

original column volume = 2.3 mm × 2.3 mm × 3.14 × column length

new column volume = 0.46 mm × 0.46 mm × 3.14 × column length

original column volume/new column volume = 25

so,

the new flow rate ~ 1/25 original flow rate = 40 μl min^{-1}

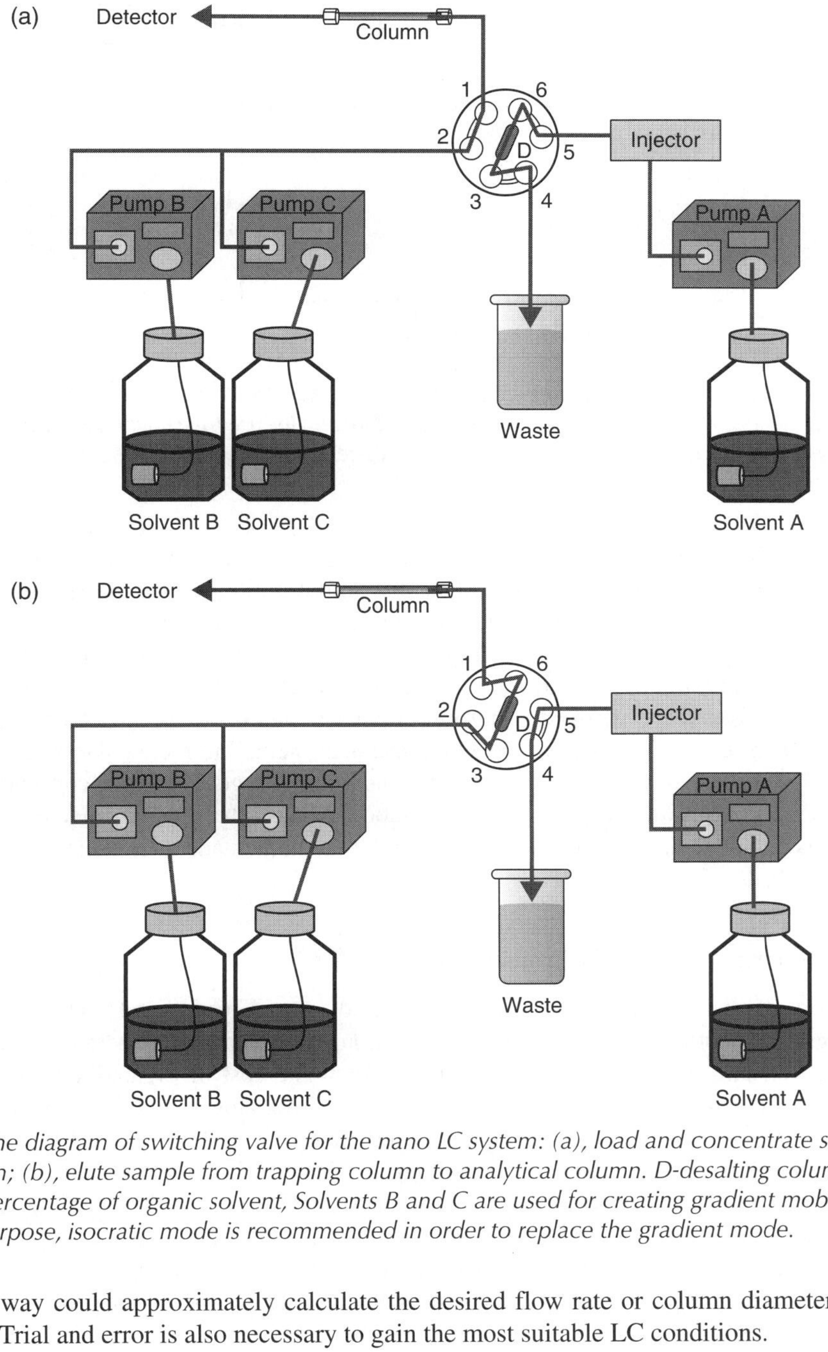

Figure 10.5 *The diagram of switching valve for the nano LC system: (a), load and concentrate sample on trapping column; (b), elute sample from trapping column to analytical column. D-desalting column. Solvent A is usually a low percentage of organic solvent, Solvents B and C are used for creating gradient mobile phase. For the green LC purpose, isocratic mode is recommended in order to replace the gradient mode.*

This simple way could approximately calculate the desired flow rate or column diameter at the similar retention time. Trial and error is also necessary to gain the most suitable LC conditions.

10.3.5 Homemade micro-scale analytical system

It is very easy to create a homemade micro-scale system from a traditional LC system. A homemade micro-scale system is very suitable for sample separation by isocratic mode. The user can slow down the flow rate

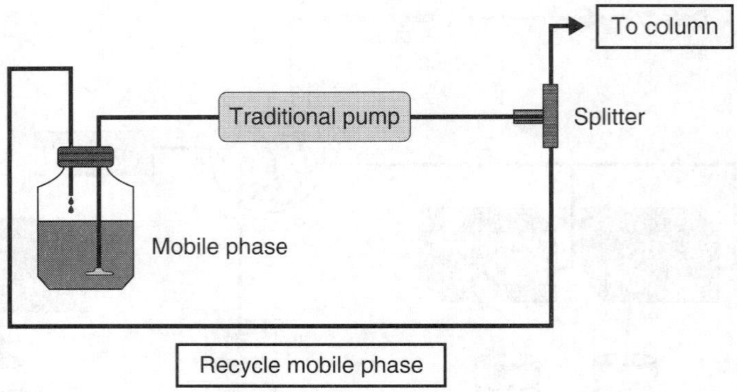

Figure 10.6 *The cartoon design of the homemade micro-scale analytical system. Mobile phase from traditional LC pump may divide into two parts easily by using a commercially inexpensive splitter. Scientists can reuse the neat mobile phase and reduce the production of organic waste.*

and choose the narrower or shorter column. A splitter is demanded to achieve this green goal. One flow direction of the splitter is to the analytical column, the other flow direction of the splitter is to the mobile phase bottle, and then the clean mobile phase can recycle until it runs out of the mobile phase. For example, an analyte is separated by a 15 cm column and the original flow rate and column diameter are 1 ml min^{-1} and 4.0 mm respectively. If your expected column diameter is 2.0 mm and column length is 15 cm, your desired flow rate is approximately 250 μl min^{-1} at the similar retention time and the same particle size of the column. According to the experimental purpose, the user can decide on the desired column diameter and flow rate by themselves. This is a simple and convenient method to reduce the production of organic waste. The mobile phase can be recycled effectively and attain the goal of green chromatography easily. However, this system is unsuitable for the gradient mode because the organic solvent percentage of the recycled eluent is unable to be determined. The diagram of the homemade micro-scale analytical system is showed in Figure 10.6.

10.3.6 Ultra Performance Liquid Chromatography (UPLC)

HPLC has been used in separation science worldwide over 30 years. Scientists and related experts are looking for new strategies to shorten the analytical (or separation) time and maintain the separation efficiency simultaneously. Reducing the analytical time could also cut the cost of organic solvents and lower the production of organic waste. Hence, ultra performance liquid chromatography (UPLC) is invented to match the requirements. In chromatography, the separation efficiency could be evaluated by the van Deemter equation. The smaller the particle size of the column, the better efficiency and resolution obtained. The difference between UPLC and traditional LC is the particle size of the column. The particle size packed in the analytical column of traditional LC is 5 or 3.5 μm, the particle size of UPLC is <2.0 μm. By using UPLC, the flow rates could be increased and the separation speed, resolution and sensitivity could also increase. The net results in the chromatogram show the shortness of the analytical time, so the reduction of the organic waste could be expected.

UPLC is similar to traditional LC except for pump pressure. 'Compared with traditional LC methods, UPLC runs at higher pressure and can reduce 95% organic solvent' says Rohit Khanna; Waters' Vice President. Transference of the analytical method from traditional LC to UPLC is easily attainable; hence, many application results are well documented, such as drug analysis [232–241], metabolite analysis [242–251], protein analysis [252–261] and biomarker analysis [262–271].

References

1. ACS Green Chemistry Institute®, Green chemistry, available at http://portal.acs.org/portal/PublicWebSite/greenchemistry/index.htm (last access 20 November 2011).
2. Kaye & Laby Online. Version 1.0 (2005) UV-visible spectroscopy, in *Tables of Physical & Chemical Constants*, 16th edn., 1995, Subsection 3.8.7., available at http://www.kayelaby.npl.co.uk (accessed at 7 November 2011).
3. Earle, M.J. and Seddon, K.R. (2000) Ionic liquids. Green solvents for the future, *Pure Appl. Chem.*, **72**, 1391–1398.
4. Mallakpour, S. and Rafiee, Z. (2007) Green methodology with ionic liquids as a media for efficient synthesis of polyamides derived from 4-(4-dimethylaminophenyl)-1,2,4-triazolidine-3,5-dione and diacid chlorides, *High Perform. Polym.*, **19**, 427–438.
5. Ge, L.; Wang, X.T.; Tan, S.N.; Tsai, H.H.; Yong, J.W.H. and Hua, L. (2010) A novel method of protein extraction from yeast using ionic liquid solution, *Talanta*, **81**, 1861–1864.
6. Sun, X.; Ji, Y.; Hu, F.; He, B.; Chen, J. and Li, D. (2010) The inner synergistic effect of bifunctional ionic liquid extractant for solvent extraction, *Talanta*, **81**, 1877–1883.
7. Liu, J.F.; Jönsson, J.Å. and Jiang, G.B. (2005) Application of ionic liquids in analytical chemistry, *Trends Anal. Chem.*, **24**, 20–27.
8. Xiaohua, X.; Liang, Z.; Xia, L. and Shengxiang, J. (2004) Ionic liquids as additives in high performance liquid chromatography analysis of amines and the interaction mechanism of ionic liquids, *Anal. Chim. Acta*, **519**, 207–211.
9. Han, D.; Tian, M.; Park, D.W.; Choi, D.K. and Row, K.H. (2009) Application of ionic liquids as mobile phase additives and surface-bonded stationary phase in liquid chromatography, *Korean J. Chem. Eng.*, **26**, 1353–1358.
10. Wang, Y.; Tian, M.; Bi, W. and Row, K.H. (2009) Application of ionic liquids in high performance reversed-phase chromatography, *Int. J. Mol. Sci.*, **10**, 2591–2610.
11. Marszałł, M.P.; Bączek, T. and Kaliszan, R. (2005) Reduction of silanophilic interactions in liquid chromatography with the use of ionic liquids, *Anal. Chim. Acta*, **547**, 172–178.
12. Kaliszan, R.; Marszałł, M.P.; Markuszewski, M.J.; Bączek, T. and Pernak, J. (2004) Suppression of deleterious effects of free silanols in liquid chromatography by imidazolium tetrafluoroborate ionic liquids, *J. Chromatogr. A*, **1030**, 263–271.
13. Marsh, K.N.; Deev, A.; Wu, A.C.T.; Tran, E. and Klamt, A. (2002) Room temperature ionic liquids as replacements for conventional solvents—A review, *Korean J. Chem. Eng.*, **19**, 357–362.
14. Anderson, J.L.; Armstrong, D.W. and Wei, G.T. (2006) Ionic liquids in analytical chemistry, *Anal. Chem.*, **78**, 2892–2902.
15. Poole, C.F. (2004) Chromatographic and spectroscopic methods for the determination of solvent properties of room temperature ionic liquids, *J. Chromatogr. A*, **1037**, 49–82.
16. Liu, R.; Liu, J.F.; Yin, Y.G.; Hu, X.L. and Jiang, G.B. (2009) Ionic liquids in sample preparation, *Anal. Bioanal. Chem.*, **393**, 871–883.
17. Baker, G.A.; Baker, S.N.; Pandey, S. and Bright, F.V. (2005) An analytical view of ionic liquids, *Analyst*, **130**, 800–808.
18. Shamsi, S.A. and Danielson, N.D. (2007) Utility of ionic liquids in analytical separations, *J. Sep. Sci.*, **30**, 1729–1750.
19. Berthod, A.; Ruiz-Ángel, M.J. and Carda-Broch, S. (2008) Ionic liquids in separation techniques, *J. Chromatogr. A*, **1184**, 6–18.
20. Vidal, L.; Chisvert, A.; Canals, A. and Salvador, A. (2010) Ionic liquid-based single-drop microextraction followed by liquid chromatography-ultraviolet spectrophotometry detection to determine typical UV filters in surface water samples, *Talanta*, **81**, 549–555.
21. He, L.; Zhang, W.; Zhao, L.; Liu, X. and Jiang, S. (2003) Effect of 1-alkyl-3-methylimidazolium-based ionic liquids as the eluent on the separation of ephedrines by liquid chromatography, *J. Chromatogr. A*, **1007**, 39–45.
22. He, C.; Li, S.; Liu, H.; Li, K. and Liu, F. (2005) Extraction of testosterone and epitestosterone in human urine using aqueous two-phase systems of ionic liquid and salt, *J. Chromatogr. A*, **1082**, 143–149.
23. Liu, J.F.; Jiang, G.B.; Chi, Y.G.; Cai, Y.Q.; Zhou, Q.X. and Hu, J.T. (2003) Use of ionic liquids for liquid-phase microextraction of polycyclic aromatic hydrocarbons, *Anal. Chem.*, **75**, 5870–5876.
24. Martín-Calero, A.; Tejral, G.; Ayala, J.H.; González, V. and Afonso, A.M. (2010) Suitability of ionic liquids as mobile-phase additives in HPLC with fluorescence and UV detection for the determination of heterocyclic aromatic amines, *J. Sep. Sci.*, **33**, 182–190.
25. Tang, F.; Tao, L.; Luo, X.; Ding, L.; Guo, M.; Nie, L. and Yao, S. (2006) Determination of octopamine, synephrine and tyramine in citrus herbs by ionic liquid improved 'green' chromatography, *J. Chromatogr. A*, **1125**, 182–188.

26. Tian, M.; Liu, J. and Row, K.H. (2009) Effect of ionic liquids as the mobile phase additives on the HPLC resolution of four active compounds from *Sophora flavescens* Ait, *Molecules*, **14**, 2127–2134.
27. Lue, B.M.; Guo, Z. and Xu, X. (2008) High-performance liquid chromatography analysis methods developed for quantifying enzymatic esterification of flavonoids in ionic liquids, *J. Chromatogr. A*, **1198–1199**, 107–114.
28. Hu, X.; Peng, J.; Huang, Y.; Yin, D. and Liu, J. (2009) Ionic liquids as mobile phase additives for high-performance liquid chromatography separation of phenoxy acid herbicides and phenols, *J. Sep. Sci.*, **32**, 4126–4132.
29. Ruiz-Angel, M.J.; Carda-Broch, S. and Berthod, A. (2006) Ionic liquids versus triethylamine as mobile phase additives in the analysis of β-blockers, *J. Chromatogr. A*, **1119**, 202–208.
30. Mizuuchi, H.; Jaitely, V.; Murdan, S. and Florence, A.T. (2008) Room temperature ionic liquids and their mixtures: Potential pharmaceutical solvents, *Eur. J. Pharm. Sci.*, **33**, 326–331.
31. Herrera-Herrera, A.V.; Hernández-Borges, J. and Rodríguez-Delgado, M.Á. (2008) Ionic liquids as mobile phase additives for the high-performance liquid chromatographic analysis of fluoroquinolone antibiotics in water samples, *Anal. Bioanal. Chem.*, **392**, 1439–1446.
32. Mezzomo, N.; Mileo, B.R.; Friedrich, M.T.; Martínez, J. and Ferreira, S.R.S. (2010) Supercritical fluid extraction of peach (*Prunus persica*) almond oil: Process yield and extract composition, *Bioresour. Technol.*, **101**, 5622–5632.
33. Sun, Y.; Wei, L.; Wang, J.; Bi, J.; Liu, Z.; Wang, Y. and Guo, Z. (2010) Optimization of supercritical fluid extraction of saikosaponins from *Bupleurum falcatum* with orthogonal array design, *J. Sep. Sci.*, **33**, 1–6.
34. Sugihara, N.; Kanda, A.; Nakano, T.; Nakamura, T.; Igusa, H. and Hara, S. (2010) Novel fractionation method for squalene and phytosterols contained in the deodorization distillate of rice bran oil, *J. Oleo Sci.*, **59**, 65–70.
35. He, S.; Zhang, Z.; Xu, F.; Zhang, S. and Lei, Z. (2010) Micronization of magnolia bark extract with enhanced dissolution behavior by rapid expansion of supercritical solution, *Chem. Pharm. Bull.*, **58**, 154–159.
36. da Silva, M.S.; Vão, E.R.; Temtem, M.; Mafra, L.; Caldeira, J.; Aguiar-Ricardo, A. and Casimiro, T. (2010) Clean Synthesis of molecular recognition polymeric materials with chiral sensing capability using supercritical fluid technology. Application as HPLC stationary phases, *Biosens. Bioelectron.*, **25**, 1742–1747.
37. Mangelings, D. and Heyden, Y.V. (2008) Chiral separations in sub- and supercritical fluid chromatography, *J. Sep. Sci.*, **31**, 1252–1273.
38. Phinney, K.W. (2005) Enantioselective separations by packed column subcritical and supercritical fluid chromatography, *Anal. Bioanal. Chem.*, **382**, 639–645.
39. Terfloth, G. (2001) Enantioseparations in super- and subcritical fluid chromatography, *J. Chromatogr. A*, **906**, 301–307.
40. West, C.; Bouet, A.; Gillaizeau, I.; Coudert, G.; Lafosse, M. and Lesellier, E. (2010) Chiral separation of phospine-containing a-amino acid derivatives using two complementary cellulosic stationary phases in supercritical fluid chromatography, *Chirality*, **22**, 242–251.
41. Stringham, R.W. and Blackwell, J.A. (1996) "Entropically driven" chiral separations in supercritical fluid chromatography. Confirmation of isoelution temperature and reversal of elution order, *Anal. Chem.*, **68**, 2179–2185.
42. Maftouh, M.; Granier-Loyaux, C.; Chavana, E.; Marini, J.; Pradines, A.; Heyden, Y.V. and Picard, C. (2005) Screening approach for chiral separation of pharmaceuticals part III. Supercritical fluid chromatography for analysis and purification in drug discovery, *J. Chromatogr. A*, **1088**, 67–81.
43. Zhao, Y.; Woo, G.; Thomas, S.; Semin, D. and Sandra, P. (2003) Rapid method development for chiral separation in drug discovery using sample pooling and supercritical fluid chromatography–mass spectrometry, *J. Chromatogr. A*, **1003**, 157–166.
44. François, I.; Pereira, A. dos S. and Sandra, P. (2010) Considerations on comprehensive and off-line supercritical fluid chromatography × reversed-phase liquid chromatography for the analysis of triacylglycerols in fish oil, *J. Sep. Sci.*, **33**, 1504–1512.
45. Mukherjee, P.S. (2007) Validation of direct assay of an aqueous formulation of a drug compound AZY by chiral supercritical fluid chromatography (SFC), *J. Pharm. Biomed. Anal.*, **43**, 464–470.
46. Mukherjee, P.S. and Cook, S.E. (2006) A feasibility study on direct assay of an aqueous formulation by chiral supercritical fluid chromatography (SFC), *J. Pharm. Biomed. Anal.*, **41**, 1287–1292.
47. Bamba, T. and Fukusaki, E. (2009) Separation of hydrophobic metabolites using monolithic silica column in high-performance liquid chromatography and supercritical fluid chromatography, *J. Sep. Sci.*, **32**, 2699–2706.
48. Matsubara, A.; Bamba, T.; Ishida, H.; Fukusaki, E. and Hirata, K. (2009) Highly sensitive and accurate profiling of carotenoids by supercritical fluid chromatography coupled with mass spectrometry, *J. Sep. Sci.*, **32**, 1459–1464.

49. Patil, S.T.; Sundaresan, M.; Bhoir, I.C. and Bhagwat, A.M. (1998) Packed column supercritical fluid chromatographic separation and estimation of acetaminophen, diclofenac sodium and methocarbamol in pharmaceutical dosage forms, *Talanta*, **47**, 3–10.
50. Kaul, N.; Agrawal, H.; Paradkar, A.R. and Mahadik, K.R. (2007) Effect of system variables involved in packed column supercritical fluid chromatography of stavudine taken as model analyte using response surface methodology along with study of thermodynamic parameters, *J. Pharm. Biomed. Anal.*, **43**, 471–480.
51. Brunelli, C.; Zhao, Y.; Brown, M.H. and Sandra, P. (2008) Pharmaceutical analysis by supercritical fluid chromatography: Optimization of the mobile phase composition on a 2-ethylpyridine column, *J. Sep. Sci.*, **31**, 1299–1306.
52. Gyllenhaal, O. and Hulthe, J. (2002) Direct injection of aqueous samples in packed column supercritical fluid chromatography of isosorbide-5-mononitrate from drug release testing, *J. Pharm. Biomed. Anal.*, **29**, 381–386.
53. Abbott, E.; Veenstra, T.D. and Issaq, H.J. (2008) Clinical and pharmaceutical applications of packed-column supercritical fluid chromatography, *J. Sep. Sci.*, **31**, 1223–1230.
54. Toribio, L.; del Nozal, M.J.; Bernal, Y.L.; Alonso, C. and Jiménez, J.J. (2008) Semipreparative chiral supercritical fluid chromatography in the fractionation of lansoprazole and two related antiulcer drugs enantiomers, *J. Sep. Sci.*, **31**, 1307–1313.
55. Brunelli, C.; Zhao, Y.; Brown, M.H. and Sandra, P. (2008) Development of a supercritical fluid chromatography high-resolution separation method suitable for pharmaceuticals using cyanopropyl silica, *J. Chromatogr. A*, **1185**, 263–272.
56. Yip, S.H.H.; Ashraf-Khorassani, M. and Taylo, L.T. (2008) Analytical scale supercritical fluid fractionation and identification of single polar lipids from deoiled soybean lecithin, *J. Sep. Sci.*, **31**, 1290–1298.
57. Bamba, T.; Shimonishi, N.; Matsubara, A.; Hirata, K.; Nakazawa, Y.; Kobayashi, A. and Fukusaki, E. (2008) High throughput and exhaustive analysis of diverse lipids by using supercritical fluid chromatography-mass spectrometry for metabolomics, *J. Biosci. Bioeng.*, **105**, 460–469.
58. Zhang, X.; Scalf, M.; Westphall, M.S. and Smith, L.M. (2008) Membrane protein separation and analysis by supercritical fluid chromatography-mass spectrometry, *Anal. Chem.*, **80**, 2590–2598.
59. White, C. (2005) Integration of supercritical fluid chromatography into drug discovery as a routine support tool. Part I. Fast chiral screening and purification, *J. Chromatogr. A*, **1074**, 163–173.
60. White, C. and Burnett, J. (2005) Integration of supercritical fluid chromatography into drug discovery as a routine support tool. Part II. Investigation and evaluation of supercritical fluid chromatography for achiral batch purification, *J. Chromatogr. A*, **1074**, 175–185.
61. Lanckmans, K.; Clinckers, R.; van Eeckhaut, A.; Sarre, S.; Smolders, I. and Michotte, Y. (2006) Use of microbore LC–MS/MS for the quantification of oxcarbazepine and its active metabolite in rat brain microdialysis samples, *J. Chromatogr. B*, **831**, 205–212.
62. Barnes, K.A.; Smith, R.A.; Williams, K.; Damant, A.P. and Shepherd, M.J. (1998) A microbore high performance liquid chromatography/electrospray ionization mass spectrometry method for the determination of the phytoestrogens genistein and daidzein in comminuted baby foods and soya flour, *Rapid Commun. Mass Spectrom.*, **12**, 130–138.
63. Wong, S.H.V. (1989) Supercritical fluid chromatography and microbore liquid chromatography for drug analysis, *Clin. Chem.*, **35**, 1293–1298.
64. Sinnaeve, B.A.; Decaestecker, T.N.; Claerhout, I.J.; Kestelyn, P.; Remon, J.P. and van Bocxlaer, J.F. (2003) Confirmation of ofloxacin precipitation in corneal deposits by microbore liquid chromatography–quadrupole time-of-flight tandem mass spectrometry, *J. Chromatogr. B*, **785**, 193–196.
65. Miao, X.S. and Metcalfe, C.D. (2003) Determination of pharmaceuticals in aqueous samples using positive and negative voltage switching microbore liquid chromatography/electrospray ionization tandem mass spectrometry, *J. Mass Spectrom.*, **38**, 27–34.
66. Majumdar, T.K.; Martin, L.L.; Melamed, D. and Tse, F.L.S. (2000) Determination of the chiral isomers of CGS 26214, a synthetic thyromimetic agent, in human plasma using microbore chiral chromatography–tandem mass spectrometry, *J. Pharm. Biomed. Anal.*, **23**, 745–755.
67. Majumdar, T.K.; Bakhtiar, R.; Melamed, D. and Tse, F.L.S. (2000) Determination of terbinafine (lamisil) in human hair by microbore liquid chromatography/tandem mass spectrometry, *Rapid Commun. Mass Spectrom.*, **14**, 1214–1219.
68. Majumdar, T.K.; Bakhtiar, R.; Melamed, D. and Tse, F.L.S. (2000) Trace-level quantitation of iralukast in human plasma by microbore liquid chromatography/tandem mass spectrometry, *Rapid Commun. Mass Spectrom.*, **14**, 476–481.
69. Prangle, A.S.; Noctor, T.A.G. and Lough, W.J. (1998) Chiral bioanalysis of warfarin using microbore LC with peak compression, *J. Pharm. Biomed. Anal.*, **16**, 1205–1212.

70. Kaltashov, I.A.; Yu, X. and Fenselau, C. (1995) Simple interface for microbore LC and electrospray ionization mass spectrometry and analysis of melphalan-alkylation sites in metallothionein, *J. Pharm. Biomed. Anal.*, **13**, 279–284.
71. Wu, N.; Webb, A.; Peck, T.L. and Sweedler, J.V. (1995) On-line NMR detection of amino acids and peptides in microbore LC, *Anal. Chem.*, **67**, 3101–3107.
72. Ji, C. and Li, L. (2005) Quantitative proteome analysis using differential stable isotopic labeling and microbore LC-MALDI MS and MS/MS, *J. Proteome Res.*, **4**, 734–742.
73. Young, J.B. and Li, L. (2007) Impulse-driven heated-droplet deposition interface for capillary and microbore LC-MALDI MS and MS/MS, *Anal. Chem.*, **79**, 5927–5934.
74. Creaser, C.S.; Lill, J.R.; Bonner, P.L.R.; Hill, S.C. and Rees, R.C. (2000) Nano-electrospray and microbore liquid chromatography-ion trap mass spectrometry studies of copper complexation with MHC restricted peptides, *Analyst*, **125**, 599–603.
75. Mills, M.J.; Maltas, J. and Lough, W.J. (1997) Automated microbore LC with peak compression for the assay of GR138950 and its active metabolite in human serum, *Chromatographia*, **45**, 275–283.
76. Lapolla, A.; Brancia, F.L.; Bereszczak, J.; Fedele, D.; Baccarin, L.; Seraglia, R. and Traldi, P. (2007) Off-line liquid chromatography-MALDI by with various matrices and tandem mass spectrometry for analysis of glycated human serum albumin tryptic peptides, *Mol. Nutr. Food Res.*, **51**, 456–461.
77. Ji, C.; Guo, N. and Li, L. (2005) Differential dimethyl labeling of n-termini of peptides after guanidination for proteome analysis, *J. Proteome Res.*, **4**, 2099–2108.
78. Hoffmann-Benning, S.; Gage, D.A.; McIntosh, L.; Kende, H. and Zeevaart, J.A.D. (2002) Comparison of peptides in the phloem sap of flowering and non-flowering perilla and lupine plants using microbore HPLC followed by matrix-assisted laser desorption/ionization time-of-flight mass spectrometry, *Planta*, **216**, 140–147.
79. Ackley, K.L.; Sutton, K.L. and Caruso, J.A. (2000) A comparison of nebulizers for microbore LC-ICP-MS with mobile phases containing methanol, *J. Anal. At. Spectrom.*, **15**, 1069–1073.
80. Pergantis, S.A.; Heahmar, E.M. and Hinners, T.A. (1995) Microscale flow injection and microbore high-performance liquid chromatography coupled with inductively coupled plasma mass spectrometry via a high-efficiency nebulizer, *Anal. Chem.*, **67**, 4530–4535.
81. Jitaru, P.; Roman, M.; Cozzi, G.; Fisicaro, P.; Cescon, P. and Barbante, C. (2009) Speciation analysis of selenoproteins in human serum by microbore affinity-HPLC hyphenated to ICP-sector field-MS using a high efficiency sample introduction system, *Microchim. Acta*, **166**, 319–327.
82. Pergantis, S.A.; Heithma, E.M. and Hinners, T.A. (1997) Speciation of arsenic animal feed additives by microbore high-performance liquid chromatography with inductively coupled plasma mass spectrometry, *Analyst*, **122**, 1063–1068.
83. Far, J.; Preud'homme, H. and Lobinski, R. (2010) Detection and identification of hydrophilic selenium compounds in selenium-rich yeast by size exclusion–microbore normal-phase HPLC with the on-line ICP–MS and electrospray Q-TOF-MS detection, *Anal. Chim. Acta*, **657**, 175–190.
84. Połeć, K.; Peréz-Calvo, M.; García-Arribas, O.; Szpunar, J.; Ribas-Ozonas, B. and Łobiński, R. (2002) Investigation of metal complexes with metallothionein in rat tissues by hyphenated techniques, *J. Inorg. Biochem.*, **88**, 197–206.
85. Chassaigne, H. and Lobinski, R. (1998) Polymorphism and identification of metallothionein isoforms by reversed-phase HPLC with on-line ion-spray mass spectrometric detection, *Anal. Chem.*, **70**, 2536–2543.
86. Castillo, A.; Roig-Navarro, A.F. and Pozo, O.J. (2007) Secondary interactions, an unexpected problem emerged between hydroxyl containing analytes and fused silica capillaries in anion-exchange micro-liquid chromatography, *J. Chromatogr. A*, **1172**, 179–185.
87. Lobinski, R.; Chassaigne, H. and Szpunar, J. (1998) Analysis for metallothioneins using coupled techniques, *Talanta*, **46**, 271–289.
88. Castillo, A.; Roig-Navarro, A.F. and Pozo, O.J. (2008) Capabilities of microbore columns coupled to inductively coupled plasma mass spectrometry in speciation of arsenic and selenium, *J. Chromatogr. A*, **1202**, 132–137.
89. Lai, C.C.; Tsai, C.H.; Tsai, F.J.; Lee, C.C. and Lin, W.D. (2001) Rapid monitoring assay of congenital adrenal hyperplasia with microbore high-performance liquid chromatography/electrospray ionization tandem mass spectrometry from dried blood spots, *Rapid Commun. Mass Spectrom.*, **15**, 2145–2151.
90. Nithipatikom, K.; Pratt, P.F. and Campbell, W.B. (2000) Determination of EETs using microbore liquid chromatography with fluorescence detection, *Am. J. Physiol. Heart Circ. Physiol.*, **279**, H857–H862.

91. Kuuranne, T.; Thomas, A.; Leinonen, A.; Delahaut, P.; Bosseloir, A.; Schänzer, W. and Thevis, W.M. (2008) Insulins in equine urine: Qualitative analysis by immunoaffinity purification and liquid chromatography/tandem mass spectrometry for doping control purposes in horse-racing, *Rapid Commun. Mass Spectrom.*, **22**, 355–362.
92. Thevis, M. and Schänzer, W. (2007) Mass spectrometric identification of peptide hormones in doping-control analysis, *Analyst*, **132**, 287–291.
93. Li, H.; Wang, H.; Schegg, K.M. and Schooley, D.A. (1997) Metabolism of an insect diuretic hormone by Malpighian tubules studied by liquid chromatography coupled with electrospray ionization mass spectrometry, *Proc. Natl. Acad. Sci. USA*, **94**, 13463–13468.
94. Chambery, A.; Severino, V.; Maro, A.D.; D'Aniello, A.; Ruvo, M. and Parente, A. (2010) Quantification of thyrotropin-releasing hormone by liquid chromatography–electrospray mass spectrometry, *Amino Acids*, **38**, 1031–1041.
95. Ghulam, A.; Kouach, M.; Racadot, A.; Boersma, A.; Vantyghem, M.C. and Briand, G. (1999) Quantitative analysis of human serum corticosterone by high-performance liquid chromatography coupled to electrospray ionization mass spectrometry, *J. Chromatogr. B*, **727**, 227–233.
96. Thevis, M.; Thomas, A.; Delahaut, P.; Bosseloir, A. and Schänzer, W. (2006) Doping control analysis of intact rapid-acting insulin analogues in human urine by liquid chromatography-tandem mass spectrometry, *Anal. Chem.*, **78**, 1897–1903.
97. Thevis, M.; Bredehöft, M.; Geyer, H.; Kamber, M.; Delahaut, P. and Schänzer, W. (2006) Determination of synacthen in human plasma using immunoaffinity purification and liquid chromatography/tandem mass spectrometry, *Rapid Commun. Mass Spectrom.*, **20**, 3551–3556.
98. Koidl, J.; Hödl, H.; Schmid, M.G.; Neubauer, B.; Konrad, M.; Petschauer, S. and Gübitz, G. (2008) Enantiorecognition of triiodothyronine and thyroxine enantiomers using different chiral selectors by HPLC and micro-HPLC, *J. Biochem. Biophys. Methods*, **70**, 1254–1260.
99. Meng, R.; Xia, W.; Sandberg, M.; Stephens, R. and Weber, S.G. (2005) Online preconcentration of thyrotropin-releasing hormone (TRH) by SDS-modified reversed phase column for microbore and capillary high-performance liquid chromatography (HPLC), *J. Chromatogr. A*, **1071**, 179–184.
100. Thevis, M.; Thomas, A.; Delahaut, P.; Bosseloir, A. and Schänzer, W. (2005) Qualitative determination of synthetic analogues of insulin in human plasma by immunoaffinity purification and liquid chromatography-tandem mass spectrometry for doping control purposes, *Anal. Chem.*, **77**, 3579–3585.
101. Wu, Y.S.; Ho, W.M.; Tsai, T.H.; Yang, L.L.; Kuo, J.S. and Cheng, F.C. (1997) Monitoring of blood catecholamines by microdialysis and microbore LC with a dual amperometric detector, *J. Pharm. Biomed. Anal.*, **16**, 77–85.
102. de Bundel, D.; Sarre, S.; van Eeckhaut, A.; Smolders, I. and Michotte, Y. (2008) Critical evaluation of acetylcholine determination in rat brain microdialysates using ion-pair liquid chromatography with amperometric detection, *Sensors*, **8**, 5171–5185.
103. Beaudry, F.; Ferland, C.E. and Vachon, P. (2009) Identification, characterization and quantification of specific neuropeptides in rat spinal cord by liquid chromatography electrospray quadrupole ion trap mass spectrometry, *Biomed. Chromatogr.*, **23**, 940–950.
104. Shao, X.M. and Feldman, J.L. (2007) Efficient measurement of endogenous neurotransmitters in small localized regions of central nervous systems *in vitro* with HPLC, *J. Neurosci. Methods*, **160**, 256–263.
105. Kehr, J.; Hu, X.J.; Yoshitake, T. and Scheller, D. (2007) Determination of the dopamine agonist rotigotine in microdialysates from the rat brain by microbore column liquid chromatography with electrochemical detection, *J. Chromatogr. B*, **845**, 109–113.
106. Yoshitake, T.; Kehr, J.; Todoroki, K.; Nohta, H. and Yamaguchi, M. (2006) Derivatization chemistries for determination of serotonin, norepinephrine and dopamine in brain microdialysis samples by liquid chromatography with fluorescence detection, *Biomed. Chromatogr.*, **20**, 267–281.
107. Clinckers, R.; Gheuens, S.; Smolders, I.; Meurs, A.; Ebinger, G. and Michotte, Y. (2005) In vivo modulatory action of extracellular glutamate on the anticonvulsant effects of hippocampal dopamine and serotonin, *Epilepsia*, **46**, 828–836.
108. Patel, B.A.; Arundell, M.; Parker, K.H.; Yeoman, M.S. and O'Hare, D. (2008) Simple and rapid determination of serotonin and catecholamines in biological tissue using high-performance liquid chromatography with electrochemical detection, *J. Chromatogr. B*, **818**, 269–276.
109. Zhang, S.; Takeda, Y.; Hagioka, S.; Takata, K.; Aoe, H.; Nakatsuka, H.; Yokoyama, M. and Morita, K. (2005) Measurement of GABA and glutamate in vivo levels with high sensitivity and frequency, *Brain Res. Protocol.*, **14**, 61–66.

110. Yoshitake, T.; Iizuka, R.; Fujino, K.; Inoue, O.; Yamagata, K.; Nohta, H. and Yamaguchi, M. (2004) Simultaneous determination of serotonin and 5-hydroxyindole-3-acetic acid in human urine by automated precolumn derivatization and semi-microbore column liquid chromatography with fluorescence detection, *Anal. Sci.*, **20**, 1687–1690.
111. Yoshitake, T.; Yamaguchi, M.; Nohta, H.; Ichinose, F.; Yoshida, H.; Yoshitake, S.; Fuxe, K. and Kehr, J. (2003) Determination of histamine in microdialysis samples from rat brain by microbore column liquid chromatography following intramolecular excimer-forming derivatization with pyrene-labeling reagent, *J. Neurosci. Methods*, **127**, 11–17.
112. Dong, Y.; Wang, L.; Shangguan, D.; Zhao, R. and Liu, G. (2003) Improved method for the routine determination of acetylcholine and choline in brain microdialysate using a horseradish peroxidase column as the immobilized enzyme reactor, *J. Chromatogr. B*, **788**, 193–198.
113. Zweigenbaum, J.A.; Henion, J.D.; Beattie, K.A.; Codd, G.A. and Poon, G.K. (2000) Direct analysis of microcystins by microbore liquid chromatography electrospray ionization ion-trap tandem mass spectrometry, *J. Pharm. Biomed. Anal.*, **23**, 723–733.
114. Kelley, W.P.; Wolters, A.M.; Sack, J.T.; Jockusch, R.A.; Jurchen, J.C.; Williams, E.R.; Sweedler, J.V. and Gilly, W.F. (2003) Characterization of a novel gastropod toxin (6-bromo-2-mercaptotryptamine) that inhibits shaker k channel activity, *J. Biol. Chem.*, **278**, 34934–34942.
115. Inceoglu, B.; Lango, J.; Wu, J.; Hawkins, P.; Southern, J. and Hammock, B.D. (2001) Isolation and characterization of a novel type of neurotoxic peptide from the venom of the South African scorpion *Parabuthus transvaalicus* (buthidae), *Eur. J. Biochem.*, **268**, 5407–5413.
116. Barco, M.; Rivera, J. and Caixach, J. (2002) Analysis of cyanobacterial hepatotoxins in water samples by microbore reversed-phase liquid chromatography–electrospray ionisation mass spectrometry, *J. Chromatogr. A*, **959**, 103–111.
117. Lin, J.; Lin, J.; Lin, X. and Xie, Z. (2009) Capillary liquid chromatography using a hydrophilic/cation-exchange monolithic column with a dynamically modified cationic surfactant, *J. Chromatogr. A*, **1216**, 7728–7731.
118. Urban, J.; Eeltink, S.; Jandera, P. and Schoenmakers, P.J. (2008) Characterization of polymer-based monolithic capillary columns by inverse size-exclusion chromatography and mercury-intrusion porosimetry, *J. Chromatogr. A*, **1182**, 161–168.
119. Horie, K.; Ikegami, T.; Hosoya, K.; Saad, N.; Fiehn, O. and Tanaka, N. (2007) Highly efficient monolithic silica capillary columns modified with poly(acrylic acid) for hydrophilic interaction chromatography, *J. Chromatogr. A*, **1164**, 198–205.
120. Zheng, M.M.; Lin, B. and Feng, Y.Q. (2007) Hybrid organic-inorganic octyl monolithic column for in-tube solid-phase microextraction coupled to capillary high-performance liquid chromatography, *J. Chromatogr. A*, **1164**, 48–55.
121. Bisjak, C.P.; Trojer, L.; Lubbad, S.H.; Wieder, W. and Bonn, G.K. (2007) Influence of different polymerisation parameters on the separation efficiency of monolithic poly(phenyl acrylate-co-1,4-phenylene diacrylate) capillary columns, *J. Chromatogr. A*, **1154**, 269–276.
122. Jandera, P.; Urban, J.; Škeříková, V.; Langmaier, P.; Kubíćková, R. and Planeta, J. (2010) Polymethacrylate monolithic and hybrid particle-monolithic columns for reversed-phase and hydrophilic interaction capillary liquid chromatography, *J. Chromatogr. A*, **1217**, 22–33.
123. Li, Y.; Gu, B.; Tolley, H.D. and Lee, M.L. (2009) Preparation of polymeric monoliths by copolymerization of acrylate monomers with amine functionalities for anion-exchange capillary liquid chromatography of proteins, *J. Chromatogr. A*, **1216**, 5525–5532.
124. Wu, R.; Hu, L.; Wang, F.; Ye, M. and Zou, H. (2008) Recent development of monolithic stationary phases with emphasis on microscale chromatographic separation, *J. Chromatogr. A*, **1184**, 369–392.
125. Nandi, P. and Lunte, S.M. (2009) Recent trends in microdialysis sampling integrated with conventional and microanalytical systems for monitoring biological events: A review, *Anal. Chim. Acta*, **651**, 1–14.
126. Soonthorntantikul, W.; Leepipatpiboon, N.; Ikegami, T.; Tanaka, N. and Nhujak, T. (2009) Selectivity comparisons of monolithic silica capillary columns modified with poly(octadecyl methacrylate) and octadecyl moieties for halogenated compounds in reversed-phase liquid chromatography, *J. Chromatogr. A*, **1216**, 5868–5874.
127. Núñez, O.; Ikegami, T.; Miyamoto, K. and Tanaka, N. (2007) Study of a monolithic silica capillary column coated with poly(octadecyl methacrylate) for the reversed-phase liquid chromatographic separation of some polar and non-polar compounds, *J. Chromatogr. A*, **1175**, 7–15.

128. Gu, C.; He, J.; Jia, J.; Fang, N.; Simmons, R. and Shamsi, S.A. (2010) Surfactant-bound monolithic columns for separation of proteins in capillary high performance liquid chromatography, *J. Chromatogr. A*, **1217**, 530–539.
129. Ren, L.; Liu, Y.; Dong, M. and Liu, Z. (2009) Synthesis of hydrophilic boronate affinity monolithic capillary for specific capture of glycoproteins by capillary liquid chromatography, *J. Chromatogr. A*, **1216**, 8421–8425.
130. Saito, K.; Uemura, E.; Ishizaki, A. and Kataok, H. (2010) Determination of perfluorooctanoic acid and perfluorooctane sulfonate by automated in-tube solid-phase microextraction coupled with liquid chromatography-mass spectrometry, *Anal. Chim. Acta*, **658**, 141–146.
131. Wang, F.; Jiang, X.; Feng, S.; Tian, R.; Jiang, X.; Han, G.; Liu, H.; Ye, M. and Zou, H. (2007) Automated injection of uncleaned samples using a ten-port switching valve and a strong cation-exchange trap column for proteome analysis, *J. Chromatogr. A*, **1171**, 56–62.
132. Pröfrock, D. and Prange, A. (2009) Compensation of gradient related effects when using capillary liquid chromatography and inductively coupled plasma mass spectrometry for the absolute quantification of phosphorylated peptides, *J. Chromatogr. A*, **1216**, 6706–6715.
133. Casanovas, A.; Carrascal, M.; Abián, J.; López-Tejero, M.D. and Llobera, M. (2009) Discovery of lipoprotein lipase pi isoforms and contributions to their characterization, *J. Proteomics*, **72**, 1031–1039.
134. Kubáň, P. and Hauser, P.C. (2007) High-performance liquid chromatography with contactless conductivity detection for the determination of peptides and proteins using a monolithic capillary column, *J. Chromatogr. A*, **1176**, 185–191.
135. Wang, F.; Dong, J.; Ye, M.; Wu, R. and Zou, H. (2009) Integration of monolithic frit into the particulate capillary (IMFPC) column in shotgun proteome analysis, *Anal. Chim. Acta*, **652**, 324–330.
136. Iwasaki, M.; Miwa, S.; Ikegami, T.; Tomita, M.; Tanaka, N. and Ishihama, Y. (2010) One-dimensional capillary liquid chromatographic separation coupled with tandem mass spectrometry unveils the *Escherichia coli* proteome on a microarray scale, *Anal. Chem.*, **82**, 2616–2620.
137. Baker, E.S.; Livesay, E.A.; Orton, D.J.; Moore, R.J.; Danielson III, W.F.; Prior, D.C.; Ibrahim, Y.M.; LaMarche, B.L.; Mayampurath, A.M.; Schepmoes, A.A.; Hopkins, D.F.; Tang, K.; Smith, R.D. and Belov, M.E. (2010) An LC-IMS-MS platform providing increased dynamic range for high-throughput proteomic studies, *J. Proteome Res.*, **9**, 997–1006.
138. Slebos, R.J.C.; Brock, J.W.C.; Winters, N.F.; Stuart, S.R.; Martinez, M.A.; Li, M.; Chambers, M.C.; Zimmerman, L.J.; Ham, A.J.; Tabb, D.L. and Liebler, D.C. (2008) Evaluation of strong cation exchange versus isoelectric focusing of peptides for multidimensional liquid chromatography-tandem mass spectrometry, *J. Proteome Res.*, **7**, 5286–5294.
139. Zheng, X.; Wu, S.L.; Hincapie, M. and Hancock, W.S. (2009) Study of the human plasma proteome of rheumatoid arthritis, *J. Chromatogr A*, **1216**, 3538–3545.
140. Ekegren, T.; Hanrieder, J.; Aquilonius, S.M. and Bergquist, J. (2006) Focused proteomics in post-mortem human spinal cord, *J. Proteome Res.*, **5**, 2364–2371.
141. Zhang, H.; Yi, E.C.; Li, X.J.; Mallick, P.; Kelly-Spratt, K.S.; Masselon, C.D.; Camp II, D.G.; Smith, R.D.; Kemp, C.J. and Aebersold, R. (2005) High throughput quantitative analysis of serum proteins using glycopeptide capture and liquid chromatography mass spectrometry, *Mol. Cell. Proteomics*, **4**, 144–155.
142. Green-Church, K.B.; Nichols, K.K.; Kleinholz, N.M.; Zhang, L. and Nichols, J.J. (2008) Investigation of the human tear film proteome using multiple proteomic approaches, *Mol. Vis.*, **14**, 456–470.
143. Metz, T.O.; Qian, W.J.; Jacobs, J.M.; Gritsenko, M.A.; Moore, R.J.; Polpitiya, A.D.; Monroe, M.E.; Camp II, D.G.; Mueller, P.W. and Smith, R.D. (2008) Application of proteomics in the discovery of candidate protein biomarkers in a diabetes autoantibody standardization program sample subset, *J. Proteome Res.*, **7**, 698–707.
144. Livesay, E.A.; Tang, K.; Taylor, B.K.; Buschbach, M.A.; Hopkins, D.F.; LaMarche, B.L.; Zhao, R.; Shen, Y.; Orton, D.J.; Moore, R.J.; Kelly, R.T.; Udseth, H.R. and Smith, R.D. (2008) Fully automated four-column capillary LC-MS system for maximizing throughput in proteomic analyses, *Anal. Chem.*, **80**, 294–302.
145. Belov, M.E.; Anderson, G.A.; Wingerd, M.A.; Udseth, H.R.; Tang, K.; Prior, D.C.; Swanson, K.R.; Buschbach, M.A.; Strittmatter, E.F.; Moore, R.J. and Smith, R.D. (2004) An automated high performance capillary liquid chromatography-Fourier transform ion cyclotron resonance mass spectrometer for high-throughput proteomics, *J. Am. Soc. Mass Spectrom.*, **15**, 212–232.
146. Schubert, B.; Pavlic, M.; Libiseller, K. and Oberacher, H. (2008) Unraveling the metabolic transformation of tetrazepam to diazepam with mass spectrometric methods, *Anal. Bioanal. Chem.*, **392**, 1299–1308.

147. Yu, H.; Straubinger, R.M.; Cao, J.; Wang, H. and Qu, J. (2008) Ultra-sensitive quantification of paclitaxel using selective solidphase extraction in conjunction with reversed-phase capillary liquid chromatography/tandem mass spectrometry, *J. Chromatogr. A*, **1210**, 160–167.
148. Bailón-Pérez, M.I.; García-Campaña, A.M.; del Olmo-Iruela, M.; Gámiz-Gracia, L. and Cruces-Blanco, C. (2009) Trace determination of 10 β-lactam antibiotics in environmental and food samples by capillary liquid chromatography, *J. Chromatogr. A*, **1216**, 8355–8361.
149. Arvidsson, B.; Allard, E.; Sjögren, E.; Lennernäs, H.; Sjöberg, P.J.R. and Bergquist, J. (2009) Online capillary solid phase extraction and liquid chromatographic separation with quantitative tandem mass spectrometric detection (SPE-LC–MS/MS) of ximelagatran and its metabolites in a complex matrix, *J. Chromatogr. B*, **877**, 291–297.
150. Lu, C.Y. and Feng, C.H. (2007) Micro-scale analysis of aminoglycoside antibiotics in human plasma by capillary liquid chromatography and nanospray tandem mass spectrometry with column switching, *J. Chromatogr. A*, **1156**, 249–253.
151. Lu, C.Y.; Chang, Y.M.; Tseng, W.L.; Feng, C.H. and Lu, C.Y. (2009) Analysis of angiotensin ii receptor antagonist and protein markers at microliter level plasma by LC–MS/MS, *J. Pharm. Biomed. Anal.*, **49**, 123–128.
152. Lu, C.Y.; Yang, C.M. and Feng, C.H. (2009) Trace analysis of glyclazide in human plasma at microscale level by mass spectrometry, *J. Sep. Sci.*, **32**, 3411–3417.
153. Ruiz-Jiménez, J. and de Castro, M.D.L. (2007) In-column micro-high-performance liquid chromatographic concentration-separation prior to ultraviolet detection for the determination of chlorophenols in water samples, *J. Chromatogr. A*, **1174**, 78–84.
154. Wang, Y. and Griffiths, W.J. (2008) Capillary liquid chromatography combined with tandem mass spectrometry for the study of neurosteroids and oxysterols in brain, *Neurochem. Int.*, **52**, 506–521.
155. Qu, J.; Qu, Y. and Straubinger, R.M. (2007) Ultra-sensitive quantification of corticosteroids in plasma samples using selective solid-phase extraction and reversed-phase capillary high-performance liquid chromatography/tandem mass spectrometry, *Anal. Chem.*, **79**, 3786–3793.
156. Neto, A.J.; Rodrigues, J.C.; Fernandes, C.; Titato, G.M.; Alves, C. and Lanças, F.M. (2006) Automated microcolumn-switching system for drug analysis by direct injection of human plasma, *J. Chromatogr. A*, **1105**, 71–76.
157. Edwards, J.L.; Edwards, R.L.; Reid, K.R. and Kennedy, R.T. (2007) Effect of decreasing column inner diameter and use of off-line two-dimensional chromatography on metabolite detection in complex mixtures, *J. Chromatogr. A*, **1172**, 127–134.
158. Wind, M.; Gebhardt, K.; Grunwald, H.; Spickermann, J.; Donzelli, M.; Kellenberger, L.; Muller, M.; Fullhardt, P.; Schmitt-Hoffmann, A. and Schleimer, M. (2007) Elucidation of the in vitro metabolic profile of stable isotope labeled BAL19403 by accurate mass capillary liquid chromatography/quadrupole time-of-flight mass spectrometry and isotope exchange, *Rapid Commun. Mass Spectrom.*, **21**,1093–1099.
159. Sundström, I.; Arts, J.; Westerlund, D. and Andrén, P.E. (2010) In vivo investigation of brain and systemic ketobemidone metabolism, *Analyst*, **135**, 405–413.
160. Wind, M.; Spickermann, J.; Schleimer, M.; Donzelli, M.; Gebhardt, K.; Sturm-Haurany, R.; Klauer, D.; Fullhardt, P. and Schmitt-Hoffmann, A. (2006) Investigation of low-abundant in vitro metabolites of stable isotope-labelled BAL4815 by accurate mass capillary-LC-ESI-qTof-MS and MS/MS, *J. Mass Spectrom.*, **41**, 903–910.
161. Wind, M.; Grunwald, H.; Gebhardt, K.; Illig, K.; Spickermann, J.; Nuoffer, C.; Roussel, P.; Klauer, D.; Fullhardt, P.; Schmitt-Hoffmann, A. and Schleimer, M. (2009) Investigation of the species-dependent in vitro metabolism of BAL30630 by stable isotope labeling and isotope exchange experiments analyzed by capillary liquid chromatography coupled to mass spectrometry, *J. Chromatogr. A*, **1216**, 3946–3953.
162. Sinnaeve, B.A.; Storme, M.L. and van Bocxlaer, J.F. (2005) Capillary liquid chromatography and tandem mass spectrometry for the quantification of enkephalins in cerebrospinal fluid, *J. Sep. Sci.*, **28**, 1779–1784.
163. Baseski, H.M.; Watson, C.J.; Cellar, N.A.; Shackman, J.G. and Kennedy, R.T. (2005) Capillary liquid chromatography with ms^3 for the determination of enkephalins in microdialysis samples from the striatum of anesthetized and freely-moving rats, *J. Mass Spectrom.*, **40**, 146–153.
164. Song, Y.; Quan, Z.; Evans, J.L.; Byrd, E.A. and Liu, Y.M. (2004) Enhancing capillary liquid chromatography/tandem mass spectrometry of biogenic amines by pre-column derivatization with 7-fluoro-4-nitrobenzoxadiazole, *Rapid Commun. Mass Spectrom.*, **18**, 989–994.
165. Song, Y.; Shenwu, M.; Dhossche, D.M. and Liu, Y.M. (2005) A capillary liquid chromatographic/tandem mass spectrometric method for the quantification of γ-aminobutyric acid in human plasma and cerebrospinal fluid, *J. Chromatogr. B*, **814**, 295–302.

166. Törnkvist, A.; Sjöberg, P.J.R.; Markides, K.E. and Bergquist, J. (2004) Analysis of catecholamines and related substances using porous graphitic carbon as separation media in liquid chromatography-tandem mass spectrometry, *J. Chromatogr. B*, **801**, 323–329.
167. Zhang, X.; Rauch, A.; Lee, H.; Xiao, H.; Rainer, G. and Logothetis, N.K. (2007) Capillary hydrophilic interaction chromatography/mass spectrometry for simultaneous determination of multiple neurotransmitters in primate cerebral cortex, *Rapid Commun. Mass Spectrom.*, **21**, 3621–3628.
168. Parrot, S.; Lambás-Señas, L.; Sentenac, S.; Denoroy, L. and Renaud, B. (2007) Highly sensitive assay for the measurement of serotonin in microdialysates using capillary high-performance liquid chromatography with electrochemical detection, *J. Chromatogr. B*, **850**, 303–309.
169. Shackman, H.M.; Shou, M.; Cellar, N.A.; Watson, C.J. and Kennedy, R.T. (2007) Microdialysis coupled on-line to capillary liquid chromatography with tandem mass spectrometry for monitoring acetylcholine in vivo, *J. Neurosci. Methods*, **159**, 86–92.
170. Amirkhani, A.; Heldin, E.; Markides, K.E. and Bergquist, J. (2002) Quantitation of tryptophan, kynurenine and kynurenic acid in human plasma by capillary liquid chromatography-electrospray ionization tandem mass spectrometry, *J. Chromatogr. B*, **780**, 381–387.
171. Jung, M.C.; Shi, G.; Borland, L.; Michael, A.C. and Weber, S.G. (2006) Simultaneous determination of biogenic monoamines in rat brain dialysates using capillary high-performance liquid chromatography with photoluminescence following electron transfer, *Anal. Chem.*, **78**, 1755–1760.
172. Perry, M.; Li, Q. and Kennedy, R.T. (2009) Review of recent advances in analytical techniques for the determination of neurotransmitters, *Anal. Chim. Acta*, **653**, 1–22.
173. Ding, S.; Dudley, E.; Chen, L.; Plummer, S.; Tang, J.; Newton, R.P. and Brenton, A.G. (2006) Determination of active components of *Ginkgo biloba* in human urine by capillary high-performance liquid chromatography/mass spectrometry with on-line column-switching purification, *Rapid Commun. Mass Spectrom.*, **20**, 3619–3624.
174. Dalluge, J.J.; Nelson, B.C.; Thomas, J.B.; Welch, M.J. and Sander, L.C. (1997) Capillary liquid chromatography/electrospray mass spectrometry for the separation and detection of catechins in green tea and human plasma, *Rapid Commun. Mass Spectrom.*, **11**, 1753–1756.
175. Fukusaki, E.; Harada, K.; Bamba, T. and Kobayashi, A. (2005) An isotope effect on the comparative quantification of flavonoids by means of methylation-based stable isotope dilution coupled with capillary liquid chromatography/mass spectrometry, *J. Biosci. Bioeng.*, **99**, 75–77.
176. Kotani, A.; Takahashi, K.; Hakamata, H.; Kojima, S. and Kusu, F. (2007) Attomole catechins determination by capillary liquid chromatography with electrochemical detection, *Anal. Sci.*, **23**, 157–163.
177. Chen, J.; Song, Y. and Li, P. (2007) Capillary high-performance liquid chromatography with mass spectrometry for simultaneous determination of major flavonoids, iridoid glucosides and saponins in *Flos lonicerae*, *J. Chromatogr. A*, **1157**, 217–226.
178. Hughey, C.A.; Wilcox, B.; Minardi, C.S.; Takehara, C.W.; Sundararaman, M. and Were, L.M. (2008) Capillary liquid chromatography–mass spectrometry for the rapid identification and quantification of almond flavonoids, *J. Chromatogr. A*, **1192**, 259–265.
179. Kotani, A.; Kojima, S.; Hayashi, Y.; Matsuda, R. and Kusu, F. (2008) Optimization of capillary liquid chromatography with electrochemical detection for determining femtogram levels of baicalin and baicalein on the basis of the FUMI theory, *J. Pharm. Biomed. Anal.*, **48**, 780–787.
180. Song, Y.; Zhao, S.; Tchounwou, P. and Liu, Y.M. (2007) A nanoparticle-based solid-phase extraction method for liquid chromatography–electrospray ionization-tandem mass spectrometric analysis, *J. Chromatogr. A*, **1166**, 79–84.
181. Rosales-Conrado, N.; León-González, M.E.; Pérez-Arribas, L.V. and Polo-Díez, L.M. (2005) Capillary liquid chromatography of chlorophenoxy acid herbicides and their esters in apple juice samples after preconcentration on a cation exchanger based on polydivinylbenzene-n-vinylpyrrolidone, *J. Chromatogr. A*, **1076**, 202–206.
182. Armenta, S. and Lendl, B. (2010) Capillary liquid chromatography with off-line mid-IR and Raman micro-spectroscopic detection: Analysis of chlorinated pesticides at ppb levels, *Anal. Bioanal. Chem.*, **397**, 297–308.
183. Holm, A.; Molander, P.; Lundanes, E. and Greibrokk, T. (2003) Determination of rotenone in river water utilizing packed capillary column switching liquid chromatography with UV and time-of-flight mass spectrometric detection, *J. Chromatogr. A*, **983**, 43–50.
184. Rosales-Conrado, N.; León-González, M.E.; Pérez-Arribas, L.V. and Polo-Díez, L.M. (2005) Effect of temperature on the separation of chlorophenoxy acids and carbamates by capillary high-performance liquid chromatography and UV (or diode array) detection, *J. Chromatogr. A*, **1081**, 114–121.

185. Campíns-Falcó, P.; Verdú-Andrés, J.; Sevillano-Cabeza, A.; Herráez-Hernández, R.; Molins-Legua, C. and Moliner-Martinez, Y. (2010) In-tube solid-phase microextraction coupled by in valve mode to capillary LC-DAD: Improving detectability to multiresidue organic pollutants analysis in several whole waters, *J. Chromatogr. A*, **1217**, 2695–2702.
186. Gou, Y. and Pawliszyn, J. (2000) In-tube solid-phase microextraction coupled to capillary LC for carbamate analysis in water samples, *Anal. Chem.*, **72**, 2774–2779.
187. Cháfer-Pericás, C.; Herráez-Hernández, R. and Campíns-Faló, P. (2007) In-tube solid-phase microextraction-capillary liquid chromatography as a solution for the screening analysis of organophosphorus pesticides in untreated environmental water samples, *J. Chromatogr. A*, **1141**, 10–21.
188. Vitta, Y.; Moliner-Martínez, Y.; Campíns-Falcó, P. and Cuervo, A.F. (2010) An in-tube SPME device for the selective determination of chlorophyll a in aquatic systems, *Talanta*, **82**, 952–956.
189. Sabatini, L.; Barbieri, A. and Violante, F.S. (2005) Development and validation of a capillary high-performance liquid chromatography/electrospray tandem mass spectrometric method for the quantification of bisphenol a in air samples, *Rapid Commun. Mass Spectrom.*, **19**, 3468–3472.
190. Pitarch, E.; Hernandez, F.; ten Hove, J.; Meiring, H.; Niesing, W.; Dijkman, E.; Stolker, L. and Hogendoorn, E. (2004) Potential of capillary-column-switching liquid chromatography-tandem mass spectrometry for the quantitative trace analysis of small molecules application to the on-line screening of drugs in water, *J. Chromatogr. A*, **1031**, 1–9.
191. Cháfer-Pericás, C.; Herráez-Hernández, R. and Campíns-Falcó, P. (2006) On-fibre solid-phase microextraction coupled to conventional liquid chromatography versus in-tube solid-phase microextraction coupled to capillary liquid chromatography for the screening analysis of triazines in water samples, *J. Chromatogr. A*, **1125**, 159–171.
192. Sedyohutomo, A.; Lim, L. and Takeuchi, T. (2008) Development of packed-column suppressor system for capillary ion chromatography and its application to environmental waters, *J. Chromatogr. A*, **1203**, 239–242.
193. Lim, L.W.; Okouchi, Y. and Takeuchi, T. (2007) On-line preconcentration of trace carcinogenic polycyclic aromatic hydrocarbons (PAHs) in microcolumn liquid chromatography via large volume injection, *Talanta*, **72**, 1600–1608.
194. Moliner-Martínez, Y.; Herráez-Hernández, R. and Campíns-Faló, P. (2007) A microanalytical method for ammonium and short-chain primary aliphatic amines using precolumn derivatization and capillary liquid chromatography, *J. Chromatogr. A*, **1164**, 329–333.
195. Olsen, R.; Ovrebø, S.; Thorud, S.; Lundanes, E.; Thomassen, Y.; Greibrokk, T. and Molander, P. (2008) Sensitive determination of a glyoxal–DNA adduct biomarker candidate by column switching capillary liquid chromatography electrospray ionization mass spectrometry, *Analyst*, **133**, 802–809.
196. Fujii, S.I.; Inagaki, K.; Takatsu, A.; Yarita, T. and Chiba, K. (2009) Determination of phosphorus using capillary electrophoresis and micro-high-performance liquid chromatography hyphenated with inductively coupled plasma mass spectrometry for the quantification of nucleotides, *J. Chromatogr. A*, **1216**, 7488–7492.
197. Wang, H.; Cao, H. and Wang, Y. (2010) Quantification of N^2-carboxymethyl-2′-deoxyguanosine in calf thymus DNA and cultured human kidney epithelial cells by capillary high-performance liquid chromatography-tandem mass spectrometry coupled with stable isotope dilution method, *Chem. Res. Toxicol.*, **23**, 74–81.
198. Matter, B.; Malejka-Giganti, D.; Csallany, A.S. and Tretyakova, N. (2006) Quantitative analysis of the oxidative DNA lesion, 2,2-diamino-4-(2-deoxy-β-D-*erythro*-pentofuranosyl) amino]-5(2*H*)-oxazolone (Oxazolone), in vitro and in vivo by isotope dilution-capillary HPLC-ESI-MS/MS, *Nucl. Acids Res.*, **34**, 5449–5460.
199. Goodenough, A.K.; Schut, H.A.J. and Turesky, R.J. (2007) Novel LC-ESI/MS/MSn method for the characterization and quantification of 2′-deoxyguanosine adducts of the dietary carcinogen 2-amino-1-methyl-6-phenylimidazo[4,5-*b*] pyridine by 2-D linear quadrupole ion trap mass spectrometry, *Chem. Res. Toxicol.*, **20**, 263–276.
200. Chen, H.J.C. and Chen, Y.C. (2009) Analysis of glyoxal-induced DNA cross-links by capillary liquid chromatography nanospray ionization tandem mass spectrometry, *Chem. Res. Toxicol.*, **22**, 1334–1341.
201. Wohlgemuth, J.; Karas, M.; Jiang, W.; Hendriks, R. and Andrecht, S. (2010) Enhanced glyco-profiling by specific glycopeptide enrichment and complementary monolithic nano-LC (ZIC-HILIC/RP18e)/ESI-MS analysis, *J. Sep. Sci.*, **33**, 1–11.
202. Snovida, S.I.; Bodnar, E.D.; Viner, R.; Saba, J. and Perreault, H. (2010) A simple cellulose column procedure for selective enrichment of glycopeptides and characterization by nano LC coupled with electron-transfer and high-energy collisional-dissociation tandem mass spectrometry, *Carbohydr. Res.*, **345**, 792–801.
203. Nittis, T.; Guittat, L.; LeDuc, R.D.; Dao, B.; Duxin, J.P.; Rohrs, H.; Townsend, R.R. and Stewart, S.A. (2010) Revealing novel telomere proteins using in vivo cross-linking, tandem affinity purification, and label-free quantitative LC-FTICR-MS, *Mol. Cell. Proteomics*, **9**, 1144–1156.

204. Wang, H.; Wong, C.H.; Chin, A.; Kennedy, J.; Zhang, Q. and Hanash, S. (2009) Quantitative serum proteomics using dual stable isotope coding and *nano* LC-MS/MSMS, *J. Proteome Res.*, **8**, 5412–5422.
205. Yu, N.H.; Ho, E.N.M.; Wan, T.S.M. and Wong, A.S.Y. (2010) Doping control analysis of recombinant human erythropoietin, darbepoetin alfa and methoxy polyethylene glycol-epoetin beta in equine plasma by nano-liquid chromatography–tandem mass spectrometry, *Anal. Bioanal. Chem.*, **396**, 2513–2521.
206. Nanni, P.; Levander, F.; Roda, G.; Caponi, A.; James, P. and Roda, A. (2009) A label-free nano-liquid chromatography-mass spectrometry approach for quantitative serum peptidomics in Crohn's disease patients, *J. Chromatogr. B*, 877, 3127–3136.
207. Lortie, M.; Bark, S.; Blantz, R. and Hook, V. (2009) Detecting low-abundance vasoactive peptides in plasma: Progress toward absolute quantitation using nano liquid chromatography-mass spectrometry, *Anal. Biochem.*, **394**, 164–170.
208. Voss, K.; Stahl, S.; Hogan, B.M.; Reinders, J.; Schleider, E.; Schulte-Merker, S. and Felbor, U. (2009) Functional analyses of human and zebrafish 18-amino acid in-frame deletion pave the way for domain mapping of the cerebral cavernous malformation 3 protein, *Hum. Mutat.*, **30**, 1003–1011.
209. Ravanat, C.; Wurtz, V.; Ohlmann, P.; Fichter, M.; Cazenave, J.P.; VanDorsselaer, A.; Lanza, F. and Gachet, C. (2009) Use of tandem Biacore–mass spectrometry to identify platelet membrane targets of novel monoclonal antibodies, *Anal. Biochem.*, **386**, 237–243.
210. Chen, Z.; Peng, I.C.; Sun, W.; Su, M.I.; Hsu, P.H.; Fu, Y.; Zhu, Y.; DeFea, K.; Pan, S.; Tsai, M.D. and Shyy, J.Y.J. (2009) AMP-activated protein kinase functionally phosphorylates endothelial nitric oxide synthase Ser633, *Circ. Res.*, **104**, 496–505.
211. Ellis, J.; Grimm, R.; Clark, J.F.; Pyne-Gaithman, G.; Wilbur, S. and Caruso, J.A. (2008) Studying protein phosphorylation in low MW CSF fractions with capLC-ICPMS and nanoLC-CHIP-ITMS for identification of phosphoproteins, *J. Proteome Res.*, **7**, 4736–4742.
212. Barnes, S.; Shonsey, E.M.; Eliuk, S.M.; Stella, D.; Barrett, K.; Srivastava, O.P.; Kim, H. and Renfrow, M.B. (2008) High-resolution mass spectrometry analysis of protein oxidations and resultant loss of function, *Biochem. Soc. Trans.*, **36**, 1037–1044.
213. Flora, J.W.; Edmiston, J.; Secrist, R.; Li, G.; Rana, G.S.J.B.; Langston, T.B. and McKinney, W. (2008) Identification of in vitro differential cell secretions due to cigarette smoke condensate exposure using nanoflow capillary liquid chromatography and high-resolution mass spectrometry, *Anal. Bioanal. Chem.*, **391**, 2845–2856.
214. Geiser, L.; Eeltink, S.; Svec, F. and Fréchet, J.M.T. (2008) In-line system containing porous polymer monoliths for protein digestion with immobilized pepsin, peptide preconcentration and nano-liquid chromatography separation coupled to electrospray ionization mass spectroscopy, *J. Chromatogr A*, **1188**, 88–96.
215. Bang, D.Y.; Ahn, E.J. and Moon, M.H. (2007) Shotgun analysis of phospholipids from mouse liver and brain by nanoflow liquid chromatography/tandem mass spectrometry, *J. Chromatogr. B*, **852**, 268–277.
216. Rocco, A. and Fanali, S. (2009) Enantiomeric separation of acidic compounds by nano-liquid chromatography with methylated-β-cyclodextrin as a mobile phase additive, *J. Sep. Sci.*, **32**, 1696–1703.
217. Fanali, S.; D'Orazio, G.; Lomsadze, K. and Chankvetadze, B. (2008) Enantioseparations with cellulose tris(3-chloro-4-methylphenylcarbamate) in nano-liquid chromatography and capillary electrochromatography, *J. Chromatogr. B*, **875**, 296–303.
218. D'Orazio, G. and Fanali, S. (2008) Enantiomeric separation by using nano-liquid chromatography with on-column focusing, *J. Sep. Sci.*, **31**, 2567–2571.
219. Fanali, S.; Aturki, Z.; Kasicka, V.; Raggi, M.A. and D'Orazio, G. (2005) Enantiomeric separation of mirtazapine and its metabolites by nano-liquid chromatography with UV-absorption and mass spectrometric detection, *J. Sep. Sci.*, **28**, 1719–1728.
220. Si-Ahmed, K.; Tazerouti, F.; Badjah-Hadj-Ahmed, A.Y.; Aturki, Z.; D'Orazio, G.; Rocco, A. and Fanali, S. (2010) Optical isomer separation of flavanones and flavanone glycosides by nano-liquid chromatography using a phenyl-carbamate-propyl-β-cyclodextrin chiral stationary phase, *J. Chromatogr. A*, **1217**, 1175–1182.
221. Fantacuzzi, M.; Bettoni, G.; D'Orazio, G. and Fanali, S. (2006) Enantiomeric separation of some demethylated analogues of clofibric acid by capillary zone electrophoresis and nano-liquid chromatography, *Electrophoresis*, **27**, 1227–1236.
222. Fanali, S.; D'Orazio, G.; Quaglia, M.G. and Rocco, A. (2004) Use of a hepta-Tyr antibiotic modified silica stationary phase for the enantiomeric resolution of D,L-loxiglumide by electrochromatography and nano-liquid chromatography, *J. Chromatogr. A*, **1051**, 247–252.

223. Healy, L.O.; Murrihy, J.P.; Tan, A.; Cocker, D.; McEnery, M. and Glennon, J.D. (2001) Enantiomeric separation of R,S-naproxen by conventional and nano-liquid chromatography with methyl-β-cyclodextrin as a mobile phase additive, *J. Chromatogr. A*, **924**, 459–464.
224. Rosales-Conrado, N.; León-González, M.E.; D'Orazio, G. and Fanali, S. (2004) Enantiomeric separation of chlorophenoxy acid herbicides by nano liquid chromatography-UV detection on a vancomycin-based chiral stationary phase, *J. Sep. Sci.*, **27**, 1303–1308.
225. Blais, C.; Blasco, T.; Maria, A.; Dauphin-Villemant, C. and Lafont, R. (2010) Characterization of ecdysteroids in *Drosophila melanogaster* by enzyme immunoassay and nano-liquid chromatography–tandem mass spectrometry, *J. Chromatogr. B*, **878**, 925–932.
226. Si-Ahmed, K.; Tazerouti, F.; Badjah-Hadj-Ahmed, A.Y.; Aturki, Z.; D'Orazio, G.; Rocco, A. and Fanali, S. (2010) Analysis of hesperetin enantiomers in human urine after ingestion of blood orange juice by using nano-liquid chromatography, *J. Pharm. Biomed. Anal.*, **51**, 225–229.
227. Hu, F.; Deng, C.; Liu, Y. and Zhang, X. (2009) Quantitative determination of chlorogenic acid in *Honeysuckle* using microwave-assisted extraction followed by nano-LC-ESI mass spectrometry, *Talanta*, **77**, 1299–1303.
228. Fanali, S.; Aturki, Z.; D'Orazio, G. and Rocco, A. (2007) Separation of basic compounds of pharmaceutical interest by using nano-liquid chromatography coupled with mass spectrometry, *J. Chromatogr. A*, **1150**, 252–258.
229. García-Villalba, R.; Carrasco-Pancorbo, A.; Zurek, G.; Behrens, M.; Bässmann, C.; Segura-Carretero, A. and Fernández-Gutiérrez, A. (2010) Nano and rapid resolution liquid chromatography–electrospray ionization–time of flight mass spectrometry to identify and quantify phenolic compounds in olive oil, *J. Sep. Sci.*, **33**, 2069–2078.
230. Randall, K.L.; Argoti, D.; Paonessa, J.D.; Ding, Y.; Oaks, Z.; Zhang, Y. and Vouros, P. (2010) An improved liquid chromatography–tandem mass spectrometry method for the quantification of 4-aminobiphenyl DNA adducts in urinary bladder cells and tissues, *J. Chromatogr. A*, **1217**, 4135–4143.
231. Hsieh, S.H.; Huang, H.Y. and Lee, S. (2009) Determination of eight penicillin antibiotics in pharmaceuticals, milk and porcine tissues by nano-liquid chromatography, *J. Chromatogr. A*, **1216**, 7186–7194.
232. Krishnaiah, Ch.; Reddy, A.R.; Kumar, R. and Mukkanti, K. (2010) Stability-indicating UPLC method for determination of Valsartan and their degradation products in active pharmaceutical ingredient and pharmaceutical dosage forms, *J. Pharm. Biomed. Anal.*, **53**, 483–489.
233. Qin, F.; Li, N.; Qin, T.; Zhang, Y. and Li, F. (2010) Simultaneous quantification of Venlafaxine and O-desmethylvenlafaxine in human plasma by ultra performance liquid chromatography–tandem mass spectrometry and its application in a pharmacokinetic study, *J. Chromatogr. B*, **878**, 689–694.
234. Phapale, P.B.; Lee, H.W.; Lim, M.S.; Kim, E.H.; Kim, S.D.; Park, J.; Lee, M.; Hwang, S.K. and Yoon, Y.R. (2010) Rapid determination of finasteride in human plasma by UPLC–MS/MS and its application to clinical pharmacokinetic study, *J. Chromatogr. B*, **878**, 1718–1723.
235. Blonk, M.I.; van der Nagel, B.C.; Smit, L.S. and Mathot, R.A.A. (2010) Quantification of levetiracetam in plasma of neonates by ultra performance liquid chromatography-tandem mass spectrometry, *J. Chromatogr. B*, **878**, 675–681.
236. Phapale, P.B.; Lee, H.W.; Lim, M.S.; Seong, S.J.; Kim, E.H.; Park, J.; Lee, M.; Hwang, S.K. and Yoon, Y.R. (2010) Liquid chromatography-tandem mass spectrometry quantification of levosulpiride in human plasma and its application to bioequivalence study, *J. Chromatogr. B*, **878**, 2280–2285.
237. Xiong, Z.; Zhang, Y.; Qin, F.; Qin, T.; Yang, S. and Li, F. (2010) Hydrophilic interaction liquid chromatography-tandem mass Spectrometry for the determination of adefovir in human plasma and its application to a pharmacokinetic study, *J. Chromatogr. B*, **878**, 2111–2116.
238. Han, M.; Chen, J.; Chen, S. and Wang, X. (2010) Development of a UPLC-ESI-MS/MS Assay for 20(S)-protopanaxadiol and pharmacokinetic application of its two formulations in rats, *Anal. Sci.*, **26**, 749–753.
239. Bhatt, M.; Shah, S. and Shivprakash (2010) Development of a high-throughput method for the determination of ethosuximide in human plasma by liquid chromatography mass spectrometry, *J. Chromatogr. B*, **878**, 1605–1610.
240. Kikura-Hanajiri, R.; Kawamura, M.; Miyajima, A.; Sunouchi, M. and Goda, Y. (2010) Determination of a new designer drug, N-hydroxy-3,4-methylenedioxymethamphetamine and its metabolites in rats using ultra-performance liquid chromatography-tandem mass spectrometry, *Forensic Sci. Int.*, **198**, 62–69.

241. Rao, D.D.; Satyanarayana, N.V.; Reddy, A.M.; Sait, S.S.; Chakole, D. and Mukkanti, K. (2010) A validated stability-indicating UPLC method for desloratadine and its impurities in pharmaceutical dosage forms, *J. Pharm. Biomed. Anal.*, **51**, 736–742.
242. Birkler, R.I.D.; Støttrup, N.B.; Hermannson, S.; Nielsen, T.T.; Gregersen, N.; Bøtker, H.E.; Andreasen, M.F. and Johannsen, M. (2010) A UPLC–MS/MS Application for profiling of intermediary energy metabolites in microdialysis samples—A method for high-throughput, *J. Pharm. Biomed. Anal.*, **53**, 983–990.
243. Erratico, C.A.; Szeitz, A. and Bandiera, S.M. (2010) Validation of a novel in vitro assay using ultra performance liquid chromatography–mass spectrometry (UPLC/MS) to detect and quantify hydroxylated metabolites of BDE-99 in rat liver microsomes, *J. Chromatogr. B*, **878**, 1562–1568.
244. Denooz, R.; Mercerolle, M.; Lachâtre, G. and Charlier, C. (2010) Ultra-performance liquid chromatography-tandem mass spectrometry method for the determination of bupropion and its main metabolites in human whole blood, *J. Anal. Toxicol.*, **34**, 280–286.
245. Wang, P.; Sun, H.; Lv, H.; Sun, W.; Yuan, Y.; Han, Y.; Wang, D.; Zhang, A. and Wang, X. (2010) Thyroxine and reserpine-induced changes in metabolic profiles of rat urine and the therapeutic effect of Liu Wei Di Huang Wan detected by UPLC-HDMS, *J. Pharm. Biomed. Anal.*, **53**, 631–645.
246. Li, N.J.; Liu, W.T.; Li, W.; Li, S.Q.; Chen, X.H.; Bi, K.S. and He, P. (2010) Plasma metabolic profiling of Alzheimer's disease by liquid chromatography/mass spectrometry, *Clin. Biochem.*, **43**, 992–997.
247. Zhao, X.; Fritsche, J.; Wang, J.; Chen, J.; Rittig, K.; Schmitt-Kopplin, P.; Fritsche, A.; Häring, H.U.; Schleicher, E.D.; Xu, G. and Lehmann, R. (2010) Metabonomic fingerprints of fasting plasma and spot urine reveal human pre-diabetic metabolic traits, *Metabolomics*, **6**, 362–374.
248. Michopoulos, F.; Theodoridis, G.; Smith, C.J. and Wilson, I.D. (2010) Metabolite profiles from dried biofluid spots for metabonomic studies using UPLC combined with oaToF-MS, *J. Proteome Res.*, **9**, 3328–3334.
249. Want, E.J.; Wilson, I.D.; Gika, H.; Theodoridis, G.; Plumb, R.S.; Shockcor, J.; Holmes, E. and Nicholson, J.K. (2010) Global metabolic profiling procedures for urine using UPLC–MS, *Nat. Protocol.*, **5**, 1005–1018.
250. Bruce, S.J.; Jonsson, P.; Antti, H.; Cloarec, O.; Trygg, J.; Marklund, S.L. and Moritz, T. (2008) Evaluation of a protocol for metabolic profiling studies on human blood plasma by combined ultra-performance liquid chromatography/mass spectrometry: From extraction to data analysis, *Anal. Biochem.*, **372**, 237–249.
251. Li, F.; Patterson, A.D.; Höfer, C.C.; Krausz, K.W.; Gonzalez, F.J. and Idle, J.R. (2010) Comparative metabolism of cyclophosphamide and ifosfamide in the mouse using UPLC–ESI-QTOFMS-based metabolomics, *Biochem. Pharmacol.*, **80**, 1063–1074.
252. Deng, X.; Liang, J.; Lin, Z.X.; Wu, F.S.; Zhang, Y.P. and Zhang, Z.W. (2010) Natural taurine promotes apoptosis of human hepatic stellate cells in proteomics analysis, *World J. Gastroenterol.*, **16**, 1916–1923.
253. Bordeerat, N.K.; Georgieva, N.I.; Klapper, D.G.; Collins, L.B.; Cross, T.J.; Borchers, C.H.; Swenberg, J.A. and Boysen, G. (2009) Accurate quantitation of standard peptides used for quantitative proteomics, *Proteomics*, **9**, 3939–3944.
254. Winter, D.; Seidler, J.; Ziv, Y.; Shiloh, Y. and Lehmann, W.D. (2009) Citrate boosts the performance of phosphopeptide analysis by UPLC-ESI-MS/MS, *J. Proteome Res.*, **8**, 418–424.
255. Yang, H.Y.; Kwon, J.; Cho, E.J.; Choi, H.I.; Park, C.; Park, H.R.; Park, S.H.; Chung, K.J.; Ryoo, Z.Y.; Cho, K.O. and Lee, T.H. (2010) Proteomic analysis of protein expression affected by peroxiredoxin V knock-down in hypoxic kidney, *J. Proteome Res.*, **9**, 4003–4015.
256. Jones, L.M.; Zhang, H.; Vidavsky, I. and Gross, M.L. (2010) Online, high-pressure digestion system for protein characterization by hydrogen/deuterium exchange and mass spectrometry, *Anal. Chem.*, **82**, 1171–1174.
257. Winter, D.; Kugelstadt, D.; Seidler, J.; Kappes, B. and Lehmann, W.D. (2009) Protein phosphorylation influences proteolytic cleavage and kinase substrate properties exemplified by analysis of in vitro phosphorylated Plasmodium falciparum glideosome-associated protein 45 by nano-ultra performance liquid chromatography-tandem mass spectrometry, *Anal. Biochem.*, **393**, 41–47.
258. Sinha, S.; Zhang, L.; Duan, S.; Williams, T.D.; Vlasak, J.; Ionescu, R. and Topp, E.M. (2009) Effect of protein structure on deamidation rate in the Fc fragment of an IgG1 monoclonal antibody, *Protein Sci.*, **18**, 1573–1584.
259. Luna, L.G.; Williams, T.L.; Pirkle, J.L. and Barr, J.R. (2008) Ultra performance liquid chromatography isotope dilution tandem mass spectrometry for the absolute quantification of proteins and peptides, *Anal. Chem.*, **80**, 2688–2693.

260. Ren, Y.; Han, Z.; Chu, X.; Zhang, J.; Cai, Z. and Wu, Y. (2010) Simultaneous determination of bovine α-lactalbumin and β-lactoglobulin in infant formulae by ultra-high-performance liquid chromatography-mass spectrometry, *Anal. Chim. Acta*, **667**, 96–102.
261. Kay, R.G.; Gregory, B.; Grace, P.B. and Pleasance, S. (2007) The application of ultra-performance liquid chromatography/tandem mass spectrometry to the detection and quantitation of apolipoproteins in human serum, *Rapid Commun. Mass Spectrom.*, **21**, 2585–2593.
262. Shiraishi, K.; Matsuzaki, K.; Matsumoto, A.; Hashimoto, Y. and Iba, K. (2010) Development of a robust LC-MS/MS method for determination of desmosine and isodesmosine in human urine, *J. Oleo Sci.*, **59**, 431–439.
263. Legido-Quigley, C.; Stella, C.; Perez-Jimenez, F.; Lopez-Miranda, J.; Ordovas, J.; Powell, J.; van-der-Ouderaa, F.; Ware, L.; Lindon, J.C.; Nicholson, J.K. and Holmes, E. (2010) Liquid chromatography-mass spectrometry methods for urinary biomarker detection in metabonomic studies with application to nutritional studies, *Biomed. Chromatogr.*, **24**, 737–743.
264. Tsutsui, H.; Maeda, T.; Toyo'oka, T.; Min, J.Z.; Inagaki, S.; Higashi, T. and Kagawa, Y. (2010) Practical analytical approach for the identification of biomarker candidates in prediabetic state based upon metabonomic study by ultraperformance liquid chromatography coupled to electrospray ionization time-of-flight mass spectrometry, *J. Proteome Res.*, **9**, 3912–3922.
265. Zhang, H.; Il'yasova, D.; Sztaray, J.; Young, S.P.; Wang, F. and Millington, D.S. (2010) Quantification of the oxidative damage biomarker 2,3-dinor-8-isoprostaglandin-F2α in human urine using liquid chromatography-tandem mass spectrometry, *Anal. Biochem.*, **399**, 302–304.
266. Bouchard, M.; Normandin, L.; Gagnon, F.; Viau, C.; Dumas, P.; Gaudreau, E. and Tremblay, C. (2009) Repeated measures of validated and novel biomarkers of exposure to polycyclic aromatic hydrocarbons in individuals living near an aluminum plant in Quebec, Canada, *J. Toxicol. Environ. Health A*, **72**, 1534–1549.
267. Misaka, S.; Sato, H.; Yamauchi, Y.; Onoue, S. and Yamada, S. (2009) Novel dry powder formulation of ovalbumin for development of COPD-like animal model: Physicochemical characterization and biomarker profiling in rats, *Eur. J. Pharm. Sci.*, **37**, 469–476.
268. Cai, S.; Huo, T.; Li, N.; Xiong, Z. and Li, F. (2009) Lysophosphatidylcholine-biomarker of metformin action: Studied using UPLC/MS/MS, *Biomed. Chromatogr.*, **23**, 782–786.
269. Nakanishi, H.; Iida, Y.; Shimizu, T. and Taguchi, R. (2009) Analysis of oxidized phosphatidylcholines as markers for oxidative stress, using multiple reaction monitoring with theoretically expanded data sets with reversed-phase liquid chromatography/tandem mass spectrometry, *J. Chromatogr. B*, **877**, 1366–1374.
270. Maher, A.D.; Patki, P.; Lindon, J.C.; Want, E.J.; Holmes, E.; Craggs, M. and Nicholson, J.K. (2008) Seminal oligouridinosis: Low uridine secretion as a biomarker for infertility in spinal neurotrauma, *Clin. Chem.*, **54**, 2063–2066.
271. Al-Dirbashi, O.Y.; Santa, T.; Rashed, M.S.; Al-Hassnan, Z.; Shimozawa, N.; Chedrawi, A.; Jacob, M. and Al-Mokhadab, M. (2008) Rapid UPLC-MS/MS method for routine analysis of plasma pristanic, phytanic, and very long chain fatty acid markers of peroxisomal disorders, *J. Lipid Res.*, **49**, 1855–1862.

11

Green Analytical Atomic Spectrometry

Martín Resano, Esperanza García-Ruiz and Miguel A. Belarra

Department of Analytical Chemistry, University of Zaragoza, Zaragoza, Spain

11.1 Atomic spectrometry in the context of Green Analytical Chemistry

Owing to the increasing public awareness on the importance of environmental issues, there is a firm and growing tendency to evaluate the environmental impact of every human activity. Of course, chemistry is particularly sensitive with this situation. In this regard, the concept of green chemistry was developed in the last decade of the twentieth century, a concept that has continued evolving in order to develop suitable patterns that should permit carrying out different chemical processes, and particularly those related with synthesis and production, in a more environmentally-friendly way [1, 2].

Obviously, the environmental impact of analytical labs is expected to be much lower than those related with the synthesis and production of large amounts of chemicals, but that does not excuse analytical chemists to try to adapt their methodologies to the main basic principles of green chemistry, which seems to be, a priori, a feasible goal [3]. In this regard, the response of the analytical community has been positive, since the number of publications dealing with Green Analytical Chemistry [4] has increased steadily during the last 15 years (from 5 to more than 20 papers per year).

This work aims at discussing the environmental characteristics of the main atomic spectrometry techniques used for elemental analysis in the UV-Vis region, together with those based on mass spectrometry for detection. In fact, some authors state that spectrometric techniques are already rather green, since the sample subjected to analysis are relatively small and sample pretreatment is often simple [5, 6]. This state holds especially true for classic molecular techniques such as IR and UV-Vis molecular absorption or Raman spectrometry. It is, however, more dubious that atomic spectrometric techniques fulfill the requisites for being considered as green techniques owing to three main reasons:

Handbook of Green Analytical Chemistry, First Edition. Edited by Miguel de la Guardia and Salvador Garrigues.

1) All these techniques show the clear inconvenience of needing a high energetic supplement that cannot been overlooked from an environmental point of view [7], and that is inherent to all of them since they require atom formation, which is usually accomplished after dissociation of a molecular species in the gaseous phase.

$$\mathrm{MA\ (sol.)} \Longleftrightarrow \mathrm{MA\ (gas)} \Longleftrightarrow \mathrm{M\ (gas)} + \mathrm{A\ (gas)}$$

   In order to achieve this goal, it is important to provide a sufficient amount of energy, in addition to an adequate working environment. Moreover, in many situations, it is also necessary to provide additional energy for the formation of excited atoms or even ions. Only the cold vapor technique, used for Hg determination, where formation of atoms is achieved by means of conventional redox chemical reaction, can be considered as a true exception to this rule.
2) Some of these techniques require the use of a high flow of gaseous species, such that it is necessary to take into account the environmental impact stemming from the production of those (e.g., air liquefaction followed by subsequent distillation for Ar production), as well as the effect deriving from the emission of these gases or their derivatives to the atmosphere (e.g., for flame techniques). Table 11.1 shows a summary of the main environmental drawbacks that can be linked to the production of atoms for analytical purposes.
3) Finally, considering the current degree of sophistication of modern atomic spectrometers, the effects arising from their production should not be underestimated either. Price may be considered as an indirect measure of the manufacturing complexity and to some extent, of its possible environmental impact. Then, it is necessary to remark that these instruments are significantly more expensive than their molecular counterparts. Moreover, traditionally atomic absorption techniques require the use of a different lamp per analyte. On the contrary, emission and mass spectrometry-based techniques are simpler in that aspect and besides, allow multi-element analysis, simplifying time and cost of analytical procedures [8] and overall, making them greener.

In any case, these basic considerations do not exclude atomic spectrometrists from their responsibility to try to develop new analytical methods characterized by a lower environmental footprint. In fact, the main steps of a typical analytical procedure can be divided into four main categories, as shown in Figure 11.1.

As discussed by Tobieszewski *et al.* [9], every one of these categories is a priory amenable of improvement concerning its environmental impact. However, these authors point out that the forth step (evaluation of the results) is of little relevance to this point of view. On the other hand, the first step (sample collection) offers few possibilities for improvement, except for those cases in which analytical techniques that permit in-line measurements are used. Unfortunately, that is often not the case for atomic spectrometric techniques.

Concerning the third step (final analysis), the main characteristics of atomic techniques have been discussed before. The development of more ecological alternatives to obtain the analytical signal is challenging and, in most cases, these alternative approaches do not trespass the academic world to penetrate into routine labs. Obviously, only if the corresponding analytical instruments are commercially available it becomes feasible to make a real impact in this regard.

In the case of electrothermal atomization, the use of short atomization programs, minimizing or even avoiding drying and pyrolysis steps, have been proposed [10]. Even though this may be seen as a step forward in the right direction, most of the energetic consumption takes place in the atomization step, which, obviously, can hardly be eliminated. The development of alternative atomizers seems to be of more ecological relevance. The use of W-based atomizers permits a very significant reduction in energy consumption for sample vaporization and atomization [11]. Moreover, instruments based on these atomizers are often of small dimensions, allowing for their easier transportation to the field in order to carry out in situ analysis (field analytical chemistry) [12], whose results are beneficial also considering the first analytical step (sample collection).

Table 11.1 Environmental disadvantages associated with the production of atoms for main atomic spectrometry techniques.

Atom formation	Main atomic techniques	Main environmental disadvantages
Flame	F-AAS, F-AES	High gas consumption (C_2H_2, N_2O) Gas emission (CO_2)
Electrothermal	GFAAS	High electric consumption Moderate gas consumption (Ar)
Plasma	ICP-AES ICPMS	High gas consumption (Ar)
Plasma	GD-AES GD-MS	High electric consumption

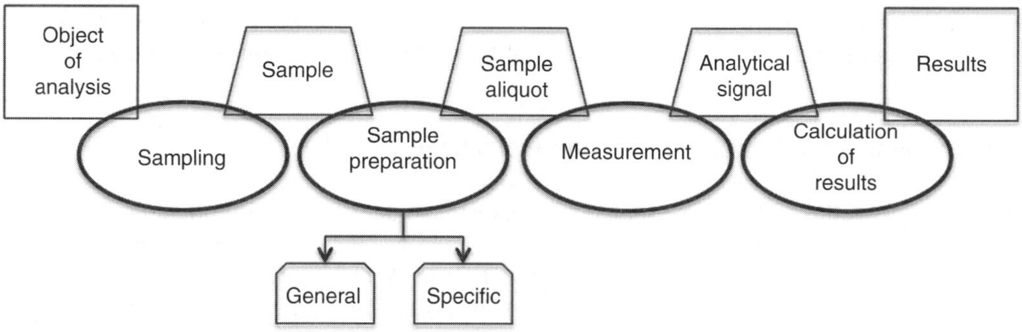

Figure 11.1 Main steps of a typical analytical procedure.

Similar advantages could apply to plasma-based techniques if the use of microplasmas spreads. Obviously, the use of such instruments represents a lower environmental impact during their construction as well as during their operation, opening new ways for miniaturization [13].

A case in which the research phase has proved successful enough to develop commercially available instruments is high-resolution continuum source atomic absorption spectrometry. This technique permits monitoring the whole spectrum (190 nm; to 900 nm) with the same lamp (high energy Xe lamp operating in hot spot mode), thus avoiding the need for using element specific lamps, and therefore reducing the environmental impact deriving from manufacturing all these lamps [14]. Moreover, this technique also allows for the monitoring of molecular species, which only need vaporization instead of atomization. Thus, in some cases, it is feasible to work at significantly lower temperatures, therefore minimizing energy consumption [15].

The second step (sample preparation) is, without question, the one that has received most attention in the development of Green Analytical Chemistry applications, either from a general point of view [16] or in the particular field of atomic spectrometry [17], which seems sensible because, on many occasions, this step represents the main risk for analyte contamination [18]. The various operations included in this step could be classified in two different areas, even if this classification is not always very simple: (1) general operations, terms referred to are those that are independent from the technique that will be subsequently deployed to obtain the analytical signal (e.g., the dissolution of a solid sample); (2) specific operations, when their execution depends on the use of a particular analytical technique (for instance, the formation of slurries, which can be directly analyzed by some atomic techniques only). This work will only discuss recent developments of the second category, since more general developments will be covered in other chapters of this book.

Concerning sample pretreatment, it is obvious that the most ecological approach would be to avoid this step at all, such that the analytical signal would be obtained directly from the sample. This is the ultimate wish of Green Analytical Chemistry [19], which, typically, can be accomplished only in practice for some very simple liquid samples. However, when the target sample is solid, there are some atomic spectrometry techniques that also permit pursuing this objective [20]. If those techniques can also deliver multi-element information, the obvious advantages of such approach can somewhat counterbalance the negative aspects discussed in the beginning of this introduction. Despite the positive environmental impact stemming from the use of direct solid sampling techniques in atomic spectrometry, these have received very little attention in Green Analytical Chemistry publications. Therefore, this work will be mainly focused on them.

Finally, it is also necessary to remark that these solid sampling techniques can be used as screening methods, one of the aspects that is considered of the utmost importance within Green Analytical Chemistry [3, 21]. As will be discussed, these solid sampling techniques are typically characterized by relatively high imprecision values, which can be a serious limitation in some cases (for instance, if the result is very close to a legal threshold). Even in these circumstances, these techniques can be used to filter out a large number of samples, such that a more complex and energy and time-consuming methodology would only be deployed for a reduced number of them, thus helping in the implementation of Green Analytical Chemistry methodologies [22].

11.2 Improvements in sample pretreatment strategies

This section discusses those alternative sample pretreatment strategies, proposed to be used in conjunction with atomic spectrometric techniques, which represent improvements from the perspective of the green chemistry policy or, most properly, follow the concept of Green Analytical Chemistry [17].

In this regard, elimination or reduction of organic solvents and/or highly toxic or ecotoxic reagents and the reduction of waste generation are obvious goals, and miniaturization approaches may help in achieving these objectives.

Miniaturization is a trend that not only affects analytical chemistry, but science in general, becoming a hot research topic during the last decade [23–25]. The range of typical sizes in most scientific fields (material science, robotics, electronics, etc.) has decreased from macro (m) to micro (μm) and, nowadays, to nanometers. There are specific journals devoted to that particular topic, such as *lab on a chip*. From the point of view of Green Analytical Chemistry, it is obvious that miniaturization allows for the use of smaller quantities of sample, with the consequent savings in the amount of reagents used and the corresponding decrease in waste materials. These aspects, until now, have been considered as positive owing to the obvious reduction in the general cost of the analysis. However, miniaturization should also be regarded as a greener approach.

Alternatively, the use of Room Temperature Ionic Liquids (RTIL), defined as any salt with a melting point below the room temperature, could also be interesting for the development of greener atomic spectrometry methods [26–29]. These liquids show a great potential in green chemistry applications due to their significantly smaller vapor pressure, which make them almost ideal solvents, although their current application in analytical chemistry is restricted for the moment to their use in gas chromatography, MALDI-MS and in preconcentration processes prior to electrothermal atomic absorption spectrometry [30–32].

Some particular improvements that have proved more common and successful in atomic spectrometry are discussed below.

11.2.1 Specific improvements

Lixiviation methods, based on the use of diluted reagents (mainly acids) do not appear to be the best candidates to start a ranking of greener analytical pretreatment methods. However, a real effort has been made in this field to become greener. These methods involve the removal of the analyte from the sample, but generally do not involve matrix dissolution, except in the case of biological samples. More efficient lixiviation schemes can be implemented with the support of ultrasounds, microwaves or increased pressure, as well as using enzymatic approaches or nanomaterials.

11.2.1.1 Ultrasounds

In recent years, use of ultrasounds has proved useful for decreasing the amount of harmful reactants involved, as well as simplifying experimental procedures. This is a clean technology with numerous analytical applications such as surface cleaning, extraction, digestion, homogenization, and so on. One work that exemplifies well the benefits that can be achieved from the point of view of the application of Green Analytical Chemistry to atomic spectrometry has been published by Gil *et al.* [33]. In this work, a new cold vapor technique is described for the determination of mercury by means of an ultrasound-assisted reduction/volatilization process, followed by atomic absorption spectrometry for Hg detection and quantification. In this method, the use of a chemical reducing agent or of any other type of chemical is avoided, consequently eliminating production of those hazardous wastes typically related to the use of such reducing agents.

11.2.1.2 Microwaves

Microwave-assisted sample pretreatment/digestion is practically the method of choice in atomic spectroscopy. From a green point of view, the rapid heating achieved as a consequence of the microwave irradiation and the use of closed vessels under controlled conditions of pressure and temperature are some of the main advantages offered. In addition to this, one of the most important improvements to date was first reported by Lorentzen and Kingston [34], who developed a method that permitted to achieve total digestion of a sample using diluted acids only, instead of concentrated reactants as usual. Recent work carried out by Flores *et al.* [35] based on digestion by microwave-induced combustion is also noteworthy. In this work, traces of metals of catalyst residues are determined in carbon nanotubes using oxygen for this combustion, and absorbing the resulting products in diluted solutions of acids, prior to flame or graphite furnace atomic spectroscopy, ICP-OES or ICP-MS monitoring [35].

11.2.1.3 Pressurized liquid extraction

Pressurized liquid extraction, a well-known technique for the extraction of organic compounds, is based on the use of solvents at a high pressure and/or high temperature without reaching the critical point [36]. In this situation, the properties of the solvents are modified and the solubility of the analytes in the extracting solution is increased. Due to this increased solubility, the extraction can be performed with smaller amounts of solvents, and the generation of waste is subsequently reduced. In addition to this, it is possible to use water as extracting solvent, which further enhances the green side of this approach [37].

This method has been proposed as an environmental friendly alternative to acid digestion procedures for metal extraction, using home-made or commercially available devices developed for this purpose. There are a high number of applications based on this strategy [38–41].

11.2.1.4 Enzymatic approaches

It is possible to combine all the methods of lixiviation described before with the use of enzymes, resulting in a group of procedures that can be applied to biological samples working under mild conditions, without the use of any toxic reagents. However, in the end, depending of the type of enzyme used, it is still necessary to deal with the residues.

As an example of the latest research in this field, Moreda-Piñeiro *et al.* proposed the use of a pressurized enzymatic-assisted hydrolysis extraction for arsenic speciation in seafood samples prior to HPLC-ICP-MS determination together with a simultaneous in situ clean up procedure that speeds up the sample treatment [42].

11.2.1.5 Extraction-preconcentration approaches with nanotubes

Carbon nanotubes have been mentioned before in another context (as potential target sample). It is also becoming trendy to evaluate the potential of these materials for the simultaneous extraction and preconcentration of trace metals and also, for the separation of one metal from another one, as performed elsewhere [43–46]. All the methods of extraction-preconcentration based on the use of nanotubes do not exempt the use of different reactants, but the required volumes are very small and, in the case of their combustion, there are no significant residues.

11.2.1.6 Cloud point extraction

A new sample pre-processing procedure with a methodology that agrees with the Green Analytical Chemistry principles is the cloud point extraction. This is a solventless solvent extraction, in which the solvent phase is replaced by a surfactant in which preconcentration and separation of the analytes from the sample takes place [47–49]. The small volume of surfactant phase used confers good characteristics to this technique, although these surfactants are not always innocuous. However, the more eco friendly tendency is based on the use of non toxic surfactants [50].

11.2.2 Slurry methods

Slurry sampling techniques are used in atomic spectrometry as a means to carry out the (quasi) direct determination of metals in solid samples, and provide some good analytical features, including some advantages over conventional procedures based on sample digestion, such as:

- increased sample throughput,
- prevention of analyte losses and/or contamination,
- the same automated device used for liquid dosing may also be applied for slurry introduction.

Most of these features are shared with truly direct solid sampling techniques that will be discussed later on this chapter. However, it should be mention that slurries techniques still require the use of reagents that can be considered as contaminants for acid leaching (e.g., HNO_3) or as surfactants (Triton X-100, for example).

Drawbacks in the past associated with sedimentation of particles or flotation of light material of the slurry are avoided with the addition of surfactant agents to the medium. With respect to the liquid media used to generated the slurry, in some occasion water can be a good choice, working in that case with an environmental friendly reactant but, as stated before, the use of acids is very frequent, which a priori causes the same problems as sample digestion related to waste generation. However, it has to be remarked that the amounts of acids used are normally much lower than those used for wet digestions.

Most of the applications of slurry sampling techniques are designed for atomic absorption spectrometry (AAS). There are two recent works of Ferreira *et al.* [51] and Resano *et al.* [52] that review the state-of-the-art of this technique. Recent trends in this context involved the use of high-resolution continuum source AAS, which has been demonstrated to offer relevant advantages in conjunction with the slurry sampling approach for the direct analysis of yogurt samples [53] or for the simultaneous determination of Cd and Fe in sewage sludge. More comments on this technique will be provided later on this chapter [54].

Next to atomic absorption spectrometry, inductively coupled plasma (ICP)-based atomic emission spectrometry (AES) has also been deployed successfully together with slurry sampling. The backbone in the application of slurry sampling to ICP techniques is the stability and homogeneity of the slurry, necessary in order to obtain a homogeneous aerosol for introduction into the plasma, but it is not an easy task to solve the problems related to particle size distribution, a fundamental parameter that controls the efficiency for transport and atomization-excitation in the plasma. All these problems can be mostly solved if slurry sampling is coupled to electrothermal vaporization prior to ICP-AES or ICP-mass spectrometry (MS) detection [55, 56].

Another sample pretreatment approach consists in the combination of slurry sampling with chemical vapor generation [57]. However, on some occasions, authors have reported the use of strong acid media, including aqua regia alone or combined with HF to prepare the slurries, as well as time-consuming approaches lasting several hours [58]. Thus, some of these approaches can hardly be looked as very green.

11.3 Direct solid sampling techniques

Despite all the improvements discussed before, when the goal is to analyze solid samples, the major breakthrough would consist in not needing any sample pretreatment at all. There are a number of solid sampling techniques that permit to achieve this goal in some situations, although in cases a minor degree of sample pretreatment may still be required.

In addition to these advantages, these techniques offer other significant analytical improvements such as a higher sensitivity (as a consequence of avoiding the dilution that accompanies sample digestion), a high sample throughput (by eliminating the digestion and preconcentration steps) and a reduced risk of suffering from contamination and/or analyte losses. Obviously, they also show some drawbacks, as their precision depends considerably on sample homogeneity and the development of suitable calibration schemes that ensure accuracy is not always straightforward.

It is the main aim of this chapter to present information on the current potential of some solid sampling techniques that, according to advances reported in the literature during the last decade, offer possibilities to implement greener strategies in atomic spectrometry. The techniques discussed comprise solid sampling-graphite furnace atomic absorption spectrometry (SS-GFAAS), solid sampling-electrothermal vaporization-inductively coupled plasma mass spectrometry (SS-ETV-ICPMS), laser ablation-inductively coupled plasma mass spectrometry (LA-ICPMS) and glow discharge-mass spectrometry (GD-MS).

11.3.1 Basic operating principles of the techniques discussed

The fundamentals of the techniques discussed in this paper are illustrated in Figure 11.2 and their main analytical properties are summarized in Table 11.2.

In SS-GFAAS and SS-ETV-ICPMS, the solid sample is weighed and introduced into a graphite furnace that is subsequently subjected to a temperature program in an Ar atmosphere. The temperature (which can reach up to 3000°C) and holding time of each individual step of the temperature program can be accurately controlled, and the total time per measurement is usually 2–3 min.

In SS-GFAAS, the goal of this temperature program is to produce atoms of the target element. These atoms interact with the radiation of a line source (traditionally a hollow cathode lamp), providing the analytical signal (atomic absorbance) [52]. This is typically a mono-element technique, which is its main disadvantage for many applications. However, this technique has been recently revitalized with the arrival of high-resolution continuum source technology [14], as will be discussed later on.

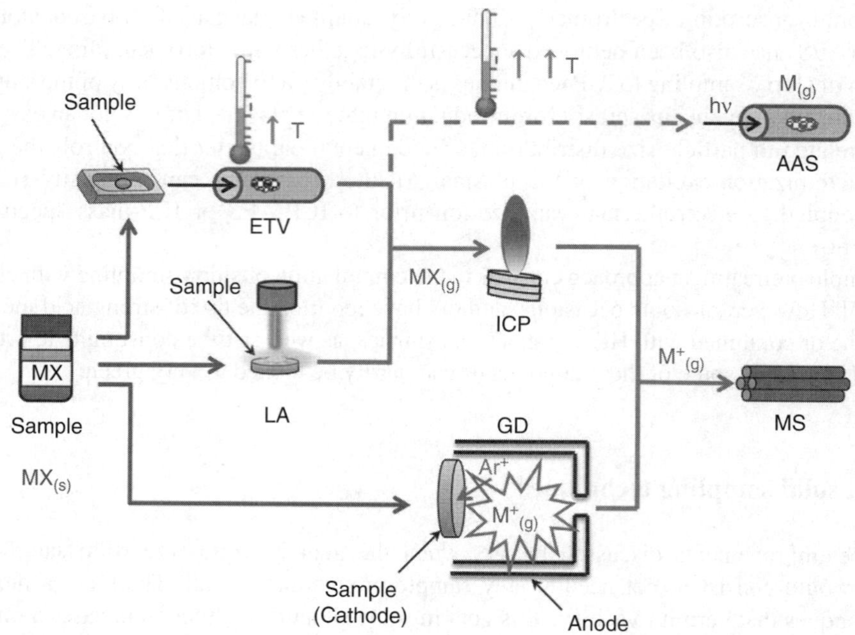

Figure 11.2 Basic operating principles of the different solid sampling techniques discussed in this work: solid sampling graphite furnace atomic absorption spectrometry, solid sampling electrothermal vaporization-inductively coupled plasma mass spectrometry, laser ablation-inductively coupled plasma mass spectrometry and glow discharge-mass spectrometry. M represents the analyte (normally a metal), while X represents an ionic species.

Table 11.2 Comparison of the main characteristics of the solid sampling techniques discussed in this paper.

	GF-AAS	ETV-ICPMS	LA-ICPMS	GD-MS
Types of samples	All (cutting into pieces of appropriate size may be required)	All (cutting into pieces of appropriate size may be required)	All (preparation of tablets for powders is required)	Conductors[a] (in the form of disks[b])
Usual working range	10 ng g^{-1} to 100 µg g^{-1}[c]	1 ng g^{-1} to 1000 µg g^{-1}	100 ng g^{-1} to %	1 ng g^{-1} to %
Precision	5–15%	5–15%	5–25%	5–10%
Selectivity	High	Medium-High[d]	Medium-High[d]	Medium-High[d]
Isotopic Analysis	No	Yes	Yes	Yes
Multi-element potential	Usually no	Yes	Yes	Yes
Sample throughput	15 min. per analyte	15 min. per sample	2–5 min. per sample	10 min. per sample
Sample Consumption	0.1–1 mg	0.1–1 mg	100 ng – 10 µg	10–100 mg
Spatial Resolution				
Lateral	Poor (mm)	Poor (mm)	Excellent (µm)	Poor (mm)
In depth	Poor (mm)	Poor (mm)	Very good (≈ 0.1 µm)	Excellent (< 10 nm)

[a]Some strategies allow for the analysis of insulators, for example, mixture with a conductive material; [b]Analysis of different sample geometries may be possible with some instruments; [c]Depends on the availability of less sensitive atomic lines; [d]These aspects greatly depend on the type of mass analyzer.

In SS-ETV-ICPMS, on the other hand, the graphite furnace is only used to vaporize the analyte(s), and the atoms or molecules of the analyte(s) formed are subsequently transported by means of an Ar flow into an ICPMS for detection and quantification. The high temperature of the ICP (up to 10 000°C) ensures efficient atomization and ionization of the analyte species, and the detection power of the mass analyzer (a quadrupole filter in most cases) permits the 'simultaneous' (i.e., during the same measurement) quantification of different isotopes of the analyte or analytes of interest [59].

A different approach, also based on the potential of ICPMS for ultratrace analysis, is known as laser ablation (LA)-ICPMS. The principle of the technique is based on the use of a very short laser pulse (a few nanoseconds or even femtoseconds) that delivers a burst of energy to the sample surface. A small amount of material is ablated and the aerosol thus formed is transported by a gas flow into the ICPMS instrument, where ionization, detection and quantification take place. When this technique is used, the amount of material that is analyzed is orders of magnitude below that of SS-GFAAS or SS-ETV-ICPMS (μg or less), so it can be considered as a *quasi* non-destructive technique. Moreover, the use of a laser system with a very small spot size (from a few μm to hundreds of μm) allows spatial information to be obtained both laterally and/or in depth [60].

Finally, in glow discharge (GD)-based techniques, a high potential is also applied between two electrodes (with the sample acting as the cathode) that are immersed in an inert gas environment (most often Ar) at reduced pressure. The application of this potential causes the gas to break down to form electrons and positively charged ions, which results in the formation of a plasma. The positive ions are accelerated towards the sample surface with sufficient energy to cause the release of surface material into the gas phase. The atoms released undergo a large number of collisional processes in the plasma, resulting in excitation and ionization. Under these conditions, it is possible to monitor the analyte atomic emission. However, for maximum sensitivity, the analyte ions thus formed can be extracted into a mass spectrometer, where they can be separated according to mass-to-charge ratio and quantified [61].

11.3.2 Sample requirements and pretreatment strategies

The requirements that a sample should fulfill for analysis with these techniques are diverse. In SS-GFAAS and SS-ETV-ICPMS, both powders and discrete particles can be directly analyzed, although cutting the sample into pieces of appropriate size (typically a few milligrams) may be required, which may be easy in some applications (e.g., polymer analysis) but more cumbersome for other samples (e.g., metals). On the other hand, for the rest of the techniques, the sample must be present in a compact form, and pelleting is thus needed if analysis of a sample in powder form is intended. Moreover, for GD-MS analysis, the sample must be electrically conductive or be mixed with a conductive binder prior to analysis, unless the so-called radio-frequency GD technique is used [61]. However, the latter is not commercially available yet.

It has to be taken into account that, depending on precision requirements and on sample characteristics, some sample pretreatment might be advisable. For instance, grinding of very inhomogeneous samples (typically environmental samples such as soils and sediments) may result in a more homogeneous target for analysis, resulting in an improved precision. Alternatively, optimum measurement conditions can be used to improve the reliability of the results. For instance, in SS-GFAAS carrying out a suitable number of replicates and using the median as representative value allows obtaining more accurate data in most situations [20, 62], while in LA-ICPMS the ablation mode known as raster or line profiling (where the laser beam is translated across the sample surface while ablation proceeds) permits achieving a more representative sampling compared with the alternative mode known as drilling, whereby the sample is continuously ablated in-depth at a fix spot [63].

In any case, it is relevant to remark that those sample pretreatment strategies typically used for these techniques do not involve the use of acids or toxic solvents.

11.3.3 Analyte monitoring: The arrival of high-resolution continuum source atomic absorption spectrometry

It is beyond the scope of this chapter to discuss in detail the different measurement protocols associated with the different techniques mentioned before, besides the basic discussion provided in Section 11.3.1. However, owing to the significant differences resulting from the recent introduction of high-resolution continuum source AAS instrumentation, some of which represent a relevant contribution to the development of greener approaches, this technique will be covered in more detail herein.

The use of continuum source atomic absorption spectrometry (CS-AAS) cannot be considered as a new topic. In fact, the first atomic absorption spectrometers, designed by Kirchhoff and others in the nineteenth century, already used a continuum source of radiation. However, the results were not satisfactory, with the consequence that atomic emission spectrometry was clearly the preferred technique for a long time [64].The main reason behind these poor results is that observing (and quantifying in a reliable way) the slight reduction in intensity obtained as a result of the absorption of atoms in the emission intensity of a continuum source is only possible with a monochromator of very high resolution that provides a spectral bandwidth corresponding to the half-width of atomic absorption lines (a few picometers). Thus, it is not surprising that when atomic absorption spectroscopy was reassessed as a feasible technique for trace elemental analysis by Sir Alan Walsh in the early 1950s, he designed his instrumentation based on the use of a line source (e.g., a hollow cathode lamp), which emits sharp radiation lines at the exact wavelengths at which atomic absorption takes place. This operation mode can provide satisfactory results even when using spectrometers of low resolution (at the nm level), since the actual analytical resolution is achieved because there is an optimal degree of correspondence between the emission (lamp) and the atomic (absorbance of analyte) lines. Even though line source (LS)-AAS has been the dominant technique during the last five decades, the idea of using CS-AAS instead was never abandoned. The reason is that the concept of LS-AAS requires the use of a specific lamp for every analyte to be determined, forcing laboratories to accumulate tens of lamps (some of which will only occasionally be put to use) and making AAS a mono-element technique. There was always the idea that the use of CS-AAS could be an elegant way to overcome this situation, but for many years it was not possible to design spectrometers of sufficient resolution, as well as to obtain a suitable continuum source able to provide sufficient energy in the far UV.

These and other technical difficulties seem to have been overcome by Becker-Ross and co-workers, who developed a continuum source-AAS instrument, based on (2) a high pressure xenon short-arc lamp operating at temperatures of approximately 10 000°C, capable of providing a high intensity in the visible and (far) UV region; (2) an optical system based on an echelle monochromator dispersing light in two steps (by using a prism first and an echelle grating afterwards) and (3) a linear CCD array for detection [14, 65]. This device, whose concept is shown in Figure 11.3, has been made commercially available by Analytik Jena in two instrumental configurations: a flame-only device (ContrAA 300), and a device that combines flame plus graphite furnace in tandem (ContrAA 700). The latter appeared in the market in late 2007. Since it is the graphite furnace atomizer the one that permits the direct analysis of solid samples, this section will focus on GFAAS only.

The main advantages of this HR-CS GFAAS setup have been summarized by Welz as follows [65]:

(a) Improved signal stability. The array detector permits the simultaneous monitoring of two hundred 'wavelength sectors' (pixels), enabling efficient correction for many unwanted spectroscopical events (e.g., lamp flicker noise, temporal changes in the transmittance of the gas phase, etc.). This aspect, in combination with the higher radiation intensity of the Xe lamp compared with that of HCL, makes it feasible to obtain better-defined signals at lower analyte levels, and results in somewhat improved limits of detections (a factor of 3 to 5).

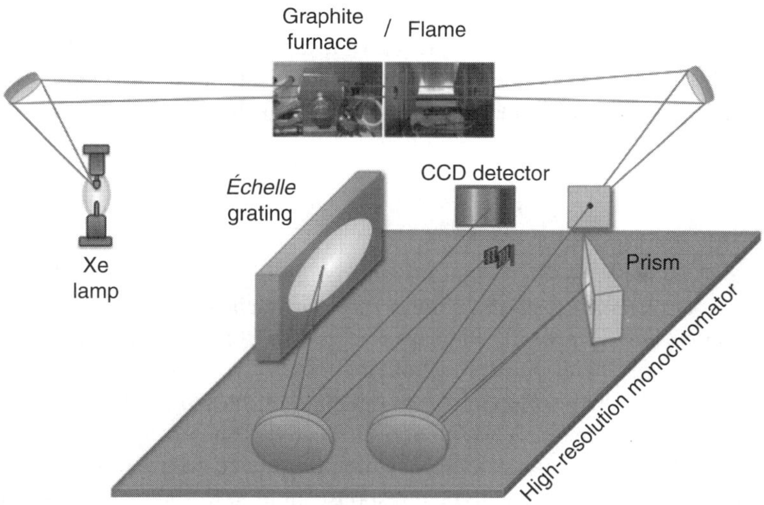

Figure 11.3 *Instrumental setup of a high-resolution continuum source atomic absorption spectrometer.*

(b) HR-CS GFAAS offers superior background correction potential. In the first place, this technique allows monitoring the spectral neighborhood of the analyte lines, thus permitting to detect potential interferences. In the case that these interferences cannot be resolved in time (by optimization of the temperature program combined with the use of chemical modifiers), the technique still offers a considerably superior approach for the correction of complex, high and even rapidly changing backgrounds, since they are monitored simultaneously with the atomic absorption lines [66].

(c) It is possible to decrease the sensitivity and therefore expand the working range by measuring the absorbance at the wings of absorption lines [67].

(d) It opens the doors to monitoring low levels of non-metals, such as Cl, F, P or S, for which using narrow molecular absorption 'lines' may be an appealing alternative since the most sensitive atomic lines for these elements are situated in the far UV region, which is not accessible for LS-AAS instrumentation [68].

(e) Even though the spectral bandwidth that is simultaneously monitored is very narrow (approx. 0.2–0.3 nm in the UV region and up to 0.5 nm in the visible region), there is still some potential to carry out multi-element monitoring either simultaneously (if the elements shows atomic lines that are very close) or sequentially, by means of a multiple atomization temperature program [69].

From the point of view of Green Analytical Chemistry, these advantages can be considered as very relevant, since they become really useful when the goal is to develop methods for the direct analysis of solid samples and other complex materials (e.g., oils). Indeed, for these challenging applications, being capable of detecting lower analyte contents, correcting the background in a more efficient way and achieving some flexibility with regard to the sensitivity (it has to be remembered that when the goal is to directly analyze a solid material it is not very simple to dilute the sample), are all key benefits. Multi-element analysis is obviously another important advantage, as discussed before.

However, a conceptual change is also possible according to (d). In fact, the use of the high-resolution continuum source technique is not only restricted to the monitoring of atomic absorption, but also molecular lines can be detected and quantified. This is of immediate importance to detect elements that could hardly be measured before by means of GFAAS, such as Cl, F or S. It makes also feasible to detect molecular species as an alternative to the production of atoms, which is a process that typically requires more temperature (meaning higher energy consumption). For instance, in the case of direct determination of P from biological solid samples, monitoring PO lines makes it feasible the use of a vaporization temperature of only 1900°C, while for efficient P atomic absorption measurement the use of an atomization step of 2600°C is necessary [15]. The same holds true for some metals as well. Instead of monitoring Al atoms, it becomes possible to monitor AlF molecules, for instance, that are easier to vaporize.

11.3.4 Calibration

This is always an important aspect when it comes to analyze solid samples. Most solid sampling techniques rely on the availability of solid materials of a composition sufficiently close to that of the target sample for achieving accurate results. This is typically the case for glow discharge or laser ablation-based techniques. However, the benefits of using multi-element techniques for detection, whether we are referring to optical emission or mass spectrometry, may soften this requirement a bit. When the signal of an element of known concentration is monitored simultaneously with that of the analytes and used as internal standard, it becomes possible to use materials of different composition for calibration, since the use of this internal standard should correct for the expected difference in the analytical response, at least to some extent. In any case, analysis of materials for which a large number of certified reference materials are available (e.g., metals or glasses) is always more straightforward with these techniques, as it becomes easier to choose calibration standards that are relatively similar to the target sample, which are sufficiently homogeneous and in which the analytes of interest are present.

The case for those techniques based on the use of the graphite furnace (GFAAS or ETV-ICPMS) is exceptional in this regard. The multi-step possibilities of these techniques make the process of measurement a bit slower in comparison with the previously mentioned ones. However, by judicious selection of the atomization/vaporization temperature, in combination with the use of chemical modifiers, it is often possible to separate in time the release of the analytes and that of the major matrix components. Under these circumstances, matrix effects are greatly reduced and the use of aqueous standards for calibration becomes feasible, as will be shown in some of the applications discussed in the next point.

11.3.5 Selected applications

The last part of this chapter will be devoted to briefly discuss some applications related with the techniques mentioned before that were considered to be of significance in the context of Green Analytical Chemistry and that have been published during the last five years. This part does not try to be comprehensive and more examples of solid samples applications can be found in other reviews [52, 70, 71].

For instance, a clear example is the implementation of solid sampling techniques for the analysis of platiniferous ores in the platinum group metal (PGM) industry [72]. Owing to the low levels at which these elements and Au are found in the target ores (below 10 µg PGM g^{-1}), their heterogeneous distribution within the matrix, the difficulties associated with bringing these elements into solution in a quantitative way and the presence of other elements at much higher levels in the samples, with the subsequent risk for spectral overlaps, analysis of platiniferous ores is one of the most challenging analytical tasks that can be undertaken in elemental analysis.

Obviously, the importance of these samples makes it compulsory to develop reliable methodologies for their analysis. PGM determination is normally preceded by isolation of the target elements from matrix components and preconcentrated by means of fire assay methods. Different types of fire assay methods can be deployed for this purpose. The Pb fire assay offers a relatively cheap and fast alternative. The NiS fire assay is more time consuming and expensive, but offers a better result in terms of analyte recoveries for all of the elements of interest.

Regardless of the fire assay method selected, the typical analytical strategy involves the dissolution of the Pb or NiS buttons in which the analytes are present. However, on top of being labor-intensive and time-consuming, there is one important additional disadvantage associated with the dissolution of these buttons when used for routine laboratory analysis at an industrial level. That is the generation of toxic wastes, such as dissolutions with high levels of Pb in the case of Pb assay. In the case of the NiS assay, the procedure results in the generation of an enormous quantity of waste HCl solution containing high levels of Ni, which is very costly to dispose of. Even worse is the production of a large amount of very toxic H_2S (g), which has to be removed by scrubbing and oxidized using H_2O_2 and NaOH. Obviously, the implementation of 'greener' approaches would represent important benefits from both economical and environmental points of view.

Thus, there is an ongoing tendency to implement direct solid sampling approaches to analyze this NiS or Pb buttons, greatly minimizing all these concerns [72]. This is a special solid sampling case because the analytes are extracted into a new matrix of known composition (practically 100% NiS or Pb), which is more homogeneous than the former ore. Thus, it is simpler to prepare matrix-matched standards for calibration and use the signal of Pb or Ni as internal standard for quantification. Despite these benefits, and the typical sensitivity and multi-element potential of mass spectrometry techniques, there are still some problems deriving from the occurrence of spectral overlaps (e.g., $^{206}Pb^{++}$ for the monitoring of Rh unique stable isotope, $^{103}Rh^{+}$, in the case of analysis of Pb buttons; ArNi and ArCu polyatomic ions for the monitoring of Ru, Rh and Pd nuclides). The first two works shown in Table 11.3 [63, 73] have evaluated different possibilities for overcoming these overlaps, such as the use of an ICPMS equipped with a reaction cell in which a gas reactant (NH_3 in this case) is added for selective chemical removal of the overlapping species. Satisfactory results were demonstrated with that approach for LA-ICPMS analysis of Pb [73] or NiS buttons [63]. It can be stated that Pb buttons can be directly analyzed with the laser, while the NiS buttons need to be ground for achieving better homogeneity and pelletized with the help of a binder (polyethylene wax) for obtaining good mechanical properties. Alternatively, the use of the highest mass resolution setting of a glow discharge-sector field-mass spectrometer proved to be sufficient to permit accurate analysis of the Pb buttons mentioned before [73].

Other field in which the use of direct solid sampling methods has resulted beneficial is analysis of polymers. There is a growing concern on the use and abuse of polymers and their effect on their environment, particularly considering that some of the additives traditionally used for these samples are very toxic. In this regard, there is a recent European Commission Directive, the Restriction of Hazardous Substances (RoHS) directive 2002/95/CE (complemented by directive 2005/618/CE), regulating the maximum amount of Cd, Cr, Hg and Pb, among other species, permitted in all types of plastics used to produce electronic equipment.

Analysis of these polymers is not always an easy task and sometimes requires the use of hazardous reagents (e.g., organic solvents, acids, etc.). In some cases, the main problem is the composition of the polymers themselves, which may result in the release of unwanted substances. Interestingly, these types of samples are amenable of relatively easy direct analysis by means of graphite furnace-based methods, since their matrix is easily removed during the pyrolysis. Moreover, these are typically very homogenous samples. Thus, several examples of polymer analysis are reported in the literature using GFAAS or ETV-ICPMS. References [74–78] are just a few of them.

Of these examples, References [74] and [75] show the potential of LA-ICPMS and ETV-ICPMS for multi-element analysis. However, while for LA-ICPMS the use of solid standards for calibration was considered a need, in ETV-ICPMS it is possible to use just aqueous standards for the reasons discussed in the previous paragraph. Moreover, [75] epitomizes very well the properties of the direct approach because it deals with the analysis of fluoropolymers that are very resistant to acids, and thus very hard to digest. Combustion approaches are typically used for analysis of these materials, but they result in the production of HF. While it may be possible that some HF is formed during the process of ablation or vaporization of these samples when using the solid sampling techniques discussed in this chapter, it has to be kept in mind that the amount will always be minimal, as these techniques work with much smaller samples amounts than conventional digestion/combustion approaches, and still get equal or better sensitivity owing to the fact that samples are not diluted.

Of the toxic elements typically controlled in polymers, probably the most difficult to determine is Hg, owing to the extreme volatility of its species, which are prone to be lost before the measurement takes place. This element is also not simple for GFAAS monitoring, but the arrival of the HR CS GFAAS technology has improved considerably the situation, as it brings a more powerful and accurate way to correct for background, as demonstrated in [77]. With this technique, it becomes possible to directly determine Hg in polymers in a straightforward way, even when most matrix components are almost simultaneously vaporized together with the analyte, because it is feasible to achieve a signal that is directly comparable with that obtained for aqueous standards.

The benefits of HR CS GFAAS have also been shown clearly for analysis of oils and biodiesel samples. While these are not solid samples, they are very complex to bring into solution and in fact, several works have explored the advantages deriving from treating them as if they were solid samples (weighing them and depositing them into the graphite furnace using a solid sampling accessory) instead of as liquids, owing to their high viscosity [79–81]. Analysis of these samples is very tedious and requires again the use of acids or of organic solvents to prepare emulsions. Instead, direct analysis using a solid sampling strategy requires just 10-minute homogenization in an ultrasonic bath and micro amounts of a surfactant (Triton X-100) and Pd solution as chemical modifier [79, 80]. Moreover, the use of HR CS AAS makes it possible to determine more than one element simultaneously, providing their atomic lines are very close, as discussed in Section 3.3. This feature is shown in the papers published by Dittert *et al.*, in which Cr and Fe [79] and Co and V [80], respectively, are determined in simultaneous mode.

Another important advantage of the HR CS GFAAS technique discussed before is the possibility of monitoring molecular lines. These typically correspond to the hyperfine rotational structure of molecular electronic transitions that, owing to the high resolution of the instrument (1 to 2 pm, depending on the wavelength), can be fully appreciated. Most of these 'lines' show the typical width of an atomic line (approximately 10 pm). This feature opens possibilities to determine elements that were not accessible before to AAS and these can be determined directly in solid samples. For instance, Ferreira *et al.* investigated the monitoring of CS lines for S determination in solid biological samples [82] with satisfactory results. As discussed before, even for elements for which atomic lines are available (e.g., P), monitoring of molecular bands (PO) may be useful to develop milder atomization procedures that result in the consumption of less energy [15].

Other types of samples that are hard to dissolve are those of environmental interest, such as soils and sludges. Again, dissolution of these samples requires the use of concentrated and dangerous reagents, such as HF. The use of solid sampling is an appealing alternative [22, 83]. Moreover, it has to be considered whether true quantitative results are always needed. In many situations, important decisions are made on the basis of whether an analyte is above or below a determined value (e.g., a legal limit), rather than on detailed, specific chemical information. Therefore, one of the current trends in Green Analytical Chemistry is to develop screening systems, defined as those systems which can provide a clear binary (Yes/No) response with the maximum simplicity and the minimum possible delay [84, 85]. These methods can be used in a two-stage system to filter out a large number of samples, such that only in those situations in which the response is considered as not totally reliable, would a confirmatory answer be required by means of an alternative technique. In this regard, it has been demonstrated that many of the solid sampling approaches discussed in this chapter, which perhaps may suffer to provide very precise results owing to the lack of homogeneity of these environmental materials, are instead suitable to be used as screening methods.

In Reference [22], Resano *et al.* proposed a method for Hg screening in soil samples to be used for agricultural purposes. This is a clear example of a situation in which only a binary response (Yes/No) is

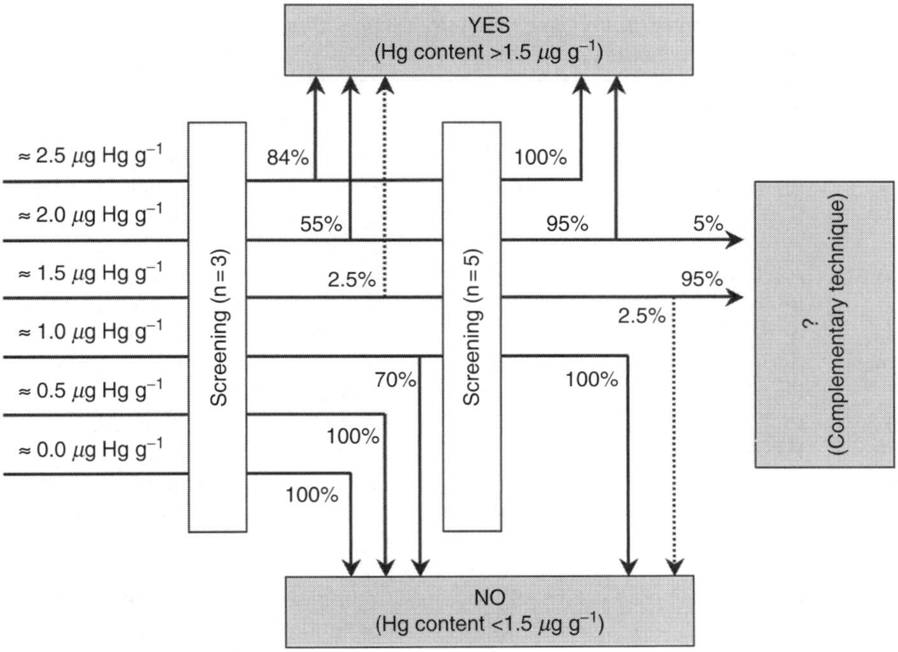

Figure 11.4 *Results obtained from the screening of Hg in real soil samples of different Hg concentrations. Reproduced by permission of the Royal Society of Chemistry (http://pubs.rsc.org/en/Content/ArticleLanding/2005/JA/b509645f). For more information, see Reference [22]. Reproduced by permission of The Royal Society of Chemistry © 2005.*

needed, as there is an EC legal limit 1.5 μg g^{-1} for Hg. The screening method proposed allows obtaining a reliable answer in 6–10 min per sample, and only for those samples with a Hg content very close to the legal value (1.15–1.95 μg g^{-1}), would the use of a complementary technique for further analysis be required. Figure 11.4 shows the results obtaining when applying the method to real soil samples of varying concentrations. It can be seen that in the majority of cases, samples can be screened correctly with only three replicates (6 min). Only for those samples with contents close to the threshold it is necessary to carry out two more replicates. Eventually, if the sample content is practically identical to that of the threshold (1.5 μg g^{-1}), it is not feasible to obtain a reliable response.

Finally, some further examples of industrial samples that are very difficult to dissolve, such as very refractory materials, and for which the use of direct solid sampling have proved advantageous are also shown in Table 11.3 [86–88].

11.4 Future for green analytical atomic spectrometry

Despite the lack of literature highlighting this aspect, it is obvious that the use of direct solid sampling techniques contributes to the development of greener atomic spectrometric methods. This issue, together with other advantageous aspects (e.g., increased sample throughput and excellent sensitivity), may help in counterbalancing the classical disadvantages of these techniques (e.g., limited precision), which can be further deploy in many situations for which they are perfectly suited, such as analysis of materials of complex dissolution or the development of screening schemes.

Table 11.3 *Selected applications featuring direct analysis of solid samples and complex matrices of particular interest for the concept of green chemistry.*

Sample	Analyte(s)	Technique	Features	Reference
Platiniferous ores	Au, Cu, Ir, Ni, Pd, Pt, Rh, Ru	LA-ICPMS GD-MS	Comparison of solid sampling techniques for direct analysis of Pb buttons obtained after Pb fire assay of samples	73
Platiniferous ores	Au, Ir, Pd, Pt, Rh, Ru	LA-ICPMS	Direct determination of platinum group metals and Au in NiS buttons obtained after fire assay of samples	63
Various polymers	Al, As, Ba, Cd, Cr, Pb, Sb, Ti	LA-ICPMS	Use of reaction cell to overcome spectral overlaps	74
Fluorocarbon polymers	Cr, Cu, Fe, K, Mn, Pb, Zn	ETV-ICPMS	Use of Pd as chemical modifier and as internal standard	75
Polyethylene, sewage sludge and biological CRMs	Cr	GFAAS ETV-ICPMS	Evaluation of strategies permitting calibration with aqueous standards depending on sample characteristics	76
Various polymers	Hg	GFAAS HR CS GFAAS	Comparison of line source and continuum source GFAAS for Hg monitoring in polymers	77
Various polymers	Cd, Pb	GFAAS	Analysis of real-life samples and monitoring of compliance with RoHS directive	78
Crude oil	Cr, Fe	HR CS GFAAS	Simultaneous determination of Cr and Fe	79
Crude oil	Co, V	HR CS GFAAS	Simultaneous determination of Co and V	80
Biodiesel	P	GFAAS	Direct analysis of biodiesel samples using solid sampling autosampler	81
Biological CRMs	P	HR CS GFAAS	Potential of HR CS GFAAS for monitoring either atomic P or molecular PO lines	15
Biological CRMs	S	HR CS GFAAS	Potential of HR CS GFAAS for monitoring or molecular CS lines	82
Soil, ore, polypropylene and bovine liver	Ag	GFAAS	Linearity expansion using 3-field mode Zeeman-effect	83
Soils	Hg	GFAAS	Performance of the method as quantitative or as a screening method	22
Charcoal and carbon black	Cr, Cu, Fe, Mn, Mo, Ni, V	HR CS GFAAS	Comparison of the potential of the technique with that of a digestion approach	86
Titanium oxide	As, Cd, Hg, Pb, Sb, Zn	ETV-ICPMS	Use of two internal standards (Pd and Te) to compensate for matrix effects	87
Boron nitride	Al, Ca, Cr, Cu, Fe, Mg, Mn, Si, Ti, Zr	GFAAS ETV-ICP-OES	Evaluation of different chemical modifiers, depending on analyte and technique	88

Acknowledgments

This work has been funded by the Spanish Ministry of Science and Innovation (Project CTQ2009–08606) and the Aragon Government ('Programa de apoyo a la I + D + I' from ARAID Foundation and 'Obra Social de IberCaja', with additional support from Inycom).

References

1. Anastas, P.T. and Warner, J.C. (1998) *Green Chemistry: Theory and Practice*, Oxford University Press, New York.
2. Clark, J.H. (1999) Green Chemistry: challenges and opportunities, *Green Chem.*, **1**, 1–8.
3. Anastas, P.T. (1999) Green Chemistry and the role of analytical methodology development, *Crit. Rev. Anal. Chem.*, **29**, 167–175.
4. de la Guardia, M. and Ruzicka, J. (1995) Towards environmentally conscientious analytical chemistry through miniaturization, containment and reagent replacement, *Analyst*, **120**, 17N.
5. Keith, L.H.; Gron, L.U. and Young, J.L. (2007) Green analytical methodologies, *Chem. Rev.*, **107**, 2695–2708.
6. He, Y.; Tang, L.; Wu, X.; Hou, X.D. and Lee, Y.I. (2007) Spectroscopy: The best way toward Green Chemistry?, *Appl. Spectrosc. Rev.*, **42**, 119–138.
7. de la Guardia, M. (2010) Green Analytical Chemistry, TrAC-*Trend. Anal. Chem.*, **29**, 577.
8. de la Guardia, M. and Armenta, S. (2011) Multianalyte determination versus one-at-a-time methodologies, in *Green Analytical Chemistry: Theory and Practice*, (eds M. de la Guardia and S. Armenta) Comprehensive Analytical Chemistry 57, Elsevier, Oxford,.
9. Tobiszewski, M.; Mechlińska, A.; Zygmunt, B. and Namieśnik, J. (2009) Green Chemistry in sample preparation for determination of trace organic pollutants, TrAC-*Trend. Anal. Chem.*, **28**, 943–951.
10. Halls, D.J. (1995) Analytical minimalism applied to the determination of trace elements by atomic spectrometry, *J. Anal. Atom. Spectrom.*, **10**, 169–175.
11. Hou, X. and Jones, B.T. (2002) Tungsten devices in analytical atomic spectrometry, *Spectrochim. Acta B*, **57**, 659–688.
12. Hou, X. and Jones, B.T. (2000) Field instrumentation in atomic spectroscopy, *Microchem. J.*, **66**, 115–145.
13. Karanassios, V. (2004) Microplasmas for chemical analysis: analytical tools or research toys?, *Spectrochim. Acta B*, **59**, 909–928.
14. Welz, B.; Becker-Ross, H.; Florek, S. and Heitmann, U. (2005) *High-Resolution Continuum Source AAS*, Wiley-VCH Verlag GmbH & Co. KGaA, Weinheim.
15. Resano, M.; Briceño, J. and Belarra, M.A. (2009) Direct determination of phosphorus in biological samples using a solid sampling-high resolution-continuum source electrothermal spectrometer: comparison of atomic and molecular absorption spectrometry, *J. Anal. Atom. Spectrom.*, **24**, 1343–1354.
16. Tobiszewski, M.; Mechlińska, A. and Namieśnik, J. (2010) Green Chemistry - theory and practice, *Chem. Soc, Rev.*, **39**, 2869–2878.
17. Moreda-Piñeiro, A.; Barciela-Alonso, M.C.; Domínguez-González, R.; Peña-Vázquez, E.; Herbello-Hermelo, P. and Bermejo-Barrera, P. (2009) Alternative solid sample pretreatment methods in green analytical atomic spectrometry, *Spectrosc. Lett.*, **42**, 394–417.
18. Koel, M. and Kaljurand, M. (2006) Application of the principles of Green Chemistry in analytical chemistry, *Pure Appl. Chem.*, **78**, 1993–2002.
19. Moros, J.; Garrigues, S. and de la Guardia, M. (2010) Vibrational spectroscopy provides a green tool for multi-component analysis, *TrAC-Trend. Anal. Chem.*, **29**, 578–591.
20. Belarra, M.A.; Resano, M.; Vanhaecke, F. and Moens, L. (2002) Direct solid sampling with electrothermal vaporization/atomization: what for and how?, *TrAC-Trend. Anal. Chem.*, **21**, 828–839.
21. Armenta, S.; Garrigues, S. and de la Guardia, M. (2008) Green Analytical Chemistry, *TrAC-Trend. Anal. Chem.*, **27**, 497–511.
22. Resano, M.; García-Ruiz, E.; Aramendía, M. and Belarra, M.A. (2005) Solid sampling-graphite furnace atomic absorption spectrometry for Hg monitoring in soils. Performance as a quantitative and as a screening method, *J. Anal. Atom. Spectrom.*, **20**, 1374–1380.

23. De Mello, A. (2002) Miniaturization, *Anal. Bioanal. Chem.*, **372**, 12–13.
24. Vilkner, T.; Janasek, D. and Mank, A. (2004) Micro total analysis systems. Recent developments, *Anal. Chem.*, **76**, 3373–3386.
25. Dittrich, P.S.; Tachikawa, K. and Mank, A. (2006) Micro total analysis systems. Latest advancements and trends, *Anal. Chem.*, **78**, 3887–3907.
26. Liu, J.F.; Jonsson, J.A. and Jiang, G.B. (2005) Application of ionic liquids in analytical chemistry, *TrAC-Trend. Anal. Chem.*, **24**, 20–27.
27. Pandey, S. (2006) Analytical applications of room-temperature ionic liquids: a review of recent efforts, *Anal. Chim. Acta*, **556**, 38–45.
28. Anderson, J.L.; Armstrong, D.W. and Wei, G.T. (2006) Ionic liquids in analytical chemistry, *Anal. Chem.*, **78**, 2892–2902.
29. Topal, S.Z.; Ertekin, K.; Topkaya, D.; Alp, A. and Yenigul, B. (2008) Emission based oxygen sensing approach with tris(2,2'-bipyridyl)-ruthenium(II)chloride in Green Chemistry reagents; room temperature ionic liquids, *Microchim. Acta*, **161**, 209–216.
30. Ding, J.; Melton, T. and Armstrong, D.W. (2004) Chiral ionic liquids as stationary phases in gas chromatography, *Anal. Chem.*, **76**, 6819–6822.
31. Mank, M.; Stahl, B. and Boehm, G. (2004) 2,5-Dihydroxybenzoic acid butylamine and other ionic liquid matrixes for enhanced MALDI-MS analysis of biomolecules, *Anal. Chem.*, **76**, 2938–2950.
32. Martinis, E.M.; Olsina, R.A.; Altamirano, J.C. and Wuilloud, R.G. (2008) Sensitive determination of cadmium in water samples by room temperature ionic liquid-based preconcentration and electrothermal atomic absorption spectrometry, *Anal. Chim. Acta*, **628**, 41–48.
33. Gil, S.; Lavilla, I. and Bendicho, C. (2006) Ultrasound-promoted cold vapor generation in the presence of formic acid for determination of mercury by atomic absorption spectrometry, *Anal. Chem.*, **78**, 6260–6264.
34. Lorentzen, E.M.L. and Kingston, H.M. (1996) Comparison of microwave-assisted and conventional leaching using EPA method 3050B, *Anal. Chem.*, **68**, 4316–4320.
35. Mortari, S.R.; Cocco, C.R.; Bartz, F.R.; Dresssler, V.L. and Flores, E.M.M. (2010) Fast digestion procedure for determination of catalyst residues in La- and Ni-based carbon nanotubes, *Anal. Chem.*, **82**, 4298–4303.
36. Carabias-Martinez, R.; Rodríguez-Gonzalo, E.; Revilla-Ruiz, P. and Hernández-Méndez, J. (2005) Pressurized liquid extraction in the analysis of food and biological samples, *J. Chromatogr. A*, **1089**, 1–17.
37. Kronholm, I.; Hartonen, K. and Rielckola, M.-L. (2007) Analytical extractions with water at elevated temperatures and pressures, *TrAC-Trend. Anal. Chem.*, **26**, 396–412.
38. Morales-Muñoz, S.; Luque-Garcia, J.L. and Luque de Castro, M.D. (2003) Acidified pressurized hot water for the continuous extraction of cadmium and lead from plant materials prior to ETAAS, *Spectrochim. Acta Part B*, **58**, 159–165.
39. Zougagh, M.; Valcárcel, M. and Ríos, A. (2004) Supercritical fluid extraction: A critical review of its analytical usefulness, *TrAC-Trend. Anal. Chem.*, **23**, 399–405.
40. Alonso-Rodriguez, E.; Moreda-Piñeiro, J.; López-Mahía, P.; Muniategui-Lorenzo, S.; Prada-Rodríguez, D.; Moreda-Piñeiro, A. and Bermejo-Barrera, P. (2007) Use of chelating solvent-based pressurized liquid extraction combined with inductively coupled plasma-optical emission spectrometry for trace element determination in atmospheric particulate matter, *J. Anal. At. Spectrom.*, **22**, 1089–1096.
41. Moreda-Piñeiro, J.; Alonso-Rodriguez, E.; López-Mahía, P.; Muniategui-Lorenzo, S.; Prada-Rodríguez, D.; Moreda-Piñeiro, A. and Bermejo-Barrera, P. (2007) Development of a new sample pre-treatment procedures based on pressurized liquid extraction for the determination of metals in edible seaweed, *Anal. Chim. Acta*, **598**, 95–102.
42. Moreda-Piñeiro, J.; Alonso-Rodriguez, E.; Moreda-Piñeiro, A.; Moscoso-Pérez, C.; Muniategui-Lorenzo, S.; López-Mahía, P.; Prada-Rodríguez, D. and Bermejo-Barrera, P. (2010) Simultaneous pressurized enzymatic hydrolysis extraction and clean up for arsenic speciation in seafood samples before high performance liquid chromatography-inductively coupled plasma-mass spectrometry determination, *Anal. Chim. Acta*, **679**, 63–73.
43. Afzali, D.; Ghaseminezhad, S. and Taher, M.A. (2010) Separation and preconcentration of trace amounts of gold(III) ions using modified multiwalled carbon nanotube sorbent prior to flame atomic absorption spectrometry determination, *J. AOAC Int.*, **93**, 1287–1292.

44. Soylak, M. and Unsal, Y.E. (2010) Chromium and iron determinations in food and herbal plant samples by atomic absorption spectrometry after solid phase extraction on single-walled carbon nanotubes (SWCNTs) disk, *Food Chem. Toxicol.*, **48**, 1511–1515.
45. Chen, S.; Liu, C.; Yang, M.; Lu, D.; Zhu, L. and Wang, Z. (2009) Solid phase extraction of Cu, Co and Pb on oxidized single-walled carbon nanotubes and their determination by inductively coupled plasma mass spectrometry, *J. Hazard. Mater.*, **170**, 247–251.
46. Zhao, X.; Song, N.; Jia, Q. and Zhou, W. (2009) Determination of Cu, Zn, Mn and Pb by microcolumn packed with multiwalled carbon nanotubes on-line coupled with flame absorption spectrometry, *Microchim. Acta*, **166**, 329–335.
47. Ojeda, C.B. and Rojas, F.S. (2009) Separation and preconcentration by a cloud point extraction procedure for determination of metals: an overview, *Anal. Bioanal. Chem.*, **394**, 759–782.
48. Burguera, J.L. and Burguera, M. (2009) Recent on-line processing procedures for biological samples for determination of trace elements by atomic spectrometric methods, *Spectrochim. Acta Part B*, **64**, 451–458.
49. Butler, O.; Evans, H.; Fisher, A.; Hill, S.; Harrington, C.; Taylor, A.; West, M. and Ellis, A. (2010) Atomic spectrometry updates: A 25-year retrospective, J. *Anal. At. Spectrom.*, **25**, 1546–1566.
50. Meeravali, N.N. and Jiang, S.-J. (2008) Interference free ultra trace determination of Pt, Pd and Au in geological and environmental samples by inductively coupled plasma quadrupole mass spectrometry after a cloud point extraction, *J. Anal. At. Spectrom.*, **23**, 854–860.
51. Ferreira, S.L.C.; Miro, M.; da Silva, E.G.P.; Matos, G.D.; dos Reis, P.S.; Brandao, G.C.; dos Santos, W.N.L.; Duarte, A.T.; Vale, M.G.R. and Araujo, R.G.O. (2010) Slurry sampling-An analytical strategy for the determination of metals and metalloids by spectroanalytical techniques, *Appl. Spectrosc. Rev.*, **45**, 44–62.
52. Resano, M.; Vanhaecke, F. and de Loos-Vollebregt, M.T.C. (2008) Electrothermal vaporization for sample introduction in atomic absorption, atomic emission and plasma mass spectrometry-a critical review with focus on solid sampling and slurry analysis, *J. Anal. At. Spectrom.*, **23**, 1450–1475.
53. Brandao, G.C.; de Jesus, R.M.; da Silva, E.G.P. and Ferreira, S.L.C. (2010) Use of slurry sampling for the direct determination of zinc in yogurt by high resolution-continuum source flame atomic absorption spectrometry, *Talanta*, **81**, 1357–1359.
54. Vignola, F.; Borges, D.L.G.; Curtius, A.J.; Welz, B. and Becker-Ross, H. (2010) Simultaneous determination of Cd and Fe in sewage sludge by high-resolution continuum source electrothermal atomic absorption spectrometry with slurry sampling, *Microchem. J.*, **95**, 333–336.
55. Huang, S.Y. and Jiang, S.J. (2010) 8-Hydroxyquinoline-5-sulfonic acid as the modifier for the determination of trace elements in cereals by slurry sampling electrothermal vaporization ICP-MS, *Anal. Methods*, **2**, 1310–1315.
56. Amberger, M.A. and Broekaert, J.A.C. (2010) Direct multielement determination of trace elements in boron carbide powders by slurry sampling ETV-ICP-OES, *J. Anal. At. Spectrom.*, **25**, 1308–1315.
57. Chen, F.Y. and Jiang, S.J. (2009) Slurry sampling flow injection chemical vapor generation inductively coupled plasma mass spectrometry for the determination of As, Cd, and Hg in cereals, *J. Agric. Food Chem.*, **57**, 6564–6569.
58. Matusiewicz, H. (2003) Chemical vapor generation with slurry sampling: A review of atomic absorption applications, *Appl. Spectrosc. Rev.*, **38**, 263–294.
59. Aramendia, M.; Resano, M. and Vanhaecke, F. (2009) Electrothermal vaporization-inductively coupled plasma-mass spectrometry: A versatile tool for tackling challenging samples. A critical review, *Anal. Chim. Acta*, **648**, 23–44.
60. Resano, M.; Garcia-Ruiz, E. and Vanhaecke, F. (2010) Laser ablation-inductively coupled plasma mass spectrometry in archaeometric research, *Mass Spectrom. Rev.*, **29**, 55–78.
61. Pisonero, J.; Fernandez, B. and Gunther, D. (2009) Critical revision of GD-MS, LA-ICP-MS and SIMS as inorganic mass spectrometric techniques for direct solid analysis, *J. Anal. At. Spectrom.*, **24**, 1145–1160.
62. Belarra, M.A.; Resano, M. and Castillo, J.R. (1999) Theoretical evaluation of solid sampling electrothermal atomic absorption spectrometry for screening purposes, *J. Anal. At. Spectrom.*, **14**, 547–552.
63. Resano, M. Garcia-Ruiz, E.; McIntosh, K.S. and Vanhaecke, F. (2008) Laser ablation-inductively coupled plasma-dynamic reaction cell-mass spectrometry for the determination of platinum group metals and gold in NiS buttons obtained by fire assay of platiniferous ores, *J. Anal. At. Spectrom.*, **23**, 1599–1609.
64. Welz, B. (2005) High-resolution continuum source AAS: the better way to perform atomic absorption spectrometry, *Anal. Bioanal. Chem.*, **381**, 69–71.

65. Becker-Ross, H.; Florek, S.; Heitmann, U.; Huang, M.D. and Okruss, M. (2006) Continuum source atomic absorption spectrometry and detector technology: A historical perspective, *Spectrochim. Acta Part B*, **61**, 1015–1030.
66. Becker-Ross, H.; Florek, S. and Heitmann, U. (2000) Observation, identification and correction of structured molecular background by means of continuum source AAS - determination of selenium and arsenic in human urine, *J. Anal. At. Spectrom.*, **15**, 137–141.
67. Heitmann, U.; Welz, B.; Borges, D.L.G. and Lepri, F.G. (2007) Feasibility of peak volume, side pixel and multiple peak registration in high-resolution continuum source atomic absorption spectrometry, *Spectrochim. Acta Part B*, **62**, 1222–1230.
68. Welz, B.; Lepri, F.G.; Araujo, R.G.O.; Ferreira, S.L.C.; Huang, M.D.; Okruss, M. and Becker-Ross, H. (2009) Determination of phosphorus, sulfur and the halogens using high-temperature molecular absorption spectrometry in flames and furnaces-A review, *Anal. Chim. Acta*, **647**, 137–148.
69. Resano, M.; Rello, L.; Flórez, M. and Belarra, M.A. (2011) On the possibilities of high-resolution continuum source graphite furnace atomic absorption spectrometry for the simultaneous or sequential monitoring of multiple atomic lines, *Spectrochim. Acta Part B*, **66**, 321–328.
70. Welz, B.; Mores, S.; Carasek, E.; Vale, M.G.R.; Okruss, M. and Becker-Ross, H. (2010) High-resolution continuum source atomic and molecular absorption spectrometry - A review, *Appl. Spectrosc. Rev.*, **45**, 327–354.
71. Beauchemin, D.; Grégoire, D.C.; Günther, D.; Karanassios, V.; Mermet, J.M. and Wood, T.J. (2000) *Discrete sample introduction techniques for inductively coupled mass spectrometry* (ed. D. Barcelo), Comprehensive Analytical Chemistry, Vol. XXXIV, Elsevier, Amsterdam.
72. Resano, M.; Garcia-Ruiz, E.; Belarra, M.A.; Vanhaecke, F. and McIntosh, K.S. (2007) Solid sampling in the determination of precious metals at ultratrace levels, *TrAC-Trend. Anal. Chem.*, **26**, 385–395.
73. Resano, M.; Garcia-Ruiz, E.; McIntosh, K.S.; Hinrichs, J.; Deconinck, I. and Vanhaecke, F. (2006) Comparison of the solid sampling techniques laser ablation-ICP-MS, glow discharge-MS and spark-OES for the determination of platinum group metals in Pb buttons obtained by fire assay of platiniferous ores, *J. Anal. At. Spectrom.*, **21**, 899–909.
74. Resano, M.; Garcia-Ruiz, E. and Vanhaecke, F. (2005) Laser ablation-inductively coupled plasma-dynamic reaction cell-mass spectrometry for the multi-element analysis of polymers, *Spectrochim. Acta Part B*, **60**, 1472–1481.
75. Resano, M.; Aramendia, M.; Devos, W. and Vanhaecke, F. (2006) Direct multi-element analysis of a fluorocarbon polymer via solid sampling-electrothermal vaporization-inductively coupled plasma mass spectrometry, *J. Anal. At. Spectrom.*, **21**, 891–898.
76. Resano, M.; Garcia-Ruiz, E.;Vanhaecke, F.; Crespo, C. and Belarra, M.A. (2004) Evaluation of solid sampling-electrothermal vaporization-inductively coupled plasma mass spectrometry and solid sampling-graphite furnace atomic absorption spectrometry for the direct determination of Cr in various materials using solution-based calibration approaches, *J. Anal. At. Spectrom.*, **19**, 958–965.
77. Resano, M.; Briceño, J. and Belarra, M.A. (2009) Direct determination of Hg in polymers by solid sampling-graphite furnace atomic absorption spectrometry A comparison of the performance of line source and continuum source instrumentation, *Spectrochim. Acta Part B*, **64**, 520–529.
78. Duarte, A.T.; Dessuy, M.B.; Silva, M.M.; Vale, M.G.R. and Welz, B. (2010) Determination of cadmium and lead in plastic material from waste electronic equipment using solid sampling graphite furnace atomic absorption spectrometry, *Microchem. J.*, **96**, 102–107.
79. Dittert, I.M.; Silva, J.S.A.; Araujo, R.G.O.; Curtius, A.J.; Welz, B. and Becker-Ross, H. (2009) Direct and simultaneous determination of Cr and Fe in crude oil using high-resolution continuum source graphite furnace atomic absorption spectrometry, *Spectrochim. Acta Part B*, **64**, 537–543.
80. Dittert, I.M.; Silva, J.S.A.; Araujo, R.G.O.; Curtius, A.J.; Welz, B. and Becker-Ross, H. (2010) Simultaneous determination of cobalt and vanadium in undiluted crude oil using high-resolution continuum source graphite furnace atomic absorption spectrometry, *J. Anal. At. Spectrom.*, **25**, 590–595.
81. Lyra, F.H.; Dias Carneiro, M.T.W.; Brandao, G.P.; Pessoa, H.M. and de Castro, E.V.R. (2009) Direct determination of phosphorus in biodiesel samples by graphite furnace atomic absorption spectrometry using a solid sampling accessory, *J. Anal. At. Spectrom.*, **24**, 1262–1266.
82. Ferreira, H.S.; Lepri, F.G.; Welz, B.; Carasek, E. and Huang, M.-D. (2010) Determination of sulfur in biological samples using high-resolution molecular absorption spectrometry in a graphite furnace with direct solid sampling, *J. Anal. At. Spectrom.*, **25**, 1039–1045.

83. Resano, M.; Aramendia, M.; Garcia-Ruiz, E.; Crespo, C. and Belarra, M.A. (2006) Solid sampling-graphite furnace atomic absorption spectrometry for the direct determination of silver at trace and ultratrace levels, *Anal. Chim. Acta*, **571**, 142–149.
84. Simonet, B.M.; Ríos, A. and Valcárcel, M. (2004) Unreliability of screening methods, *Anal. Chim. Acta*, **516**, 67–74.
85. Song, R.; Schlecht, P.C. and Ashley, K. (2001) Field screening test methods: performance criteria and performance characteristics, *J. Hazardous Mater.*, **83**, 29–39.
86. Lepri, F.G.; Borges, D.L.G.; Araujo, R.G.O.; Welz, B.; Wendler, F.; Krieg, M. and Becker-Ross, H. (2010) Determination of heavy metals in activated charcoals and carbon black for Lyocell fiber production using direct solid sampling high-resolution continuum source graphite furnace atomic absorption and inductively coupled plasma optical emission spectrometry, *Talanta*, **81**, 980–987.
87. Aramendia, M.; Resano, M. and Vanhaecke, F. (2009) Determination of toxic trace impurities in titanium dioxide by solid sampling-electrothermal vaporization-inductively coupled plasma mass spectrometry, *J. Anal. At. Spectrom.*, **24**, 41–50.
88. Barth, P.; Hassler, J.; Kudrik, I. and Krivan, V. (2007) Determination of trace impurities in boron nitride by graphite furnace atomic absorption spectrometry and electrothermal vaporization inductively coupled plasma optical emission spectrometry using solid sampling, *Spectrochim. Acta Part B*, **62**, 924–932.

12
Solid Phase Molecular Spectroscopy

Antonio Molina-Díaz, Juan Francisco García-Reyes and Natividad Ramos-Martos

Analytical Chemistry Research Group, Department of Physical and Analytical Chemistry, University of Jaén, Jaén, Spain

In this chapter, the principles of solid phase molecular spectroscopy (SPMS) are presented, together with its main contributions to the Green Analytical Chemistry field in both batch mode and flow-based methodologies. Solid phase molecular absorption (spectrophotometry) and emission (fluorescence) strategies and procedures will be presented and described from the point of view of the Green Analytical Chemistry concept. Procedures based on the measurement of absorption or emission that is, solid phase spectrophotometric and spectrofluorimetric procedures are discussed, paying particular attention to those flow procedures based on the detection of an intrinsic analyte property, which exhibit features that map well against Green Analytical Chemistry requirements. The main features of Flow Injection Analysis, Multicommutation, Bead Injection and Sequential Injection Analysis concepts applied to SPMS are addressed in detail. Finally, selected examples of flow SPMS application are presented and the potential of this group of methodologies critically discussed and envisaged from the point of view of Green Analytical Chemistry.

12.1 Solid phase molecular spectroscopy: an approach to Green Analytical Chemistry

Molecular spectroscopy comprises a suite of diverse analytical techniques, whose requirements usually map well against the features of Green Analytical Chemistry [1]. Amongst them, solid phase molecular spectroscopy (SPMS) is a single approach that has led to the development of several green analytical methodologies in last few years. It is based on the use of solid phase microbeads from an appropriate material to retain onto them the target species by establishing an equilibrium between the active sites of the sorbent and the target species to be analyzed in the solution. The retained (and concentrated) species is directly

Handbook of Green Analytical Chemistry, First Edition. Edited by Miguel de la Guardia and Salvador Garrigues.

determined by measuring a molecular spectroscopic property (typically absorbance or fluorescence) directly on the solid. The analytical signal obtained is then related to the analyte concentration in the sample. Therefore, no previous elution step is performed before measuring, thus obtaining interesting analytical features in terms of sensitivity and selectivity despite using a very single and inexpensive apparatus (easy to perform in any routine laboratory). Determinations in the $\mu g\ l^{-1}$ range can be achieved with SPMS using either UV-Vis molecular absorption spectroscopy or molecular emission techniques (fluorescence) [2]. Besides this, as a consequence of the selective conditions provided by the solid phase under optimized operating conditions, straightforward green analytical molecular (absorption or emission) spectroscopic procedures can be developed without needing derivative reagents, simply based on the native absorption (typically in the UV region) or on the native fluorescence of the analytes.

The concept of SPMS was first introduced in batch mode [3]. A few years later, SPMS was combined and implemented with Flow Injection Analysis (FIA) systems. Interestingly, the term Green (Analytical) Chemistry is not mentioned as keywords in papers published on this methodology, despite these methods ideally fulfill the GAC requirements [4–6].

In this chapter, SPMS procedures based on the measurement of native absorption or emission that is, solid phase spectrophotometric and spectrofluorimetric procedures are discussed in detail. These procedures represent the more valuable contribution of SPMS to Green Analytical Chemistry. Both batch mode and flow-based SPMS methodologies will be considered and envisaged from the point of view of GAC concept.

12.2 Fundamentals of solid phase molecular spectroscopy

Solid phase molecular spectroscopy is based on the immobilization by equilibration of the target species (analyte or a suitable reaction product) on an appropriate solid support (usually microbeads from a polymeric or a non polar sorbent material or ion-exchanger resins). Next, beads are collected by filtration and transferred to an appropriate measurement cell (typically 1 mm optical path length cell) as a suspension with a relatively small volume of equilibrated solution (ca. 200 µl). The analytical spectroscopic property from the target species, directly retained on the solid support is measured against a microbead layer blank and related to the sample analyte concentration [7, 8].

12.2.1 Solid phase absorption (spectrophotometric) procedures

The experimental value of the measured absorbance, A, from a microbead thin layer containing the target species retained comprises three main contributions [7]:

$$A = A_A + A_R + A_S \tag{12.1}$$

where A_A is the absorbance of the analyte (or its derivative product) sorbed on the microbeads, A_R is the absorbance of the background (solid support plus reagent, if any), and A_S, that of the interstitial solution among the microbeads, which is, usually, negligible.

The absorbance of interest species is:

$$A_A = A - (A_R + A_S) \tag{12.2}$$

The values of A, A_R and A_S depend on the microbeads' packing degree. Therefore, packing of the microbeads in the cell may affect the values of A_A. Since the solid support packing in the cell is not reproducible, its background signal affects the value measured by the instrument, so rendering no reproducible measurements. To overcome this issue, two measurements have to be performed, one in the maximum signal wavelength, and another one in a region where only the support has contribution to the signal, but not the analyte. By calculating the difference, the analyte net signal is obtained.

The net absorbance of the sample species on the solid microbead, A_A, is given [7] by

$$A_A = \varepsilon_S l_S C_S \tag{12.3}$$

where ε_S is the apparent molar absorptivity of the species on the solid phase as observed in the measurement cell (kg mol^{-1} cm^{-1}), l_S the mean light path length through the resin layer which can be supposed equal to 0.1 cm and, C_S, the analyte concentration in the solid phase (mol kg^{-1}). When a volume V (l) of a sample at concentration C_O (mol l^{-1}) of analyte is equilibrated with m (mg) of solid microbeads, the concentration on the solid phase (C_S, mol kg^{-1}) will be:

$$C_S = C_O V1000 /(m + V/D) \tag{12.4}$$

where D is the distribution ratio of the sample component (l/Kg). Therefore,

$$A_A = \varepsilon_S l_S C_O V1000/(m + V/D) \tag{12.5}$$

This equation shows the dependence of A_A on the amount of support used to equilibrate *(m)* and the sample solution volume (*V*) used in the procedure.

From this equation, the influence of m and V on sensitivity of a SPMS procedure can be inferred. *D* shows usually a high value, so for low *V* values (typically, less than 1 l), the fraction *V/D* can be neglected as compared with *m* and then Equation (12.4) in these conditions keeps as follows:

$$A_A = \varepsilon_S l_S C_O V1000/m = K/m \tag{12.6}$$

This equation displays an inversely proportional relation between the signal measured on solid microbeads and the amount of solid phase used in a SPMS procedure (*K* being constant). Miniaturization is favored in SPMS, so that sensitivity can be increased by reducing the mass of solid support as much as possible. However, this feature is somewhat limited by cell volume and light pathway instrumental geometry, which will affect the minimum required amount of solid support.

Figure 12.1 shows the influence of solid support amount on sensitivity in the determination of thiamin by UV-SPMS which is in accordance with Equation (12.5) [9]. A 10-fold increase of molar absorptivity coefficient (one order of magnitude) was reached when 30 mg of solid is used compared to the use of 300 mg.

In contrast, sensitivity in SPMS methodology can be enhanced by increasing the sample volume. In our case, the fraction *V/D* can be neglected as compared with *m* for *V* values lower than 1liter. This leads to an Equation (12.7) which shows a linear relationship between analytical signal and sample volume:

$$A_A = \varepsilon_S l_S C_O V1000/m = K'V \tag{12.7}$$

where K' is constant. Therefore, sensitivity can be enhanced by increasing the sample volume used. As V increases, the equilibration time also does, so that a compromise solution has to be selected.

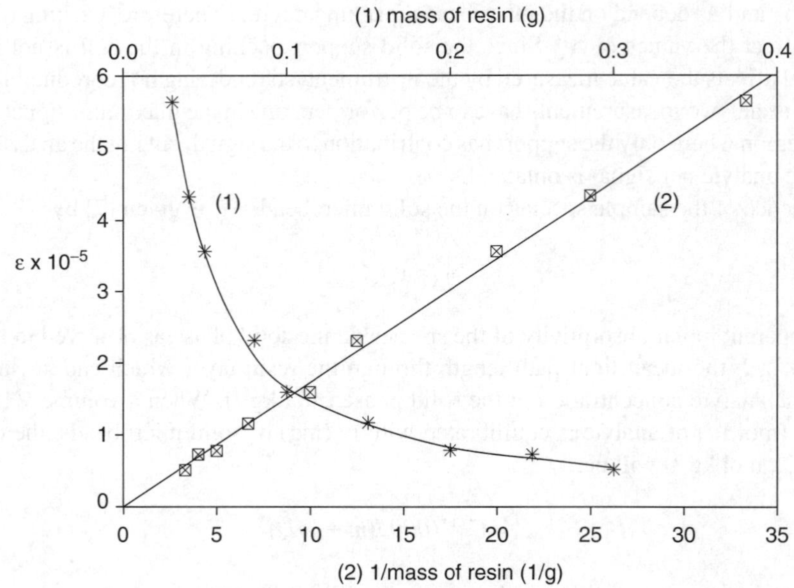

Figure 12.1 *UV-SPMS batch procedure for thiamine. Dependence of molar absorptivity coefficient on amount of resin: (1) mass of solid phase; (2) 1/mass of solid phase. Reprinted from [9]. © 1998 American Chemical Society.*

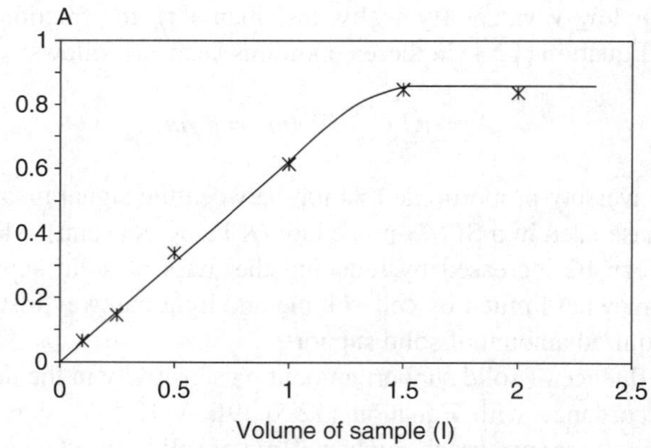

Figure 12.2 *Influence of the sample volume on signal development in a Vis-SPMS batch procedure for vanadium (V). Reproduced from [10] with permission from Elsevier, © 1995.*

An example of the effect of the sample volume on sensitivity can be observed in Figure 12.2 [10]. This experimental data is consistent with Equation (12.6). Sensitivity is increased about one order of magnitude when sample volume increases from 10 to 1000 ml. Note that the observed plateau is due to the need of drastically increasing equilibration time with sample volumes higher than 1 l.

12.2.2 Solid phase emission (fluorescence) procedures

In solid phase spectrofluorimetric procedures, a similar relationship between the analytical signal and the initial concentration of analyte in solution can be established. The fluorescence signal of the analyte sorbed on the solid support, *I*, is given by the following expression:

$$I = \Phi_F I_0 \varepsilon_S l_S Cs \tag{12.8}$$

where ϕ_F is the fluorescence efficiency in the solid phase, *Io* is the intensity of the excitation light beam and $\varepsilon_{S,}$ $l_{S,}$ and *Cs* are as above described. From Equations (12.4) and (12.8), it follows:

$$I = \Phi_F I_0 \varepsilon_S l_S C_O V1000/(m + V/D) \tag{12.9}$$

and if the fraction V/D is neglected, as it usually occurs:

$$I = \Phi_F I_0 \varepsilon_S l_S C_O V1000/m \tag{12.10}$$

indicating a dependence of the fluorescence signal on *m* and *V* similar to that found for solid phase molecular spectrophotometry.

In flow mode, similar relations to previous described equations can be obtained. However, there is an additional advantage: when the system is flowing for a few seconds, the packing keeps constant and the baseline shows a constant signal, so the spectrophotometric solid phase based measurements can be performed at only one wavelength [11].

12.3 Batch mode procedures

SPMS performed in the UV region is an easy approach used when the analyte shows intrinsic absorption in the UV region and can be selectively retained on the solid microbeads. No reagent is needed for the procedure except a buffer solution which is often required to set the optimum sample solution pH value to allow the optimized analyte retention. Therefore, the waste generated does not contain any potential hazardous reagent. Moreover, the solid support can be regenerated and reutilized, although this advantage usually is not exploited by the authors.

Interestingly, a key feature in UV-SPMS is the compatibility of the solid support with the nature of the detector. While Sephadex exchanger gels or Silica C_{18} are, in practical terms, transparent enough in the UV region to allow measurements compatibility [9], other support such as aromatic polymers (type Dowex or Amberlite) cannot be used for UV-SPMS [11]. In batch mode, UV-SPMS has been applied for the determination of both cationic and anionic species in a wide variety of matrices. It must be noted that the main applications reported are limited to active pharmaceutical formulations [12, 13]. The developed procedures represent a simple GAC alternative to other procedures for quality control which include the use of more sophisticated and/or less environmental friendly methods, even official ones. In these cases, relatively low sample volume is used (typically 10 ml) and, consequently, low equilibration time is needed.

An illustrative example of UV-SPMS application is the microdetermination of thiamine (vitamin B_1) in the presence of other B-group vitamins [9]. Its cationic nature allows the selective retention on Sephadex CMC-25 cation-exchange gel microbeads. The effect of solid phase amount is shown in Figure 12.1. Sensitivity (expressed as apparent molar absorptivity) increases from 5.63×10^5 to 2.19×10^7 l mol^{-1}cm^{-1} for a 100-fold sample volume increase.

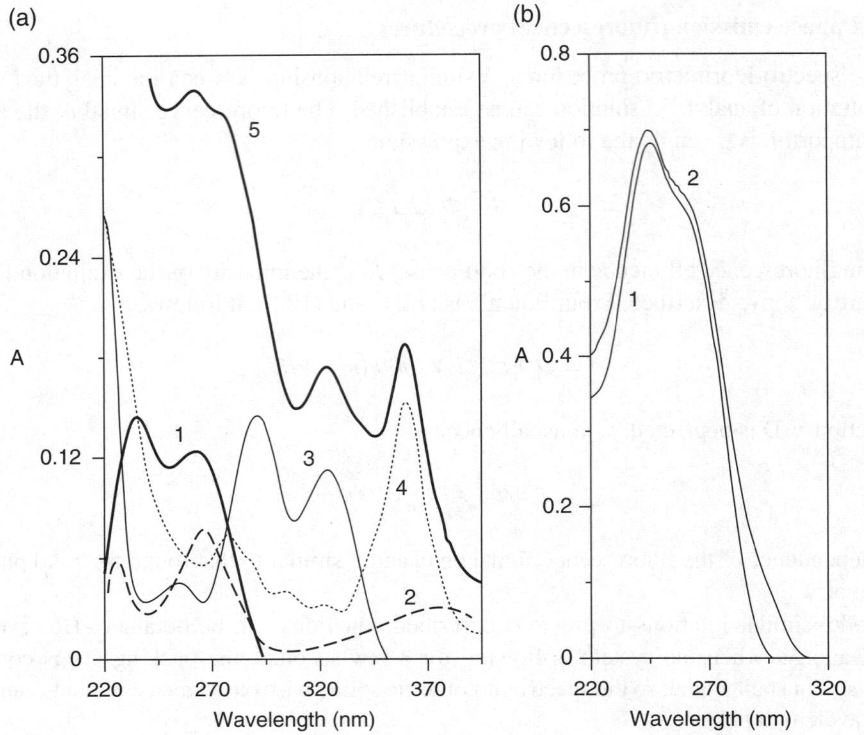

Figure 12.3 *Effect of the presence of other B-group vitamins on the absorption spectrum of thiamine. (a) In solution: (1) thiamine, (2) B_2, (3) B_6, (4) B_{12}, and (5) mixture of thiamine, B_2, B_6, and B_{12}. (b) On resin: (1) thiamine and (2) mixture of thiamine, B_2, B_6, and B_{12}. [Thiamine] = 1.48×10^{-5} M; [B_2] = 1.33×10^{-6} M; [B_6] = 2.92×10^{-5} M; [B_{12}] = 5.90×10^{-6} M; pH 4.8. The spectra on resin were obtained with 30 mg of resin and 10 mL sample volume. Reprinted from [9]. © 1998 American Chemical Society.*

Figure 12.3 successfully illustrates selectivity performance for vitamin B_1 microdetermination by this procedure: Figure 12.3(a) shows the UV spectra of the B-group vitamins in solution and their mixture; and Figure 12.3(b) shows the solid phase UV spectra of both thiamine and vitamins B mixture. The excellent selectivity shown by the UV procedure is self-evident as well as its potential for determining selectively thiamine in the presence of other vitamins B.

Native fluorescence SPMS procedures show similar features regarding sensitivity and selectivity compared with the respective conventional methods based on solution measurements. Besides, the inherent increased selectivity shown by fluorescence compared to spectrophotometry should be noticed. Examples of fluorescence SPMS are the determination of acetylsalicylic acid and pyridoxine [14]

It should be pointed out that 1 mm quartz cells are used to perform the measurements and the usual fluorescence signal measured is the diffuse transmitted fluorescence emitted from the solid microbeads with the retained species, the angle formed between the cell plane and the excitation beam being 45° [15].

12.4 Flow mode procedures

Batch SPMS exhibits very attractive analytical performance. Amongst the main features, one should highlight the sensitivity increase up to four orders of magnitude higher (than the respective classical homogeneous solution procedure) and remarkable selectivity. On one hand, in order to exploit sensitivity, high sample volumes should be used [2]. On the other hand, high selectivity can be reached by appropriately choosing working conditions and using a few ml of sample volume [9] in one step procedure without derivative reagents. Nevertheless, SPMS in batch mode is time and labor consuming due to bead manipulation stages (load, unload, filtration, transferring to cell, etc.). With a set of six cells, about 10–12 analysis can be performed per hour working with 10 ml sample volume. After measurement, the beads are usually discarded, thus involving the use of large amounts of solid support. In this sense, the implementation of SPMS with flow analysis methodology became one of the most relevant advances in SPMS. The solid microbeads are placed inside an appropriate flow cell, so that successive determinations are performed on the same beads, as the solid phase can be regenerated after each measurement. Therefore, important saving of reagents, sample solution and waste generation is achieved. Moreover, relevant improvements in accuracy, reliability, precision, sample throughput and versatility are other remarkable features of this implementation.

This concept, first introduced by Ruzicka and Hansen [16] as the 'flow-through optosensoŕ, has miniaturized the SPMS methodology and its analytical potential has been widely demonstrated, including easy-to-perform determinations on a same sample in very different fields of the chemical analysis [17, 18, 19]. It works similarly to a liquid chromatographic process in which the molecular spectroscopic detector is continuously monitoring the signal on the separation column itself: separation and detection processes are integrated and they occur at the same place and at the same time.

Taking into account these features, flow-based SPMS is a more environmental friendly principle than SPMS in batch mode. A key aspect of flow SPMS to be pointed out, is the reutilization of the solid microbeads, which allows to perform a high number of successive determinations, that is, the need to achieve the retention of the analyte to be reversible, so that the active sites of the sensing surface keep ready for the next determination. Thus, a few mg (typically 30–50 mg) of beads can be reused hundreds of times, drastically reducing the amount of solid phase required.

There are two ways of performing the regeneration (elution step) of the solid surface: (1) after the signal reaches its maximum value, an eluent solution (either via injection of a known volume of eluent or using a selection valve) is passed through it and (2) to choose an appropriate carrier able to also elute by itself the retained species [11], so originating a transitory signal. The first option shows a lower sample throughput and in addition, a more reduced lifetime of the solid support. Both drawbacks are overcome by using the second option.

The most usual assembly for implementing SPMS with flow analysis is the conventional flow injection analysis based approach (FIA), and, in a minor extension, the multicommutation principle (MC). These two implemented flow-SPMS methodologies are particularly attractive from the GAC point of view when detection is based on monitoring of an intrinsic property. In the following section, these two approaches will be presented and discussed along with alternative flow SPMS systems such as Sequential Injection Analysis (SIA) and Bead Injection Spectroscopy (BIS).

12.4.1 Monitoring an intrinsic property

When an intrinsic analyte property is monitored, flow-SPMS is a valuable contribution to GAC. For instance, measurements in UV region in conventional homogeneous solution exhibit scarce selectivity. Nevertheless, to perform the on-line spectroscopic measurements directly on the solid phase, increases drastically the selectivity towards the targeted species compared to respective homogeneous solution measurements. The selected operating conditions (nature of solid support, pH, polarity, retention behavior, etc.) provide appropriate selectivity conditions, contributing to the development of single and rapid procedures fulfilling the GAC requirements [20].

For this purpose, the more common approach is FIA-SPMS. The first attempt was the flow solid phase spectrophotometric determination of Cu(II) ion based on the use of a cation-exchange resin as solid phase and

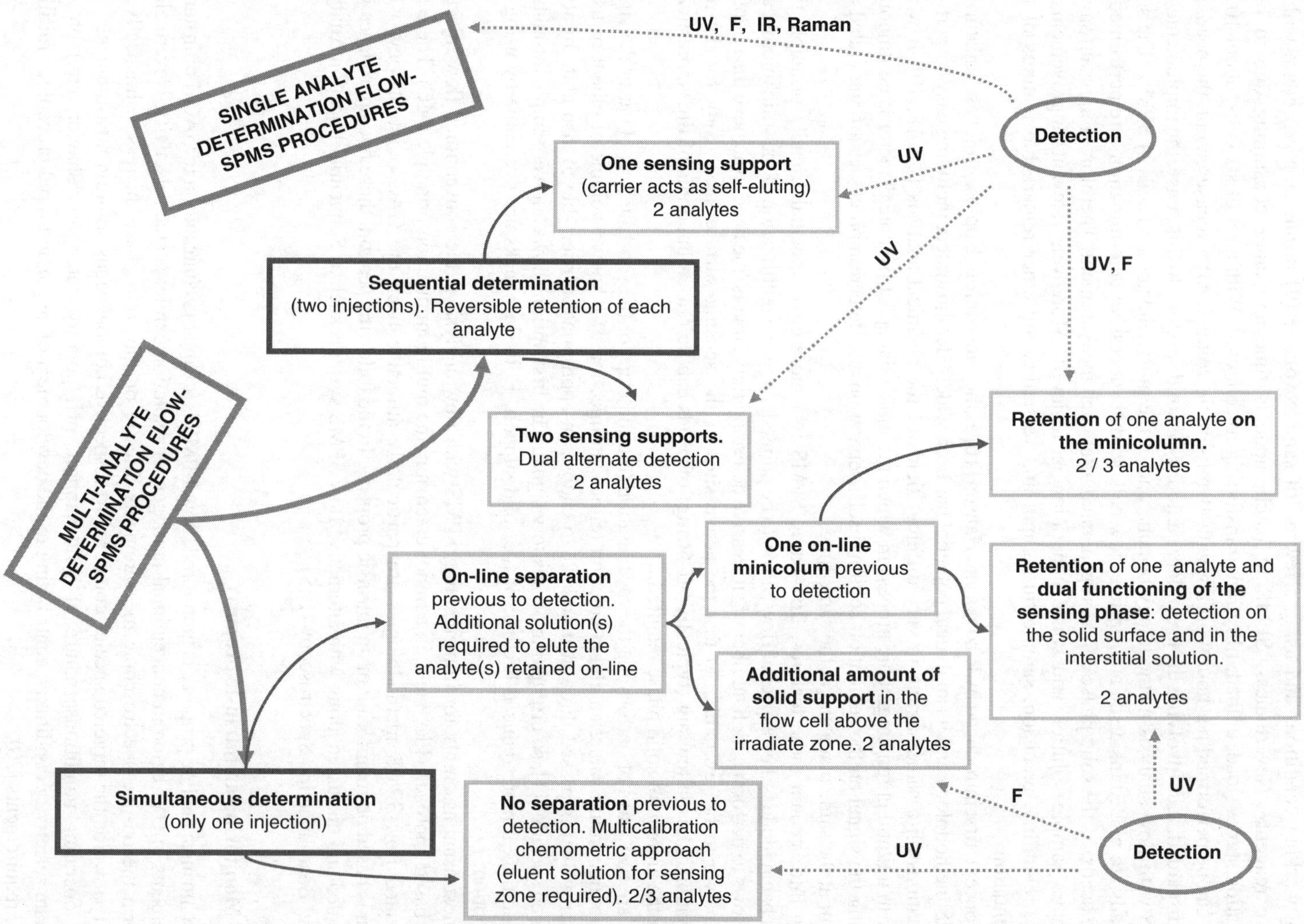

Figure 12.4 Flow–SPMS procedures monitoring an intrinsic analyte property. UV: Ultraviolet; F: fluorescence; IR: Infrared.

the monitorization of the analyte native absorbance [21]. Cumulative measurements were performed on the solid phase from several samples analysis and, thus, regeneration was achieved by using a concentrate HNO_3 solution. These conditions reduce drastically the lifetime of the support. A better alternative is the use of a carrier self-eluting solution.

Figure 12.4 summarizes the more commonly employed flow-SPMS strategies based on the monitoring of an intrinsic analyte property, along with the respective detection technique for each case and the number of analytes that can be determined.

UV, Fluorescence, Infrared and Raman Spectroscopy have been described for the development of single-analyte flow-SPMS systems. These methods are based on the use of single manifolds, consisting of a single channel manifold with the carrier solution also acting as eluting, or including an additional injection or selection valve for introducing the eluent in the system [11]. On the other hand, described multi-analyte flow-SPMS systems (able to determine up to three analytes) are based on the measurement of either UV or fluorescence signals. They require in some cases the same single channel manifold above mentioned (when no separation is performed previous to detection); but in most cases, more complex manifolds are required in order to perform on-line separations, which are attained by using either (1) an on-line minicolumn: or (2) and extra-amount of sensing microbeads in the flow cell to provide the separation of the analytes before the detection area. Thus, other additional channels are needed to introduce the respective eluent solution(s) [5].

12.4.1.1 Solid phase molecular spectroscopy: FIA procedures

Flow-through sensors for monitoring an intrinsic property have been described for both single-analyte and multi-analyte determinations.

Single determination procedures The simplest flow-through optosensor configuration for single-analyte determination, and without involving derivatization reaction of the analyte has two main features: (1) the use of a monochannel configuration in which the carrier solution also acts as eluent; and (2) the analytical signal is an analyte intrinsic property. From the point of view of GAC the more remarkable features of these systems are:

1) Drastic reduction of sample volume required (from ca. 100–1500 ml in batch mode to around 100–1000 µl in flow-through sensors);
2) No other chemical is used except the carrier self-eluting solution;
3) The possibility of reusing the sensing zone (solid support). This saves a significant amount of solid phase when compared to batch mode procedures. The typical average lifetime of these sensing supports is around 200–300 determinations.

This type of single-analyte flow through optosensors have been applied to different types of samples such as foodstuffs and cosmetics (butylated hydroxyanisole, butylated hydroxytoluene polyphenols in grape and n-propyl gallate) [22, 23, 24], pharmaceutical preparations (riboflavin, thiamine) [25] and waters (cyanide and polycyclic aromatic hydrocarbons) [26, 27]. Although the most used molecular spectroscopic detection techniques in these systems are UV and fluorescence, there are also flow-through optosensors for single-analyte determination have been developed based on solid surface Room Temperature Phosphorescence (RTP) [28]. Besides, vibrational spectroscopic techniques (NIR, MR and Raman) has also been explored [29, 30, 31]. Molecularly imprinted polymers (MIP) are novel sensing materials with higher selectivity and have been applied in flow SPMS with RTP based transduction [32], chemiluminometry [33] and bioluminometry [34].

Multi-analyte procedures Multi-analyte flow-through optosensors are able to determine more than one analyte (typically two or three analytes) per sample. Figure 12.4 summarizes the described strategies for these sensors. As can be seen, there are two main groups:

1) Systems with simultaneous determination of two or three analytes from only one injection.
2) Systems performing sequential determination of two analytes (requiring two successive injection of sample).

Simultaneous determinations There are two main options for performing simultaneous determinations:

1) Without performing separation previous to detection. These are systems which combine SPMS-FIA with multicalibration chemometric approaches. Up to three analytes can easily be simultaneously determined with only a single injection in a carrier stream using a diode array detector to acquire the solid phase UV spectral data [35] or a spectrofluorimeter when the analytes pass through the sensing zone.
2) On-line separation previous to detection. Several options have been described:

   a) A minicolumn placed on-line retains one (or two) of the analytes, while another one passes to the sensing zone generating its transitory signal. The further elution of the second (and, if it is the case, the third one) analyte from the minicolumn by appropriately chosen eluting solution(s) allows the simultaneous multicomponent analysis with only one sample injection in a manifold which provides the temporary sequentiation of the analyte arrival to the sensing microbeads, by means of the separation performed in the minicolumn [36].
   b) The use of photo induced fluorescence (PIF) has also been implemented in the development of sensors for simultaneous determination of two analytes. Similar strategy to that just above described was used. Although a photoreactor on line is included before the detector [37], no chemical reagent is used, so these multisensors can be also considered as belonging to the same class that those ones included in this section.

      The multidetermination in these three described cases is performed on the sensing support for all the analytes. This design imposes an additional selectivity condition: the retention of the analytes in the minicolumn, so making the multisensor more versatile and selective. In this easy way, multideterminations are performed with a single GAC SPMS system.
   c) The described set up in point (a) can still be simplified without using the minicolumn and substituting it by a little additional amount of sensing microbeads inside the flow cell above the detection area. This strategy works in a similar way to that from on-line minicolumn separation, but it saves solid phase, and simplifies the manifold, with favorable subsequent advantages [19]. In this system, the analytes have to show a very different affinity behavior towards the sensing phase in order to get the separation inside the flow cell.
   d) Another interesting strategy in multi-determination sensors is to perform a dual functioning of the detection zone containing the sensing support: (i) detection in the interstitial (homogeneous) solution between solid microbeads, which offers a very low effective path length (first analyte) and (ii) the solid phase transitory measurement of the second analyte as it passes through the flow cell, after eluting it form the online minicolumn, being eluted by the carrier itself. This strategy is used when the first analyte cannot be retained on the sensing support and its concentration is much higher than the second one [38].

Sequential determinations Unlike simultaneous determinations, in sequential determinations, one injection per each analyte is needed. Different systems for the determination of two analytes have been described. The described approaches are based on the following strategies:

1) Two successive injections of sample are performed, using two different appropriate carrier-eluting solutions, one for each of them. The sensing surface responds each time to only one analyte in the sample or, in some cases, to the sum of both analytes in one of the injections. High selectivity is achieved just by the interaction of the analyte with the solid phase in the different surface microenvironment appropriately provided by the respective carrier. They really are very single environment-friendly systems (and, hence, simple GAC methods) which have proved their utility [39].
2) Two successive injections of sample are performed, combined with the use of two sensing supports exhibiting each of them different chemical nature. Such as system has been described using as analytical property the intrinsic UV absorbance of the analytes. The supports were placed, respectively, in the measurement and the reference cells in a double beam spectrophotometer, so obtaining, respectively, positive (measurement cell) and negative peaks (reference cell) depending on the cell where the analyte was measured [40].

12.4.1.2 Solid phase molecular spectroscopy: multicommutated procedures

Multicommutation (MC) allows the design of dynamic manifolds based on the use of discrete commutation devices (three way solenoid valves) so that each analytical step of the flow system procedure can be implemented independently, and the systems can easily be reconfigured by means of software [41]. Only the reagent necessary for sample analysis is introduced in the system, thus being in compliance with the GAC principles more closely than conventional FIA approaches.

The combination of MC with SPMS (SPMS-MC) has become a very fruitful implementation in the field of optosensors. Both, single- and multi-analyte SPMS-MC systems based on analyte native properties (using UV and fluorescent detection techniques) have been described [42, 43], using the same strategies applied for SPMS-FIA multi-optosensors. SPMS-MC implementation exhibits interesting properties in line with the main principles of GAC concept [43]: (1) minimization of both sample and reagents consumption (inherent to multicommutation principle) is achieved by re-cycling solutions to their respective vessels while they are not being used in the flow system; (2) no derivatizing reagent is used; and, (3) one single injection allows multicomponent determination together with enhanced sensitivity and selectivity provided by the solid phase.

12.4.2 Monitoring derivative species

Flow-SPMS procedures based on the use of derivative reactions involves a less ideal GAC approach than those earlier described. Nevertheless, it is worthwhile mentioning them because they still keep very interesting features related with the GAC principles. The number of described flow-through optosensors monitoring derivative product is lower than those measuring a native analytical signal from the analyte(s). It is due, probably, to the difficulty of regenerating the solid sensing zone when retaining derivatized species.

12.4.2.1 FIA-solid phase molecular spectroscopy with derivatizing reactions

There are two different designs for sensors which monitor a reaction product: (1) on-line derivatization, before the sample plug reaches the solid phase area and retention of the derivative species on the sensing zone; or (2) to immobilize the reagent on the sensing support and to generate the reaction product just when the analyte in the sample plug reaches the sensing microbeads. In this second option, successive reutilization of the immobilized reagent onto the solid phase is performed. Sensors based on this option integrate the reaction, the retention (separation/preconcentration) and the detection processes on the solid phase, so these three processes occur at the same time and in the same place of the flow system, just on the solid phase, while the former design integrates only the retention and detection processes.

Optosensors with on-line reaction development In these cases, the reagent is injected in the flow system to perform an on-line reaction before the derivative species is monitored [44, 45]. The most common design uses one merging zone with synchronized injections of both sample and reagent. Therefore, the amount of reagent is drastically minimized. Spectrophotometric and luminescence, including lanthanide sensitized luminescence and RTP SPMS-FIA based methods with derivatizing reactions have been described [18, 46, 47, 48, 49, 50].

Optosensors with immobilization of the derivative reagent The permanent immobilization of the derivatizing reagent on the sensing microbeads simplifies the manifold of those SPMS-FIA systems with derivative reaction and offers a remarkable saving of reagent. A relatively small amount of immobilized reagent can be reutilized for a large number of injections, when an appropriate eluent is chosen, which is able to selectively elute the analyte (or reagent product) from the microbeads, keeping the reagent retained after each determination on the solid sensing phase for the next measurement. From the point of view of GAC, this option is preferable than performing an on-line reaction. This design is applied usually for metal-ion optosensors based on chelating reactions with chromogenic of fluorogenic reagents [51, 52, 53] although chemiluminescence optosensors have also been described including reagent, enzyme or analyte inmobilization [54, 55].

12.4.3 Recent flow-SPMS based approaches

12.4.3.1 Solid phase molecular spectroscopy combined with sequential injection analysis (SIA) and multisyringe flow injection analysis (MSFIA)

The implementation of SIA with SPMS is a recently described approach. Two single-analyte SIA-SPMS optosensors have been described for fluorimetric detection of thiamine and paracetamol, respectively [56, 57]. Besides, a multi-analyte method using dual spectroscopic detection has been proposed for determination of vitamins B_2, B_6 and C [58]. The system combines simultaneous native fluorescence determination of vitamins B_2 and B_6 (using only one sample injection) with chemiluminescence detection for vitamin C based on its direct oxidation with $KMnO_4$ in H_2SO_4 medium in a second sample injection [58]. Therefore, this multisensor combines the two essential operation modes of flow SPMS: native intrinsic signals (vitamins B_2 and B_6) and derivative reaction (vitamin C). The combination of SIA principle with SPMS shows a remarkable saving of reagents and minor waste generation, thus being an attractive SPMS approach from the point of view of GAC.

With regards to MSFIA, a chemiluminometric MSFIA-SPMS sensor has been described based on reagent immobilization (luminol) for determination of orthophosphate in waters [59]. Combination of MSFIA with the use of optical fiber and a disk-based solid phase preconcentration as sensing phase [60] for determination of sulfur in waters is another example of contribution to GAC methods showing a drastic reduction of reagents and sample compared to the similar methods, including those in homogeneous solution FIA procedures.

12.4.3.2 Solid phase molecular spectroscopy combined with bead injection

Bead Injection Spectroscopy (BIS) uses minute amounts of beads on which the reagent is adsorbed and the analyte preconcentrated and monitored via optical fibers in specially designed (jet-ring-configured) flow-through cells. Once the bead suspension is pumped into the cell, beads are retained in the cell and solution goes through a circular gap narrower than the diameter of the beads [61]. After measurement, beads are automatically discarded. BIS is a similar approach to SPMS, with the only difference of the step of injection and renovation of the solid support [62].

BIS was first associated to SIA. Then, simpler assemblies based on FIA have been proposed in the literature for both single and multi-analyte determination [63, 64] using available commercial flow cells. As a consequence of renewable surface principle of SPMS-BIS, three important advantages can be highlighted: (1) there is no deactivation or contamination of sensing surface; (2) microbeads compaction due to continued stream circulation does not take place; and (3) no reversibility of the sorption/elution process is required. It should be noted that BIS-SPMS is an interesting alternative to SPMS procedures when high ionic strength eluent solutions are required in order to elute completely those species strongly retained on the support. Note that BIS operates in a somewhat similar way to SPMS in batch mode, considering that in both approaches the beads are for single use. Nevertheless, the automation of the procedure benefits from the additional advantages of SPMS-FIA (or SPMS-SIA) compared to batch SPMS mode, including saving in both sample and derivatizing reagent [6]. Recently, magnetic beads have been used as alternative to other more common supports in the development of renewable surface optosensors and a few BIS-SIA immusensors have been described based on an immunoreaction performed on the magnetic beads surface with immobilized antibodies [65].

12.5 Selected examples of application of solid phase molecular spectroscopy

Representative examples of optosensors are discussed, focused on particular cases able to perform multideterminations without derivatizing reagent, matching the main requirements of GAC procedures. For instance, the simultaneous determination of carbendazim and benomyl in environmental water samples by means of a fluorimetric flow-SPMS optosensor is an interesting example of SPMS-FIA system. The continuous-flow assembly used is depicted schematically in Figure 12.5(a). The sample solution containing the binary mixture is inserted by means of the injection valve into the carrier/eluent stream and pumped to the flow-through cell, which contain certain amount of solid support above the detection area. Then carbendazim, because of its faster retention-elution kinetic process, reaches the beads in the detection zone and is eluted by the carrier itself, originating a transitory signal; meanwhile, benomyl interacts strongly with the support on the top part of the solid support, just above the irradiated zone, being retained there. When carbendazim was totally eluted by the carrier itself, by turning the selection valve, an appropriate solution elutes benomyl, carrying it to the irradiated zone, and performing its transitory analytical signal. The fluorescence emission was continuously and simultaneously monitored at two pairs of excitation/emission wavelengths, so that a double fiagram is obtained [19]. The amount of sensing support placed inside the flow cell is a key parameter in order to obtain complete separation of the analytes in the cell before both of them reach the detection area. Figure 12.5(b) shows the influence of the solid amount on the binary mixture separation.

An interesting and single SPMS-FIA multioptosensor for the determination in the UV region of two analytes is that based on the use of two appropriately alternate carrier/self-eluting streams. Figure 12.6(a) shows the single designed manifold for UV determination of ascorbic acid and paracetamol. The sample with the binary mixture is injected alternately in one of the respective carrier/self-eluting solutions (acetate buffer for ascorbic acid and alkaline NaCl solution for paracetamol), selected by means of the selection valve, SV. Each carrier/self-eluting stream allows to selectively retain one of the analytes and, simultaneously, exclude the other one. Figure 12.6(b) shows the fiagram corresponding to the calibration of this single optosensor able to solve an analytical problem which cannot be solved in a similar way with the respective manifold in homogeneous solution (without using a sensing support). It was successfully applied to pharmaceutical analysis [39].

Another example of UV-SPMS-FIA for determining two analytes is that shown in Figure 12.7. In this case, the manifold (Figure 12.7a) also uses two appropriate alternate carrier/self-eluting solutions, and, besides, two

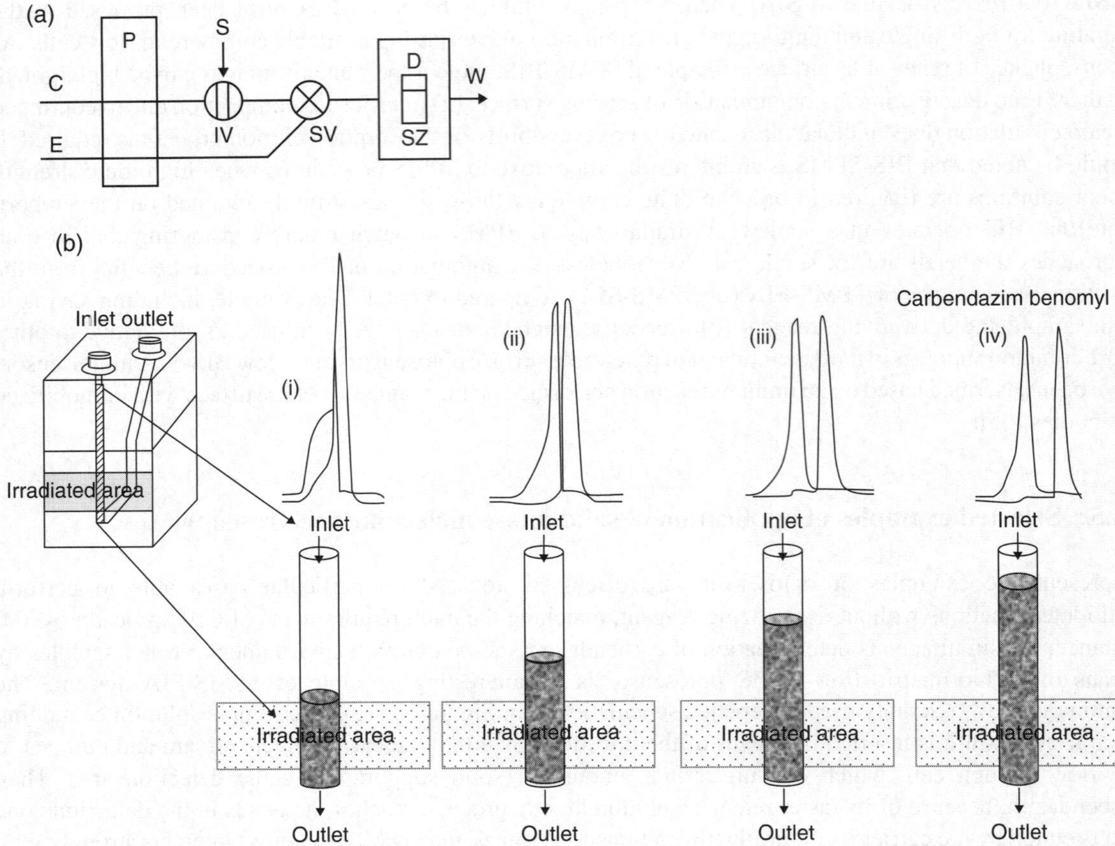

Figure 12.5 (a) Manifold used for Fluorimetric-SPMS optosensor for resolution of the mixture carbendazim and benomyl using additional solid support in the flow cell. P: peristaltic pump. IV/SV: injection/selection valve. C: carrier; E: eluting solution; S: sample; D: UV-detector. SZ: sensing zone. W: waste. (b) Scheme showing the optimization of the separation which occurs in the flow-cell just above the detection area as a function of the solid microbeads amount: (i) with 41.6 mg of C_{18}; (ii) with 47.2 mg of C_{18}; (iii) with 52.9 mg of C_{18} and (iv) with 59.3 mg of C_{18} (22 mm height above irradiated zone). Reproduced from [19] with permission from Elsevier, © 2003.

different sensing supports (SZ_1 and SZ_2), each placed inside a quartz flow cell in a double beam spectrophotometer. The carrier solution is selected by SV_1 and pumped alternately to SZ_1 or SZ_2 by means SV_2). The system was designed to determine binary mixture of vitamins B_1 (Sephadex SPC-25 cation exchange gel as solid support in the sample beam) and C (Sephadex anion-exchange gel as sensing support in the reference beam). The sample is successively injected in the respective carrier solution. With an acetate buffer as carrier solution vitamin B_1 is selectively retained on the cation exchange gel, while vitamin C cannot be retained, but excluded. A transitory positive signal is thus obtained, as the carrier itself elutes the analyte. With a phosphate buffer is selected as second carrier solution, vitamin C is selectively retained, given transitory negative signals in this case (Figure 12.6b). Once again, analytical problem is thus solved in a very single way working in UV region. The working wavelength corresponds to absorption maximum of vitamin B_1, which is the component in much lower concentration in the samples (up to 40 times lower than vitamin C). In this way

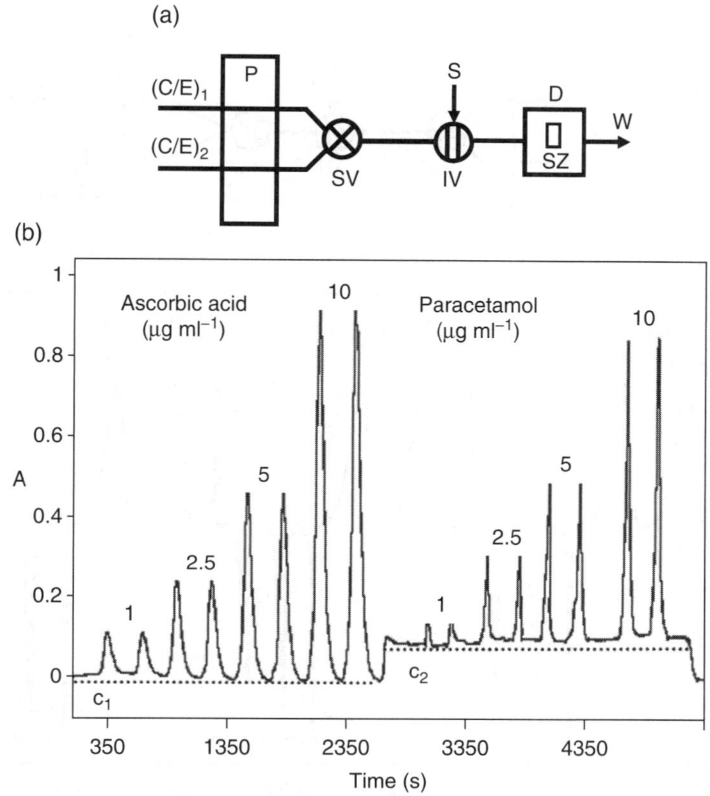

Figure 12.6 *(a) Manifold of the UV-SPMS optosensor for sequential determination of ascorbic acid (AA) and paracetamol (PCT). P: peristaltic pump. IV/SV: injection/selection valve. C/E: carrier self-eluting solution; S: sample; D: UV-detector. SZ: sensing zone. W: waste. (b) Fiagram obtained in the calibration of the sensor. Note the higher baseline in the measurement of PCT signals with respect to AA, due to the compaction of the anion-exchange gel Sephadex QAE-A25 solid support beads when the second carrier (alkaline solution) flows through the system. Reproduced from [39] with permission from Elsevier, © 2000.*

the need of performing a previous dilution when determining vitamin C is avoided. The sensor was successfully applied to pharmaceuticals analysis [40]. These two latter examples correspond to sequential multidetermination FIA-SPMS systems; meanwhile, the former is an example of simultaneous one. Other interesting UV and fluorescence flow SPMS procedures for flow multideterminations have been described based in different strategies. The reader is submitted to literature for additional information [35–38, 66, 67, 68, 69].

12.6 The potential of flow solid phase envisaged from the point of view of Green Analytical Chemistry

Very simple and reliable methodologies that contribute to a greener analytical chemistry have been provided by SPMS in both batch and different flow modes (FIA, SIA, BIS, MC…) described in this chapter, saving sample, reagents and waste generation. In general, they exhibit very attractive analytical features in terms of

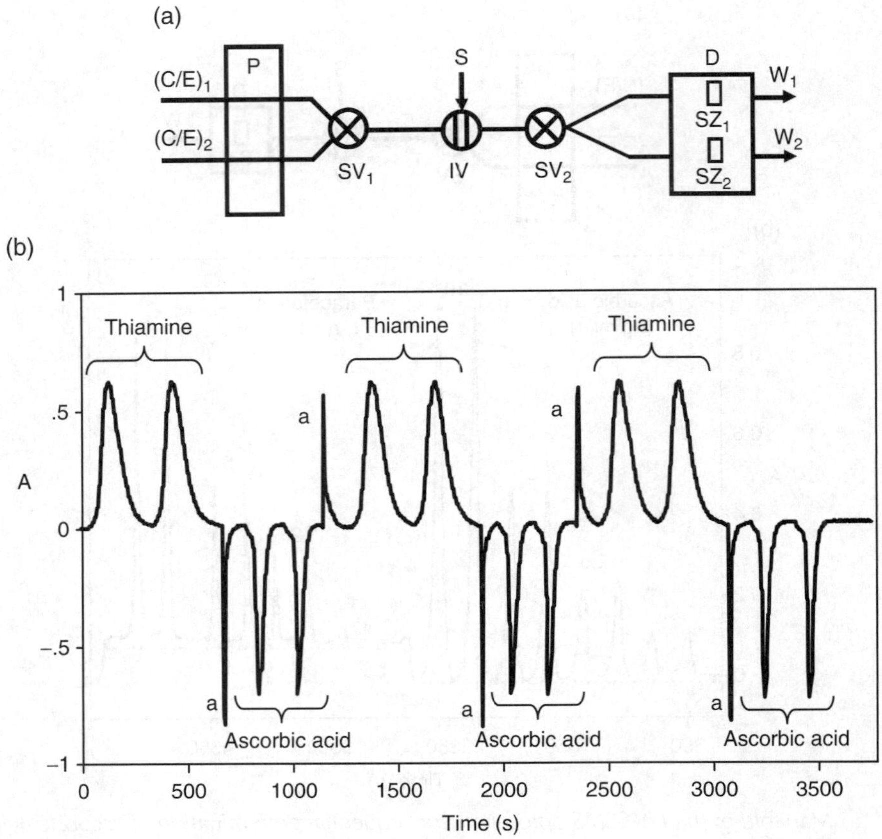

Figure 12.7 (a) Manifold of the UV-SPMS optosensor for sequential determination of thiamine and ascorbic acid using two sensing zones strategy. P: peristaltic pump. IV/SV: injection/selection valve. C/E: carrier self-eluting solution; S: sample; D: UV-detector. SZ: sensing zones. W: waste. (b) Fiagram obtained with the sequential UV-SPMS-FIA optosensor. Peaks 'a' correspond to transitory compaction when alternatively changing the carrier solution. For details, see [40].

sensitivity and selectivity both provided by the analyte(s) retention on the solid sensing support, together with appropriately optimized working conditions. Table 12.1 shows the consumption of reagent(s) and solid phase amounts in SPMS (batch and automatic modes) and the respective conventional solution FIA procedures (for single determinations). Several conclusions can be drawn from these data:

- in general, SPMS in any flow mode shows lower consumption of reagents. The larger volume of reagent necessary in flow-SPMS is usually only that of the carrier solution. SPMS-BIS shows the minimum reagent consumption followed by SIA- and MC-SPMS;
- in all cases, except SPMS-BIS, flow-SPMS saves an extraordinary amount of solid support due to the reuse of the solid phase (about 300 times less amount than SPMS in batch mode; and
- available conventional FIA solution methods show very much larger reagent consumption that SPMS procedures.

Table 12.1 *Comparative values of solution(s) volume and solid phase amount used in SPS (batch and automatic modes) and conventional homogeneous solution FIA for various analytes. Reproduced from [6] with permission from Elsevier © 2010.*

Analyte	SPMS (Batch)	SPMS-FIA	SPMS-SIA	SPMS-MC	SPMS-BIS	FIA (Conventional)	References
Thiamine	1–100 ml acetate buffer 0.1 M 30 mg ss (UV)	2.5 ml acetate buffer 0.15 M 0.1 mg ss (UV) 13.2 ml (acetonitrile 60% + acetonitrile 4 % + Na_2SO_3 0.1 M + NaOH 0.1 M) 0.1 mg ss (PIF)	—	—	—	—	[9, 71, 72]
Ascorbic Acid	1 ml $Na_2B_4O_7$ buffer 0.03 M 30 mg ss (UV)	2.14 ml acetate buffer 0.04 M 0.1 mg ss (UV)	1.2 ml H_2SO_4 0.1 mg ss (CL)	—	—	—	[58,73,74]
Pyridoxine	1–5 ml succinate buffer 0.02 M 100 mg ss (Fluor.)	2.3 ml NaCl 0.031 M 0.1 mg ss (Fluor.) 1.5 ml NaCl 0,08 M 0.1 mg ss (UV)	1 ml H_2SO_4 0.005 M (Fluor.)	1.5 ml HCl 0.01 M (Fluor.)	—	—	[42, 75–78]
Fe (total)	0.5–5 ml AA 1% 0.4–6 ml Fz 0.019 M	—	—	—	0.1 ml AA 10 mg/l 0.33 mg Fz	3.45 ml AA 1% 3.20 ml (Fz 0.01 M + HTMA 0.14 M)	[79, 80, 81]

(continued)

Table 12.1 (*continued*)

Analyte	SPMS (Batch)	SPMS-FIA	SPMS-SIA	SPMS-MC	SPMS-BIS	FIA (Conventional)	References
	1–100 ml tartrate buffer 0.22 M 40 mg ss (Vis)				33 mg ss (Vis)	(Vis)	
Ascorbic Acid	1–4 ml Fe(III)1.26 × 10^{-4} M 1–6 ml Fz 1.17 × 10^{-3} M 1–10 ml HMTA 0.25 M 30 mg ss (Vis)	—	—	—	0.1 ml Fe(III) 8 mg l^{-1} 0.33 mg Fz 33 mg ss (Vis)	3.45 ml Fe(III) 1.8 × 10^{-4} M 3.20 ml (Fz 0.01 M + HTMA 0.14 M) (Vis)	[80, 82, 83]
Amiloride	1–50 ml acetate buffer 0.01 M 30 mg ss (Fluor.)	3.08 ml acetate buffer 0.4 M 0.1 mg ss (Fluor.)	—	—	—	—	[84, 85]

UV: UV detection; Vis: Spectrophotometric Visible detection; Fluor.: fluorimetric detection; PIF: photo induced fluorescence detection; CL: Chemiluminescence detection; ss: solid support; HMTA: Hexamethylenetetramine; Fz: Ferrozine (3-(2-pyridyl)-5,6-diphenyl-1,2,4-triazine-4′,4″-disulfonic acid disodium salt); AA: Ascorbic acid

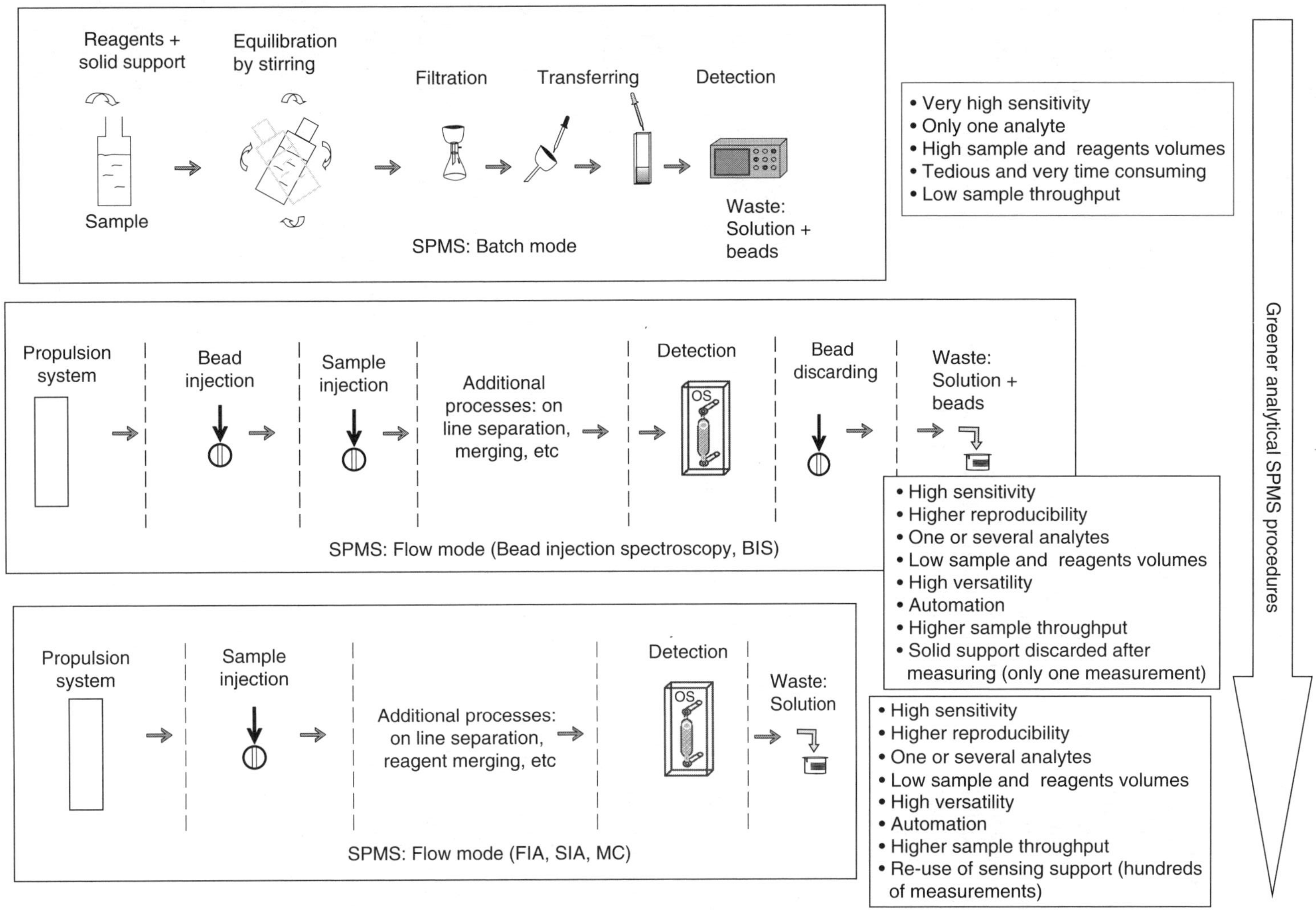

Figure 12.8 *Schematic comparison of SPMS procedures from the GAC point of view.*

In SPMS-MC implementation, the potentiality offered by the easy reconfiguration of the systems (just by software), makes the SPMS concept an attractive, interesting and cost effective analytical tool from a GAC point of view, able to solve real analytical problems.

It is worth to mention the GAC potential of multi-determinations by flow-SPMS without using derivative reagents and the different single ways to perform strategies. Nevertheless, we must also mention the drawback related with these systems when applied to complex sample matrices. In these cases, high matrix effects and significant sensitivity decreases are expected [70]. On the other hand, it has to be considered the disadvantage of deactivation of the sensing surface with repeated use, so needing replacement of the solid support in the cell from time to time. This deactivation is more quickly observed when analyzing complex samples. The implementation SPMS-BIS overcomes this problem, while for samples such as waters or active principles in pharmaceuticals, flow SPMS is more robust and of practical cost-effective use. Finally, Figure 12.8 shows a schematic comparison of the gradual approximation to GAC concept of the most significant SPMS methodologies presented in this chapter, summarizing, besides, the respective more relevant features of these methodologies which contribute to the development of GAC methods.

References

1. He, Y.; Tang, L.; Wu, X.; Hou, X. and Lee, Y.I. (2007) Spectroscopy: The best way toward green analytical chemistry?, *Appl. Spec. Reviews*, **42**, 119–138.
2. Ortega-Barrales, P.; Molina-Díaz, A.; Pascual-Reguera, M.I. and Capitán-Vallvey, L.F. (1997) Solid-phase spectrophotometric determination of trace amounts of hydrazine at sub-ng ml^{-1} level, *Anal. Chim. Acta*, **353**, 115–122.
3. Yoshimura, K.; Waki, H. and Ohashi, S. (1976) Ion-exchanger colorimetry—I: Micro determination of chromium, iron, copper and cobalt in water, *Talanta*, **23**, 449–454.
4. Armenta, S. and de la Guardia, M. (2009) Green Spectroscopy: A scientometric picture, *Spectrosc. Lett.*, **42**, 277–283.
5. García–Reyes, J.F.; Gilbert-López, B. and Molina-Díaz, A. (2009) Flow-through solid phase spectroscopy: A contribution to Green Analytical Chemistry, *Spectrosc. Lett.*, **42**, 383–393.
6. Molina-Díaz, A.; García Reyes, J.F. and Gilbert López, B. (2009) Solid Phase spectroscopy from the point of view of green analytical chemistry, *Trends Anal. Chem.*, **29**, 654–666.
7. Yoshimura, K. and Waki, H. (1985) Ion-exchanger phase absorptiometry for trace analysis, *Talanta*, **32**, 345–352.
8. Capitán, F.; Sánchez-Palencia, G.; Navalón, A.; Capitán-Vallvey, L.F. and Vilchez, J.L. (1992) Simultaneous determination of molybdenum and tungsten by first-derivative synchronous solid-phase spectrofluorimetry, *Anal. Chim. Acta*, **259**, 345–353.
9. Ortega Barrales, P.; Fernández de Córdova, M.L. and Molina Díaz, A. (1998) Microdetermination of Vitamin B_1 in the Presence of Vitamins B_2, B_6, and B_{12} by Solid-Phase UV Spectrophotometry, *Anal. Chem.*, **70**, 271– 275.
10. Fernández de Córdova, M.L.; Molina Díaz, A.; Pascual-Reguera, M.I. and Capitán-Vallvey, L.F. (1995) Solid-phase spectrophotometric determination of trace amounts of vanadium at sub-ng/ml level with 4-(2-pyridylazo)resorcinol, *Talanta*, **42**, 1057–1065.
11. Molina Díaz, A.; Ruiz Medina, A. and Fernández de Córdova, M.L. (2002) The potential of flow-through optosensors in pharmaceutical analysis. A review. *J. Pharm and Biomed. Anal.*, **28**, 399–419.
12. Capitán-Vallvey, L.F.; de Orbe, I.; Valencia, M.C. and Berzas-Nevado, J.J. (1994) A solid- phase spectrophometric method for the determination of sulphatiazole, *Talanta*, **4**, 1327–1333.
13. Fernández de Córdova, M.L.; Ortega Barrales, P. and Molina Díaz, A. (1998) Sensitive and selective determination of diclofenac sodium in pharmaceutical preparations by solid phase ultraviolet absorptiometry, *Anal. Chim. Acta*, **369**, 263–268.
14. Ortega Algar, S.; Ramos Martos, N. and Molina Díaz, A. (2003) Fast and single solid phase fluorescence spectroscopic batch procedure for (acetyl) salicylic acid determination in drug formulations, *J. Pharm. Biomed. Anal.*, **31**, 439–446.

15. Capitán, F.; Manzano, E.; Vilchez, J.L. and Capitán-Vallvey, L.F. (1989) Determination of trace aluminum in natural waters by ion exchanger fluorometry, *Anal. Sci.*, **5**, 549–555.
16. Ruzicka, J. and Hansen, E.H. (1985) Optosensing at active surfaces: a new detection principle in flow injection analysis, *Anal. Chim. Acta*, **173**, 3–21.
17. Ortega-Algar, S.; Ramos-Martos, N. and Molina-Díaz, A. (2003) A flow-through fluorimetric sensing device for determination of α- and β-naphthol mixtures using a partial least-squares multivariate calibration approach, *Talanta*, **60**, 313–323.
18. Llorent-Martínez, E.J.; Ortega-Barrales, P. and Molina-Díaz, A. (2005) Multicommuted optosensor for the determination of pipemidic acid in biological fluids, *Anal. Biochem.*, **347**, 330–332.
19. García Reyes, J.F.; Ortega Barrales, P. and Molina Díaz, A. (2003) Gel-surface enhanced fluorescence sensing system coupled to a continuous-flow assembly for simultaneous monitoring of benomyl and carbendazim, *Anal. Chim. Acta*, **493**, 35–45.
20. Ruiz Medina, A.; Fernández de Córdova, M.L. and Molina Díaz, A. (1999) A very simple resolution of the mixture paracetamol and salicylamide by flow injection–solid phase spectrophotometry, *Anal. Chim. Acta*, **394**, 149–158.
21. Yoshimura, K. (1987) Implementation of ion-exchanger absorptiometric detection in flow analysis systems, *Anal. Chem.*, **59**, 2922–2924.
22. Capitán-Vallvey, L.F.; Valencia, M.C. and Arana Nicolás, E. (2001) Monoparameter sensors for the determination of the antioxidants butylated hydroxyanisole and n-propyl gallate in foods and cosmetics by flow injection spectrophotometry, *Analyst*, **126**, 897–902.
23. Capitán-Vallvey, L.F.; Valencia, M.C. and Arana Nicolás, E. (2002) Flow-through sensor for determination of butylated hydroxytoluene in cosmetics, *Anal. Lett.*, **35**, 65–81.
24. Arce, L.; Lista, A.G.; Ríos, A. and Valcárcel, M. (2001) Screening of polyphenols in grape marc by on-line supercritical fluid extraction-flow through sensor, *Anal. Lett.*, **34**, 1461–1476.
25. Gong, Z. and Zhang, Z. (1996) Cyclodextrin-based optosensor for the determination of riboflavin in pharmaceutical preparations, *Analyst*, **121**, 1119–1122.
26. Recalde-Ruiz, D.L.; Andrés-García, E. and Díaz-García, M.E. (2000) Fluorimetric flow injection and flow-through sensing systems for cyanide control in waste water, *Analyst*, **125**, 2100–2105.
27. Fernández-Sánchez, J.F.; Segura-Carretero, A.; Costa-Fernández, J.M.; Bordel, N.; Pereiro, R.; Cruces-Blanco, C.; Sanz-Medel, A. and Fernández-Gutierrez, A. (2003) Fluorescence optosensors based on different transducers for the determination of polycyclic aromatic hydrocarbons in water, *Anal. Bioanal. Chem.*, **377**, 614–623,
28. Sánchez-Barragán, I.; Fernández-Sánchez, J.F.; Pereiro, R.; Sanz-Medel, A.; Salinas, A.; Segura, A.; Fernández-Gutierrez, A. and Ballesteros González, J.M. (2005) Molecularly imprinted polymers based on iodinated monomers for selective room temperature phosphorescence optosensing of fluoranthene in water, *Anal. Chem.*, **77**, 7005–7011.
29. Ortega-Barrales, P.; Ayora Cañada, M.J.; Molina Díaz, A.; Garrigues, S. and de la Guardia, M. (1999) Solid phase Fourier transform near infrared spectroscopy, *Analyst*, **124**, 579–582.
30. Haberkorn, M.; Hinsmann, P. and Lendl, B. (2002) A mid-IR flow-through sensor for direct monitoring of enzyme catalysed reactions. Case study: Measurement of carbohydrates in beer, *Analyst*, **127**, 109–113.
31. Ruedas Rama, M.J.; López-Sánchez, M.; Ruiz-Medina, A.; Molina-Díaz, A. and Ayora-Cañada, M.J. (2005) Flow-Through Sensor with Fourier Transform Raman Detection for Determination of Sulfonamides, *Analyst*, **130**, 1617–1623.
32. Dias, A.C.B.; Figueiredo, E.C.; Grassi, V.; Zagatto, E.A.G. and Arruda, M.A.Z. (2008) Molecularly imprinted polymer as a solid phase extractor in flow analysis, *Talanta*, **76**, 988–996.
33. Xiong, Y.; Zhou, H.; Zhang, Z.; He, D. and He, C. (2006) Determination of hydralazine with flow injection chemiluminescence sensor using molecularly imprinted polymer as recognition element, *J. Pharm. Biomed. Anal.*, **41**, 694–700.
34. He, D.; Zhang, Z.; Zhou, H. and Huang, Y. (2006) Micro flow sensor on a chip for the determination of terbutaline in human serum based on chemiluminescence and a molecularly imprinted polymer, *Talanta*, **69**, 1215–1220.
35. Ayora Cañada, M.J.; Pascual Reguera, M.I.; Molina Díaz, A. and Capitán-Vallvey, L.F. (1999) Solid-phase UV spectroscopic multisensor for the simultaneous determination of caffeine, dimenhydrinate and acetaminophen by using partial least squares multicalibration, *Talanta*, **49**, 691–701.
36. García Reyes, J.F.; Llorent Martínez, E.J.; Ortega Barrales, P. and Molina Díaz, A. (2004) Multiwavelength fluorescence based optosensor for simultaneous determination of fuberidazole, carbaryl and benomyl, *Talanta*, **64**, 742–749.

37. López Flores, J.; Fernández de Córdova, M.L. and Molina Díaz, A. (2007) Flow-through optosensor combined with photochemically induced fluorescence for simultaneous determination of binary mixtures of sulfonamides in pharmaceuticals, milk and urine, *Anal. Chim. Acta*, **600**, 164–171.
38. Fernández de Córdova, M.L.; Ortega Barrales, P.; Rodríguez Torné, G. and Molina Díaz, A. (2003) A flow injection sensor for simultaneous determination of sulfamethoxazole and trimethoprim by using Sephadex SP C-25 for continuous on-line separation and solid phase UV transduction, *J. Pharm. Biomed. Anal.*, **31**, 669–677.
39. Ruiz-Medina, A.; Fernández-de Córdova, M.L.; Ayora-Cañada, M.J.; Pascual-Reguera, M.I. and Molina-Díaz, A. (2000) A flow-through solid phase UV spectrophotometric biparameter sensor for the sequential determination of ascorbic acid and paracetamol. *Anal. Chim. Acta*, **404**, 131–139.
40. Ruiz-Medina, A.; Ortega Barrales, P.; Fernández-de Córdova, M.L. and Molina-Díaz, A. (2002) A continuous flow solid phase spectroscopic sensor using two sensing zones: determination of thiamine and ascorbic acid. *J. AOAC Intern.*, **85**, 369–374.
41. Reis, B.F.; Giné, M.F.; Zagatto, E.A.G.; Lima, J.L.F.C. and Lapa, R.A. (1994) Multicommutation in flow analysis. Part 1. Binary sampling: concepts, instrumentation and spectrophotometric determination of iron in plant digests. *Anal. Chim. Acta*, **293**, 129–138.
42. Llorent-Martinez, E.J.; Dominguez-Vidal, A.; Ortega-Barrales, P.; de la Guardia, M. and Molina-Diaz, A. (2005) Implementation of multicommutation principle with flow-through multioptosensors, *Anal. Chim. Acta*, **545**, 113–118.
43. García-Reyes, J.F.; Llorent-Martínez, E.J.; Ortega-Barrales, P. and Molina-Díaz, A. (2006) Determination of thiabendazole residues in citrus fruits using a multicommuted fluorescence-based optosensor, *Anal. Chim. Acta*, **557**, 95–100.
44. García-Reyes, J.F.; Ortega-Barrales, P. and Molina-Díaz, A. (2005) Development of a solid surface fluorescence-based sensing system for aluminium monitoring in drinking water, *Talanta*, **65**, 1203–1208.
45. García-Reyes, J.F.; Ortega-Barrales, P. and Molina-Díaz, A. (2006) Sensing of trace amounts of cadmium in drinking water using a single fluorescence-based optosensor, *Microchem. J.*, **82**, 94–99.
46. Ayora-Cañada, M.J.; Pascual-Reguera, M.I. and Molina-Díaz, A. (2000) Determination of vanadium by solid phase spectrophotometry in a continuous flow system, *Intern. J. Environ. Anal. Chem.*, **76**, 319–330.
47. Gusmao Portela, J.; Spínola Costa, A.C. and Gomes Teixeira, L.S. (2004) Determination of Vitamin B_6 in pharmaceutical formulations by flow injection-solid phase spectrophotometry, *J. Pharm. Biomed. Anal.*, **34**, 543–549.
48. Chen, D.; Luque de Castro, M.D. and Valcárcel, M. (1992) Flow-through sensing device based on derivative synchronous fluorescence measurements for the multi-determination of B_6 vitamers, *Anal. Chim. Acta*, **26**, 269–274.
49. Alava-Moreno, F.; Díaz-García, M.E. and Sanz-Medel, A. (1993) Room temperature phosphorescence optosensor for tetracyclines, *Anal. Chim. Acta*, **281**, 637–644.
50. Traviesa-Alvarez, J.M.; Costa-Fernández, J.M.; Pereiro, R. and Sanz-Medel, A. (2007) Direct screening of tetracyclines in water and bovine milk using room temperature phosphorescence detection, *Analytica Chimica Acta*, **589**, 51–58.
51. Ayora-Cañada, M.J.; Pascual-Reguera, M.I. and Molina-Díaz, A. (1998) Continuous flow-through solid phase spectrophotometric determination of trace amounts of zinc, *Anal. Chim. Acta*, **375**, 71–80.
52. Lenarczuck, T.; Glab, S. and Koncki, R. (2001) Application of Prussian blue-based optical sensor in pharmaceutical analysis, *J. Pharm. Biomed. Anal.*, **26**, 163–169.
53. De la Torre, M.; Fernández-Gámez, F.; Lázaro, F.; Luque de Castro, M.D. and Valcárcel, M. (1991) Spectrofluorimetric flow-through sensor for the determination of beryllium in alloys, *Analyst*, **116**, 81–83.
54. Li, B.; Zhang, Z. and Lin, Y. (2001) Chemiluminescence flow biosensor for hydrogen peroxide with immobilized reagents, *Sens. Actuators B*, **72**, 115–119.
55. Du, J.; Sehn, L. and Lu, J. (2003) Flow injection chemiluminescence determination of epinephrine using epinephrine-imprinted polymer as recognition material, *Anal. Chim. Acta*, **489**, 183–189.
56. Zhu, H.; Chen, H. and Zhou, Y. (2003) Determination of thiamine in pharmaceutical preparations by sequential injection renewable surface solid-phase spectrofluorimetry, *Anal. Sci.*, **19**, 289–294.
57. Llorent-Martínez, E.J.; Šatínský, D.; Solich, P.; Ortega-Barrales, P. and Molina-Díaz, A. (2007) Fluorimetric SIA optosensing in pharmaceutical analysis: Determination of paracetamol. *J. Pharm. Biomed. Anal.*, **45**, 318–31.

58. Llorent-Martínez, E.J.; Ortega-Barrales, P. and Molina-Díaz, A. (2008) Sequential injection multi-optosensor based on a dual-luminescence system using two sensing zones: application to multivitamin determination, *Microchim. Acta*, **162**, 199–204.
59. Morais, I.P.A.; Miró, M.; Manera, M.; Estela, J.M.; Cerdá, V.; Souto, V.M.R.S. and Rangel, A.O.S.S. (2004) Flow-through solid-phase based optical sensor for the multisyringe flow injection trace determination of orthophosphate in waters with chemiluminescence detection, *Anal. Chim. Acta.*, **506**, 17–24.
60. Ferrer, L.; de Armas, G.; Miró, M.; Estela, J.M. and Cerdà, V. (2005) Flow-through optical fiber sensor for automatic sulfide determination in waters by multisyringe flow injection analysis using solid-phase reflectometry, *Analyst*, **130**, 644–651.
61. Ruzicka, J. and Scampivia, L. (1999) From flow injection to bead injection, *Anal. Chem.*, **71**, 257A–263A.
62. Miró, M. and Frenzel, W. (2004) Flow-through sorptive preconcentration with direct optosensing at solid surfaces for trace-ion analysis, *Trends Anal. Chem.*, **23**, 11–20.
63. Jitmanee, K.; Kradtap Hartwell, S.; Jakmunee, J.; Jayasvasti, S.; Ruzicka, J. and Grudpan, K. (2002) A simple flow injection system with bead injection for trace iron determination, *Talanta*, **57**, 187–192.
64. Ruedas Rama, M.J.; Ruiz Medina, A. and Molina Díaz, A. (2004) Implementation of flow-through multi-sensors with bead injection spectroscopy: fluorimetric renewable surface biparameter sensor for determination of berillium and aluminium, *Talanta*, **62**, 879–886.
65. Tanaka, K. and Imagawa, H. (2005) Rapid and reagent-saving immunoassay using innovative stirring actions of magnetic beads in microreactors in the sequential injection mode, *Talanta*, **68**, 437–441.
66. Ruedas Rama, M.J.; Ruiz Medina, A. and Molina Diaz, A. (2002) Use of a solid sensing zone implemented with unsegmented flow analysis for simultaneous determination of thiabendazole and warfarin, *Anal. Chim. Acta*, **459**, 235–243.
67. Garcia-Reyes, J.F.; Ortega-Barrales, P. and Molina Diaz, A. (2004) Development of a single fluorescence-based optosensor for rapid simultaneous determination of fungicides benomyl and thiabendazole in waters and commercial formulations, *J. Agric. Food Chem.*, **52**, 2197–2202.
68. Ortega-Barrales, P.; Ruiz Medina, A.; Fernández-de Córdova, M.L.; Molina Diaz, A. (2002) A flow-through solid-phase spectroscopic sensing device implemented with FIA solution measurements in the same flow cell: determination of binary mixtures of thiamine with ascorbic acid or acetylsalicylic acid, *Anal. Bioanal. Chem.*, **373**, 227–232.
69. Dominguez-Vidal, A.; Ortega-Barrales, P. and Molina Diaz, A. (2007) Environmental water samples analysis of pesticides by means of chemometrics combined with fluorimetric multioptosensing. *J. Fluor.*, **17**, 271–277.
70. Llorent-Martínez, E.J.; García-Reyes, J.F.; Ortega-Barrales, P. and Molina Diaz, A. (2007) Multicommuted fluorescence based optosensor for the screening of bitertanol residues in banana samples, *Food Chem.*, **102**, 676–682.
71. Ortega-Barrales, P.; Fernández de Córdova, M.L. and Molina Diaz, A. (1998) A selective optosensor for UV spectrophotometric determination of thiamine in the presence of other vitamins B, *Anal. Chim. Acta*, **376**, 227–233.
72. López-Flores, J.; Fernández de Córdova, M.L. and Molina Diaz, A. (2005) Implementation of flow-through solid phase spectroscopic transduction with photochemically induced fluorescence: determination of thiamine, *Anal. Chim. Acta*, **535**, 161–168.
73. Ruíz-Medina, A.; Fernández de Córdova, M.L. and Molina Diaz, A. (1999) A rapid and selective solid-phase UV spectrophotometric method for determination of ascorbic acid in pharmaceutical preparations and urine, *J. Pharm. Biomed. Anal.*, **20**, 247–254.
74. Molina Diaz, A.; Ruíz-Medina, A. and Fernández de Córdova, M.L. (1999) Determination of ascorbic acid by use of a flow-through solid phase UV spectrophotometric system, *Fresenius J. Anal. Chem.*, **363**(1), 92–97.
75. Ortega Algar, S.; Ramos Martos, N. and Molina Diaz, A. (2003) Native fluorescence determination of pyridoxine hydrochloride (vitamin B_6) in pharmaceutical preparations after sorption on Sephadex SP C-25, *Spectrosc. Lett.*, **36**, 133–149.
76. Ruíz-Medina, A.; Fernández de Córdova, M.L. and Molina Diaz, A. (1999) Flow injection-solid phase spectrofluorimetric determination of pyridoxine in presence group B-vitamins, *Fresenius J. Anal. Chem.*, **363**(3), 265–269.
77. Ayora Cañada, M.J.; Pascual Reguera, M.I. and Molina Diaz, A. (2000) Selective determination of pyridoxine in the presence of hydrosoluble vitamins using a continuous-flow solid phase sensing device with UV detection, *Intern. J. Pharm.*, **202**, 113–120.

78. García-Reyes, J.F.; Llorent-Martínez, E.J.; Ortega-Barrales, P. and Molina Diaz, A. (2006) A multicommuted fluorescence-based sensing system for simultaneous determination of Vitamins B_2 and B_6, *Anal. Chim. Acta*, **555** 128–133.
79. Fernández de Córdova, M.L.; Ruíz-Medina, A. and Molina Diaz, A. (1997) Solid-phase spectrophotometric microdetermination of iron with ascorbic acid and ferrozine, *Fresenius' J. Anal. Chem.*, **357**, 44–49.
80. Ruedas Rama, M.J.; Ruíz-Medina, A. and Molina Diaz, A. (2003) Bead injection spectroscopic flow-through renewable surface sensors with commercial flow cells as an alternative to reusable flow-through sensors, *Anal. Chim. Acta*, **482**, 209–217.
81. Pascual-Reguera, M.I.; Ortega-Carmona, I. and Molina Diaz, A. (1997) Spectrophotometric determination of iron with ferrozine by flow-injection analysis, *Talanta*, **44**, 1793–1801.
82. Molina Diaz, A.; Ortega-Carmona, I. and Pascual-Reguera, M.I. (1998) Indirect spectrophotometric determination of ascorbic acid with ferrozine by flow-injection analysis, *Talanta*, **47**, 531–536.
83. Ortega Barrales, P.; Fernández de Córdova, M.L. and Molina Diaz, A. (1998) Indirect determination of ascorbic acid by solid-phase spectrophotometry, *Anal. Chim. Acta*, **360**, 143–152.
84. Ortega-Barrales, P.; Pellerano, G.; Vázquez, F.A. and Molina Diaz, A. (2002) Rapid and sensitive determination of amiloride by cation exchange preconcentration and direct solid-phase UV detection, *Anal. Lett.*, **35**(9), 1491–1504.
85. Domínguez-Vidal, A.; Ortega-Barrales, P. and Molina-Díaz A. (2002) Fast flow-injection fluorimetric determination of amiloride by using a solid sensing zone, *Talanta*, **56**, 1005–1013.

13

Derivative Techniques in Molecular Absorption, Fluorimetry and Liquid Chromatography as Tools for Green Analytical Chemistry

José Manuel Cano Pavón, Amparo García de Torres, Catalina Bosch Ojeda, Fuensanta Sánchez Rojas and Elisa I. Vereda Alonso

Department of Analytical Chemistry, University of Málaga, Málaga, Spain

13.1 The derivative technique as a tool for Green Analytical Chemistry

Derivative spectroscopy is an analytical technique of great utility for extracting both qualitative and quantitative information from spectra or chromatograms composed of unresolved bands. This makes unnecessary the preparation of other samples to obtain spectra or chromatograms with a higher level of resolution and thus can achieve significant savings on products and solvents. Although derivative techniques were introduced more 50 years ago and have demonstrable advantages for the solution of specific analytical problems, this technique has been accepted only hesitantly, because of the initial lack of reasonably priced instrumentation and original limitation to the first derivative. However, the introduction firstly of electronic differentiation by a microcomputer interfaced with the spectrophotometer and more recently the use of a adequate computer for the mathematical treatment of derivative signal by means of an adequate algorithm makes possible the plotting of the first, second or higher order derivatives of a spectrum with respect to wavelength or a chromatogram with respect to time.

The derivative method has found application not only in ultraviolet-visible region spectrophotometry, but also in fluorimetry (normal [1] and synchronous scanning [2]). The use of derivative techniques is not restricted to special cases, but may be of advantage whenever quantitative study of normal spectrum or

Handbook of Green Analytical Chemistry, First Edition. Edited by Miguel de la Guardia and Salvador Garrigues.

chromatogram is difficult. Its disadvantage is that the differentiation degrades the signal-to-noise ratio, so that some form of smoothing is required in conjunction with differentiation [3].

13.1.1 Theoretical aspects

13.1.1.1 Basic properties of derivative spectra

For a single-peak spectrum, the first-derivative is a plot of the gradient $dS/d\lambda$ (S being absorbance or fluorescence or another type of signal) envelope versus wavelength and features a maximum and a minimum; the vertical distance between these is the amplitude, which is proportional to the analyte concentrations; theoretically, $dS/d\lambda$ is zero at λ_{max} for the band in the normal spectrum. The second-derivative spectrum, $d^2S/d\lambda^2$ vs. λ, has two maxima with a minimum between them, at λ_{max} of the normal absorption (or emission) band [4]. In principle, both peak heights (measured from $d^2A/d\lambda^2 = 0$) are proportional to the analyte concentration but the amplitude can also be measured by the so-called tangents method, in which a tangent is drawn to the maxima and the amplitude is measured vertically from the tangent to the minimum [5]. Other possibilities have been considered [6], and background absorbance of fluorescence cam be eliminated.

The differentiation discriminates against broad bands, emphasizing sharper features to an extent that increases with increasing derivative order, because for bands (Gaussian or Lorentzian) the amplitude D_n of the n-th derivative is related to the n-th power of the inverse of the band-width, w, of the normal spectrum [7]:

$$D_n \propto \left(\frac{1}{w}\right)^n$$

Thus, for two bands A and B of equal absorbance but different width, the derivative amplitude of the sharper band (A, for example) is greater than that of the broader (B) by a factor that increases with increasing derivative order [8,9]:

$$\frac{D_{n,A}}{D_{n,B}} \propto \left(\frac{w_B}{w_A}\right)^n$$

For this reason, the use of derivative spectra can increase the detection sensitivity [10,11] of minor spectral features.

For quantitative analysis in UV-visible spectrophotometry, if Beer's law is obeyed for the normal spectrum, the following equation can be obtained:

$$\frac{d^n y}{d\lambda^n} = \frac{d^n \varepsilon}{d\lambda^n}.l.c$$

where A = absorbance, ε = molar absorptivity, l = cell path-length and c = concentration of the analyte, and this form the basis of analytical determinations [12].

In fluorimetry, the basic equation is $I_f = kP_o c$ (at diluted dissolution). The n order derivative gives:

$$\frac{d^n I_f}{d\lambda^n} = \frac{d^n k}{d\lambda^n}.P_0.c$$

Where I_f is the intensity of the fluorescence signal, P_o the power of the radiation and c the concentration.

13.1.1.2 Effect of differentiation on signal-to-noise ratio

An inconvenience of the derivative technique is that the signal-to-noise ratio (SNR) becomes worse for progressive higher orders. The noise of a normal spectrum may be expressed as the standard deviation σ_o of the absorbance (or fluorescence intensity) of a blank; for the n-th order derivative, the standard deviation σ_n can be calculated from σ_o by the usual rules for error propagation [13].

The SNR of derivative signals depends on the shape of the spectrum, and has been evaluated for different bands. In Gaussian bands, for example, if we suppose as SNR of 1 in the normal spectrum, the SNR is given by 2.02/M in the first derivative, $3.26/M^2$ in the second, $8.10/M^3$ in the third, $17.8/M^4$ in the fourth, and so on, where M is the number of points in the peak full-width at half maximum. For example, if M = 12, the SNR values for the first four derivatives are 0.168, 0.022, 0.0047 and 0.00086; that is to say, if n is large, the SNR of the derivatives will be very small.

For this reason, practical derivative techniques include some degree of low-pass filtering or smoothing to control the increase in noise. The process of smoothing involves a convolution of the data series with a smoothing function consisting of a set of weighing coefficients. The various smoothing methods differ only in the way that the coefficients are calculated.

The effect of smoothing depends on two variables: (1) the smoothing ratio, which is the ratio of the width of the smoothed peak to the number M of data points corresponding to the peak full-width of half maximum, and (2) the number of times that the smoothing is done. Moreover, the attenuation factor is determined by the soothing ratio. A detailed study of the influences of these variables on the derivative spectra has been reported by O'Haver [14].

The ratio of the SNR of the unsmoothed n-th derivative $(SNR)_n$ to that of the unsmoothed normal SNR $(SNR)_o$ is given by [3]:

$$\frac{(SNR)_n}{(SNR)_0} = \alpha_n c_n r^{(n+0.5)} \sqrt{M}$$

where C_n is a constant which increased with derivative order n, and α_n is the attenuation factor of the n-th derivative for a smoothing ratio r.

13.2 Derivative absorption spectrometry in the UV-visible region

The derivative spectra are always more complex than the zero-order spectrum. The first derivative is the rate of change of absorbance against wavelength. It starts and finished at zero, passes through zero at the same wavelength as λmax of the absorbance band with first a positive and then a negative band, with the maximum and minimum at the same wavelengths as the inflection points in the absorbance band. This bipolar function is characteristic of all the odd-order derivatives. The most characteristic feature of the second-order derivative is a negative band with the minimum at the same wavelength as the maximum on the zero-order band. It also shows two additional positive satellite bands on either side of the main band. The fourth derivative shows a positive band. The presence of a strong negative or positive band, with the minimum or maximum at the same wavelength as λ_{max} of the absorbance band, is characteristic of the even-order derivatives. Note that the number of bands observed is equal to the derivative order plus one.

Derivative UV spectroscopy has been widely used as a tool for quantitative analysis, characterization; and quality control in the agricultural, pharmaceutical and biomedical fields. This outstanding feature coupled with zero-crossing, least-square deconvolution, or Fourier transform data-processing techniques, has received increasing attention in single and multi-component quantitative analysis, especially in UV-absorbing matrices.

A wide study has been published in a book by Talsky [15], in which the instrumentation used was described, as well as numerous methodologies and applications. Diverse reviews published from 1986 to 2009 [16–19] exposed the different aspects of derivative spectrometry: theoretical, instrumental devices and analytical applications. Other interesting reviews are published in 2004 by Karpinska about various aspects of application in chemical analysis [20] and in 2005 by El-Sayed and El-Salem about developments and analytical applications [21].

13.2.1 Strategies to greener derivative spectrophotometry

Miniaturization of analytical methods and instrumentation has resulted in precisely controlled trace level analyses, and their high environmental compatibility owing to insignificant amount of solvent and reagent consumption, low waste production, and low operational cost. Most of the current analytical procedures consist of a sample preparation step, and quantitation by chromatography and spectrometry.

Solvent extraction has been most widely used method for sample preparation. However, the classical mode of solvent extraction performed in a separatory funnel has inconveniences of unfavourable partition equilibria, formation of emulsions, necessity to use large volume of hazardous and ozone depleting organic solvents and difficulty in waste disposal. Miniaturization of extraction process by using microlitre volumes of the organic solvent has avoided many of these problems, allowed new ways in which the sample pre-treatment can be performed [22–24].

UV-Vis spectrophotometry is a mature analytical technique applied to many thousands of determinations owing to its simplicity, flexibility, low cost and convenience. Due to the widespread use of UV-Vis spectrophotometers for routine analysis and as a result of the great demand to decrease the sample volume needed to perform a measurement, UV-Vis micro-spectrophotometry have been developed.

13.2.1.1 Micro-scale derivative spectrophotometry batch procedures by use of internal standard

The feasibility of UV-Vis spectrophotometry for multi-component analysis has encouraged us to use the method of internal standard in micro-scale multi-step spectrophotometric procedures and thus, enhance their analytical performance and reduce adverse environmental impact.

The analytical signal is defined as the ratio between the signals of analyte and of internal standard. The most important feature of internal standard (IS) in UV-Vis spectrophotometry is that less rigorous procedures become possible and, from the green chemistry point of view, the micro-scale batch procedures can be used without sacrificing the results quality.

The well-known limitation of separation/preconcentration procedures is the increasing risk of analytical errors. Such source of errors can be controlled by using a method of IS, which is common in chromatographic and in some spectrometric techniques.

Recently, the method of IS was also applied in spectrophotometry UV-Vis and the improved analytical performance was achieved both for many-fold dilution and for preconcentration procedures. In the case of many-fold dilution (about 103), the important advantage was the use of micro volumes, and in the preconcentration procedure, the enrichment factor about 400 was achieved by evaporation of organic phase after extraction and dissolution of the residue in micro volume (200 μl). Considering the general requirements for an IS in different analytical techniques, in spectrophotometry the following parameters should be taken into account:

1. Similar chemical properties of the IS and of the analyte;
2. No interactions between the IS and the analyte;
3. Low absorbance of the IS in the region of the absorption band of the analyte;
4. High absorbance of the IS in the region of low analyte absorption signal (in the UV-Vis region);
5. The IS should not be a natural component of the sample.

The serious limitation of separation/preconcentration, or a many fold dilution procedure, is the risk of analytical errors which can cause the poor analytical performance of the method. The above mentioned facts encouraged us to introduce the method of IS in the multistage spectrophotometric procedures.

Also, a number of official analytical methods still in use in pharmaceutical and food industries rely on traditional, batch UV-Vis spectrophotometric procedures. Down-scaling seems to be attractive, owing to lower quantities of reagents and solvents employed, in agreement with the principles of green chemistry. With the aim to enhance the analytical performance of spectrophotometric micro-assays, the use of IS method had been proposed for UV-Vis spectrophotometry in general, and for derivative spectrophotometry in particular. In Table 13.1, some applications of IS in micro-scale derivative spectrophotometry are presented.

The general requirements for an IS in UV-Vis spectrophotometry, including derivative spectrophotometry and its applications in the procedures involving many-fold dilution or extraction/preconcentration of the analyte are presented recently by Wrobel and Wrobel [30].

13.2.1.2 Fibre optic-linear array detection spectrophotometry

Currently, there is an increasing interest in reducing the sample volume to the micro-scale level in many fields such as food, environment, clinical, and so on. Micro-sample quantitation is nowadays possible owing to the introduction of micro-volume cuvettes and capillaries, as well as micro-volume UV-Vis spectrophotometers.

Table 13.1 *Applications of internal standard in micro-scale derivative spectrophotometry.*

Analyte	Internal standard	Analytical signal	Detection limit	Samples	Ref.
Azinphos-methyl	Erioglaucine A and acetophenone in methanol	${}^1D_{analyte(219.59\,nm)}/{}^1D_{IS(268.93\,nm)}$	0.043 mg	Pharmaceutical formulations	[25]
Chlorpromazine hydrochloride	1,10-phenantroline	${}^2D_{analyte(256\,nm)}/{}^2D_{IS(236\,nm)}$	3.97 ng ml^{-1}	Pharmaceutical formulations	[26]
Chromium	Erioglaucine A	${}^1D_{analyte(591.3\,nm)}/{}^1D_{IS(653.5\,nm)}$	0.2 μg l^{-1}		[27]
Malondialdehyde	Erioglaucine A	${}^1D_{analyte(543.1\,nm)}/{}^1D_{IS(644.4\,nm)}$	0.19 μM	Human serum	[28]
Miconazole	Methylene blue	${}^2D_{analyte(236.9\,nm)}/{}^2D_{IS(663.2\,nm)}$	50 mg l^{-1}	Pharmaceutical creams	[29]

With recent advances in sensitive array detectors, fibre optic wave guides, high speed electronics and powerful softwares, many new generations of spectrometers have been developed. These new spectrometers use charged couple devices (CCDs) or photodiode arrays to replace the photomultipliers and avalanche photodiodes used in conventional spectrometers. They offer excellent performance for a wide range of spectroscopic applications from the UV to the near IR. Because of their unique combination of outstanding sensitivity, high speed, low noise, compactness, instantaneous capture of full spectra, low cost and robustness, these detectors have revolutionized spectroscopic detection. A quick glance at today's instrumentation market indicates the popularity of the CCD as the detector of choice. The overwhelming benefits of array detectors are simultaneous and multi-wavelength data acquisition.

On the other hand, the use of fibre optics as light guidance allows a great modularity and flexibility in setting up an optical measurement system. The fibre optic-linear array detection spectrometers (FOLADS) based on CCD-detector have made a major impact on simultaneous and real-time data collection in analytical spectroscopy from the UV to NIR. Simultaneous determination of analytes using multi-wavelength data acquisition using CCD linear array spectrophotometry is very powerful technique in this area.

Recent applications of this system are presented by Shokoufi *et al.* [31,32]. The authors described the design and construction of the FOLADS in combination with cloud point extraction (CPE) using a cylindrical micro-cell. The simultaneous preconcentration and determination of different analytes of a mixture were achieved by using the newly developed CPE/FOLADS technique. Cobalt and nickel were used as an example for evaluation of proposed method. Also, simultaneous preconcentration and determination of palladium and cobalt have been established by dispersive liquid–liquid microextraction (DLLME) in combination with fibre optic-(CCD) linear array detection spectrophotometry. In this combined method, the remained phase after DLLME was determined by FOLADS using a 50 µl cylindrical micro-cell. The overlapping spectra of the complexes were collected by FOLADS and the multi-wavelength absorbance data were processed using Matlab 7 software for first-derivative spectrophotometry.

13.3 Derivative fluorescence spectrometry

The ability to analyse complex multicomponent mixtures without resorting to tedious separation procedures is extremely useful for routine analysis. Fluorescence spectroscopy is one example of an analytical technique suitable for multicomponent analysis due to its inherent sensitivity, selectivity and versatility. In addition, the low limits of detection make molecular fluorescence the basis of many analytical methods. Fluorimetry is more selective than UV-Vis absorption spectrometry for two reasons. First, many molecules absorb in the UV or visible range but no exhibit detectable fluorescence. Second, two wavelengths (excitation and emission) are available in fluorimetry, but only one wavelength is available in absorptiometry. If two sample constituents with similar absorption spectra fluoresce at different wavelengths, they may be distinguished from one another by appropriate choice of emission wavelength. Similarly, two compounds that have similar fluorescence spectra but absorb strongly at different wavelengths may be distinguished by proper choice of excitation wavelength (selective excitation). However, the selectivity of fluorimetry is limited by the broad, featureless nature of the absorption and fluorescence spectra of most molecules. In UV-Vis absorption and fluorescence spectra, bandwidths of 25 nm or more are common, especially for polar or hydrogen-bonding molecules in polar solvents. Thus, the absorption and fluorescence spectra of a mixture of fluorescent compounds may be a jumble of overlapping bands. The derivative fluorescence technique can be used to enhance minor spectral features and reduce the effect of many of the spectral interferences. The first paper published about derivative luminescence spectrometry [33] demonstrated the useful of the derivative technique in luminescence spectrometry for both qualitative and quantitative applications. However, SNR generally worsens with increasing derivative orders, although in some cases the SNR of the derivatives exceeds that of the normal spectrum. Such samples must be subjected to preliminary cleanup and separation (which may be quite time-consuming) or specialized sample-preparation and measurement techniques may be used; these may be instrumentally complex or time-consuming.

13.3.1 Derivative synchronous fluorescence spectrometry

In conventional molecular fluorescence spectroscopy, two types of spectra are generally discerned: emission and excitation spectra. An emission spectrum is obtained by scanning the emission wavelength, λ_{em}, as the sample is irradiated at a single wavelength, λ_{ex}. Similarly, an excitation spectrum is obtained by scanning the excitation wavelength while recording the emission signal at a single wavelength. The other possibility is to scan both the monochromators simultaneously, which is called synchronous fluorescence spectrometry (SFS). The concept of SFS was first put forward by Lloyd [34] and further developed by Vo-Dinh [35]. This technique has several variants, depending on the scan-rates of the two monochromators:

- If the scan rate is constant for both the monochromators and therefore a constant wavelength interval, λ, is kept between λ_{em} and λ_{ex}, the technique is known as constant-wavelength SFS.
- The excitation and emission wavelengths may be varied simultaneously in such manner that a constant frequency difference, ν, is maintained between them. This technique, constant-energy SFS has scarcely been used.
- The excitation and emission wavelengths may be varied simultaneously but at different rates. These different rates allow the construction of planes at angles between 45° and 90° to the excitation x-axis throughout the whole spectrum. Known variously as variable-angle synchronous fluorimetry, variable separation synchronous fluorimetry or variable offset synchronous fluorimetry, this technique also has not been used in many applications.

13.3.1.1 Derivative constant-wavelength SFS

The constant-wavelength SFS technique is very simple and is the most frequently used for all synchronous modes. Therefore generally, SFS refers to this technique. The synchronous fluorescence spectra are simpler, narrower and hence more distinct than conventional excitation and emission fluorescence spectra. These features make this technique especially useful for the elimination of matrix effects and the resolution of mixtures. However, although synchronous spectra are often sufficiently resolved for analytical purposes, their derivation to the first or second order spectra is also useful to resolve closely spaced bands. Thus, when conventional spectra are strongly overlapped, the synchronous technique reduces the extent of overlap but does not allow the mixture to be adequately resolved. However, complete resolution is still possible if the derivative synchronous spectrum is recorded. The combination of synchronous and derivative fluorimetry enhances minor spectral features and allows more reliable identification of chemical species. In addition, the SNR values obtained by differentiation of the synchronous spectrum are greater than those obtained by differentiation of the conventional spectrum. This is so because the amplitude of the derivative signals is inversely proportional to the bandwidth of the original spectrum and, as stated above, a salient feature of synchronous fluorimetry is the narrowing of the spectral bandwidth in relation to conventional fluorimetry. When adequate instrumental conditions are used, this increase in amplitude results in higher SNR values for the derivative synchronous spectrum than those obtained in the zero-order spectrum recorded by conventional fluorimetry. Therefore, the use of the synchronous scan reduces the degradation of the SNR in the derivative spectrum, one of the most significant disadvantages of derivative spectrometry. Thus, derivative SFS is a relevant tool in the Green Analytical Chemistry field because it permits the analysis of multi-component with a single sample treatment and with a significant reduction in samples, reagents and waste generation.

The combination of derivative and SFS was first suggested by John and Soutar [36] as a mean of improving the selectivity of analytical methods. The advantages as well as the limitations of this method have been the topic of several reports [37,38]. The applicability of derivative synchronous fluorescence has been shown in the determination of mixtures [39–41], some of which could only be satisfactorily resolved by time-consuming or expensive techniques. Recent development of this technique has been reviewed by Patra and Mishra [42].

Specificity is a particular problem in the determination of fluorescent drugs [43]. Derivative SFS technique has been used to develop simple, rapid and sensitive spectrofluorimetric methods for the simultaneous determination of drugs and nowadays this technique is the most widely used in pharmaceutical analysis of several mixtures in their co-formulated dosage forms and biological fluids. Because of the need to monitor the drug level in human fluids, a variety of methods was developed for the quantitative estimation; these methods are not free from disadvantages [44]. The biochemical assays are faster, more specific and generally more sensitive than microbiological techniques. However, they require the use of radioactive isotopes and in addition, they require expensive reagents and sophisticated equipment. Several chromatographic methods have been also reported for the assay of the drug. These methods offer a high degree of specificity, but most of these are time consuming, requiring sample pretreatment prior to the measurement, different internal standards, when applied to serum or urine, and the use of organic solvents. These drawbacks, in addition to the instrumental limitations, preclude their use in routine clinical analysis and in Green Analytical Chemistry.

The second-derivative constant-wavelength SFS technique is the most frequently used for the simultaneous determination of drugs. The suitability of this technique has been widely shown since the 1980s, for example, for the simultaneous determination of epinephrine and norepinephrine in urine, with analytical recovery about 94% for epinephrine and 91% for norepinephrine [45]. First and second derivative SFS was employed in the determination of the alkaloid berberine [46]. By using second-derivative SFS, the simple resolution of the anti-coagulant rodenticides, warfarin and bromadiolone, in the presence of β-cyclodextrin was accomplished through a differential effect on the fluorescence intensity of these compounds; mixtures of warfarin and bromadiolone in the ratios 4:1 and 1:10 were satisfactorily resolved [47]. Also, the simultaneous determination of nonsteroidal anti-inflammatory drugs [48] and of salicylic acid and its derivatives or its metabolites in human serum [49,50], urine [51] and aspirin formulations [52] has been reported. Lianidou *et al.* [53] reported an improved spectrofluorimetric method for the simultaneous determination of diflunisal and salicylic acid in serum and urine samples not requiring the use of organic solvents. The method was based on the formation of ternary Tb-EDTA-Diflunisal and Tb-EDTA-salicylic acid complexes and second derivative SFS technique, allowing the simultaneous determination in a single scan, and without removal of proteins. In the last decade, second-derivative constant-wavelength SFS has been utilized for the determination of different drugs mixtures in their co formulated dosage forms and biological fluids, such as oxytetracycline in medical premixes and feeds [54], cinnarizine and domperidone in pharmaceutical preparations and biological fluids [55], fluphenazine hydrochloride and nortriptyline hydrochloride in pharmaceutical preparations [56], sulpiride and its alkaline degradation product [57].

This derivative technique has been also used to analyse mixture of several hydrocarbon compounds. Lloyd [58], made evident the enhancement of relatively minor spectral features by second-derivative synchronous fluorimetry comparing the results obtained for a motor oil by direct and derivative SFS. Vo-Dinh [59] employed this technique to analyse a synthetic mixture of several aromatic compounds. Later, it was applied to an extract from an atmospheric sample (collected in a workplace environment) which had previously been analysed by direct synchronous fluorimetry [60]. The second-derivative peaks were narrower than those found by direct synchronous scanning. Benzo(a)pyrene (in smoke-flavour agents) has also been determined by second-derivative, constant-wavelength SFS [61]. Similar methods have been reported for the determination of PAH pollutants in drinking water [62,63]: even in methods capable of determining 11 PAHs [64] in a mixture of 18.

Simultaneous analysis of 10 PAHs in a mixture of 18 has also been reported, but by second derivative constant-energy SFS [65].

There have been few applications of derivative synchronous fluorimetry in inorganic analysis. First-derivative SFS has been used for the determination of magnesium in drinking water samples [66], the method was based on the formation of a fluorescent chelate with salicylaldehyde 2-pyridylhydrazone. A similar method was developed for the determination of cadmium through cadmium-benzyl 2-pyridyl ketone 2-quinolyl-hydrazone chelate [67], with similar results. Salicylaldehyde thiocarbohydrazone has been used for the simultaneous determination of Ga and Zn in biological samples, wine, drinking water and wastewater by using second-derivative synchronous fluorimetry [68]. Also, salicylaldehyde carbohydrazone has been used for the simultaneous determination of Ga and Al in biological samples by using second derivative synchronous fluorimetry [69].

Although there are many individual determinations that benefit from the use of derivative SFS, the main interest in this technique lies in its ability to resolve complex mixtures. This has been shown in the analysis of binary and ternary mixtures of titanium, zirconium and hafnium, by formation of fluorescent chelates with biacetylmonoxime nicotinylhydrazone in an acidic medium [70]. With second-derivative SFS, it is possible to analyse binary mixtures in a single scan, and ternary mixtures in two scans. A method for determination of Al and Be in a mixture by first-derivative synchronous solid-phase spectrofluorimetry has also been developed [71]. Aluminium and beryllium reacted with Morin to give fluorescent complexes with fairly broad and overlapping spectra. These complexes were fixed on a dextran-type resin packed in a 1 mm silica cell, and the fluorescence was measured directly with a solid-surface attachment. By using second-derivative SFS the determination of samarium, europium and terbium was accomplished by mean of the formation of their complexes with quinaldic acid and phenantroline in aqueous solution [72]. The cationic surfactant hexadecyl-trimethylammonium bromide enhances the fluorescence intensity of both molybdenum and tungsten complexes with Alizarin Red S. This characteristic has been used for sensitive simultaneous SFS determinations of both ions [73].

13.3.1.2 Derivative variable-angle SFS

1-naphthol and 2-naphthol were employed to evaluate the combination of variable-angle SFS and derivative techniques. Second-derivative variable-angle SFS permitted the rapid simultaneous identification and quantitative determination of 1-naphthol and 2-naphthol in a mixture from a single spectrum [74]. Although, the derivative variable-angle SFS, in its pure concept, has not been utilized ever. The resulting spectra represent the intensity profile of a cut through the excitation-emission matrix, the trajectory of which is a continuous function. It is not possible to take the derivative because of the different values of the limits in the point where the trajectory angle is varied. Therefore an appropriate empirical equation involving the effect of each compound or a further treatment of the data by a multilinear regression program, for resolving complex mixtures may be needed [75]. A first derivative non-linear variable-angle synchronous fluorescence method by means of a program (Ftotal program) has been developed [76] for analysing a mixture of fluorescent doping substances with broad and highly overlapped excitation and emission spectra (atenolol, propanolol, dipyridamole and amiloride) with satisfactory results.

13.3.1.3 Derivative matrix isopotential SFS

In the two last decades, a fluorimetric technique called matrix isopotential synchronous fluorescence (MISF) has been applied to the determination of drugs. This technique consists on performing synchronous scans through a trajectory joining points of equal intensity on a fluorescence matrix three-dimensional spectrum. Provided the matrix used exhibits an almost invariable composition, a constant background signal can be maintained even if the fluorescence intensity changes. This is possible if a portion of the total fluorescence spectrum along a trajectory connecting points of identical intensity (isopotential trajectory) is selected from an initial point to final excitation and emission wavelengths. In order to ensure a sensitivity level similar to that of a direct determination in absence of background fluorescence, trajectories must pass by the fluorescence intensity peaks of both analytes. The performance of this technique can be improved by using derivative methods (DMISF). It was found that the simultaneous detection of spectrally closely overlapping compounds is possible using MISF method combined with derivative techniques. The DMISF method has high sensitivity and selectivity, but at higher noise levels the sensitivity of detection decreases more rapidly. DMISF is capable to discriminate fluorophores of similar quantum yield being in not more than 1% concentration compared to interfering high concentration components at low noise level. In practically noiseless circumstances, MISF is capable to discriminate compounds differing in their concentrations in more than two orders of magnitude [77]. This method has been applied to the simultaneous direct fluorimetric determination of diflunisal and salicylic acid in serum [78]. The simultaneous determination is accomplished by using a single isopotential trajectory that passes by the maxima in the fluorescence bands of diflunisal and salicylic acid in order to ensure a sensitivity level similar to that of a direct determination in absence of background fluorescence. Although the MSIF technique suppressed the fluorescence background of serum, the overlap between the diflunisal and salicylic acid MISF spectra precluded the simultaneous determination of the two compounds. Satisfactory results could be obtained by measuring the analytical signals at the zero-crossing points in the first derivative of matrix isopotential synchronous scan. The main advantages of this method include: aqueous solutions are used over organic solvents required in others methods; it can be directly applied to untreated serum samples; and limits of detection are similar or lower than obtained in the other described methods [78].

13.3.1.4 Combination of multivariate statistical methods with derivative SFS

Multivariate statistical methods, such as principal component regression (PCR) and partial least-squares (PLS), have been applied in the analysis of mixtures, using mainly infrared or UV-Visible absorption spectroscopy. Only a small number of methods have been described in which fluorescence data are used for multicomponent analysis, by multivariate calibration methods [79]. Jones *et al.* [80] reported the use of excitation, emission and synchronous spectra for the determination of acyclovir and guanine, applying several multivariate calibration methods. In addition, they investigated the utilization of second-derivative synchronous spectra in the multivariate calibration methods. They found that the use of synchronous spectral data was the best choice for the determination of acyclovir, whereas the use of second-derivative synchronous spectra was the best choice for the determination of guanine. Aghamohammadi *et al.* [81] evaluated the potential of combining normal, synchronous and derivative fluorimetry with multivariate methods to determine aflatoxin B_1 (AFB1) in complex matrix of pistachio samples. Normal fluorescence spectra and their derivatives and synchronous spectra and its first derivative were used separately as analytical signal to make the calibration models (multiple linear regression (MLR), PCR and PLS1). They concluded that AFB1 determination by synchronous fluorimetry in combination with MLR or PLS calibration showed a better performance when compared with derivative spectral and normal fluorescence spectra.

13.4 Use of derivative signal techniques in liquid chromatography

One of the most serious problems with chromatography workers are confronted is the occurrence of only partially resolved peaks arising from coelution of solutes in the samples or from similarities between their retention times. Traditionally, this type of problem was addressed by modifying the experimental conditions by trial and error until the aforesaid errors were minimized. Thus, different mobile or stationary phases (columns) or even working techniques (e.g. isocratic or gradient elution) were tested, which was time-consuming and involved consumption of expensive solvents.

The advent of multi-dimensional detection systems and the affordability of personal computers provides with software that allows storage and subsequent processing of chromatographic data has fostered developments of new experimental procedure for the characterization of unresolved peaks. There are four basic alternatives to computer-assisted resolution of chromatographic peaks, namely [82]:

1. Fitting the chromatogram peak to known functions. There are some precedents to the use of this alternative in gas chromatography ranging for the use of Gaussian and non-Gaussian models, to convoluted Gaussian curves with exponential decay and fast Fourier transform techniques. Solutions involving comparison of logarithmic spectra or the use of chemometric methods [83, 84] have been tested in liquid chromatography as implemented with diode-array detectors.
2. Integration by tracing a line perpendicular to the baseline from the valley between two peaks or one joining the valley and the end of the second peak (skimming) by computing the area of each separately from the two zones thus established.
3. Use of derivative techniques. As far as practical applications, the information supplied by derivative spectroscopy used as detection system in liquid chromatography has been exploited in two different ways, namely:
   a. By using the complete derivative obtained using the first derivative of the elution profile obtained at the wavelength of the absorption maximum. In theory, the derivative should be zero at this point and therefore the disappearance of the main peak may reveal the presence of other constituents with different absorption features. This procedure is called 'null spectral derivative technique';
   b. By using the complete derivative obtained by recording the elution profile. This procedure, known as 'spectral derivative mapping technique' was studied theoretically by Grant *et al.* [83], who discussed specific cases where the spectral curves of potential impurities lay within the spectral band of the major components. This procedure has been applied in the resolution of diverse mixtures.
4. A procedure has been developed by using the second derivative spectra of the components obtained around the maxima signal of the overlapped chromatographic peaks. A diode-array spectrophotometer is used with this purpose. This procedure, which has been used firstly to the analysis of active components in insecticide formulations, makes possible the easy transformation of a chromatographic problem in a spectrophotometric problem [84].

References

1. O'Haver, T.C. (1976) Modulation and derivative techniques in luminescence spectroscopy: Approaches to increased analytical selectivity, in *Modern Fluorescence Spectroscopy*, Vol. 1 (ed. E.L. Wehry), pp. 65–81, Plenum Press, New York.
2. John, P. and Soutar, I. (1976) Identification of crude oils by synchronous excitation spectrofluorimetry, *Anal. Chem.*, **48**, 520–524.
3. O'Haver, T.C. and Bagley, T. (1981) Signal-to-noise ratio in higher order derivative spectrometry, *Anal. Chem.*, **53**, 1876–1878.
4. Green, G.L. and O'Haver, T.C. (1974) Derivative luminescence spectrometry, *Anal. Chem.*, **46**, 2191–2196.
5. Schmitt, A. (1978) Derivative spectroscopy. An introduction with practical examples, *Tech. Lab.*, **5**, 1207–1215.
6. Osipov, V.M. and Borisova, N.F. (1985) Parametric reduction of a spectrum in derivative spectroscopy, *Zh. Prikl. Spektrok.*, **42**, 603–606.
7. Fell, A.F. and O'Haver, T.C. (1982) Bioanalytical applications of derivative and multichannel spectroscopy, *Anal. Proc.*, **19**, 398–403.

8. Butler, W.L. and Hopkins, D.W. (1970) Higher derivative analysis of complex absorption spectra, *Photochem. Photobiol.*, **12**, 439–450.
9. Gulyayev, B. A. and Litvin, F.F. (1970) First and second derivatives of absorption spectrum of chlorophyll and associate pigments in cells of higher plants and algae at 20°C, *Biofizika*, **15**, 670–680.
10. Talsky, G.; Mayring, L. and Kreuzer, H. (1978) Derivative spectrophotometry of higher order for the fine resolution, *Angew. Chem.*, **90**, 563–564.
11. Fell, A.F.; Jarvie, D.R. and Stewart, M.J. (1981) Analysis of paraquat by second derivative spectroscopy, *Clin. Chem.*, **22**, 286–292.
12. Fell, A.F. (1983) Biomedical applications of derivative spectroscopy, *TrAC- Trend Anal. Chem.*, **2**, 13–66.
13. Cahill, J.E. (1979) Derivative spectroscopy: understanding its application, *Am. Lab.*, **11**, 79–85.
14. O'Haver, T.C. (1982) Derivative spectroscopy: theoretical aspects, *Anal. Proc.*, **19**, 22–28.
15. Talsky, G. (1994) *Derivative Spectrophotometry*, Wiley-VCH Verlag, Weinheim.
16. Sánchez Rojas, F.; Bosch Ojeda, C. and Cano Pavón, J.M. (1988) Derivative ultraviolet-visible region absorption spectrophotometry and its analytical applications, *Talanta*, **35**, 753–761.
17. Bosch Ojeda, C.; Sánchez Rojas, F. and Cano Pavón, J.M. (1995) Recent developments in derivative ultraviolet/visible absorption spectrophotometry, *Talanta*, **42**, 1195–1214.
18. Bosch Ojeda, C. and Sánchez Rojas, F. (2004) Recent developments in derivative ultraviolet/visible absorption spectrophotometry, *Anal. Chim. Acta*, **518**, 1–24.
19. Sánchez Rojas, F. and Bosch Ojeda, C. (2009) Recent developments in derivative ultraviolet/visible absorption spectrophotometry: 2004–2008, *Anal. Chim. Acta*, **635**, 22–44.
20. Karpinska, J. (2004) Derivative spectrophotometry-recent applications and directions of developments, *Talanta*, **64**, 801–822.
21. El-Sayed, A.A.Y. and El-Salem, N.A. (2005) Recent developments of derivative spectrophotometry and their analytical applications, *Anal. Sci.*, **21**, 595–612.
22. Pena-Pereira, F.; Lavilla, I. and Bendicho, C. (2009) Miniaturized preconcentration methods basegd on liquid-liquid extraction and their application in inorganic ultratrace analysis and speciation: A review, *Spectrochim. Acta Part B*, **64**, 1–15.
23. Xu, L.; Basheer, C. and Lee, H.K. (2007) Development in single-drop microextraction, *J. Chromatogr. A*, **1152**, 184–192.
24. Xu, L.; Basheer, C. and Lee, H.K. (2009)Chemical reactions in liquid-phase microextraction, *J. Chromatogr. A*, **1216**, 701–707.
25. Wróbel-Zasada, K.; Wróbel-Kaczmarczyk, K.; López de Alba, P.L.; López-Martínez, L. and García-López, M.A. (1996) Application of internal standard for derivative-spectrophotometric determination of azinphos-methyl in commercial formulations, *Talanta*, **43**, 1055–1060.
26. Karpinska, J. (2000) The use of internal standard method for derivative-spectrophotometric determination of chlorpromazine hydrochloride, *Anal. Lett.*, **33**, 1555–1566.
27. Wróbel-Zasada, K.; Wróbel-Kaczmarczyk, K.; López de Alba, P.L.; López-Martínez, L. and García-López, M.A. (1997) Enhanced spectrophotometric determination of chromium(VI) with diphenylcarbazide using internal standard and derivative spectrophotometry, *Talanta*, **44**, 2129–2136.
28. Serafín Muñoz, A.H.; Preciado Puga, M.; Wróbel-Zasada, K.; Garay Sevilla, M.E. and Wróbel-Kaczmarczyk, K. (2004) Micro assay for malondialdehyde in human serum by extraction-spectrophotometry using an internal standard, *Microchim. Acta*, **148**, 285–291.
29. Wróbel-Zasada, K.; Wróbel-Kaczmarczyk, K.; de la Garza Rodríguez, I.M.; López de Alba, P.L. and López-Martínez, L. (1999) Determination of miconazole in pharmaceutical creams using internal standard and second derivative spectrophotometry, *J. Pharm. Biomed. Anal.*, **20**, 99–105.
30. Wróbel-Zasada, K. and Wróbel-Kaczmarczyk, K. (2009) Micro-scale UV/Vis spectrometric batch procedures by use of an internal standard-A Green Chemistry approach, *Spectrosc. Lett.*, **42**, 327–333.
31. Shokoufi, N.; Shemirani, F and Memarzadeh, F. (2007) Fiber optic-linear array detection spectrophotometry in combination with cloud point extraction for simultaneous preconcentration and determination of cobalt and nickel, *Anal. Chim. Acta*, **601**, 204–211.
32. Shokoufi, N.; Shemirani, F and Assadi, Y. (2007) Fiber optic-linear array detection spectrophotometry in combination with dispersive liquid-liquid microextraction for simultaneous preconcentration and determination of palladium and cobalt, *Anal. Chim. Acta*, **597**, 349–356.

33. Green, G.L. and O'Haver, T.C. (1974) Derivative luminescence spectrometry, *Anal. Chem.*, **46**, 2191–2196.
34. Lloyd, J.B.F. (1971) Synchronized excitation of fluorescence emission spectra, *Nature Phys. Sci.*, **231**, 64–65.
35. Vo-Dinh, T. (1981) Synchronous excitation spectroscopy, in *Modern Fluorescence Spectroscopy*, Vol. 4 (ed.E.L. Wehry), pp 167–192, Plenum Press, New York.
36. John, P. and Soutar, I. (1976) Identification of crude oils by synchronous excitation spectrofluorimetry, *Anal. Chem.*, **48**, 520–524.
37. Latz, H.W.; Ullman, A.H. and Winefordner, J.D. (1978) Limitations of synchronous luminescence spectrometry in multicomponent analysis, *Anal. Chem.*, **50**, 2148–2149.
38. Lloyd, J.B.F. (1980) Limitations of synchronous luminescence spectrometry in multicomponent analysis, *Comments, Anal. Chem.*, **52**, 189–191.
39. Rubio, S.; Gómez-Henz, A. and Valcárcel, M. (1986) Analytical applications of synchronous fluorescence spectroscopy, *Talanta*, **33**, 633–640.
40. García Sánchez, F.; Cruces, C. and Ramos Rubio, A.L. (1986) Synchronously scanned derivative spectrofluorimetry: A new approach for the analysis of multicomponent fluorescent simples, *J. Mol. Struct.*, **143**, 473–476.
41. Gutiérrez, C.; Rubio, S.; Gómez-Henz, A. and Valcárcel, M. (1987) Determination of histamine by derivative synchronous fluorescence spectrometry, *Anal. Chem.*, **59**, 769–773.
42. Patra, D. and Mishra, A.K. (2002) Recent developments in multi-component synchronous fluorescence scan analysis, *TrAC-Trend Anal. Chem.*, **21**, 787–798.
43. Jiang, C. and He, J. (2002) Simultaneous determination of Aloe-emodin and Rhein by synchronous fluorescence spectroscopy, *J. Pharm. and Biomed. Anal.*, **29**, 737–742.
44. Al-Majed, A.A.; Belal, F.; Abounassif, M.A. and Khalil, N. (2003) Fluorimetric determination of gentamicin in dosage forms and biological fluids through derivatization with 4-chloro-7-nitrobenzo-2-oxa-1,3-diazole (NBD-Cl), *Microchim. Acta*, **141**, 1–6.
45. Valcárcel, M.; Gómez-Henz, A. and Rubio, S. (1985) Simultaneous determination of epinephrine and norepinephrine in urine by derivative synchronous fluorescence spectroscopy, *Clin. Chem.*, **31**, 1790–1794.
46. Ramos Rubio, A.L.; Cruces Blanco, C. and García Sánchez, F. (1986) Synchronous scanning first and second derivative spectrofluorimetric determination of the alkaloid berberine, *Fresenius' J. Anal. Chem.*, **323**, 153–156.
47. Panadero, S.A.; Gómez-Henz, A. and Pérez-Bendito, D. (1993) Simultaneous determination of warfarin and bromadiolone by derivative synchronous fluorescence spectrometry, *Talanta*, **40**, 225–230.
48. Pérez Ruiz, T.; Martínez Lozano, C.; Tomás, V. and Carpena, J. (1998) Analysis of binary mixtures of flufenamic, meclofenamic and mefenamic acids by derivative synchronous fluorescence spectrometry, *Talanta*, **47**, 537–545.
49. Konstantinos, D.G. and Ioannou, P.C. (1992) Simultaneous determination of acetylsalicylic acid and its major metabolites in human serum by second-derivative synchronous fluorescence spectrometry, *Analyst*, **117**, 877–882.
50. Konstantinos, D.G. and Ioannou, P.C. (1994) Simultaneous determination of diflunisal and salicylic acid in human serum by second-derivative synchronous fluorescence spectroscopy, *Eur. J. of Pharm. Sci.*, **1**, 209–217.
51. Salinas, F.; Muñoz de la Peña, A. and Duran-Meras, I. (1990) Determination of salicylic acid and its metabolites in urine by derivative synchronous spectrofluorimetry, *Analyst*, **115**, 1007–1011.
52. Konstantinos, D.G.; Ioannou, P.C. and Efstathiou, C.E. (1991) Simultaneous determination of acetylsalicylic and salicylic acids in human serum and aspirin formulations by second-derivative synchronous fluorescence spectrometry, *Analyst*, **116**, 373–378.
53. Lianidou, E.S.; Ioannou, P.C.; Polydorou, C.K. and Efstathiou, C.E. (1996) Synchronous scanning second derivative spectrofluorimetry for the simultaneous determination of diflunisal and salicylic acid added to serum and urine as ternary complexes with terbium and EDTA, *Anal. Chim. Acta*, **320**, 107–114.
54. Fernández-González, R.; García-Falcón, M.S. and Simal-Gándara, J. (2002) Quantitative analysis for oxytetracycline in medicated premixes and feeds by second-derivative synchronous spectrofluorimetry, *Anal. Chim. Acta*, **455**, 143–148.
55. Walash, M.I.; Belal, F.; El-Enany, N. and Abdelal, A.A. (2008) Second-derivative synchronous fluorometric method for the simultaneous determination of cinnarizine and domperidone in pharmaceutical Preparations. Application to biological fluids, *J. Fluoresc.*, **18**, 61–74.
56. Walash, M.I.; El-Brashy, A.; El-Enany, N. and Kamel, M.E. (2009) Second-derivative synchronous fluorescence spectroscopy for the simultaneous determination of fluphenazine hydrochloride and nortriptyline hydrochloride in pharmaceutical preparations, *J. Fluoresc.*, **19**, 891–904.

57. Abdelal, A.; El-Enany, N. and Belal, F. (2009) Simultaneous determination of sulpiride and its alkaline degradation product by second derivative synchronous fluorescence spectroscopy, *Talanta*, **80**, 880–888.
58. Lloyd, J.B.F. (1980) Examination of petroleum products of high relative molecular mass for forensic purposes by synchronous fluorescence spectroscopy. Part I. Appraisal of experimental factors, *Analyst*, **105**, 97–109.
59. Vo-Dinh, T. (1981) Synchronous excitation spectroscopy, in *Modern Fluorescence Spectroscopy*, Vol. 4 (ed. E.L. Wehry), pp. 167–191, Plenum Press, New York.
60. Vo-Dinh, T. (1982) Synchronous luminescence spectroscopy: Methodology and applicability, *Appl. Spectrosc.*, **36**, 576–581.
61. Falcon, S.G.; Amigo, S.G.; Yusty, M.A.L.; López de Alda, M.J. and Lozano, J.S. (1996) Determination of benzo[a] pyrene in smoking-flavour agents (water-soluble liquid smoke) by second derivative constant-wavelength synchronous spectrofluorimetry, *Talanta*, **43**, 659–664.
62. Villaizan, M.J.L.; Falcon, M.S.G.; Amigo, S.G.; Lozano, J.S. and Yusty, M.A.L. (1996) Second-derivative constant-wavelength synchronous scan spectrofluorimetry for determination of benzo[b]fluoranthene, benzo[a]pyrene and indeno[1,2,3-cd]pyrene in drinking water, *Talanta*, **43**, 1405–1412.
63. Villaizan, M.J.L.; Lozano, J.S. and Yusty, M.A.L. (1995) Determination of polycyclic aromatic hydrocarbons in drinking and surface waters from Galicia (N.W. Spain) by constant-wavelength synchronous spectrofluorimetry, *Talanta*, **42**, 967–970.
64. Eiroa, A.A.; Vazquez-Blanco, E.; Lopez-Mahia, P.; Muniaegui-Lorenzo, S. and Prada-Rodríguez, D. (1998) Simultaneous determination of 11 polycyclic aromatic hydrocarbons (PAHs) by second-derivative synchronous spectrofluorimetry considering the possibility of quenching by some PAHs in the mixture, *Analyst*, **123**, 2113–2117.
65. Eiroa, A.A.; Vazquez-Blanco, E.; Lopez-Mahia, P.; Muniaegui-Lorenzo, S. Prada-Rodríguez, D. and Fernández, E.F. (2000) Determination of polycyclic aromatic hydrocarbons (PAHs) in a complex mixture by second-derivative constant-energy synchronous spectrofluorimetry, *Talanta*, **51**, 677–684.
66. Cruces Blanco, C. and García Sánchez, F. (1984) Spectrofluorimetric and synchronous scanning first derivative spectrofluorimetry for determination of magnesium with salicylaldehyde 2-pyridylhydrazone, *Anal. Chem.*, **56**, 2035–2038.
67. García Sánchez, F.; Navas, A. and Santiago, M. (1985) Synchronous scanning derivative spectrofluorimetry for the determination of cadmium with benzyl-2-pyridylketone 2-quinolylhydrazone, *Anal. Chim. Acta*, **167**, 217–223.
68. Ureña Pozo. M.E.; García de Torres, A. and Cano Pavón, J.M. (1987) Simultaneous determination of gallium and zinc in biological samples, wine, drinking water and waste water by derivative synchronous fluorescence spectrometry, *Anal. Chem.*, **59**, 1129–1133.
69. Cano Pavón, J.M.; García de Torres, A. Ureña Pozo, M.E. (1990) Simultaneous determination of gallium and aluminium in biological samples by conventional luminescence and derivative synchronous fluorescence, *Talanta*, **37**, 385–391.
70. Rubio, S.; Gómez-Hens, A. and Valcárcel, M. (1985) Analysis of binary and ternary mixtures of titanium, zirconium, and hafnium by derivative synchronous fluorescence spectrometry, *Anal. Chem.*, **57**, 1101–1106.
71. Capitán, F.; Manzano, E.; Navalón, A.; Vilchez, J.L. and Capitán-Vallvey, L.F. (1992) Simultaneous determination of aluminium and beryllium by first-derivative synchronous solid-phase spectrofluorimetry, *Talanta*, **39**, 21–27.
72. Du, X.; Gao, J.; Xie, Q. and Kang, J. (1994) Simultaneous determination of samarium, europium and terbium with quinaldic acid and phenanthroline by synchronous derivative fluorimetry, *Talanta*, **41**, 201–204.
73. Blanco, C.C.; García Campana, A.M.; Barrero, F.A. and Ceba, M.R. (1995) Micellar medium for the analysis of complex mixtures of molybdenum and tungsten by derivative synchronous spectrofluorimetry in steels, *Talanta*, **42**, 1037–1044.
74. Li, Y.Q.; Huang, X.Z.; Xu, J.G. and Chen, G.Z. (1994) Rapid simultaneous analysis of 1-naphthol and 2-naphthol in water samples by derivative variable-offset synchronous fluorescence spectroscopy, *Talanta*, **41**, 695–701.
75. Murillo Pulgarín, J.A.; Alañón Molina, A. and Fernández López, P. (1998) Simultaneous determination of atenolol, propranolol, dipyridamole and amiloride by means of non-linear variable-angle synchronous fluorescence spectrometry, *Anal. Chim. Acta*, **370**, 9–18.
76. Murillo, J.A. and Alañón, A. (1993) FTOTAL, A program to process total luminescence spectra, *Comp. Chem.*, **17**, 341–354.

77. Makkai, G.; Buzády, A. and Erostyák, J. (2010) Sensitivity test of derivative matrix isopotential synchronous fluorimetry and least squares fitting methods, *J. Fluoresc.*, **20**, 87–94.
78. Murillo Pulgarín, J.A.; Alañón Molina, A.; Fernández López, P. and Sánchez-Ferrer Robles, I. (2007) Direct determination of closely overlapping drug mixtures of diflunisal and salicylic acid in serum by means of derivative matrix isopotential synchronous fluorescence spectrometry, *Anal. Chim. Acta*, **583**, 55–62.
79. Nørgaard, L. (1995) A multivariate chemometric approach to fluorescence spectroscopy, *Talanta*, **42**, 1305–1324.
80. Jones, R.; Coomber, T.J.; McCormick, J.P.; Fell, A.F.; Clark, B.J.; Hill, S.; Ebdon, L.; Williams, J.G. and Gray, A.L. (1988) Recent advances in atomic and molecular spectroscopy, *Anal. Proc.*, **25**, 381–388.
81. Aghamohammadi, M.; Hashemi, J.; Kram, G.A. and Alizadeh, N. (2007) Enhanced synchronous spectrofluorimetric determination of aflatoxin B1 in pistachio samples using multivariate analysis, *Anal. Chim. Acta*, **582**, 288–294.
82. Cano Pavon, J.M. and Gallardo Melgarejo, A. (1990) Applications of derivative spectrophotometry todetection in liquid chromatography, *Quim. Analitica*, **9**, 299–303.
83 Grant, A. and Battacharyya, P.K. (1985) Application of derivative spectroscopy to the determination of chromatographic peak purity, *J. Chromatogr.*, **347**, 219–235.
84 Jimena García, J.A.J.; Plaza, J.G. and Cano Pavón, J.M.C. (1994) Use of derivative spectrophotometry in the resolution of overlapped peaks in liquid chromatography and its application in the analysis of active components in insecticide formulations, *J. Liquid Chromatogr.*, **17**, 277–285.

14
Greening Electroanalytical Methods

Paloma Yáñez-Sedeño[1], José M. Pingarrón[1] and Lucas Hernández[2]

[1] *Department of Analytical Chemistry, Faculty of Chemistry, University Complutense of Madrid, Madrid, Spain*
[2] *Department of Analytical and Instrumental Analysis, Universidad Autónoma de Madrid, Madrid, Spain*

14.1 Towards a more environmentally friendly electroanalysis

Nowadays, modern electroanalysis uses materials, solvents and techniques in accordance with the principles that characterize Green Analytical Chemistry. Electrodes are fabricated from non toxic materials, green solvents are employed and techniques and methodologies that make possible the reduction of the sample size and the amount of waste products are currently used. Although these characteristics result from the constant evolution towards a more environmentally friendly electroanalysis, it should be remarked, however, that their achievement is due undoubtedly to the especial properties of the electrochemical detection approach. On the one hand, there is a high versatility of the detection element itself, the working electrode, which can be fabricated and modified with many diverse materials and procedures; on the other hand, the high sensitivity of the electrochemical measurements and the inherent miniaturization capability of electrodes and electrochemical cells make possible easy adaptation of this detection mode to analytical microsystems. Finally, the possibility of obtaining analytical signals without sample treatment or derivatization removes the need for organic solvents.

These aspects are commented on in this chapter with the aim of describing briefly the state of the art of electroanalytical chemistry in connection with the green chemistry. In the course of the sections in which this chapter has been divided, the relevant aspects related to the different items are commented upon and some illustrative examples from the recent literature are given.

Handbook of Green Analytical Chemistry, First Edition. Edited by Miguel de la Guardia and Salvador Garrigues.

14.2 Electrode materials

14.2.1 Alternatives to mercury electrodes

Stripping analysis is a high sensitivity technique widely used in environmental analysis due to its usefulness to determine traces of metal ions and organic compounds. Since the invention of polarography by Heyrovsky, the mercury electrode was established as the preferred electrode for use in this technique due to the high overpotential for the hydrogen evolution, which makes it possible to work at moderately negative potential values, as well as the renewability of the electrode surface. The current trend of green chemistry that encourages avoiding the use and generation of toxic and pollutant substances, as well the recent European regulations that have prohibited exporting and storing of metallic mercury, have led to the search of competitive electrodes fabricated with a minimal amount of mercury or definitely, with other materials.

14.2.1.1 Amalgam electrodes

Electrodes composed of solid and liquid amalgams, in which the content of mercury is minimum, have been developed as alternatives to the different configurations available as mercury electrodes: dropping (DME), hanging drop (HMDE) and film mercury (MFE) electrodes. Numerous works described the methodology of construction of solid amalgam electrodes (MeSAEs), polished solid amalgam discs (p-MeSAE) or metal or carbon surfaces modified with a deposited mercury film (MF-MeSAEs) as well as their application to the determination of trace elements by means of stripping techniques. A novel configuration is the cylindrical silver amalgam film electrode (Hg/Ag)FE, which is renewable although it is fragile and cannot be used in polarography. Also, silver liquid amalgam film-modified silver solid amalgam annular band electrodes (AgLAF-AgSAE) have been recently reported [1]. These electrodes are simply and rapidly renewable and have been used both in voltammetric stripping and polarography. Figure 14.1 shows a scheme of this configuration. Its functioning is based on the cyclic renewal of the electrode surface by forming a fresh film of the liquid amalgam prior to each measurement. The mechanic system of film renewal is simple and provides an excellent repeatability and reproducibility as well as a long stability; better than that of mercury film electrodes.

Solid amalgam electrodes are characterized by possessing a solid and homogeneous surface that can be regenerated also by applying surface treatments and a simple electrochemical regeneration, this leading to reproducible voltammograms to be achieved. They can be easily prepared by mixing fine metal powder (Ag,Cu,Au,Bi,Sn or Zn) with liquid mercury. After the amalgamated process, the electrodes behave in a similar way to traditional hanging drop mercury electrodes, exhibiting a wide range of cathodic potentials and a high sensitivity. The use of these electrodes is compatible with the environmentally friendly analytical procedures since the amount of mercury employed is very small. Among them, the polished silver solid amalgam electrode (p-AgSAE) possesses the advantage of not forming inter-metallic products with other metals present in the working medium [2].

14.2.1.2 Bismuth electrodes

In the 2000 year, Professor Joseph Wang's group introduced the bismuth carbon electrode (BiCE), consisted of a thin bismuth film deposited onto a glassy carbon or carbon paste substrate for voltammetric stripping analysis [3,4]. Due to its low toxicity, its ability to form amalgams with many metals, its partial insensitivity to dissolved oxygen and the wide accessible potential window, the BiCE was rapidly accepted as an attractive alternative to mercury electrodes. Currently, the use of the bismuth electrode is very widespread. One of the most reported applications is the determination of lead in different materials, which have been recently

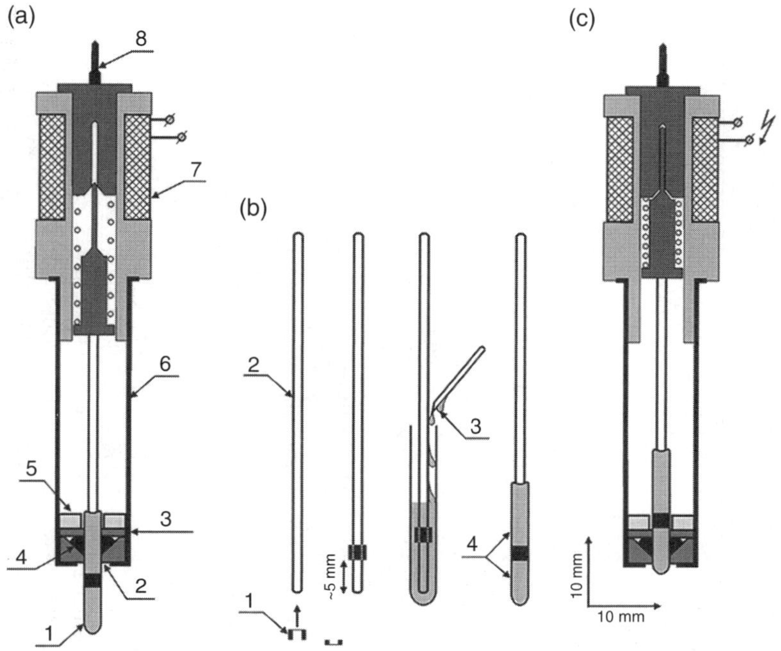

Figure 14.1 *(a) The AgLAF-AgSAE-ready-for-measurement configuration: (1) silver solid amalgam annular band electrode (AgSAE); (2) O-ring, 3,5 mm; (3) O-ring, 1 mm; (4) 1% (in wt%) silver liquid amalgam (AgLA); (5) PTFE centring element; (6) electrode body; (7) solenoid; (8) electric contact pin; (b) Construction of the AgSAE: (1) silver tube; (2) stainless steel wire, (3,4) resin. (c) The AgLAF-AgSAE-standby configuration. Reproduced from [1] with permission from Elsevier, © 2010.*

reviewed [5]. This work reveals that at present only 20% of lead determinations are carried out with mercury electrodes whereas 37% are performed with bismuth electrodes. Table 14.1 summarizes some applications of this kind of electrode to the determination of metal traces and also of some organic compounds [6–45].

Recent examples of determination of other metals include the use of a bismuth film electrode for the environmentally friendly determination of Tl(I) traces in non deoxygenated solutions. The experimental procedure involved anodic stripping voltammetry on an in situ plated rotating bismuth film electrode, and allowed a linear calibration graph in the sub-micromolar concentration range to be obtained with a detection limit of 10.8 nM. The electrode was applied to the analysis of soil and river water samples [6]. Bismuth film microelectrodes (BiFMEs) were used for the determination of trace Co(II) by adsorptive stripping voltammetry (AdCSV). A comparison of the stripping performance between bismuth and mercury film microelectrodes revealed distinct practical advantages of the BiFME [8]. Using square-wave voltammetry, a detection limit of 70 ng l^{-1} was achieved as well as an excellent reproducibility with 2.4% RSD at the 1 μg l^{-1} level (n = 10) with only 2 min pre-concentration time in the presence of dissolved oxygen.

Bismuth electrodes, similarly to other electrodes used in stripping voltammetry for the determination of metals, can suffer interference from other metal ions due to the formation of inter-metallic compounds or by peaks overlapping. In these cases, electrode modification with ionophores like Nafion minimizes such drawbacks since this kind of polymers exhibit a defined selectivity. For instance, it has been demonstrated that the ionophore–Nafion modified bismuth film electrode can offer largely improved sensitivity and selectivity for stripping measurement of cadmium [10].

Table 14.1 *Analytical methods involving bismuth electrodes.*

Electrode	Technique	Analytes/Sample	Analytical characteristics	Ref.
Bi/poly(p-ABSA)/GCE	DPASV	Pb(II), Cd(II), Zn(II)/river water	LOD, μg l^{-1}: 0.80 (Pb); 0.63 (Cd); 0.62 (Zn)	6
Bi/Nf/GCE	DPASV	Pb(II), Cd(II), Zn(II)/vegetables	LOD, μg l^{-1}: 0.30 (Zn); 0.17 (Cd, Pb)	7
Bi/COC chip electrode	SWASV	Pb(II), Cd(II)/soil pore, ground water	LOD: 8 μg l^{-1}	8
Bi/GCE	SWASV	Pb(II), Cd(II), Zn(II), Tl(I), In(III), Cu(II)	LOD, μg l^{-1}: 4 (Tl); 2 (In); 5 (Cu)	9
Bi/CPE	SWASV	Pb(II), Cd(II)	LOD, μg l^{-1}: 0.41 (Pb); 0.49 (Cd)	10
Bi/TBDOT/Nf/GCE	SWASV	Cd(II)	LOD, 1.3×10^{-3} M	11
Bi/silicon wafer	SWASV	Pb(II), Cd(II)	LOD, μg l^{-1}: 5 (Cd), 3 (Pb)	12
Bi/Nf/2,2′-bipy/GCE	SWASV	Zn(II), Cd(II), Pb(II)/waters, wine	LOD, nM: 8.6 (Zn); 1.1 (Cd); 0.37 (Pb)	13
Bi/GCE	DPASV	Sn(II)/sea water	LOD, 0.26 ng l^{-1}	14
Bi/CuE	SWASV	Ni(II)/ground water, electrolytic bath	LOD, 0.1 μM	15
Bi/GCE or Bi/SPCE	SWAdSV with DTPA	Cr(VI)/river water	LOD, 0.3 nM	16
Bi_2O_3/SPCE	Chronopo-tentiometry	Pb(II)/soil, waste waters	LOD, 8 μM	17
Bi/CFMEs	SWAdCSV with Cp	U(VI)/sea water	LOD, 3 μM	18
Bi/SPCE	SWASV	Pb(II), Cd(II), Zn(II)/river water	LOD, μg l^{-1}: 1.3 (Pb), 0.7 (Cd), 12 (Zn)	19
Bi/GCE	SI-MFSA/ SWASV	Cd(II), Pb(II)/water	LOD, μg l^{-1}: 1.4 (Cd), 6.9 Pb(II)	20
Bi/Cnanofibres/Nf/GCE	DPASV	Cd(II), Pb(II)/river water, human blood	LOD, μg l^{-1}: 0.09 (Cd(II); 0.02 (Pb(II)	21
Bi/GCE	SWAsSV with PAN	Fe(III)/sea water, reference materials	LOD: 0.1 μg l^{-1}	22
Bi/GCE	DPCSV	Se (IV)/multi-vitamin tablets, hair	L.R. 2–30 μg l^{-1}; LOD: 0.1 μg l^{-1} (300s)	23
Bi/GCE	SWASV	In(III), Cd(II), Pb(II)	L.R. 10–90 μg l^{-1} In(III)	24
Bi/GCE	SWAdSV with Cp	Al(III)/water, dow metal, bronze	LOD: 0.5 μg l^{-1} (240 s)	25
Bi (Se)/C electrode	SWCSV	As(III)/reference material, water	L.R. 0.01–1.0 μg l^{-1} (30 s); LOD: 0.7 ng l^{-1}	26
Bi/rotating disc GCE	SWAdSV with DTPA	Cr(III), Cr(VI)/river water	LOD: 0336 nM (Cr VI)	27

Bi/GCE	SWAdSV with ChA/bromate	V(V)/ground water	L.R. 5–25 μg l^{-1} (2 min); LOD: 0.20 μg l^{-1}	28
Bi/GCE	SWASV with Ga(III)	Cu(II)/tap water	L.R. 2–500 μg l^{-1}; LOD: 1.4 μg l^{-1}	29
Bi/CNTs-PSS/GCE	DPASV	Pb(II), Cd(II)/water	LOD, μg l^{-1}: 0.04 (Pb); 0.02 (Cd)	30
Bi/GCE	DPV with oxalic acid	Ti(IV)/waters, mussel tissue	LOD: 0.01 μM	31
Bi nanopowder thick-film C electrode	SWASV	Tl (I)	LOD: 003 μg l^{-1} (10 min)	32
Bi/poly(p-ABSA)/GCE	DPAdSV	PAN, PAR, azobenzene/beverages	LOD, nM: 41 (PAN); 33 (PAR); 38(azobenzene)	33
Bi/GCE	DPCSV	sulfadiazine/pharmaceuticals	LOD, 2.1 μM	34
Bi/GCE	DPASV	metallothionein	LOD: 3.86×10^{-8} M	35
Bi/GCE	SWCSV	methyl parathion/garlic	LOD, 1.2 ng ml^{-1}	36
Bi/GCE	SWV	aminosalycilate drugs/pharmaceuticals	5.0–350 μM (sulfasalazine)	37
Bi/CPE or Bi/GCE	SWV	herbicides	LOD: 6 μM (metribuzin)	38
Bi/CPE	SWV with CTAB	nitrobenzene	L.R. 1 – 100 μM; LOD: 0.83 μM	39
Bi/CuE	DPV	Paraquat/river water	L.R. 0.66 –48 μM; LOD: 0.093 μM	40
Bi-powder CPE	SWCSV	cysteine	L.R. 1–50 μM; LOD: 0.3 μM (600 s)	41
Bi/GCE	FIA	*Escherichia coli*	LOD: 100 cfu ml^{-1}	42
				43
				44
				45

Key: CETAB: cetyltrimethylammonium bromide; ChA: chloranilic acid; CFME: carbon fibre microelectrode; COC: cyclic olefin copolymer; Cp: cupferron; CTAB, cetyltrimethylammonium bromide; DTPA: diethylenetriammine pentaacetic acid, Nf: Nafion; pABSA: poly(*p*-aminobenzene sulfonic acid; PAN: 1-(2-pyridylazo)-2-naphthol; PAR: 4-(2-pyridylazo)-resorcinol; PSS: poly(sodium 4-styrenesulfonate); SI-MSFA: sequential injection monosegmented flow analysis; TBDOT: N,N,N′,N′-tetrabutyl-3,6-dioxaoctanedi(thioamide).

Microfabrication and screen-printed techniques allow preparation of planar bismuth electrodes with well defined patterns. These fabrication techniques provide ready-to-use Bi electrodes along with the integrated Ag/AgCl reference electrode and gold counter electrode. This kind of devices can be used for fast in situ heavy metal measurements in a natural- or bio-environment in an easy manner. [13]. For example, a screen-printed microband electrode prepared by ex situ bismuth deposition was reported for the sensitive determination of cadmium in undeaerated and unstirred solutions. A detection limit of 1.3 $\mu g\ l^{-1}$ was achieved using square-wave anodic stripping voltammetry and a deposition time of 120 s. The method was successfully applied to a non treated river water sample [14].

The fabrication of electrodes modified with nanomaterials is applicable also to this area. A bismuth-modified carbon nanotube electrode (Bi-CNT electrode) was prepared and applied to the determination of trace lead, cadmium and zinc. Bismuth film was prepared by in situ plating of bismuth onto the screen-printed CNT electrode. The simultaneous determination of lead, cadmium and zinc was performed by square wave anodic stripping voltammetry. The Bi-CNT electrode exhibited well-defined, reproducible and sharp stripping signals. The peak current response increased linearly with the metal concentration in a range of 2–100 $\mu g\ l^{-1}$, and the detection limits were 1.3 $\mu g\ l^{-1}$ for lead, 0.7 $\mu g\ l^{-1}$ for cadmium and 12 $\mu g\ l^{-1}$ for zinc (S/N = 3). This nanostructured electrode was successfully applied to the analysis of trace metals in real environments [15].

Good results were also obtained when bismuth film electrodes were applied to the determination of organic compounds, such as sulfadiazine pharmaceuticals [36] or aminosalycilate drugs [38]. Adsorptive stripping analysis is usually employed for this kind of determinations. In this case, in order to avoid interference from other species that could be adsorbed on the electrode surface as, for example, surfactants, Nafion or other polymers such as p-aminobenzene sulfonic acid (pABSA) are also used. For example, this polymer has been recently employed for the construction of a composite film-coated glassy carbon electrode (Bi/poly(p-ABSA), which was applied to the determination of azo-compounds in drinks [39].

Due to the easy preparation and regeneration of bismuth film electrodes together with their good reproducibility and stability, the combination of this kind of electrode and cathodic voltammetry performances opens new opportunities for a fast, simple and sensitive analysis of pesticides [41]. For example, the herbicide paraquat can be determined in a concentration range between 0.66 and 48 μM using a bismuth film-modified copper electrode [43]. Differential pulse voltammetry between −1.1 to −0.5 V versus Ag/AgCl was used and the developed method was applied to the analysis of river water samples. Also the pesticide methyl parathion was determined using a bismuth-film electrode prepared by ex situ bismuth deposition onto a glassy carbon electrode. As in the previous case, the sufficiently wide negative potential window available made the bismuth film electrode suitable for this application. Square wave voltammetry between −0.3 and −0.9 V versus SCE was employed, the cathodic voltammetric response being proportional to the concentration of methyl parathion between 3.0 and 100 ng ml^{-1} with a detection limit of 1.2 ng ml^{-1} [41].

14.2.1.3 Other metallic film electrodes

Other metals different to bismuth have been proposed for the construction of film electrodes. For example, gold films have been employed with this purpose although the usefulness of the electrodes fabricated with this metal is limited to the detection of species with rather positive oxidation potentials, such as mercury or antimonium. Lead film electrodes have been also reported although the toxicity of this metal and the limited accessible potential window are important disadvantages. Tin film electrodes can be used for anodic stripping although they possess a narrow range of anodic polarization. Also, antimonium electrodes have been proposed due to much lower toxicity of this metal compared to mercury. The preparation of antimonium film electrodes, deposited on glassy carbon, was described for the first time in 2007 for the determination of heavy metals by anodic stripping voltammetry [46]. More recently, electrodes of electrodeposited antimonium on carbon paste have been used for the determination of trace metals by potentiometric stripping [47]. Moreover, a Sb film electrode constructed by a photolithographic microfabrication process was prepared. This fabrication procedure provides small size, cost effective, reproducible, mass produced and disposable devices, and have been applied to the determination of trace nickel by adsorptive stripping voltammetry [48].

14.2.1.4 Other electrode materials

Boron-doped diamond film electrodes (BDDFE) can be also considered as another environmentally friendly alternative for electroanalytical chemistry. BDDFE has attracted the attention of researchers because it exhibits a stable background current, wide potential window for aqueous and non-aqueous electrolytes, high thermal conductivity, high hardness, high electrochemical stability, low organic molecule adsorption and high sensitivity for analytical purposes [49]. There are numerous examples of applications of BDDFE in the literature especially for the determination of pharmaceutical compounds such as nitrofurantoin [49], lyncomicin [50] or fluvastatin [51], as well as in food analysis for the determination of antioxidant mixtures [52], among others.

Other environmentally friendly materials have also been used as electrode modifiers for the development of stripping methods for metal traces and organic compounds. This is the case of β-cyclodextrin (β-CD), a versatile oligosaccharide acting as a host molecule for a wide variety of possible guests. Different CD-derivatives have been employed for this purpose. For example, a highly sensitive and reproducible lead sensor based on a cyclodextrin-modified gold electrode was described. A self-assembled monolayer (SAM) of thiolated β-cyclodextrin (6-(2-mercapto-ethylamino)-6-deoxy-β-cyclodextrin (MEA-ß-CD)) was prepared on the gold electrode to construct a mercury-free sensor for Pb^{2+} with a linear response over the 1.7×10^{-8} – 9.3×10^{-7} M concentration range and a detection limit of 7.1×10^{-9} M [54]. Similar configurations were also used for the determination of organic compounds. An example is the development of a sensor to detect azobenzene, a toxic industrial pollutant, by means of a gold electrode modified with a 6-O-toluenesulfonyl-β-CD (β-CD-6-OTs) layer. The inclusion complex formed at the modified electrode allowed azobenzene determination in solution by square wave voltammetry with a detection limit of 1.0×10^{-10} M. An additional advantage is the reusability of the β-CD-modified electrode. When azobenzene on the β-CD layer is irradiated by UV light (360 nm) in stirred solution, it is detached from the β-CD layer [54].

The determination of uranyl in environmental samples is a challenging task that has given rise to numerous procedures. Against methods using expensive instrumentation such neutron activation analysis or inductively coupled plasma emission spectrometry, electrochemical methods based on stripping analysis are faster, easier and cheaper. Although the majority of these methods have been performed using mercury electrodes, other electrode materials can be also used. For instance, a carbon paste electrode containing benzo-15-crown-5 (B15C5) has been reported recently for trace determination of uranyl ion using differential pulse voltammetry. Due to the good compatibility between the ionic radius of uranyl and the cavity size of the crown ether, modification of carbon paste using B15C5 resulted in facile electrochemical reduction of UO_2^{2+} with enhanced current. The proposed method demonstrated a good sensitivity, with a limit of detection of 0.03 μg l^{-1}. Furthermore, it was also selective and, so, it was successfully applied to the determination of uranyl in the presence of Pb^{2+} and Cd^{2+} as well to the determination of UO_2^{2+} in ores and industrial effluents [55].

Metal electrodes such as copper electrodes are also employed for some especial applications. So, a clean alternative method for the determination of chemical oxygen demand (COD) using a copper electrode as an electrocatalytic sensor was described. The measuring principle is based on the oxidation current of organic compounds in the wastewater. An analytical linear range of 53.0–2 801.4 mg l^{-1} COD with a detection limit of 20.3 mg l^{-1} COD was achieved. The procedure was successfully applied to the COD determination in wastewater from soft industries. The results were in good agreement with those obtained using the conventional (i.e. dichromate) COD method. Accordingly, the COD value of a sample could be determined in a simple, rapid and accurate manner and the end products did not contain toxic metals [56].

14.2.2 Nanomaterial-based electrodes

The preparation of electrodes modified with nanomaterials has open new avenues in Electroanalysis. Nanostructured surfaces possess properties that are highly beneficial for the construction of electrochemical sensors and biosensors. From a general point of view, they have the ability of enhancing the electron transfer reactions, show electrocatalytic effects for many relevant compounds with the consequent reduction of the corresponding detection potential values and exhibit high resistance to surface fouling by the products of the electrochemical reaction. These properties improve the electroanalytical performance of the resulting sensors in terms of selectivity and sensitivity as well as the reproducibility of the measurements. In the particular case of biosensors, the use of nanomaterials allows optimization of biomolecule orientation, their distribution and proximity both in connection with the electrode surface and the reaction substrate, preserving at the same time their stability and biological activity. From the green chemistry point of view, the attention paid in recent years to the development of green methods for the synthesis or preparation of nanomaterials with electrochemical applications is particularly interesting. In this field, the 12 principles of green chemistry have provided a framework for the designing of new materials, products, processes and systems in nanoscience and nanotechnology.

14.2.2.1 Carbon nanotubes and graphene

Modification of electrode surfaces with carbon nanotubes is being a common practice to achieve sensitive and selective sensing electroanalytical devices for specific applications since several years. Lately, functionalization methods have been developed with the aim of improving the properties of this nanomaterial and attaining a higher versatility of their applications. Nowadays, functionalization of carbon nanotubes is a common strategy in electroanalysis laboratories. Usually, acids such as sulfuric or nitric (or their mixtures) are used with the objective of forming carboxylic groups that can serve for enhancing nanotubes solubility or dispersability as well as for the further formation of covalent bindings by amidation, esterification, thiolation and other. However, these methodologies show some serious drawbacks such as the structural damage produced or the low process performance and the high pollution [57]. Green methods to functionalize carbon nanotubes have been described in recent literature. For instance, the incorporation of hydroxyl groups to the surface of multiwalled carbon nanotubes (MWNTs) by one step free radical addition of 4,4-azobis (4-cyanopentanol) (ACP) in water has been recently reported. Once the MWCNT-OH obtained, polymeric materials can be incorporated. For example, poly(ε-caprolactone)-grafted MWNTs (MWNT-g-PCL) were synthesized by in situ ring-opening polymerization of ε-caprolactone in the presence of MWNT–OH in room temperature ionic liquids (RTILs). The use of 1-n-butyl-3-methylimidazolium tetrafluoroborate ($BMIMBF_4$) as solvent cannot only improve the dispersability and stability of nanotubes during polymerization, but also enhance the functionalization effectiveness as a result of higher polymerization rate and amount of the grafted-polymer compared to general organic solvents [58]. Wrapping is another useful functionalization alternative without modifying the sp^2 structure of nanotubes. For instance, using β-cyclodextrins, β-CD-MWCNT composites can be obtained with a good solubility/dispersion in water. Furthermore, owing to the excellent biocompatibility nature of β-CD, the obtained supramolecular composites were expected to have potential applications in the preparation of biosensing devices [59]. The combination of polymer wrapping technique and layer-by-layer self-assembly allowed the non-covalent attachment of β-CD on MWCNTs through van der Waals forces between CNTs and β-CD and hydrogen-bonding interaction between the adjacent β-CD molecules as the principal driving forces. Figure 14.2 displays a scheme of the possible interaction mechanism. High resolution transmission electron microscopy (HRTEM) studies revealed that the β-CD cavities are perpendicular to the MWCNT surface. Since a β-CD molecule has two different hydroxyl groups on the two ends of its cavity, it is possible that either end of the β-CD molecule can interact with MWCNTs.

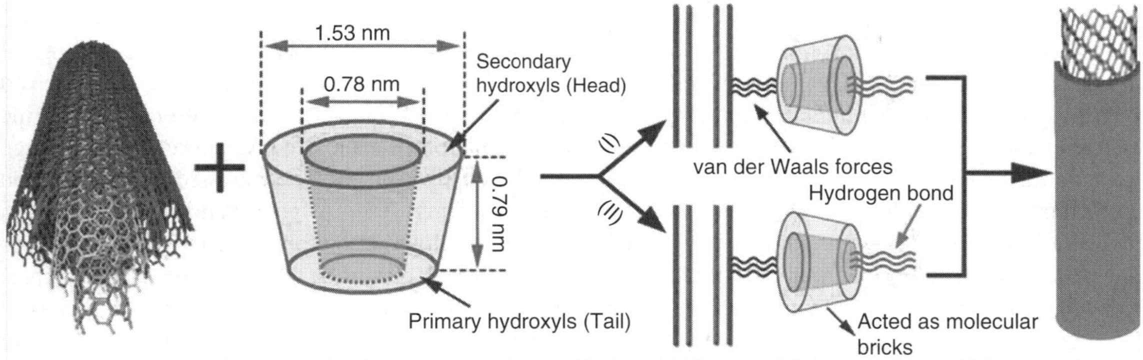

Figure 14.2 *Schematic picture of the MWCNTs-β-CD composites formation mechanism. Adapted from [60].* *© 2008 American Chemical Society.*

Novel nanocomposites of β-cyclodextrin and MWCNTs have also been prepared and deposited on glassy carbon substrates to prepare β-CyDex/MWNT modified electrodes. These novel modified electrodes offer a new strategy for the simultaneous determination of guanine (G), adenine (A) and thymine (T). Well separated voltammetric peaks are obtained between G and A (330 mV), A and T (170 mV) mixtures. The detection limits for individual analysis of A, T and G were 0.75, 6.768 and 33.67 nmol l^{-1}, respectively. The functional properties of the composite films when differential pulse voltammetry was used showed that the β-CyDex/MWNT nanocomposites modified electrodes were sensitive and could be readily applied in biosensor devices [60].

Due to its extraordinary electronic, thermal and mechanical properties, graphene has emerged as a fascinating material in electronic, optoelectronic and sensing applications. Chemical reduction of graphene oxide (GO) is considered to be an efficient approach to produce processable graphene sheets on a large scale. Different reducing agents are used for the chemical reduction of GO but, unfortunately, the reducing agents involved are hazardous. Furthermore, most chemical reduction processes are time-consuming and complicated. There are a few reports on environmentally-friendly processes to reduce graphene oxide to graphene either by chemical or electrochemical methods. However, a 'unified green approach' to the overall process of graphene synthesis has not been achieved yet. Among others, a method based on the use of aluminium powder in acid medium has been recently reported. Though the reduction mechanism is still unclear, this procedure provides a new green route to prepare graphene sheets with high purity and high electrical conductivity on a large scale without toxicity [61]. A microwave-based approach towards eco-friendly synthesis of graphene nanosheets, with advantages such as simplicity, high productivity, low cost and short processing times has also been described [62]. Together with such procedures, green methods for the preparation of stable dispersions of graphene without the need of stabilizers are also being developed. For example, an environmentally friendly freeze drying process which allows the preparation of solvent-free graphene nanosheets, has been recently described. [63].

14.2.2.2 Metal nanoparticles

Green methods for the preparation of gold nanoparticles have appeared in the literature in the last years. An interesting example of the fabrication of gold nanostructured films is based on the prior oxidation of a gold electrode surface at a high potential of 10 V followed by the reduction of the gold oxides using β-D-glucose as a non-toxic reducing agent. A fractal gold nanofilm atop the gold electrode was produced by this method in a way of completely green chemistry [64]. The growing trend of exploring bacteria, fungi, actinomycetes and plant materials for the biosynthesis of nanoparticles is considered as eco-friendly and a green technological approach. For example, the synthesis of gold nanoparticles from gold precursor using the extract derived from the marine sponge, *Acanthella elongata*, has been reported. Water-soluble organics present in the marine sponge extract were mainly responsible for the reduction of gold ions to nanosized Au particles. The sponge extract added to 10^{-3} M $HAuCl_4$ aqueous solution at 45°C changed to pinkish ruby red colour solution confirming the bio-reduction within 4 h with continuous stirring [65]. A gold nanoparticles-chitosan nanocomposite gel, for the immobilization of K562 leukaemia cells, was also prepared using a whole process that did not introduce any environmental toxicity or biological hazards according to the fundamental principles of green chemistry. Gold nanoparticles were produced in situ in a solution of chitosan, which is a non toxic natural polysaccharide. Moreover, the protocols for the gel preparation and the immobilization of cells were simple and environmentally friendly. Further modification of a glassy carbon electrode with the prepared gel produced an irreversible voltammetric response and an increased electron transfer resistance upon cell immobilization with a good correlation to the logarithmic value of their concentration [66]. Also, a green colloidal method was described for the fabrication of polymer stabilized gold nanoparticles and nanofilms using ascorbic acid as reducing agent of Au^{3+} ions and gum arabic as dispersing agent. Ascorbic acid is a nontoxic reducing species that originates the rapid formation of gold atoms. Furthermore, gum arabic, which is a mixture of saccharides and glycoproteins, acts as a nontoxic dispersing agent preventing particles aggregation [67]. Using a related approach, a green method for the preparation of platinum nanoparticles by reduction of H_2PtCl_6 with nanocrystalline cellulose from cotton as the reducing agent, was also described [68]. A simple and effective green chemistry route for the facile synthesis of nanowire-like Pt nanostructures at one step was also developed. In the reaction, dextran acted as a reductive agent as well as a protective agent for the Pt nanostructures. Simple mixing of precursor aqueous solutions of dextran and K_2PtCl_4 at 80°C resulted in spontaneous formation of the Pt nanostructures. The as-prepared nanowire-like Pt nanostructures were immobilized on glassy carbon electrodes using an electrochemical coupling strategy, and the resulting nanostructured electrode exhibited an excellent electrocatalytic activity for the reduction of oxygen and the oxidation of NADH [69].

The preparation of silver nanoparticles by different green synthesis approaches involving the use of mixed-valence polyoxometallates, polysaccharides, irradiation and biological methods, among others, has been reviewed. Regarding biological methods, different bio-organisms may act both as reducing and capping agents for Ag nanoparticles synthesis. The reduction of Ag^+ ions by combinations of biomolecules found in their extracts such as enzymes/proteins, amino acids, polysaccharides, and vitamins is environmentally benign. The extract of unicellular green algae *Chlorella vulgaris* was used to synthesize single-crystalline Ag nanoplates at room temperature. Proteins in the extract provide dual function of Ag^+ reduction and shape-control in the nanosilver synthesis. The carboxyl groups in aspartic and/or glutamine residues and the hydroxyl groups in tyrosine residues of the proteins were suggested to be responsible for the Ag^+ ion reduction. Plant extracts from live alfalfa, the broths of lemongrass, geranium leaves and others have also served as green reactants in the synthesis of Ag nanoparticles [70].

14.3 Solvents

An important concern in the implementation of greener analytical methods is the substitution of the organic solvents commonly used in many protocols of sample preparation and analytes determination for other more environmentally friendly solvents. Nowadays the search for other alternative solvents also implies the use of their characteristics to introduce additional improvements such as a reduction in the time of analysis or an enhancement in sensitivity or selectivity. Regarding electroanalytical approaches, alternative solvents can provide wider potential windows that allow performing studies, synthesis or experiments unable to be carried out in aqueous or conventional hydro-organic media. Furthermore, enhanced diffusion coefficients of electroactive solutes can be achieved.

14.3.1 Ionic liquids

The use of ionic liquids (ILs) in electroanalysis is important from two fundamental sides. On the one hand, they have been favourably explored as electrolytes and working media while, on the other hand, they have been shown to be useful for the fabrication of electrode materials. From an electrochemical point of view, ILs offer enormous advantages since they possess excellent conductivity, are thermally and chemically stable and exhibit a wide accessible potential window. As an example of application, a recent description of a solubilization and determination method of metal oxides pollutant particles in a single step, based on the use of a specific IL, the protonated betaine bis(trifluoromethylsulfonyl)imide, [Hbet][Tf2N], can be commented. This IL solubilizes metal oxide particles due to the presence of acid functional groups in its structure. In this way, the application of the conventional electrochemical method involving a strong acid treatment step to dissolve the metal oxide particulates and an intricate detection process that makes in situ monitoring of metal oxide particulate pollutants in soil, water and airborne particulates extremely challenging, can be avoided.

The detection step is carried out with an electrochemical sensor consisting of indium tin oxide (ITO) printed electrodes (working, counter and pseudo reference electrode) and a thin layer of [Hbet] [Tf2N] coated onto the ITO electrode as the selective solubilization medium and electrolyte (Figure 14.3). This figure displays the voltammograms corresponding to the simultaneous detection of PbO, CdO and CuO present at the 8 ng l^{-1} level at room temperature. An insignificant stripping peak was observed after the addition of PbO, CdO and CuO into pure [Hbet][Tf2N] upon a 2 min deposition time (black line). Conversely, larger, sharper, and more defined separated peaks were obtained with the addition of 10 ng l^{-1} bismuth oxide into [Hbet] [Tf2N] and the same deposition time (grey line). The results clearly demonstrated that (1) the [Hbet][Tf2N] can work as electrolyte as well as a solubilization medium to quickly dissolve the targeted metal oxides; (2) the addition of bismuth greatly improved the stripping signals of PbO, CdO and CuO [71].

Another relevant application of ILs as working media is their use for polymer electro-synthesis. The monomer is dissolved in IL medium and is electropolymerized at an appropriate electrode. In this way, the use of organic solvents is avoided in many cases and, moreover, it can be achieved that the grown polymer has especial characteristics. So, poly(pyrrol) (PPy) electrode coatings prepared in $BMIMPF_6$ exhibited better electrocatalytic activity and higher selectivity for dopamine detection than those prepared in aqueous media as a consequence of enhancing the oxidation current and decreasing the oxidation overpotential for this compound [72]. The substitution of conventional media for ILs to perform the synthesis of electronic conducting polymers can be also beneficial for the electro-mechanic properties of the polymer and its lifetime. In these applications, the IL acts as electrolyte and as source and drain of dopants depending on the polymer is in its oxidized or reduced form. Accordingly, the choice of the IL has a great influence on the polymer redox chemistry, affecting the oxidation and reduction potential values as well as the number of voltammetric signals. These differences were observed, for instance, in the synthesis of polythiophenes, where the cation of the IL affects also the coating morphology [73].

Concerning the incorporation of ILs to the electrode materials, their use as agglutinants in the preparation of carbon paste electrodes should be emphasized. The more relevant advantage of ILs with respect to the other organic agglutinants conventionally employed is their conducting nature. The carbon ionic liquid electrodes (CILEs) posses a high conductivity and are able to provide rapid electron transfer and low overpotential values for diverse electroactive analytes. For example, Wang *et al.* developed a carbon paste microelectrode by mixing of graphite and 1-butyl-3-methyl-imidazolium hexafluorophosphate ($BMIM\text{-}PF_6$).

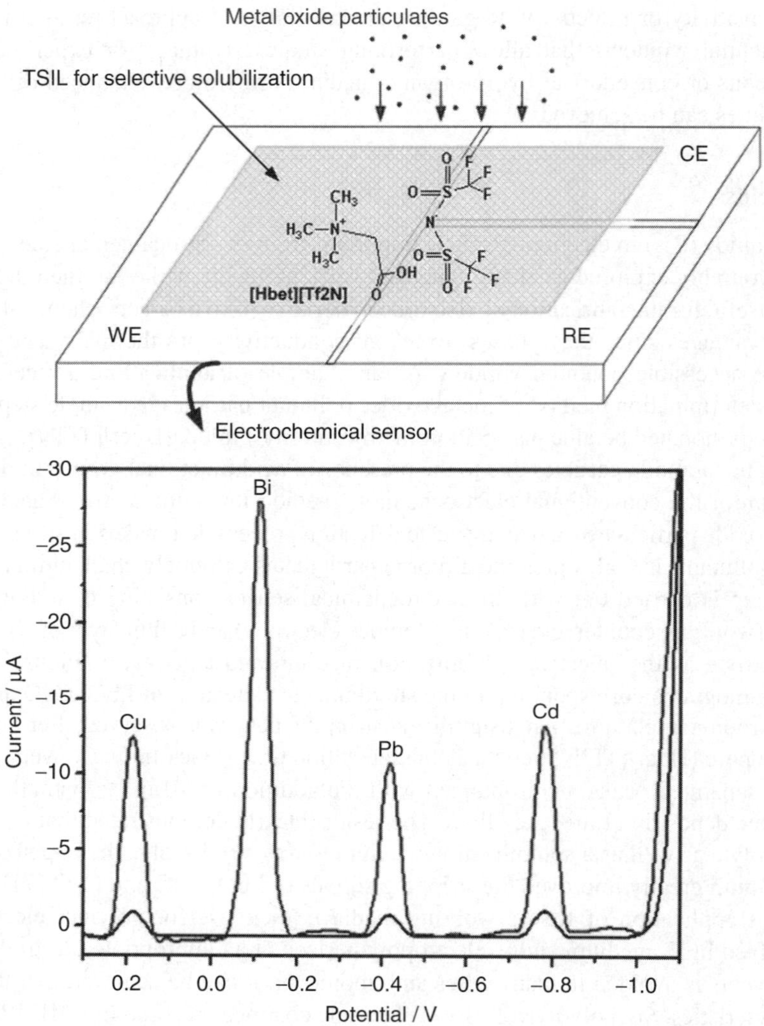

Figure 14.3 *Electrochemical sensor fabricated with a thin layer of a task specific ionic liquid (TSIL) as a selective solubilization medium for metal oxide particulates. Square wave stripping voltammograms from 8 ng l^{-1} PbO, CdO and CuO in the absence (black line) and presence (grey line) of 10 ng l^{-1} Bi_2O_3. SW ASV parameters were: 120 s deposition at −1.1 V, Frequency and amplitude are 25 Hz and 25 mV. Reproduced from [71] with permission from Elsevier, © 2010.*

This microelectrode improved the sensitivity, stability, reproducibility and response linearity for NADH compared with the equivalent electrode constructed with carbon paste and mineral oil. Moreover, the surface passivation of CILE as a result of the NADH oxidation was much lower [74]. Also, it is possible to observe direct electron transfer of proteins on the surface of these electrodes, as occurred for haemoglobin (Hb) entrapped in a sodium alginate hydrogel film on the surface of a $BMIMPF_6$ carbon paste electrode. The presence of the IL provides a biocompatible interface that improves the electron transfer and retains the Hb bioactivity on the electrode exhibiting a pair of well defined and quasi-reversible peaks by cyclic voltammetry corresponding to the Hb Fe(III)/Fe(II) redox couple [75].

Another interesting application of ILs is that related with the synthesis of metal nanoparticles. It has been demonstrated that imidazolium ionic liquids constitute a suitable working medium for the preparation of catalytically active transition-metal nanoparticles (NPs). So, platinum, rhodium and iridium NPs were prepared in 1-butyl-3-methyl-imidazolium with control of size, near-monodispersity, shape and stabilization [76]. Recently, the first synthesis of anisotropic gold nanocrystals (AuNC) by means of a seeding approach using the greener ionic liquid, 1-butyl-3-methylimidazolium lauryl sulfate, [BMIM] [$C_{12}H_{25}OSO_3$] has been reported. Crystal growth was successfully achieved by successive addition of the growth solution containing $HAuCl_4$ and ascorbic acid to 2.5-nm Au seeds. These nanocrystals were stabilized by the ionic liquid that served as solvent and capping agent via a two layer micellar formation by the butyl chain of the organic cation and the lauryl chain of the anion. [77].

14.3.2 Supercritical fluids

Since the beginning of the use of microelectrodes in electroanalysis, supercritical fluids (SCFs) were conceived as appropriate media to carry out electrochemical measurements. These fluids are considered important from the point of view of Green Analytical Chemistry as they can be employed instead of some organic solvents. Despite this and because of the high electrical resistivity of these media, their use was unfeasible for electrodes of conventional size. Microelectrodes have a critical surface of size approaching the diffusion layer. The current magnitude they provide is low and this minimizes the ohmic drop effects, thus being possible to use them in supercritical fluids such as the carbon dioxide which, moreover, is non-toxic, relatively inert and easy to be purified. Some separation methods by supercritical fluids chromatography using microelectrodes to perform electrochemical detection of compounds of environmental significance were described. For instance, a platinum microelectrode 25 μm in diameter allowed the determination of components in a mixture of phenols and nitroaromatic compounds working with both oxidative and reductive amperometric detection modes, using carbon dioxide modified with a low concentration (1–5%) of methanol as the mobile phase. The comparison with the results obtained with an UV detector revealed among other advantages the significantly less pronounced baseline drift and higher signal-to-noise ratio values (between 20 and 40 versus 4–10 obtained with UV detection) [78].

More recent interesting applications of SCFs have to do with the preparation of materials employed in electrochemistry and, in particular, with nanomaterials. SFCs have been used to deposit thin films of metal on a variety of surfaces and to incorporate metal nanoparticles on several organic and inorganic substrates. Different approaches have been reported to achieve these preparations. Among them, supercritical fluid deposition, especially using supercritical carbon dioxide, is particularly attractive because no liquid wastes are generated, no solvent residues remain on the substrate and the mass transfer rates are faster [79]. The synthesis of carbon nanotubes is another area where the use of SCFs is of high interest. The synthesis of this nanomaterial in gas phase produces low yields and poor crystallinity, and high temperatures are required. As an alternative, SCFs have been used to prepare well-crystallized CNTs using carbon dioxide as the carbon source in a relatively low temperature process and lithium metal as a reducing agent [80]. Compared with CO_2 gas, $scCO_2$ enhanced the adsorption of carbon dioxide on the surface of lithium and therefore facilitated and accelerated the electron transfer process. Large amounts of CNTs were produced with a diameter of 55 nm and more than one micrometer in length, without the use of a catalyst and in relatively mild and cost-effective reaction conditions. A similar approach was also reported to produce multiwalled carbon nanotubes (MWNTs) in supercritical toluene [81]. The concept was to take advantage of pressure effects to lower the high temperatures usually required to synthesize CNTs. In this process, supercritical toluene served both as the carbon source for nanotube growth, and as the reaction medium, and ferrocene, iron or platinum iron nanocrystals were used as catalysts.

14.4 Electrochemical detection in flowing solutions

14.4.1 Injection techniques

Flow injection analysis (FIA) has been recognized as an attractive tool for the development of green methodologies. In particular, methods based on flow injection coupled with electrochemical detection have proved their analytical utility, mainly because of their high sensitivity independently of the cell miniaturization, possibility of use turbid samples and low cost. FIA has demonstrated to be a useful technique for the rapid acquisition of preliminary chemical information about the occurrence of organic compounds in environmental samples. Very recently, the use of FIA methodologies to determine chemical oxygen demand (COD) or total concentration indices of halogenated organic compounds or phenols in water samples, have been reviewed [82]. Electrochemical detection is probably the most environmentally friendly alternative for COD measurements. On the contrary of the classical determination, which implies the use and generation of big amounts of extremely hazardous wastes, some recent electrochemical methods for direct COD determination have been proposed which do not need sample pre-treatment and are characterized by a limited reagent consumption, easy automation and low environmental impact. Methods based on the use of UV/TiO_2 packed bead reactors for sample digestion followed by the amperometric measurement of the O_2 concentration changes have been proposed [83]. Other improvements in the COD determination using EC detection are based on F-PbO2 modified electrode, and are useful within a linear range between 100 and 1200 mg l^{-1} [84]. Another electrochemical alternative involving a Ti/TiO_2/PbO_2 electrode for the flow injection with amperometric detection allows us to determine COD by applying a detection potential of +1.35 V [85].

Total organic fluorine (TOF) determination was also performed as an amount of inorganic fluoride obtained after defluorination reaction and analysis of recovered fluoride by using flow-injection system with potentiometric detection [86]. Defluorination reactions took place in a reaction coil in conditions of constant flow using sodium biphenyl. Figure 14.4 shows a scheme of the used manifold. An in-flow phase separator of small volume was needed for the well functioning of the potentiometric detector, which is based on a fluoride ion selective electrode incompatible with organic solvents. This system was simpler and easier to optimize than that based on fluorimetric detection and allows minimization of manual operations, thus providing improvement in precision, elimination of aggressive chemical handling as well as reduction of possible sources of contamination.

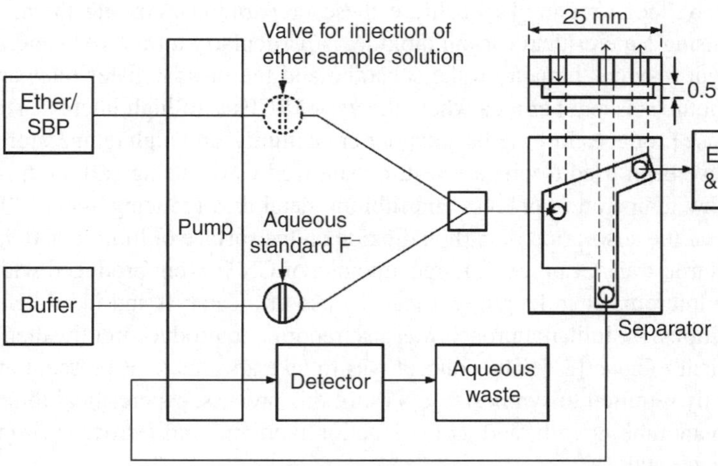

Figure 14.4 FIA system design with the use of an on-line defluorination reaction with sodium biphenyl, phase separator and potentiometric detection. Reproduced from [86] with permission from Elsevier, © 2007.

Sequential injection analysis (SIA) was proposed in 1990 by Ruzicka and Marshall [87] as the second generation of flow injection analysis (FIA). This technique is based on the formation of well defined sample and reagent zones. The combination of axial and radial dispersion modes gives rise to penetration on each zone with the other. The propulsion device is stopped between the aspiration of small volumes of sample and reagents, and then the flow direction is reversed and the zones in the holding coil are driven to the detector. So, the reaction products are formed in areas of well defined concentration gradient generating transient signals that provide good analytical reproducibility [88].

Applications of electrochemical detection in SIA were reviewed by Pérez-Olmos *et al.* [89]. Among others, several methods for the analysis of environmental samples have been described. For instance, amperometric detection was employed for the determination of primary explosive azides in ground and lake waters [90]. Also, SIA-anodic stripping voltammetric methods (SIA-ASV) for the determination of trace metals were reported. A recent example is the use of a carbon nanotubes screen-printed electrode to determine lead, cadmium and zinc, achieving detection limits of 0.2 μg l^{-1} for Pb(II), 0.8 μg l^{-1} for Cd(II) and 11 μg l^{-1} for Zn(II). Besides a high sensitivity, SIA allowed a high throughput to be obtained for on-line monitoring of trace heavy metals in herb samples [91]. Another configuration for Cd(II) and Pb(II) determination by ASV involved a sequential injection mono-segmented flow analysis (SI-MSFA) system with a Bi film electrode. The SI-MSFA device (Figure 14.5) provided an appropriate platform for the preparation of a homogeneous solution

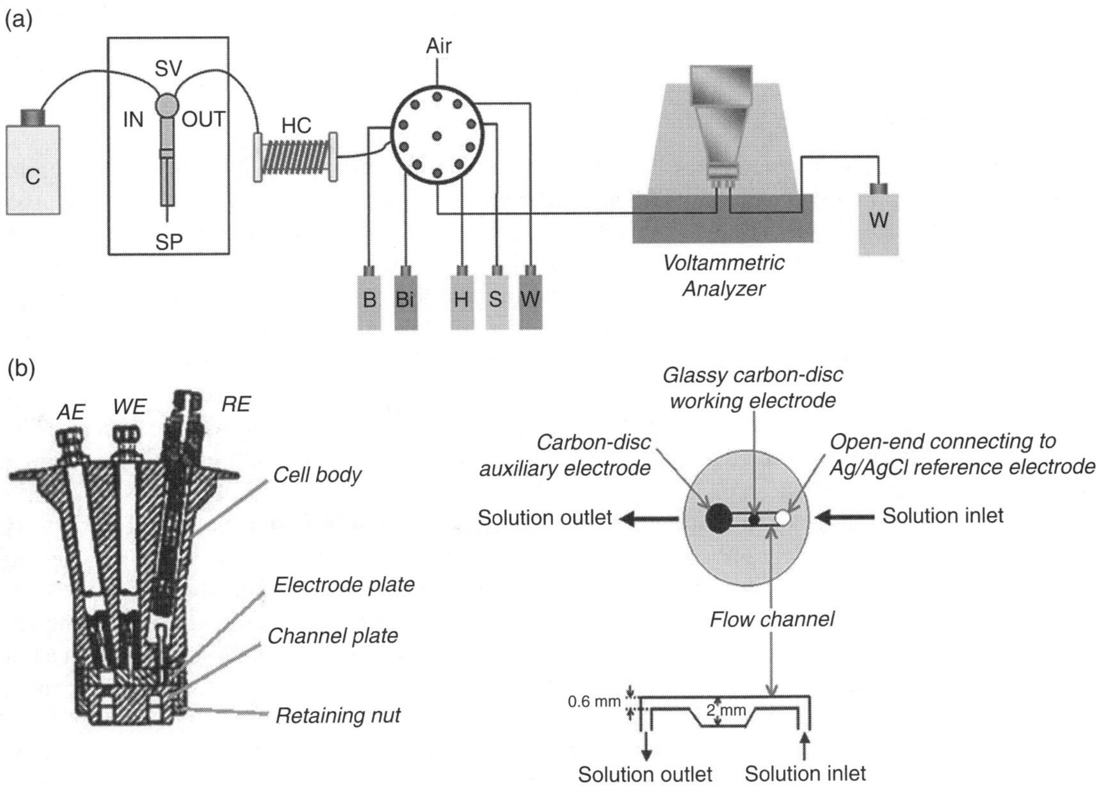

Figure 14.5 *(a) Schematic diagram of the SI-MSFA system. C: carrier (deionized water), SP: syringe pump, SV: switching valve, HC: holding coil, B: 0.2 M acetate buffer, Bi: 40 mg l^{-1} Bi(III) in 3M acetate buffer, H: cleaning solution (0.1 M HNO_3), S: mixed metals standard/sample and W: waste. (b) A thin layer electrochemical flow-cell and the flow channel, AE: auxiliary electrode, WE: working electrode and RE: reference electrode. Reproduced from [92] with permission from Elsevier, © 2009.*

zone of both Bi(III) for the in situ plating of the Bi film electrode and sample for ASV detection at a flow through thin layer electrochemical cell. The ranges of linearity of the calibration plots ranged between 10 and 100 μg l^{-1} and the detection limits were 1.4 and 6.9 μg l^{-1} for Cd(II) and Pb(II), respectively, allowing the application of this system to the analysis of water samples collected from a draining pond of zinc mining [92].

Another injection technique is batch-injection analysis (BIA). It was developed in 1991 [93] and involves the injection of a small volume of analyte sample with an automatic micropipette above the sensing surface of the detector which is immersed in a large volume of blank solution. A transient signal during the flow of sample over the detector surface is obtained. The main characteristics of this technique are short response times and virtually zero dispersion. Moreover, when electrochemical detection is used, BIA allows the analysis of small sample volumes with little dilution in supporting electrolyte. An illustrative example of a BIA methodology using electrochemical detection is the amperometric determination of phosphate based on the reduction of phosphomolybdate complex at a carbon paste electrode [94]. This method shows as relevant characteristics the simplicity of the equipment, the low reagents consumption and the low detection limit (0.3 μM). The good analytical performance made possible the application to the ortophosphate determination in seawater. Furthermore, stripping voltammetry using BIA relies on performing the electrodeposition or adsorption step during sample injection. Then, a potential scan is applied to oxidize or reduce the accumulated species. An example is the simultaneous determination of lead and cadmium using BIA-SWASV with fibrinogen-coated bismuth films deposited on carbon paste electrodes. The BIA system provided the possibility of rapid and simple electroanalysis of microlitre samples with detection limits of 0.2 μg l^{-1} and 0.1 μg l^{-1} for Cd and Pb, respectively [95].

14.4.2 Miniaturized systems

Miniaturization of analytical systems has been a matter of concern during recent years with the ultimate objective of achieving 'lab-on-a-chip' or μTAS in which miniaturization affects the overall analytical process [96]. These systems are closely related with the principles of green chemistry due to the extremely low volumes of sample and reagents required and the possibility of performing all the analysis steps in the same device. Electrochemical detection is particularly useful in these systems because of its inherent ability for miniaturization with no losses in sensitivity. In this context, various microsystem designs were reported in which the electrochemical cell was partially or totally integrated in the separation system. The technology employed for microfabrication allows the design of unique structures that can be tailored for specific purposes [97]. Multiple channel networks permit the introduction and mixing of reagents with zero dead volume connections. Also, theoretically, the reduced dimensions should provide improved separation performance over that of conventional systems.

The determination of oxygen consumption is usually employed to test biological activity. The Clark-type oxygen sensor is the most widely utilized electrode for this type of applications not only for fermentation monitoring but also in clinical chemistry and biosensor development. Various miniaturized Clark sensors have been proposed. For example, a novel micromachining technique was used to integrate a miniature Clark-type oxygen sensor with a microstructure. The oxygen chip consisted of a glass substrate with a three-electrode configuration, separated and connected by a groove and a poly(dimethylsiloxane)(PDMS) container with an immobilized PDMS oxygen-permeable membrane. Good linearity, a lifetime of at least 60 h and a fast response time of 6.8 s were the most relevant analytical characteristics of this configuration. Moreover, the miniature Clark oxygen sensor could be integrated with a microfluidic system to form a μ-TAS for measuring single cells such as an embryo or a low-volume biological sample [98].

Another interesting example is the development of an environmentally friendly disposable heavy metal ion sensor composed of a microfabricated bismuth (Bi) working electrode to perform anodic stripping voltammetry of Pb (II) and Cd (II) inside the microchannels of a microchip. A scheme of the fabricated

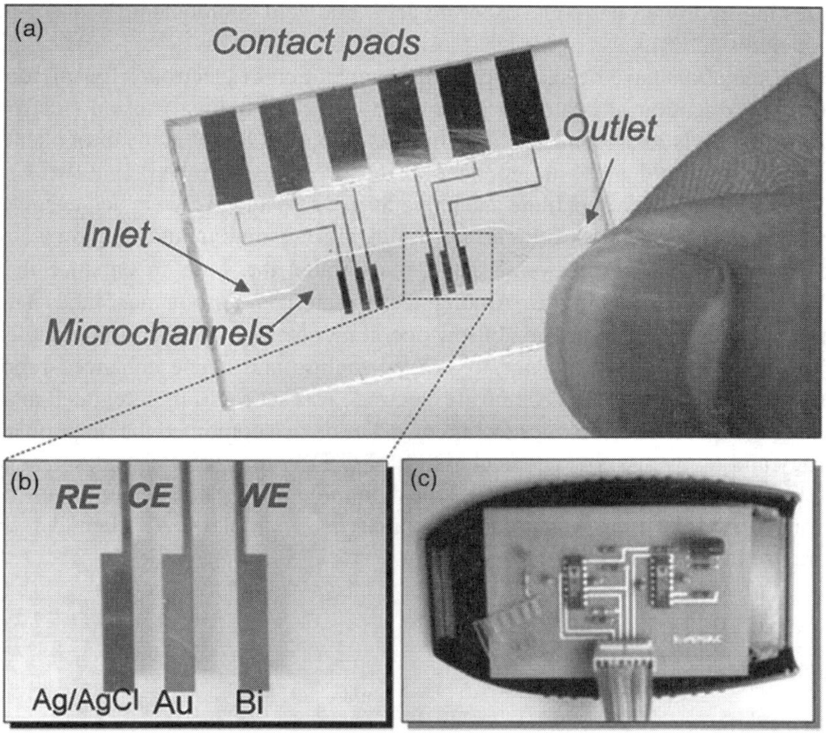

Figure 14.6 *Scheme of the disposable heavy metal microchip. (a) The entire chip including the sensor array and the microfluidic channels; (b) the magnified picture of the three-electrode sensor, and (c) the customized detection chips along with the chip. Reproduced from [99] with permission from Elsevier, © 2008.*

device is depicted in Figure 14.6. Low values of detection limits of 8 ppb for Pb(II) using 60 s deposition time, and 9.3 ppb for Cd (II) for 90 s deposition time were achieved. The sensor chip was applied to the in situ monitoring of Cd (II) concentration in environmental samples of soil pore and ground water, and to the online direct measurement of Cd (II) concentration in biological and clinical samples, such as cell culture media in its native environment [99].

Microfabrication techniques nowadays allow the design of unique structures that can be tailor-made for specific application [98]. Furthermore, multiple channel networks make possible the introduction and mixing of reagents with zero dead volume connections. Also, theoretically, the small dimensions of these systems allow enhancements in the separation performance with respect to conventional systems. In particular, the development of micro-fabricated capillary electrophoresis (CE) chips and their use for rapid biochemical analysis have demonstrated that this separation technique is also enhanced through miniaturization. Moreover, electrochemical detection is ideally suited for miniaturization to the microchip format. Photolithographic techniques used to construct the microchips can also be used to fabricate reproducible and inexpensive microelectrodes for these applications. A major challenge for electrochemical detection on microchips is the ability to isolate the detection potential from the separation field. Furthermore, microfabrication technology offers the possibility for creating multiple electrode detectors and arrays on a very small scale which can be used for monitoring various analytes simultaneously or for selective detection of analytes via chemically reversible redox reactions [98]. An illustrative example of a capillary electrophoresis micro-analytical chip with integrated electrodes for sample injection, separation and electrochemical detection is the one designed and applied for testing hydrogen peroxide, ascorbic acid, uric acid simultaneously and glucose showing a very good reproducibility [100].

A wide variety of materials have been applied to the preparation of electrodes for microchip CE. Carbon ink was used for the fabrication of coated-metallic wires to get cylindrical carbon electrodes which were integrated in a poly(dimethylsiloxane) microchip. This was employed for the analysis of phenolic compounds: dopamine, epinephrine, catechol and 4-aminophenol. Linear responses in the 0.1–190 μM range, and limits of detection of 140 and 105 nM for dopamine and epinephrine, respectively, were achieved. Moreover, a high signal stability without significant electrode surface fouling during continuous injection was observed [101]. On the other hand, the use of nanomaterials has demonstrated the ways to enhance the capabilities of electrochemical sensing in microchip-CE. Among these materials, carbon nanotubes have been widely employed because they allow the application of low operating potentials with substantially enhanced signal-to-noise ratios together with high resistance to surface fouling and hence enhanced reproducibility. For example, a multiwalled carbon nanotube composite electrode was successfully integrated in an amperometric microanalyser using the green tape ceramics technology. The device combined the properties provided by the ceramic technology and the well-known characteristics of CNT composite electrodes, and was used for the analysis of free chlorine at very low concentrations in swimming pool waters. Continuous monitoring of the analyte during long periods of time was possible by applying a simple electrochemical pretreatment step application before use in new analysis [102].

14.5 Biosensors

14.5.1 Greening biosurface preparation

The preparation of electrochemical biosensors nowadays shares most of green alternatives described for other types of configurations in the precedent sections. For instance, one of the recent developments in the research with enzyme biosensors involves the possibility of using ionic liquids as working media in which enzymes retain their biological activity. It has been demonstrated that the use of these media provides relevant advantages such as high conversion rates, high enantioselectivity, better enzyme stability, as well as better recoverability and recyclability. [103]. Enzyme electrochemical biosensors prepared in ionic liquid medium exhibit the additional advantage of a relative wide potential window of electrochemical stability and conductivity [104]. For example, it was reported that ionic liquid sol–gel enzyme electrodes retained the high activity of horseradish peroxidase (HRP) and provided long-term stability of HRP in storage [104]. These characteristics combined with the already commented green properties of ILs have made these media to be converted in an interesting material for enzyme immobilization and the development of a new generation of enzyme bioelectrodes. Mushroom tyrosinase was also demonstrated to be active in ionic liquids such as [BMIm][PF_6], [BMIm][BF_4], and [BMIm][$MeSO_4$] [105]. The increasing number of published articles on this subject makes predictable that ILs turn into green solvents for biocatalytic processes, capable of exhibiting also modified selectivity [106], since the enzymatic performance may vary using ionic liquids with different structure and composition whose ionic interaction with the enzyme affect its catalytic behaviour.

The combination of ILs with nanomaterials has driven the development of new biosensing surfaces with enhanced analytical performance. It is well known that metal nanoparticles, and in particular gold nanoparticles, have served as efficient support for biomolecules immobilization, exhibiting a high capability for extending biomolecules' stability. At present, the synergetic effects that arise from coupling these nanoparticles with ILs allow the improvement of not only the stability, but also for fastening the enzyme reaction [107]. An example is the construction of a biosensor for the determination of adrenaline based on the use of platinum nanoparticles dispersed in the ionic liquid 1-butyl-3-methylimidazolium hexafluorophosphate

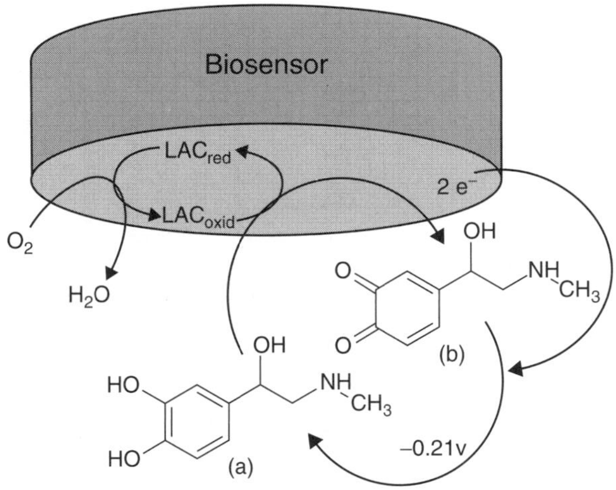

***Figure 14.7** Schematic representation of the enzymatic process between (a) adrenaline, oxygen and laccase, and electrochemical reduction of the (b) adrenalinequinone at the biosensor surface. Reproduced from [108] with permission from Elsevier, © 2009.*

(Pt-BMI.PF_6) and square wave voltammetry. This bioelectrode makes use of enzyme laccase from *Aspergillus oryzae*, which is considered as a 'green enzyme' since it is a phenol oxidase that catalyses phenols oxidation generating water as the reaction product (see Figure 14.7).

The calibration plot exhibited linearity in the 9.99×10^{-7} to 2.13×10^{-4} mol l^{-1} ($r = 0.9998$) concentration range, with a detection limit of 2.93×10^{-7} mol l^{-1}. Furthermore, the biosensor showed a nice stability (ca. 90 days; 300 determinations) and good repeatability and reproducibility [108].

RTILs have been also widely used for the preparation of DNA biosensors. In these applications, other materials such as chitosan are often used in combination with ionic liquids to improve both the homogeneity of the sensor surface and the biomolecules' immobilization. Furthermore, recent designs involve also conducting polymers and/or metal nanoparticles in order to enhance the conductivity of the sensor surface. For example, Figure 14.8 shows a scheme displaying the composition and functioning of a screen-printed DNA sensor, in which polyaniline nanotubes (PANINTs) and the ionic liquid (IL) [C12mim][PF_6] were used as dopants of the commercial conductive ink. The electrode surface was covered with chitosan and gold nanoparticles, thus obtaining a homogeneous, stable and highly conductive electrode. A DNA assay was developed using sandwich hybridization and chronocoulometry was employed to quantify hexamine ruthenium, RuHex, as electrochemical probe. This configuration allowed the concentration of the target DNA to be detected in the range from 1.0×10^{-16} M to 1.0×10^{-14} M, with and a detection limit of 8.0×10^{-17} M [109].

On the other hand, the synergic effect between carbon nanotubes and oxide-based nanomaterials was shown to increase the amount of ssDNA immobilized on the electrode surface this giving rise to an increase of the electrochemical responses. For instance, a recent configuration that implies a modified carbon ionic electrode (CILE) containing V_2O_5 nanobelts, multiwalled carbon nanotubes and chitosan can be mentioned. The CILE was prepared by using N-hexylpyridinium hexafluorophosphate (HPPF_6) as the binder with the graphite powder. This biosensor design was employed to immobilize ssDNA probe using methylene blue (MB) as the electrochemical indicator of the hybridization reaction. The ssDNA sequence could be detected in the 1.0×10^{-11} -1.0×10^{-6} mol l^{-1} concentration range and the biosensor was applied to the detection of the loop-mediated isothermal amplification (LAMP) product of *Yersinia enterocolitica* gene sequence in pork meat [110].

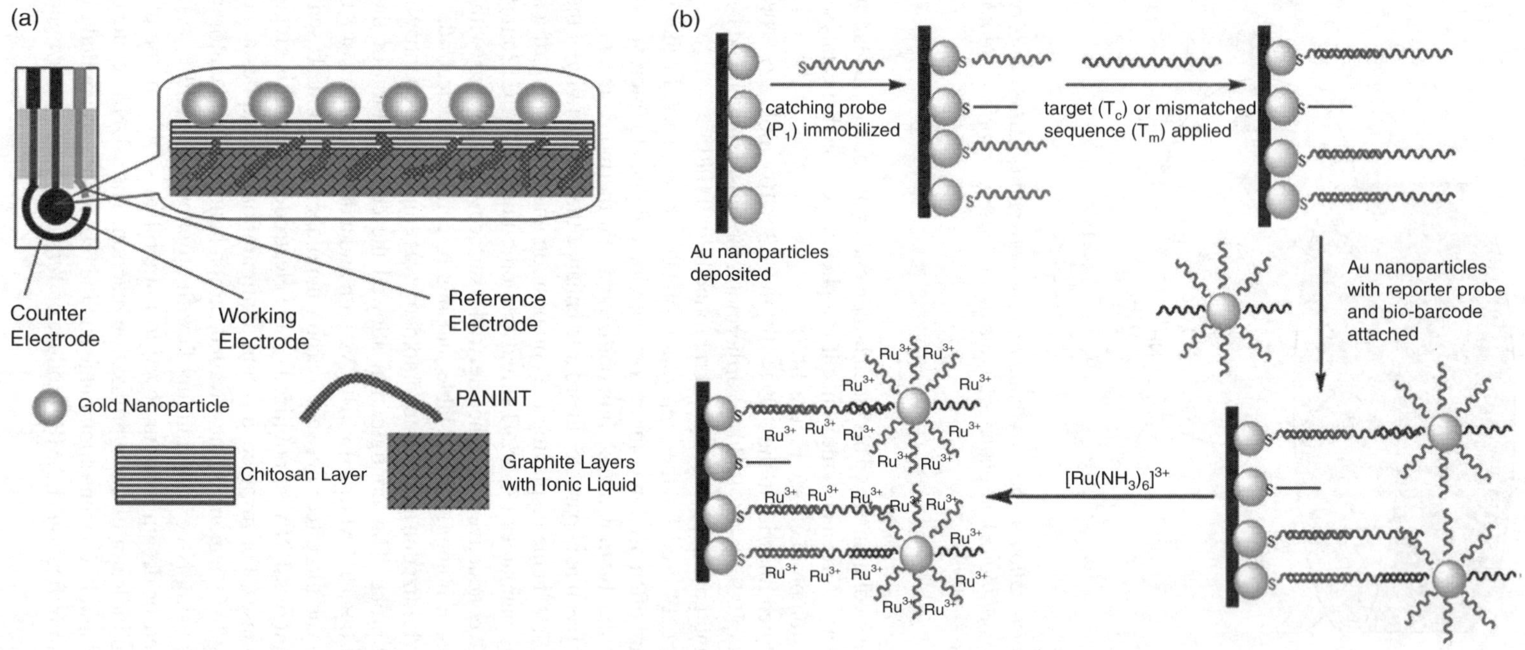

Figure 14.8 *(a) The pattern of the PANINT-IL/chitosan/AuNP SPE and the structure of the working electrode. (b) Procedure for the detection of DNA with RuHex as electrochemical probe. Reproduced from [109] with permission from Elsevier, © 2010.*

The search for environmentally more acceptable and user friendly systems for field biosensing detection has long been pursued. Various green approaches based on enzyme biosensors have been described. A recent example is the development of a highly sensitive amperometric biosensor for organophosphate (OPs) pesticides based on the immobilization of acetylcholinesterase (AChE) on multiwalled carbon nanotubes (MWCNTs)-β-cyclodextrin (β-CD) composite-modified glassy carbon electrodes. The composite was prepared using polymer wrapping strategy. The good dispersability and porous structure of MWCNTs-β-CD composite provided a favourable microenvironment for maintaining AChE bioactivity for screening of OPs exposure. MWCNTs promoted electron-transfer reactions at a lower potential and catalysed the electro-oxidation of thiocholine, thus increasing detection sensitivity. Based on the inhibition of OPs on the AChE activity, a detection limit of 2 nM (S/N = 3) was achieved using dimethoate as a model pesticide [111].

The attractive green properties of microsystems have been also profited for the development of biosensing methods. Nowadays, analytical microdevices have found application in the preparation of electrochemical biosensors due to their versatility, low cost and minimum energy consumption as well as their inherent characteristics of room saving and low volumes of sample, reagents and wastes [112]. In the particular case of clinical applications, the use of protein chips is the best choice to analyse real samples with small volume. Protein chip-based electrochemical sensors can be readily miniaturized and incorporated into microfluidic components for the development of lab-on-a-chip devices. The smaller sample volumes and reduced reaction distances in such devices would be expected to significantly increase the efficiency of immunological reactions. In particular, immunosensors coupled with electrochemical detectors using microelectrodes arrays have attracted extensive interest due to their powerful potential in simultaneously probing disease relevant markers [113]. On-chip integration of immunological and enzymatic reactions, amperometric detection and microchip operation allow immunoassays to perform more rapidly, easily and economically. An illustrative example is a microchip-based electrochemical immunoassay prepared to determine toxicity due to pesticides in a glass lab-on-a-chip based on the enzymatic inhibition of acetylcholinesterase immobilized on magnetic beads. The reproducible insertion of a controlled amount of enzyme-coupled magnetic beads inside the chip channel and their immobilization in a capture region with the aid of a magnetic field allowed the easy renewal of the biosensing material after each determination in a highly reproducible manner. The experimental conditions for the detection were optimized using thiocholine (TCh) as the target compound. Furthermore, the method was applied to the determination of carbofuran down to the nanomolar level [114].

14.5.2 Direct electrochemical transfer of proteins

In recent years, studies on protein film electrochemistry have been widely reported. Direct electron transfer of proteins with electrodes is a versatile and powerful strategy to construct third generation biosensors and also to investigate the chemistry of redox proteins. However, achieving an efficient direct electron transfer is not an easy task because the redox active prosthetic groups are usually buried in the protein molecule structure. On one hand, adsorption of the protein on the electrode surface induces deactivation and denaturation. Moreover, unfavourable orientations on the electrode surface difficult protein molecules to perform redox reactions. In spite of these difficulties, various electrodes modified with diverse materials have demonstrated to provide a suitable microenvironment to enhance direct electron transfer rate of proteins with electrodes. Ionic liquids are such materials [75]. Furthermore, carbon nanotubes in combination or not with ILs have been also used. For example, the direct electrochemistry of myoglobin (Mb) entrapped in a Nafion film on a multiwalled carbon nanotubes (MWCNTs) modified carbon ionic liquid electrode (CILE) was investigated using 1-ethyl-3-methylimidazolium tetrafluoroborate ($EMIMBF_4$) as the modifier. Spectroscopic results revealed that the Mb molecule on the surface of MWCNTs/ CILE retained its native structure. Cyclic voltammetry showed that a pair of well-defined quasi-reversible redox peaks appeared in the pH 7.0 phosphate buffer solution (PBS), which was attributed to the direct electron transfer of Mb heme Fe(III)/Fe(II) redox couples with the modified electrode. This Nafion/Mb/MWCNTs/CILE exhibited excellent electrocatalytic activity towards different substrates including trichloroacetic acid (TCA), hydrogen peroxide (H_2O_2) and sodium nitrite ($NaNO_2$) [115]. Another nice application involved single-walled carbon nanotubes (SWCNTs) selectively wrapped by a water-soluble, environmentally friendly, biocompatible polymer chitosan (CHI) for the construction of a bioelectrochemical platform for biosensing purposes making use of the direct electron transfer (DET) of glucose oxidase (GOD). The DET between GOD and the electrode surface was observed with a formal potential of about ca. −460mV versus SCE in phosphate buffer solution. The heterogeneous electron transfer rate constant and the surface coverage of GOD were estimated to be $3.0\,s^{-1}$ and 1.3×10^{-10} mol cm^2, respectively. The experimental results demonstrated that the immobilized GOD retained its catalytic activity towards the oxidation of glucose [116].

14.6 Future trends in green electroanalysis

Similar to that which has occurred with other scientific disciplines, modern electroanalytical chemistry has faced the challenge of developing new strategies able to fulfil the green chemistry principles. Current research is more and more involved in the strict observance of these principles. In particular it is oriented to the search for alternative green solvents and electrode materials, and the development of electroanalytical on-line methodologies with miniaturized systems making use of the recent advances in microfabrication technologies. The enormous potentials of electroanalytical chemistry in these kind of developments combined with the more and more widely used biological recognition systems pave the way for a new green generation of electrochemical biodevices with numerous potential applications in the clinical, agro-food and environmental fields.

References

1. Baś, B. Baś,S. (2010) Rapidly renewable silver amalgam annular band electrode for voltammetry and polarography, *Electrochem. Commun.,* **12**, 816–819.
2. De Sousa Dantas, A.N.; De Souza, D.; Soares de Lima, J.E.; de Lima-Neto, P. and Nunes Correia, A. (2010) Voltammetric determination of ketoconazole using a polished silver solid amalgam electrode, *Electrochim. Acta*, **55**, 9083–9089.
3. Wang, J.; Lu, J.; Hocevar, S.B.; Farias, P.A.M. and Ogorevc, B. (2000) Bismuth-Coated Carbon Electrodes for Anodic Stripping Voltammetry, *Anal. Chem.*, **72**, 3218–3222.
4. Wang, J. (2005) Stripping Analysis at Bismuth Electrodes: A Review, *Electroanalysis,* **17**, 1341–1346.
5. Arduini, F.; Quintana Calvo, J.; Amine, A.; Palleschi, G. and Moscone, D. (2010) Bismuth-modified electrodes for lead detection, *TrAC Trends Anal. Chem.*, **29**, 1295–1304.
6. Jorge, E.O.; Neto, M.M.M. and Rocha, M.M. (2007) A mercury-free electrochemical sensor for the determination of thallium(I) based on the rotating-disc bismuth film electrode, *Talanta*, **72**, 1392–1399.
7. Lee, G.H.; Lee, H.M.; Uhm, Y.R.; Lee, M.K. and Rhee, C.K. (2008) Square-wave voltammetric determination of thallium using surface modified thick-film graphite electrode with Bi nanopowder, *Electrochem. Commun.*, **10**, 1920–1923.
8. Hutton, E.A.; Hocevar, S.B. and Ogorevc, M. (2005) Ex situ preparation of bismuth film microelectrode for use in electrochemical stripping microanalysis, *Anal. Chim, Acta*, **537**, 285–292.
9. Lin, L.; Thongngamdee, S.; Wang, J.; Lin, Y.; Sadik, O.A. and Lye, S.Y. (2005) Adsorptive stripping voltammetric measurements of trace uranium at the bismuth film electrode, *Anal. Chim. Acta*, **535**, 9–13.
10. Jiang, L.; Wang, Y.; Ding, J.; Lou, T. and Qin, W. (2010) An ionophore–Nafion modified bismuth electrode for the analysis of cadmium(II), *Electrochem. Commun.*, **12**, 202–205.

11. Xu, H.; Zeng, L.; Huang, D.; Xian, Y. and Jin, L. (2008) A Nafion-coated bismuth film electrode for the determination of heavy metals in vegetable using differential pulse anodic stripping voltammetry: An alternative to mercury-based electrodes, *Food Chem.*, **109**, 834–839.
12. Li, D.; Jia, J. and Wang, J. (2010) Simultaneous determination of Cd(II) and Pb(II) by differential pulse anodic stripping voltammetry based on graphite nanofibers–Nafion composite modified bismuth film electrode, *Talanta*, **83**, 332–336.
13. Zou, Z.; Jang, A.; MacKnight, E.; Wu, P.-M.; Do, J.; Bishop, P.L. and Ahn, C.H. (2008) Environmentally friendly disposable sensors with microfabricated on-chip planar bismuth electrode for *in situ* heavy metal ions measurement, *Sens. Actuators B*, **134**, 18–24.
14. Zaouak, O.; Authier, L.; Cugnet, C.; Castetbon, A. and Potin-Gautier, M. (2009) Bismuth-Coated Screen-Printed Microband Electrodes for On-Field Labile Cadmium Determination, *Electroanalysis*, **21**, 689–695.
15. Hwang, G.H.; Han, W.K.; Park, J.S. and Kang, G.S. (2008) Determination of trace metals by anodic stripping voltammetry using a bismuth-modified carbon nanotube electrode, *Talanta*, **76**, 301–308.
16. Wu, Y.; Li, N.B. and Luo, H.Q. (2008) Simultaneous measurement of Pb, Cd and Zn using differential pulse anodic stripping voltammetry at a bismuth/poly(*p*-aminobenzene sulfonic acid) film electrode, *Sens. Actuators B*, **133**, 677–681.
17. Wang, J.; Lu, J.; Kirgöz, U.A.; Hocevar, S.B. and Ogorevc, B. (2001) Insights into the anodic stripping voltammetric behavior of bismuth film electrodes, *Anal. Chim. Acta*, **434**, 29–34.
18. Hwang, G.-H.; Han, W.-K.; Hong, S.-J.; Park, J-S. and Kang, S.-G. (2009) Determination of trace amounts of lead and cadmium using a bismuth/glassy carbon composite electrode, *Talanta*, **77**, 1432–1436.
19. Kokkinos, C.; Economou, A.; Raptis, I.; Efstathiou, C.E. and Speliotis, T. (2007) Novel disposable bismuth-sputtered electrodes for the determination of trace metals by stripping voltammetry, *Electrochem. Commun.*, **9**, 2795–2800.
20. Torma, F.; Kádár, M.; Tóth, K. and Tatár, E. (2008) Nafion®/2,2′-bipyridyl-modified bismuth film electrode for anodic stripping voltammetry, *Anal. Chim. Acta*, **619**, 173–182.
21. Hutton, E.A.; Hocevar, S.B.; Mauko, L. and Ogorevc, B. (2006) Bismuth film electrode for anodic stripping voltammetric determination of tin, *Anal. Chim. Acta*, **580**, 244–250.
22. Legeai, S.; Bois, S. and Vittori, O. (2006) A copper bismuth film electrode for adsorptive cathodic stripping analysis of trace nickel using square wave voltammetry, *J. Electroanal. Chem.*, **591**, 93–98.
23. Lin, L.; Lawrence, N.S.; Thongngamdee, S.; Wang, J. and Lin, Y. (2005) Catalytic adsorptive stripping determination of trace chromium (VI) at the bismuth film electrode, *Talanta*, **65**, 144–148.
24. Kadara, R.O.and Tothill, I.E. (2008) Development of disposable bulk-modified screen-printed electrode based on bismuth oxide for stripping chronopotentiometric analysis of lead (II) and cadmium (II) in soil and water samples, *Anal. Chim. Acta*, **623**, 76–81.
25. Siriangkhawut, W.; Pencharee, S.; Grudpan, K. and Jakmunee, J. (2009) Sequential injection monosegmented flow voltammetric determination of cadmium and lead using a bismuth film working electrode, *Talanta*, **79**, 1118–1124.
26. Segura, R.; Toral, M.I. and Arancibia, V. (2008) Determination of iron in water samples by adsorptive stripping voltammetry with a bismuth film electrode in the presence of 1-(2-piridylazo)-2-naphthol, *Talanta*, **75**, 973–977.
27. Zhang, Q.; Li, X.; Hongzhou, H.S. and Yuan, Z. (2010) Determination of selenium by differential pulse adsorptive stripping voltammetry at a bismuth film electrode, *Electrochim, Acta*, **55**, 4717–4721.
28. Charalambous, A. and Economou, A. (2005) A study on the utility of bismuth-film electrodes for the determination of In(III) in the presence of Pb(II) and Cd(II) by square wave anodic stripping voltammetry, *Anal. Chim. Acta*, **547**, 53–58.
29. Kefala, G.; Economou, A. and Sofoniou, M. (2006) Determination of trace aluminium by adsorptive stripping voltammetry on a preplated bismuth-film electrode in the presence of cupferron, *Talanta*, **68**, 1013–1019.
30. Jiajie, L. and Nagaosa, Y. (2007) Cathodic stripping voltammetric determination of As(III) with in situ plated bismuth-film electrode using the catalytic hydrogen wave, *Anal. Chim. Acta*, **593**, 1–6.
31. Jorge, E.O.; Rocha, M.M.; Fonseca, I.T.E. and Neto, M.M.M. (2010) Studies on the stripping voltammetric determination and speciation of chromium at a rotating-disc bismuth film electrode, *Talanta*, **81**, 556–564.
32. Wang, J.; Lu, D.; Thongngamdee, S.; Lin, Y. and Sadik, O.A. (2006) Catalytic adsorptive stripping voltammetric measurements of trace vanadium at bismuth film electrodes, *Talanta*, **69**, 914–917.

33. Prior, C.; Lenehan, C.E. and Walker, G.S. (2007) Utilising gallium for enhanced electrochemical copper analysis at the bismuth film electrode, *Anal. Chim. Acta*, **598**, 65–73.
34. Jia, X.; Li, J. and Wang, E. (2010) High-sensitivity determination of lead(II) and cadmium(II) based on the CNTs-PSS/Bi composite film electrode, *Electroanalysis*, **22**, 1682–1687.
35. Bobrowski, A.; Królicka, A.; Pacan, K. and Zarębski, J. (2009) Application of the bismuth Film Electrode for Voltammetric Determination of titanium using Ti(IV)-Oxalate-Chlorate Catalytic System, *Electroanalysis*, **21**, 2415–2419.
36. Campestrini, I.; de Braga, O.C.; Vieira, I.C. and Spinelli, A. (2010) Application of bismuth-film electrode for cathodic electroanalytical determination of sulfadiazine, *Electrochim. Acta*, **55**, 4970–4975.
37. Yang, M.; Zhang, Z.; Hu, Z. and Li, J. (2006) Differential pulse anodic stripping voltammetry detection of metallothionein at bismuth film electrodes, *Talanta*, **69**, 1162–1165.
38. Nigović, B.; Śimunić, B. and Hocevar, S. (2009) Voltammetric measurements of aminosalicylate drugs using bismuth film electrode, *Electrochim. Acta*, **54**, 5678–5683.
39. Xie, X.Y.; Luo, H.Q. and Li, N.B. (2010) Determination of azo compounds by differential pulse voltammetry at a bismuth/poly(p-aminobenzene sulfonic acid) film electrode and application for detection in food stuffs, *J. Electroanal. Chem.*, **639**, 175–180.
40. Moreno, M.; Bermejo, E.; Chicharro, M.; Zapardiel, A. and Sánchez Arribas, A. (2009) Cathodic Electrochemical Determination of Herbicides in Acid Media Using a Bismuth Film Electrode, *Electroanalysis*, **21**, 415–421.
41. Du, D.; Ye, X.; Zhang, J. and Liu, D. (2008) Cathodic electrochemical analysis of methyl parathion at bismuth-film-modified glassy carbon electrode, *Electrochim. Acta*, **53**, 4478–4484.
42. Luo, L.; Wang, X.; Ding, Y.; Li, Q.; Jia, J. and Deng, D. (2010) Electrochemical determination of nitrobenzene using bismuth-film modified carbon paste electrode in the presence of cetyltrimethylammonium bromide, *Anal. Methods*, **2**, 1095–1100.
43. de Figueiredo-Filho, L.C.S.; dos Santos, V.B.; Campos Janegitz, B.; Brito Guerreiro, T.; Fatibello-Filho, O.; Censi Faria, R. and Marcolino-Junior, L.H. (2010) Differential pulse voltammetric determination of paraquat using a bismuth-film electrode, *Electroanalysis*, **22**, 1260–1266.
44. Baldrianova, L.; Agrafiotou, P.; Svancara, I.; Vytras, K. and Sotiropoulos, S. (2008) The determination of cysteine at Bi-powder carbon paste electrodes by cathodic stripping voltammetry, *Electrochem. Commun.*, **10**, 918–921.
45. Zhang, W.; Tang, H.; Geng, P.; Wang, Q.; Jin, L. and Wu, Z. (2007) Amperometric method for rapid detection of Escherichia coli by flow injection analysis using a bismuth nano-film modified glassy carbon electrode, *Electrochem. Commun.*, **9**, 833–838.
46. Hocevar, S.B.; Sÿvancara, I.; Ogorevc, B. and Vytr&acaron;s, K. (2007) Antimony film electrode for electrochemical stripping analysis, *Anal. Chem.*, **79**, 8639–8643.
47. Tesarŏva, E. and Vytrăs, K. (2009) Potentiometric stripping analysis at antimony film electrodes, *Electroanalysis*, **21**, 1075–1080.
48. Kokkinos, C.; Economou, A.; Raptis, I. and Speliotis, T. (2009) Novel disposable microfabricated antimony-film electrodes for adsorptive stripping analysis of trace Ni(II), *Electrochem. Commun.*, **11**, 250–253.
49. de Lima-Neto, P.; Correia, A.N.; Portela, R.R.; da Silva Juliao, M.; Linhares-Junior, G.F. and de Lima, J.E.S. (2010) Square wave voltammetric determination of nitrofurantoin in pharmaceutical formulations on highly boron-doped diamond electrodes at different boron-doping contents, *Talanta*, **80**, 1730–1736.
50. Boonsong, K.; Chuanuwatanakul, S.; Wangfuengkanagul, N. and Chailapakul, O. (2005) Electroanalysis of lincomycin using boron-doped diamond thin film electrode applied to flow injection system, *Sens. Actuators B*, **108**, 627–632.
51. Dogan, B.; Tuncel, S.; Uslu, B. and Özkan, S.A. (2007) Selective electrochemical behavior of highly conductive boron-doped diamond electrodes for fluvastatin sodium oxidation, *Diamond Rel. Mater.*, **16**, 1695–1704.
52. Antigo Medeiros, R.; Rocha-Filho, R.C. and Fatibello-Filho, O. (2010) Simultaneous voltammetric determination of phenolic antioxidants in food using a boron-doped diamond electrode, *Food Chem.*, **123**, 886–891.
53. Li, W.; Jin, G.; Chen, H. and Kong, J. (2009) Highly sensitive and reproducible cyclodextrin-modified gold electrodes for probing trace lead in blood, *Talanta*, **78**, 717–722.
54. Li, Y.Y.; Jiang, Z.Y.; Chen, J.Q.; Li, H.L. and Zhang, H.Q. (2008) Environment-Friendly Determination of Low Concentration Azobenzene β-Cyclodextrin-Modified Electrode, *Biomed. Environ. Sci.*, **21**, 479–484.

55. Agrahari, S.K.; Kumar, S.D. and Srivastava, A.K. (2009) Development of a carbon paste electrode containing benzo-15-crown-5 for trace determination of the uranyl ion by using a voltammetric technique, *J. AOAC Int.*, **92**, 241–247.
56. Silva, C.R.; Conceição, C.D.C.; Bonifácio, V.G.; Fatibello Filho, O. and Teixeira, M.F.S. (2009) Determination of the chemical oxygen demand (COD) using a copper electrode: a clean alternative method, *J. Solid State Electrochem.*, **13**, 665–670.
57 Yang, Y.; Qiu, S.; Xie, X.; Wang, X.; Kwok, R. and Li, Y. (2010) A facile, green, and tunable method to functionalize carbon nanotubes with water-soluble azo initiators by one-step free radical addition, *Appl. Surf. Sci.*, **256**, 3286–3292.
58. Yang, Y.; Qiu, S.; Hea, C.; He, W.; Yu, L. and Xie, X. (2010) Green chemical functionalization of multiwalled carbon nanotubes with poly(epsilon-caprolactone) in ionic liquids, *Appl. Surf. Sci.*, **257**, 1010–1014.
59. Liu, K.S.; Fu, H.G.; Xie, Y.; Zhang, L.L.; Pan, K. and Zhou, W. (2008) Assembly of β-cyclodextrins acting as molecular bricks onto multiwall carbon nanotubes, *J. Phys. Chem. C*, **112**, 951–957.
60. Shen, Q. and Wang, X. (2009) Simultaneous determination of adenine, guanine and thymine based on β-cyclodextrin/MWNTs modified electrode, *J. Electroanal. Chem.*, **632,** 149–153.
61. Fan, Z.; Wang, K.; Wei, T.; Yan, J.; Song, L. and Shao, B. (2010) An environmentally friendly and efficient route for the reduction of graphene oxide by aluminum powder, *Carbon*, **48**, 1686–1689.
62. Sridhar, V.; Jeon, J.H. and Oh, I.K. (2010) Synthesis of graphene nano-sheets using eco-friendly chemicals and microwave radiation, *Carbon*, **48**, 2953–2857.
63. Cao, Y.; Feng, J. and Wu, P. (2010) Preparation of organically dispersible graphene nanosheet powders through a lyophilization method and their poly(lactic acid) composites, *Carbon*, **48**, 3834–3839.
64. Zhao, W.; Xu, J.J.; Shi, C.G. and Chen, H.Y. (2006) Fabrication, characterization and application of gold nano-structured film, *Electrochem. Commun.*, **8**, 773–778.
65. Inbakandan, D.; Venkatesan, R. and Khan, S.A. (2010) Biosynthesis of gold nanoparticles utilizing marine sponge Acanthella elongata (Dendy, 1905), *Colloid Surf. B*, **81**, 634–639.
66. Ding, L.; Hao, C.; Xue, Y. and Ju, H. (2007) A bio-inspired support of gold nanoparticles-chitosan nanocomposites gel for immobilization and electrochemical study of K562 leukemia cells, *Biomacromolecules*, **8**, 1341–1346.
67. Medina-Ramírez, I.; González-García, M. and Louise Liu, J. (2009) Nanostructure characterization of polymer-stabilized gold nanoparticles and nanofilms derived from green synthesis, *J. Mater. Sci.*, **44**, 6325–6332.
68. Benaissi, K.; Johnson, L.; Walsh, D.A. and Thielemans, W. (2010) Synthesis of platinum nanoparticles using cellulosic reducing agents, *Green Chem.*, **12**, 220–222.
69. Yang, W.; Yang, C.; Sun, M.; Yang, F.; Ma, Y.; Zhang, Z. and Yang, X. (2009) Green synthesis of nanowire-like Pt nanostructures and their catalytic properties, *Talanta*, **78**, 557–564.
70. Sharma, V.K.; Yngard, R.A. and Lin, Y. (2009) Silver nanoparticles: Green synthesis and their antimicrobial activities, *Adv. Coll. Interf. Sci.*, **145**, 83–96.
71. Lu, D.; Shomali, N. and Shen, A. (2010) Task specific ionic liquid for direct electrochemistry of metal oxides, *Electrochem. Commun.*, **12**, 1214–1217.
72. Zhang, A.; Chen, J.; Niu, D.; Wallace, G.G. and Lua, J. (2009) Electrochemical polymerization of pyrrole in $BMIMPF_6$ ionic liquid and its electrochemical response to dopamine in the presence of ascorbic acid, *Synth. Met.*, **159**, 1542–1545.
73. Pringle, J.M.; Forsyth, M.; MacFarlane, D.R.; Wagner, K.; Hall, S.B. and Officer, D.L. (2005) The influence of the monomer and the ionic liquid on the electrochemical preparation of polythiophene, *Polymer*, **46**, 2047–2052.
74. Musameh, M. and Wang, J. (2008) Sensitive and stable amperometric measurements at ionic liquid–carbon paste microelectrodes, *Anal. Chim. Acta*, **606**, 45–49.
75. Wei, S.; Dandan, W.; Ruifang, G. and Kui, J. (2007) Direct electrochemistry and electrocatalysis of hemoglobin in sodium alginate film on a $BMIMPF_6$ modified carbon paste electrode, *Electrochem. Commun.*, **9**, 1159–1164.
76. Schrekker, H.S.; Gelesky, M.A.; Stracke, M.P.; Schrekker, C.M.L.; Machado, G.; Teixeira, S.R.; Rubim, J.C. and Dupont, J. (2007) Disclosure of the imidazolium cation coordination and stabilization mode in ionic liquid stabilized gold(0) nanoparticles, *J. Colloid Interf. Sci.*, **316**, 189–195.
77. Obliosca, J.M.; Arellano, I.H.J.; Huang, M.H. and Arco, S.D. (2010) Double layer micellar stabilization of gold nanocrystals by greener ionic liquid 1-butyl-3-methylimidazolium lauryl sulfate, *Mat. Letters*, **64**, 1109–1112.

78. Wallenborg, S.R.; Markides, K.E. and Nyholm, L. (1997) Oxidative and reductive amperometric detection of phenolic and nitroaromatic compounds in packed capillary column supercritical fluid chromatography, *J. Chromatogr. A*, **785**, 121–128.
79. Erkey, C.J. (2009) Preparation of metallic supported nanoparticles and films using supercritical fluid deposition, *J. Supercrit. Fluids*, **47**, 517–522.
80. Lou, Z.; Chen, Q.; Wang, W. and Zhang, Y. (2003) Synthesis of carbon nanotubes by reduction of carbon dioxide with metallic lithium, *Carbon*, **41**, 3063–3067.
81. Smith, D.K.; Lee, D.C. and Korgel, B.A. (2006) High yield multiwall carbon nanotube synthesis in supercritical fluids, *Chem. Mater.*, **18**, 3356–3364.
82. Maya, F.; Estela, J.M. and Cerdà, V. (2010) Flow analysis techniques as effective tools for the improved environmental analysis of organic compounds expressed as total indices, *Talanta*, **81**, 1–8.
83. Kim, Y.C.; Sasaki, S.; Yano, K.; Ikebukuro, K.; Hashimoto, K. and Karube, I. (2001) Photocatalytic sensor for the determination of chemical oxygen demand using flow injection analysis, *Anal. Chim. Acta*, **432**, 59–66.
84. Li, J.; Li, L.; Zheng, L.; Xian, Y.; Ai, S. and Jin, L. (2005) Amperometric determination of chemical oxygen demand with flow injection analysis using F-PbO_2 modified electrode, *Anal. Chim. Acta*, **548,** 199–204.
85. Li, J.; Zheng, L.; Li, L.; Shi, G.; Xian, Y. and Jin, L. (2007) Photoelectro-synergistic catalysis combined with a FIA system application on determination of chemical oxygen demand, *Talanta*, **72**, 1752–1756.
86. Musijowski, J.; Trojanowicz, M.; Szostek, B.; Lima, J.L.F.C.; Lapa, R.; Yamashita, H.; Takayanagi, T. and Motomizu, S. (2007) Flow-injection determination of total organic fluorine with off-line defluorination reaction on a solid sorbent bed, *Anal. Chim. Acta*, **600**, 147–154.
87. Ruzicka, J. and Marshall, G.D. (1990) Sequential injection: a new concept for chemical sensors, process analysis and laboratory assays, *Anal. Chim. Acta*, **237**, 329–343.
88. Ruzicka, J. (1992) The second coming of flow-injection analysis, *Anal. Chim. Acta*, **261**, 3–10.
89. Pérez-Olmos, R.; Soto, J.C.; Zárate, N.; Araújo, A.N. and Montenegro, M.C.B.S.M. (2005) Sequential injection analysis using electrochemical detection: A review, *Anal. Chim. Acta*, **554**, 1–16.
90. Echols, R.T.; James, R.R. and Aldstadt, J.H. (1997) Determination of primary explosive azides in environmental samples by sequential injection amperometry, *Analyst*, **122**, 315–319.
91. Injang, U.; Noyrod, P.; Siangprohb, W.; Dungchai, W.R.; Motomizu, S. and Chailapakul, O. (2010) Determination of trace heavy metals in herbs by sequential injection analysis-anodic stripping voltammetry using screen-printed carbon nanotubes electrodes *Anal. Chim. Acta*, **668**, 54–60.
92. Siriangkhawut, W.; Pencharee, S.; Grudpan, K. and Jakmunee, J. (2009) Sequential injection monosegmented flow voltammetric determination of cadmium and lead using a bismuth film working electrode, *Talanta*, **79**, 1118–1124.
93. Wang, J. and Taha, Z. (1991) Batch injection analysis, *Anal. Chem.*, **63**, 1053–1056.
94. Calvo Quintana, J.; Idrissi, L.; Palleschi, G.; Albertano, P.; Amine, A.; Rhazi, M.E. and Moscone, D. (2004) Investigation of amperometric detection of phosphate: Application in seawater and cyanobacterial biofilm samples, *Talanta*, **63**, 567–574.
95. Adraoui, I.; Rhazi, M.E. and Amine, A. (2007) Fibrinogen-coated bismuth film electrodes for voltammetric analysis of lead and cadmium using the batch injection analysis, *Anal. Lett.*, **40**, 349–354.
96. Alegret, S. (2003) *Integrated Analytical Systems,* Ch. 1, Elsevier, Amsterdam.
97. Gawron, A.J.; Martin, R.S. and Lunte, S.M. (2001) Microchip electrophoretic separation systems for biomedical and pharmaceutical analysis, *Eur. J. Pharm. Sci.*, **14**, 1–12.
98. Wu, C.C.; Yasukawa, T.; Shiku, H. and Matsue, T. (2005) Fabrication of miniature Clark oxygen sensor integrated with microstructure, *Sens. Actuators B*, **110**, 342–349.
99. Zou, Z.; Jang, A.; MacKnight, E.; Wu, P.M.; Do, J.; Bishop, P.L. and Ahn, C.H. (2008) Environmentally friendly disposable sensors with microfabricated on-chip planar bismuth electrode for *in situ* heavy metal ions measurement, *Sens. Actuators B*, **134**, 18–24.
100. Virdi, G.S.; Chutani, R.K.; Rao, P.K. and Kumar, S. (2008) Fabrication of low cost integrated micro-capillary electrophoresis analytical chip for chemical analysis, *Sens. Actuators B*, **128**, 422–426.
101. Ding, Y.; Ayon, A. and García, C.D. (2007) Electrochemical detection of phenolic compounds using cylindrical carbon-ink electrodes and microchip capillary electrophoresis, *Anal. Chim. Acta*, **584**, 244–251.

102. Olivé-Monllau, R.; Martínez-Cisneros, C.S.; Bartrolí, J.; Baeza, M. and Céspedes, F. (2011) Integration of a sensitive carbon nanotube composite electrode in a ceramic microanalyzer for the amperometric determination of free chlorine, *Sens. Actuators B*, **151**, 416–422.
103. Moniruzzaman, M.; Nakashima, K.; Kamiya, N. and Goto, M. (2010) Recent advances of enzymatic reactions in ionic liquids, *Biochem. Eng. J.*, **48**, 295–314.
104. Liu, Y.; Shi, L.; Wang, M.; Li, Z.; Liu, H. and Li, J. (2005) A novel room temperature ionic liquid sol-gel matrix for amperometric biosensor application, *Green Chem.*, **7**, 655–659.
105. Yang, Z. Yue, Y.J. and Xing, M. (2008) Tyrosinase activity in ionic liquids, *Biotechnol. Lett.*, **30**, 153–158.
106. Yang, Z. and Pan, W. (2005) Ionic liquids: Green solvents for nonaqueous biocatalysis, *Enzyme Microb. Technol.*, **37**, 19–28.
107. Jiang, Y.; Guo, C.; Xia, H.; Mahmood, I.; Liu, C. and Liu, H. (2009) Magnetic nanoparticles supported ionic liquids for lipase immobilization: Enzyme activity in catalyzing esterification, *J. Molec. Cat. B*, **58**, 103–109.
108. Brondani, D.; Weber Scheeren, C.; Dupont, J. and Cruz Vieira, I. (2009) Biosensor based on platinum nanoparticles dispersed in ionic liquid and laccase for determination of adrenaline, *Sens. Actuators B*, **140**, 252–259.
109. Ren, R.; Leng, C. and Zhang, S. (2010) A chronocoulometric DNA sensor based on screen-printed electrode doped with ionic liquid and polyaniline nanotubes, *Biosens. Bioelectron.*, **25**, 2089–2094.
110. Sun, W.; Qin, P.; Gao, H.; Li, G. and Jiao, K. (2010) Electrochemical DNA biosensor based on chitosan/nano-V_2O_5/MWCNTs composite film modified carbon ionic liquid electrode and its application to the LAMP product of Yersinia enterocolitica gene sequence, *Biosens. Bioelectron.*, **25**, 1264–1270.
111. Dua, D.; Wang, M.; Cai, J. and Zhang, A. (2010) Sensitive acetylcholinesterase biosensor based on assembly of β-cyclodextrins onto multiwall carbon nanotubes for detection of organophosphates pesticide, *Sens. Actuators B*, **146**, 337–341.
112. Carvalhal, R.F.; Machado, D.S.; Mendes, R.K.; Almeida, A.L.J.; Moreira, N.H.; Piazetta, M.H.O.; Gobbi, A.L. and Kubota, L.T. (2010) Development of a disposable amperometric biosensor for salicylate based on a plastic electrochemical microcell, *Biosens. Bioelectron.*, **25**, 2200–2004.
113. Chen, H.; Jiang, C.; Yu, C.; Zhang, S.; Liu, B. and Kong, J. (2009) Protein chips and nanomaterials for application in tumor marker immunoassays, *Biosens. Bioelectron.*, **24**, 3399–3411.
114. Llopis, X.; Pumera, M.; Alegret, S. and Merkoçi, A. (2009) Lab-on-a-chip for ultrasensitive detection of carbofuran by enzymatic inhibition with replacement of enzyme using magnetic beads, *Lab-on-a-Chip*, **9**, 213–218.
115. Sun, W.; Li, X.; Wang, Y.; Li, X.; Zhao, C. and Jiao, K. (2009) Electrochemistry of myoglobin in Nafion and multi-walled carbon nanotubes modified carbon ionic liquid electrode, *Bioelectrochem*, **75**, 170–175.
116. Zhou, Y.; Yang, H. and Chen, H.Y. (2008) Direct electrochemistry and reagentless biosensing of glucose oxidase immobilized on chitosan wrapped single-walled carbon nanotubes *Talanta*, **76**, 419–423.

Section III

Strategies

15

Energy Savings in Analytical Chemistry

Mihkel Koel

Institute of Chemistry, Faculty of Science, Tallinn University of Technology, Tallinn, Estonia

15.1 Energy consumption in the analytical methods

The chemical sciences and the chemical industry are at the heart of the most productive multidisciplinary initiatives. Chemistry contributes to our well being, high life expectancy and economic prosperity, satisfying the desire for better materials for everyday life and accommodation, for drugs to cure illnesses and improve health, for the provision of pure water in addition to a host of other human activities. New environmental regulations and a growing social consciousness of the need to protect nature, have nudged the chemical sciences and industry toward a new framework in which pollution prevention and energy saving are the central considerations. Energy and chemistry are key themes in ongoing discussions about the sustainable future. This situation presents an enormous technological and scientific challenge that requires the development of innovative products and tools to minimize the impact on the environment.

In addition to reducing pollution and the production of waste, conserving energy has become a crucial issue. The principles of green chemistry [1] provide good guidelines for further development and are increasingly being adopted by both academia and industry.

These principles can be applied to analytical chemistry and other fields of chemistry as well as to chemical technology. They mainly concern reducing waste and curtailing the use of organic solvents and toxic chemicals; but saving energy is equally important and has a major impact on development in analytical chemistry. This is aptly expressed in the principles of Green Engineering proposed by Anastas and Zimmerman [2]:

Design for Energy Efficiency – The energy requirements of chemical processes should be decreased. These are largely, influenced by sample preparation and analysis conditions. Processes should be designed for ambient temperature and pressure, in order to minimize the energy costs associated with extremes of temperature and pressure.

Handbook of Green Analytical Chemistry, First Edition. Edited by Miguel de la Guardia and Salvador Garrigues.

Maximize Mass-, Space- and Time-Efficiency – Processes and systems should be designed to maximize mass-, space-, and time-efficiency. If a system is designed for, used, or applied at less than maximum efficiency, resources are being wasted throughout the lifecycle.

Conserve Complexity – The choices and complexity of the design must take into account recycling, reuse, or beneficial disposition across all design scales.

Meet the Need, Minimize the Excess – Minimize the use of underutilized and unnecessary materials to increase efficiency and conserve energy technologies that target the specific needs and demands of end users provide alternatives to devising solutions for extreme or unrealistic conditions.

Integrate Local Material and Energy Flows – Products, processes and systems must integrate and interconnect with available energy and material flows. By taking advantage of existing energy and material flows, the need to generate energy and/or acquire and process raw materials is minimized. This strategy can be employed in processes to use heat generated by exothermic reactions to drive other reactions with high activation energies.

Design for a commercial 'afterlife' – Design decisions on the end-of-life of a product should be based on the material and energy invested. The instrumentation used in analytical chemistry could be recycled. This would extend the life cycle of an analytical instrument. To reduce waste, components that remain functional and valuable could be recovered for reuse and/or reconfiguration. The design of next-generation products, processes and systems must take into account the reuse of the valuable properties of recovered components.

The US Environmental Protection Agency EPA has enunciated a similar policy: green engineering is the design, commercialization, and use of processes and products that are feasible and economical while minimizing; (1) the generation of pollution at the source and (2) the risk to human health and the environment. Green engineering embraces the concept that decisions to protect human health and the environment can have the greatest impact and be most cost-effective if they are applied at the outset to the design and development of a process or product [3].

Strategies to decrease the use of energy in laboratories are now common and the principles outlined above are shaping the trends in analytical chemistry and analytical laboratories.

There are at least three options for development:

1. The practical approach: conserving or reducing use (short- to medium-term);
2. The administrative approach: education and training (medium-term);
3. The benign technologies approach: substituting critical components (long-term).

The traditional short- to medium-term approach to reducing the consumption of energy in a process is minimizing and prevention of waste. This can yield substantial savings and can be viewed as a way to buy time while developing more sustainable approaches. However, if a sizeable investment has been made in a particular solution, for example, advanced thermal oven, this technology has a tendency to persist even if the situation has changed, the technology has advanced, or it is no longer an adequate solution to a particular problem.

Some progress has been made with regard to medium-term strategies and savings in laboratories, such as the computerized control of lab buildings. However, many of these improvements are only fully effective if the people working in the labs – the ' end- users' – are involved in the energy saving efforts. There are ways that the laboratory energy consumer can reduce the overall consumption of energy in the lab and this can be achieved by means of intensive education programs and practical training. It is a significant task for laboratory managers to educate lab researchers in the ways they can save energy, and to involve them in reducing energy consumption.

The third and long-term approach is the objective of green chemistry – that chemistry itself would generate no waste and become more effective in its use of energy. The key to sustainability is to design chemical processes, equipment and instruments that produce a minimum amount of waste and use energy only at the point of need and in only the amount necessary to achieve the reaction between the components. One approach is to design equipment and processes that operate at close to ambient conditions (room temperature and normal pressure). It also includes biomimetics; modelling products and processes after biological systems. The type of thermal energy commonly used is non-specific to the process; that is it is not directed at the chemical bond. Much of the energy is 'wasted' in heating up reactors, solvent and even the surrounding environment, instead of targeting the molecules undergoing reaction. This brings us to the second approach; using other sources of energy, such as light, microwaves, and possibly, sound.

Analytical chemistry is somewhat different from synthetic chemistry, the aim of which is to produce (new) materials. The analyst must supply information about the existence and amount of chemicals based on certain signals. Analytical signals are obtained as a result of interactions between relevant forms of energy and the levels of constituents of the sample (e.g. atoms, ions, molecules). Usually, in order to perform this interaction most effectively, an external energy system (e.g. chemical energy, radiation, heat) must be applied via some medium (solvents). The signals are measured in physical quantities, such as volume, mass and electrical charges, and in differences of temperature and radiation energy, described qualitatively and quantitatively by wavelengths or frequencies and intensities.

Analytical instruments receive signals and present them in a usable form (e.g. a data file of values, spectra, chromatograms, or pictures) for further information processing.

Many developments in analytical chemistry coincide with green chemistry and engineering, and result in energy savings:

- Replacing wet-chemistry in sample preparation and treatment;
- Searching for alternative solvents, and using solventless techniques;
- Reducing the need for derivatization by means of different methods of detection and signal processing;
- Treating samples efficiently by replacing thermal energy with alternatives (ultrasonic irradiation, microwaves, laser irradiation);
- Using photochemical activation and mechanical activation;
- Analysing samples directly without treatment, by using spectroscopy or, surface analysis, mainly using lasers as the light source;
- Decreasing the volume of the sample; miniaturization using laboratory-on-a-chip for analysis, and biosensors and immunoassays;
- Automating and using hyphenated techniques.

In general, the development of instrumental methods produces a more efficient use of energy, especially when the method is highly automated and uses a minimal amount of sample. Hyphenating several methods of sample treatment, and separating components or integrating separation and complex methods of detection can lead to a more efficient use of energy. Other forms of energy, for example, ultrasound and microwave, can replace wasteful thermal energy.

In analytical chemistry, these developments are directed towards increasing sensitivity and specificity. This can be accomplished more effectively with specialized instruments than with universal ones.

There is a rapidly growing demand for new analytical instruments in field analytical chemistry (FAC), in which, unlike traditional laboratory-based methods, measurement is performed on the site where the analyte or sample is located. The most important criterion for an ideal FAC method is a minimum requirement for power and consumables, which leads to the development of effective and sensitive portable instrumentation.

Increasing interest in process monitoring is motivated by advances in technology and the use of new materials and also as a result of challenges associated with new monitoring devices and legislation aimed at reducing pollution. To be efficient and effective, analysis should be carried out in real time, that is, the results should be obtained 'instantaneously'.

A closely related topic is that of chemical screening, which is the examination of a large number of objects (for instance, goods or even people) in order to identify those with a particular chemistry-related problem or feature. Because time and cost are more important issues for screening methods than obtaining a complete analysis, compromise is required. A simple extraction process might be more suitable than a complex system with higher recoveries, or a set of selective detectors might be more appropriate than a universal model. Seeking economical alternatives would be beneficial for monitoring programs, and effective solutions can make the monitoring process more energy efficient.

Measurement science continues to build on advances in computing and communication industry technology and on sensor-related research conducted in university and research laboratories. These developments in miniaturization, electro-optics, new materials, and information/data handling are showing promise for measurement breakthroughs using emerging nanomaterials, photonics, improved bioassay techniques, and the like. The field of nanoscience is already having an impact; nanomaterials are making significant improvements in separation science and sensing and providing potential building material for new sensors. The future of analytical technology will be influenced significantly by innovations in micro-instrumentation, and versatile microanalytical systems will provide the flexibility needed to improve and make laboratory and process control operations more economical and energy efficient.

It is necessary to consider the principles of green chemistry when designing new methods and technologies; automation and miniaturization are not ends in themselves, but rather means to increase environmental friendliness and the quality of analytical metrics. This is an important aspect of analytical chemistry. Analytical data-gathering procedures are comprised of several steps: field sampling and sample handling, laboratory sample preparation and separation, detection, quantization, and identification. All of these steps must be performed with high metrological quality, meaning that the earlier-mentioned parameters must be provided with measurement results. Whenever changes, enhancements or replacements are made in analytical methods, the aim should be to improve the metrological quality of the procedure. The application of green chemistry principles pertaining to waste reduction and the efficient use of energy must increase the quality of analytical procedures.

15.2 Economy and saving energy in laboratory practice

Laboratories are energy-intensive facilities, consuming many times more energy than the average academic building. Human comfort is not the main consideration as it is in residences and most commercial buildings. Health and safety requirements and environmental conditions for research are equally important. Laboratories are complex operations with multiple service requirements. They use large quantities of heated and cooled, one-pass air for ventilation and fume hoods, electricity to operate fans, lighting, and an enormous variety of specialized lab equipment, as well as water and process chilled water. Some lab facilities also use substantial quantities of natural gas. Due to the high air change rates, multiple mechanical and electrical service requirements, and the levels of heat gain generated by the equipment, the energy efficiency strategies commonly used in commercial buildings cannot be readily or easily applied in laboratories. Laboratory buildings are always going to be expensive to operate. Labs use four to five times more energy than classrooms or offices per square metre.

Organic, inorganic and synthetic chemistry research laboratories make extensive use of fume cupboards. These facilities have high ventilation requirements to remove solvent vapours and gaseous fumes. Analytical laboratories have less need of ventilation for containment and safety. Their research is more instrumental than practical; it involves materials testing or electronics, and generates high power loads from a variety of electrical instruments that also produce relatively high requirements for air conditioning and process water cooling. Temperature and humidity often need to be closely controlled; and low levels of airborne pollutants must be maintained.

Rapid and substantial increases in energy costs and a growing consciousness of the pending scarcity of natural resources have led to more responsible management of resources and prevention of pollution, where a reduction in energy use is most effective. Facility design improvements, including computerized control, can lead to substantial energy savings in lab buildings. Renewable technologies such as solar hot water heating, localized wind power or photovoltaic systems should also be considered. These would require careful analysis, because laboratories need reliable energy on demand, rather than the intermittent and diffuse supply that is often associated with these technologies. An auxiliary power source would be required to meet the functional requirements of the lab. These measures must be accompanied by good laboratory housekeeping, which also contributes significantly to reducing energy consumption.

15.2.1 Good housekeeping, control and maintenance

Energy conservation measures related to the economical use of equipment time and optimized procedures are closely connected with good laboratory practice. An important place to begin is with control and maintenance. The equipment must operate at design specifications, and temperature controls have to be optimized. This can easily be achieved by modern computerized digital monitoring and control systems which are designed to give early warning of malfunction or inefficient operation.

Good housekeeping is also essential. Regular maintenance and lubrication can improve the energy efficiency and extend the life of equipment. High efficiency lubricants and lightweight moving parts are good ways to reduce energy consumption. Leaking valves and pipelines should be repaired. Air and water flows must be adjusted to match demand. Freezers need to be defrosted regularly to eliminate frost build-up, and checked to dispose of old samples. Automatic glassware washers or autoclaves should be fully loaded.

Efficiency strategies for laboratories must cover three main areas, which are closely interconnected [4]:

1. ventilation systems;
2. heating and cooling systems;
3. lighting systems.

The following activities support these strategies:

- Optimizing ventilation and minimizing the areas that require high ventilation rates; using high-performance fume cupboards/safety cabinets.
- Making economical use of hoods – significant reductions in energy consumption can be achieved just by shutting their sashes, which reduces the air flow rate and energy requirements of variable air volume (VAV) fume hoods.
- Limiting the heating and cooling of lab rooms to the minimum time required; using optimum start controllers and occupancy detectors to avoid excessive out-of-hours operation.
- Promptly reporting abnormal room temperatures – room temperatures that are too hot or cool may be caused by faulty thermostats or other controls that are malfunctioning or have drifted from their set-points.
- Using shades and blinds can help keep interiors cool on sunny days.
- Matching the type of energy to process requirements and having a variety of energy sources to meet the specific demands of processes performed in the lab.
- Efficiently insulating processes that must maintain internal temperature differences.

- Using heat exchanges to harvest heat produced in one process for other processes, or for optimizing the temperature in the workplace.
- Using energy-efficient equipment with timers to automatically shut them off when not in use (i.e. at night).
- Selecting electrical devices with the Energy Star logo or other statements attesting that they have been designed to conserve energy.
- Installing occupancy and daylight sensors to control lighting; turning off lights and electrical equipment whenever they are not needed.
- Using small area lamps rather than overhead lights that illuminate the entire laboratory.
- Maximizing the use of sunlight if the design of the building allows.
- Identifying analytical equipment and processes that can be shut down when not in use; batch processing when constant operation is not necessary.
- Using timers to turn equipment on and off. Computers and associated equipment such as monitors consume substantial amounts of electricity and should be turned off overnight or placed in standby mode for shorter periods of inactivity, such as when the users are attending meetings or seminars.

Safety is the highest priority in laboratory design and operation. Energy efficiency measures should always maintain or improve safety relative to standard practice. Energy efficiency, on the other hand, should have more to do with the laboratory quality assurance program than with green chemistry. The green approach could be emphasized when undertaking challenging and measurable energy savings by introducing new, less energy-intensive analysis methods and procedures.

Programs are available to help improve the environmental and energy performance of chemical laboratories, such as Labs21®. This is a voluntary partnership program sponsored by the US Environmental Protection Agency and the US Department of Energy [5] that is dedicated to improving the environmental performance of US laboratories. Adopting the Labs21 approach will encourage laboratories to: (1) make capital investment decisions based on life cycle cost savings, (2) pursue advanced, energy-efficient HVAC (Heating, Ventilating and Air Conditioning) technologies, (3) design systems that recover and exchange waste heat and other forms of free energy, and (4) incorporate renewable energy systems.

The energy consumed by 'plug-in' equipment (freezers, autoclaves, centrifuges, etc.) constitutes from 10 to as much as 50% of the total energy use in a laboratory (not including energy associated with cooling). However, scant attention has been paid to improving their efficiency, partly because of the real and perceived lack of energy-efficient choices in laboratory equipment, and partly because of the lack of energy-consumption data on laboratory equipment. Many scientists, laboratory managers and laboratory design consultants are beginning to use energy efficiency as a selection criterion for lab equipment, and some manufacturers are starting to advertise the 'green features' of their products. Labs21 has developed a toolkit – Energy-Efficient Laboratory Equipment Wiki [6] – to share information about laboratory equipment efficiency among users, and encourage manufacturers to provide more data on the energy consumption use characteristics of their products.

15.3 Alternative sources of energy for processes

Analytical signals are the result of interactions between constituents of the sample (e.g. atoms, ions, molecules) and some form of energy. This interaction usually takes place in a gas or liquid medium. The signals are measured in electrical charges, differences of temperatures, of wavelengths or frequencies, and intensities of radiation energy. Contemporary analytical instruments generate and measure signals and present them in usable form for further information processing.

To be efficient, energy must be applied precisely at the point of need and in only the amount necessary to produce the reaction between the components. Because this reaction depends on temperature, careful temperature control is needed to generate effective analytical signals. The thermal energy that is commonly used is non-specific to the process, that is, it is not targeted at the chemical bond. Much of the energy is 'wasted' in heating up reactors, solvent and even the surrounding environment, instead of targeting only the molecules undergoing reaction. In many cases, it is possible to use other sources of energy, such as light, microwaves, or sound, on the place of action.

15.3.1 Using microwaves in place of thermal heating

The most common way to prevent heating the surrounding environment is to incorporate microwave heating into the process.

Microwaves are 1 m to 1 mm (0.3–300 GHz) in length electromagnetic radiation, and therefore have frequencies similar to radar and telecommunication devices. Household and industrial appliances are regulated and operate on fixed frequencies (commonly 2.45 GHz). Microwave energy is a non-ionizing type of electromagnetic radiation that causes molecular motion through the migration of ions and the rotation of dipoles without altering the molecular structure. Because the mechanism by which microwave energy is absorbed is complex and varies for different substances, it is not a universal substitute. The absorbed energy is transferred to molecular kinetic energy and the sample heats up almost instantaneously. Substances that do not have a dipole moment (or in which one cannot be induced) cannot be directly heated by microwaves. Because of the large distances between molecules, gases also cannot be heated by microwave radiation. Nonetheless, microwaves are a more efficient source of heating than conventional thermal heating because energy is imparted directly to the medium in which the process takes place, rather than through the walls of a vessel.

The use of microwave ovens in laboratories is common and microwave–assisted sample preparation techniques are used in analytical laboratories all over the world [7, 8]. The latest advances in the application of microwave techniques to various fields of analytical chemistry include: sample digestion for elemental analysis, solvent extraction, sample drying, measurement of moisture, analyte desorption and adsorption, sample clean up, chromogenic reaction, speciation and nebulization of sample solutions [9].

15.3.1.1 Accelerated extraction

Elevated temperatures produce high efficiency extraction as a result of increased diffusivity of the solvent into the interior of the matrix and enhanced desorption of the components from its active areas. The effect of microwave energy is greatly dependent on the nature of both the solvent and the matrix. Usually, a solvent is chosen that strongly absorbs microwave energy. However, in some cases (for thermolabile compounds), the microwaves are absorbed only by the matrix, which heats the sample and releases the solutes into the cold solvent.

Analytical equipment for microwave-assisted sample preparation can be classified according to: the type of microwave energy (multi-mode in cavity or focused by waveguide) applied to the sample, the subjection or lack of subjection to overpressure (from open atmospheric to closed pressure vessels of more than 80 atm), or the approach – dynamic (a flowing stream of mixture passes through a microwave field) or static (the stream of mixture is stopped when it is contained within a microwave field). Closed vessel systems (under controlled pressure and temperature) are generally of the multi-mode type (the microwave radiation is dispersed into a cavity where the sample is located, so that it is randomly irradiated) and open-vessel systems (under atmospheric pressure, open or closed to the atmosphere) are of the focused type (the microwave radiation is focused on a restricted zone where the sample is placed).

The unique feature of MW extraction is that components can be extracted selectively either by selective heating of a phase or single component of a system (for example, free water molecules in vegetable cells), or by choosing an organic solvent which targets a particular component. Pressure is an important variable in closed systems; however, it is directly dependent on temperature. The high temperatures MW heating can produce in closed systems may significantly change the properties of many solvents and they can even reach supercritical conditions.

Microwave accelerated extraction (MAE) uses organic solvents or mixtures of solvents for extraction. The most typical are methanol, isopropanol, and mixtures of hexane and acetone. Using aqueous solutions and non-ionic surfactants as extractants is also possible. Ionic liquids are attractive media for MW radiation processes because their ability to adsorb MW radiation energy is extremely high. MAE (mainly in its focused energy form) is currently considered a good alternative to traditional Soxhlet extraction for different analytes in solid samples because it reduces extraction time (e.g., 20–30 min per batch of as many as 12 samples), uses small amounts of solvents (30 ml in MAE versus 300 ml in Soxhlet extraction), and improves extraction yields. The EPA has approved MAE (Method 3546) for extracting organic compounds from solid environmental samples [10].

The benefits of microwave-assisted solvent extraction include:

- Fast extractions that can be completed within minutes;
- Lower solvent usage, which reduces the cost of purchasing and disposing of solvents;
- Higher recovery of analytes;
- Simplicity, which precludes the need to modify the existing chemistry;
- Precise computerized control of all process parameters, which enables reproducibility;
- Using microwave ovens for organic extraction elevates sample preparation to the same level of sophistication as sample analysis.

However, MAE does have several drawbacks: the extract must be filtered after extraction and almost always requires clean-up (because MAE is very efficient), polar solvents are needed and the equipment is moderately expensive. Applying microwave energy to inflammable organic compounds can pose serious hazards; therefore, it is strongly recommended that only equipment approved for laboratories be used.

15.3.1.2 Assisted digestion

Conventional wet-sample preparation methods for decomposing solid samples to analyse the content of metals are usually carried out in vessels containing the sample and a large volume of digestion reagent(s), typically 15 to 100 ml of nitric acid, pure or combined with hydrochloric or hydrofluoric acid, which is heated for a long time using a hot plate, heating mantle, or oven. This type of open-vessel digestion has many drawbacks, including the use of large volumes (and multiple additions) of reagents, a significant potential for contamination of the sample by other materials and the laboratory environment, and the exposure of the analyst and the laboratory to corrosive fumes.

Closed-vessel microwave digestion uses a substantially different technology to decompose samples [11]. Most solid samples can be decomposed using near stoichiometric quantities of reagents, typically 10 ml and can usually be completed in 10 to 15 minutes. This decrease in sample preparation time can be attributed to the higher pressure in closed vessels and rapid heating of the sample mixture. The higher temperatures, which are achieved almost instantaneously, give microwave digestion a kinetic advantage over hot plate digestion. In a closed vessel system, it is possible to retain even volatile elements. Digestion is more complete because many acids have greater oxidation potential at elevated temperatures. Effective dissolution procedures depend on the specific temperature profile and the reagents. This is important not only for complete dissolution, but also for reproducible extraction (leaching), analyte solubility, analyte volatility, species stability, and most importantly, safety. The most commonly used digestion reagents are nitric acid, hydrochloric acid, hydrofluoric acid, sulfuric acid, perchloric acid and hydrogen peroxide. The analyst must evaluate the reactivity of the acids with each specific matrix because every matrix has unique chemical interactions.

A similarly effective modification of procedure using microwaves is employed in the sample introduction step of atomic spectrometry. Thermospray (TSP) has recently attracted more attention because it generates a hot and very fine primary aerosol, which enables a high percentage of analyte to reach the atomization cell. TSP thus performs much better than conventional pneumatic nebulizers in terms of sensitivity and limits of detection for most elements. When TSP with adapted thermal conduction is used for heating a quartz or stainless steel capillary to generate aerosols, a heat gradient occurs from the capillary wall to the centre of the liquid stream. If the liquid contains a relatively large amount of analyte, or the temperature is not controlled accurately, the analyte will be deposited on the capillary wall. For this reason, it is difficult to use a thermospray technique to analyse slurries or sample solutions with a high salt content. If the quartz capillary is heated with microwave radiation, there is no heat gradient from the wall to the centre of the liquid stream and a minimum amount of analyte will be deposited on the wall. Moreover, the solvent vapour will combine with the remaining liquid, and theoretically, the droplet diameter distribution of the aerosols will be much narrower. In most studies of TSP, the capillary diameter must be very small (25–200 mm) to avoid localized overheating, whereas with microwaves, the capillary diameter can be up to 1 mm. Therefore, microwave thermospray nebulization (MWTN) is effective at a lower liquid pressure, and can be used to analyse sample solutions with high salt content or even slurries [12].

15.3.1.3 Drying, assisted desorption and adsorption

Microwave drying is a simple method for removing volatile compounds from a sample which is faster and consumes less energy than hot air drying. Water is usually the main volatile compound that has to be removed from a sample. Because microwave radiation energy is directly absorbed by water, heating and evaporation occurs much more quickly than under standard conditions. As the moisture is removed, heating decreases and the temperature of the sample drops, preventing overheating. Microwave-assisted drying can also be done at reduced pressure; water (or other microwave-absorbing solvents) evaporates at lower temperatures, so there is less decomposition or loss of volatile organic compounds and formation of localized hot spots. Microwave drying has been also used to dry precipitates in gravimetric analysis. The advantages of microwave drying under vacuum conditions include the fact that it is a milder thermal mode of drying, vapours are removed more rapidly, reducing the analysis time, and the degradation or oxidation of samples is minimized. Traditional oven- or air-drying techniques are time- consuming (8 h to a few days), whereas microwave vacuum drying of sediments takes less time and processes more samples. The same type of procedure can be performed after digestion, to evaporate the solvent and concentrate the sample for further analysis. In recent years, microwave drying has gained popularity as an alternative method for drying marine sediments and a variety of food products, such as fruits, vegetables, snack foods and dairy products. Some precautions must be taken with microwave drying because of possible interactions between microwave energy and the sample.

Microwave radiation is not extensively used for preparing samples for bio-analysis and it has some drawbacks. Until the mechanism whereby microwave radiation mediates an alternative catalytic reaction is better understood, it is difficult to predict which biochemical and analytical processes will benefit from microwave energy.

15.3.2 Using ultrasound in sample treatment

Ultrasonic irradiation can also save significant amounts of time and energy in several sample treatment procedures. Ultrasound (US) is mainly used in analytical laboratories to improve or accelerate the preliminary steps of the analytical process. US aids in conditioning the sample (e.g. mixing or solubilizing) and transporting it to detectors (e.g. with US nebulization, levitation or slurry formation) [13]. Ultrasonic equipment operating in the range of 20–100 kHz is relatively inexpensive and readily available. Some high frequency (500 kHz) equipment has also been used for more fundamental mechanistic investigations.

Ultrasound energy causes acoustic cavitation – bubble formation and implosion – in solutions. The collapse of these bubbles generates extremely high pressures and temperatures at the interface between the bubble and another phase. Moreover, the extremely high effective temperatures and pressures at the interface between an aqueous solution and a solid matrix, combined with the oxidative energy of hydroxyl radicals and hydrogen peroxide created during the sonolysis of water, enhances chemical reactivity. Acoustic cavitation thus provides a unique interaction between energy and matter and more extractive power is obtained from the solvent.

15.3.2.1 Accelerated extraction and leaching

Ultrasonic radiation is another alternative for accelerating some steps in solid sample pre-treatment, such as dissolution, fusion and leaching. Cavitation from acoustic vibrations facilitates the removal of analytes from the matrix surface and creates microenvironments with high temperatures and pressures. Sonication-assisted extraction (SAE) is faster (5–30 min per sample) than the Soxhlet mode and allows large amounts of sample to be extracted with lower energy consumption. The same range of solvents is suitable for SAE as for Soxhlet extraction. SAE is usually conducted statically in discontinuous batch mode; however, SAE can also be dynamic (DSAE). DSAE has several advantages over conventional batch mode, such as continuous exposure of the sample to fresh solvent, which improves extraction kinetics. Furthermore, post-extraction filtration and rinsing are eliminated and solvent consumption and the risk of loss and/or contamination of the extracted species during manipulation is reduced. Ultrasonic leaching of metals from sediments, although not yet sufficiently exploited, could be an attractive alternative to conventional, acid bomb and microwave digestion; except for the time required for digestion, unloading reactors is faster in the absence of high temperature and pressure. The type and concentration of acid in the liquid extractant is the most critical parameter for ultrasound leaching. Unfortunately, SAE uses as much solvent as Soxhlet extraction and filtration is required after extraction. Moreover, it is labour-intensive because apart from the polarity of the solvent, the efficiency of the extraction is dependent upon the nature and homogeneity of the sample matrix, the ultrasound frequency and the sonication time. The EPA has approved sonication-assisted extraction as Method 3550C [14].

Ultrasound has been used to overcome the effects of electrode passivation in electrochemical systems, allowing sensitive electroanalysis to be carried out in hostile media such as blood, effluents, foodstuffs and fuels [15]. Mass transport is greatly increased via acoustic streaming and micro-jetting that results from cavitational collapse on the surface of the electrode. In addition, surface activation via cavitational erosion can be used to activate otherwise passivated electrodes. Therefore, the coupling of power-ultrasound with electroanalytical techniques dramatically improves analysis times, detection limits and electrode lifespan. Sono-electroanalysis methods have been developed using readily available and relatively inexpensive equipment that can be tailored to specific analytical situations [16].

15.3.2.2 Assisted matrix solid-phase dispersion

Matrix solid-phase dispersion (MSPD) is a form of microextraction for both liquids and solids. For liquid samples, the sorbent is dispersed in the sample to absorb the analyte. The sorbent is then separated from the liquid and transferred to a column and then the analyte is desorbed thermally for further analysis. For solid or semi-solid samples, a suitable solid-phase material (e.g. derivatized silica, silica gel, sand or Florisil) is manually blended with the sample. The solid support and sample are then transferred to a column and eluted with an appropriate solvent. In MSPD, sample disruption and dispersal onto the particles of the support material takes place in a single step. The most important factors for this kind of dispersion extraction are the characteristics of the solid-phase material (e.g. particle size), the ratio of sample to support material, the use of chemical modifiers (e.g. acids, bases, or chelating agents), the elution solvents and volume. The MSPD technique is particularly well suited to biological and plant-derived samples. Its main advantages are the ability to process very small amounts of sample, and low solvent consumption.

Ultrasonic-assisted matrix solid-phase dispersion (UA-MSPD), is a method that has been developed for extracting and removing pesticides (OPPs) and triazines from a variety of fruits [17]. Ultrasonic energy has been found to accelerate and improve extraction efficiency during solid–liquid extraction of pesticides. Compared with conventional MSPD, the proposed method improves extraction efficiency, decreases the dispersion of results and completes sample treatment within a few minutes. These results, characterized with low limit of detection (LODs), indicate that the procedure is suitable for accurate determination of target analytes at levels specified in current legislation, for amounts of sample as small as 100 mg [18].

15.3.3 Light as a source of energy

Light is a clean reagent that leaves no traces; therefore, organic photochemistry is considered a green chemical approach [19]. The most abundant supply of energy on Earth comes from solar radiation – for all practical purposes, a perennial source that will continue to exceed every conceivable need of mankind (25 000–75 000 kWh per day and hectare). Human beings directly use only a small fraction of this energy. Green plants use solar energy and photocatalysis occupies a central place in ecological equilibria. A variety of artificial synthetic methods have been developed in which the photon is used as a chemical reagent, but these have no direct connection with processes occurring in nature. Photons play the same role as activating reagents and/or catalysts in organic synthesis; they introduce alternative paths so that the reaction can occur in cold temperatures under mild conditions. Furthermore, contrary to chemical activating agents, light leaves no spent reagent or residue in the reaction mixture and thus simplifies the work-up. Despite this advantage, synthetic organic photochemistry has been widely ignored by the chemical industry, mainly because most artificial light sources require large amounts of energy to generate light. However, light-emitting diodes (LED) could improve this situation and they are available in a wide range of emission wavelengths, facilitating the selection of the best match between the source and reagent absorption. Researchers are currently studying the late nineteenth century origins of organic photochemistry, when sunlight was the only available source of radiation [20]. In contrast to the early, often improvised equipment, solar photochemical reactions are now performed with modern solar reactors and moderately to highly concentrated sunlight, which produce high time-space yields [21]. This can therefore be considered an energy-saving process. However, solar energy is not used directly in analytical methods because of its rather low intensity and wide spectral window.

Light and other sources of electromagnetic radiation are the main means of producing signals in analytical chemistry. Other sources of light than solar are widely used in analytical chemistry; spectroscopical methods are the most commonly selected for detecting and identifying substances. Laser-based light has affected a major revolution in spectroscopy. It is giving rise to new analytical techniques based on laser-induced plasma generation and related techniques of spectroscopic analysis. Energy is conserved by heating very small amounts of sample in a very short time. When a short-duration laser pulse with efficient energy density is focused on the surface of a target material, the surface temperature rises to 5000–6000 K in an extremely short time (~7 ns), resulting in the formation of very small volumes of plasma (microplasma) in which the breaking of chemical bonds produce electronically excited atoms and ions, regardless of the type and shape of the material [22]. These excited species in the plasma can almost all be used for analysis purposes. Laser-induced plasma spectrometry, or laser-induced breakdown spectrometry (LIBS), as it is more commonly known, has the following advantages: (1) simple to nonexistent sample preparation and lack of contact with the sample, (2) minimal consumption and destruction of the sample, (3) high sensitivity and selectivity for trace analysis, (4) rapid analysis, (5) multi-element analysis, and (6) the potential for process analysis. In typical LIBS measurements, multiple laser pulses are formed and collected, and an ensemble-averaged method is applied to the spectra. LIBS is performed using a single laser shot, making it ideal for applications that require rapid chemical analysis and/or non-destructive measurements. In single-shot LIBS, one laser pulse forms a single plasma for spectral analysis. By capturing the light from the microplasma within the 200–980 nm range, elements are identified by their specific wavelengths, quantities of chemicals are measured by the intensity of light at specific wavelengths and all matter is detected, because every element emits somewhere in the 200–980 nm range. LIBS is inherently sensitive and because of the short response time, it is considered a real-time method. Furthermore, LIBS detects small particles, especially when the size of the particle is similar to the 20–30 micron diameter of the focused laser beam.

Single-shot LIBS shows great promise for the chemical analysis and identification of a wide range of types of sample. Single-shot LIBS has proven potential for analysing cultural heritage, biological, environmental and forensic samples, as well as explosives, military applications, metals and metal alloys, combustion analysis and catalytic converters. The advantages of single-shot LIBS include rapid chemical analysis and the preservation and mapping of samples. The most significant drawback of this technique is the reproducibility of the spectral line intensity, which can be affected by numerous factors. Shots may vary as a result of the laser source (e.g. pulse instability), interactions (e.g. laser pulse–plasma, plasma–sample and laser pulse–sample), the amount of sample ablated, characteristics of the sample (e.g. roughness, chemical composition and homogeneity), matrix effects, environmental factors (e.g. presence of particulates or moisture) and light scattering. These shot-to-shot variations make it difficult to use single-shot LIBS for quantitative work.

Matrix-assisted laser desorption/ionization (MALDI) is a similar, soft ionization technique used in mass spectrometry. It facilitates the analysis of biomolecules (such as proteins, peptides and sugars) and large organic molecules (such as polymers, dendrimers and other macromolecules), which tend to be fragile and fragment when ionized by more conventional methods. Ionization is triggered by a laser beam (normally a nitrogen laser). A matrix is used to protect the biomolecule from being destroyed by a direct laser beam and to facilitate vaporization and ionization. MALDI techniques typically employ the use of UV lasers (337 nm, 355 nm and 266 nm). Infrared lasers are also sometimes used because of their softer mode of ionization.

When the laser is fired, various processes take place. The matrix absorbs and, it is believed, is primarily ionized by the laser energy. In theory, the matrix then transfers part of its charge to the analyte molecules, thus ionizing them while protecting them from the disruptive energy of the laser [23]. Ions resulting from this process consist of a neutral molecule [M] and an added or subtracted ion. MALDI is capable of producing single-charged ions, but multiple-charged ions ($[M^+nH]n^+$) can also be created, as a function of the matrix, the laser intensity and/or the voltage.

Despite their high efficiency, these instruments are complicated and expensive, and not every laboratory can afford them.

15.4 Using alternative solvents for energy savings

More regulation of industrial solvents and waste treatment, and increasingly stringent pollution-control legislation have increased the attractiveness of alternative, non-toxic, environmentally acceptable industrial solvents. Energy-intensive separation techniques, such as distillation, are also becoming more expensive because of rising energy costs.

There are two classes of solvents – supercritical fluids and ionic liquids – that have the potential to make processes environmentally safer and to reduce energy consumption by replacing organic solvents in liquid/liquid extractions, catalysis and separations. Using them would reduce or eliminate the costs, disposal requirements, and hazards associated with volatile organic compounds (VOCs). Their main advantage, compared with common solvents, is an ability to fine-tune the properties of the solvent media, which lets them replace specific solvents in a variety of different processes. Improving process efficiency is the way to reduce energy consumption. Furthermore, the unique properties of both liquids have led to new fields of research: supercritical CO_2 and ionic liquids operate together because ionic liquids do not dissolve in supercritical fluid; however, the fluid is miscible with ionic liquid and this allows supercritical fluid to be used to extract compounds from ionic liquid [24].

15.4.1 Advantages of ionic liquids

Ionic liquids have two main advantages: non-volatility and thermal robustness. Most of them operate in a wide thermal range (typically –40°C to 200°C) with no measurable vapour pressure. This permits a high level of kinetic control and eliminates solvent vapours. Negligible vapour pressure is the main support for the claim that ILs are a 'green' solvent.

The properties of ILs (fluidity over a broad temperature range, extremely low vapour pressure, miscibility or non-miscibility with water and other solvents, ability to dissolve many organic and inorganic substances, electrical conductivity, electrochemical stability, lack of inflammability, explosiveness, non-coordination and lack of corrosivity) can be tuned by an appropriate choice of anion and cation. This attribute can be called the 'chemical tunability' of ionic liquids – a class of solvents that possess similar physical properties but different chemical behaviour. There are ILs with specific functional groups, which has given rise to the identification of task-specific ionic liquids (TSILs) and gained wide attention.

In addition to the advantages provided by their physical properties, ILs usually produce higher reaction rates, higher yields, and better selectivities than conventional organic solvents when used in organic synthesis. They have been used in a large number of reactions, especially in catalysis, and particularly in biocatalysis and biochemical technology because many enzymes retain their activity in these solvents. The application of ILs – often motivated by the perception that they are 'green' solvents – has been tested in the chemical, petroleum and allied fields as reaction media, used for electroplating, acid-gas scrubbing and desulfurization of transportation fuels. Replacing volatile solvents with ILs is expected to prevent the emission of VOCs.

The environmental impact of the life cycle phases of ILs relative to alternative methods has not yet been studied. A life cycle assessment (LCA) is essential to legitimatize any claims of 'greenness'. The same statement can be made with regard to energy consumption. More proof is required that analytical processes using ionic liquids as media, reagents or catalysts will be faster and consume less energy to generate measurable analytical signals. Research is also being carried out to find an effective use for ionic liquids in analytical applications [25].

15.4.2 Using subcritical and supercritical fluids

Tunability is important aspect of alternative solvents like supercritical fluids that makes them attractive to researchers. Small changes in temperature and/or pressure around the critical point of supercritical fluids cause significant changes in density and other physical parameters that make it possible to tune the solubility and other properties of the solvent. Therefore, using solvents in their supercritical condition expands the overall spectrum of solubility, polarity and volatility of solvents and mobile phases. The ability to fine-tune the properties of the solvent medium makes it a good candidate to replace other solvents in a variety of different processes, or to create new methods for analysing samples. The same solvent can be used for different applications and procedures. However, appropriate instrumentation that can generate the pressure and temperature required to reach the supercritical state (for example, for CO_2, the critical temperature is 31°C and the pressure is 73 bar) is needed for working with supercritical fluids. Their pumps and heating systems consume energy, making these instruments somewhat energy-intensive. The physical properties of the fluids also limit their application. The polarity and solvating power of carbon dioxide, the most popular fluid, are too low for many common analytes. This substantially limits the use of supercritical CO_2 as a method that could universally replace Soxhlet extraction. However, supercritical extraction (SFE) of non-polar analytes is very rapid and efficient, reducing energy consumption and the use of volatile organic solvents.

Supercritical fluid chromatography (SFC) using liquid carbon dioxide as an eluent is the most popular green alternative to traditional liquid chromatography (LC). It is used mainly as a preparative mode in chromatography. Supercritical fluid chromatography typically uses carbon dioxide above or near its critical temperature and pressure bar, combined with an organic modifier such as methanol or ethanol. Most of the available SFC analytical instrumentation is also LC-compatible. The pumps and mixing chambers are already at high pressure; however, the pump heads must be cooled. The major alteration is the need to keep the UV detector at high pressure. SFC provides faster separations than LC and uses minimal amounts of organic solvents. SFC mobile phases are carbon dioxide-based, but may contain up to 15% ethanol, methanol or acetonitrile. Theoretically, mixed-phase SFC solvents are less environmentally benign than single-phase CO_2, but they are significantly easier to dispose of or recycle than mixed organic-aqueous LC solvents. Capillary SFC provides high-resolution chromatography at much lower temperatures than gas chromatography and facilitates fast analysis of thermolabile compounds [26]. This type of SFC has been viewed as an extension of gas chromatography (GC) in which some of the thermal energy required for mobilizing the solutes is replaced by solvation energy. The use of capillary SFC grew swiftly, mainly because of the novel combination of supercritical mobile phases and open tubular fused silica column technology. GC detectors such as flame ionization, electron capture, nitrogen phosphorus, and sulfur chemiluminescence were popular with SFC.

SFE and SFC are particularly advantageous when they use a carbon dioxide-based phase. Carbon dioxide gas does cause global warming but the carbon dioxide used in these procedures contributes no new chemicals to the environment, because it is usually a by-product of other chemical reactions or is obtained directly from the atmosphere, and the gas is usually recycled when it is used in large quantities.

In liquid chromatography, when elevated temperatures are required for efficiency, using only aqueous streams is often proposed. S. Heinisch and J-L. Rocca recently reviewed elevated temperature liquid chromatography [27]. They found that, in addition to the non-toxicity advantage of superheated water, as discussed above, it is also possible to use flame ionization, low wavelength UV, inductively coupled plasma mass spectrometry and even nuclear magnetic resonance spectroscopy detectors, with hot deuterium oxide as the mobile phase [28, 29, 30]. Temperature programming instead of gradient elution could be practical in some situations. For example, for separations using micro- or nanocolumns, when gradient elution is difficult to operate at very low flow-rates in a particular system, temperature programming is generally recommended with the capillary columns because heat transfer is more effective with small inner diameters. However, temperature programming is sometimes problematic. A wider temperature interval than ambient to 200°C is usually required for programming. Commercially available equipment is not yet able to cover such a wide range of temperatures. The required temperature ramp is often too steep (>20°C min^{-1}) to be applied to HPLC separations because heat transfer in liquids is very slow compared with heat transfer in gas. Heating a steel HPLC column (a material with very low thermal conductivity) can be too slow as well. Temperature programming is mainly used to improve isocratic isothermal analysis rather than to replace gradient elution.

A frequently overlooked aspect of elevated temperature chromatography is the fact that it consumes more energy. G. van der Vorst *et al.* [31] performed an *exergetic* life cycle analysis of a chromatographic separation of enantiomers of a racemic mixture of phenyl acetic acid derivatives in order to compare preparative HPLC with SFC. Exergy refers to the maximum work that is required to bring a system into equilibrium during a process [32]. The energy required to produce a chromatogram can be substantial for both HPLC and SFC. Considering instrumentation alone (in their terminology – 'α system boundary'), the exergy consumption of a preparative HPLC technique is about 25% higher than that of SFC, due to its inherently higher use of organic solvents. To determine the exergy consumption of a plant one must consider its capacity ('β system boundary') and perform calculations taking into account the physical boundaries of the plant and the resources passing those boundaries. These include the resources purchased by the pharmaceutical company to perform the separation and produce the racemic mixture, the cooling and industrial water, the cooling and heating medium, steam and the cost of storing the product. Preparative SFC appears to be more advantageous because it requires about 30% less resources than preparative HPLC as quantified in exergy. However, the cumulative exergy extracted from the natural environment to deliver all the mass and energy flows to the α and β system boundaries via the overall industrial network (γ system boundary) reveals that preparative SFC actually requires about 34% more resources than preparative HPLC. The astonishing conclusion of the work [31] is that the most sustainable process in terms of integral resource consumption is preparative HPLC. The authors reason that the electricity required for heating and cooling and the production of liquid CO_2 argues against the use of preparative SFC. Therefore, the actual greenness of elevated temperature chromatography cannot be proven until a thorough exergy life cycle analysis has been performed. It is hoped that exergy calculations will progress from academic exercises to organizational policy issues.

15.5 Efficient laboratory equipment

D. Raynie [33] has divided laboratory equipment into three categories, based on the energy it consumes per sample (see Table 15.1)

The energy assessment Table 15.1 indicates that wet chemistry methods such as titration and those based on biochemical assays will be in the 'green' area. Simple instrumental methods such as GC and, HPLC are in the 'yellow' zone. More complicated and combined instrumental methods are energy intensive and they are therefore in the 'red' column. High sample throughput is clearly important because it reduces energy consumption per sample. Solvents create additional energy requirements. Solvent evaporation is directly related to energy use, and a method that requires the evaporation of high volumes of solvent is unfavourable.

Spectroscopic methods require remarkably little energy. Many of them are reagentless, non-destructive, fast and able to determine several analytes in the same run. For these reasons, they are environmentally friendly and can be considered green analytical methods [34]. Unique analytical characteristics, such as high sensitivity,

Table 15.1 *Instrument ranking according to energy consumption.*

≤ 0.1 kWh per sample *green*	≤ 1.5 kWh per sample *yellow*	> 1.5 kWh per sample *red*
SPE (vacuum assistance)	Hot plate solvent evaporation is less than 2.5 h with a green instrument	Hot plate solvent evaporation is more than 2.5 h with a green instrument
Rotavap	ASE	
SPME	SFE	Hot plate solvent evaporation is more than 1 h with a yellow instrument
Sonicator	Ultrasound probe	
FTIR	Microwave	Soxhlet
UV-Vis spectrometer	AA spectrometer (flame or furnace)	
Fluorescence spectrometer	ICP-MS	NMR
UPLC	GC	GC-MS
Titration	LC	LC-MS
Immunoassay		X-ray diffractometer
Microbiological assay		
Hot plate solvent evaporation is less than 10 min	Hot plate solvent evaporation is less than 1 h with a yellow instrument	

high spectral resolution, and very low limits of detection, can be achieved by using lasers. However, their major drawbacks include the cost of the instrument, increased complexity and the fact that not all wavelengths are available, especially simultaneously. However, miniaturizing the detectors and light source (e.g. diode lasers) has increased the use of spectroscopic methods in various sensors and lab-on-a-chip applications.

15.5.1 Trends in sample treatment

Green chemists are paying close attention to sampling, especially sample preparation, which often involves clean-up and pre-concentration steps because it requires solvents and other chemicals that generate large amounts of waste. The handling of solvents also consumes considerable energy.

Typical sample preparation processes may include homogenization and filtration, centrifugation, distillation, extraction, fractionation and concentration. The successful execution of these processes will ensure that the analyte is in a form compatible with the analytical system. Sampling and sample pre-treatment usually account for over 60% of the total analysis time and the quality of these steps largely determines the success of an analysis from a complex matrix. Therefore, selecting and optimizing a sample preparation plan greatly affects the reliability, accuracy and material and energy consumption of an analysis.

According to the principles of green chemistry, new methods of sample preparation must:

- Significantly reduce the use of organic solvents;
- Reduce vapour emissions and the generation of liquid and solid wastes;
- Eliminate the use of highly toxic or ecotoxic reagents; and
- Reduce labour (time) and energy consumption.

Preparation of the sample must be tailored to the analysis and is dependent on the sample matrix and the type and amount of analytes. The selectivity of the sample preparation and exhaustiveness of the extraction are often key factors in target-compound analysis. The speed of extraction, complexity of the instrumentation, simplicity and flexibility of the method and ruggedness of the process are crucial factors for energy consumption.

At-line or on-line coupling of the extraction and clean-up steps with the separation process is an important consideration in the development of sample preparation methods; miniaturization has been a key factor in designing integrated analytical systems to provide higher sample throughput and/or automated operation, to make more economical use of solvents and chemicals and to conserve energy.

The possible approaches include solventless methods, microextraction processes and new solvent systems. Solventless methods of sample preparation can be categorized as gas, membrane-based, and solid-phase extraction [35].

The main analysis and identification of volatile analytes in gaseous media is performed on a gas chromatograph. Extractions using supercritical gases like CO_2 were discussed under the heading of alternative solvents. Membrane extraction is a solventless method when the analytes are volatile and the accepting media is a gas phase; otherwise, it is a form of liquid-liquid extraction. The solid-phase extraction (SPE) technique is derived from liquid chromatography and in recent years, SPE has gradually replaced most liquid-liquid extraction (LLE) methods as the preferred technique, especially for biological samples prior to quantitative analysis [36]. The analyte is then desorbed from the sorbent into the analyser inlet by means of an appropriate solvent or thermally (the true solventless approach). SPE reduces, solvent consumption and costs and thus the process can be automated. In the most popular microextraction system, a fused silica fibre is coated with sorbent and the analytes are partitioned between the sorbent rod and the matrix phase [37]. The method is simple and can be readily automated. It was the starting point for the development of a series of new sample preparation methods.

SPE, in all its modes and formats, is arguably the preferred method for preparing liquid samples, especially those containing more polar analytes. On-line and automated solid-phase extraction–liquid chromatography (SPE–HPLC) is a fully mature technique that can easily be miniaturized. In addition to the effectiveness of the extraction technique, many other factors can affect the analytical plan [38]. New alternative solvents (fluorous, polymeric, ionic liquids etc.) can make conventional liquid–liquid extraction systems more efficient by reducing the volume of solvent required. However, the equipment and operating costs, complexity of the method, volume of organic solvent and level of automation must also be considered.

All of the above processes consume additional energy to conduct the analysis, for such requirements as using additional equipment or processes, evaporating solvent, drying the sample, and so on. Traditional sample preparation techniques such as liquid–liquid extraction (LLE) and Soxhlet extraction (SOX) rely on solvents for extraction. They often use large amounts of organic and chlorinated solvents, and have long extraction times (usually 6–24 h), which makes them unfavourable from an environmental and energetic perspective. Pressurized–liquid extraction (PLE) in its static (SPLE) and dynamic (DPLE) forms was developed to replace traditional sample pre-treatment techniques. It is evident that applying high temperatures to conserve energy shortens the time and increases the efficiency of the process. PLE, also known as pressurized fluid extraction (PFE), pressurized solvent extraction (PSE), accelerated solvent extraction (ASE) and enhanced solvent extraction, generates faster extractions because of higher diffusivity, improved solubilization capability and more efficient analyte interactions in liquid solvents at temperatures above their boiling points. PLE provides quantitative extractions with less solvent consumption and substantially shorter extraction times than the Soxhlet method. For a number of organic trace pollutants in various solid and semi-solid matrices, recoveries with PLE are comparable or even better than those using Soxhlet extraction. Solvents similar to those used for Soxhlet extraction usually yield good results, so it is relatively straightforward to replace old methods with PLE. Mixtures of low-polar and high-polar solvents generally provide more efficient extractions of analytes than single solvents. The temperatures are usually in the range of 60–200°C. Except for labile analytes or samples, higher extraction temperatures will increase the efficiency of PLE as a result of enhanced sample wetting and better penetration of the extraction solvent, and also because of higher diffusion and desorption of the analytes from the matrix to the solvent. It is possible to use purely aqueous solvents in PLE, (i.e. pressurized hot water extraction (PHWE), also called sub-critical water extraction (SWE), hot-water extraction, high-temperature water extraction, superheated water extraction and hot liquid-water extraction). This technique uses water as an extraction solvent at temperatures of 100–350°C and at pressures high enough to keep it liquid. Typical extraction times for PLE are 10–30 min, which substantially reduces the energy requirement per sample. However, PLE instruments can seldom be used for PHWE because the sealing materials do not withstand the high temperatures.

15.6 Effects of automation and micronization on energy consumption

As mentioned above, the demand for chemical analysis is growing rapidly, especially for environmental purposes. Real-time field measurement capability is needed for continuous environmental monitoring on land and – even more importantly – in oceans and seas. Process analytical chemistry, for obtaining analytical data on large-scale production operations, is another swiftly expanding discipline [39,40]. These requirements are putting pressure on labs to increase their throughput and shorten response times to produce data for decision-making. In addition, laboratories must be able to determine whether the planned procedure will be unique or used routinely. There is growing demand for miniaturized measurement devices, scanning sensor devices and rapid data handling with high-throughput and intensive experimentation platforms. The objectives are changing from simply proving the feasibility of a chemical reaction to undertaking more in-depth scientific studies and industrial piloting. The quality of the data must simultaneously be maintained or improved.

The major factors for laboratory efficiency are equipment costs, operating costs including energy, solvents, chemicals and consumables, complexity of the methods, and level of automation. These factors determine the cost per analysis. To achieve the highest possible gains in productivity, the laboratory process flow must be carefully studied to identify the points at which automation would have a measurable impact. Sensitive instruments with high productivity will improve the overall efficiency of the laboratory. Laboratory automation can eliminate defects and errors in laboratory processes, and it can often be a way to minimize human errors.

The high costs of instrumental analysis may prohibit more extended studies of environmental screening and field monitoring. Screening chemical residues is not easy because the methods must be capable of achieving extremely low limits of detection (LOD). The task is also complex, because even when low LOD values can be achieved, they vary widely depending on the analyte, the sampling technique, and the sample preparation and matrix, which is especially important for food samples. It can also be difficult to achieve the levels of sensitivity required for all the analytes with one pre-treatment method.

Legislation requires many different compounds and matrices to be monitored. Despite considerable activity in this area of research, there are still many gaps in government-mandated laboratory monitoring plans for certain matrices and residues. For this reason, multi-residue ‘catch-all’ methods, or even combinations of methods for different drug residues using definitive techniques such as LC–MS, are highly appealing in terms of their high throughput and sensitivity, and ability to meet legislated requirements. However, they are expensive, and they require solvents and chemicals for sample pre-treatment and qualified specialists to operate the laboratories.

A reasonable balance must be found between laboratory and field experiments. Field instrumentation in particular needs serious improvement. Ideal analytical field instruments should (1) be able to produce the required information on a real-time or near real-time basis, (2) be capable of in situ/at-site analysis and have little or no need for sample preparation, (3) be portable for field use with a minimum requirement for power (battery power is desirable), consumables such as gases or/solvents, and clean space for handling samples and (4) be sufficiently sensitive and selective.

15.6.1 Miniaturization in sample treatment

Reducing the use of solvents is the main force driving the development of sample treatment methods. Fortunately, solvent reduction is almost always accompanied by decreased energy consumption. Miniaturization is one way to reduce solvent and energy use. Several attempts have been made to miniaturize liquid-liquid extraction (LLE). Three of them will be discussed: single-drop microextraction (SDME), liquid-phase microextractions (LPME) and electro membrane extraction (EME).

In SDME, a microdrop of solvent (ca. 1–3 μl) is suspended from the tip of a conventional microsyringe and then exposed to the headspace of the sample or immersed in a sample solution in which it is immiscible [41, 42]. The analytes diffuse into this droplet. Once the extraction is complete, the drop is retracted into the syringe and injected into the chromatographic system for analysis. The main advantage of SDME is its high enrichment factor, which is due to the very small volume of the acceptor phase. The other advantages of the single-drop technique include suitability to a wide range of solvents, minimal solvent consumption and low cost. In addition, no preconditioning is required and memory effects are avoided.

Hollow fibre systems operate similarly to solvent drops, but the solvent is immobilized in the pores of the fibre. The fibre can also function as a supported liquid membrane, and an acceptor can be placed inside the lumen [43]. This system is called liquid phase microextraction (LPME). The advantage of LPME is the presence of a fibre that supports organic solvent and hinders the dissolution or evaporation of the solvent that occurs in SDME. High enrichment factors can also be achieved with this technique because of the relatively small volume of the acceptor phase. The main disadvantage of all membrane techniques is that the extraction tends not to be exhaustive. The recoveries are usually at a level similar to those in SPME, and precise calibration is required. Selective extraction can be achieved by a careful choice of the solvent, as well as the material and pore size of the membrane. Pore size is important because in addition to the extraction process, exclusion by size takes place, which increases selectivity. Membrane extraction systems based on hollow fibres and membrane bags can be used for on-site sampling.

It is well known from chemical engineering literature that LLE processes can be manipulated and enhanced by the application of electrical fields across the liquid-liquid interface [44, 45].

This process is called electro membrane extraction (EME). The electrical field aids the movement of charged compounds, so that the target analytes are charged during extraction. Differently charged analytes have different mobilities that can be controlled by electrical potentials and the applied potential controls the selectivity of the extraction. It is important that the solvent remain immobilized in the pores of the hollow fibre during extraction, so that it serves as a stable, supported liquid membrane (SLM). EME has already been shown to be feasible for complex samples and it has been optimized for different types of analytes (bases, acids, zwitterionic compounds, non-polar compounds, and polar compounds). It is faster than LPME with similar equipment, but more research needs to be conducted to check and validate the reliability of the quantitative procedures [46].

Integrating processes and combining analytical methods result in substantial reductions in energy consumption. The energy efficiency of flow-though methods is much greater than those employing batch processes. It is not unusual for both heating and cooling to be required for any given step or stage of a batch process. Although this can be avoided during research, it is not routinely achieved in practice for a variety of reasons. It is generally true that the presence of large instrument installations and their supporting equipment discourages the adoption of newer technologies. The increases in efficiency or cost savings of the new methods and technologies need to be impressive to offset this reality.

Microfluidics can reduce the consumption of toxic eluents in instrumental processes. 'Micro' refers to at least one of the following features: minimal volumes of eluent, small size, or low energy consumption. Microfluidics is the science of designing, manufacturing, and formulating devices and processes capable of handling volumes of fluid on the order of nanolitres (10^{-9} litre) or picolitres (10^{-12} litre). The dimensions of the devices range from millimetres to micrometres. Microfluidics hardware is constructed and designed differently from that of its conventional macroscale analogues. As the scale diminishes and the dimensions of a device reach a certain size, the particles of fluid, or particles suspended in the fluid, become comparable in size with the apparatus itself, dramatically altering the behaviour of the system. Capillary force changes the way in which fluids pass through micrometre-diameter tubes from that in which they flow through macroscale channels.

Microfluidics and related microanalysis systems have the potential to reduce energy consumption; however, in their present stage of development, the typical set-up consists of a small microfluidic chip surrounded by desktop-sized instruments and power supplies. To solve this 'world-to-chip' interface problem, detectors and analysers will need to be further miniaturized. Integrating miniaturized supporting elements will result in true micro- total analysis systems (μTAS) and produce additional energy savings. μTAS combines all the components of the system on one chip. Pumps, valves, mixers and so on must be miniaturized in order to integrate the system, which necessitates a choice between active methods – efficient, but difficult to miniaturize, and requiring energy sources – and passive methods – easier to integrate, but less efficient. For many processes or tasks, especially in biotechnology, macroscopic volumes of molecules and/or particles have to be concentrated in micro-volumes in order to use chips. Miniaturization of the whole treatment chain remains a huge challenge. Additional hurdles are created by the physical behaviour of miniaturized systems, such as solid walls absorbing the target molecules or, capillary force preventing the fluid from entering the microchannels. According to some researchers, the world-to-chip interface is the main bottleneck in the development of μTAS [47]. Solving this problem is especially critical for high-throughput applications. Some solutions that have appeared in the literature have been reviewed and discussed in an article [48].

Many elements of these systems can be replaced by passive components that operate without external power, using gravity, air pressure, or simple manual actions to manipulate the fluids. The development of such simple and possibly disposable devices is primarily motivated by the need for simple point-of-care (POC) tests for underdeveloped countries, but they could also be useful in developed countries where most medical diagnostics are performed in centralized, well-equipped laboratories.

Some rather ingenious solutions have been proposed. George Whitesides and his colleagues at Harvard University have recently demonstrated that three-dimensional microfluidic devices can be made from stacked layers of ordinary paper and tape [49]. This is the cheapest and most robust microfluidic device to date. Because of the wicking ability of the paper, the devices don't require external pumps for the liquids. To define the pathways, each layer of paper is patterned with a common photoresist and treated with UV light. The layers of patterned paper are then alternated with layers of double-sided tape and holes are cut in the tape to connect the channels in adjacent layers of paper. A complex routing can be achieved that can direct liquid horizontally and vertically to an array of 1024 detection zones on the bottom. Reagents or antibodies are placed in the detection zones prior to assembly. Such devices could provide highly parallel, independent assays.

Paper-based microfluidic devices are especially useful for distributed healthcare in the developing world and for environmental monitoring and water testing. Paper-based 3D microfluidic devices have capabilities that are difficult to achieve with conventional open-channel glass or polymer microsystems. 3D paper-based devices wick fluids and distribute microlitre volumes of sample from single inlet points to arrays of detection zones that number in the thousands. This capability makes it possible to carry out a range of new analytical protocols simply and inexpensively on a piece of paper and without external pumps.

The wicking property of paper is already routinely exploited in medical tests such as those for blood glucose, pregnancy, and HIV. The immuno-chromatographic strip (ICS) assay is a good example. Within a few minutes of a sample being applied to the test strip, an indicator line reveals the presence or absence of the relevant analyte. The test sample migrates through the filter paper matrix, which contains both fixed and impregnated antibodies. As the sample migrates, it solubilizes the tagged antibodies and binds them to the target. The analyte then encounters the second set of antibodies and forms an antibody-analyte-antibody sandwich matrix with a visible signal. Because ICS contains inexpensive reagents and components, they cost the end user less than US$2 in many cases. ICS strips require relatively little and sometimes no sample processing, and they do not require external instruments.

Simple physical chemical phenomena, such as capillary action, evaporation, endo/exothermic reaction, gravity and laminar flow in microchannels are suitable for microfluidics-based disposable diagnostic devices without instrumentation. Because a power supply is probably not available, some of the energy can be supplied by the analyst. In their review, Weigl *et al.* give many examples of such designs [50].

Microfluidics has aroused much hope and excitement among analytical chemists. As of 2009, there were more than 5000 publications on microfluidics applications in chemistry and biochemistry [51]. The period of disillusionment is past and a search has begun for a 'killer' application. Becker defines a killer application as one that has such desirable properties that it is capable of generating very large revenues with attractive margins in a comparatively short period of time [52]. Becker is optimistic. He expects that a killer application in microfluidics will appear within 5–10 years and predicts that such an application will probably be in the domain of diagnostics; perhaps a point-of-care device or diagnostics apparatus for a physician's office, or it might be associated with veterinary medicine or food inspection. The energy savings from this kind of instrument is obvious.

15.6.2 Using sensors

Because resource consumption – energy or chemical – and waste generation are central issues, portability is an important attribute of a green analytical instrument. Instruments that can be taken to the analysis site or point-of-care (POC) are in increasing demand. The POC might be a hospital, home or crime scene. Of all the requirements, low power consumption is the most crucial. Smaller analytical instruments consume much less energy and produce faster analyses for higher sample throughput.

The current trend in analysers, especially for on-line use, is toward sensors that are amenable to miniaturization. The new generation of sensors offers tremendous potential for obtaining the desired analytical information in a manner that is faster, simpler and cheaper than that of traditional laboratory-based instruments. A chemical sensor can be defined as a device that, as a result of a process of chemical interaction, transforms chemical or biochemical information into a signal (usually electrical or optical), in a way that responds to changes in sample concentration in the chemical environment. A sensor contains three parts: a recognition element, a transducer, and a signal processor capable of continuously and reversibly reporting a chemical concentration [53, 54]. The transducer transforms a signal obtained by the sensor element into an electrical signal. In addition to rapidly transforming information about the concentration of a specific compound into a signal, and maintaining its activity over a long period of time, a sensor is usually small and inexpensive. Because of its size, the energy requirements of sensors are rather low.

The following groups of sensors have been developed:

- Optical sensors based on absorbance, reflectance, fluorescence, refractive index, optothermal effects and light scattering;
- Electrochemical sensors based on voltammetry and potentiometry;
- Electrical sensors based on metal oxides, organic semiconductors and electrolytic conductivity;
- Mass sensitive detectors based on piezoelectric devices and surface acoustic waves;
- Paramagnetic sensors for oxygen; and
- Thermometric sensors based on the heat effects of chemical reactions.

Biosensors, which offer prospects for combining the recognition of biological events with electronic signal transduction and for designing a new generation of bio-electronic devices that can perform novel functions, are especially promising. From an energy perspective, their operating temperature is close to room temperature, so the requirement for heating or cooling is minimal; the temperature merely needs to be stabilized.

Progress in electronics and instrumentation has made it possible to measure extremely low electrical signals. Wireless technology enables, responses from individual sensors to be received automatically, the signals transformed into analytical information and changes over time assessed from a remote location. This is far more economical than conventional process control and environmental monitoring, in which samples are collected and transported to a central laboratory to determine their chemical composition. A high degree of integration, efficiency, speed, and disposability makes these systems attractive for on-site environmental or industrial applications.

Periodic calibration or completely removing the need for calibration is a particular challenge. If it can be guaranteed that all the sensors within a batch are exactly equal within stipulated limits of uncertainty, then only one of them needs to be calibrated. Another challenge is producing a real-time response without kinetic effects, an impediment to the use of flow systems. By providing timely, safe and effective analytical information such devices offer direct and reliable assessment of the production process or of the effects and gradient of contaminants on site, greatly reducing analytical costs. Recent developments in wireless communication technology and miniaturizing electronics have created the possibility to easily form sensor networks that will greatly enhance monitoring the natural environment (which might be a home or hospital) and, in some cases, lead to new measurement techniques or the deployment of sensors that was previously impossible. As mentioned previously, the quality assurance of the data from instrumentation deployed remotely and for long periods of time is a major issue that must be addressed. However, according to Lieberzeit and Dickert [55], despite the great potential of sensors, studies have mainly been confined to research laboratories and have often not yet reached the stage of technical development.

Although the cost and complexity of analytical instruments has decreased, the implementation of various methods, especially in less developed countries where the need to monitor the environment, food and water is higher than elsewhere, has been hindered by the relatively high cost of instrumentation. Some very interesting proposals have been advanced, using consumer electronics, including low-cost optical equipment such as digital scanners, video cameras, and video screens for routine chemical analysis. Suslick and co-workers provided the first illustration of this method [56] when they showed how a simple flatbed digital scanner could be converted into a powerful analytical instrument for identifying subtle colour changes in chemical dyes to detect volatile compounds. More recently, scientists from Sweden's Lindköpping University have demonstrated the usefulness of personal computers for various colorimentric assays. Filippini *et al.* showed how a computer monitor and an inexpensive webcam could be used as a spectrophotometer [57]. This computer screen photoassisted technique (CSPT) is based on the fact that a computer screen can be easily programmed to display different colours by, combining three (red, green and blue) narrow band emission profiles. Various aspects of the use of computer screens as light sources combined with chemometric data processing have been enthusiastically studied by the Lindköpping group [58]. At first sight, CSPT technology looks like a school science project, but the value lies in its accessibility to an extremely wide audience. Using computer screens as controlled light sources and web cameras as image detectors, is an obvious option for performing quick colorimetric tests in homes or primary care units [59, 60]. It has significant potential to support global monitoring of environmental and sanitary parameters, using internet browsers [61].

Other consumer products with optoelectronic parts can be used as optical sensors in analytical instruments. Compact disc-based, high-density microarrays were studied many years ago by H. Kido *et al.* [62]. Their research indicates that compact disk-based microarray technology can produce qualitative and quantitative results and has the potential for simultaneous multi-analyte analyses. The use of conventional computer optical disk drives for chemical and biological sensing has been reported. Several groups of researchers like S. Lange [63] *et al.* and R. Potyrailo *et al.* [64, 65, 66, 67] have demonstrated that an analogue signal can be extracted from a drive for quantitatively detecting optical changes in sensor films deposited on conventional CD and DVD optical disks. The use slightly modified common CD readers highlights the huge potential of this method, especially for assays using standard audio–video disk surfaces for environmental or clinical analysis, drug detection, or high-throughput multi-residue screening applications.

Methods using CSPT and CD readers are cost-effective, sensitive, and suitable for point-of-need applications, or for diagnostics in settings with meagre resources that require low-priced, low-energy and uncomplicated technology.

Instrumentation is an essential part of the analysis process and a thorough life cycle assessment of these instruments accounted into the greenness of the method. The functional life span of an instrument is a design consideration that affects its EOL (end-of-life) options. As a general rule, an instrument that requires a large input of resources should be durable and reusable. Equipment that has a short life span should be designed in such a way that the minimum amount of resources is expended to obtain results, especially if reuse or recycling is not feasible. This also applies to consumables, especially those related to bioanalysis, in which single-use consumables are very common. The design of these products must ensure that their disposal is easy and energy efficient, or that the energy can be recovered, for example, by using plastics that can be burned without producing toxic emissions.

15.7 Assessment of energy efficiency

Products and processes are often labelled 'green' for advertising purposes without any solid evidence, because the term garners a positive response. However, a more objective reason for this practice is the lack of measurable characteristics to verify the greenness of a product or process. This situation also applies to analytical chemistry.

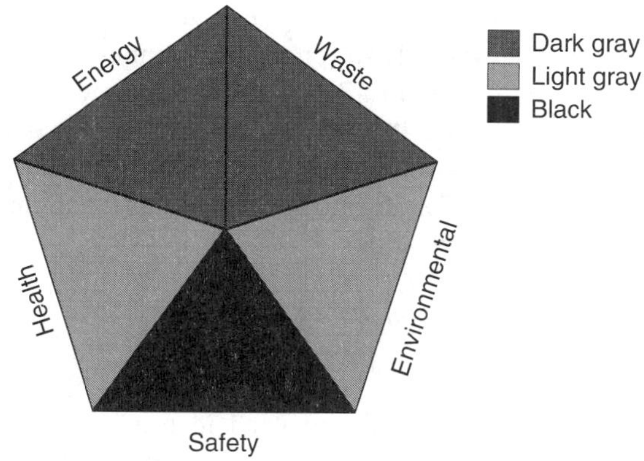

Figure 15.1 *Setup of the green assessment profile symbol.*

The main emphasis is on the metrological quality of the data, and only recently has attention been directed toward the environmental friendliness of the methods by which the data are obtained. Scientists and engineers are aware of this problem and some useful concepts have already been developed [68].

Assessing the greenness of analytical methods is difficult because of their complexity and the abundance of parameters. Databases of analytical methods illustrate this situation well. One of the most comprehensive lists of analytical methods is the National Environmental Methods Index (NEMI) [69], which is a free, internet-searchable database. It is continually expanding to include a growing number of methods and types of methods. Within the database, methods can be easily searched, sorted and compared. It is hoped that users will begin to demand 'greener' methods. However, meeting performance requirements (e.g. level of detection, degree of bias, precision, etc.) is the user's primary consideration in selecting a method. NEMI is the first database to include greenness profiles of the methods. The greenness profile is one of the parameters that have been developed for the assessment of analytical methodologies [70]. The criteria were defined and developed collaboratively by a large number of experts in environmental methods. The profile is oriented towards toxicity, bioaccumulation, reactivity, waste generation, corrosivity, safety and related factors, which are divided into four risk categories; health, safety, environment and waste. A statistical survey of analytical methods indicates that many methods generate waste because they use a large amount of solvents and hazardous chemicals. One disadvantage of this approach is that an energy criterion is not included in the greenness profile. All the steps in the analysis contribute to laboratory energy consumption and must be taken into account to calculate an energy score for the method. Douglas Raynie has developed energy considerations as an additional assessment criterion [33]. The assessment is comprised of five categories of potential risk; health, safety, environment, energy and waste. It also specifies three different levels to score chemical methods from 1–3 on each attribute, using readily available chemical data and represented by colours green, yellow and red respectively. This assessment tool enables individual researchers to judge between conflicting criteria.

This assessment criterion can be represented by a figure of a pentagon consisting of five coloured triangles according to the level of risk (see Figure 15.1).

The energy score is calculated according to the total amount of energy in kWh required to analyse one sample: for ≤ 0.1 kWh, the score is 1 (dark gray), for >0.1 to ≤ 1.5 kWh, the score is 2 (light gray), and for > 1.5 kWh, the score is 3 (black) (look also Table 15.1). All of the energy consumed in the analysis process, including that required for sample preparation and the analyser, is included in this score. This three-tiered system gives more flexibility in assessing methods, but requires more data to create the profile. The number of parameters required to profile the greenness of a method can be debated, but it is certain that the availability of this kind of information helps in the selection of a method that performs best in terms of environmental friendliness and metrological quality.

With the ever-increasing information in databases of analytical methods, users have the data they need to support more informed selections of methods and to improve the greenness of analytical laboratory operations.

The most common parameter – the E-factor [71] – is in principle very simple and well suited to characterizing chemical processes. Although it does not consider material life cycle stages apart from production, the E-factor is a measure of environmental impact and sustainability that is often employed by chemists. The E-factor consists of the ratio of product to the total inputs (or all materials used in the production process) and is expressed by the following equation:

$$E = \frac{input\ material[kg]}{product[kg]}.$$

The E-factor takes into account all the chemicals involved in production. Energy and water inputs are generally not included in E-factor calculations, nor are products of combustion, such as water vapour or carbon dioxide. Because of its simplicity and despite the challenges of including recycled compounds in formulas, this parameter is attracting attention and is even being used to analyse complex processes [72]. However, the E-factor is not directly applicable to analytical chemistry because the 'product' of an analytical laboratory is not quantifiable in kilograms, so the equation cannot be used to calculate its E-factor. All chemicals and solvents could in principle become waste, even after careful recycling. However, the ratio of the requisite amount of chemicals and solvents to the amount of sample required to obtain a measurable analytical signal can be used to compare different analytical methods.

Atom economy is another important metric of material efficiency in green chemistry [73]. Atom economy calculates the efficiency with which atoms that are used as feedstocks in chemical transformations are incorporated into the final product. Unlike the E-factor, which is based on the specific conditions and circumstances of a process, atom economy is an intrinsic metric that measures the theoretical efficiency of a process under perfect conditions. This metric is most frequently applied to chemical transformations in which substances of discrete molecular structure are transformed into new, homogeneous target products.

In analytical chemistry, an efficiency parameter could be developed based on analytical signals. To apply this approach to analytical measurements, it must be possible to assess how many molecules of analyte will give a measurable analytical signal where the theoretical limits of one molecule are known. Optical spectroscopy, especially fluorescence, electrochemistry and bioanalysis are areas of analytical chemistry with highly sensitive methods, due to the extremely selective nature of the reactions. According to this approach, the sensitivity of the method correlates with its greenness.

Unfortunately, it is rare that a single analyte can be analysed without an interfering matrix; the sample preparation is usually complex, and the calculations are therefore more complicated.

Energy metrics are usually similar to those for mass; one can take into account all forms of energy, such as that used for heating or cooling per kg of product.

Some potential metrics might include [74]:

$$Energy\ Intensity = \frac{Total\ \ process\ energy[MJ]}{product[kg]},$$

and

$$Life\ Cycle\ Energy = \frac{Life\ Cycle\ energy\ requirements[MJ]}{product[kg]}$$

in which all energy requirements for every step are included into the life cycle (process, material manufacturing, recovery, treatment).

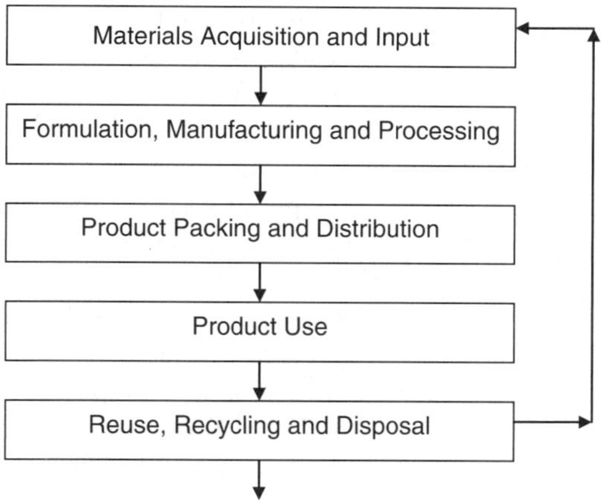

Figure 15.2 *Schematic of a product life cycle.*

As mentioned earlier, it is important to account for the energy requirements of the complete life cycle, which would include not only the energy required for signal generation (in interaction), but also that needed for operation of the instrument, recycling of materials and treatment of waste.

Clark *et al.* [75] proposed reaction input energy per mole of target product in units of kWh mol^{-1} as an energy metric for chemical reactions. Developing an energy metric for analytical chemistry is more complicated because the product of the process is an analytical signal from different sources – chemical or electrochemical reactions or physical processes – and the signals are measured in physical quantities such as volume, mass, electrical charges, differences of temperatures, or radiation energy as expressed qualitatively and quantitatively in wavelengths or frequencies and intensities. An energy metric for analytical chemistry must also take into account the amount of additional energy that is required to generate a measurable analytical signal. It is possible to compare the energy (Energy Intensity) needed to generate a signal at particular levels of detection.

A life cycle analysis (LCA) is a technique for quantifying and assessing the inputs and outputs affecting the environmental performance of a product throughout its life cycle, from production, through use, to disposal (Figure 15.2).

LCA includes all aspects of production, such as capital and maintenance costs, efficiency, safety and hazard risk, and waste management. The life cycle parameters of the product are associated with the flexibility of its design and operation. LCA can be a powerful tool for systematically studying the environmental influence of a product and assist in identifying opportunities to improve its environmental performance.

Chemical analysis is a process that follows a particular method, but its output is an analytical signal rather than a tangible product. In analytical chemistry, one must consider the complete life cycle of all the components of the method, that is, the instruments and all the materials and consumables required to generate the analytical signal, from probe sampling up to the presentation of the data. In principle, the life cycle of the method continues until an analysis is performed on the same instrument following the same procedure; therefore, the number of samples analysed according to this method is an important parameter.

LCA is an approach that can greatly assist in determining the greenness of product. Standardized life cycle assessment (LCA) protocols already exist, such as the ISO 14040 series. However, standards and data for toxicity assessments of materials are lacking. The LCA should also be supported by information on the ways packaging and distribution affect the environmental impact of the product [76]. For those reasons, some comprehensive, quantitative analyses of chemical products have been conducted, but none have been undertaken for analytical methods. LCA is not directly applicable to information obtained from the measurement process, but it is applicable to the components (chemicals, solvents, instrumentation, data processing equipment, etc.) of the process. Therefore, LCA can be used to assess the environmental impact of analytical methods.

LCA approach is important to apply also on analytical instrumentation what is directly related to analysis on methods. This position is according with a principle of green engineering that concerns designing for commercial 'after-life'; products, processes and systems should be designed for functionality in a commercial 'after-life'. To reduce waste, components that remain functional and valuable can be recovered for reuse and/or reconfiguration. The next-generation design of products, processes and systems must take into account the functional use of the valuable properties of recovered components.

In case of matured techniques where big changes in operating principles have not happened during decades like chromatography and instruments made in the 1970s still perform well. New models do not improve the quality of analysis, and there is no urgent need to replace instrumentation often. However, it is not popular among practitioners and manufacturers. Whereas the attitude of the latter is understandable, the desire of the analyst to replace instrumentation every second year is hardly logical, even if the laboratory budget can accommodate it. The appearance of new instruments on the market can be welcomed if this is a reflection of a scientific breakthrough. Nevertheless, a plethora of new models of standard scientific instruments appears on the market every year even though no new scientific discoveries have been implemented. The instrument markets are rather saturated, and only the info-technological revolution can be followed from interfacing with new computers and installed new version of software.

References

1. Anastas, P.T. and Warner, J.C. (1998) *Green Chemistry: Theory and Practice*, Oxford University Press, New York.
2. Anastas, P.T. and Zimmerman, J.B. (2003) Design through the 12 principles of green engineering, *Env. Sci. & Tech.*, **37**, 94A–101A.
3. US Environmental Protection Agency (n.d.) *Green Engineering Homepage*. Available at http://www.epa.gov/opptintr/greenengineering/ (accessed 8 November, 2011).
4. Mathew, P.A.; Sartor, D.A.; Bell, G.C. and Drummond, D. (2007) Major energy efficiency opportunities in laboratories—Implications for health and safety, *J. Chem. Health & Safety*, **5**, 31–39.
5. US Department of Energy (n.d.) Available at http://www.epa.gov/lab21gov/ (accessed 7 November, 2011).
6. Labs21 (n.d.) *Energy_Efficient_Laboratory_Equipment_Wiki*. Available at http://labs21.lbl.gov/wiki/equipment/index.php/ (accessed 7 November, 2010).
7. Loupy, A. (ed.) (2006) *Microwaves in Organic Synthesis,* 2nd edn. Wiley-VCH, Weinheim.
8. Kingston, H.M. and Haswell, S.J. (ed.) (1997) *Microwave-Enhanced Chemistry: Fundamentals, Sample Preparation, and Applications*, ACS,.
9. Srogi, K. (2006) A review: Application of microwave techniques for environmental Analytical Chemistry, *Anal. Letters,* **39,** 1261–1288.
10. US Environmental Protection Agency (2007) *Method 3546: Microwave Extraction*. Available at http://www.epa.gov/epawaste/hazard/testmethods (accessed 8 November, 2010).
11. Richter, R. (2003) *Clean Chemistry: Techniques for the Modern Laboratory*, Milestone Press.
12. Jin, Q.; Liang, F.; Zhang, H.; Zhao, L.; Huan, Y. and Song, D. (1999) Application of microwave techniques in Analytical Chemistry, *TRAC-Trend. Anal. Chem.*, **18**, 479–485.
13. Priego-Capote, F. and Luque de Castro, M.D. (2004) Analytical uses of ultrasound - I. Sample preparation, *TRAC-Trend. Anal. Chem.*, **23**, 829–838.
14. US Environmental Protection Agency (2007) *Method 3550C: Ultrasonic Extraction*. Available at:http://www.epa.gov/epawaste/hazard/testmethods (accessed 8 November, 2010).

15. Saterlay, A.J. and Compton, R.G. (2000) Sonoelectroanalysis - an overview, *Fresenius J. Anal. Chem.*, **367**, 308–313.
16. Priego-Capote, F. and Luque de Castro, M.D. (2004) Analytical uses of ultrasound II. Detectors and detection techniques, *TRAC-Trend. Anal. Chem.*, **23**, 829–838.
17. Fidalgo-Used, N.; Blanco-Gonźalez, E. and Sanz-Medel, A. (2007) Sample handling strategies for the determination of persistent trace organic contaminants from biota simples, *Anal. Chim. Acta*, **590**, 1–16.
18. Ramos, J.J.; Rial-Otero, R.; Ramos, L. and Capelo, J.L. (2008) Ultrasonic-assisted matrix solid-phase dispersion as an improved methodology for the determination of pesticides in fruits, *J. Chromatogr.* A, **1212**, 145–149.
19. Albini, A. and Fagnoni, M. (2004) Green Chemistry and photochemistry were born at the same time, *Green Chem.*, **6**, 1–6.
20. Roth, H. D. (2001) Twentieth century developments in photochemistry. Brief historical sketches, *Pure Appl. Chem.*, **73**, 395–403.
21. For overviews on different reactor types, see: (a) Malato, S. Blanco, J. Vidal, A. and Richter, C. (2002) Photocatalysis with solar energy at a pilot-plant scale: an overview, *Appl. Catal. B*, **37**, 1–15; (b) Funken, K.H. and Ortner, J. (1999) Technologies for the solar photochemical and photocatalytic manufacture of specialities and commodities: A review, *Z. Phys. Chem.*, **213**, 99–105.
22. Stchur, P.; Yang, K.X.; Hou, D.; Sun, T. and Michel, R.G. (2001) Laser exited atomic fluorescence spectrometry—A review, *Spectrochim. Acta B*, **56**, 1565–1592.
23. Peter-Katalinic, J. and Hillenkamp, F. (2007) *MALDI MS: A Practical Guide to Instrumentation, Methods and Applications*, Wiley-VCH, Weinheim.
24. Blanchard, L.A.; Hancu, D.; Beckman, E.J. and Brennecke, J.F. (1999) Green processing using ionic liquids and CO2, *Nature*, **399**, 28–29.
25. Koel M. (ed.) (2008) *Ionic Liquids in Chemical Analysis*, CRC Press, Boca Baton.
26. Griffiths, P.R. (1988) Contemporary SFC – Accomplishments and limitations, *Anal. Chem.*, **60**, 593A-597A.
27. Heinisch, S. and Rocca, J.L. (2009) Sense and nonsense of high-temperature liquid chromatography, *J. Chromatogr. A*, **1216**, 642–658.
28. Smith, R.M.; Chienthavorn, O.; Wilson, I.D.; Wright, B. and Taylor, S.D. (1999) Superheated heavy water as the eluent for HPLC-NMR and HPLC-NMR-MS of model drugs, *Anal. Chem.*, **71**, 4493–4497.
29. Smith, R.M. (1999) Buffered superheated water as an fluent for reversed-phase high performance liquid chromatography, *Chromatographia*, **50**, 485–489.
30. Louden, D.; Handley, A.; Taylor, S.; Sinclair, I.; Lenz, E. and Wilson, I.D. (2001) High temperature reversed-phase HPLC using deuterium oxide as a mobile phase for the separation of model pharmaceuticals with multiple on-line spectroscopic analysis (UV, IR, H-1-NMR and MS) *Analyst*, **126**, 1625–1629.
31. van der Vorst, G.; van Langenhove, H.; de Paep, F.; Aelterman, W.; Dingenen, J. and Dewulf, J. (2009) Exergetic life cycle analysis for the selection of chromatographic separation processes in the pharmaceutical industry: preparative HPLC versus preparative SFC, *Green Chem.*, **11**, 1007.
32. Perrot, P. (1998) *A to Z of Thermodynamics*, Oxford University Press.
33. Raynie. D. and Driver, J.L. (2009) Green Assessment of Chemical methods, *Presentation on to the 13th Green Chem & Eng Conference*, Maryland.
34. He, Y.; Tang, L.; Wu, X.; Hou, X. and Lee, Y. (2007) Spectroscopy: The best way toward Green Analytical Chemistry? *Appl. Spectrosc. Rev.*, **42**, 119–138.
35. Ioffe, B.V.; Vitenberg, A.G. and Manatov, I.A. 1984 *Head-Space Analysis and Related Methods in Gas Chromatography*, John Wiley& Sons, Inc., New York.
36. Thurman, E.M. and Mills, M.S. (1998) *Solid-Phase Extraction, Principles and Practice*, John Wiley & Sons, Inc., New York.
37. Pawliszyn, J. (1997) *Solid Phase Microextraction: Theory and Practice*, Wiley-VCH, Weinheim.
38. Dean, J.R. and Xiong, G. (2000) Extraction of organic pollutants from environmental matrices: selection of extraction technique, *TRAC-Trend. Anal. Chem.*, **19**, 553.
39. Workman, J.; Creasy, K.E.; Dohetry, S.; Bond, L.; Koch, M.; Ullman, A. and Veltkamp, D. (2001) Process Analytical Chemistry, *Anal. Chem.*, **73**, 2705–2718.

40. Workman, Jr. J.; Koch, M.; Lavine, B. and Chrisman, R. (2009) Process Analytical Chemistry, *Anal. Chem.*, **81**, 4623–4643.
41. Xu, L.; Basheer, C. and Lee, H.K. (2007) Developments in single-drop microextraction, *J. Chromatogr. A*, **1152**, 84–192.
42. Ouyang, G.; Zhao, W. and Pawliszyn, J. (2005) Kinetic calibration for automated headspace liquid-phase microextraction, *Anal. Chem.*, **77**, 8122–8128.
43. Lee, J.; Lee, H.K.; Rasmussen, K.E. and Pedersen-Bjergaard, S. (2008) Environmental and bioanalytical applications of hollow fiber membrane liquid-phase microextraction: A review, *Anal. Chim. Acta*, **624**, 253–268.
44. Stichlmair, J.; Schmidt, J. and Proplesch, R. (1992) Electroextraction – a novel separation technique, *Chem. Eng. Sci.*, **47**, 3015–3022.
45. Pedersen-Bjergaard, S. and Rasmussen, K.E. (2008) Electrical potential can drive liquid-liquid extraction for sample preparation in chromatography, *TRAC-Trend. Anal. Chem.*, **27**, 934–941.
46. Østegaard Kjelsen, I.J.; Gjelstad, A.; Rasmussen, K.E. and Pedersen-Bjergaard, S. (2008) Low-voltage electromembrane extraction of basic drugs from biological samples, *J. Chromatogr., A*, **1180**, 1–9.
47. Wang, J.H. and Fang, Z.L. (2004) The third generation of flow injection analysis: Current situation and perspectives of Lab-on-Valve scheme, *Chin J., Anal. Chem.*, **32**, 1401–1406.
48. Jesson, G.; Kylberg, G. and Andersson, P. (2003) A versatile macro-to-micro dispensing system, *Micro Total Analysis Systems* (eds. M.A. Nothrup, K.F. Jensen, D.J. Harrison), p. 155–158.
49. Martinez, A.W.; Phillips, S.T. and Whitesides, G.M. (2008) Simple telemedicine for developing regions: Camera phones and paper-based microfluidic devices for real-time, off-site diagnosis, *Anal. Chem.*, **80**, 3699–3707.
50. Weigl, B.; Domingo, G.; LaBarre, P. and Gerlach, J. (2008) Towards non- and minimally instrumented, microfluidics-based diagnostic devices, *Lab. Chip*, **8**, 1999–2014.
51. Ohno, K.; Tachikawa, K. and Manz, A. (2008) Microfluidics: Applications for analytical purposes in chemistry and biochemistry, *Electrophoresis*, **29**, 4443.
52. Becker, H. (2009) Hype, hope and hubris: the quest for the killer application in microfluidics, *Lab. Chip*, **9**, 2119–2122.
53. Gründler, P. (2007) *Chemical Sensors: An Introduction for Scientists and Engineers*, Springer, Heidelberg.
54. Cattrall, R.W. (1997) *Chemical Sensors,* Oxford University Press, Oxford.
55. Lieberzeit, P.A. and Dickert, F.L. (2007) Sensor technology and its application in environmental analysis, *Anal. Bioanal. Chem.*, **387**, 237–247.
56. Rakow, N. and Suslick, K. (2000) A colorimetric sensor array for odour visualization, *Nature*, **406**, 710–713.
57. Filippini, D.; Svensson, S. and Lundström, I. (2003) Computer screen as a programmable light source for visible absorption characterization of (bio)chemical assays, *Chem. Commun.*, **2**, 240–241.
58. Bakker, J.W.P.; Filippini, D. and Lundström, I. (2005) Computer screen photo-assisted reflectance fingerprinting, *Sensors Actuat. B-Chemical*, **107**, 580–586.
59. Filippini, D. and Lundstrom, I. (2006) Measurement strategy and instrumental performance of a computer screen photo-assisted technique for the evaluation of a multi-parameter colorimetric test strip, *Analyst*, **131**, 111.
60. Filippini, D.; Alimelli, A.; Di Natale, C.; Paolesse, R.; D'Amico, A. and Lundstrom, I. (2006) Chemical sensing with familiar devices, *Angew. Chem.- Int.Ed.*, **45**, 3800–3803.
61. Filippini, D.; Di Natale, C.; Paolesse, R.; D'Amico, A. and Lundstrom, I. (2007) Computer screen photo-assisted techniques for global monitoring of environmental and sanitary parameters, *Sensors Actuat. B-Chemical*, **121**, 93–102.
62. Kido, H.; Maquieira, A. and Hammock, B.D. (2000) Disc-based immunoassay microarrays, *Anal. Chim. Acta*, **411**, 1–11.
63. Lange, S.A.; Roth, G.; Wittermann, S.; Lacoste, T.; Vetter, A.; Grassle, J.; Kopta, S.; Kolleck, M.; Breitinger, B.; Wick, M.; Horber, J.K.H.; Dubel, S.and Bernard, A. (2006) Measuring biomolecular binding events with a compact disc player device, *Angew. Chem., Int. Ed.*, **45**, 270–273.
64. Potyrailo, R.A.; Morris, W.G.; Leach, A.M.; Hassib, L.; Krishnan, K.; Surman, C.; Wroczynski, R.; Boyette, S.; Xiao, C.; Shrikhande, P.; Agree, A. and Cecconie, T. (2007) Theory and practice of ubiquitous quantitative chemical analysis using conventional computer optical disk drives, *Appl.Optics*, **46**, 7007–7017.

65. Potyrailo, R.A. and Mirsky, V.M. (2008) Combinatorial and high-throughput development of sensing materials: The first 10 years, *Chem. Rev.*, **108**, 770–813.
66. Potyrailo, R.A.; Morris, W.G.; Leach, A.M.; Sivavec, T.M.; Wisnudel, M.B. and Boyette, S. (2006) Analog signal acquisition from computer optical disk drives for quantitative chemical sensing, *Anal. Chem.*, **78**, 5893.
67 Potyrailo, R.A.; Morris, W.G.; Wroczynski, R.; Hassib, L.; Miller, P.; Dworken, B.; Leach, A.M.; Boyette, S. and Xiao, C. (2009) Multi-wavelength operation of optical disk drives for chemical and biological analysis, *Sensors Actuat. B.*, **136**, 203–208.
68. Koel, M. and Kaljurand, M. (2010) *Green Analytical Chemistry*, RSC.
69. National Environmental Methods Index (n.d.) *NEMI Homepage*. Available at http://www.nemi.gov (accessed 7 November, 2011).
70. Keith, L.H.; Gron, L.U. and Young, J.L. (2007) Green analytical methodologies, *Chem. Rev.*, **107**, 2695–2708.
71. Sheldon, R.A. (1992) Organic synthesis – past, present and future, *Chem. Ind. (London)*, **23**, 903–906.
72. Sheldon, R.A. (2007) The E-factor: Fifteen years on, *Green Chem.*, **9**, 1273–1283.
73. Trost, B.M. (1991) The atom economy - A search for synthetic efficiency, *Science,* **254**, 1471–1477.
74. Constable, D.J.C.; Jimenez-Conzalez, C. and Lapkin, A. (2009) Process Metrics, in *Green Chemistry Metrics. Measuring and Monitoring Sustainable Processes* (eds. A. Lapkin and D. Constable), John Wiley & Sons, Ltd, Chichester.
75. Gronnow, M.J.; White, R.J.; Clark, J.H. and Macquarrie, D.J. (2005) Energy efficiency in chemical reactions: comparative study of different reaction techniques, *Org. Proc. R & D.*, **9**, 516–518.
76. Anastas, P.T. and Lankey, R.L. (2000) Life cycle assessment and Green Chemistry: the yin and yang of industrial ecology, *Green Chem.*, **2**, 289–295.

16

Green Analytical Chemistry and Flow Injection Methodologies

Luis Dante Martínez, Soledad Cerutti and Raúl Andrés Gil

Instituto de Química de San Luis, Universidad Nacional de San Luis-CONICET, San Luis, Argentina

16.1 Progress of automated techniques for Green Analytical Chemistry

The term 'Green Chemistry' was first used in 1991 [1, 2]. Green chemistry gives us new tools that permit chemists to perform chemistry in a more environmentally benign manner. Therefore, philosophy and ideas on green chemistry are those previously developed in analytical laboratories (GAC). This concept refers to the utilization of chemistry techniques and methodologies that reduce or eliminate the use or generation of feedstocks, products, by-products, solvents, reagents, and so on that are hazardous to human health or the environment as a whole. Taking into account current public concern on environmental matters, environmental analytical studies and the consequent use of toxic reagents and solvents have increased to a point at which they became unsustainable to continue without an environmentally friendly perspective [3]. The adverse environmental impact of analytical methodologies has been reduced in three different ways: (1) reduction of the amount of solvents required in sample pretreatment; (2) reduction in the amount and the toxicity of solvents and reagents employed in the measurement step, especially by automation and miniaturization; and (3) development of alternative direct analytical methodologies not requiring solvents or reagents. Milestones of both concepts and methodologies leading to the present state of GAC occurred in the twentieth century; J. Namieśnik *et al.* resumed them all in a recent article [1]. Among these concepts and methods, the development of flow injection analysis (FIA), sequential injection analysis (SIA), and miniaturization of analytical systems are highlighted.

The analytical process is comprised of several steps such as field sampling, field sample handling, laboratory sample preparation, separation and quantitation, and statistical evaluation. Though each step can affect the environment, they make different contributions towards pollution and have different potentials for

Handbook of Green Analytical Chemistry, First Edition. Edited by Miguel de la Guardia and Salvador Garrigues.

being made greener [4]. For these reasons, analytical chemists tried to change the application of existing analytical methodologies and looked for new ones that could use smaller amounts of harmful solvents and reagents, and less energy. The process accelerated when ecological consciousness increased considerably, resulting in 'greener' analytical procedures, miniaturized analytical instruments and the lab-on-valve concept. This chapter attempts to examine the state-of-the-art of Green Analytical Chemistry within the most popular automated and flow injection techniques of analysis.

16.2 Flow injection analysis

The International Union of Pure and Applied Chemistry (IUPAC) defines automation as

> *the use of combinations of mechanical and instrumental devices to replace, refine, extend, or supplement human effort and facilities in the performance of a given process, in which at least one major operation is controlled without human intervention, by a feedback mechanism* [5].

Thus, automation extends beyond the simple instrumentation aspects, and transgresses into management techniques, so it is difficult to look at it in terms of a single scientific subject area. Automation also crosses disciplinary boundaries and clinical, industrial, and process chemists have much to gain from each other's experience of automation in routine use. Automatic methods of analysis have become well established and common features in most analytical laboratories; this required a fundamental change in the role of the analytical chemist. Research into automation of analytical techniques is prompted by the need to produce cost-effective solutions to an ever increasing demand for chemical analysis. The demand comes from increased legislative control in many areas, as well as a growing concern for the quality of the environment, both in general and in the workplace.

Automation improves measurements through (1) fast sampling and cleaning of the measurement devices, (2) fast analytical control of the baseline/background, (3) an easy and rapid standardization, (4) improved repeatability, (5) a considerable reduction of reagent, solvent consumption, and waste generation, and (6) a possible enhancement of the analytical sensitivity through the online coupling of detection with preconcentration/separation techniques. In the development of automation, the main milestones correspond with the use of segmented flow, flow injection analysis (FIA) developments, advances regarding monosegmented flow and those for sequential injection analysis (SIA) and multicommutation [6].

As one of the important events in GAC, the considerable number of publications of FIA is evidence of its growth and wide applicability. First described over 30 years ago [7], FIA developments have been spectacular, leading to more than 20 000 articles have appeared in various international analytical chemical journals in a multitude of languages, to which should be added hundreds of monographs and PhD theses. The ready adoption of FIA is not the least due to the fact that it broke with the centuries long concept of making chemical analysis via steady-state conditions (sample and reagent(s) homogenization, chemical equilibrium), but indeed that it devised new and novel ways to make use of non-equilibrium conditions to exploit hereto unknown procedures. During its existence, FIA has, however, undergone certain changes and modifications, which can be classified as the 'Three Generations of FIA' [8]. In the early 1990s a variant of FIA was introduced, that is, Sequential Injection Analysis [9]. Termed the second generation of FIA, it was at the end of that decade supplemented by the third generation of FIA, also named the Bead Injection Lab-on-valve (BI-LOV) [10]. Both these approaches have, in their own right, proven to entail a number of specific advantages. Thus, for instance, miniaturization of the manifolds drastically reduced the consumption of sample and reagent solutions, and hence led to generation of minute amounts of waste. On the other hand, it allowed complex sample manipulations to be facilitated in simple fashions. In addition, it readily permitted

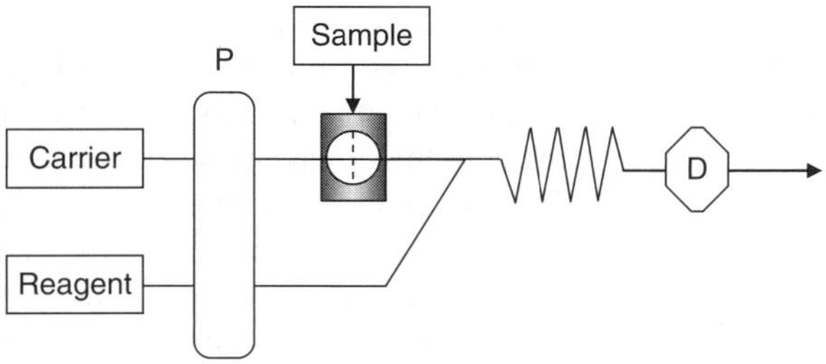

Figure 16.1 *Representation of a FIA system. A defined volume of sample is injected into a continuously flowing carrier stream which is merged with a reagent stream. The resulting transient generation of product is monitored by a suitable detector (D).*

the integration of sequential unit operations. The most important characteristics of each type of approach will be discussed in the following sections.

A typical FIA manifold is depicted in Figure 16.1. As is well known, FIA is a continuous flow method where the sample is injected into a carrier stream, usually containing reagents. As the sample disperses, it reacts with reagents. Mixing continues during downstream flow until the mixture reaches a detector, typically an electrochemical or spectrophotometric device. The physical mixing of the chemicals and the extent of reactions between the chemicals are incomplete, typically giving a throughput advantage. Precisely timed injection events and controlled flow rates in the manifold reduce noise to allow high reproducibility of these transient signals [11].

The fundamentals of FIA were already mentioned in the first publication [6], that is, (1) injection of a well-defined volume of sample; (2) reproducible and precise timing of the manipulations it is subjected to in the system, from the point of injection to the point of detection (controllable dispersion); and (3) the creation of a concentration gradient of the injected sample, providing a transient, but strictly reproducible readout of the recorded signal. The combination of these features, as recorded by the detector, which may continuously observe an absorbance, an electrode potential, or any other physical parameter as it changes on passage of the sample material through the detector device, makes it unnecessary to achieve chemical equilibrium. Any point on the path toward the steady-state signal is as good a measure as the steady-state itself, provided that this point can be reproduced repeatedly, and this is certainly feasible in FIA with its inherently exact timing. Consequently, this feature has not only allowed to perform chemical assays much faster, and hence facilitate higher sampling rates, than in conventional procedures, but more importantly it has permitted to implement procedures which are difficult, or, in fact, impossible, to effect by traditional means, many of which are entirely unique. Just to mention of few, one can point to examples such as: (1) use of bio- and chemiluminescence for analytical purposes [12]. These very sensitive procedures were virtually nonexistent prior to the introduction of FIA. (2) Exploitation of kinetic discrimination schemes where, even subtle, differences in the reaction rates of occurring chemical reactions are judiciously used. Another example is the elimination of interferences due to the presence of transition metal ions, such as Cu, Ni, and Co, in hydride generation schemes [13]. (3) Utilizing intermediary/metastable constituents of specific analytical characteristics for the analytical readout [14]. (4) Drawing on enzymatic degradation procedures either for measurements of substrates or enzymatic activities (which traditionally have been very difficult to execute) [15] not the least via stopped-flow measurements [16] or (5) to perform appropriate on-line sample pretreatments (separation and/or preconcentration) procedures [17, 18].

The evolution of flow systems comprises the use of solid-phase reagents (in all its possible configurations), and especially immobilized reagents in solid-phase spectrophotometry (SPS), which Yoshimura *et al.* introduced in 1976 [19]. In SPS, the solid support is placed inside the flow cell and analyte retention, preconcentration, and detection are performed simultaneously, increasing the sensitivity and the selectivity of the analytical procedures and reducing the reagents consumed. The advantages of using automated solid phase extraction (SPE, in- and on-line approaches) or microextraction (SPME) in combination with FIA (and all of the rest FI techniques) as a way of replacing toxic reagents have been experimentally demonstrated [20–25]. The extraction procedure accomplishes two goals: separation of the analyte from interfering species in the sample; and, preconcentration of the analyte to increase the sensitivity. The great advantage of SPE is that both organic compounds and inorganic species can be extracted. Among multiple examples, a time-based multi-syringe flow injection (MSFI) approach was proposed for automated disk-based sorbent extraction of three nitro-substituted phenol isomers followed by on-line simultaneous determination of individual species by diode-array spectrophotometry [26]. The MSFI manifold was coupled to a multi-position selection valve to an analyzer that ensured minimum operational maintenance via exploitation of membrane switching protocols. As compared with other previously reported methods, the proposed flow-through disk-based system could be considered as an environmentally friendly approach because the use of harmful organic solvents was avoided. Circumventing the generally used cadmium filings and carcinogenic reagents or by reaction with phenolic compounds in highly concentrated sulfuric acid medium for the determination of nitrate in waters, a green analytical procedure for nitrate determination in natural waters based on direct spectrophotometric measurements in ultraviolet, using a flow-injection system with an anion-exchange column for separation of nitrate from interfering species, allowing selective nitrate determination without employing toxic reagents [27]. Also, an environmentally friendly methodology was developed for quality control analysis of commercial pesticide formulations containing Malathion, using flow injection analysis (FIA) – Fourier transform infrared (FTIR) spectrometry, without any pre-treatment of samples, saving time and reducing the volume of harmful chlorinated solvents employed in the analysis (2 ml) [28]. A flow injection spectrophotometric method was proposed for the determination of chloride ion in natural waters by reaction with $Hg(SCN)_2$ immobilized in an epoxy resin bead [29]. The manifold allowed a throughput sample of 100 determinations per hour. The main advantage of the developed method was the 400% reduction of the Hg waste solution generated when compared to conventional methods for chloride determination based on the same spectrophotometric reaction. In addition, the retention of reagents in solid supports was proposed for the monitoring of sub-nanogram amounts of chlorpyrifos in fruit based on immobilizing luminol or periodate on an anion-exchange column [30]. A reagentless process and reduced background noise was achieved. Low reagent consumption and the use of water as eluent are some of the 'green' advantages of the proposed method. In this 'green' sense, a procedure for determination of carbaryl pesticide in natural waters was recently developed [31]. The flow system was designed with solenoid micro-pumps in order to improve mixing conditions and minimize reagent consumption as well as waste generation. The degradation of the generated waste yielded a colorless residue with 94% reduction in total organic carbon after 5 min of exposure to UV radiation. Toxicity tests with marine bacteria *Vibrio fischeri* classified the waste as non-toxic and the cost of the waste degradation as very low. This research group also developed a methodology to determine free chlorine in natural waters using similar approaches and pursuing the same environmentally-friendly aims [32].

Increasing numbers of publications reveal researchers interest in waste management [33]. A highly concern situation is laboratory wastes. They are not discarded directly into the environment but are treated off-line. However, this practice increases the running costs of analytical laboratories and creates a problem through the accumulation of toxic residues. A green alternative to waste storage is therefore on-line recovery of solvents and toxic or expensive reagents [3]. A useful way of recovering reagents on-line is to incorporate a cation exchange mini-column after a flow-through cell. In this way, the main reagent can be regenerated on-line and heavy-metal ions can be removed from the waste, thus reducing both reagent consumption and waste generation [34]. The method involved spectrophotometric determination of Pb with chromogenic reagent arsenazo(III), the accumulation of lead onto the cationic exchanger, and subsequent regeneration of the chromogenic reagent, made the system reversible and the reagents reusable. The method was successively applied to the assay of lead in high-octane gasoline. Simple FIA has also been adapted for post-analysis in-line waste detoxification, such as the TiO_2 catalyzed UV mineralization of aromatics [35]. On-line solid-phase extraction coupled to different detectors using chelating resins and other low-cost adsorbents have been used to recover metals at trace levels in different matrices [36–42]. The FI-based manifolds have the advantages of reducing solvent usage and exposure, disposal costs and extraction time for sample preparation. Consequently, SPE has been successfully used for the separation, preconcentration, and sensitive determination of metal ions in numerous types of samples.

Despite of the significant advantages of FIA compared to batch assays, adaptation of FI to industrial environment for process analytical purposes was hindered by the following drawbacks: (1) complicated flow manifolds, (2) lack of flow stability due to peristaltic pumps do not provide steady flow on 24 h basis operation and therefore require frequent maintenance increasing this way the cost of the analysis, and (3) the continuous flow of reagents produces considerable amounts of waste material in the case of 24 h process control applications. These disadvantages were more or less overcome by the introduction of Sequential Injection Analysis, which is considered as the second generation of FIA-based techniques.

16.3 Sequential injection analysis

While most FIA-procedures employ continuous, unidirectional pumping of carrier and reagent streams, SIA is based on using programmable, bidirectional discontinuous flow as precisely coordinated and controlled by a computer. Sequential injection has shown the ability of performing different determinations without system reconfiguration (placing different reagents on the ports of the selection valve); as a consequence of the non-continuous consumption, a reagent saving is promoted [9, 13, 43–45]. A sketch of a typical SIA-manifold is reproduced in Figure 16.2.

The core of the system is a multi-position selection valve with six ports as it enables the sequential selection of the various solutions and the subsequent redirection towards the detection system. Most of the characteristics

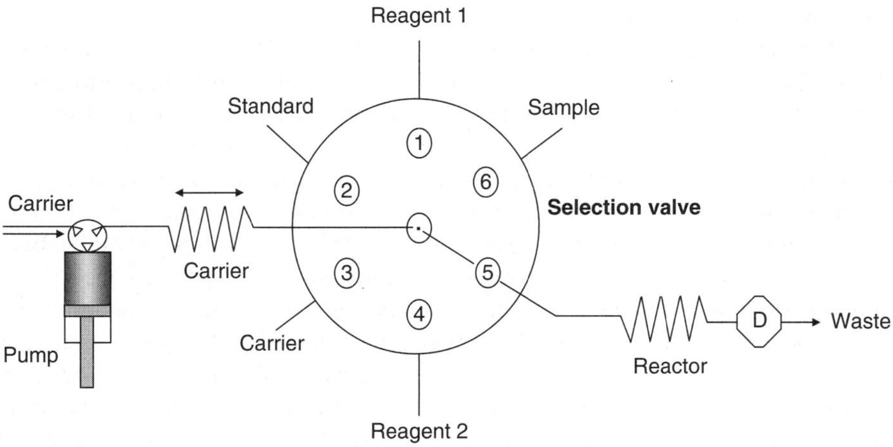

Figure 16.2 *Sequential injection analysis system. Selection valve (SV) and central CC of which randomly can address each of the six individual ports. Initially, sample and reagent are by means of the syringe pump (SP), and via the central CL, aspirated into a HC and stacked there as individual zones. Thereafter the zones are through Port 5 and the reaction coil forwarded to D, which monitors the product formed as the result of the dispersion of the zones into each other during the transport.*

attributed to sequential injection are due to the selection valve. In fact, the placement of different reagents on the ports has allowed different determinations with the same manifold [13, 45] being the structure of the selection valve itself that gives robustness to the sequential injection technique. This valve is equipped with a central communication channel (CC) that can be made to address each of the peripheral ports (1–6), and a central communication line (CL) which, via a holding coil (HC) that prevents the contamination of the carrier, is connected to a syringe pump. By directing the central CC to the individual ports, well-defined sample and reagent zones are initially time-based aspirated sequentially into the HC where they are stacked one after the other. Afterwards, the sample zone can be manipulated in different ways within the SIA manifold by taking advantage of the stopped-flow, bi-directional nature of fluid handling in SIA (as opposed to FIA), the result of which is monitored by the detector.

SIA allows the exact metering of even small volumetric volumes (of the order of a few tenths of a microliter or even less), and that it, thanks to the use of a syringe pump, readily and reproducibly permits flow reversals. Generally, the use of reagents and carrier in SIA is reduced to about a tenth of that used with FIA. Besides, it is extremely economical as to consumption of sample and reagents, and hence in waste generation. And since all manipulations are computer controlled, it is easy and simple to reprogram the system from one application to another one. However, it is generally difficult to accommodate (stack) more than two reagents along with the sample, although additional reagents might be added further downstream, that is, by making an FIA/SIA-hybrid. And due to the use of a syringe pump, SIA has a somewhat limited operating capacity, although this in practice rarely is a constricting factor. As with FIA, high sample throughputs are achievable, thus providing the potential of pseudo-continuous process monitoring [46].

From the all stated, it is not surprising that SIA has got an increasing number of applications in analytical chemistry, it has proven itself especially useful for various separation and preconcentration schemes relying on, for instance, on-line liquid-liquid extraction, precipitation/co-precipitation in knotted reactors, or solid-phase extractions in column reactors with either hydrophilic or hydrophobic packing materials. This advantage of easily coupling separation devices to the sequential injection valve without overall manifold reconfiguration has been fully exploited and reported in the literature. In fact, about 45% of all the listed papers described an on-line treatment of the sample. Within these papers, different types of in-line treatment are considered, some aiming for the separation and/or preconcentration of the analyte (~70%) and others aiming for the change of the oxidation state of the analyte (~30%) [43].

SIA-based methods have been applied to the analysis of a wide variety of analytes in matrices as diverse as water [43], food [47], beverages [48], bioprocesses [49], environmental [50], pharmaceutical [51], and industrial processes [52], the latter use having proven to be especially attractive. Published methods have incorporated sample treatment steps such as on-line dilution, preconcentration, separation on miniature columns, pH adjustment, phase transfer with membrane sampling devices, and immunoassays.

SIA permits great reduction in the volumes of waste, and several metal ions were determined in waters using a double-line sequential injection spectrophotometric system [53]. The proposed configuration added sample and chromogenic reagents as merging zones. The methodology was applied to the spectrophotometric determination of copper, iron, manganese, and zinc in samples of diverse origin using different chromogenic reagents and detection wavelengths. FI-Solid-Phase Spectrophotometry was proposed for iron determination [54]. Iron(II) was reversibly retained on 1-(2-thiazolylazo)-2-naphthol immobilized on C_{18}-bonded silica. The metal ion was eluted as iron(II) with a small volume of a dilute-acid solution without removing the immobilized reagent, which could be used for at least 100 determinations. The proposed procedure reduced effluent generation and consumed micro amounts of reagents. SIA has also been used to construct a potentiometric electronic tongue for the determination of cadmium, copper, and lead [55]. For the building of the complete response model, a set of 81 samples was prepared automatically by the SIA system. A method for lactic acid quantification was proposed in an elegant sequential injection system. In comparison with some enzymatic methods for spectrophotometric determination of lactic acid exploiting flow systems, simplicity in operational procedure, robust and easy control of temperature and solutions handling, and low cost of reagents can be highlighted [56]. A sequential injection system was also recently developed for the determination of sulfates by using spectrophotometric detection and multivariate curve resolution-alternating least squares [57]. The proposed flow methodology allowed in-line preparation of the samples, with the corresponding increase in sampling rate and simplicity analysis. Additionally, the reagent consumption and waste generation was minimal. In addition, to these excellent SIA applications, this topic has been reviewed in multiple recently published articles [58–68].

16.4 Lab-on-valve

All current procedures could become environmentally friendly by reducing the amounts of reagents consumed and it can easily be achieved by downscaling the manifold components and arranging them in a single device. This concept is known as the micro-total analytical system (μ-TAS), which involves arranging all steps of sample processing in a single device of a few square centimeters [69]. Micro FIA (μ-FIA) systems, which exploit microelectronic techniques to integrate pumps, mixing and reaction chambers as well as detectors in a single chip provide the so-called lab-on-valve (LOV) concept.

As the third generation of FIA analysis, in the LOV (see Figure 16.3) configuration an integrated microconduit is placed on top of the selection valve. The microconduit is potentially designed to incorporate and handle all the necessary unit operations required for a given assay, that is, act as a small laboratory; hence the name LOV.

Thus, it may contain facilities such as mixing points for the analyte and reagents, appropriate column reactors packed for instance with immobilized enzymes, or small beads furnished with active groups such as ion-exchangers, which in themselves might be manipulated within the LOV in exactly the same manner as liquids, and even detection facilities. For optical assays (e.g., UV/Vis or fluorometry), this can readily be achieved by use of optical fibers, the ends of which, furthermore, can be used to define the optical path length

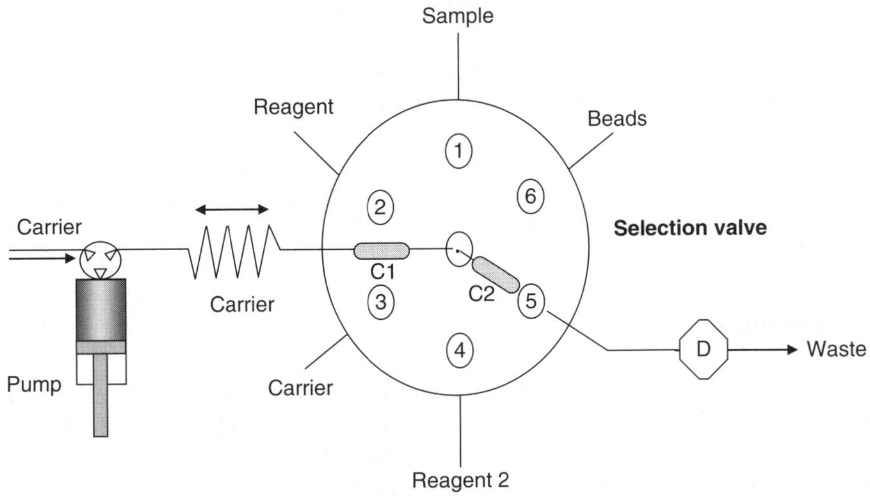

Figure 16.3 LOV system. The microconduit placed atop the selection valve should ideally act as a small laboratory. However, when large instrumental detector devices are to be used, it is necessary to employ external detection as shown in the figure. Besides aspirating liquids, it is also possible to handle small beads (furnished with active functional groups), which can be used to integrate small packed column reactors.

to yield optimal measurement conditions. Thus, one of the fibers is used to direct the light from a power source into the LOV, while the other one serves to guide the transmitted light to an appropriate detection device. For other detectors, such as AAS or ICPMS, it is, of course, necessary to make use of external detection devices, in which instances the LOV can be used as an effective front-end means for introducing the analyte into the detector in an intelligent and optimal fashion. In SIA and LOV all flow programming is computer controlled, which implies that it is readily possible, via random access to reagents and appropriate manipulations, to devise different assay protocols in the microsystems.

The LOV is an ideal facility for performing on-line sample pre-treatments (such as separations and pre-concentrations) of samples with the aim to determine metals at trace levels [70, 71]. Thus, not only sample can be aspirated, but packed columns reactors can also be generated by aspirating beads, with special surface characteristics, from a reservoir (the syringe on top). The reactors are realized by fitting the column positions with appropriate stoppers, which will retain the beads, yet allow solutions to flow freely. Furthermore, beads are commercially available with various surface groups/properties and can be manipulated and directed between different column positions within the LOV. Appropriate eluents can be aspirated, and the eluate propelled to an external detection device, as sandwiched by air or liquid segments in order to preserve its integrity. And after the assay, the beads can be reused or they can be discarded and new ones aspirated (the so called renewable approach), depending on the circumstances. The SI-LOV renewable column approach with BI (bead injection) is a very attractive methodology in many contexts. Thus, immunoassays can take advantage of the fact that since the beads upon completion of measurement are discarded, no restrictions are placed on the binding characteristics between antibody and antigen [72].

In the so-called μSI-BI-LOV mode [73], the flow cell is configured into the jet-ring-cell approach, in order to perform bioligand interactions assays. Thus, an appropriate suspension of beads is injected into and entrapped within the flow cell; then, the beads are perfused with analyte solution. Afterwards, the loaded beads are exposed to various stimuli, and the (bio)chemical reactions taking place on the bead surface are recorded in real time.

Waste generation, reagents and sample consumption can also be minimized by reducing the size of manifolds developed. A clear example was the micro-flow system comprising a planar glass chip with a PDMS top plate fitted with a fiber-optic probe, employed as optical sensor, for monitoring Cu(II) [74]. By miniaturizing the manifold, a simple, precise, environmentally friendly (small amounts of reagents), and readily adaptable to the analysis of large sample series microextraction-spectrophotometric assay was developed for the determination of malondialdehyde (MDA) in blood after formation of MDA-thiobarbituric acid (TBA) adduct [75].

Miniaturized systems with low or insignificant waste production are a promise for the development of methodologies meeting the requirements of GAC. Therefore, the concept of miniaturizing both separation and detection systems has an essential role to play.

16.5 Multicommutation

Multicommutation is a fully computer-controlled modern way to mechanize analytical procedures based on using an array of three-way valves for sample and reagent handling [76]. This is a methodology of increasing use for the development of chemical analysis systems [77–84]. In the development of multicommutation, the main milestone corresponds with the first paper published by Reis *et al.* in 1994 [85]. The multicommutation technique has had an exponential growth in the past 10 years indicating the current importance of multicommutated studies. Multicommutation is an adaptation of flow injection (multicommutated flow system, MCFIA) that utilizes multiple solenoid valves as separate switching devices to create a more flexible flow path that is able to use significantly less reagents than FIA. These computer-controlled valves can be

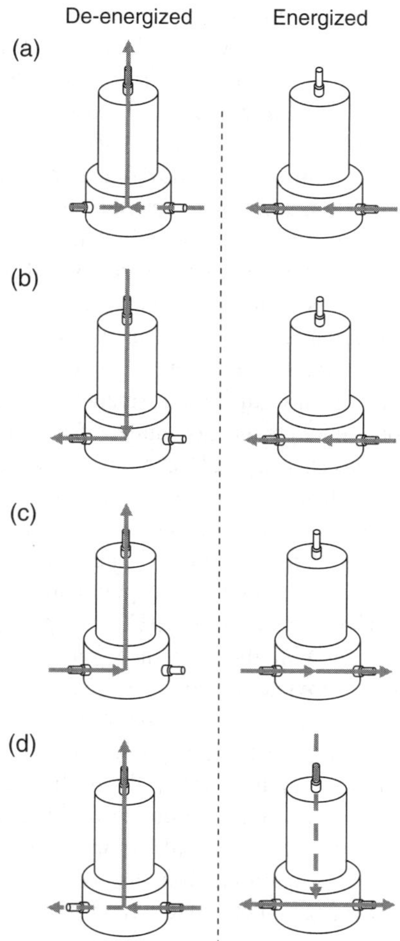

Figure 16.4 *Multicommutation. Three-way solenoid valves. (a) Three-way normally closed; (b) three-way normally open; (c) three-way directional control; and (d) three-way multipurpose.*

compared to an electronic circuit that presents a variable number of active nodes that can adopt two possible states: ON and OFF (see Figure 16.4).

Multicommutation includes different configurations or mechanical commutations in flow analysis [86]; development of commutating devices has led to a continuous enhancement of the system versatility thus system performance. In addition, flow rate modifications and stream reversals are usually accomplished without manifold reconfigurations. Stream redirecting, washing stream, confluent stream, stopped-flow approach, recycling sample carrier stream of a single-line flow system, processing sample stopped inside the analytical path, stacks of different zones feeding the analytical path, establishment of tandem streams, achievement of segmented flows, exchange of manifold components, trapping the processing sample inside an incubation coil, and two inlet and two outlet streams are some of the multiple potentialities available when using commutation in its different configurations. In this way, the uncertainty associated with introducing volume aliquots of sample can be reduced by controlling the time of sampling and the reaction speed, and yields can be improved through the interfaces created [87].

The use of solenoid pumps for propelling the solutions has been recently introduced and is known as Multipumping Flow Systems (MPFS). In addition, Multisyringe Flow Injection Analysis (MSFIA) has been developed implementing multicommutation principles in a modified Sequential Injection Analysis (SIA) system [88, 89]. Main components of multicommutation-based set-ups are depicted in Figure 16.5 (a to e) [90]. The propulsion system used to move each of the reagents involved or the entire reagents together is shown in Figures 16.5(a) and Figure 16.5(b). The evolution of multisyringe burette coupled with an additional injection system to simply using a multisyringe piston pump with solenoid commutation valves is depicted in Figures 16.5(c) and 16.5(d). In addition, minipumps (Figure 16.5e) allow the miniaturization of the propulsion system and the reduction of energy, integrating mixing and merging zones and providing truly portable set-ups.

Applications of multicommutation flow systems are a current research area in analytical chemistry and pharmaceutical journals. Some multicommuted flow systems involve processing of a single sample in such a way that two or more analytical signals are obtained. As different analytical results are obtained for the same sample under different diluting conditions, an additional accuracy assessment is readily available [91]. Detection techniques including chemiluminescence and fluorescence are the most utilized for multicommuted studies [88]. Applications relying on UV/Vis spectrophotometry or electroanalytical techniques are scarce, perhaps in view of the unavoidable formation of undesirable concentration gradients that might impair proper sample monitoring [88].

Recent literature includes a flow system based for fast and clean determination of cyclamate by reaction of nitrite and subsequent spectrophotometric determination of excess nitrite by iodometry [92], the system was designed with a set of solenoid micropumps to minimize reagent consumption and waste generation, as well as spectrophotometric determination of phenols in water [93]. The approach allowed a 200-fold reduction of the reagent consumption in comparison with the reference batch method. In addition, the chloroform extraction for analyte concentration was obviated. Another example is application of multicommutation to vibrational spectroscopy by the infrared determination of benzene in gasoline; the method permitted direct determination of the analyte without any pre-treatment of samples. Advantages of the method were low solvent consumption and a very high throughput sample [94]. In this sense, recent developments regarding the use of vibrational spectroscopy techniques for detection in flow-injection analysis systems, from classical FIA to modern (sequential injection analysis (SIA) or multicommutation) have been summarized by M. de la Guardia and co-workers [95]. Besides, multicommutation has been applied for Hg determination in milk by hydride generation atomic fluorescence spectrometry (HG-AFS) [96]. The method reduces four-fold reagent consumption and effluent generation, and it also increases sample throughput. In addition, analytical characteristics of multicommutation have been compared with those obtained by classical FIA and SIA, in improving the automated spectrophotometric determination of carbaryl with p-aminophenol (PAP) [97]. Multicommutation provided a limit of detection (LOD) comparable to that obtained using FIA and lower than that found by SIA, but generating a lower total waste volume, which was comparable to that found in SIA and six times lower than that obtained by classical FIA. This comparative study showed that SIA is the best strategy for reducing reagent consumption but multicommutation provides a faster, more sensitive alternative to SIA. Regarding to this, a comparative and illustrative table highlighting the performance of the different FI-based approaches regarding the most desirable green characteristics is presented in Table 16.1. In this sense, and considering the sample manipulation features of the two most popular FI methodologies in the academic labs, the following aspects can be mentioned: computer controlled and automated SIA single-channel manifolds and pumps – syringe types – are more robust that those multiple-channels configurations and peristaltic pumps required for FIA. As a consequence, sample and reagents consumptions, and waste generation are considerably reduced. In addition, SIA manifolds enable to implement a wide range of assays. On other note, a recent article describes how a multicommuted flow injection analysis setup for photometric determination of iodate in table salt can be downscaled [98]. The setup consisted of a flow system module and a LED based photometer, with a total internal volume of a few microliters. Profitable features, such as a low reagent consumption high sampling throughput, and low waste generation were achieved. In general and

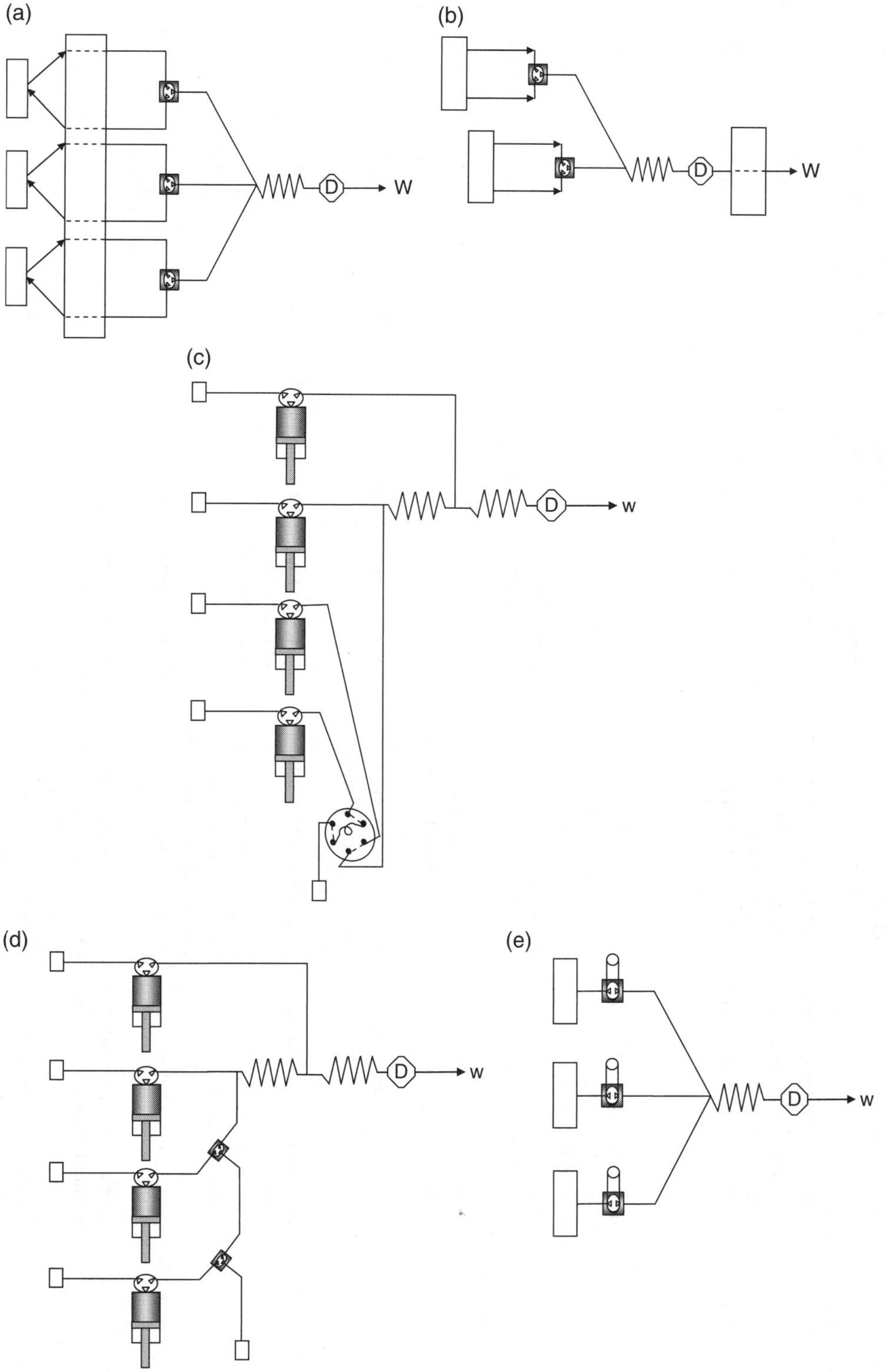

Figure 16.5 *Regular multicommutation set-ups: (a, b) solenoid-valves based; (c, d) multi-syringes based, and (e) mini-pump based set-ups. D: Detector; W: Waste. Adapted from [86] with permission from Elsevier, © 2002.*

Table 16.1 *Some examples of the flow-based techniques and greener strategies related to replacement of toxic reagents, miniaturization of reagents and wastes, recovery of reagents, on-line decontamination of wastes, reagent-free methodologies, among others.*

Flow-based procedure	Analyte/Sample	Sample/Reagent consumption	Waste generation	Sampling frequency (h^{-1})	Precision (% R.S.D.)	Limit of detection	Reference
Multi-Syringe FIA-UV	Nitro-substituted phenols in mineral, tap, and sea water	1.5 ml of sample. The use of organic solvents is circumvented	Not informed	135	≤ 5.0	0.3–3.2 μM l^{-1}	[26]
Closed FIA-FTIR system	Malathion in commercial formulations	5 ml of sample injected in a closed manifold in which 2 ml of $CHCL_3$ was recirculated	2 ml of $CHCL_3$	30	0.2–0.8	12 μg l^{-1}	[28]
FIA-UV	Chloride in natural waters utilizing $Hg(SCN)_2$ immobilized in epoxy resin	Sample-500 μl; Fe(III)-50 mg/determination; Hg(II)-12 mg/det.-; SCN-0.07 mg/det.-	400% reduction of waste generation	100	2.2	1.4×10^{-5} mol l^{-1}	[29]
FIA-Chemiluminescence	Chlorpyrifos in fruits	~ 1 ml	Not informed	120	< 3.0	0.18 ng ml^{-1}	[30]
FIA-UV	Carbaryl in natural waters	1.9 μg of p-aminophenol and 5.7 μg of potassium metaperiodate	2.6 ml	55	0.7	1.7 μg l^{-1}	[31]
FIA-UV	Free-chlorine in waters	The consumption of reagents was reduced 20 000 fold	Not informed	60	0.9	6.8 μg l^{-1}	[32]
Cyclic FIA-UV/Vis	Lead in gasoline using Arsenazo (III)	20 μl injected volume	Recirculation of reagents	Not informed	0.88–1.85	0.04 mg l^{-1}	[34]
SIA-UV	Calcium, iron and magnesium in food	150 μl of sample, 75 μl of HNO_3, and 30 and 112 μg of complexing reagents	Not informed	7	≥ 4.0	Not informed	[47]
SIA-Potentiometric titration	Acidity in soft drinks	110 μl of sample	Not informed	45	≥ 0.5	Not informed	[48]
SIA-UV	Lactic acid in yogurt and fermented mash samples	100 μl of sample, 50 μl of reagent (Ce^{4+}) and 100 μl of carrier (H_2SO_4)	Not informed	22	1.5	0.16 mg l^{-1}	[56]
SIA-UV	Sulfate in water and biodiesel samples	83 μl of sample and 166 μl of reagents	Not informed	6	0.32[a]	1.42 mg l^{-1}	[57]
μSI	Ammonia, glycerol, glucose and free iron for monitoring of fermentations	μl amounts	~30 ml for 40 assays	Not informed	1.5	ppm range	[70]

[illegible]	Trace metals	Not informed	Beads can or cannot be discarded according to the approach used	Not informed	1.5 (ETAAS), 2.9 (ICPMS)	10.2 (ETAAS) – 13 (ICPMS) ng ml^{-1}	[71]
BIS (Bead Injection Spectroscopy)/FIA	Muscarinic receptor antagonism on living cells	4.5–5.5 ml	Not informed	11	Not informed	Not informed	[72]
μFI with Optical Sensor	Cu(II) in wastewaters	2 μl	200% reduction	30	2.5	0.1 μg ml^{-1}	[74]
μFI-UV	Malondialdehyde in human serum of diabetic patients	20 ml of serum sample and 420 ml of aqueous phase	Not informed	Not informed	0.7–0.8	0.19 mmol l^{-1}	[75]
MSFIA-UV	Sulfide in waters	450 or 200 μl of sample, 160 μl of reagents (DMPD and Fe^{3+}) and 1200 or 900 μl of carrier (HCl)	3.6-fold reduction in comparison with typical FI-manifolds)	13–27	1.38–1.44	0.09 and 0.15 mg l^{-1}	[91]
Multicommuted-FI-UV	Cyclamate in table sweeteners	3 mg of KI and 1.3 μg of NaNO2	2.0 ml of effluent per determination	60	1.7[b]	30 μM l^{-1}	[92]
Multicommuted-FI-AFS	Mercury in milk	Sample volume of 32.5 ml and a reagent volume consumption of 102.5 ml for 100 analysis	94.5 ml h^{-1} with on-line waste treatment	70	3.4	0.011 ng g^{-1}	[96]
Multicommuted-UV	carbaryl determination with p-aminophenol (PAP)	2, 0.092 and 0.005 g/1000 determinations of NaOH, KIO_4 and PAP respectively	Total waste volume per sample of 1.7 ml comparable to that found in SIA and 6 times lower than that obtained by conventional FIA	70	0.5[b]	26 mg l^{-1}	[97]
Multicommuted-FIA-UV	Iodate in table salt	7.3 μg of N,N-diethyl-phenylenediamine (DPD) per determination	600 μl per determination	117	0.9	17 μg l^{-1}	[98]

[a] Relative root mean squared error, [b] Coefficient of variations.

similar to SIA, multicommuted systems use significantly lower volumes of sample and reagents than those required in classical FIA [76, 77, 87]. However, other analytical figures of merit (e.g., sampling rate and sensitivity) could be better than those of the SIA system [77, 87, 99].

16.6 Conclusions and remarks

The use of GAC methods has proved to be a smart strategy to provide both environmental and economic benefits, so we would like to propose for these methods the term 'Sustainable Analytical Procedures' (SAPs). The progress of flow methodologies has contributed to GAC, but we are convinced that their potential has not yet been exploited fully. Diversity of flow manipulation combined with the possibilities in detection systems have evolved extraordinarily in diversity and size. Therefore, the combination of flow handling techniques and detection systems has led to a countless number of possibilities.

Flow systems have been developed for nearly all types of samples ranging from pharmaceutical preparations to complex solid samples such as soil and food. Multipumping flow systems have proved to be a cost-effective, valuable alternative to fully automated analytical methods for pharmaceutical, food, environmental, agro-industrial and large-scale routine clinical analysis. Moreover, they are fast, precise and accurate, and require less operator intervention or maintenance than classical FIA. Miniaturization is one way to avoid side effects of analytical methods, and has been the subject of a significant number of research efforts. In this respect, combination of modern analytical techniques with breakthroughs in microelectronics and miniaturization allows development of powerful analytical devices for effective control of processes and pollution. Combining miniaturization in analytical systems with advances in chemometrics is very important. Of course, development and improvement of new components for instrumentation is critical in GAC. Using examples, we have illustrated the power and the versatility of modern analytical systems and their potential for minimizing the consumption of hazardous substances and the amounts of waste generated during assays.

References

[1] Tobiszewski, M.; Mechlinska, A.; Zygmunt, B. and Namiesnik, J. (2009) Green Analytical Chemistry in sample preparation for determination of trace organic pollutants, *TrAC-Trend Anal. Chem.*, **28**, 943–951.

[2] Anastas, P.T. (1999) Green Chemistry and the role of analytical methodology development, *Crit. Rev. Anal. Chem.*, **29**, 167–175.

[3] Armenta, S.; Garrigues, S. and de la Guardia, M. (2008) Green Analytical Chemistry, *TrAC-Trend Anal. Chem.*, **27**, 497–511.

[4] Wardencki, W.; Curylo, J. and Namiesnik, J. (2007) Trends in solventless sample preparation techniques for environmental analysis, *J. Biochem. Biophys. Meth.*, **70**, 275–288.

[5] Stockwell, P.B. (1990) The role of flow injection analysis within the framework of an automated laboratory, *J. Autom. Chem.*, **12**, 95–103.

[6] Rodenas-Torralba, E.; Morales-Rubio, A. and de la Guardia, M. (2006) Scientometric picture of the evolution of the literature of automation in spectroscopy and its current state, *Spectrosc. Lett.*, **39**, 513–532.

[7] Ruzicka, J. and Hansen, E.H. (1975) Flow injection analysis. Part I. A new concept of fast continuous flow analysis, *Anal. Chim. Acta*, **78**, 145–157.

[8] Hansen, H. and Wang, J. (2005) The three generations of flow injection analysis, *Anal. Lett.*, **37**, 345–359.

[9] Ruzicka, J. and Marshall, G.D. (1990) Sequential injection: a new concept for chemical sensors, process analysis and laboratory assays, *Anal. Chim. Acta*, **237**, 329–343.

[10] Ruzicka, J. (2000) Lab-on-valve: universal microflow analyzer based on sequential and bead injection, *Analyst*, **125**, 1053–1060.

[11] Keith, L.H.; Gron, L.U. and Young, J.L. (2007) Green analytical methodologies, *Chem. Rev.*, **107**, 2695–2708.
[12] Ruzicka, J. and Hansen, E.H. (1988) *Flow Injection Analysis*, John Wiley & Sons, Inc., New York.
[13] Hansen, E.H. (2004) The impact of flow injection on modern chemical analysis: has it fulfilled our expectations? And where are we going? *Talanta*, **64**, 1076–1083.
[14] Bendtsen, A.B. and Hansen, E.H. (1991) Spectrophotometric flow injection determination of trace amounts of thiocyanate based on its reaction with 2-(5-bromo-2-pyridylazo)-5-diethylaminophenol and dichromate: assay of the thiocyanate level in saliva from smokers and non-smokers, *Analyst*, **116**, 647–651.
[15] Hansen, E.H. (1994) Flow injection analysis: a complementary or alternative concept to biosensors, *Talanta*, **41**, 939–948.
[16] Hansen, E.H. (1988)Exploitation of gradient techniques in flow injection analysis, *Fresenius J. Anal. Chem.*, **329**, 656–659.
[17] Fang, Z.L. and Burguera, J.L. (1989) *Analytical methods and techniques. Flow Injection Atomic Spectroscopy, Practical Spectroscopy Series*, Marcel Dekker, New York.
[18] Fang, Z.L. (1995) *On-Line Preconcentration for Flame and Hydride Generation of Atomic Absorption Spectrometry; and Flow Injection Technique for Electrothermal Atomic Absorption Spectrometry. Flow Injection Atomic Spectrometry*, John Wiley & Sons, Inc., New York.
[19] Yoshimura, K.; Waki, H. and Ohashi, S. (1976) Ion-exchanger colorimetry—I: Microdetermination of chromium, iron, copper and cobalt in water, *Talanta*, **23**, 449–454.
[20] Appelqvist, R. Beecher, G.R. Bergamin, H. *et al.* (1986) Flow injection analysis (FIA) – a personal view, *Anal. Chim. Acta*, **180**, 1–67.
[21] Molina-Díaz, A.; García-Reyes, J.F. and Gilbert-López, B. (2010) Solid-phase spectroscopy from the point of view of Green Analytical Chemistry, *TrAC-Trend Anal. Chem.*, **29**, 654–666.
[22] Garrigues, S.; Armenta, S. and de la Guardia, M. (2010) Green strategies for decontamination of analytical wastes, *TrAC-Trend Anal. Chem.*, **29**, 592–601.
[23] Hansen, E.H. and Miró, M. (2007) How flow-injection analysis (FIA) over the past 25 years has changed our way of performing chemical analyses, *TrAC-Trend Anal. Chem.*, **26**, 18–26.
[24] Gómez, V. and Callao, M.P. (2007) Multicomponent analysis using flow systems, *TrAC-Trend Anal. Chem.*, **26**, 767–774.
[25] Luque de Castro, M.D. (1992) Solid phase reactors in flow injection analysis, *TrAC-Trend Anal. Chem.*, **11**, 149–155.
[26] Manera, M.; Miró, M.; Estela, J.M. and Cerdá, V. (2007) Multi-syringe flow injection solid-phase extraction system for on-line simultaneous spectrophotometric determination of nitro-substituted phenol isomers, *Anal. Chim. Acta*, **582**, 41–49.
[27] Melchert, W.R. and Rocha, F.R.P. (2005) A green analytical procedure for flow-injection determination of nitrate in natural waters, *Talanta*, **65**, 461–465.
[28] Quintas, G.; Morales-Noé, A.; Armenta, S.; Garrigues, S. and de la Guardia, M. (2004) Fourier transform infrared spectrometric determination of malathion in pesticide formulations, *Anal. Chim. Acta*, **502**, 213–220.
[29] Silva, C.R.; Vieira, H.J.; Canaes, L.S.; Nobrega, J.A. and Fatibello-Filho, O. (2005) Flow injection spectrophotometric method for chloride determination in natural waters using $Hg(SCN)_2$ immobilized in epoxy resin, *Talanta*, **65**, 965–970.
[30] Song, Z.; Hou, S. and Zhang, N. (2002) A New Green Analytical Procedure for Monitoring Sub-nanogram Amounts of Chlorpyrifos on Fruits Using Flow Injection Chemiluminescence with Immobilized Reagents, *J. Agr. Food Chem.*, **50**, 4468–4474.
[31] Melchert, W.R. and Rocha, F.R.P. (2010) A greener and highly sensitive flow-based procedure for carbaryl determination exploiting long pathlength spectrophotometry and photochemical waste degradation, *Talanta*, **81**, 327–333.
[32] Melchert, W.R.; Oliveira, D.R. and Rocha, F.R.P. (2010) An environmentally friendly flow system for high-sensitivity spectrophotometric determination of free chlorine in natural waters, *Microchem. J.*, **96**, 77–81.
[33] Dutta, S. and Das, A.K. (2010) Analytical perspective on waste management for environmental Remediation, *TrAC-Trend Anal. Chem.*, **29**, 636–644.
[34] Zenki, M.; Minamisawa, K. and Yokoyama, T. (2005) Clean analytical methodology for the determination of lead with Arsenazo III by cyclic flow-injection analysis, *Talanta*, **68**, 281–286.

[35] de la Guardia, M.; Khalaf, K.D.; Hasan, B.A.; Morales-Rubio, A. and Carbonell, V. (1995) In-line, titanium dioxide-catalysed, ultraviolet mineralization of toxic aromatic compounds in the waste stream from a flow injection-based resorcinol analyzer, *Analyst*, **120**, 231–235.
[36] Camel, V. (2003) Solid-phase extraction of trace elements, *Spectrochim. Acta Pt. B*, **58**, 1177–1233.
[37] Pacheco, P.H.; Spisso, A.; Cerutti, S.; Smichowski, P. and Martinez, L.D. (2010) Non-chromatographic screening method for the determination of mercury species. Application to the monitoring of mercury levels in Antarctic samples, *Talanta*, **82**, 1505–1510.
[38] Ferrúa, N.; Cerutti, S.; Salonia, J.A.; Olsina, R.A. and Martinez, L.D. (2007) On-line preconcentration and determination of mercury in biological and environmental samples by cold vapor-atomic absorption spectrometry, *J. Hazard. Mater.*, **141**, 693–699.
[39] Gil, R.A.; Cerutti, S.; Gásquez, J.A.; Olsina, R.A. andMartinez, L.D. (2006) Preconcentration and speciation of chromium in drinking water samples by coupling of on-line sorption on activated carbon to ETAAS determination, *Talanta*, **68**, 1065–1070.
[40] Cerutti, S.: Salonia, J.A.; Gásquez, J.A.; Olsina, R.A. and Martinez, L.D. (2003) Determination of scandium in river water by ICP-OES with flow injection on-line preconcentration using knotted reactor and ultrasonic nebulization, *J. Anal. Atom. Spectrom.*, **18**, 1198–1201.
[41] Cerutti, S.: Wuilloud, R.G. and Martinez, L.D. (2005) The knotted Reactors and their role in flow-injection on-line preconcentration systems coupled to atomic spectrometry-based detectors, *Appl. Spectrosc. Rev.*, **40**, 71–101.
[42] Silva, M.F.; Cerutti, S. and Martinez, L.D. (2006) Coupling cloud point extraction to instrumental detection systems for metal analysis, *Microchim. Acta*, **155**, 349–364.
[43] Mesquita, R.B.R. and Rangel, A.O.S.S. (2009) A review on sequential injection methods for water analysis, *Anal. Chim. Acta*, **648**, 7–22.
[44] Pimenta, A.M.; Montenegro, M.C.B.S.M.; Araújo, A.N. and Martínez-Calatayud, J. (2006) Application of sequential injection analysis to pharmaceutical analysis, *J. Pharmaceut. Biomed. Anal.*, **40**, 16–34.
[45] Wang, J. and Hansen, E.H. (2005) Trends and perspectives of flow injection/sequential injection on-line sample-pretreatment schemes coupled to ETAAS, *TrAC-Trend Anal. Chem.*, **24**, 1–8.
[46] Barnett, N.W.; Lenehan, C.E.; Lewis, S.W. and Barnett, N.W. (1999) Sequential injection analysis: an alternative approach to process analytical chemistry, *TrAC-Trend Anal. Chem.*, **18**, 346–353.
[47] Oliveira, C.C.; Sartini, R.P. and Zagatto, E.A.G. (2000) Microwave-assisted sample preparation in sequential injection: spectrophotometric determination of magnesium, calcium, and iron in food, *Anal. Chim. Acta*, **413**, 41–48.
[48] van Staden, J.F.; Mashamba, M.G. and Stefan, R.I. (2002) Determination of the total acidity in soft drinks using potentiometric sequential injection titration, *Talanta*, **58**, 1109–1114.
[49] Baxter, P.J. and Christian, G.D. (1996) Sequential injection analysis—a versatile technique for bioprocess monitoring, *Account. Chem. Res.*, **29**, 515–521.
[50] Cerda, V.; Cerda, A.; Cladera, A. (2001) *et al.* Monitoring of environmental parameters by sequential injection analysis, *TrAC-Trend Anal. Chem.*, **20**, 407–418.
[51] Liu, X.Z. and Fang, Z.L. (2000) The automation of pharmaceutical analysis using sequential-injection techniques, *Lab. Robotics Automation*, **12**, 60–66.
[52] Barnett, N.W.; Lenehan, C.E. and Lewis, S.W. (1999) Sequential injection analysis: an alternative approach to process analytical chemistry, *TrAC-Trend Anal. Chem.*, **18**, 346–353.
[53] Morais, I.P.A.; Souto, M.R.S. and Rangel, A.O.S.S. (2005) A double-line sequential injection system for the spectrophotometric determination of copper, iron, manganese, and zinc in waters, *J. AOAC Int.*, **88**, 639–644.
[54] Teixeira, L.S.G. and Rocha, F.R.P. (2007) A green analytical procedure for sensitive and selective determination of iron in water samples by flow-injection solid-phase spectrophotometry, *Talanta*, **71**, 1507–1511.
[55] Mimendia, A.; Legin, A.; Merkoci, A. and del Valle, M. (2010) Use of sequential injection analysis to construct a potentiometric electronic tongue: application to the multidetermination of heavy metals, *Sensor. Actuator. B*, **146**, 420–426.
[56] Dias, A.C.B.; Silva, R.A.O. and Arruda, M.A.Z. (2010) A sequential injection system for indirect spectrophotometric determination of lactic acid in yogurt and fermented mash samples, *Microchem. J.*, **96**, 151–156.
[57] del Río, V.; Larrechi, M.S. and Callao, M.P. (2010) Determination of sulphate in water and biodiesel samples by a sequential injection analysis-Multivariate curve resolution method, *Anal. Chim. Acta*, **676**, 28–33.

[58] Wang, Y.; Chen, M.-L. and Wang, J.-H. (2007) New Developments in Flow Injection/Sequential Injection On-line Separation and Preconcentration Coupled with Electrothermal Atomic Absorption Spectrometry for Trace Metal Analysis, *Appl. Spectrosc. Rev.*, **42**, 103–118.

[59] Pasamontes, A. and Callao, M.P. (2006) Sequential injection analysis linked to multivariate curve resolution with alternating least squares, *TrAC-Trend Anal. Chem.*, **25**, 77–85.

[60] Segundo, M.A. and Magalhaes, L.M. (2006) Multisyringe Flow Injection Analysis: State-of-the-Art and Perspectives, *Anal. Sci.*, **22**, 3–8.

[61] Economou, A.; Tzanavaras, P.D. and Themelis, D.G. (2005) Sequential-Injection Analysis: Principles, Instrument Construction, and Demonstration by a Simple Experiment, The entity from which ERIC acquires the content, including journal, organization, and conference names, or by means of online submission from the author.*J. Chem. Educ.*, **82**, 1820–1822.

[62] Segundo, M.A. and Rangel, O.S.S. (2002) Flow analysis: a critical view of its evolution and perspectives, *J. Flow Inj. Anal.*, **19**, 3–8.

[63] Allan, C.V. and Jorge, C.M. (2009) Applying sequential injection analysis (SIA) and response surface methodology for optimization of Fenton-based processes, *Talanta*, **77**, 1081–1086.

[64] Kikas, T. and Ivaska, A. (2007) Potentiometric measurements in sequential injection analysis lab-on-valve (SIA-LOV) flow-system, *Talanta*, **71**, 160–164.

[65] Amorim, C.G.; Araujo, A.N. and Montenegro, M.C.B.S.M. (2007) Exploiting sequential injection analysis with lab-on-valve and miniaturized potentiometric detection, *Talanta*, **72**, 1255–1260.

[66] van Staden, J.F. and Stefan, R.I. (2004) Chemical speciation by sequential injection analysis: an overview, *Talanta*, **64**, 1109–1113.

[67] Economou, A. (2005) Sequential-injection analysis (SIA): A useful tool for on-line sample-handling and pre-treatment, *TrAC-Trend Anal. Chem.*, **24**, 416–425.

[68] Perez-Olmos, R.; Soto, J.C.; Zarate, N.; Araujo, A.N.; Lima, J.L.F.C. and Saraiva, M.L.M.F.S. (2005) Application of sequential injection analysis (SIA) to food analysis, *Food Chem.*, **90**, 471–490.

[69] Manz, A.; Graber, N. and Widmer, A. Manz, N. Graber and H.M. Widmer, H.M. (1990) Miniaturized total chemical analysis systems: A novel concept for chemical sensing, *Sensor. Actuator.*, **B1**, 244–248.

[70] Wu, C.H.; Scampavia, L.; Ruzicka, J. and Zamost, B. (2001) Microsequential injection: fermentation monitoring of ammonia, glycerol, glucose, and free iron using the novel lab-on-valve system, *Analyst*, **126**, 291–297.

[71] Hansen, E.H. and Wang, J.H. (2002) Implementation of suitable FI/SI-sample separation/preconcentration schemes for determination of trace metal concentrations using detection by ETAAS and ICPMS, *Anal. Chim. Acta*, **467**, 3–12.

[72] Hodder, P.S.; Beeson, C. and Ruzicka, J. (2000) Equilibrium and kinetic measurements of muscarinic receptor antagonism on living cells using bead injection spectroscopy, *Anal. Chem.*, **72**, 3109–3115.

[73] Ruzicka, J. and Scampavia, L. (1999) From flow injection to bead injection, *Anal. Chem.*, **71**, 257A–263A.

[74] Leelasattarathkul, T.; Liawruangrath, S.; Rayanakorn, M.; Liawruangrath, B.; Oungpipat, W. and Youngvises, N. (2007) Greener analytical method for the determination of copper(II) in wastewater by micro flow system with optical sensor, *Talanta*, **72**, 126–131.

[75] Serafin Muñoz, A.H.; Preciado Puga, M.; Wrobel, K.; Garay Sevilla, M.E. and Wrobel, K. (2004) Microassay for malondialdehyde in human serum by extraction-spectrophotometry using an internal standard, *Microchim. Acta*, **148**, 285–291.

[76] Zagatto, E.A.G.; Reis, B.F.; Oliveira, C.C.; Sartini, R.P. and Arruda, M.A.Z. (1999) Evolution of the commutation concept associated with the development of flow analysis, *Anal. Chim. Acta*, **400**, 249–256.

[77] Giné, M.F.; Bergamin, H.; Zagatto, E.A.G. and Reis, B.F. (1980) Simultaneous determination of nitrate and nitrite by flow injection analysis, *Anal. Chim. Acta*, **114**, 191–197.

[78] Bergamin, H.; Zagatto, E.A.G.; Krug, F.J. and Reis, B.F. (1978) Merging zones in flow injection analysis: Part 1. Double proportional injector and reagent consumption, *Anal. Chim. Acta*, **101**, 17–23.

[79] Wang, X.D.; Cardwell, T.J.; Cattrall, R.W. and Jenkins, G.E. (1998) Pulsed flow chemistry. A new approach to the generation of concentration profiles in flow analysis, *Anal. Commun.*, **35**, 97–101.

[80] Ruzicka, J.; Marshall, G.D. and Christian, G.D. (1990) Variable flow rates and a sinusoidal flow pump for flow injection analysis, *Anal. Chem.*, **62**, 1861–1866.

[81] Brienza, S.M.B.; Zagatto, E.A.G. and Matsumoto, O.M. (1991) A simplified split zone flow injection system, *Lab. Robotics Automation*, **3**, 115–118.

[82] Toei, J. (1988) Determination of glucose in clinical samples by flow reversal flow injection analysis, *Analyst*, **113**, 475–478.

[83] Arruda, M.A.Z. and Zagatto, E.A.G. (1987) A simple stopped-flow method with continuous pumping for the spectrophotometric flow-injection determination of boron in plants, *Anal. Chim. Acta*, **199**, 137–145.

[84] Cardwell, T.J.; Cattrall, R.W.; Cross, J.G.; O'Connell, G.R.; Petty, J.D. and Scollary, G.R. (1991) Determination of titratable acidity of wines and total acidity of vinegars by discontinuous flow analysis using photometric end-point detection, *Analyst*, **116**, 1051–1054.

[85] Reis, B.F.; Gine, M.F.; Zagatto, E.A.G.; Lima, J.L.F.C. and Lapa, R.A.S. (1994) Multicommutation in flow analysis. 1. Binary sampling. Concepts, instrumentation and spectrophotometric determination of iron in plant digests, *Anal. Chim. Acta*, **293**, 129–138.

[86] Rocha, F.R.P.; Reis, B.F.; Zagatto, E.A.G.; Lima, J.L.F.C.; Lapa, R.A.S. and Santos, J.L.M. (2002) Multicommutation in flow analysis: concepts, applications and trends, *Anal. Chim. Acta*, **468**, 119–131.

[87] Segundo, M.A.; Oliveira, H.M.; Lima, J.L.F.C.; Almeida, M.G.S. and Rangel, A.O.S.S. (2005) Sample introduction in multi-syringe flow injection systems: comparison between time-based and volume-based strategies, *Anal. Chim. Acta*, **537**, 207–214.

[88] Feres, M.A.; Fortes, P.R.; Zagatto, E.A.G.; Santos, J.L.M. and Lima, J.L.F.C. (2008) Multi-commutation in flow analysis: Recent developments and applications, *Anal. Chim. Acta*, **618**, 1–17.

[89] Cerda, V.; Estela, J.M.; Forteza, R.; Cladera, A.; Becerra, E.; Altimira, P. and Sitjar, P. (1999) Flow techniques in water analysis, *Talanta*, **50**, 695–705.

[90] Morales-Rubio, A.; Freire dos Reis, B. and de la Guardia, M. (2009) Multi-commutation in spectrometry, *TrAC-Trend Anal. Chem.*, **28**, 903–913.

[91] Ferrer, L.; Armas, G.; Miro, M.; Estela, J.M. and Cerda, V. (2004) A multisyringe flow injection method for the automated determination of sulfide in waters using a miniaturised optical fiber spectrophotometer, *Talanta*, **64**, 1119–1126.

[92] Rocha, F.R.P.; Rodenas-Torralba, E.; Morales-Rubio, A. and de la Guardia, M. (2005) A clean method for flow injection spectrophotometric determination of cyclamate in table sweeteners, *Anal. Chim. Acta*, **547**, 204–208.

[93] Lupetti, K.O.; Rocha, F.R.P. and Fatibello-Filho, O. (2004) An improved flow system for phenols determination exploiting multicommutation and long pathlength spectrophotometry, *Talanta*, **62**, 463–467.

[94] Ródenas-Torralba, E.; Ventura-Gayete, J.; Morales-Rubio, A.; Garrigues, S. and de la Guardia, M. (2004) Multicommutation fourier transform infrared determination of benzene in gasoline, *Anal. Chim. Acta*, **512**, 215–221.

[95] Armenta, S.; Garrigues, S. and de la Guardia, M. (2007) Recent developments in flow-analysis vibrational spectroscopy, *TrAC-Trend Anal. Chem.*, **26**, 775–787.

[96] Cava-Montesinos, P.; Ródenas-Torralba, E.; Morales-Rubio, Á.; Cervera, M.L. and de la Guardia, M. (2004) Cold vapour atomic fluorescence determination of mercury in milk by slurry sampling using multicommutation, *Anal. Chim. Acta*, **506**, 145–153.

[97] Reis, B.F.; Morales Rubio, A. and de la Guardia, M. (1999) Environmentally friendly analytical chemistry through automation: comparative study of strategies for carbaryl determination with p-aminophenol, *Anal. Chim. Acta*, **392**, 265–272.

[98] Borges, S.S.; de Souza Peixoto, J.; Feres, M.A. and Reis, B.F. (2010) Downscaling a multicommuted flow injection analysis system for the photometric determination of iodate in table salt, *Anal. Chim. Acta*, **668**, 3–7.

[99] Santos, J.L.M.; Ribeiro, M.F.T.; Dias, A.C.B.; Lima, J.L.F.C. and Zagatto, E.A.G. (2007) Multi-pumping flow systems: The potential of simplicity, *Anal. Chim. Acta*, **600**, 21–28.

17
Miniaturization

Alberto Escarpa[1], Miguel Ángel López[1] and Lourdes Ramos[2]

[1]*Department of Analytical Chemistry and Chemical Engineering, University of Alcala, Madrid, Spain*
[2]*Department of Instrumental Analysis and Environmental Chemistry, Institute of Organic Chemistry, CSIC, Madrid, Spain*

This chapter highlights miniaturization in sample preparation as a valuable alternative for Green Analytical Chemistry in two broad parts: non-integrated approaches and integrated ones or 'lab-on-a-chip' technology. Regarding the first part, state-of-the-art will be discussed on the basis of examples selected from representative application areas and involving conventional instrumental techniques for final determination of the target compounds. Emphasis will be on those techniques and approaches that have already demonstrated their practicality by the analysis of real-life samples, and in particular on those dealing with the accurate determination of minor organic components. The second part of this chapter will describe selected approaches to lab-on-a-chip scenarios for integration of sample treatment based on the microfluidic behaviour as well as microfabrication and (electro-)focusing techniques.

The potential of latest developments in this field for sample treatment simplification and complete hyphenation and integration of analytical process will be discussed and the most pressing remaining limitations evaluated. Forces and weakness will be critically evaluated and the main future trends highlighted.

Handbook of Green Analytical Chemistry, First Edition. Edited by Miguel de la Guardia and Salvador Garrigues.

17.1 Current needs and pitfalls in sample preparation

Analytical procedures typically involve a number of equally relevant steps for sampling, sample treatment, isolation of the target compounds, identification, quantification and data handling. Most modern instrumental techniques are sufficiently mature enough to enable the hyphenation of different separation techniques with each other and with detectors that often provide a high selectivity and sensitivity. The use of such a sophisticated and powerful hyphenated systems for final instrumental analysis contrasts sharply with the often large-scale and highly manipulative traditional methodologies still in use for sample treatment (e.g. liquid–liquid extraction (LLE) or Soxhlet extraction). Due to the essentially non-selective character of the initial extraction steps, especially in the case of trace analysis, a number of subsequent treatments are usually required for isolation of the analytes of interest from primarily coextracted matrix components and then from other structurally related compounds frequently present in the sample and that could interfere in their final determination (unless separation-plus-detection is highly selective). The several analytical treatments involved in such procedures are usually carried out off-line, which makes them tedious, expensive in terms of time and reagents consumption, prone to contamination and degradation of the analytes because of the continual manipulation of the extracts, and often result in the generation of a relatively high amount of waste. Thereby, apart from their inherent analytical shortcomings, they can also be considered the less environmentally friendly part of the analytical process. These limitations explain the many efforts carried out in recent years to develop novel techniques that contribute to solve the most pressing drawbacks of such traditional methodologies and, when possible, allow increasing integration and/or simplification of the several treatment steps. The studies in this field have led to the development of new faster and more powerful and/or versatile extraction techniques [1] and, in many instances, also to partial and even full hyphenation and automation of the analytical process. In most instances, the proposed approaches fulfil the goals of Green Analytical Chemistry (GAC), that is, contribute to reduce, or virtually eliminate, the use of hazardous chemicals as well as the amount of wastes generated [2].

For obvious reasons, the ideal situation would be the complete elimination of sample pre-treatment from the analytical process. However, and despite the actual degree of development of the analytical instrumentation used for final determination, in most cases, this is not feasible. In this context, miniaturization is a valuable analytical tool that, per se, efficiently contributes to greening the analytical process by reducing the amount of solvents and reagents involved in the analysis. But miniaturization is also a worthy alternative that contributes to solve some of the most pressing drawbacks of conventional sample preparation methods by favouring the hyphenation, or at least simplification and integration of the analytical treatments; the development of complete on-line and/or (semi-)automated systems, and the effective reduction of energy cost and wastes, as will be shown next. In addition, miniaturization is probably the best alternative when dealing with the analysis of size-limited samples, when studying processes that take place in times shorter than those involved by traditional methodologies and when using separation techniques with limited sensitivity due to their reduced loading capacity, such as narrow-bore chromatography or electrophoretic separation systems.

On the other hand, total miniaturization or 'lab-on-a-chip' technology offers unique advantages for GAC because of the extremely low sample and reagents consumption as well as negligible wastes becomes as very environmentally friendly in the GAC landscape. Because of its inherent complexity, the sample treatment step is less developed than those of separation and detection on lab-on-a-chip devices being its integration one of the main challenges into the scene. From an analytical point of view, the analysis of real samples is often the ultimate goal of a target study. Although the volume of research in this area is impressive, because of the difficult to integrate sample treatment on lab-on-a-chip; real sample analysis is still in its infancy and constitutes one of the major challenges in the field. However, for these types of microdevice, a very unique role in the future for analysis of real samples is highly expected because of the inherent possibilities of microfabrication technology, the unique behaviour of *microfluidics*, and the possibilities of accurate fluidic manipulation.

In two broad parts, this chapter reviews the current state-of-the-art regarding miniaturization in sample preparation and discusses most recent developments and achievements in the field on the basis of selected examples. In particular, attention will focus on the analysis of trace organic compounds due to the difficulty associated to this type of determination. Nevertheless, if relevant, examples will also be taken from other application areas. In all cases, emphasis will be on techniques that have already demonstrated their practicality by the analysis of real-life samples.

17.2 Non-integrated approaches for miniaturized sample preparation

In this section, the different analytical approaches will be revised on the base of sample preparation techniques best suited for each type of matrix. Nevertheless, many of these techniques can be used at different stages of sample pre-treatment for most matrices.

17.2.1 Gaseous and liquid samples

17.2.1.1 Thermal desorption-based techniques

Purge-based techniques have successfully been used for years in combination with gas chromatography (GC) for the analysis of volatile components in liquid, semi-solid and solid samples. There are solvent-free approaches among which the most significant and widely used is the so-called Purge-and-Trap (P&T). During the last decades, several automatic systems have been commercialized for this already mature and well established technique. Thereby, research in this field remains essentially in the development of novel sorbents for improved analyte retention and of new electronics allowing faster heating and cooling of the different parts of the instrumentation. Probably one of the most interesting recent achievements regarding this technique has been the introduction of a direct thermal desorption (DTD) liner-exchange unit for determination of (semi-)volatile analytes in solid samples with subsequent GC analysis. The system, introduced by de Koning *et al.* [3], enables complete sample preparation to be performed in an automated, on-line and miniaturized fashion. The feasibility of the approach for obtaining information from sample sizes which could hardly be handled with other extraction techniques has been illustrated, for example, by determination of wood preservatives using only 10 mg of pine sapwood powder, or by obtaining paleontogical information from 1–20 pollen grains (pollen size ca. 20 µm). The possibility of performing reactions directly in the liner and/or using a fresh liner-plus-vial for each sample with automatic exchange of both opens new application possibilities to be exploited in future and that certainly will contribute to further simplify sample handling.

17.2.1.2 Solvent-based extraction techniques

In principle, the simplest strategy to miniaturize a procedure is reducing the dimensions of the systems used in earlier approaches. This approach has been used in the case of liquid-liquid extraction (LLE) yielding different modern miniaturized techniques.

When the volumes of the aqueous sample and the extractant are small enough, LLE can be performed in a chromatographic vial and the analytical approach is called in-vial LLE. The experimental parameters affecting the analytes partition process, and so to be optimized, are similar to those of LLE. Salting out of the mixture and in-vial derivatization of the analytes can also be used to improve the extraction efficiency. However, apart from its simplicity, probably the most interesting features of the techniques include the significant reduction

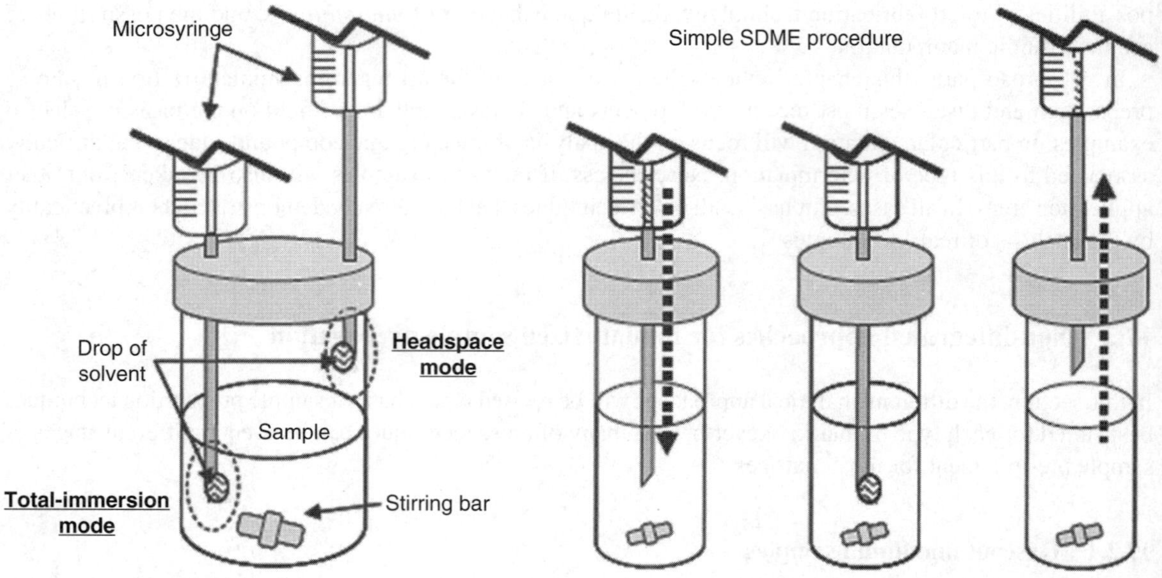

Figure 17.1 *Basic configurations for SDME and steps in the SDME immersion mode [7].*

of the amount of organic solvent(s) used, the faster separation of phases due to the (virtual) elimination of the emulsions, and the very favourable phases' ratio. All together, these contribute to increase the efficiency of the extraction process as compared to conventional LLE and to green this classical and still widely used extraction technique. In addition, when the resulting extractant is clean enough to avoid compromising the final instrumental determination, the organic phase can directly sampled by conventional autosamplers and injected in the separation-plus-detection system without any further treatment or concentration.

Up to now, the in-vial LLE has provided satisfactory results for the fast and green (semi-)automated extraction of analytes with medium and low polarity from relatively clean aqueous samples. Typical experiments involve sample volumes of 1–2 ml and ca. 500 µl of an organic solvent for which the target compounds showed a high affinity. The use of membranes is recommended in the case of dirty matrices to reduce the amount of matrix components coextrated. However, this approach does not look to be efficient enough in the analysis of non-aqueous samples, for which additional clean-up before instrumental analysis is usually mandatory.

Miniaturization can also be accomplished by developing new set-ups and techniques on the base of already known concepts or by developing completely novel concepts. Several new micro-LLE-based techniques, generically named as solvent microextraction (SME) techniques, have been developed during the last 15 years using these two approaches [4]. These techniques can be classified according to different criteria. For simplicity, in this chapter we will follow a terminology and classification essentially based on the number of phases involved in the extraction process and the two basic working modes: direct immersion sampling and headspace (HS) sampling.

The simplest technique belonging to the SME group is single-drop microextraction (SDME) [5, 6], in which a single microdrop of a water-insoluble solvent suspended at the tip of a GC syringe is either immersed in an aqueous sample or exposed to the HS of a sample contained in a vial (Figure 17.1). Typical extractant and aqueous volumes are 1–8 µl and 1–10 ml, respectively. Although SDME is an equilibrium technique, it allows enrichment factors as large as 300 with extraction times as short as 1–15 min. Stirring of the sample (up to ca. 600 rpm to prevent drop dislodgment), salting-out, application of temperature and analyte

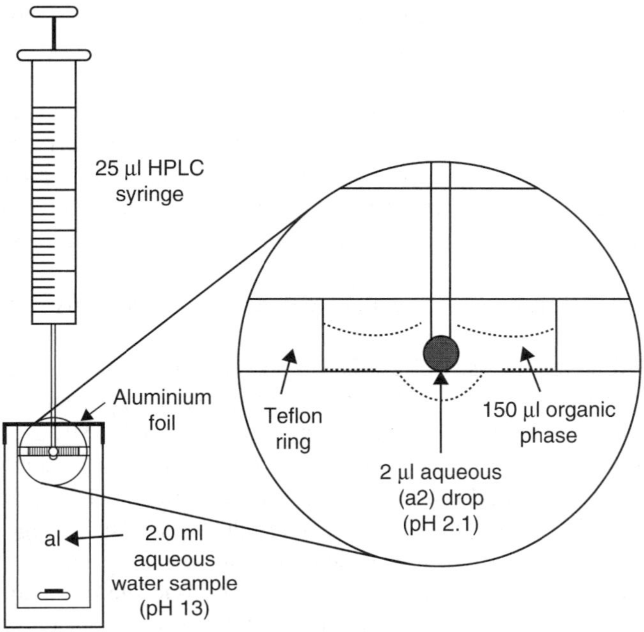

Figure 17.2 *Basic configuration for LLLME [11].*

derivatization (to reduce its polarity or increase its volatility) are common practices that, in general, contribute to increase the extraction efficiency and reduce the analysis time. The simplicity of the analytical procedure, the possibility to perform it manually or (semi-)automatic using an autosampler, and the feasibility of obtaining ready-to-analyse extracts have probably been additional factors contributing to the rapid development and acceptation of this environmental-friendly technique in different research fields.

Direct-immersion SDME has demonstrated to be useful for the extraction of relatively non-polar and semivolatile analytes from water samples that contain little or no particulate or dissolved matter. However, the analysis of more complex matrices, such as urine, requires a previous filtration of the sample [8]. Due to its characteristics, the technique is particularly suited for the treatment of size-limited samples, as recently demonstrated by Wu *et al.* [9], who used it for the simple, fast and efficient extraction of drugs from relatively complex samples (i.e. biological fluids). The analysis required only 10 µl of sample and 0.5–1.0 µl of organic extractant, and was named drop-to-drop microextraction (DDME). Despite the relatively high limits of detection (LODs), the technique allowed proper detection of the investigated drugs in blood, serum and urine.

Application of SDME to the analysis of polar compounds required a modification that resulted in a three-phase SDME system named liquid-liquid-liquid microextraction (LLLME) [10]. In this approach, the deionized polar analytes were preconcentrated from the aqueous sample in a few microlitres of organic phase placed in a PTFE ring and subsequently back-extracted in an aqueous micro-drop that acted as receiving phase (Figure 17.2). Next, this micro-drop was withdrawn into the syringe and directly subjected to liquid chromatography (LC) or capillary electrophoresis (CE) analysis. Therefore, the organic phase acts as an organic liquid membrane allowing the simultaneous enrichment and purification of the analytes. Its higher stability as compared to the organic drop of the two-phase SDME format allows higher stirring rates, something that combined with the small volume of receiving organic phase, resulted in fast extraction processes (ca. 15 min) with higher enrichment factors (in the 200–500 range). The complete renewal of the phases in between extractions also contributes to reduce the risk of cross-contamination. Although the efficiency of this approach has conveniently been proved through the analysis of model buffered solutions, to the best of our knowledge, real-life applications are still needed.

SDME can also be accomplished by direct exposure of the drop to the headspace of the investigated sample. In this case, the technique is named headspace single-drop microextraction (HS-SDME) and can be applied to gaseous, aqueous and solid samples. The technique performs very efficiently for the preconcentration of volatile non-polar analytes, and has the advantage over direct-immersion SDME of providing cleaner extracts in shorter analytical times due to the possibility of using higher stirring rates. For the rest, experimental parameters affecting the efficiency of the process are essentially the same as for the immersion mode.

All previously described SDME-based techniques are static and consequently, the main factor determining both the extraction efficiency and the extraction time is the diffusion of the extracted analytes from the drop surface to its inner part. Although the use of less viscous solvents and higher stirring rates and temperatures can contribute to increase this diffusion rate, the constant renovation of the solvent surface by using a dynamic approach is probably a more effective approach. Two type of dynamic SDME are possible: in-syringe and in-needle SME. In the former approach, the aqueous sample or headspace is withdraw into the syringe needle or lumen and ejected repeatedly to perform the desired solvent enrichment [12]. In the in-needle dynamic approach [13, 14], around 90% of the extraction drop is withdrawn into the syringe needle and then pushed out again repeatedly for sample exposure. For obvious reasons, the in-syringe approach is more effective when dealing with relatively pristine samples. Meanwhile, the in-needle one may be more useful for the analysis of relatively 'dirty' samples, that is, samples containing relatively high amount of matrix components that could affect the subsequent instrumental analysis.

17.2.1.3 Hollow fibre-protected two/three-phase solvent microextraction

Hollow fiber-protected two-phase solvent microextraction (HF(2)ME) was introduced by He and Lee in 1997 [6] with the name of liquid-phase microextraction. In its simplest version, the technique involves a small-diameter microporous polypropylene tube (the hollow fibre), usually sealed at one end, to contain the organic extracting solvent. The open end of the hollow fibre is attached to a syringe needle used to fill the fibre with the organic solvent. Once filled, the fibre is immersed in the vial containing the investigated aqueous sample to allow analytes migration through its walls. After a preselected extraction time, the solvent is withdrawal with the syringe and transferred to the instrument selected for analytes determination, typically GC. HF(2)ME can consequently be considered a liquid-liquid membrane extraction [4] and so it is more appropriate than SDME for the analysis of 'dirty' aqueous samples. The use of larger extractant volumes (typically in the 4–20 µl range) and the possibility of using higher stirring rates are other advantages of HFME over SDME. On the other hand, HF(2)ME usually involves longer extraction times than SDME (20–60 min versus 5–15 min with SDME), and, at least LVI was used, only a fraction of the organic extractant is transferred to the instrument selected for final determination. In addition, and although it can be adapted for use with an autosampler [15], probably its main limitation is that each individual hollow fibre should carefully be sized and prepared before use [4].

The three phases involved in HF(3)ME are the aqueous sample investigated, the water-immiscible organic solvent that fills the pores of the hollow fibre polymer before this is attached to the syringe needle, and an aqueous acceptor phase that is placed in the lumen of the fibre with the help of the syringe [16]. HF(3)ME is operated in a way similar to HF(2)ME but, since the final acceptor solution is aqueous, the technique is used to extract water-soluble analytes from aqueous matrices, and LC and CE are usually preferred for final instrumental determination of the tested analytes. Similarly to that explained for three-phases SDME, the pH of the aqueous sample and the acceptor phase are key parameters controlling the efficiently of the HF(3)ME process.

HF(2)ME and HF(3)ME, which can be used in a static or dynamic mode similarly to that described for SDME, share their most pressing shortcomings, namely relatively long extraction times, difficulty of complete automation and intensive manual preparation of the fibre before use. However, a number of examples can be found in the literature proving the practicality of the approach for the preconcentration of analytes of divergent polarity from size-limited aqueous samples. Interestingly, this process can be favoured by the application of a difference of potential between the two phases [17]. In this case, the technique is referred as electromembrane extraction (EME).

In an attractive modificatio the fibre, filled with solvent, is sealed at both ends, which allows to place it directly into the stirred solution for extraction. In this case, after a preselected extraction time, the fiber it retrieved form water and the enriched solvent is removed by polymer puncture with a chromatographic syringe. This technique is called solvent bar microextraction [18] and can also be use as a two- or three-phase system.

17.2.1.4 Dispersive liquid-liquid microextraction

The dispersive liquid-liquid microextraction (DLLME) was introduced in 2006 by Assadi's group [19] and can be considered a modification of the miniaturized LLE. In this technique, a relatively small amount of a water-immiscible extraction solvent (typically 10–50 μl) is dissolved in 0.5–2 ml of a water-soluble solvent and rapidly injected with a syringe into the investigated aqueous sample (up to 10 ml). The fast injection of the mixture of organic solvents into the water makes the water-immiscible solvent to be dispersed in the aqueous mass as small micro-drops in which the target analytes are rapidly extracted. The enriched organic phase is then separated from the aqueous sample by centrifugation or frozen (depending on its density) and directly subjected to instrumental analysis, typically by GC. Application to polar analytes requires previous pH adjustment and/or in situ derivatization, which can be accomplished by direct addition of the derivatization agent to the sample or by dispersion together with the extraction solvent.

The several manual manipulations involved in DLLME made the technique difficult to automate and the use of internal standards and surrogates even more necessary than for previously revised SME-based techniques. Despite these shortcomings, many examples of application of DLLME to water analysis can be found in the literature, which is considered approval of the rapid acceptation of this green, fast and efficient (enrichment factors in the 200–900 range) extraction technique for analytes of different polarity.

17.2.1.5 Sorption-based extraction techniques

Many techniques currently in use for pretreatment of gaseous, fluid or liquid samples are based on trapping the investigated analytes on, or in, a suitable sorbent. The preconcentrated test compounds are subsequently desorbed, in a more or less selective fashion, by elution with a relatively small amount of solvent, either in a vial or in an appropriated interface; or by thermal desorption, typically in the injection port of the instrument selected for final determination. The later approach avoids dilution but, for obvious reasons, is limited to (semi-)volatile thermally stable compounds. Greening of these methodologies has been accomplished by either scaling down the original conventional size technique or by developing novel miniaturized techniques, as it will be shown in this section.

Solid phase extraction Solid phase extraction (SPE) is probably the most widely accepted technique for preconcentration and clean-up of analytes from fluids and aqueous samples. Nowadays, the large variety of sorbents commercially available makes this technique suitable for the determination of analytes with divergent chemical structures and polarities. Thereby, research regarding sorbents is mainly focused in the development of novel sorbents allowing higher loading capacities, higher efficiency for retention of highly polar analytes from aqueous matrices, and in the synthesis of class-selective immuno-sorbents (ISPEs) and molecular imprinted polymers (MIPs) allowing an increased selectivity of the retention process and so contribute to simplify the subsequent clean-up and/or detection steps [20, 21]. As an example, in a recently published study, five ionic liquid-modified porous polymers with different imidazolium-based functional groups were obtained and a new molecular imprinting technique was introduced to form the ordered functional groups in the porous structure [22]. The sorbent finally selected allowed selective SPE of tanshinone commercial functional groups.

The progress achieved in past decades in the synthesis of sorbent materials allowed obtaining very pure small-size sorbent particles in a very reproducible way. These small particles (ca. 40 μm) provided higher retention capacities than conventional-size ones, an interesting feature that promoted the use of smaller SPE cartridges without a significant lost of retention efficiency and with adequate breakthrough volumes. In fact, reducing the size of the conventional 1–6 ml SPE syringe barrels to the 10 mm × 1–2 mm i.d. of the so-called Prospeckt-type cartridges used in the hyphenated systems led to a reduction of sample volumes from 0.5–1.01 to less than 50–100 ml. Actually, quite often, even 5– 10 ml suffices to obtain similar LOD of 0.01–0.1 μg l^{-1} with SPE-LC and 1000-fold lower with SPE-GC that previously required 100-fold larger volumes. More importantly, quantitative elution of the analytes can be achieved with 50–100 μl of the appropriate solvent, that is, with a volume small enough to allow complete transfer to the instrument selected for final determination. This fact promoted the development of hyphenated and automated systems for (unattended) and fast (ca. up 20 min) treatment of aqueous samples with minimal solvent consumption and wastes generation [23]. A scheme of an on-line SPE-GC-MS system is shown in Figure 17.3.

In principle, the small volumes and short analytical times involved in these valve-based hyphenated miniaturized SPE systems should made method development faster than in conventional approaches. The experimental parameters to consider during method optimization are the same as for conventional SPE, namely the nature and amount of sorbent, the nature of the solvents used in the different SPE steps and, in particular, their flow rates. The main reasons for low analyte recoveries are in general also similar: reduced sorbent capacity or too strong retention, slow kinetic of the sorption process (or, in other words, too high sample and/or solvent flow rates), and a possible adsorption of the analytes in the tube used to connect the different parts of the system. On the other hand, in these close systems the risk of analyte degradation and oxidation is greatly reduced as compared to the (open) conventional approaches. Finally, the exposition of operator to hazardous solvents is greatly reduced.

Accepting that the hyphenation of SPE with LC and GC can nowadays be considered achieved goals, at present, development in this area is mainly orientated to the progressive reduction of the sample amount required for accurate determination of trace compounds and to the further simplification of the sample treatment methodologies. Both aspects usually rely on the use of any of the high-capacity or highly selective sorbents nowadays available and/or the use of very powerful MS- [25] or, preferably, MS/MS-based [26] detectors. Such approaches have been demonstrated to be useful for extremely fast determinations that could hardly be carried out with analysis times similar to those involved by conventional sample treatment procedures. For example, replacing the SPE–LC part for a single short column, SSC (1–2 cm length), and using MS–MS as detection system facilitated the real-time study of analyte degradation at the trace level [26, 27], with LC run times of frequently, only some 3 min. A slight modification of the conventional valves configuration typically adopted for miniaturized SPE to introduce a filter before the SPE cartridge allowed the direct injection of soil and sediment slurries in the SPE-GC-MS system. This set-up allowed in-deep evaluation of the so-called fast adsorption of pesticides in these complex matrices by simultaneous analysis of both phases (i.e. water and soil/sediment) in less than 45 min and with a single injection of the mixture in the system [28]. Finally, the progressive reduction of the sample size required to perform these types of hyphenated analyses has made possible, in some cases, the direct injection of the aqueous sample [29] or of the aqueous extract obtained from fruits and vegetables [30], with LODs low enough to consider the methods

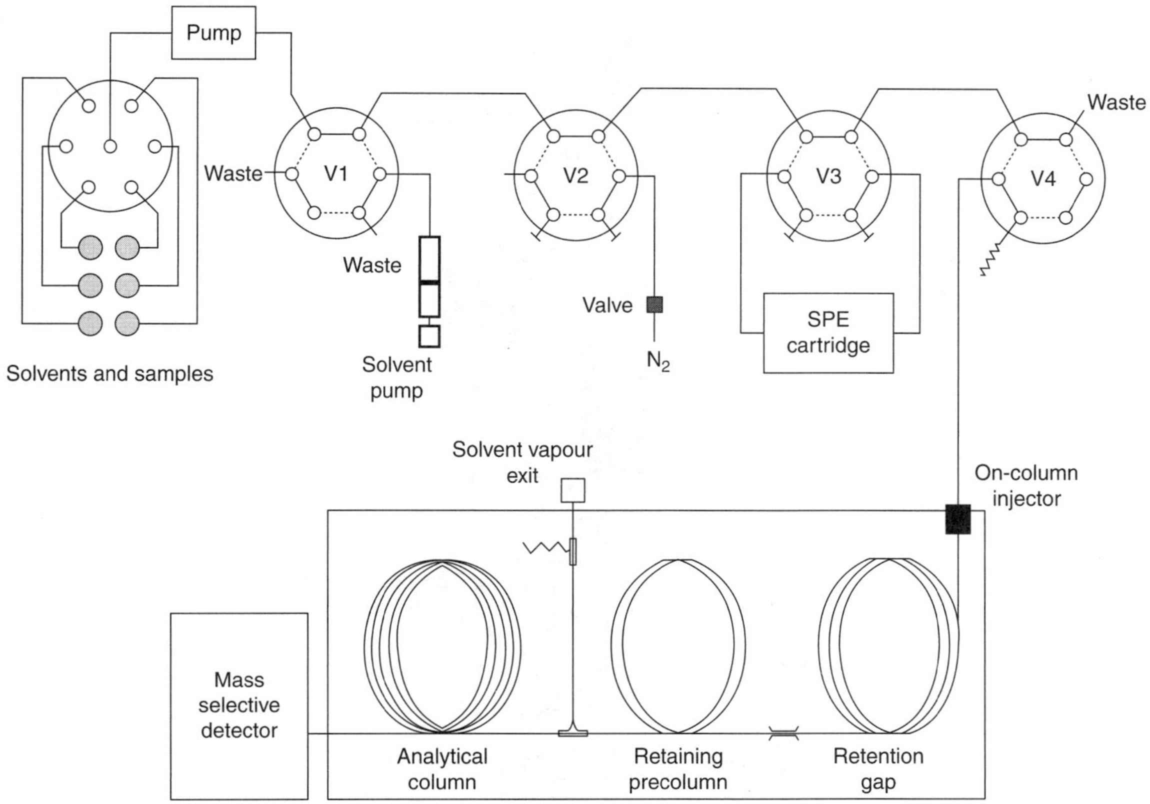

Figure 17.3 *Scheme of on-line SPE-GC-MS system [24].*

appropriate even for the fast screening of selected relevant pesticides. However, in such applications, the use of a highly selective and sensitive detector, such as MS-MS, becomes newly mandatory. In the case of the GC coupling, these trends have promoted an important development in the field of large volume injection (LVI). Some of the novel LVI interfaces [31, 32] allow direct water injection in GC, and the introduction of up to 10 ml of solvent using the slow injection mode or multiple fast injections in a packed liner. Nevertheless, the real practicability of these approaches in the different application areas is strongly depending on the type and concentration of the interferences present in the sample.

Up to now, and somehow surprisingly due to their many attractive features, the number of studies involving the use of SPE disks in hyphenated SPE-LC and SPE-GC has been rather limited [33]. Resin disks (0.7 mm diameter) have also been mounted inside the removable needle chamber of a 50 µl Hamilton gas-tight syringe. This set-up was demonstrated to be a valuable miniaturized automated alternative that enabled the efficient preconcentration of substituted benzenes from a volume of water as small as 2.5 ml. The method provided recoveries higher than 90% at the 10 ng ml^{-1} level with GC-FID and required only 5 µl of acetonitrile for desorption [34]. This study can be considered the precedent of a recently introduced modification, the so-called microextraction in packed syringe (MEPS). MEPS is a miniaturized SPE technique in which a sorbent packing material (ca. 1 mg) is placed at the top of the syringe needle. For extraction, 10–250 µl of an aqueous sample is successively withdrawn and ejected so allowing preconcentration of the analytes on the

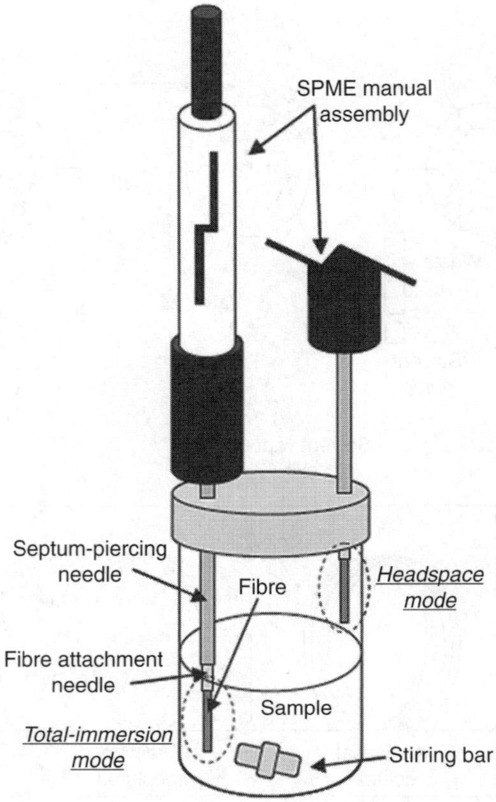

Figure 17.4 Basic configurations for SPME [7].

selected sorbent. Analyte elution is typically accomplished with 20–50 µl of an appropriate solvent, which can directly be transferred to the GC/LC port. MEPS applications include, for instance, the determination of PAHs in water [35] and of drugs in blood [36].

Other SPE formats, such as packing in conventional micropipettes or 96-well-plate, although popular in certain application areas such as high throughput clinical analysis, are still scarcely used in other research areas, in which they are far from considered really established and accepted configurations.

Solid phase microextraction Solid phase microextraction (SPME) was introduced in 1990 by Pawliszyn's group [37] as a (virtually) solvent-free preconcentration technique in which the analyte(s) is (are) adsorbed onto a fused-silica fibre coated with an appropriate sorbent layer by simple exposure of the fibre for a pre-selected time to the headspace of the sample or by direct immersion in a liquid sample (Figure 17.4).

Despite been an equilibrium (i.e. non-exhaustive) technique and the initial limitations regarding the nature of the commercialized sorbent coating, SPME was rapidly accepted as a simple, reproducible, miniaturized and green technique, and its feasibility for fast and accurate analysis of (semi-)volatile compounds was illustrated through a number of application studies. Today, on-line coupling of SPME with LC and GC are achieved goals and a number of systems (e.g. autosamplers) allowing complete automation of the process are commercially available. Alternative SPME-formats have also been proposed [38, 39] although, in general, they have achieved only a rather limited success.

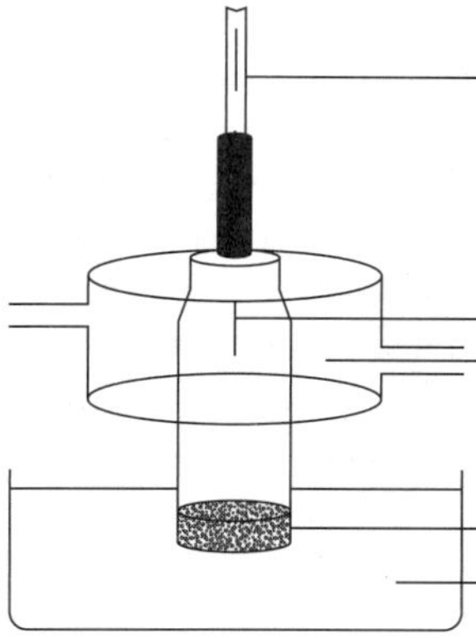

Figure 17.5 *Scheme of the SPME-based system [49].*

SPME can be considered a well established and widely accepted technique. Thereby, at present, research in this field is mainly focused on the development of new coatings and analytical strategies that contribute to improve the sensitivity of the technique.

The variety of commercially available fibre coatings has increased significantly during the last year, something that have contributed to expand the range of analyte classes that can be successfully analysed. Today, in addition to the originally introduced non-polar PDMS, semi-polar polydimethyl siloxane–divinylbenzene (PDMS–DVB), polar polyacrylate (PA), Carbowax–divinylbenzene (CW–DVB) liquid-like phases, coated porous particle phases such a polydimethyl siloxane–Carboxen (PDMS– Carboxen), poly(3-methylthiophene) and Nafion are commercialized. Other less frequently used coatings include carbon nanotubes [40], several crown ethers [41], MIPs [42] or anodized metals [43]. The use of selective sorbents, such a MIPs, contributes to increase the selectivity of the process, so significantly reducing the need of previous fractionation and purification steps in the analysis of complex extracts and the risk of instrument contamination [7].

Strategies involving derivatization of the analytes in the aqueous phase have extended the range of application of SPME to very polar [44] or ionic substances [45, 46]. Applications involving on-fibre derivatization require conversion of the analytes after extraction by applying the reagent as a gas and are, as far as we know, still scarce in the literature. The relatively high RSD values typically associated to this approach (e.g. 10–35% for chemical warfare agents at 1–20 μg ml^{-1} levels, n = 6 [47]) can be regarded as an indicator of conditions which are difficult to control. Generally speaking, although the analysis of aqueous samples can be accomplished without (or with little) pre-treatment, SPME of target compounds from more complex (solid) matrices typically requires a previous separation of the analytes from the main matrix components [48], involves longer extraction times and is frequently less exhaustive than for liquid samples because of the less favourable extraction conditions. Interestingly, a new SPME-based system has been designed to allow cooling on the upper part of the sample vial and simultaneous heating and ultrasonic extraction of the bottom part of the vial (Figure 17.5). The system was applied to the HS-SPME with a 100 μm PDMS fibre of compounds with medium volatility, such as polychlorinated dibenzo-*p*-dioxins and furans (PCDD/Fs) polychlorinated dibenzo (PCDFs), from soil slurries. Although the use of GC-MS/MS and appropriated internal standards were mandatory, the feasibility of the approach for fast screening of some types of heavily contaminated samples was demonstrated [49].

Stir-bar-sorptive extraction In a typical stir-bar-sorptive extraction (SBSE) experiment, a magnetic stir bar coated with 55 or 219 μl PDMS (corresponding to magnets 10 and 40 mm long, respectively) is rotated into an aqueous sample (or extract) for a selected and often fairly long extraction time [50]. SBSE of the headspace of a gas, liquid or solid sample contained in a sealed vial is also possible, although less frequently used. The magnetic stir-bar can also be inserted into a short length of PDMS or silicon tubing. Irrespective of the format, the surface area of the stir bar is greater than that of the SPME fibre and the volume of the adsorbent is increased at least by a factor of 100, which results in a higher phase ratio than in SPME and hence, a higher extraction efficiency and lower LODs. Once the extraction step is completed, the stir bar is removed, often manually, and transferred to either the injection port of a GC for thermal desorption [51], or into a solvent for LC analysis [52, 53]. As for SPME, in the former approach, all preconcentrated analytes are (virtually) transferred to the instrument selected for final determination; in the latter, only a fraction of the concentrated extract is frequently introduced in the LC system. Nonetheless, a novel desorption unit enables fully automated analysis of 98 or 196 PDMS-coated stir bars [54].

A number of studies have demonstrated the feasibility of SBSE for the preconcentration of analytes with medium to low polarity and divergent volatility and the several advantages of SBSE as compared to SPME in most of these applications [55]. However, the technique has not been as widely accepted as could be anticipated, probably due to the limited number of coatings materials commercially available. At present, efforts in this field focus on the development of 'dual-phase twisters' in which the conventional PDMS phase is combined with another sorbent to increase the selectivity and/or efficiency of the extraction process [56].

Finally, it is worth mentioning that due to its particular features, and similarly to SPME, SBSE can be use as a convenient sampling system which can be easily transported for subsequent laboratory analysis or, even better, for on-site determination of the target compounds when combined with portable and miniaturized instrumentation [57].

17.2.2 Solid samples

The first step in the analysis of (semi-)solid samples is usually the exhaustive extraction of the target compounds from the matrix in which they are entrapped. The essentially non-selective character of this initial treatment makes mandatory the subsequent purification of the obtained extract, firstly by the rough elimination of chemically non-related main matrix components (e.g. organic matter, lipids, proteins…) and then, if required, by removal of other chemically related analytes that can interfere in the final instrumental determination of the investigated compounds. Pre-treatment techniques revised in previous sections can be used for these purification steps. However, the initial extraction treatment requires the use of different analytical techniques for which miniaturization (i.e. use of sample size smaller than 0.5 g) is more the exception than the rule.

17.2.2.1 Matrix-solid phase dispersion

The main limitation encountered when developing miniaturized and hyphenated sample preparation methodologies for the treatment of (semi-)solid samples is the lack of commercial (small size) instrumentation corresponding to the conventional (i.e. large scale) technique. In other words, miniaturization of such procedures frequently demands the previous design and development of tailored instrumentation. Otherwise, when conventional instruments are used for the treatment of small samples, amounts of reagent and volumes of solvents in the range of those used for the treatment of regular size samples are typically involved. Therefore, neither the approach nor the methods can really be considered as green and the main goal of miniaturization remains unachieved.

Probably the only exception to this general consideration is matrix-solid phase dispersion (MSPD). In MSPD, a (semi-)solid sample is blended with an appropriate sorbent until a homogenous distribution of the matrix components on the sorbent surface is achieved. The dried and homogeneous mixture obtained then packed into a column (or syringe barrel) from which the analytes of interest are eluted with a suitable solvent. MSPD can also be applied to liquid or viscous matrices by simply mixing the sample and the sorbent; and

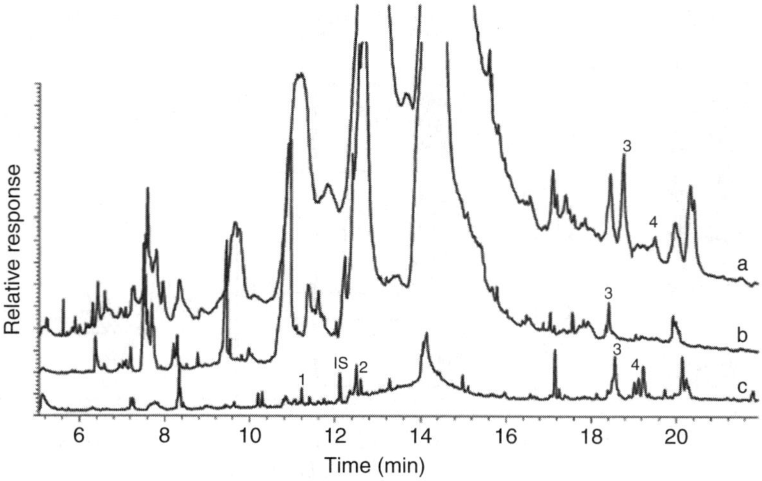

Figure 17.6 *Comparison of GC-MS chromatograms obtained by MSPD of 40 mg Porcellio scaber with 100 μl of (a) ethyl acetate from a C8-bonded silica/sample mixture, (b) ethyl acetate from a C8-bonded silica/sample mixture and washing before extraction, and (c) n-hexane from silica/sample mixture. Peak identification: (1) diazinon, (2) malathion, (3) permethrin, (4) cyfuthrin and (IS) parathion-methyl [59].*

subsequent sedimentation and homogenous packing of the slurry into a column. The two main parameters affecting the efficiency of the extraction process are the sorbent and the eluent. Proper selection of their nature and amount can result on specific retention of impurities on the sorbent and selective elution of the target compounds, and so yield (virtually) ready-to-analyse extracts. When the collected extracts are not cleaned enough (or the separation-plus-detection technique selective enough), some extra purification step(s) should be incorporated to the sample preparation protocol. These treatments can be carried out off-line [58] or, more interestingly, on-line or in-line with the MSPD process. Regarding the latter approach, additional clean-up of the MSPD eluates can be accomplished by, for example, washing of the packed matrix-sorbent mixture with a selected solvent for selective removal of interferences before analyte collection [59, 60] or by packing of an extra sorbent layer at the bottom of the MSPD column [61, 62]. In this case, the chosen sorbent should perform the selective retention of the interfering components washed up from the MSPD mixture with the eluent without affecting the target analytes which should pass through this phase unaltered. Figure 17.6 shows an illustrative example of the different degree of clean-up achieved when applying some of these different strategies to the analysis of pesticides in single insects, for example, 40 mg *Porcellio scaber* [59].

MSPD can be a particularly suitable analytical alternative when dealing with the analysis of limited-size samples [59, 61]. Other interesting features are its simplicity, the fact it does not require any special equipment and that it is typically completed in less than 1 h. In addition, when combined with GC-MS or LC-MS for final determination, LODs in the low-μg g^{-1} range are easily obtained even although no extra treatment of the collected extracts was carried out [59, 63, 64]. In other words, MSPD is not only feasible for the analysis of relatively abundant components, but also for the determination of trace analytes even if as a small amount of sample as 25 mg were used [65].

17.2.2.2 Enhanced solvent extraction techniques

The application of enhanced solvent extraction techniques to the miniaturized treatment of (semi-)solid samples has been, up to now, rather rare in the literature. The lack of dedicated commercial instrumentation adequate for this type of determination is probably the main reason. Nevertheless, some research groups have

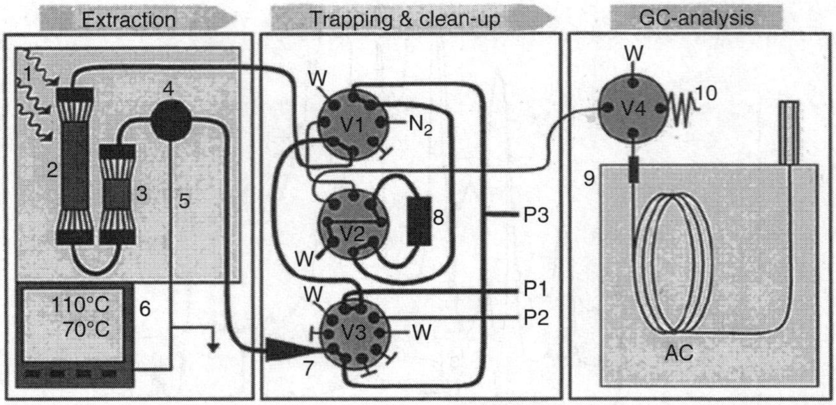

Figure 17.7 Scheme of the DMAE-SPE-LVI-GC system: (1) microwave oven, (2) preheater, (3) extraction vessel, (4) mixing tee, (5) thermocouple, (6) temperature regulator, (7) restrictor, (8) SPE cartridge, (9) PTV-GC-NPD and (10) fused-silica leak.

developed their own instruments or adapted conventional ones in order to perform pressurized liquid extraction (PLE), microwave-assisted extraction (MAE) or ultrasonic-assisted extraction of selected analytes from solid and semisolid matrices with minimum consumption of sample, solvent and chemical reagents.

The first miniaturized PLE systems were described in the literature 10 years after the introduction of the technique. In some of these systems, a GC oven was used to heat up the extraction cell and, when water was used as extractant, the coil installed before the cell [66, 67]. In other cases, a dedicated instrument with a miniaturized oven was designed [68, 69]. In one of these systems [68], a heatable 10 × 3.0 mm i.d. stainless-steel extraction cell was used for quantitative extraction of the 16 EPA PAHs (recoveries, 90–110%) from 50 mg soil with only 100 µl of toluene. Direct injection of 50 µl of this raw extract into a PTV system containing an appropriated sorbent, enabled in-liner clean-up of the extracts before GC-MS analysis and ensured LODs as low as 2–9 ng g^{-1} soil for a large majority of the target compounds. Despite the very heterogeneous samples investigated, satisfactory RSDs in the 2–15% range were reported. Replacing the original small size ceramic oven by a larger one resulted in an increased flexibility regarding the extraction cell size used [69]. This contributed to wider the range of possible application studies as well as that of the analytical strategies to be used. As an example, the system was employed for the selective pressurized extraction of endogenous PCBs from foodstuffs of animal origin, including complex fatty matrices such as fish, meat and eggs with satisfactory results.

The studies involving miniaturized MAE system are still extremely scarce in the literature. After the first attempt reported by Cresswell and Haswell [70], Ericsson and Colmsjo [71] proposed inserting a preheating column in front of the extraction cell in the microwave cavity and the back-elution of the target compounds from the 10 ×2 mm PLRP-S SPE disposable cartridge used for on-line SPE of the extracted PAHs. The method involved only 60 mg of sample and 400 µl of MTDE as an eluent and demonstrated a satisfactory performance for the determination of the target compounds in a reference sediment (recoveries 88–104%, RSDs 1–10%). The system was subsequently hyphenated with a GC equipped with a nitrogen-phosphor detector (GC-NPD) for on-line analysis of organophosphate esters in air samples [72]. A schematic diagram of the set-up used in this study is shown in Figure 17.7.

Despite its many positive features, the use of ultrasound for analytical applications is relatively recent. Nevertheless, some applications have already demonstrated the potential of sonication for the miniaturized, rapid, relatively inexpensive and quantitative extraction of several types of analytes, ranging from metals to different classes of organic compounds. These applications have involved the use of ultrasonic baths, water baths in which an ultrasonic probe was installed close to the extraction cell, sonoreactors and, more recently, ultrasonic probes. As expected, the extraction time decreased from baths to sonoreactors and from this to probes. However, in most of these applications subsequent clean-up of the obtained extracts was mandatory. Off-line and on-line approaches have been proposed to perform these purification treatments and, in general, satisfactory figures of merit have been reported for these types of approaches proving the feasibility of the technique for the development of hyphenated systems [73].

Interestingly, ultrasound energy has recently been used to speed up other extraction or purification processes, such as SPE [74] or MSPD [75]. The first approach was used by Albero *et al.* to increase the rapidity and efficiency of SPE of pesticides from juice. In the second one, the MSPD mixture, packed in a close SPE cartridge, was wetted with the selected extraction solvent and placed in a sonoreactor for a preselected time. Then, the extracted analytes were eluted from the column and directly subjected to instrumental analysis. The feasibility of this so-called ultrasonic-assisted matrix solid phase dispersion (UA-MSPD) was illustrated for the analysis of selected triazines and organophosphorous pesticides from fruits contaminated at levels similar to the MRL set in EU legislations. Complete sample preparation was accomplished in only 1 min and involved only 100 mg of fruit peel, a similar amount of dispersant sorbent and a few ml of organic solvent.

In our opinion, all these promising preliminary results indicate that enhanced solvent extraction techniques, alone or in combination with other previously mentioned techniques, might become interesting and efficient miniaturized analytical alternatives to other more conventional leaching procedures. However, to achieve a level of development and maturity similar to that achieved at present for other previously described sorbent- or solvent-based techniques, more work is still demanded from both academia and especially companies who should support and promote the development of appropriate analytical instrumentation for these types of determinations.

17.3 Integrated approaches for sample preparation on microfluidic platforms

17.3.1 Microfluidic platforms in sample preparation process

The micro total analysis system concept (μTAS) or 'lab-on-a-chip', was developed from the modification of the total analysis system (TAS) approach by downsizing and integrating its multiple steps onto a single device, yielding a sensor-like system with a fast response time, low sample consumption, on-site operation, and high stability [76, 77]. Apart from its small size, the relevant/key feature derived from μTAS concept is the possibility of handling fluidics on the nanolitre and even picolitre scale, which has widened the scope of μTAS to now be called *microfluidics*. Microfluidics is the science and technology of systems that process or manipulate small amounts of fluidics (10^{-9} to 10^{-18} l), using channels measuring from tens to hundreds of micrometres [78]. However, miniaturization is more than simply the scaling down of well-known systems since the relative importance of forces and processes changes with scale. One of the most relevant characteristics of microfluidics is the omnipresence of laminar flow, in which viscous forces dominate over inertia. This means that turbulence is often unattainable and that molecule transportation only occurs through diffusion, which has direct consequences on the designs of these types of microsystems [79]. Also, small volumes reduce the time taken to analyse a product; the unique behaviour of liquids at the microscale allows greater control of molecular concentrations and interactions; and reagent costs and the amount of chemical waste can be greatly reduced. In consequence, microfluidics exploits both its most obvious characteristic -small size- and less obvious characteristics of fluids in microchannels. It offers fundamentally new capabilities in the control of concentrations of molecules in space and time.

On the other hand, analytical microsystems (AM) can be seen as the 'main product' still in its infancy of successful systems and applications in the field of μTAS or 'lab-on-a-chip' systems which have evolved both innovative analytical chemistry methods and microfluidics cleverly combined. They are the *new* systems, in the *new* labs becoming a *new* picture on analytical landscape. One of the most representative formats of analytical microsystems is microchip technology. A microchip is a closed analytical microsystem constituted at least by one microchannel where a sample plug is critically loaded and other microchannel(s) where sample pre-treatment and separation of analytes is performed interfaced suitably to reservoirs where different solutions and samples are deposited. Microchannels and reservoirs are fabricated in microchips using photolithography and etching (Glass), micromoulding (PMMA) or casting (PDMS) to form channels for sample injection, sample pre-treatment, analyte separation or reaction, and detection. Once all solutions, including those of the samples, are loaded, the samples are typically transferred hydrodynamically or electrokinetically into an injector region. Then their components are filtered/preconcentrated/separated/ subjected to derivatization schemes by application of pressure or a high voltage, and afterwards detected with a suitable detection system. The system integration concept on the microchip platform eliminates the necessity of most fluidic connections that otherwise link microfluidic components. Avoiding such connections greatly reduces sample dispersion, delay times and dead volumes between the different microchip compartments, therefore significantly increasing the separation power of such integrated miniaturized systems being the monolithic approach the most commonly used. In a general sense, the design of microchips has undergone significant development from simple single-channel structures to increasingly complex ones. As we will see in the following sections, current designs of microchips allow reactions on-chip and separations in multiple channels. Micro reactors can also be added to perform on-line reactions such as (bio)-chemical reactions and post column derivatization.

The development of practical analytical microsystems or microchips [78] (especially those for bioanalysis) continues rapidly, although, given its early focus, this area has been slower than expected to reach widespread routine use. Part of the problem is that there is limited technology in two parts of the analytical process: sample preparation (including analytical separation of the mixtures into single analyte bands) and analyte detection. Biological samples, particularly clinical samples (such as blood or faeces), or those obtained by environmental and food sampling, are often dilute or complicated. Before these samples can be analysed by microfluidic devices, they must be converted to a form that is compatible with the intended analysis and then introduced into the analytical device. The procedures required to complete these tasks are surprisingly sample-dependent, and not necessarily 'micro' in scale. After a sample has been prepared, introduced into the analytical device and processed, it must then be detected. Having the microfluidic chips as just a small part of a system in which sample preparation and detection are much more complicated than the chip's operation may be appropriate in some circumstances, but does detract from the potential advantages of microfluidic devices.

Furthermore, it is important to point out in the scope of this chapter that not all analytical steps and/or microsystems are equally incorporated in the analytical scene of the labs and the incidence level of miniaturization is not homogeneous. In fact, the detection and separation steps are the ones most investigated, whereas sample preparation is notably the least developed [80]. These purposes involve two important hot-points: the integration of all analytical steps and dealing with real samples. The philosophy of these statements discussed above is depicted in Figure 17.8. This figure shows the monolithic approach (microchip technology) as well as the intensity of integration of the main analytical steps in microfluidic platforms represented by an analytical microsystem as the final product.

From the beginning, detection has been one of the main challenges for analytical microsystems, since very sensitive techniques are needed as a consequence of the ultra-small sample volumes used in micron-sized environments. In principle, a wide variety of detection alternatives can be used in microfluidic systems, but laser-induced fluorescence (LIF) and electrochemical detection (ED) primarily, along with mass spectrometry

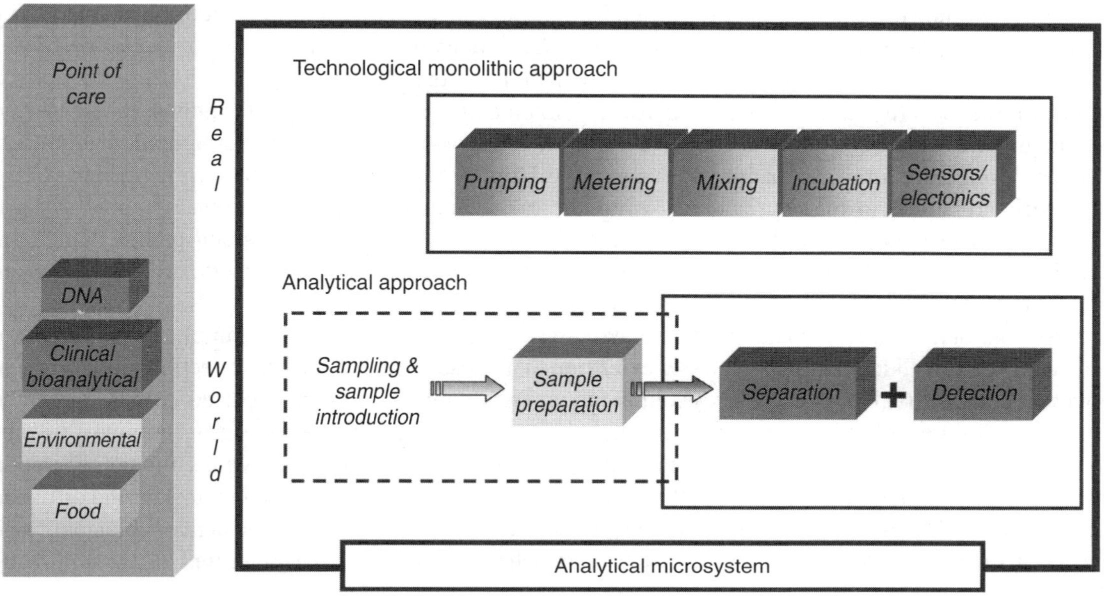

Figure 17.8 *Philosophy of the integrated analytical process on analytical microsystems.*

(MS) are the most commonly used routes. LIF was the original detection technique and it is the most often used detection scheme because of its inherent sensitivity [81]. However, the high cost and large size of the instrumental set-up for LIF has been sometimes incompatible with the concept of μTAS. In addition, tedious derivatization schemes are needed to use LIF with non-fluorescent compounds. The principal alternative to LIF detection is ED. The great importance of ED lies in its inherent miniaturization without loss of performance and its high compatibility with microfabrication techniques. Likewise, it possesses high sensitivity, its responses are not dependent on the optical path length or sample turbidity and it has low power supply requirements which are additional advantages. Proof of the predominant role of ED can be easily found in selected literature [82–84]. Apart from LIF and ED, there are other detection approaches for analytical microsystems, but from a realistic point of view, they have been developed to a lesser extent [85].

With respect to separation techniques, microchip electrophoresis (MEC) was one of the earliest examples of μTAS, and it constitutes one of the most representative examples of analytical microsystems [86]. This technology emerged as an important new analytical technique in the early 1990s with the introduction of the μTAS concept [76] and the seminal work of Manz *et al.* [87]. This new technology was a result of the marriage of the ability of conventional CE to analyse ultra-small volumes (nl) and microfabrication techniques perfected in the semiconductor industry to produce very small structures in silicon. Using CE microchips, analysis times can be reduced to seconds and extremely high separation efficiencies can be achieved. The easy microfabrication of a network of channels using materials of well-known chemistry, which themselves have good electroosmotic flows and the possibility of using the electrokinetic phenomena to move fluids, are among the most important factors to understand the relevance of CE microchips to miniaturization. Since electrokinetics is easy to apply (only a pair of electrodes are needed), electroosmotic-driven flow (EOF) has been successfully implemented using different types of materials to manufacture the channels, being glass the most commonly used. Microfabrication on polymers is faster and cheaper than on glass, so these materials have great potential for mass production. However, glass chips present the best EOF and the chemical modification on the surfaces of its micro channels is better known and easier compared to polymers. The accurate and deep description of the microfabrication techniques is obviously outside of the scope of this chapter, but readers can find more information in excellent literature [88–94].

Many of the benefits already discussed for CE microchips could equally be applied to downsized chromatographic techniques; however, relatively few examples of chip-based chromatographic instruments are covered in the literature compared with chip-based CE devices. This fact is not surprising since CE is almost perfectly suited to miniaturization. Nevertheless, the miniaturization of chromatographic systems involves a number of technical challenges, such as the microfabrication of valves and pumps, which are generally not encountered in CE [80]. In other words, an in a general sense, electrokinetic driven systems are dominant to the pressure system.

On the contrary, sample preparation steps involving emerging technology for sample preparation/pre-treatment have been addressed in much lesser extension than those of separation and detection, mainly due to its inherent complexity. However, a very exciting unique opportunities could be expected because of (1) the intrinsic possibilities of microfabrication technology to create sophisticated designs (complex layouts according to sample requirements) and microstructures (filters) in connection with sample preparation requirements; (2) the natural possibilities offered by microfluidics: the presence of laminar flow and diffusion in action; and (3) the ease of using electrokinetic phenomena to move fluidics into the network channels with accuracy (focusing on flow techniques). These approaches will properly be presented in the following sections.

In addition and as noted previously, an analyte may be present in real samples at extremely low concentrations. This combined with the ultra-small detection volumes encountered in microfluidic systems (pl–nl) makes sensitive detection methods a prerequisite for most analyses. Although fluorescence and electrochemical detection methods provide for extremely low detection limits, it is often thus desirable to incorporate sample preconcentration prior to detection within microfluidic systems.

There are three main sub-categories of sample pre-treatment: isolation/clean-up, sample preconcentration and sample derivatization. A selection of microfluidic-based approaches within each category will be discussed in the following sections towards selected examples. In addition, a sections dealing about sample pre-treatment for cell analysis and some relevant immunoassay strategies on microfluidic platforms are also given. Excellent literature is proposed by further reading [80, 95–100].

17.3.2 The isolation of analyte from the sample matrix: filtering approaches

A challenging task faced by the analytical chemist when dealing with raw samples is extracting/isolating the analyte of interest from the sample matrix. It is fair to say that the majority of non-ideal samples arrive in a format incompatible with most analytical instrumentation, and requires some degree of clean-up. Even well-defined samples (e.g. aqueous solutions) require basic filtration prior to analysis. Other desirable techniques may include dilution, liquid/liquid extraction and solid phase extraction or even cell lysis in biological treatment as we will discuss in Section 17.3.6.

Possibly the most essential step when performing analysis in microfluidic systems is the filtration of sample prior to processing. Due to the typical small dimensions in microstructures, particulates can cause serious operational problems, providing sites for nucleation or blockage. The simplest solution is to filter all reagents and sample prior to introduction. Unfortunately, most conventional filtration methods require fluidic volumes far greater than those required for analytical processing. Consequently, it is desirable to integrate sample filtration on-chip prior to analysis.

In this way, two approaches have been employed: structurally-based filters (filtering and retention by integrated flow restrictions and controlled by manufacturing process) and diffusion based filtration (filtering by diffusion in laminar flows controlled with diffusion). Structurally-based or microfabricated filters have been proposed in popular architectures like frits, pillar structures or flow restrictions within fluidic channels to mimic conventional filters. Besides, like H-filters structures have been the most used in diffusion based filtration process. Filtration can be induced by allowing the analytes of interest to migrate across a laminar boundary (between a sample and solvent stream) while unwanted heavier particulates are retained in the original fluid stream (namely H-filter). It is very important to underline that, both approaches are initially linked to the heart of microfabrication and microfluidic technologies itself. The first one because of the well-known integration technology of multiple micropieces into a single microdevice; whereas the second one works on the basis of an inherent property in microfluidics: presence of laminar flow and diffusion 'in action'. However, the ultimate limit of the structurally-based filters is the resolution of manufacturing process, while the great advantage of the second one is that all the process is controlled by physics, although a very sophisticated technology is also required.

17.3.2.1 Physical particle filtering and retention by integrated flow restrictions

Microfabricated filters have been described either for collection of particles and clean-up of sample for downstream processing. An approach has been to create microfabricated frits, pillar structures or flow restrictions within fluidic channels to mimic conventional filters. Particulates can be physically retained within the structure as long as the diameter of the particulate is larger than the feature dimensions of the microstructure, and solution flow is unimpeded. However, such designs have typically been used as bead traps, or stationary phase reactors, in which a chemical reaction can be performed without causing blockage to the rest of the fluidic network. Microfabricated filters based on this concept have been described [101–104]. In the work reported by Anderson *et al.* [104], different designs for a particle trapping component in a flow-through device based on small pillars have been creatively illustrated. The variable number of pillars composing the filter can have different shape and pitch. The geometrical placement of the pillars has been constrained by three demands: (1) the flow resistance of the filter must remain below certain limits; (2) the filter should not be sensitive to clogging and, (3) the collected beads should be held within a relatively small confined area to facilitate detection. Some basic microfilter designs were initially studied and assayed. In the first one, the filter was placed inside a flow channel where the number of pillars is limited by the channel width. This means that a narrow channel (50 μm) having a small number of pillars is highly sensitive to clogging. In a second configuration, channel was widened at the entrapment location. This design reduces the risk for clogging and results in homogeneous flow characteristics over the filter; although the beads are confined to a long strip instead of a small square or round area entailing a more difficult detection. In the last design, the pillars defined a square reaction chamber in which the beads were successfully trapped. A larger number of pillars constitute the filter which results in a more clogging, insensitive filter. This last design fulfilled the design criteria mentioned above and was therefore chosen for fabrication and evaluation. Figure 17.9 shows a schematic of the micromachined filter-chamber device (a) and the filter chamber (b) (side view). The pillars are 3 μm wide and 50 μm high with a spacing of 2 μm [104].

An alternative to the necessity of filtering within microfluidic systems has been proposed making use of the concept of lateral percolation [105]. In lateral percolation filters, the sample penetrates a filter bed of posts along the plane/face of the microstructure, continuing perpendicular through the filter bed (lateral to the point of entry). Particles are then retained in the bed as fluid flows laterally through the structure. First generation lateral percolation filters were fabricated in quartz using deep reactive ion etching techniques, with the filter element comprising a network of intersecting channels (1.5 × 10 μm) situated at the bottom of the fluidic reservoirs. Particulates with dimensions larger than the minimum filter features (1.5 μm) were successfully restricted and the robustness of the device was examined with a variety of particulates (including dust particles and bacterial cells).

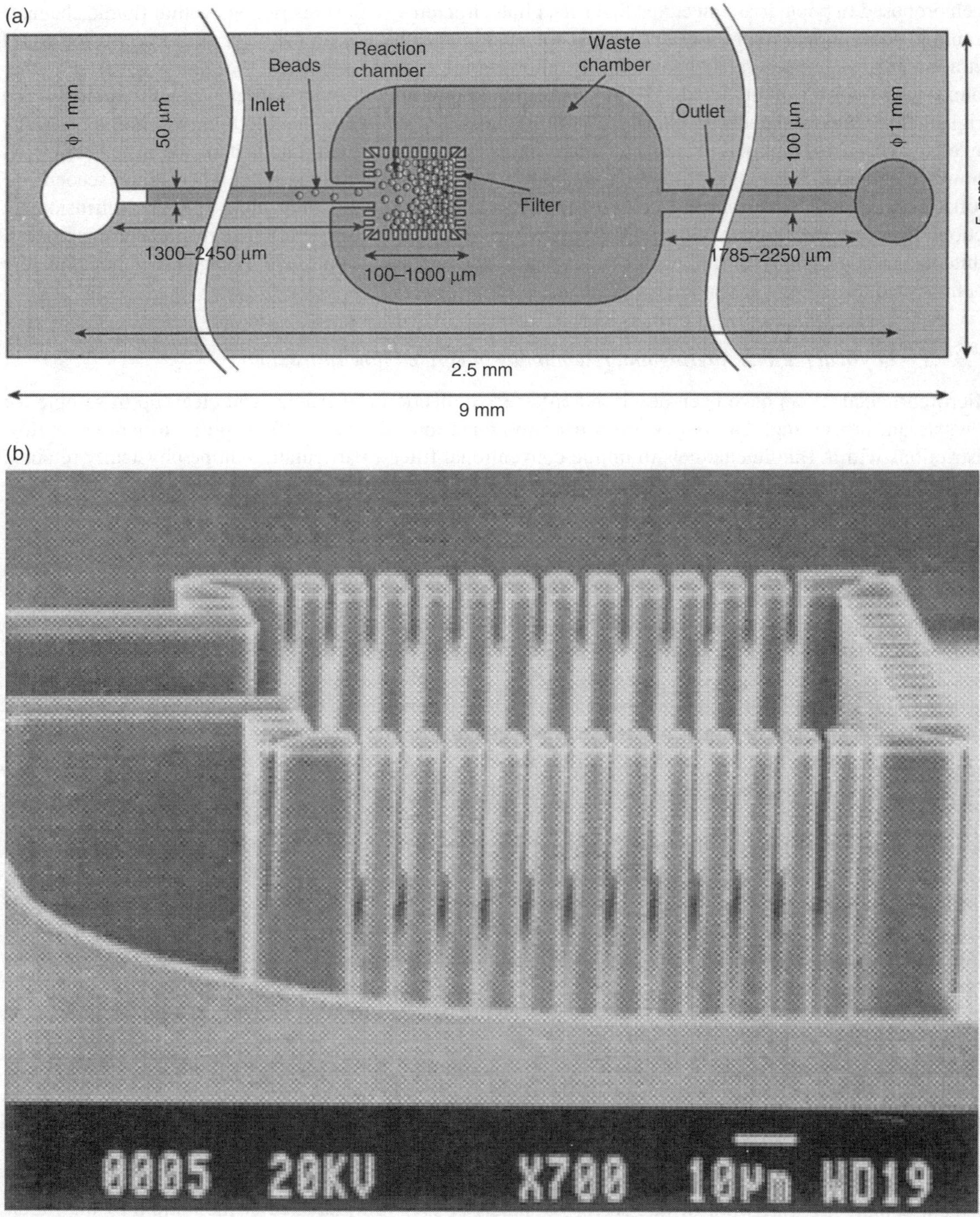

Figure 17.9 *(a) A schematic of the micromachined filter-chamber device. (b) SEM image of the filter-chamber. Adapted from [104] with permission from Wiley-VCH Verlag GmbH & Co. KGaA © 2001.*

In previous discussed approaches, conceptually speaking, the effectiveness and application of the filter structure is determined by the resolution limits of the manufacturing process. Consequently, filtering of sub-micron sized particulates using physical structures puts stringent demands on device fabrication. To this aim, sacrificial layer technology has been used to create flow filters for particles as small as 10 microns [106, 107]. A detailed discussion of such approaches may be found elsewhere [96].

To fabricate microfiltering devices more easily, sandwiching a thin polycarbonate track-etched membrane between cover and bottom plates or the photopolymerization of a membrane into a channel have been introduced [108, 109]. Indeed, the potential integration of functions in microfluidic chips has been demonstrated by implementing on-chip preconcentration of proteins prior to on-chip protein sizing by sodium docecyl sulfate-polyacrylamide gel electrophoresis (SDS-PAGE) [109]. Two polymeric elements, a thin (50 μm) size exclusion membrane for preconcentration and a longer (cm) porous monolith for protein sizing, were fabricated in situ using photopolymerization. Contiguous placement of the two polymeric elements in the channels of a microchip enabled simple and zero dead volume integration of the preconcentration with SDS-PAGE. The size exclusion membrane was polymerized in the injection channel using a shaped laser beam, and the sizing monolith was cast by photolithography using a mask and UV lamp. Proteins injected electrophoretically were trapped on the upstream side of the size exclusion membrane (cut off 10 kDa) and eluted off the membrane by reversing the electric field. Subsequently, the concentrated proteins were separated in a cross-linked polyacrylamide monolith that was patterned contiguous to the size exclusion membrane. The extent of protein preconcentration is easily tuned by varying the voltage during injection or by controlling the sample volume loaded.

17.3.2.2 Filtering by diffusion in laminar flows

A unique feature of microfluidic systems is the dominance of laminar flow conditions at the flow rates normally used (nl s^{-1}-μl s^{-1}), due to the small dimensions of the channels. Reynolds number (Re) informs about the behaviour of the fluidic in terms of chaotic flows or steadier, better-behaved flows. The dimensionless Reynolds number is given in the following equation, Re:

$$\mathrm{Re} = \frac{\rho d v}{\eta}$$

where d is the typical length scale (e.g. the diameter or the channel depth), and v is the average velocity of the moving liquid. With this number we can obtain an impression of the flow behaviour in microsystems. From empirical observations, physicists and engineers have found that Reynolds numbers larger than about 2300 correspond to what is called turbulent flow. Under this regime inertial forces are dominant. As we can see from the inspection of this equation, large Reynolds numbers are attained at higher liquid densities, higher flow velocities, larger typical length scales, or lower viscosities. The region in which the Reynolds number is between about 2000 and 3000 is called the regime of transitional flow and the region in which it is below about 2000 is referred to as the laminar flow regime. Again, by examining the equation given above, we can find that low Reynolds numbers are attained at lower velocities, smaller dimensions, smaller densities, or higher viscosities. In consequence, the decrease of scale drives inherently to obtain a laminar flow and turbulence is not allowed.

An interesting quality of laminar flow is that material transport between two adjacent streamlines only takes place by diffusion, and not by turbulent mixing as it is done in a test tube. This inherent feature of microfluidics is conceptually present in Figure 17.10 towards two classic designs: T and H-designs. Interestingly, this property can be exploited to allow for the discrimination (or filtering) of molecular or particular species. Simply put, species of low molecular mass have greater mobility (larger diffusion coefficients) than large molecular species. Consequently, filtration (or spatial discrimination) can be induced

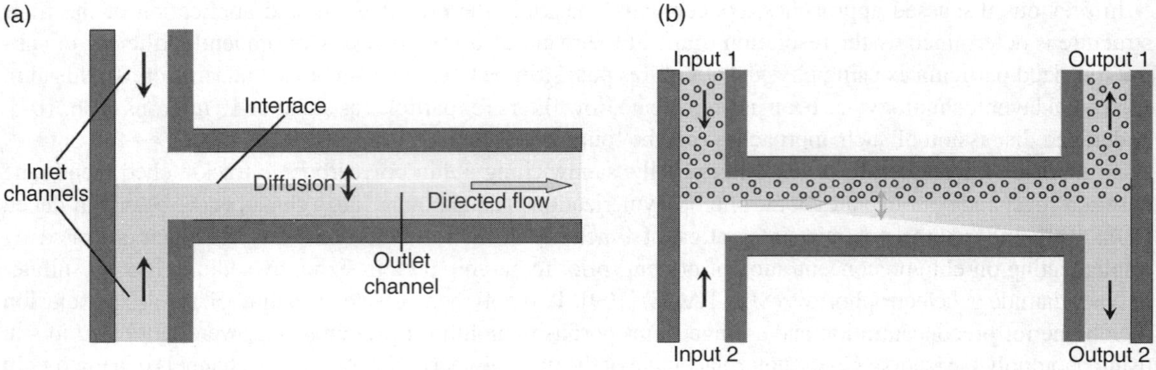

Figure 17.10 Conceptual drawing of two filtering microdevices based on laminar flow and diffusion principles. (a) A junction where two liquids meet and flow downstream side by side. (b) H-filter. Reprinted from [100] with permission from John Wiley & Sons © 2009.

by allowing analytes of interest to migrate across a laminar boundary (between a sample and solvent stream) whilst retaining unwanted heavier particulates in the original fluid stream. The process can be further controlled by altering the time in which the two fluids come into contact. Importantly, this approach addresses many of the problems associated with structurally-based filters, since its operation is reliant on the control of molecular diffusion rather that the resolution of the manufacturing process.

Microstructures based on this concept were first demonstrated by James Brody, Paul Yager and co-workers [110]. Initial devices incorporated both structural (etched barriers) and diffusion based filtration, and they were successful at discriminating and isolating 16 and 2.6 μm spheres. Due to the relative complexity of the fabrication process (a three mask process incorporating anisotropic etching of silicon) the authors developed their ideas to create microfabricated filters based solely on diffusional discrimination[111, 112]. Such H-filters function by bringing two laminar flows (a carrier stream and eluent stream) together in a central channel. Molecules (in the carrier stream) may then diffuse across the fluid barrier created at the boundary between the two streams. Highly mobile analyte molecules can cross between streams quickly, whilst heavier particles remain in the carrier stream. Consequently, only particles which have crossed the boundary before reaching the output channels will end up in the filtered output (see Figure 17.10b). Importantly, the approach is highly configurable since the time allowed for diffusional transfer between streams is directly controlled by fluid velocity and the length of the channel.

More examples exploiting this unique behaviour in microfluidics will be shown along different sections of this chapter.

17.3.3 The isolation of analytes from the sample matrix: extraction approaches

17.3.3.1 Liquid/liquid extraction

Liquid/liquid extraction (LLE) is a technique widely used in conventional sample pre-treatment methodologies and describes the physical process by which a compound (or a mixture of compounds) is transferred from one liquid phase to another. LLE could play itself a predominant role in miniaturized systems. Indeed, the high surface-to-volume ratio and the short diffusion distances, typically within microfluidic environments combined with laminar flow conditions, offer the possibility of performing liquid-liquid extraction within microchannels without shaking. The main challenge is to induce appropriate electroosmotic flows when common organic solvents are used.

The first microfluidic system for solvent extraction was a variation of an H-filter design, fabricated in quartz and comprising 250 μm wide channels. Introduction of Fe(II) in an aqueous stream and trioctylmethylammonium chloride in an organic (chloroform) stream allowed extraction of the ion-pair product in the organic phase. Extraction was shown to occur in less than 45 s, representing an order of magnitude improvement over conventional extraction times in separation funnels [113]. The authors have also demonstrated the extraction of Ni(II) complexes in microchannels [114], integration of neutral ionophore-based ion-pair extraction on-chip [115], and sequential ion sensing via 'slug' flow in microchannel environments [116]. The last one is of particular interest since the approach allows the determination of multiple ions in a single sample by pumping aqueous and organic phases intermittently through a fluidic network.

A potential drawback when using microfluidic systems for solvent extraction is the low unit throughput (normally between 1 and 100 μl min^{-1}). This problem can be obviated by operating arrays of parallel channels concurrently. To this end, the fabrication of silicon/glass micro-contactor arrays for the extraction of single feeds at rates of 250 ml h^{-1} have been reported [117]. LLE is achieved by contacting fluidic streams at constricted openings between distinct channels. The approach is attractive since flows can be separated naturally as the channels diverge.

17.3.3.2 Solid phase extraction (SPE)

Solid phase extraction is a broadly used technique in which a target molecule is retained by a chromatographic stationary phase material and subsequently eluted in an appropriate (and selective) solvent. SPE functions either as a sample clean-up method and a preconcentration method. This fact is due not only that the target analyte is retained within the stationary phase and the unwanted components of the sample matrix flow to waste, but also after elution in the desired volume entails preconcentration of the analyte.

Several and creative approaches for performing solid phase extraction in microfluidics have been proposed. One of them entails to coat channel walls with a high affinity stationary phase, although this methodology is less common. The other one involves packing the microchannels with the stationary phase material. The advantages and disadvantages are clearly defined into miniaturized environments. Coating channel approaches depend on the available surface area for interaction and, consequently, into micron-size channels the contact surface is very small. A simpler way to increase the surface area is to pack the microchannels with stationary phase; however, the packing process is not easy and sometimes this alternative is avoided. A very attractive possibility, compatible with miniaturized dimensions, is to replace conventional stationary phase materials by a continuous porous bed in situ formed from polymerization of organic monomers. The process of bed formation is easy, since a low-viscosity monomer solution can be introduced by vacuum or pressure into the microfluidic channel prior to initiation.

Selected examples of relevant approaches will be given next in order to provide the grade of implementation of these approaches on microfluidics.

Octadecyltrimethoxysilane coated walls channels were firstly used for SPE of neutral coumarin dye [118]. A simple fluidic network allowed for both enrichment (80-fold increase in concentration) and elution of the dye within 4 min. The problem of limited surface areas in open-channel devices can be improved to some degree by utilizing sophisticated fabrication techniques. Taking in account the idea of increasing surface area by packing microchannels with stationary phase material, fabrication and testing of a 330 pl chromatographic bed integrated within an electroosmotically pumped microsystem was carried out [119, 120]. The authors utilized weirs within a microfabricated channel to trap coated silica beads (1.5–4 μm diameter). These are then used to perform both solid phase extraction and electrochromatography of small molecules. Concentration enhancements of up to 500 times were demonstrated for two fluorescent dyes. Reversed-phase beads have been used in the SPE device because of their extensive use for the chromatography of proteins and peptides.

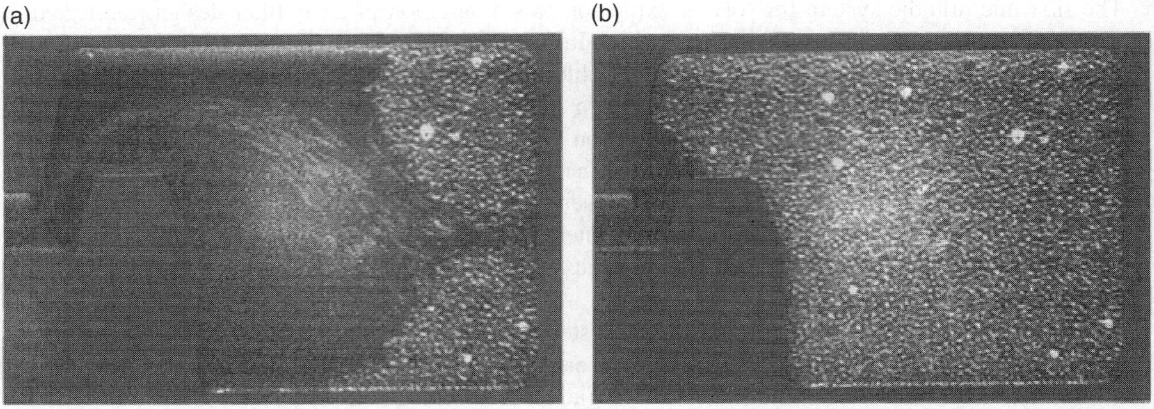

Figure 17.11 *Images of the chamber (a) at an initial stage of electrokinetic packing and (b) after it is completely filled with beads. Reprinted from [119] © 2000 American Chemical Society.*

In this work, electrokinetic packing of conventional capillaries has been demonstrated, and adapted for the microchip format. The packing procedure involved applying a suitable voltage program. The applied voltage induced EOF to flow down the bead channel, carrying the beads into the cavity. An organic solvent was required to suspend the chromatographic beads and prevent them from aggregating and plugging the relatively narrow bead channel. At the early stages of packing, the beads entering the cavity contacted the weir structures on either side of the cavity. The beads were unable to traverse the weir because the distance from the top of the weir to the bottom of the cover plate (1.0 μm) was less than the diameter of the individual particles of the packing material (1.5–4.0 μm). Figure 17.11 shows illustratively an image of the bead cavity midway through the packing procedure (a), and the cavity 10 s later (b) [119].

Similar in-stream SPE microdevices have been proposed for sample clean-up and enrichment of protein and peptide samples prior to MALDI-TOF MS analysis [121]. Silicon/glass devices incorporating a weir structure facilitate the packing of reverse-phase chromatographic beads. These beads are then used to successfully purify and enrich a 10 nM peptide mixture containing 2 M urea in 0.1 M phosphate-buffered saline prior to MS analysis. Subsequent modelling of the fluid dynamics in this system has allowed an improved grid-SPE device to be fabricated and tested for on-line proteomic sample preparation [122].

Other approaches based on pillar-like frits [123, 124] and photopolymerized frit-structure [125] have been fabricated for packing the SPE beads. Regarding the first ones, in a very interesting work [123], the channel pattern of a CEC microchip employed is shown in Figure 17.12(a), where the expanded view shows the bead chamber with frit structures (Figure 17.12b). The structures of a microchip include an injector and a bead chamber with integrated frits, where the particles of the stationary phase were completely retained. Dimensions of the frit structures are 25 × 20 μm, and the space between the structures is 3 μm. PDMS channels were coated with polyelectrolyte multilayers to avoid adsorption of hydrophobic analytes to PDMS and chromatographic beads are packed in a bead chamber with integrated frits. Silica beads (5 μm diameter, 2% in acetonitrile) were loaded and packed into the bead chamber by applying different pressures from the reservoirs. Neutral compounds and coumarins, have been successfully preconcentrated and separated.

Besides, in a photopolymerized frit-structure approach [125], a porous plug of polymethacrylate polymer (200 μm in length) was fabricated by ultraviolet irradiation in microchannels. Microcolumns of hydrophobic beads packed against the polymethacrylate plugs were utilized for the quantitative extraction of rhodamine B using a microchip layout shown in Figure 17.13(a). This work demonstrates the first utilization of

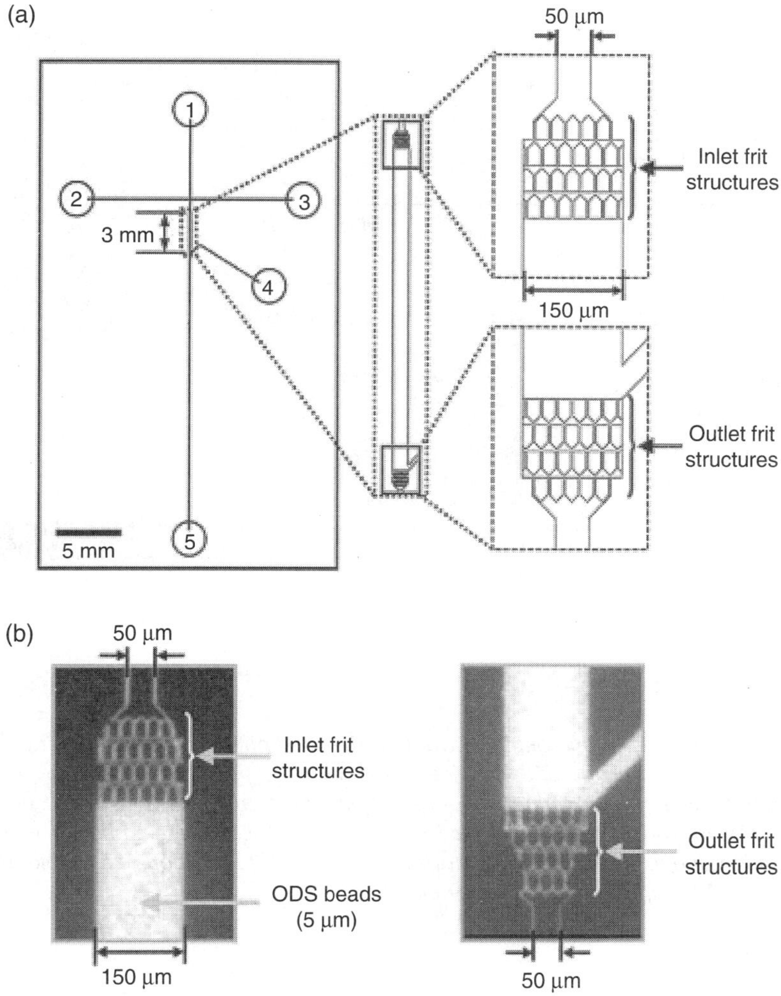

Figure 17.12 (a) Schematic of the PDMS microchip used for CEC and preconcentration: reservoirs: (1) sample; (2) buffer; (3) sample waste; (4) bead introduction; (5) waste; expanded view of a bead chamber and detailed structures of inlet and outlet frits. (b) Images of bead (5 μm diameter)-packed chambers inlet and outlet frit structures. Adapted from [123] with permission from Wiley-VCH Verlag GmbH & Co. KGaA © 2003.

photopolymerized frits for retaining beads in microfluidic channels for solid phase extraction. Although intricate fabrication of methacrylate polymers has been previously performed, fabrication of the small frits necessary for microfluidic frits was challenging. Several factors become critical to success in the photopolymerization of these methyacrylate monomers. The first one was the purity of the monomers. The second factor was the need for efficient cooling of the fluidic device during exposure. Cooling becomes necessary to eliminate undesirable polymerization. The third and most important factor was the elimination of localized fluid flow in the microchannels due to siphoning effects. These results in irregular, elongated masses of polymer that is not appropriate for holding beads as it is shown in Figure 17.13(b). Once the frit was in place, a simple procedure was used to pack the solid phase extraction column. Pulling slurry of beads in acetonitrile through the frit produced a well packed column in a short period of time (10 s).

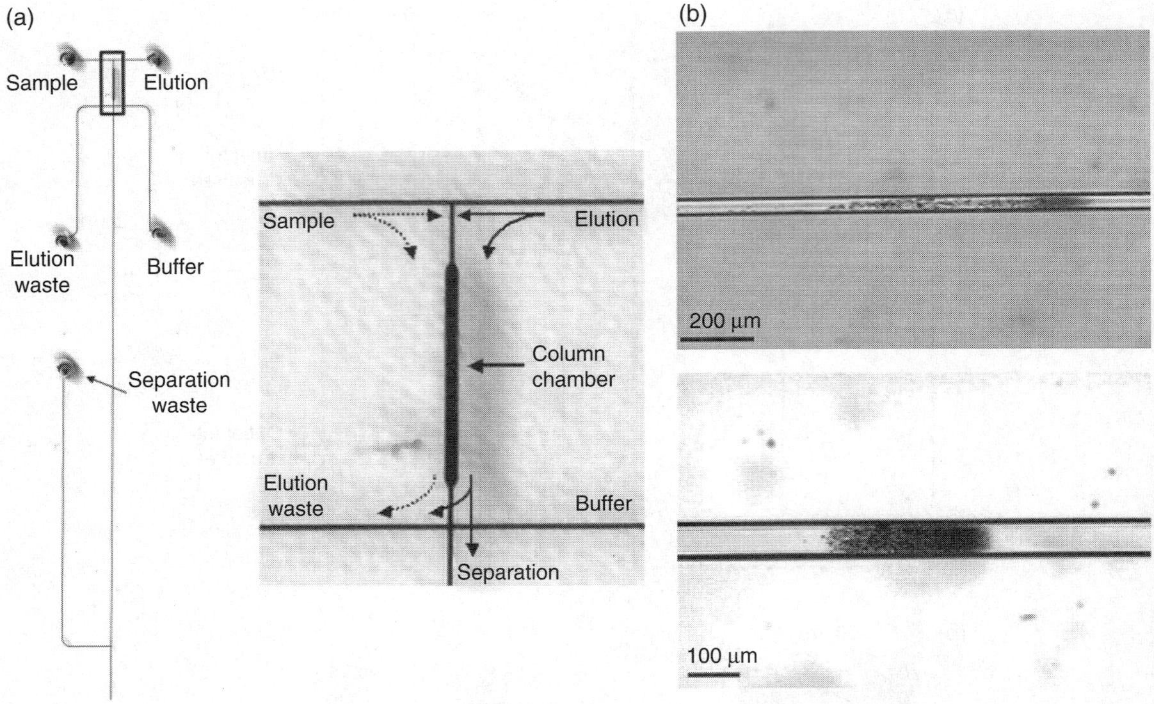

Figure 17.13 (a) Schematic of the microfluidic layout for the solid phase extraction/separation device with expanded view of the extraction region of the device. The dotted line signifies the direction of fluid flow during extraction; the solid line signifies flow during elution/injection. The narrow channels are 55 μm wide; the column chamber is 210 μm wide. All channels are 15 μm deep. (b) Fabrication of photopolymerized frits: sporadic polymerization due to siphoning effect (top) and successful fabrication of a usable frit (bottom). Adapted from [125] © 2005 American Chemical Society.

Photopolymerized frit structure has a main advantage based on the fact that it can be formed anywhere in a microchannel. For online coupling of SPE with zone electrophoresis on microchip (MCZE), a microfluidic device with a nanoporous membrane sandwiched between two PDMS substrate has been developed by Long *et al.* [126]. SPE and MCZE can be carried out in upper and lower fluidic layers, respectively. In addition, Broyles *et al.* [127] have reported a representative work showing the power of the integration filtration, concentration and separation on microchip has been reported which design is shown in Figure 17.14. Interestingly, filtration consists of an array of seven thin channels (1 μm deep and 18 μm wide) which comes together into one channel (5 μm deep and 25 μm wide). The input of the thin channels is communicated with sample reservoir and the sample loading was electrokinetically carried out. Sample concentration was performed by solid phase extraction. The separation channel was coated with C18 (1.5–4 μm) particles. Then, simply by increasing injection times, the analytes were retained and concentrated in the separation channel, being afterwards eluted with the appropriate solvent.

Moreover, a monolithic stationary phase synthesized by the polymerization of acrylate monomers for SPE-MCEC has been reported by Augustin *et al.* [128]. In the monolithic-SPE-MCEC analysis, the 200-fold enrichment and successive separation of polyaromatic hydrocarbons are accomplished at the sample injection time of 60 s.

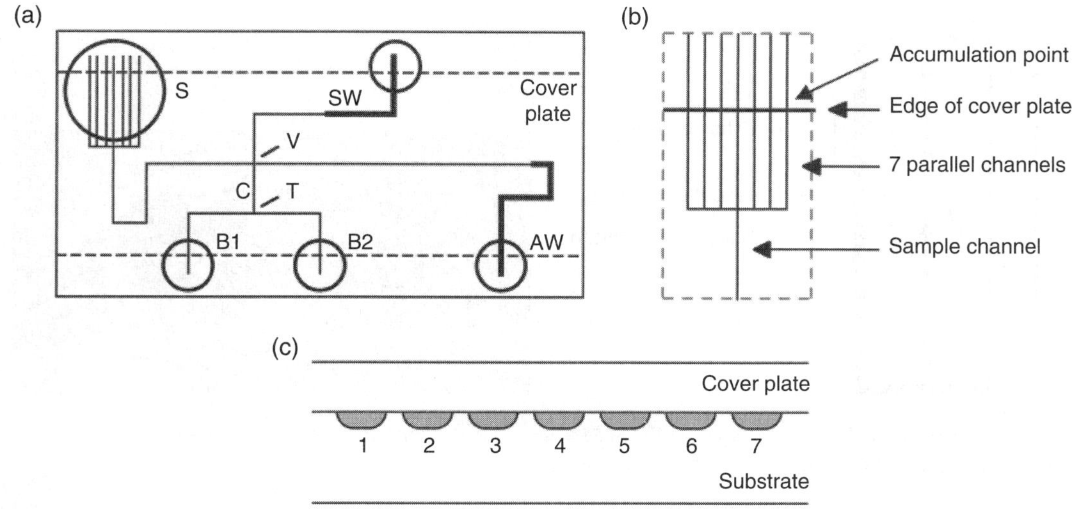

Figure 17.14 *(a) Schematic of the microchip for sample filtration, concentration and separation. The labels are (B1) buffer with 20% acetonitrile, (B2) buffer with 60% acetonitrile, (S) sample, (SW) sample waste, (AW) analysis waste, (T) T-intersection, and (V) valve. The circles represent fluid reservoirs. (b) An expanded view and (c) cross section of the filter element. The filer channels are numbered 1–7. Reprinted from [127] © 2003 American Chemical Society.*

Because of packing columns are often complex and ill-defined, an alternative is to replace conventional stationary phase materials with a continuous, porous bed of support (a porous monolith) formed by in situ polymerization of organic monomers [129]. The process of bed formation is facile, since a low-viscosity monomer solution can be introduced by vacuum or pressure into the microfluidic channel prior to initiation. Afterwards, the continuous polymer bed is attached to the channel walls, making a retaining frit or weir redundant. To this end, the preparation of microfluidic devices for SPE using porous monoliths with hydrophobic and ionisable surface chemistries has been described [130, 131]. In this work reported by Yu *et al.*, monolithic porous polymers have been prepared by photoinitiated polymerization within the channels of a microfluidic device and used for on-chip solid phase extraction and preconcentration. The preparation of the monolithic material with hydrophobic and ionizable surface chemistries was easily achieved by copolymerization of butyl methacrylate with ethylene dimethacrylate. Monoliths were prepared with two different surface chemistries at the desired location in the channel of the microchip shown in Figure 17.15(a). The large pore size of the monolith materials minimizes the flow resistance enabling use of very high flow rates of up to 10 μl min^{-1} without causing mechanical breakage of the monolith (Figure 17.15b) [130].

17.3.4 Preconcentration approaches using electrokinetics

As we have already mentioned in previous sections, in the miniaturization scene electrokinetic flow-driven systems have a relevant role because of their inherent and easy miniaturization, simplicity of fabrication and high-controlled fluidic manipulation and motivation. In addition, the analytical potency of electrokinetics has been enhanced since this principle has also been demonstrated to be very useful to analyte pre-concentration on microchips because their high versatility and control for fluidic handling offering a wide group of strategies with this purposes. Indeed, on-line sample preconcentration techniques can easily enhance the sensitivity compared to the development of the sensitive detection methods or the application of off-line preconcentration techniques. Employing on-line sample preconcentration, analytes can be rapidly and easily enriched and

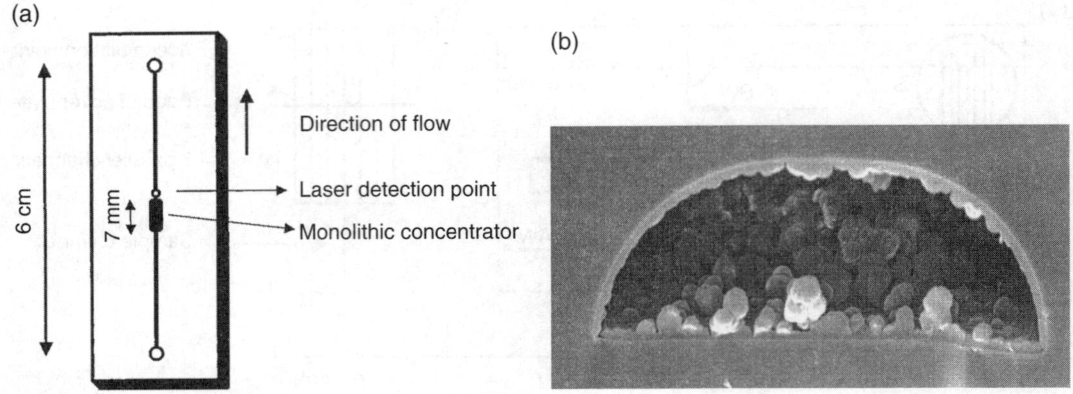

Figure 17.15 *(a) Microfluidic chip layout. (b) SEM image of the monolithic ion-exchange concentrator. Adapted from [130] © 2001 American Chemical Society.*

separated with a simple procedure by the variation in the migration velocity between the sample and running solution zones, which can be adjusted by several factors such as the conductivity, buffer components, pH of the solutions and so on. Although these methods have been developed in conventional scale (CE), several researchers have transferred these techniques from conventional format to the microchip one.

In the following sections, selected strategies and progress of integrated sample preconcentration electrokinetic techniques on microchip platforms will be studied. Firstly, preconcentration strategies based on the velocity change of the analytes between the sample and separation solution zones will be presented. In a second stage, 'focusing' techniques where the analytes are focused at the points where the migrating velocities become zero will be given toward selected examples. Some excellent general literature is proposed for further reading [132, 133].

17.3.4.1 Preconcentration based on the velocity change of the analytes between the sample and separation solution zones

Figure 17.16 shows conceptually the preconcentration mechanisms most common used on microchips based on the velocity change of the analytes between the sample and separation solution zones: field amplified stacking (FAS) and (t-) isotachophoresis (ITP). These approaches and the termed sweeping will be studied in this section.

Firstly, in the field amplified stacking (FAS), a background solution (BGS) with a high conductivity and a sample solution (S) with a low conductivity are prepared. The S is introduced into the capillary filled with the BGS, and then an appropriate voltage is applied to both ends of a capillary. The local electric field in the sample zone is higher than that in the BGS as the electric current in the capillary is constant. Therefore, the electrophoretic velocity of the analytes in the sample zone is faster than that in the BGS. This difference in the electrophoretic velocities between the S and BGS zones generates the 'stacking' effect at the S/BGS boundary, so that the analytes are concentrated around the boundary.

In MCE, the first applications of FAS have been reported by Jacobson and Ramsay [134, 135]. Preconcentration of the sample was performed using gated injection and compared with the pinched injection protocol [134]. Both principles are shown in Figure 17.17. Chip operation consisted of two modes: sample loading and separation. In the gated injection protocol, the sample migrates through the injection cross toward

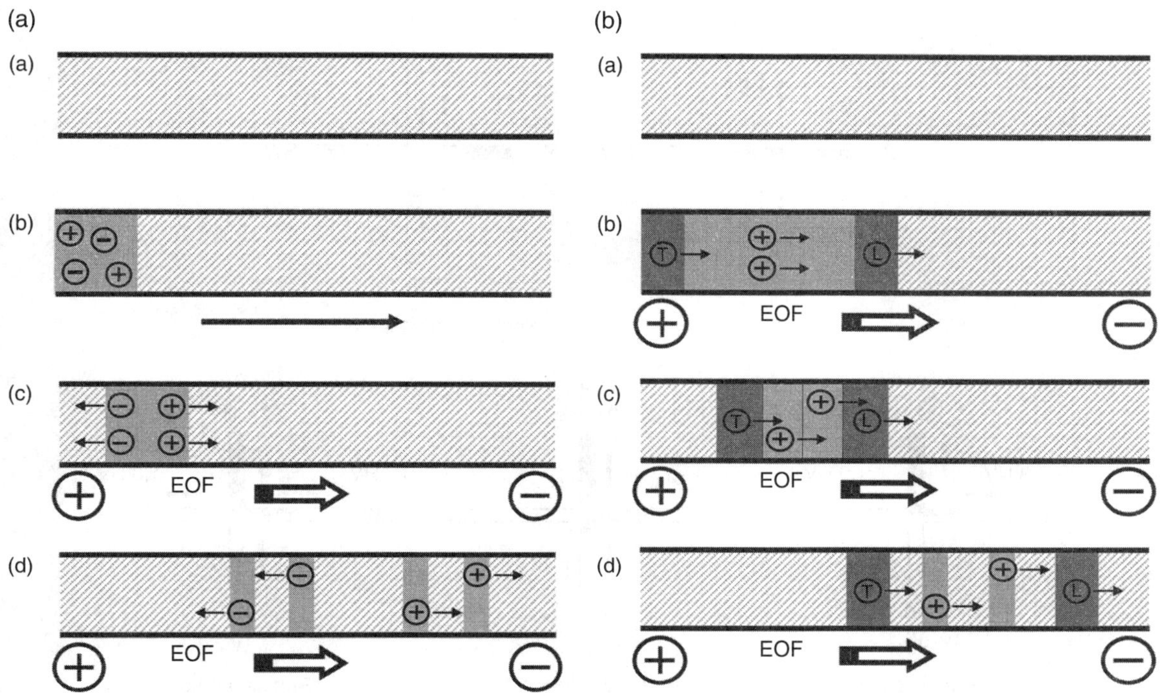

Figure 17.16 Operational procedures of the more common electrophoretic preconcentration methods. (A) Schematic diagrams of a field-amplified sample stacking (FASS) model. (a) The capillary is conditioned with BGE (a high conductivity buffer); (b) the sample, prepared in a low-conductivity matrix, is the injected to a certain length; (c) a high positive voltage is applied (EOF moves towards the outlet) that causes focusing of the analytes near boundaries between the sample zone and the BGE by effect of mobility changes; (d) stacked analytes migrate and are separated by CE. (B) Schematic diagrams of a transient isotachophoresis model for cations. (a) The capillary is conditioned with BGE; (b) the leading electrolyte, sample solution and terminating electrolyte are then injected in turn, (c) a high positive voltage is applied; (d) concentration of the analytes occurs between the leading and terminating ions during t-ITP migration and the concentrated analyte zones are separated by CE. Adapted from [133] © 2010 Bentham Science Publishers.

the sample waste reservoir. A buffer is pumped by electroosmosis from the buffer reservoir towards both the waste and sample waste reservoirs. This buffer stream prevents the sample from bleeding into the separation column. To make an injection of the sample onto the separation column, the potential at the buffer reservoir is simply floated (opening of the high voltage switch for a brief period of time (0.1–10 s), and sample migrates into the separation column as in an electrokinetic injection.

It is well known that the sensitive enhancement factor (SEF) obtained by FAS is increased to some extent with increase in the sample injection volume. However, a large volume injection of the sample also reduces the effective separation length, which may cause a poor resolution. To overcome this drawback, a fabrication of long channel for FAS on a microchip has been proposed [136]. In this excellent work, the sample filled in the long stacking channel is concentrated well and then the stacked sample is introduced into the separation channel without the loss of the effective separation length. A chip layout for full-column stacking combining a dedicated preconcentration column and a separation channel into a single system is depicted in Figure 17.18 where the stacking and separation procedure requires three steps. First,

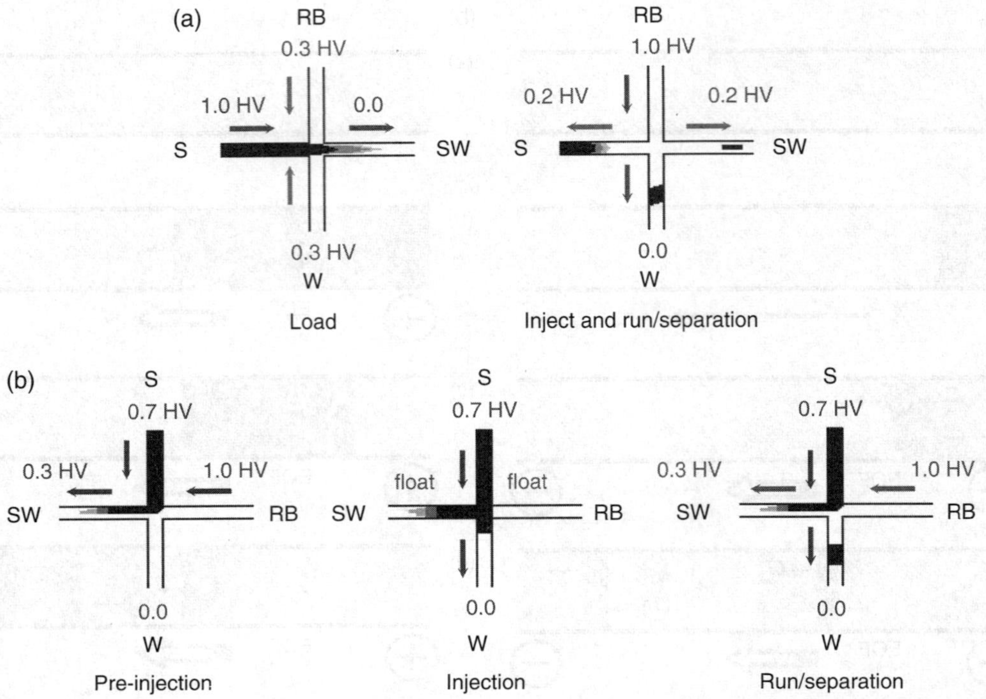

Figure 17.17 *Schematics of pinched (a) and gated (b) injection protocols. Arrows depict direction of flow of the sample and separation buffers. Adapted from [92] with permission from Wiley-VCH Verlag GmbH & Co. KGaA © 2004.*

the preconcentration column (69 mm long, 30 nl) is completely filled with the sample by applying a potential between the sample and stacking waste reservoirs. Then, the electric field is reversed by applying ground to the stacking buffer reservoir and a negative voltage to the sample waste. A high-conductivity running buffer is dragged into the capillary and sweeps the analyte anions back, thereby increasing their concentration. Finally, the stacked sample plug reaches the 9 mm long (3.9 nl) intersection of preconcentration and CE channel (38 mm long), and can be separated rapidly by applying an electrical field between buffer and buffer waste reservoir.

Several novel injection techniques for the large volume sample injection have been also reported, for example, the use of narrow channels [137]. In FAS, there is a major problem that neutral analytes cannot be concentrated. To improve this low applicability in FAS, electrokinetic chromatography (EKC) conditions on a microchip have been introduced for concentrating neutral analytes [138]. Also, the adsorption of analytes onto the inner PDMS microchannel wall was prevented using ionic liquids (ILs), while the EOF is increased due to the charge of the ILs [139].

Another difficulty usually observed in FAS is to concentrate a sample solution containing high-concentration salts, for example, sea water and blood, because a high conductivity in the sample matrix provides a smaller difference in the electric field at the BGS/S boundary. Trying to solve this trouble ITP can be used where an ionic sample solution is introduced between leading (LE) and terminating electrolyte (TE) solutions. The electrophoretic mobilities of the electrolytes are in the order of LE > S > TE. When the voltage is applied to a separation column, the sample ions are separated in the order of the electrophoretic mobilities, and then all the divided zones migrate with a uniform velocity toward a detection point (see Figure 17.16b). Next we describe selected approaches.

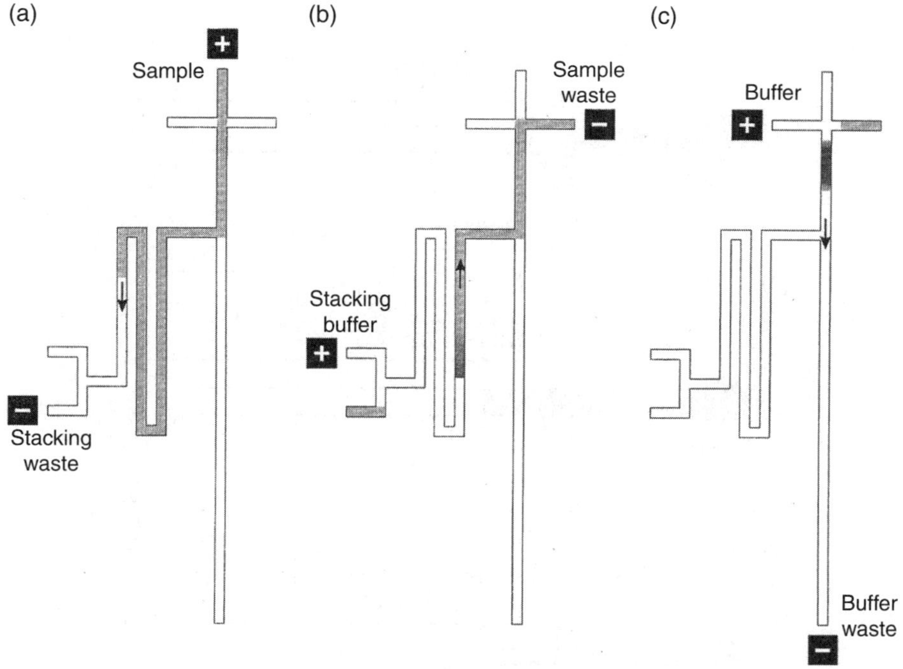

Figure 17.18 *The three process steps for full-column stacking in a coupled column configuration: (a) filling the stacking channel, (b) sweeping anions back to the separation channel, and (c) separation of stacked sample plug. Adapted from [136] with permission from Wiley-VCH Verlag GmbH & Co. KGaA © 2001.*

In MCE, the first application of the ITP technique was reported by Walker *et al.* [140]. The ITP preconcentration was carried out on a microchip with Raman spectroscopic detection system. As the samples only line along the separation channel in the order of their electrophoretic mobilities, a baseline separation cannot be obtained in the ITP condition. Also, in MCE, the first report of the application of tITP to MCE-MS has been given by Zhang *et al.* [141]. Improved resolution and 10-fold signal enhancement for peptides are obtained by the tITP preconcentration. For further improvement of the preconcentration efficiency in tITP–MCE, Williams and co-workers have developed a 'floating injection' method [142–145]. As a result, a 13-fold enhanced sensitivity is obtained for the tITP–MCZE analysis of tyrosine kinase compared to the conventional injection scheme.

When a highly diluted sample solution is introduced into a microchannel filled with a highly concentrated LE solution by the floating injection in tITP, a million-fold increase in the SEF is obtained on the conventional cross channel chip [146]. In this work, a simple and robust isotachophoresis (ITP) method was integrated on a CE microchip-base to achieve million-fold sample stacking. Figure 17.19 shows the strategy proposed: (a) the north and the south reservoirs are filled with LE, and the west reservoir is filled with a mixture of TE sample. TE/LE boundary is formed by applying vacuum at the south reservoir (while arrows show the direction of pressure-driven flows). (b) ITP stacking is initiated by applying high voltage and ground at the east and west reservoirs, respectively. The black arrows denote the direction of electric filed. Sample anions electromigrate toward the anode as EOF is suppressed. The early stage of ITP stacking results in a partial separation (i.e. moving boundary electrophoresis). (c) The field is switched toward the north reservoir to inject LE ions behind the sample and initiates C.

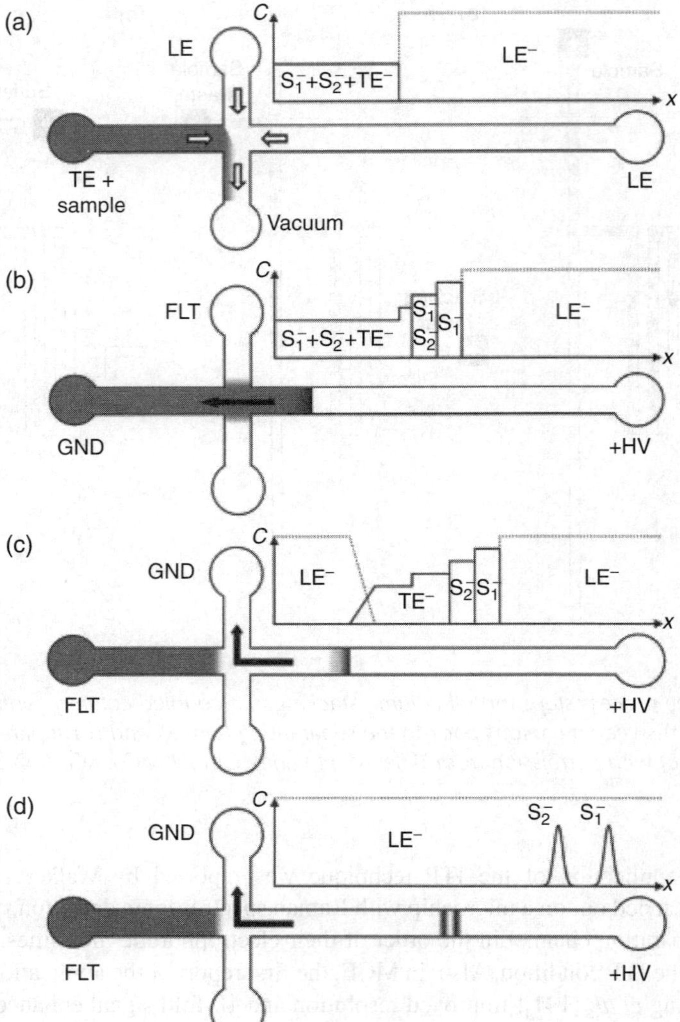

Figure 17.19 *Schematic of ITP/CE assay protocol. Adapted from [146] © 2006 American Chemical Society.*

Although it is necessary to determine the electrophoretic mobilities of sample ions prior to the ITP and tITP analyses, which may be time-consuming and troublesome, ITP is one of the most effective preconcentration techniques for ionic analytes in MCE as demonstrated above. ITP preconcentration has a wide applicability to ionic analytes mainly. As low-concentration biogenic analytes dissolved in high-conductivity matrices are well concentrated, the applications of ITP to MCE should contribute the progress of the bioanalysis.

Sweeping is another interesting alternative, less often integrated on lab-on-a-chip platforms. Previously, the online preconcentration of ionic analytes are mainly described. For neutral analytes, micellar eletrokinetic-chromatography (MEKC)-based preconcentration techniques are apparently effective. In MEKC, the difference in the interactions between the sample and micelle allows the separation of neutral compounds. Consequently, the effective velocities of the analytes are changed by the degree of the incorporation into the micelle. Quirino and Terabe developed this novel online sample preconcentration technique named 'sweeping', which utilizes the variation of the effective velocity in MEKC with ionic surfactant micelle [147–150]. In a typical sweeping condition, S and BGS are prepared with buffer solutions without and with anionic micelles such as SDS, respectively. In the case of the normal EOF condition, sample molecules migrate toward the cathode by the EOF in the S zone injected as a long plug for the preconcentration. When the sample molecules reach the S/BGS boundary, they are incorporated into the anionic micelle. Thus, the effective electrophoretic velocities of the samples are reduced. Finally, the analytes are concentrated at the S/BGS boundary. In MCE, on the other hand, only a few applications have been reported being the first approach reported by Sera *et al.* [151] an interesting example.

Recently, a novel preconcentration technique termed 'transient-trapping (tr-trapping)', which utilizes the sweeping phenomenon, has been developed in MCMEKC [152]. This technique allows not only enriching but also separating the analytes by 'trap' and 'release' mechanisms which occurred at the boundary between a sample and a partially injected micellar plug.

17.3.4.2 Preconcentration by flow focusing

Other less common approaches used in MCE, all of them based on focusing techniques, have been also proposed. In the electrophoretic analysis, several 'focusing' techniques are available for the preconcentration and separation. In these techniques, the migration direction of analytes is inversed in a separation channel. Thus, the analytes are 'focused' at the points where the migrating velocities become zero. The focusing techniques are roughly classified into three methods, that is, electric field gradient focusing (EFGF), temperature gradient focusing (TGF) and IEF.

In EFGF, sample molecules migrate in a channel applied as an electric field gradient. In the microchannel, a hydrodynamic counterflow is also introduced with a constant velocity. When the electric field at the inlet is higher than that at the outlet and the counterflow is introduced from the outlet, a faster electrophoretic migration of the S overcomes the hydrodynamic flow. Thus, the S moves toward the outlet. With decrease in the electric field along the microchannel, the electrophoretic velocity of the S is gradually decreased and finally inversed due to the counterflow, so that the S is focused and separated in the order of their electrophoretic mobilities. EFGF on planar microchips have been first reported by Humble *et al.* [153]. The focusing of green fluorescent protein (GFP) and R-phycoerythrin is carried out on a micro-EFGF device, resulting in the SEF of 104.

Ross *et al.* [154] have developed a novel focusing technique using a temperature gradient (TGF). The electrophoretic velocity of the S is gradually changed by the variation of the ionic strength of the BGS which is dependent on the temperature. TGF is more advantageous than EFGF in respect of easier operation and the applicability to a wide range of analytes. Various analytes, for example, fluorescent dyes, amino acids, GFP, DNA, labelled polystyrene particles and so on, are concentrated by TGF.

Finally, regarding focusing approaches, IEF is a powerful separation and preconcentration technique to analyse amphoteric compounds on the basis of their pIs. Generally, the IEF technique is applied to the characterization of peptides or proteins. In this case, a sample solution containing ampholytes is filled with the capillary. Then, both ends of the capillary are immersed into the catholyte and anolyte solutions, and then the electric field is applied. The S migrates in the pH gradient formed in the capillary. When the S reaches a point in the capillary based on their pIs, the charge of analytes becomes zero. Thus, the electrophoretic migration of the S is stopped. As a result, the S is focused and separated by the difference in the pIs. The miniaturization of IEF for a microfluidic format (microchip IEF; MCIEF) has been reported [155–160]. In the first report, MCIEF is carried out on a glass microchip with a simple straight channel pattern [155]. As the length of a focusing channel is reduced from 27 to 7 cm, the focusing time is decreased from 40–50 min to 5 min. To improve the analytical performance, the fabrication of a MCIEF device equipped with an on-chip diaphragm pumping [156], the combination with ESI-MS detection [157] and the application of chemiluminescence (CL) detection [158], have all been demonstrated. Apparently, MCIEF can shorten the focusing time, but the requirement of the mobilization step for the detection causes the reduction of the resolution and the increase of experimental procedures.

17.3.4.3 Electrokinetic concentration using nanochannels

As alternative to physical filtration, electrokinetics has also been used creatively for preconcentration using the double-layer properties. This novel approach has been termed as 'electrokinetic trapped'. Dai and Song have developed an electrokinetic filtration (trapping) [161, 162]. On a glass substrate with negative charges, electrical double layer (EDL) nearby the surface is overlapped in a nanochannel. As anionic molecules cannot penetrate into this overlapped EDL region, they can be trapped around the entrance of the nanochannel. Han and co-workers [163] have reported the electrokinetic preconcentration using a nanochannel fabricated onto a silicon wafer. The trapping condition can be maintained for several hours without current breakdown, which provides the SEF as high as 106–108 for small peptides. However, the difficulty in the nanochannel fabrication may be disadvantageous in spite of the extremely high enrichment ability. To achieve a highly effective preconcentration with simple fabrication procedures Kim *et al.* [164] have developed an electrokinetic trapping PDMS device. The fabricated device allows the preconcentration of proteins with the SEF of 106. The fabrication of a PDMS device for electrokinetic trapping using a nanogap has also been reported by Han and co-workers [165]. When the electric field of 1 kV is applied to a thin-walled section of the PDMS device, a nanogap is formed around the bonded surface between the PDMS and glass plate. After the formation, ion depletion and electrokinetic trapping occurs around the entrance of the nanogap.

In conclusion, both filtration and electrokinetic trapping are effective preconcentration techniques especially for large size molecules. Furthermore, the progress of the microfabrication techniques such as photopolymerization and nanogap allows on-chip filtration easily. However, the long filtrating time is one of the most significant problems in microfiltration, so that faster chip-based filtration analysis should be developed.

17.3.5 Derivatization schemes on microfluidic platforms

Derivatization on microchip has been well established in analytical microsystems. The reason of this development could be found in the traditional use of LIF as a sensitive detection system since the earliest times because it was highly required. Again, the role of microfabrication in the design of complex microcircuits offers a unique route in sample derivatization schemes, where the process is carried out before or after analyte separation and before analyte detection. The high degree of functional integration (reagent mixing, product separation and post column labelling) provides an elegant indication of the potential benefits of microfluidic systems. Derivatization schemes carried out prior to the separation (pre-column) or immediately previous to the detection (post-column), have been proposed using LIF and ED detections. Thus, derivatization in fluorescence has been well implemented into microchip in connecting with LIF detection [166]. Also, a group of works using an very attractive strategy based on the combination of suitable pre- and post-channel layouts with bioreagents such enzymes and antibodies and electrochemical detection have been reported. These strategies are shown in Figure 17.20.

An interesting approach where the incorporation of a pre-column reactor into the chip design, an integrated microfabricated device that performs the on-line derivatization/biocatalytic reactions of multiple enzymes and substrates, the electrophoretic separation of the involved analytes, and the electrochemical detection process on a chip platform, has been illustrated for the electrophoretic separation of eight amino acids in just 6 min [167] (Figure 17.20a). A new biochip strategy for performing post-column reaction of a 'class' enzyme (amino acid oxidase) on microchips platforms has also been described for the rapid and sensitive measurements

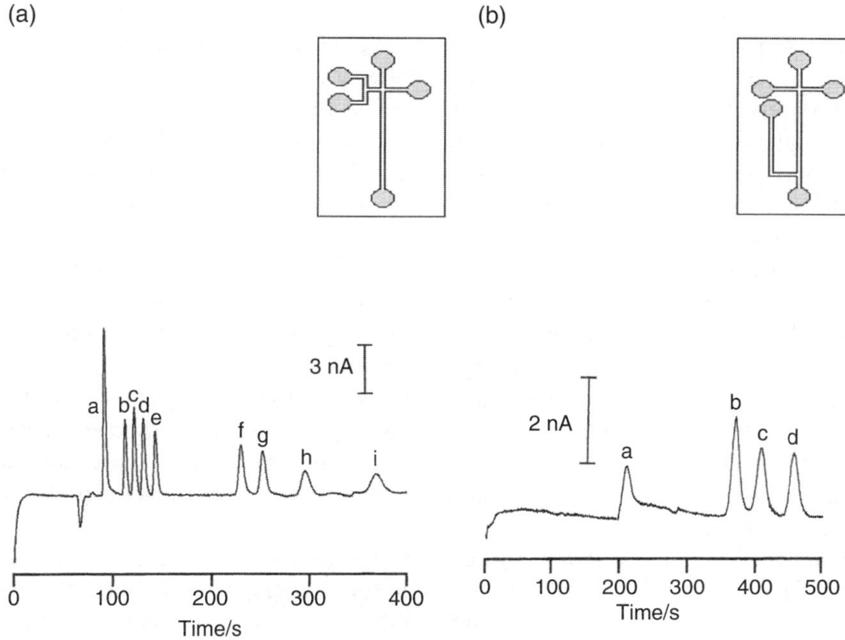

Figure 17.20 *Microchip layouts for derivatization of aminoacids (a) pre-column Adapted from [167]. © 2006 American Chemical Society, (b) post-column approaches. Adapted from [168] with permission from Wiley-VCH Verlag GmbH & Co. KGaA © 2002.*

of amino acids (arginine, isoleucine, alanine, phenylalanine) in connection to a post-column biogeneration of electroactive hydrogen peroxide [168]. The microchip layout is shown in Figure 17.20(b).

17.3.6 Sample preparation in cell analysis

One specific and exciting challenge is devoted to cell analysis. On the contrary of previous matter, this section is focused on a specific sample pre-treatment in biological analysis where different filtering, extraction and pre concentration strategies have been performed creatively on microfluidic devices. An interesting, specific and detailed review has been recently published [98].

Each analysis or detection scheme may require different sample preparation steps to obtain purified analytical targets from a variety of starting samples. The sample processing will depend to a large extent on the initial sample provided as input to the microfluidic device. This is the case of whole cells, where cell separation may be necessary, followed by cell lysis to release the proteins, nucleic acids or specific analyte into solution. The microfluidic devices have the advantages of the speed and cost to integrate various microstructures to achieve typical sample pre-treatment in cell analysis involving cell/particle filtrating and sorting, cell lysis, sample extraction and purification and so on. In this section, just selected examples of sample pre-processing steps in biological treatments on microdevices are described in order to show the potential of microfluidic technology in this field.

Firstly, regarding cell/particle filtrating and sorting, it is strongly necessary to sort, trap or concentrate the cells or exogenous material (virus) of interest from complex biological sample for obtaining accurate information in the downstream analytical step. As an example of genomic analysis, the large population of blood red cells needs to be separated from the nucleated white cells in order to remove the interfering substances. In this sense, Brody *et al.* [169] firstly designed and fabricated a novel microchip to separate cells based on the relative rigidity of the cell membrane, where cells were moving through a microarray via hydrodynamic flow. In a different approach focused on trapping white blood cells (WBCs) from whole blood, Wilding *et al.* [170] reported a silicon-glass hybrid microchips for cell isolation via their different sizes. The silicon substrate contains micropillars or 'weir-type' filters. Red blood cells (RBCs) assumed a biconcave discoid shape with a diameter of ~8 µm and a thickness of ~2 µm passed through the microstructure barrier, while larger WBC with a diameter of more than 10 µm were trapped by them. The white blood cells were then used as DNA source for microchip-based PCR.

Moreover, other microstructures such as micropillar array, microweir structure, membrane with circular, hexagonal and rectangular through holes, which was micromachined by a composite silicon nitride/parylene membrane method based on micro electronically-mechanical systems (MEMS) technology, have been used as particle filters for the gaseous sample [171]. Separation cells or particles by using microstructure barriers is not only a simple and non-destructive method but also easily permits integration of other different processes. However, problems of clogging or jamming in most of these separation microchips due to dead-end filtration are sometime presented. For dead end filtration, the fluidic flow is perpendicular to the filtration structures so that smaller particles pass through the filtration barriers along the fluid, while larger particles are stopped and clogging or jamming take place [172].

The idea of cross flow filtration with a glass-PDMS hybrid microchip for cells sorting was also reported by Sethu *et al.* [173]. The microchip fabricated by simple soft lithographic techniques employs microsieves that exploit the size and shape difference between the different cell types to obtain depletion of leukocytes from whole blood. Using this design, isolation of ~50% of the inlet RBCs, along with depletion of >97% of the inlet WBCs was performed by optimizing flow rate.

Following with the same principle, Chen *et al.* [174] have developed two kinds of microfluidic chips based on the cross flow filtration which can be more effective than conventional types in the area of avoiding clogging or jamming. Pillar-type and weir-type filtration microchips were designed and fabricated by MEMS technology, in which parallel micropillar-array and parallel microweirs were used to separate cells via their different sizes. The design of crossflow filtration microfluidic chips consisted of a PDMS-glass compounded cover and a silicon substrate. Shown in Figure 17.21(a), the design of silicon substrate includes an inlet for introducing blood sample, filtration barriers within tortuous channel, a WBC outlet for collecting WBCs, and an RBC outlet for collecting RBCs. The tortuous channel is divided into three sub-channels spaced by two-row filtration barriers in parallel. Pillar-type filtration barrier is shown in Figure 17.21(b), in which two-row pillars are about 20 µm in diameter and spaced by 6.5 µm. Weir-type filtration barrier is shown in Figure 17.21(c) in which two integral weirs are 26.5 µm high and 20 µm wide.

Moving on the second aspect in cell processing, cell lysis is a main step in proteomic analysis or genomic analysis to release proteins or gene from the cells of interest. Typical laboratory protocols for off-chip lysis steps include the use of enzymes (lysozyme), chemical lytic agents (detergents), and mechanical forces (sonication, bead milling). Cell lysis on microdevices by chemical method with sodium dodecyl sulfate (SDS), along with the movement of cells in glass chips using electrokinetic transport has been reported [175]. In this article, the initial steps made in on-chip cell transport and manipulations have been nicely addressed. The injection, transport, and manipulation of three types of cells (yeast, *Escherichia coli*, and erythrocytes) were studied. These cells were selected due to their relevance to applications in cell and molecular biology, DNA cloning and hematology. The ability to direct cell flow in a network of channels and to create packets of cells, using applied potentials to control the direction of flow, has been demonstrated. A simple lysing reaction, initiated in a controlled fashion by on-chip mixing of cells and reagents has also been shown. Potentials were applied to the various reservoirs to direct solvent flow within the four intersecting channels. As example of this cell transport and manipulations, Figure 17.22 illustrates the manipulation of *E. coli* cells

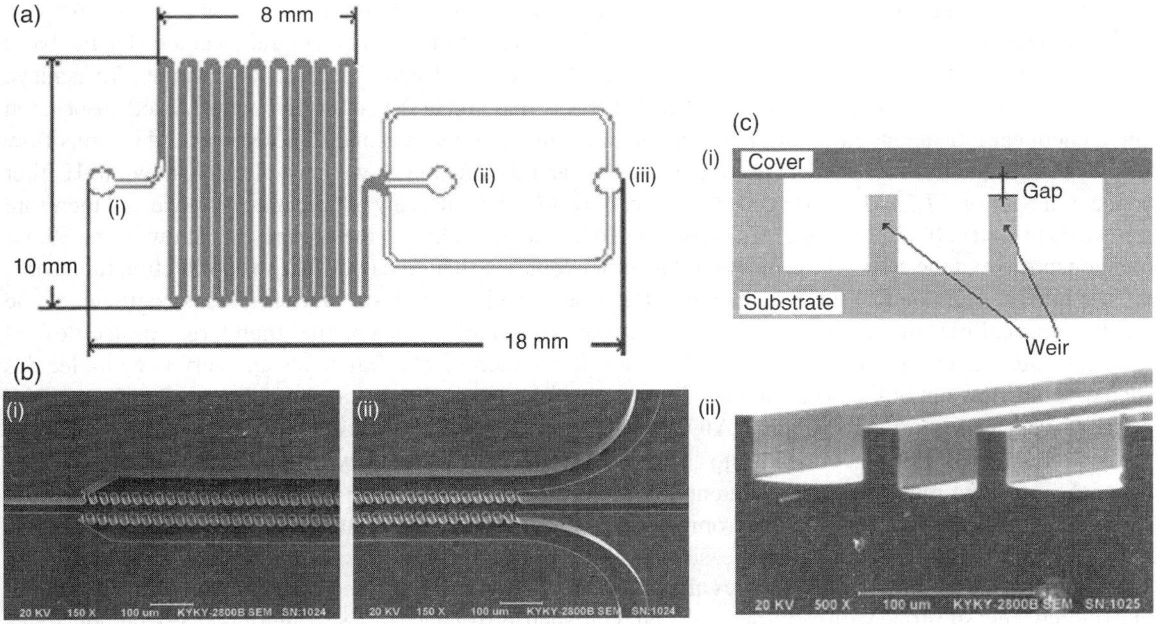

Figure 17.21 *(A) Schematic view of the silicon substrate. (a)Inlet port for introducing blood samples, (b) outlet port for collecting WBC (c) and outlet port for collecting RBC. (B) SEM micrographs of pillar-type barrier: (a) inlet, (b) outlet. (C) Weir-type filtration barrier: (a) sketch map of weir section, (b) SEM micrograph. Adapted from [174] with permission from Elsevier, © 2008.*

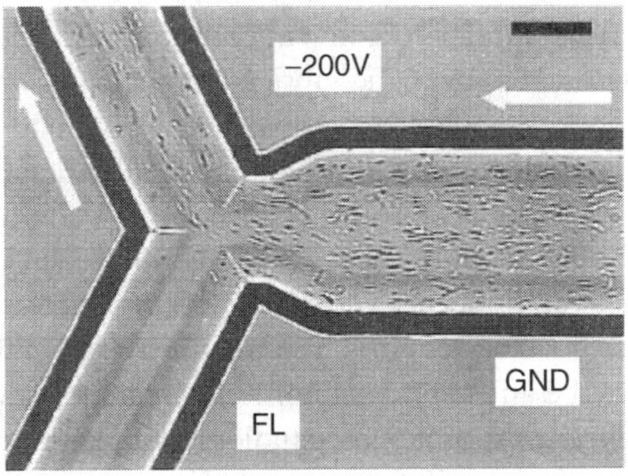

Figure 17.22 *Photomicrograph showing E. coli cell transport. White arrows show the flow direction, and the black bar shows the scales (40 μm). The –200 V and floating (FL) potentials could be toggled between the two branches of the Y to switch the flow of cells between branches. Adapted from [175] © 1997 American Chemical Society.*

at the Y-intersection of the device, using electric fields to control the direction of flow. Higher voltages give higher flow rates without cell lysis.

Unfortunately, the use of SDS might not be appropriate for enzymatic, immuno, or affinity based assays because the SDS binds to the proteins. In this sense, Schilling *et al.* [176] reported a T-type microfluidic chip that allowed the continuous lysis of bacterial cells using a mild detergent that did not affect the proteins with complete lysis of cells in less than 1 s. Moreover for the given chip, fluorescent detection of the enzymatic reaction showed that cell lysis and protein separation can be sequentially implemented on a single chip. The device showed in Figure 17.23(a) consists of three inlets, two outlets and two central channels. On the basis of their function, the main channels are named the lysis channel and the detection channel. To achieve predictable flow patterns pump rates are controlled at all inlets and at the controlled outlet. A cell suspension and a chemical lytic agent enter through separate inlets into the lysis channel. These two fluid streams flow side by side in the channel with no mixing except by lateral diffusion as we have already shown in H-filter concept in Section 17.3.2. Because cells are large, they do not diffuse any appreciable distance and therefore remain on the left half of the channel. The lytic agent permeabilizes the cell membranes and allows intracellular components to exit the cells. These intracellular components are then free to diffuse in all directions and some diffuse into the right half of the lysis channel. Because of their relative size and diffusions coefficients, the smaller intracellular molecules travel further into the right half of the channel than larger molecules. At the controlled outlet, the flow rate is such that all of the remaining cell fragments and very large molecules such as DNA flow out of the lysis channel. These components can then be transported to waste or to the inlet of another device for further separation. All molecular species that have diffused into the right half of the lysis channel are carried around the bend into the detection channel, where they occupy the left portion of the channel. The large molecules are preferentially located near the centreline of the detection channel while smaller molecules have diffused away from the centreline. Entering the right half of the detection channel is a stream containing a detection molecule, such as a fluorogenic substrate. As the two streams in the detection channel interdiffuse, the detection species allows the production of a fluorescent product through the reaction of a fluoregeneic substrate with an intracellular enzyme such as B-galactosidase. Figure 17.23(b) demonstrates the concept of the detection principle used in this impress work. The flow rates and device dimensions (including the length of the two main channels) can be varied to control the average residence time in the device, thereby allowing tuning of the system to any particular cell constituent.

Mechanical cell lysis is an alternative method if detergents interfere with downstream analysis [177]. A mechanical cell lysis device with nanostructural barbs which was used to disrupt sheep blood cells has been proposed. The microscale shear lysis similar to conventional macroscopic techniques can be achieved by forcing cells through nanoscale barbs. However, the fabricated process of nanostructures is complex. Further miniaturization cell lysis methods have included thermal, ultrasonic and electrical. Although the heating method is not applicable to proteins analysis due to protein molecules denature in the course of heating, it is simple method for releasing DNA for analysis using the heater of PCR. By other hand, although these approaches have been shown to be moderately successful, they all depend on the use of an external power supply and the devices may be quite complicated and costly to fabricate.

Extraction and purification are further required pre-treatments in cell analysis. For most of genetic analytical processes, the DNA must be amenable to amplification at a reasonably high concentration without endogenous PCR inhibitors. Also for most proteomic studies, a single protein or group of protein interested must often be isolated from the remaining cellular material. Thus once cell lysis has occurred, extraction of DNA or purification of a particular protein must take place. Microfluidic methods for extraction and purification have also been widely explored by a number of researchers as well.

As it has been seen in Section 17.3.3.2, solid phase extraction (SPE) liable to integration into a microfluidic system is one of the predominant methods currently as a chip-based methodology. Kutter *et al.* [118] reported a SPE on microdevice for enriching coumarin C460 with C18 wall coating method, but it was operated by exterior high voltage power, and again the limited surface area creates a capacity problem as we have stated in previous sections. Although particles have been a more common choice in solid phase extraction, monolithic porous polymers were prepared by photoinitiated polymerization within the channels of a microfluidic device and used for on-chip solid phase extraction and preconcentration reported by Yu *et al.* [130].

As another example of SPE on chips, the group of Liu *et al.* [178, 179] used a porous silicon with huge surface area as the solid phase matrix for adsorption DNA. Pore size of porous rectangle channel anodized at

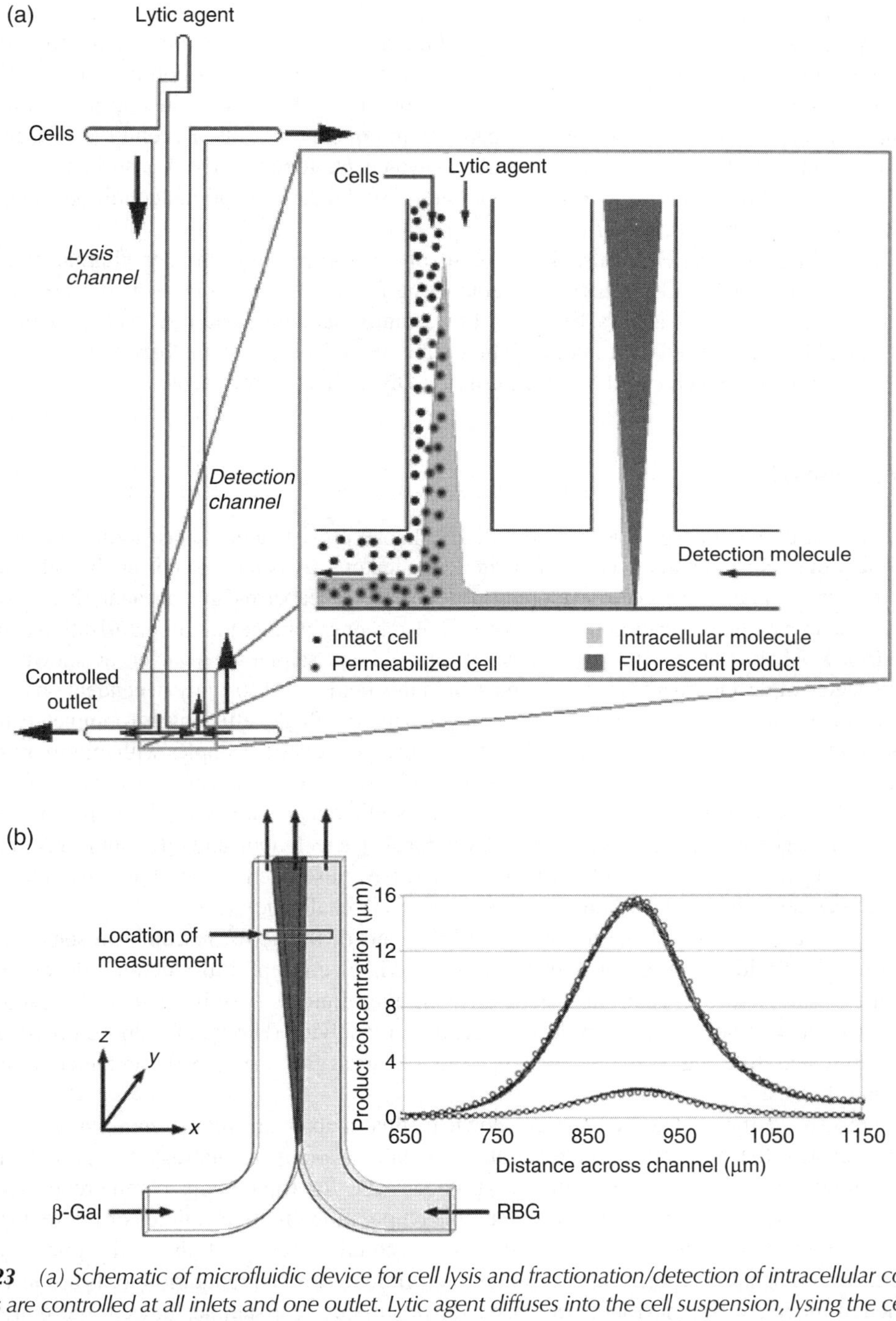

Figure 17.23 *(a) Schematic of microfluidic device for cell lysis and fractionation/detection of intracellular components. Pump rates are controlled at all inlets and one outlet. Lytic agent diffuses into the cell suspension, lysing the cells. Intracellular components then diffuse away from the cell stream and some are brought around the corner into the detection channel, where their presence can be detected by the production of a fluorescent species from a fluorogenic substrate. (b) Validation enzyme experiments in a simplified T-sensor. B-Gal is introduced on one side of the T-sensor while fluorogenic substrate, RBG, is introduced on the other side. As flow proceeds downstream, the two components interdiffuse and fluorescent resorufin is produced. Adapted from [176] © 2002 American Chemical Society.*

the optimal conditions has been determined in the range of 20 to 30 nm and surface area is approximately 400 m^2 g^{-1} by using BET technology. Thus, the surface area to volume of porous microfluidic chip is approximate 300 m^2 cm^3, which is thousands of times more than that of non-porous one. For the optimal μSPE chip, 49.5 ng PCR-amplifiable DNA was extracted per μl whole blood at the optimal condition within 15 min, which was approximately two-fold compared with commercial kits. Recently, a SPE microfluidic device for removing bovine serum albumin (BSA) in which photopolymerized adsorbent as a stationary phase of SPE was in situ polymerized has been reported [180]. Under the optimal conditions, the adsorption capacity was 36 mg BSA g^{-1} adsorbent.

Liquid/liquid extraction (LLE) is also in the case of cell processing, another alternative method for extraction and purification. LLE is achieved by contacting fluidic streams at constricted openings between distinct channels. The approach is attractive since flows can be separated naturally as the channels diverge. Reddy and Zahn [181] generated dual inlet and three inlet microfluidic systems based on organic– aqueous liquid (phenol) extraction which could be used purifying DNA directly from cells.

17.4 Final remarks

The many examples discussed in previous sections of this chapter prove that miniaturization is an active research field in a variety of application areas. The several attractive features associated to this analytical strategy made to consider it a valuable alternative when developing more efficient, cheaper and greener analytical methodologies than those in use in routine and monitoring analyses. It also seems evident that miniaturization should be the strategy to follow when dealing with the analysis of size-limited samples. However, as shown in previous sections, the degree of development achieved at present in this field is still strongly dependent on both the level of miniaturization intended and the nature of the investigated matrix. Really efficient and completely hyphenated analytical systems are already commercialized for the treatment of aqueous samples with minimum sample and reagent consumption. On the contrary, the extraction of solid and semi-solid samples remains as an (almost) unachieved goal. Various examples have illustrated successful strategies allowing the miniaturization of this step using 'small' versions of the corresponding conventional size technique and different levels of integration of these treatments with the subsequent purification steps have also been reported as prove of the concept. However, commercialization of these home-made set-ups is still challenge.

Similar considerations can be done regarding μTAS. Apart of the active research on sensitive detection principles on microfluidics since the early times of μTAS concept introduction, developments and improvements of sample preconcentration techniques on microfluidics have been also addressed as second alternative for the detection of extremely low-concentration analytes. The main proposed approaches have been microfabricated filtering systems, preconcentration strategies using SPE and electrokinetics and derivatization schemes.

Although mostly of the selected examples shown in this chapter are really impressive and they have demonstrated clearly the potency of microfluidics and lab-on-a-chip technology for performing sample preparation under microscale; however, these approaches are far from being commercial. The possible reasons could be drawn as follows. Firstly most sample preparation on microfluidics or microchips required exterior power units which are expensive and huge. Secondly, mostly of the real samples are highly complicated. Among what is reported above, some microfluidic sample preparation tools are just new ideas or experimental model, which have not been used for processing real sample. Indeed, the majority of the reported analytes are a mixture of model fluorescent dyes as LIF detection where the fluorescent derivatization is consequently also required. Last but not the least, it is not easy for the aimed analyte prepared by miniaturized sample preparation techniques to be detected by the conventional and commercial method or technology due to the fact that the present instruments usually requires more target analyte.

Because of extremely novel sensitive detection schemes cannot be commercially available soon, sample preconcentration approaches will remain for the highly sensitive analysis on microfluidic platforms. In the next step to use microfluidics for real sample analysis, therefore, not only further enhancement of the detectability but also the shorter analysis time, higher reproducibility, easier fabrication and further integration with different functional operations will be expected in sample preconcentration approaches integrated on these miniaturized platforms. The other alternative is that the sample preparation on microfluidic platforms should be designed to combine with the traditional instruments in order to transit these miniaturized sample preparation tools to commodities.

Acknowledgements

Authors thank MICINN (AGL2009–11909) and CM and FEDER program (project S2009/AGR-1464, ANALISYC-II) for financial support.

References

1. Dean, J.R. (2009) *Extraction Techniques in Analytical Sciences*, John Wiley & Sons, Ltd, Chichester.
2. Armenta, S.; Garrigues, S. and de la Guardia, M. (2008) Green Analytical Chemistry, *TrAC-Trend. Anal. Chem.*, **27**, 497–511.
3. de Koning, J.A. Blokker, P.; Jüngel, P.; Alkema, G. and Brinkman, U.A.T. (2002) Automated liner exchange - A novel approach in direct thermal desorption-gas chromatography, *Chromatographia*, **56**, 185–190.
4. Kokosa, J.M.; Przyjazny, A. and Jennot, M.A. (2009) *Solvent Microextraction. Theory and Practice*, John Wiley & Sons, Inc., Hoboken.
5. Jeannot, M.A. and Cantwell, F.F. (1997) Mass transfer characteristics of solvent extraction into a single drop at the tip of a syringe needle, *Anal. Chem.*, **69**, 235–239.
6. He, Y. and Lee, H.K. (1997) Liquid-phase microextraction in a single drop of organic solvent by using a conventional microsyringe, *Anal. Chem.*, **69**, 4634–4640.
7. Nerín, C.; Salafranca, J.; Aznar, M. and Batlle, R. (2009) Critical review on recent developments in solventless techniques for extraction of analytes, *Anal. Bioanal. Chem.*, **393**, 809–833.
8. de Jager, L.S. and Andrews, A.R.J. (2001) Development of a screening method for cocaine and cocaine metabolites in urine using solvent microextraction in conjunction with gas chromatography *J. Chromatogr. A*, **911**, 97–105.
9. Wu, H.F.; Yen, J.H. and Chin, C.C. (2006) Combining drop-to-drop solvent microextraction with gas chromatography/mass spectrometry using electronic ionization and self-ion/molecule reaction method to determine methoxyacetophenone isomers in one drop of water, *Anal. Chem.*, **78**, 1707–1712.
10. Ma, M. and Cantwell, F.F. (1999) Solvent microextraction with simultaneous back-extraction for sample cleanup and preconcentration: preconcentration into a single microdrop, *Anal. Chem.*, **71**, 388–393.
11. Zhu, L.; Tay, C.B. and Lee, H.K. (2002) Liquid–liquid–liquid microextraction of aromatic amines from water samples combined with high-performance liquid chromatography, *J. Chromatogr. A*, **963**, 231–237.
12. Shen, G. and Lee, H.K. (2003) Headspace liquid-phase microextraction of chlorobenzenes in soil with gas chromatography-electron capture detection, *Anal. Chem.*, **75**, 98–103.
13. Ouyang, G.; Zhao, W. and Pawliszyn, J. (2007) Automation and optimization of liquid-phase microextraction by gas chromatography, *J. Chromatogr. A*, **1138**, 47–54.
14. Kokosa, J.M.; Przyjazny, A. and Jones, R. (2007) *Paper 1680–4, presented at PittCon 2007*, Chicago.
15. Ouyang, G. and Pawliszyn, J. (2006) Kinetic calibration for automated hollow fiber-protected liquid-phase microextraction, *Anal. Chem.*, **78**, 5783–5788.
16. Pedersen-Bjergaard, S. and Rasmussen, K.E. (1999) Liquid–liquid–liquid microextraction for sample preparation of biological fluids prior to capillary electrophoresis, *Anal. Chem.*, **71**, 2650–2656.

17. Gjelsad, A.; Andersen, S.; Rasmussen, K. E. and Pedersen-Bjergaard, S. (2007) Microextraction across supported liquid membranes forced by pH gradients and electrical fields, *J. Chromatogr. A*, **1157**, 38–45.
18. Jiang, G. and Lee, H.K. (2004) Solvent bar microextraction, *Anal. Chem.*, **76**, 5591–5596.
19. Rezaee, M.; Assadi, Y.; Milani-Hosseini, M.R.; Aghaee, E.; Ahmadi, F. and Berijani, S. (2006) Determination of organic compounds in water using dispersive liquid–liquid microextraction, *J. Chromatogr. A*, **1116**, 1–9.
20. Fontanals, N.; Marcé, R. M. and Borrull, F. (2007) New materials in sorptive extraction techniques for polar compounds, *J. Chromatogr. A*, **1152**, 14–31.
21. Tamayo, F.G.; Turiel E. and Martin-Esteban, A. (2007) Molecularly imprinted polymers for solid-phase extraction and solid-phase microextraction: Recent developments and future trends, *J. Chromatogr. A*, **1152**, 32–40.
22. Tian, M.; Bi, W. and Row, K.H. (2011) Molecular imprinting in ionic liquid-modified porous polymer for recognitive separation of three tanshinones from Salvia miltiorrhiza Bunge, *Anal. Bioanal. Chem.*, **399**, 2495–2502.
23. Bagheri, H.; Brouwer, E.R.; Ghijsen, R.T. and Brinkman, U.A.T. (1993) On-line low-level screening of polar pesticides in drinking and surface waters by liquid chromatography—thermospray mass spectrometry, *J. Chromatogr. A*, **647**, 121–129.
24. Hankemeier, T.; van Leeuwen, S.P.J.; Vreuls, J.J. and Brinkman, U.A.T. (1998) Use of a presolvent to include volatile organic analytes in the application range of on-line solid-phase extraction–gas chromatography–mass spectrometry, *J. Chromatogr. A*, **811**, 117–133.
25. Capiello, A.; Berloni, A.; Famiglini, G.; Mangani, F. and Palma, P. (2001) Micro-SPE method for sample introduction in capillary HPLC/MS, *Anal. Chem.*, **73**, 298–302.
26. Hogenboom, A.C.; Steen, R.J.C.A.; Niessen, W.M.A. and Brinkman, U.A.T. (1998) Potential of short-column liquid chromatography with tandem mass spectrometric detection for the rapid study of pesticide degradation, *Chromatographia*, **48**, 475–480.
27. Hogenboom, A.C.; Niessen, W.M.A. and Brinkman, U.A.T. (2000) Characterization of photodegradation products of alachlor in water by on-line solid-phase extraction liquid chromatography combined with tandem mass spectrometry and orthogonal-acceleration time-of-flight mass spectrometry, *Rapid Commun. Mass Spectrom.*, **14**, (1914–1924).
28. Ramos, L.; Vreuls, J.J. Brinkman, U.A.T. and Sojo, L.E. (1999) Simultaneous filtration and liquid chromatographic microextraction with subsequent GC-MS analysis to study adsorption equilibria of pesticides in soil, *Environ. Sci. Technol.*, **33**, 3254–3259.
29. Hogenboom, A.C.; Speksnijder, P.; Vreeken, R.J.; Niessen, W.M.A. and Brinkman, U.A.T. (1997) Rapid target analysis of microcontaminants in water by on-line single-short-column liquid chromatography combined with atmospheric pressure chemical ionization tandem mass spectrometry, *J. Chromatogr. A*, **777**, 81–90.
30. Hogenboom, A.C.; Hofman, M.P.; Kok, S.J.; Niessen, W.M.A. and Brinkman, U.A.T. (2000) Determination of pesticides in vegetables using large-volume injection column liquid chromatography–electrospray tandem mass spectrometry, *J. Chromatogr. A*, **892**, 379–390.
31. Pérez, M.; Alario, J.; Vázquez, A. and Villén, J. (1999) On-line reversed phase LC-GC by using the new TOTAD (Through Oven Transfer Adsorption Desorption) interface: Application to parathion residue analysis, *J. Microcolumn Sep.*, **11**, 582–589.
32. Cortés, J.M.; Toledano, R.M.; Villén, J. and Vázquez, A. (2008) Analysis of pesticides in nuts by online reversed-phase liquid chromatography–gas chromatography using the through-oven transfer adsorption/desorption interface, *J. Agric. Food Chem.*, **56**, 5544–5549.
33. Brouwer, E.R.; Lingeman, H. and Brinkman, U.A.T. (1990) Use of membrane extraction disks for on-line trace enrichment of organic compounds from aqueous samples, *Chromatographia*, **29**, 415–418.
34. Fritz, J.S. and Masso, J.J. (2001) Miniaturized solid-phase extraction with resin disks, *J. Chromatogr. A*, **909**, 79–85.
35. El-Beaqqali, A.; Kussak, A. and Abdel-Rehim, M. (2006) Fast and sensitive environmental analysis utilizing microextraction in packed syringe online with gas chromatography–mass spectrometry: Determination of polycyclic aromatic hydrocarbons in water, *J. Chromatogr. A*, **1114**, 234–238.
36. Abdel-Rehim, M. (2009) Current advances in microextraction by packed sorbent (MEPS) for bioanalysis applications, *LC-GC Eur.*, **22**, 8–19.
37. Arthur, C. and Pawliszyn, J. (1990) Solid phase microextraction with thermal desorption using fused silica optical fibers, *Anal. Chem.*, **62**, 2145–2148.

38. Gou, Y.; Eisert, R. and Pawliszyn, J. (2000) Automated in-tube solid-phase microextraction–high-performance liquid chromatography for carbamate pesticide analysis, *J. Chromatogr. A*, **873**, 137–147.
39. Bagheri, H. and Salemi, A. (2004) Coupling of a modified in-tube solid phase microextraction technique with high performance liquid chromatography-fluorescence detection for the ultra-trace determination of polycyclic aromatic hydrocarbons in water samples, *Chromatographia*, **59**, 501–505.
40. Li, Q.; Ma, X.; Yuan, D. and Chen, J. (2010) Evaluation of the solid-phase microextraction fiber coated with single walled carbon nanotubes for the determination of benzene, toluene, ethylbenzene, xylenes in aqueous samples, *J. Chromatogr. A*, **1217**, 2191–2196.
41. Cai, L.S.; Gong, S.L.; Chen, M. and Wu, C.Y. (2006) Vinyl crown ether as a novel radical crosslinked sol-gel SPME fiber for determination of organophosphorus pesticides in food samples, *Anal. Chim. Acta*, **559**, 89–96.
42. Koster, E.H.M.; Crescenzi, C.; den Hoedt, W.; Ensing, K. and de Jong, G.J. (2001) Fibers coated with molecularly imprinted polymers for solid-phase microextraction, *Anal. Chem.*, **73**, 3140–3145.
43. Djozan, D.J. and Abdollahi, L. (2003) Anodized zinc wire as a solid-phase microextraction fiber, *Chromatographia*, **57**, 799–804.
44. Sarrión, M.N.; Santos, F.J. and Galcerán, M.T. (2000) In situ derivatization/solid-phase microextraction for the determination of haloacetic acids in water, *Anal. Chem.*, **72**, 4865–4873.
45. Cai, Y. and Bayona, J.M. (1995) Determination of methylmercury in fish and river water samples using in situ sodium tetraethylborate derivatization following by solid-phase microextraction and gas chromatography-mass spectrometry, *J. Chromatogr. A*, **696**, 113–122.
46. Mester, Z. and Pawliszyn, J. (2000) Speciation of dimethylarsinic acid and monomethylarsonic acid by solid-phase microextraction–gas chromatography–ion trap mass spectrometry, *J. Chromatogr. A*, **873**, 129–135.
47. Sng, M.T. and Ng, W.F. (1999) In-situ derivatisation of degradation products of chemical warfare agents in water by solid-phase microextraction and gas chromatographic–mass spectrometric analysis, *J. Chromatogr. A*, **832**, 173–182.
48. Bouaid, A.; Ramos, L.; González, M.J.; Fernández, P. and Cámara, C. (2001) Solid-phase microextraction method for the determination of atrazine and four organophosphorus pesticides in soil samples by gas chromatography, *J. Chromatogr. A*, **939**, 13–21.
49. Chia, K.J.; Lee, T.Y. and Huang, S.D. (2004) Simple device for the solid-phase microextraction screening of polychlorodibenzo-p-dioxins and polychlorodibenzofurans in heavily contaminated soil samples, *Anal. Chim. Acta*, **527**, 157–162.
50. Baltussen, E.; Sandra, P.; David, F. and Cramers, C. (1999) Stir bar sorptive extraction (SBSE), a novel extraction technique for aqueous samples: Theory and principles, *J. Microcolumn Sep.*, **11**, 737–747.
51. Baltussen, E.; David, F.; Sandra, P. and Cramers, C. (1999) Study into the equilibrium mechanism between water and poly(dimethylsiloxane) for very apolar solutes: adsorption or sorption?, *Anal. Chem.*, **71**, 5213–5216.
52. Popp, P.; Bauer, C. and Wennrich, L. (2001) Application of stir bar sorptive extraction in combination with column liquid chromatography for the determination of polycyclic aromatic hydrocarbons in water samples, *Anal. Chim. Acta*, **436**, 1–9.
53. Silva, A.R.M. and Nogueira, J.M.F. (2010) Stir-bar-sorptive extraction and liquid desorption combined with large-volume injection gas chromatography–mass spectrometry for ultra-trace analysis of musk compounds in environmental water matrices, *Anal. Bioanal. Chem.*, **396**, 1853–1862.
54. Sandra, P.; Tienpont B. and David, F. (2003) Multi-residue screening of pesticides in vegetables, fruits and baby food by stir bar sorptive extraction–thermal desorption–capillary gas chromatography–mass spectrometry, *J. Chromatogr. A*, **1000**, 299–309.
55. Sanchez-Avila, J.; Quintana, J.; Ventura, F.; Tauler, R.; Duarte, C.M. and Lacorte, S. (2010) Stir bar sorptive extraction-thermal desorption-gas chromatography-mass spectrometry: An effective tool for determining persistent organic pollutants and nonylphenol in coastal waters in compliance with existing Directives, *Marine Poll. Bull.*, **60**, 103–112.
56. Yu, C.H.; Yao, Z.M. and Hu, B. (2009) Preparation of polydimethylsiloxane/beta-cyclodextrin/divinylbenzene coated 'dumbbell-shaped' stir bar and its application to the analysis of polycyclic aromatic hydrocarbons and polycyclic aromatic sulfur heterocycles compounds in lake water and soil by high performance liquid chromatography, *Anal. Chim. Acta*, **641**, 75–82.

57. Roy, G.; Vuillemin, R. and Guyomarch, J. (2005) On-site determination of polynuclear aromatic hydrocarbons in seawater by stir bar sorptive extraction (SBSE) and thermal desorption GC-MS, *Talanta*, **66**, 540–546.
58. Zhao, M.; van der Wielen, F. and de Vooght, P. (1999) Optimization of a matrix solid-phase dispersion method with sequential clean-up for the determination of alkylphenol ethoxylates in biological tissues, *J. Chromatogr. A*, **837**, 129–138.
59. Kristenson, E.M.; Shahmiri, S.; Slooten, C.J.; Vreuls, J.J. and Brinkman, U.A.T. (2004) Matrix solid-phase dispersion micro-extraction of pesticides from single insects with subsequent GC–MS analysis, *Chromatographia*, **59**, 315–320.
60. Garcinuño, R.M.; Ramos, L.; Fernández-Hernando, P. and Cámara, C. (2004) Optimization of a matrix solid-phase dispersion method with subsequent clean-up for the determination of ethylene bisdithiocarbamate residues in almond samples, *J. Chromatogr. A*, **1041**, 35–41.
61. Morzycka, B. (2002) Simple method for the determination of trace levels of pesticides in honeybees using matrix solid-phase dispersion and gas chromatography, *J. Chromatogr. A*, **982**, 267–273.
62. Ramos, J.J.; González, M.J. and Ramos, L. (2004) Miniaturised sample preparation of fatty foodstuffs for the determination of polychlorinated biphenyls, *J. Sep. Sci.*, **27**, 595–601.
63. Albero, B.; Sánchez-Brunete, C.; Donoso, A. and Tadeo, J. (2004) Determination of herbicide residues in juice by matrix solid-phase dispersion and gas chromatography–mass spectrometry, *J. Chromatogr. A*, **1043**, 127–133.
64. Blasco, C.; Font, G. and Picó, Y. (2004) Determination of dithiocarbamates and metabolites in plants by liquid chromatography–mass spectrometry, *J. Chromatogr. A*, **1028**, 267–276.
65. Kristenson, E.M.; Haverkate, E.G.J.; Slooten, C.J.; Ramos, L.; Vreuls, R.J.J. and Brinkman, U.A.T. (2001) Miniaturized automated matrix solid-phase dispersion extraction of pesticides in fruit followed by gas chromatographic-mass spectrometric analysis, *J. Chromatogr. A*, **917**, 277–286.
66. Hyötyläinen, T.; Andersson, T.; Hartonen, K.; Kuosmanen, K. and Riekkola, M.L. (2000) Pressurized hot water extraction coupled on-line with LC–GC: Determination of polyaromatic hydrocarbons in sediment, *Anal. Chem.*, **72**, 3070–3076.
67. Kuosmanen, K.; Hyötyläinen, T.; Hartonen, K. and Riekkola, M.L. (2001) Pressurised hot water extraction coupled on-line with liquid chromatography–gas chromatography for the determination of brominated flame retardants in sediment samples, *J. Chromatogr. A*, **943**, 113–122.
68. Ramos, L.; Vreuls, J.J. and Brinkman, U.A.T. (2000) Miniaturised pressurised liquid extraction of polycyclic aromatic hydrocarbons from soil and sediment with subsequent large-volume injection–gas chromatography, *J. Chromatogr. A*, **891**, 275–286.
69. Ramos, J.J.; Dietz, C.; González, M. J. and Ramos, L. (2007) Miniaturised selective pressurized liquid extraction of polychlorinated biphenyls, *J. Chromatogr. A*, **1152**, 254–261.
70. Cresswell, S.L. and Haswell, S.J. (1999) Evaluation of on-line methodology for microwave-assisted extraction of polycyclic aromatic hydrocarbons (PAHs) from sediment samples, *Analyst*, **124**, 1361–1366.
71. Ericsson, M. and Colmsjö, A. (2002) Dynamic microwave-assisted extraction coupled on-line with solid-phase extraction: determination of polycyclic aromatic hydrocarbons in sediment and soil, *J. Chromatogr. A*, **964**, 11–20.
72. Ericsson, M. and Colmsjö, A. (2003) Dynamic microwave-assisted extraction coupled on-line with solid-phase extraction and large-volume injection gas chromatography: determination of organophosphate esters in air samples, *Anal. Chem.*, **75**, 1713–1719.
73. Luque-García, J.L. and Luque de Castro, M.D. (2002) Continuous ultrasound-assisted extraction of hexavalent chromium from soil with or without on-line preconcentration prior to photometric monitoring, *Analyst*, **127**, 1115–1120.
74. Albero, B.; Sanchez-Brunete, C. and Tadeo, J.L. (2003) Determination of organophosphorus pesticides in fruit juices by matrix solid-phase dispersion and gas chromatography, *J. Agric. Food Chem.*, **51**, 6915–6921.
75. Ramos, J.J. Rial-Otero, R. Ramos, L. and Capelo, J.L. (2008) Ultrasonic-assisted matrix solid-phase dispersion as an improved methodology for the determination of pesticides in fruits, *J. Chromatogr. A*, **1212**, 145–149.
76. Manz, A.; Grabner, N. and Widmer, H.M. (1990) Miniaturized total chemical analysis systems: A novel concept for chemical sensing, *Sens. Actuator. B.*, **1**, 244–248.
77. Janasek, D.; Frazke, J. and Manz, A. (2006) Scaling and the design of miniaturized chemical-analysis systems, *Nature*, **442**, 374–380.

78. Whitesides, G.M. (2006) The origins and the future of microfluidics, *Nature*, **442**, 368–373.
79. Daw, R. and Finkelstein, J. (2006) Lab on a chip, *Nature*, **442**, 367.
80. Rios, A.; Escarpa, A.; Gonzalez, M.C. and Crevillen, A.G. (2006) Challenges of analytical microsystems, *TrAC-Trend. Anal. Chem.*, **25**, 467–479.
81. Schwarz, M.A. and Hauser, P.C. (2001) Recent developments in detection methods for microfabricated analytical devices, *Lab. Chip*, **1**, 1–6.
82. Vandaveer, W.R.; Pasas, S.A.; Martin, R.S. and Lunte, S.M. (2002) Recent developments in amperometric detection for microchip capillary electrophoresis, Recent developments in amperometric detection for microchip capillary electrophoresis, *Electrophoresis*, **23**, 3667–3677.
83. Wang, J. (2005) Electrochemical detection for capillary electrophoresis microchips: A review, *Electroanalysis*, **17**, 1133–1140.
84. Pumera, M.; Merkosi, A. and Alegret, S. (2006) New materials for electrochemical sensing VII. Microfluidic chip platforms, *TrAC-Trend. Anal. Chem.*, **25**, 219–235.
85. Viskari, P.J. and Landers, J.P. (2006) Unconventional detection methods for microfluidic devices, *Electrophoresis*, **27**, 1797–1810.
86. van der Berg, A. and Bergveld, P. (2006) Labs-on-a-Chip: origin, highlights and future perspectives on the occasion of the 10th μTAS conference, *Lab-on-a-Chip*, **6**, 1266–1273.
87. Manz, A.; Harrison, D.J.; Verpoorte, E.M.J.; Fetlinger, J.C.; Paulus, A.; Lüdi, H. and Wildmer, H.M. (1992) Planar chips technology for miniaturization and integration of separation techniques into monitoring systems: Capillary electrophoresis on a chip, *J. Chromatogr. A*, **593**, 253–258.
88. Madou, M.J. (ed.) (2000) *Fundamentals of Microfabrication* (The Science of Miniaturization), CRC Press LLC, Boca Raton.
89. Becker, H. and Gärtener, C. (2003) Microreplication technologies for polymer-based μ-TAS applications, *Lab-on-a-Chip*, (eds R.E. Oosterbroek and A. van der Berg), Elsevier, Amsterdam, pp. 21–35.
90. Gardieneers, J.G. and Oosterbroek, R.E. (2003) Silicon and glass micromachining for μTAS, *Lab-on-a-Chip*, (eds R.E. Oosterbroek and A. van der Berg), Elsevier, Amsterdam, pp. 37–64.
91. Binyamin, G.; Boone, T.D.; Lackritz, H.S.; Ricco, A.J.; Sassi, A.P. and Williams, S.J. (2003) Plastic microfluidic devices: Electrokinetic manipulations, life science applications, and production technologies, *Lab-on-a-Chip*, (eds R.E. Oosterbroek and A. van der Berg), Elsevier, Amsterdam, pp. 83–112.
92. Kutter, J.P.; Mogensen, K.B.; Klank, H. and Geshke, O. (2004) Microfluidics – Components, *Microsystem Engineering of Lab-on-a-Chip Devices* (eds O. Geschke, H. Klank and P. Telleman), Wiley-VCH Verlag, Weinheim, pp. 39–77.
93. Petersen, D.; Mogensen, K.B. and H. Klank, H. (2004) Glass Micromachining, *Microsystem Engineering of Lab-on-a-Chip Devices* (eds O. Geschke, H. Klank and P. Telleman), Wiley-VCH Verlag, Weinheim, pp. 161–168.
94. Klank, H. (2004) Polymer Micromachining, *Microsystem Engineering of Lab-on-a-Chip Devices* (eds O. Geschke, H. Klank and P. Telleman), Wiley-VCH Verlag, Weinheim, pp. 169–182.
95. de Mello, A.J. and Beard, N. (2003) Dealing with 'real' samples: sample pre-treatment in microfluidic systems, *Lab. Chip*, **3**, 11N–19N.
96. Lichtenberg, J.; de Rooij, N.F. and Veerporte, E. (2002) Sample pretreatment on microfabricated devices, *Talanta*, **56**, 233–266.
97. Sueyoshi, K.; Kitagawa, F. and Otsuka, K. (2008) Recent progress of online sample preconcentration techniques in microchip electrophoresis, *J. Sep. Sci.*, **31**, 2650–2666.
98. Chen, X. and Cui, D.F. (2009) Microfluidic devices for sample pretreatment and applications, *Mycrosyst. Technol.*, **15**, 667–676.
99. Crevillén, A.G.; Hervás, M.; López, M.A.; González, M.C. and Escarpa, A. (2007) Real samples analysis on microfluidic devices, *Talanta*, **74**, 342–357.
100. Ríos, A.; Escarpa, A. and Simonet, B. (2009) *Miniaturization of analytical systems*, John Wiley & Sons, Ltd., Chichester.
101. Stemme, G. and Kittilsland, G. (1988) New fluid filter structure in silicon fabricated using a self-aligning technique, *Appl. Phys. Lett.*, **53**, 1566–1569.
102. Kittilsland, G.; Stemme, G. and Norden, B. (1990) A sub-micron particle filter in silicon, *Sens. Actuator. A*, **23**, 904–907.

103. Andersson, H.; Wijngaart, W.; Enoksson, P. and Stemme, G. (2000) Micromachined flow-through filter-chamber for chemical reactions on beads, *Sens. Actuator. B*, **67**, 203–208.
104. Andersson, H.; Wijngaart, W. and Stemme, G. (2001) Micromachined filter-chamber array with passive valves for biochemical assays on beads, *Electrophoresis*, **22**, 249–257.
105. He, B.; Tan, L. and Regnier, F. (1999) Microfabricated filters for microfluidic analytical systems, *Anal. Chem.*, **71**, 1464–1468.
106. Chu, W.H.; Chin, R.; Huen, T. and Ferrari, M. (1999) Silicon membrane nanofilters from sacrificial oxide removal, *J. Microelectromech. Syst.*, **8**, 34–42.
107. Tu, J.K.; Huen, T.; Szema, R. and Ferrari, M. (1999) Filtration of sub-100 nm particles using a bulk-micromachined, direct-bonded silicon filter, *Biome. Microdev.*, **1**, 113–119.
108. Long, Z.; Liu, D.; Ye, N.; Qin, J. and Lin, B. (2006) Integration of nanoporous membranes for sample filtration/preconcentration in microchip electrophoresis, *Electrophoresis*, **27**, 4927–4934.
109. Hatch, A.V.; Herr, A.E.; Throckmorton, J.; Brennan, J.S. and Singh, A.K. (2006) Integrated preconcentration SDS-PAGE of proteins inmicrochips using photopatterned cross-linked polyacrylamide gels, *Anal. Chem.*, **78**, 4976–4984.
110. Brody, J.P.; Osborn, T.D.; Forster, F.K. and Yager, P. (1996) A planar microfabricated fluid filter, *Sens. Actuator. A*, **54**, 704–708.
111. Brody, J.P. and Yager, P. (1997) Diffusion-based extraction in a microfabricated device, *Sens. Actuator. A*, **58**, 13–18.
112. Weigl, B.H. and Yager, P. (1999) Microfluidic diffusion-based separation and detection, *Science*, **283**, 346–347.
113. Tokeshi, M.; Minagawa, T. and Kitamori, T. (2000) Integration of a microextraction system on a glass chip: Ion-pair solvent extraction of Fe(II) with 4,7-Diphenyl-1,10-phenantrolinedisulfonic acid and tri-n-octylmethylammonium choride, *Anal. Chem.*, **72**, 1711–1714.
114. Sato, K.; Tokeshi, M.; Sawada, T. and Kitamori, T. (2000) Molecular transport between two phases in a microchannel, *Anal. Sci.*, **16**, 455–456.
115. Hisamoto, H.; Horiuchi, T.; Tokeshi, M.; Hibara, A. and Kitamori, T. (2001) On-Chip integration of neutral Ionophore-based ion pair extraction reaction, *Anal. Chem.*, **73**, 1382–1386.
116. Hisamoto, H.;T. Horiuchi, K. Uchiyama, Tokeshi, M.; Hibara, A. and Kitamori, T. (2001) On-chip integration of sequential ion-sensing system based on intermittent reagent pumping and formation of two-layer flow, *Anal. Chem.*, **73**, 5551–5556.
117. Shaw, J.; Nudd, R.; Naik, C.; Turner, C.; Rudge, D.; Benson, M. and Garman, A. (2000) Liquid-liquid extraction systems using micro-contactor arrays, *Proceedings of Micro-TAS 2000*, Kluwer Academic, Dordrecht, p. 371.
118. Kutter, J.P.; Jacobson, S.C. and Ramsey, J.M. (2000) Solid phase extraction on microfluidic devices, *J. Microcolumn Sep.*, **12**, 93–97.
119. Oleschuk, R.D.; Shultz-Lockyear, L.L.; Ning, Y.B. and Harrison, D.J. (2000) Trapping of bead-based reagents within microfluidic systems: On-chip solid-phase extraction and electrochromatography, *Anal. Chem.*, **72**, 585–590.
120. Jemere, A.B.; Oleschuk, R.D.; Ouchen, F.; Fajuyigbe, F. and Harrison, D.J. (2002) An integrated solid-phase extraction system for sub-picomolar detection, *Electrophoresis*, **23**, 3537–3544.
121. Ekstrom, S.; Malmstrom, J.; Wallman, L.; Löfgren, M.; Nilsson, J.; Laurell, T. and Marko-Varga, G. (2002) On-chip microextraction for proteomic sample preparation of in-gel digests, *Proteomics*, **2**, 413–421 (2002).
122. J. Bergkvist, Ekstrom, S.;Wallman, L.; Löfgren, M.; Marko-Varga, G. Nilsson, J. and Laurell, T. (2002) Improved chip design for integrated solid-phase microextraction in on-line proteomic sample preparation, *Proteomics*, **2**, 422–429.
123. Ro, K.W.; Chang, W.J.; Kim, H.; Koo, Y.M. and Hahn, J.H. (2003) Capillary electrochromatography and preconcentration of neutral compounds on poly(dimethylsiloxane) microchips, *Electrophoresis*, **24**, 3253–3259.
124. Tuomikoski, S.; Virkkala, N.; Rovio, S.; Hokkanen, A.; Sirn, H. and Franssila, S. (2006) Design and fabrication of integrated solid-phase extraction-zone electrophoresis microchip, *J. Chromatgr. A*, **1111**, 258–266.
125. Ramsey, J.D. and Collins, G.E. (2005) Integrated microfluidic device for solid-phase extraction coupled to micellar electrokinetic chromatography separation, *Anal. Chem.*, **77**, 6664–6670.
126. Long, Z.; Shen, Z.; Wu, D.; Qin, J. and Lin, B. (2007) Integrated multilayer microfluidic device with a nanoporous membrane interconnect for online coupling of solid-phase extraction to microchip electrophoresis, *Lab. Chip*, **7**, 1819–1824.

127. Broyles, B.S.; Jacobson, S.C. and Ramsey, J.M. (2003) Sample filtration, concentration, and separation integrated on microfluidic devices, *Anal. Chem.*, **75**, 2761–2767.
128. Augustin, V.; Proczek, G.; Dugay, J.; Descroix, S. and Hennion, M.C. (2007) Online preconcentration using monoliths in electrochromatography capillary format and microchips, *J. Sep. Sci.*, **30**, 2858–2865.
129. de Mello, A.J. (2002) On-chip chromatography: the last twenty years, *Lab. Chip*, **2**, 48N-54N.
130. Yu, C.; Davey, M.H.; Svec, F. and Frechet, J.M.J. (2001) Monolithic porous polymer for on-chip solid-phase extraction and preconcentration prepared by photoionitiated in situ polymerization within a microfluidic device, *Anal. Chem.*, **73**, 5088–5096.
131. Yu, C.; Xu, M.C.; Svec, F. and Frechet, J.M.J. (2002) Preparation of monolithic polymers with controlled porous properties for microfluidic chip applications using photoinitiated free-radical polymerization, *J. Polym. Sci. Part A: Polym. Chem.*, **40**, 755–769.
132. Chen, H. and Terabe, S. (2005) On-line sample preconcentration techniques in capillary electrophoresis, *Analysis and Detection by Capillary Electrophoresis* (eds M.L. Marina, A. Ríos and M. Valcárcel), Elsevier, Amsterdam, pp. 135–172.
133. Almeda, S.; Arce, L. and Valcárcel, M. (2010) The more and less common approaches to enhancing sensitivity in capillary electrophoresis, *Curr. Anal. Chem.*, **6**, 126–143.
134. Jacobson, S.C. and Ramsey, J.M. (1995) Microchip electrophoresis with sample stacking, *Electrophoresis*, **16**, 481–486.
135. Jacobson, S.C.; Koutny, L.B.; Hergenrder, R.; Moore, A.W.; Ramsey, J.M. (1994) Microchip capillary electrophoresis with an integrated postocolumn reactor, *Anal. Chem.*, **66**, 3472–3476.
136. Lichtenberg, J.; Verpoorte, E. and de Rooij, N.F. (2001) Sample preconcentration by field amplification stacking for microchip-based capillary electrophoresis, *Electrophoresis*, **22**, 258–271.
137. Beard, N.P.; Zhang, C.X. and de Mello, A.J. (2003) In-column field-amplified sample stacking of biogenic amines on microfabricated electrophoresis devices, *Electrophoresis*, **24**, 732–739.
138. Palmer, J.; Burgi, D.S.; Munro, N.J. and Landers, J.P. (2001) Electrokinetic injection for stacking neutral analytes in capillary and microchip electrophoresis, *Anal. Chem.*, **73**, 725–731.
139. Xu, Y.; Jiang, H. and Wang, E. (2007) Ionic liquid-assisted PDMS microchannel modification for efficiently resolving fluorescent dye and protein adsorption, *Electrophoresis*, **28**, 4597–4605.
140. Walker, P.A.; Morris, M.D.; Burns, M.A. and Johnson, B.N. (1998) Isotachophoresis separation son a microchip. Normal Raman spectroscopy detection, *Anal. Chem.*, **70**, 3766–3769.
141. Zhang, B.; Liu, H.; Karger, B.L. and Foret, F. (1999) Microfabricated devices for capillary electrophoresis-electrospray mass spectrometry, *Anal. Chem.*, **71**, 3258–3264.
142. Wainright, A.; Williams, S.J.; Ciambrone, G.; Xue, Q.; Wei, J. and Harris, D. (2002) Sample re-concentration by isotachophoreis in microfluidic devices, *J. Chromatgr. A*, **979**, 69–80.
143. Wainright, A.; Nguyen, U.T.; Bjornson, T. and Boone, T.D. (2003) Preconcentration and separation of double-stranded DNA fragments by electrophoresis in plastic microfluidic devices, *Electrophoresis*, **24**, 3784–3792.
144. Vreeland, W.N.; Williams, S.J.; Barron, A.E. and Sassi, A.P. (2003) Tandem isotachophoresis-zone electrophoresis via based-mediated stacking for increased detection sensitivity in microfluidic systems, *Anal. Chem.*, **75**, 3059–3065.
145. Kurnik, R.T.; Boone, T.D.; Nguyen, U.; Ricco, A.J. and Williams, S.J. (2003) Use of floating electrodes in transient isotachophoresis to increase the sensitivity of detection, *Lab. Chip*, **3**, 86–92.
146. Jung, B.; Bharadwaj, R. and Santiago, J.G. (2006) On-chip million fold sample stacking using transient isotachophoresis, *Anal. Chem.*, **78**, 2319–2327.
147. Quirino, J.P. and Terabe, S. (1998) Exceeding 5000-fold concentration of dilute analytes in micellar electrokinetic chromatography, *Science*, **282**, 465–468.
148. Quirino, J.P. and Terabe, S. (1999) Sweeping of analyte zones in electrokinetic chromatography, *Anal. Chem.*, **71**, 1638–1644.
149. Isoo, K. and Terabe, S. (2003) Analysis of metal ions by sweeping via dynamic complexation and cation-selective exhaustive injection in capillary electrophoresis, *Anal. Chem.*, **75**, 6789–6798.
150. Otsuka, K.; Matsumura, M.; Kim, J.B. and Terabe, S. (2003) On-line preconcentration and enantioselective separation of triadimenol by electrokinetic chromatography using cyclodextrins as chiral selectors, *J. Pharm. Biomed. Anal.*, **30**, 1861–1867.

151. Sera, Y.; Matsubara, N.; Otsuka, K. and Terabe, S. (2001) Sweeping on a microchip: Concentration profiles of the focused zone in micellar electrokinetic chromatography, *Electrophoresis*, **22**, 3509–3513.
152. Sueyoshi, K.; Kitagawa, F. and Otsuka, K. (2008) On-line sample preconcentration and separation technique based on transient trapping in microchip micellar electrokinetic chromatography, *Anal. Chem.*, **80**, 1255–1262.
153. Humble, P.H.;Kelly, R.T.; Woolley, A.T.; Tolley, H.D. and Lee, M.L. (2004) Electric field gradient focusing of proteins based on shaped ionically conductive acrylic polymer, *Anal. Chem.*, **76**, 5641–5648.
154. Ross, D. and Locascio, L.E. (2002) Microfluidic temperature gradient focusing, *Anal. Chem.*, **74**, 2556–2564.
155. Hofmann, O.; Che, D.; Cruickshank, K.A. and Mller, U.R. (1999) Adaptation of capillary isoelectric focusing to microchannels on a glass chip, *Anal. Chem.*, **71**, 678–686.
156. Guillo, C.; Karlinsey, J.M. and Landers, J.P. (2007) On-chip pumping for pressure mobilization of the focused zones following microchip isoelectric focusing, *Lab. Chip*, **7**, 112–118.
157. Wen, J.; Lin, Y.; Xiang, F.; Matson, D.W.; Udseth, H.R. and Smith, R.D. (2000) Microfabricated isoelectric focusing device for direct electrospray ionization-mass spectrometry, *Electrophoresis*, **21**, 191–197.
158. Huang, X. and Ren, J. (2005) On-line chemiluminescence detection for isoelectric focusing of heme proteins on microchips, *Electrophoresis*, **26**, 3595–3601.
159. Stoyanov, A.V.; Das, C.; Fredrickson, C.K. and Fan, Z.H. (2005) Conductivity properties of carrier ampholyte pH gradients in isoelectric focusing, *Electrophoresis*, **26**, 473–479.
160. Thormann, W.; Caslavska, J. and Mosher, R.A. (2007) Modeling of electroosmotic and electrophoretic mobilization in capillary and microchip isoelectric focusing, *J. Chromatgr. A*, **1155**, 154–163.
161. Dai, J.; Ito, T.; Sun, L. and Crooks, R.M. (2003) Electrokinetic trapping and concentration enrichment of DNA in a microfluidic channel, *J. Am. Chem. Soc.*, **125**, 13026–13027.
162. Song, S.; Singh, A.K. and Kirby, B.J. (2004) Electrophoretic concentration of proteins at laser-patterned nanoporous membranes in microchips, *Anal. Chem.*, **76**, 4589–4592.
163. Wang, Y.C.; Stevens, A.L. and Han, J. (2005) Million-fold preconcentration of proteins and peptides by nanofluidic filter, *Anal. Chem.*, **77**, 4293–4299.
164. Kim, S.M.; Burns, M.A. and Hasselbrink, E.F. (2006) Electrokinetic protein preconcentration using a simple glass/pol(dimethylsiloxane) microfluidic chip, *Anal. Chem.*, **78**, 4779–4785.
165. Lee, J.H.; Chung, S.; Kim, S.J. and Han, J. (2007) Poly(dimethylsiloxne)-based protein preconcentration using a nanogap generated by junction gap break-down, *Anal. Chem.*, **79**, 6868–6873.
166. Hadd, A.G.; Jacobson, S.C. and Ramsey, J.M. (1999) Microfluidic assays of acetylcholinesterase inhibitors, *Anal. Chem.*, **71**, 5206–5212.
167. Wang, J.; Chatrathi, M.P. and Tian, B. (2000) Micromachined separation chips with a precolumn reactor and end-column electrochemical detector, *Anal. Chem.*, **72**, 5774–5778.
168. Wang, J.; Chatrathi, M.P.; Ibañez, A. and Escarpa, A. (2002) Micromachined separation chips with post-column enzymatic reactions of 'class' enzymes and end-column electrochemical detection: Assays of amino acids, *Electroanalysis*, **14**, 400–404.
169. Brody, J.P.; Han, Y.; Austin, R.H. and Bitensky, M. (1995) Deformation and flow of red blood cells in a synthetic lattice: evidence for an active cytoskeleton, *Biophys. J.*, **68**, 2224–2232.
170. Wilding, P.; Kricka, L.J.; Cheng, J.; Hvichia, G.; Shoffner, M.A. and Fortina, P. (1998) Integrated cell isolation and polymerase chain reaction analysis using silicon microfilter chambers, *Anal. Biochem.*, **257**, 95–100.
171. Yang, X.; Yang, J.M.; Tai, Y.C. and Ho, C.M. (1999) Micromachined membrane particle filters, *Sens. Actuator. A*, **73**, 184–191.
172. Kim, A.S. and Hoek, E.M.V. (2002) Cake structure in dead-end membrane filtration: Monte carlo simulations, *Environ. Eng. Sci.*, **19**, 373–386.
173. Sethu, P.; Sin, A. and Toner, M. (2006) Microfluidic diffusive filter for apheresis (leukapheresis), *Lab. Chip*, **6**, 83–89.
174. Chen, X.; Cui, D.F.; Liu, C.C. and Li, H. (2008) Microfluidic chip for blood cell separation and collection based on crossflow filtration, *Sens. Actuator. B*, **130**, 216–221.
175. Li, P.C. and Harrison, D.J. (1997) Transport, manipulation, and reaction of biological cells on chip using electrokinetic effects, *Anal. Chem.*, **69**, 1564–1568.

176. Schilling, E.A.; Kamholz, A.E. and Yager, P. (2002) Cell lysis and protein extraction in a microfluidic device with detection by a fluorogenic enzyme assay, *Anal. Chem.*, **74**, 1798–1804.
177. Carlo, D.D.; Jeong, K.H. and Lee, L.P. (2003) Reagentless mechanical cell lysis by nanoscale barbs in microchannels for sample preparation, *Lab. Chip*, **3**, 287–291.
178. Chen, X.; Cui, D.F. and Liu, C.C. (2007) Fabrication of DNA purification microchip integrated with mesoporous matrix based on MEMS technology, *Microsyst. Technol.*, **14**, 51–57.
179. Chen, X.; Cui, D.F. and Liu, C.C. (2007) Microfabrication and characterization of porous channels for DNA purification, *J. Micromech. Microeng.*, **17**, 68–75.
180. Lee, E.Z.; Huh, Y.S.; Jun, Y.S.; Won, H.J.; Hong, Y.K.; Park, T.J.; Lee, S.Y. and Hong, W.H. (2008) Removal of bovine serum albumin using solid-phaseextraction with in situ polymerized stationary phase in amicrofluidic device, *J. Chromatogr. A*, **1187**, 11–17.
181. Reddy, V. and Zhan, J.D. (2005) Interfacial stabilization of organic–aqueous two-phase microflows for a miniaturized DNA extraction module, *J. Colloid Interface Sci.*, **286**, 158–165.

18

Micro- and Nanomaterials Based Detection Systems Applied in Lab-on-a-Chip Technology

Mariana Medina-Sánchez[1] and Arben Merkoçi[1,2]

[1]*Nanobioelectronics and Biosensors Group, Institut Català de Nanotecnologia: Universitat Autònoma de Barcelona, Bellaterra, Barcelona, Spain*
[2]*Nanobioelectronics and Biosensors Group, Institut Català de Nanotechnologia & ICREA, Barcelona, Spain*

18.1 Micro- and nanotechnology in Green Analytical Chemistry

Green Analytical Chemistry is an approach to the synthesis, processing and use of chemicals that reduce risks to humans and environment [1]. These approaches include new processes as well as new tools to minimize the effect of the pollution agents in the human health and nature. One of the main trends to achieve this goal during the design and fabrication of (bio)analytical systems is the introduction of miniaturized methods that can be disposable, cheap, with reusable materials, and that allow a reducing of the use of reagents and materials. That is the case of the lab-on-a-chip (LOC) or microfluidic platforms, where their features are in concordance with the green chemistry philosophy, therefore, the use of this technology allows preventing waste generation, by facilitating the safe use of solvents and other materials, and the optimal use of energy, making chemistry in general, and particularly bioanalytical processes, more sustainable.

Simple and miniaturized micro/nanofluidic platforms are especially interesting due to their advantages, like the use of reduced sample volume, low cost, portability and the possibility of being integrated into conventional analytical techniques. They represent promising alternative tools for point-of-care diagnostics and drug delivery between other applications.

The objective of this chapter is to present the latest trends in the use of nanomaterials (i.e. nanoparticles, quantum dots, nanotubes etc.) in combination with miniaturization and lab-on-a-chip technologies for several

Handbook of Green Analytical Chemistry, First Edition. Edited by Miguel de la Guardia and Salvador Garrigues.

analytical chemistry applications of devices that can integrate multiple laboratory functions (separation, identification and quantification of the chemical and biological analytes) on a single chip able to handle very small sample fluid volumes.

Although LOC technology already represents an established research field the efforts of scientists and engineers are facing a continuous growth toward the synergy of this technology with nanotechnology and material science in general; particularly nanomaterials. The development of novel sensors and biosensors based on nanomaterials and nanostructures – called also nanobiosensors – is opening the way to cost-effective devices that would be the perfect candidates to be integrated into LOC systems. The recent trends in the development of nanomaterials-based biosensors, followed by the LOC technology and the applications of the resulting synergic systems in diagnostics as well as other industries will be shown in the next sections.

18.2 Nanomaterials-based (bio)sensors

Nanotechnology is an emerging field, parallel to many knowledge areas where limits are established from 1 to 100 nm. Different applications can be distinguished such as: nanobiotechnology; nanomedicine; nanomaterials; nanoelectronics; and nanosensors/nanodevices, nanotechnological instrumentation and nanometrology [2].

The application of nanomaterials in the development of sensors and biosensors represents one of the main focuses of the current nanotechnology. The interesting optical and electrical properties are making nanomaterials new building blocks for the design of various kinds of sensing and biosensing devices with improved performance as well as cost-efficienct in comparison to conventional materials.

In order to quantify and analyse changing the operation of nanosensors is necessary to use transducers that convert the chemical/biological recognition events to electrical/optical/piezoelectrical signals. Several kinds of nanomaterials such as nanoparticles, nanotubes and other nanostructured materials are being used to design various transducing platforms that lead to interesting nano(bio)sensing systems or nano(bio)sensors.

Transducing platforms based on optical detections such as light absorption and scattering induced by nanoparticles used as labels have proved to be very interesting and have been extended not only to DNA [3] and protein [4] analysis fields but also to other fields (i.e. heavy metal detection [5]. Surface plasmon resonance enhancement by using nanoparticles is also an interesting optical transducing alternative that is bringing advantages to this technique in terms of sensitivity. Between the different optical techniques, those related to fluorescence measurements, including its quench monitoring while using nanoparticles as tracers of biomolecules, are very interesting not only for the shown sensitivity but the possibility to achieve multidetection through multiplexing alternatives [6].

Other kinds of transducers are the *electrochemical transducers* with two classes of biosensors; the biocatalytic devices and the affinity sensors [7]. These transducers can be based on voltammetric/amperometric measurements during which a potential is applied in the working electrode versus a reference electrode and the current related to redox reactions occurring at the working electrode is measured. Amperometric detection is commonly used with biocatalytic and affinity sensors because of its simplicity and low detection limit. Transducers based on potentiometric measurements are related with the determination of the potential difference between either an indicator or a reference electrode or two reference electrodes separated by a perm-selective membrane, when there is no significant current flowing between them. Most common are ion-selective electrodes (ISEs) based on thin films or selective membranes as recognition elements for pH, F^-, I^-, CN^-, Na^+, K^+, Ca^{2+} and so on. or even gases (i.e. CO_2, NH_3) [8]. Impedance measurements - during which resistive and capacitive properties of materials, while excited with a small amplitude sinusoidal ac typically of 2–10 mV - are also being used. On other hand, the frequency is varied over a wide range to obtain the impedance spectrum. The impedance based methods are very useful in affinity biosensors [9]; *Conductimetry*, is another technique that measures the change of the conductivity, commonly used in applications such as environmental monitoring and clinical analysis.

Field effect transistors represent other interesting transducer ISFETs (ion sensitive field effect transistors) or FETs with open gates, with the purpose of using the sum of external gate tension with its own electroactive potential or ionic solution (sample to detect and quantify) have been studied. Most of the reported devices using FET based detectors are fabricated by using conventional photolithography process [10,11,12].

An important transducer used in mass sensors as cantilever, is the *piezoelectric based transducer*. The piezoelectric effect is related to a linear electromechanical interaction between the mechanical and electrical state in crystalline materials. This process is reversible, and is used to transform the variation in the deflection in a micro/nanocantilever where the response toward an analyte reacting with a biorecognizer is measured. These kinds of measurements can also take the advantages of surface acoustic wave detection by means of an interdigitated transducer [13].

Some of the most reported optical and electrochemical nano(bio)systems/nano(bio)sensors based on the various transducing mechanisms already mentioned in this section will be shown and discussed in the following part. The operation of these devices will be discussed in relation also to various kinds of nanomaterials (i.e. nanoparticles) due to their particularities in advantages that these bring in terms of detection performance as well as the variability of assays/detection systems.

18.2.1 Optical nano(bio)sensors

The intrinsic optical properties (UV-Vis light absorption and auto-fluorescence properties) of nanoparticles (NPs) in addition to their ability to change optical properties of sensor surfaces (i.e. in surface plasmon resonance and scattering light based devices) have been approached for their sensible detection in many analytical applications (i.e. detection of DNA, proteins, cells, heavy metals etc.). Various kinds of nanoparticles and other nanomaterials (i.e. nanotubes) are being used in the fabrication of optical sensors.

Core-shell nanoparticles such as magnetic/luminescent Fe_3O_4/Eu:Gd_2O_3 synthesized by spray pyrolysis were used by Son *et al.* as both platforms for the hybridization reactions and as fluorescence labels. The developed DNA hybridization sensor is capable of perfectly distinguishing of matching targets from two-base pair mismatching, allowing the discrimination of different bacterial species [14] (Figure 18.1a).

Some nanostructures are and optimally active have great interest for their potential applications thanks to their optical properties; these are the nanoshells [15]. They consist of small silica spheres covered by a thin gold layer of approximately 100 nm in diameter. Besides their composition, two fundamental factors exist for the optical response of nanoshell: its geometric structure and its dimensions. Modifying these two parameters makes it possible to control the wavelength of the emitted light when the particle scatters the light or when it absorbs radiation and consequently temperature increases. These structures have provided hopeful results in experiments carried out in animal weaves for diagnosis and treatment in cancer patients [16].

Also, simple *gold nanoparticles* (AuNPs) are a promising approach for the cancer cells detection. The University of San Francisco showed that by conjugating gold nanoparticles with specific antibodies, it is possible to detect cancer cells and further remove them by applying ordinary light using a simple optical microscope [17].

It is well known, for example, that the NP plasmon band shifts when AuNPs are aggregated, due to a decrease in the interparticle's distance. The maximum of absorbance shifts from 520 nm to 580 nm (red to blue colour) when the aggregation takes place and the monitoring of this phenomenon can be used for biosensing purposes. For example, AuNPs suspension previously modified with ssDNA probe has been hybridized with a DNA target (complementary in its two ends with ssDNA) producing NPs aggregation with the consequent change in colour in this way. This principle was pioneered by Mirkin's group and applied for the detection of DNA characteristic of anthrax, a biological warfare agent [18] (Figure 18.1b).

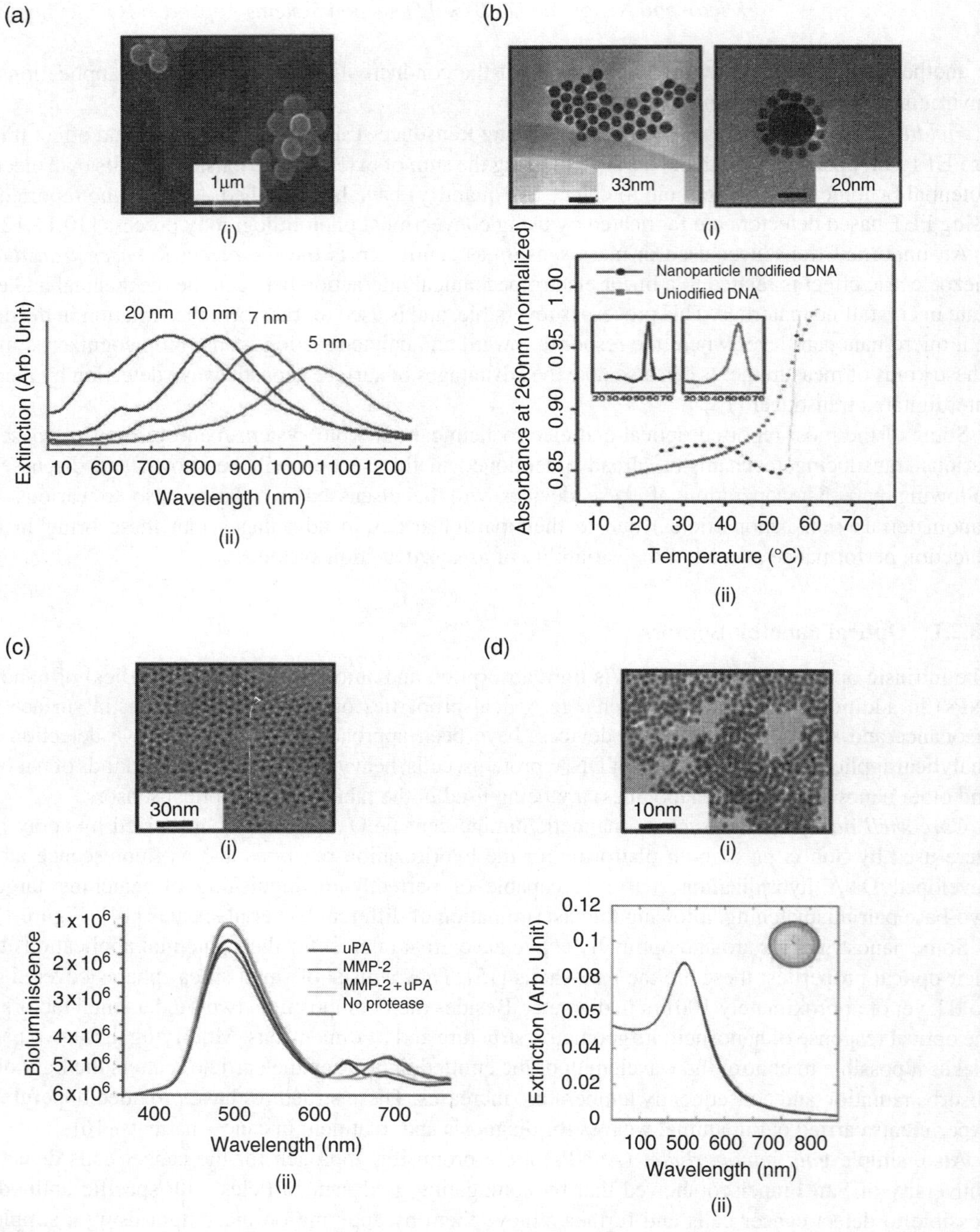

Figure 18.1 Optical nanosensors. *(a) TEM image of core (gold)-shell (silicon) nanoparticles (i) and optical resonances of gold shell-silica core nanoshells as a function of their core/shell ratio (ii) Adapted with permission from [16] © 2004. (b) TEM image of gold nanoparticles modified with a polypeptide-base (i) and absorbencies of gold nanoparticles modified or not with DNA (ii). Adapted from [18] © 2000 American Chemical Society. (c) TEM image of quantum dots (QDs) (i) and two-plex detection of matrix metalloprotinease-2 (MMP-2), 1 μg ml^{-1} and/or urokinase-type plasminogen activator (uPA) 10 μg ml^{-1} via the bioluminescence resonance energy transfer (BRET) sensitized photoluminescence (PL) of QDs with peak emission at 655 and 705 nm, respectively (ii). Adapted from [23] with kind permission from Springer Science + Business Media © 2010. (d) TEM images of hybrid nanoparticles (i), absorbance of the binding of the enzyme to the particles sterically (b) [27] Adapted by permission of The Royal Society of Chemistry © 2010.*

An interesting recent work demonstrates that by using a colourimetric method, a proteinase assay could be developed by exploiting the colour shift upon aggregation of functionalized AuNPs. This aggregation is observed by means of a colour change from red to purple. Researchers used UV spectroscopy to quantify proteinase in future biosensing platforms [19].

Another example is the use of AuNPs to elucidate the mechanism using a unique structure of the monoazacrown ether moieties, useful to detect Pb(II) in aqueous solution at ambient temperatures. These modified AuNPs were shown to be selective to Pb^{2+} ions as the colour changes from brown to purple. This method doesn't use enzymatic reaction or light-sensitive dye molecules and it doesn't require time consuming protocols or sophisticated instrumentation [20].

On the other hand, a silver nanoparticles-on-plastic sensor was recently developed. This sensor is based in polyethylene terephthalate (PET) films chemically modified with *silver nanoparticles* (AgNPs) and it represents a very sensitive platform to monitor protein-protein interactions and biotin-avidin binding by simply changing the surface functionality [21].

Quantum Dots (QDs) are also attractive nanoparticles with a size of a few nanometres that when appropriately excited, emit light with a colour depending on their size [22]. These nanostructures seem to have a great potential in the cancer cells detection. Companies like the Quantum Dots Corporation produce commercially these crystalline structures, which are used for cancer diagnosis [23] (Figure 18.1c).

QDs are highly sensitive and their versatility in changing their optical properties according to the size and surface modification is especially interesting. A recent application is the QDs-dopamine-peptide bioconjugates coupled pH sensors [24].

New nanostructures used for optical detection are *magnetic fluorescent microspheres* in the evanescent field [25], *upconversion nanoparticles* in biological labelling, imaging and therapy where their operation principle refers to non-linear optical processes that convert two or more low-energy pump photons to a higher-energy output photon [26] and some other hybrid nanoparticles with optical properties [27] (Figure 18.1d).

Bioconjugations of *silicon nanoparticles* (SiNPs) with DNA/RNA molecules can provide unique biofunctionalities. These nanoparticles usually exhibit very high luminescense and photostability, within a broad size range from 5 to 400 nm. Furthermore, dye-doped silica NPs have been extensively used in bioimaging and biochemical analysis due to their advantages such as signal enhancement, photostability and surface modification availability for the immobilization of biomolecules [28].

18.2.2 Electrochemical nano(bio)sensors

Nanotechnology has a great impact on the development of electrochemical biosensors bringing new alternatives for their designs/constructions and modifications. Nanomaterials have been useful to promote electrochemical reactions, improving the sensibility in the measurement thus the specific recognition events [29]. Several electrochemical routes (voltammetric, potentiometric, conductometric, impedimetric and scanning electrochemical microscope-based methods) have been exploited for the sensitive detection of these NP tags in bioassays [30] (Figure 18.2a).

There are many nanoparticles with electrical properties that in combination with micro and nanotechnologies are useful for developing different kinds of electrical based biosensing systems. For example, *metallic nanoparticles* (NPs) modified by bioconjugation present excellent electrochemical activity. It has given rise to their extensive use as labels in nanobiosensors in recent years.

NP-based amplification schemes have improved the sensitivity of bioelectronic assays by several orders of magnitude. In 2001, Wang's [31] and Limoges' [32]groups both pioneered the use of *AuNPs* tags for electronic detection of DNA hybridization. Their protocols are based on capturing the NPs to the hybridized target, followed by highly sensitive anodic stripping electrochemical measurement of the metal tracer.

Other strategies based on the use of *inorganic nanocrystals* offering an electrodiverse population of electrical tags as needed for designing electronic coding have been reported. Three encoding NPs (ZnS, CdS and PbS) have been used in this way to differentiate the signals of three ssDNA target strands in connection with a sandwich hybridization assays and stripping voltammetry of the corresponding metals [33].

Nanostructured metal oxides are known for their high mechanical, chemical, physical, thermal, electrical, optical, magnetic and also specific surface area properties, which make them really interesting nanostructures for nanoelectronics, nanophotonics, nanobiomaterials, nanobiolabels, and so on [34] (Figure 18.2b). Some examples of these nanostructures are ZnO, TiO_2, ZrO_2, SnO_2, CeO_2, MnO_2, Fe_3O_4 and SiO_2 with interest in applications such as optics, optoelectronics, catalysis, as sensors and actuators [35].

Inorganic-organic hybrid nanocomposites represent other kinds of nanomaterials with electrical properties. They have attracted many scientists over the last few years because of their potential for combining the different physical properties of their organic and inorganic components within a single molecular composite application in the catalysis field, as well as electronics and optics [36]. An example of the use of this material is in the field of electrochemical immunosensors. In the mentioned work, the developed sensor was fabricated by using a CNPs-doped chitosan organic–inorganic hybrid composite membrane as a matrix. This nanocomposite has low toxicity, high conductivity due to the presence of carbon nanoparticles and high compatibility due to the presence of chitosan as a matrix. These nanocomposites used as bionanolabels displayed high electrochemical responses in comparison to the convectional label methods. Various tumour markers; including carcinoma antigen 125 (CA125), carcinoembryonic antigen (CEA), cancer antigen 19-9 (CA 19-9) and prostate-specific antigen (PSA), obtaining detection limits of 2 ng ml^{-1} to 0.05 ng ml^{-1} respectively without any interference from other agents, have been studied [37].

Other nanomaterials such as *carbon nanotubes* (CNTs) being used as carriers for several thousand enzyme tags has been reported allowing the detection of DNA and proteins down to 80 copies (5.4 aM) and 2000 protein molecules (67 aM), respectively [38] with future interest in environmental monitoring.

The electrochemical qualities of carbon-based materials are significantly dependent on the surface properties, such as the creation of specific functional groups on the surface (especially oxygen-containing groups) that can considerably increase the rate of electron transfer. All these properties are the reason which makes this nanomaterial so attractive and versatile to be implemented in sensors and biosensors [39].

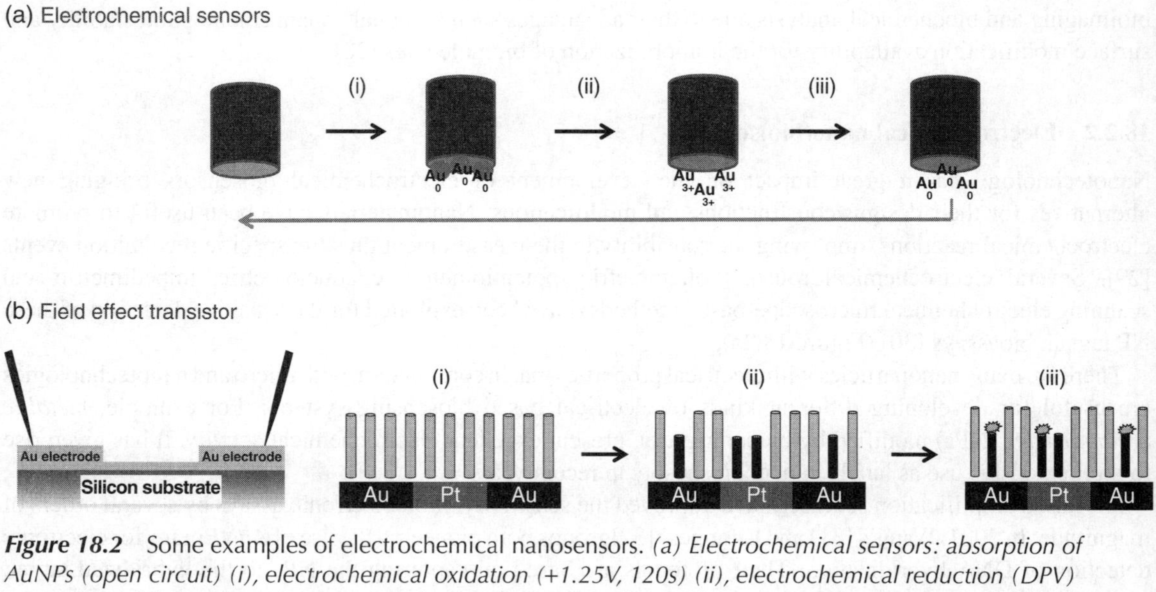

Figure 18.2 Some examples of electrochemical nanosensors. *(a) Electrochemical sensors: absorption of AuNPs (open circuit) (i), electrochemical oxidation (+1.25V, 120s) (ii), electrochemical reduction (DPV) (iii). Adapted from [34] with permission from Wiley-VCH Verlag GmbH & Co. KGaA © 2006. (b) Field effect transistor: nanowires (i), nanowires + MEA (ii), MEA molecules coupled with rhodamine B isothiocyanate (iii). Adapted from [30] with permission from Elsevier © 2008.*

Recent studies have demonstrated that CNTs exhibit strong electrocatalytic activity for a wide range of compounds, such as neurotransmitters, hydrogen peroxide, ascorbic and uric acid, cytochrome c, hydrazines, hydrogen sulfide, amino acids and deoxyribonucleic acid (DNA). Their potential applications in electrochemical sensors and biosensors due to their high conductivity are also reported [40].

Finally, an emergent nanomaterial is *graphene*, constituted by a single-atom-thick and two-dimensional carbon material. It has attracted great attention for being a novel alternative to carbon nanotubes in biosensing, as it had demonstrated its capabilities in many applications. An example of its application is the glucose sensor where the glucose or glutamate molecules were detected by the change of graphene conductance in a transistor platform showing exceptional electronic properties [41]. Some variation in the sensing platform such as a transistor with suspended gate modified with graphene could improve the signal and decrease the noise [42].

18.2.3 Other detection principles

Electrochemiluminescence (ECL) is an optical and electrical phenomenon in which a material emits light in response to an electric current or electrical field. One important application of ECL is in enzyme biosensors based on cadmium sulfide nanocrystals (CdS NCs) in situ formed onto the surface of multi-walled carbon nanotubes (MWCNTs). The MWCNT-CdS can react with H_2O_2 generating strong and stable ECL emission in neutral solution [43] (See Figure 18.3a). By following the same procedure,

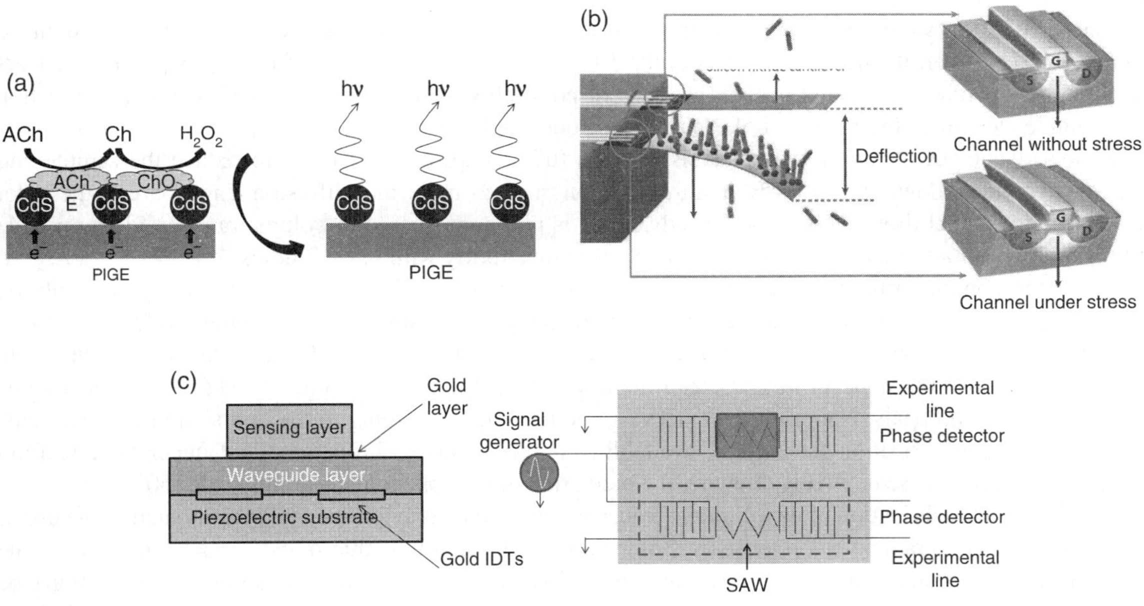

Figure 18.3 Some examples of other detection alternatives. *(a) Electroluminiscence: Schematic diagram of the multi-wall carbon nanotubes (MWCNT): Cadmium sulfide (CdS), acetylcholine esterase (AChE), ECL biosensor for detection of neurotransmitter acetylcholine (Ach). Adapted from [43] with permission from Wiley-Blackwell © 2009; (b) schematic of an embedded metal-oxide semiconductor field-effect transistor (MOSFET) cantilever system. Adapted from [46] © Annual Reviews 2009; and (c) surface acoustic wave sensor (SAW): a sinusoidal voltage is applied to the input (input interdigitated transducer - IDT), which develops an alternating electric field that is translated into a mechanical SAW by the piezoelectric effect. The velocity of the SAW is affected by mass loading, fluid viscosity and temperature on the surface of the substrate. Adapted from [45] © Expert Reviews 2009.*

the biosensors based on CdS-graphene, had shown promising results to detect H_2O_2 with acceptable linear response from 5 μM up to 1 mM with a detection limit of 1.7 μM [44].

Electromechanical detection is another interesting detection technique. It is especially interesting for *surface acoustic wave sensors* (SAW), where the mechanical vibration that propagates just below the surface of piezoelectric solids is excited by an electrical signal at the resonant frequency. The velocity of a SAW is sensitive to mass changes applied to the active area, the viscosity of the material applied to the active area and the surface temperature [45] (see Figure 18.3c).

Finally another interesting detection alternative is based on the use of nanowires or nanotubes as mass detectors. One example is the development of a nanocantilever-based sensor with a field effect transistor as a transducer (see Figure 18.3b) where the signal on the gate changes when the cantilever is deflected due to specific binding of biomolecules on its surface [46].

18.3 Lab-on-a-chip (LOC) technology

18.3.1 Miniaturization and nano-/microfluidics

One of the trends in the analytical chemistry field is oriented to the miniaturization, integration and automation of all the analytical steps into a single system. The integration of analytical operations such as filtration, extraction, mixing, separation and detection lead to lab-on-a-chip systems (LOC) or total analysis microsystems (μTAS) 47.

The first LOCs were gas chromatography systems, created in 1975 by S.C. Terry [48], of Stanford University. However, its high growth was in the 1990s with the development of micropumps, flow sensors and the concept of the integrated systems probing the possibility to integrate all steps of treatment, separation and sample cleaning, obtaining a whole analysis in a one single device [49].

Major advantages of micro-total analysis systems (μTAS) are the possibility to reduce the sample and reagents volumes (down to picolitres) and the reaction times as molecular diffusion lengths are of the order of the microchannel dimensions. Another advantage is the large surface-to-volume ratio providing a wide range of possibilities in surface-based assays due its compatibility with microfluidics.

Most common microfluidic system devices consist on microchannel networks made in flat galls or plastic substrates, in which solutions are displaced using pneumatic micropumps or by applying an electric field.

Nanofluidics involves the fluid motion studies in channels with less than 100 nm in one or more direction. The introduction of this concept was raised due to necessity to replace polymer gels used for DNA separations with solid-state materials. Nanofluidic devices can be designed according to the application requirements, advances in nanofabrication techniques, availability of new tools to investigate and understanding fluid behaviour on the nanoscale finally allowing the prediction of new phenomena at this scale [50].

The design and fabrication of nanofluidic devices requires some previous considerations such as the use of very high pressure gradient to drive fluids. Some applications combine micro and nanochannels inside the geometry [51]. In addition, some other parameters like the introduction of the liquid sample must be considered. It is not an easy process; therefore it had been widely studied. *Capillarity* considers the Young-Laplace relationship and viscous pressure drop [52]. Hibara's group fabricated nanometre-sized channels and performed fluorescence measurements of aqueous solutions [53]. Fluid transport in fused silica nanochannels with diameters between 330 nm and 850 nm, by capillary action, results in fluid velocity four times lower than the theoretical model, maybe because of the increase of viscosity through the channel. Other alternative to introduce the fluid inside the nanochannel is *dissolution of gas in liquid* based on the high solubility of CO_2 in water, reducing the pressure inside during the fluid introduction [54]. Finally, another studied option is the pressure-driven flow, demonstrating that the Navier Stokes approach is still available while local shear rate at the interface is lower than the critical shear rate [55].

18.3.2 Micro- and nanofabrication techniques

18.3.2.1 Chip fabrications

Several chip fabrication methodologies that allow the obtaining of different geometries according to the micro- and nanofluidics have been reported. A brief description of the most important technologies is given in the next sections.

Photolithography is an optical methodology that allows transferring of patterns onto certain substrates. It is essentially the same process used in lithographic printing. This technique consists in transferring a landlord from a photomask to the surface of a substrate. Three-dimensional structures fabricated using this process have potential applications in MEMS sensors/actuators, optical devices and microfluidics [56].

Soft Lithography is a manufacturing technique using elastomers, moulds and conformable photomasks. 'Soft' is called because elastomeric materials are used, especially poly(dimethilsiloxane) (PDMS) and recently poly[(3-mercaptopropyl)-methylsiloxane] (PMMS), demonstrating the successful pattern replication from the micrometre to sub-100 nm scale [57]. Generally, it is used to build devices at micro- and nanoscale. The process includes technology of impression by microcontact, moulding by replica, moulding by micro transference and micromoulding in capillaries [58].

Micro manufacturing is based on the use of a cutting tool to obtain predefined material geometries [59]. Although the sizes obtained by this technique are not as small as those obtained using lithography, it allows the fabrication of platforms for the integration with smaller devices.

Ink-printing technology was initially used for computer science and decorative objects. Now it is widely used to print on different surfaces, such as aluminium, glass, plastic and paper. Nowadays, special inks like graphite inks, silver, carbon nanotubes inks and polymer inks, among others are used. These allow printing geometries with certain thicknesses and roughness to create three-dimensional structures, without using masks [60].

Polymerization based prototyping is a layer-by-layer fabrication technique through a computer aided design (CAD). Resin is deposited in crud and then using laser traces or an agglutinant cartridge it is polymerized and solidified onto the surface. It allows the construction of 3D structures and assembled pieces [61].

Other processes used to fabricate microfluidic systems are: X-ray lithography [62], moulding by high precision, hot embossing [63], micro-injection moulding[64], and roll-to-roll embossing [65]. This technology allows building three-dimensional complex structures with a great variety of materials.

18.3.2.2 Electrode/detector fabrication methods

Screen-printed technology is based on the sequential transfer of different conductive or non-conductive inks on a variety of inert substrates through a mask with the desired geometry. Nowadays, screen-printing microfabrication technology is a well-established technology for mass production of thick film electrodes and it is widely applied to build biological or chemical sensors [66]. Screen-printed electrodes (SPE) represent one of the most important products of this technology because these are high mass production and low cost.

Generally, inert substrates used in screen-printed processes are ceramic or polymeric ones. In case of polymeric substrates, polyester substrate is the most commonly used for its durability, thermal and hydroscopic stability, clarity and stiffness. Exact composition of printing inks is uncertain, as they are commercial and patented.

Ink-jet printing is also being used as an alternative to screen-printing technology. Other micro and nanofabrication techniques based on photolithography and other 'micro' and 'nano' imprinting techniques are also extended for electrode/detector fabrication.

18.4 LOC applications

Systems based on nanomaterials-based sensors, lab-on-a-chip technology and micro/nanofabrication techniques have been used to detect different analytes of interests as: cells, proteins, DNA fragments, and so on. Lab-on-a-chip platforms for optical, electrochemical and other detection systems have been reported.

18.4.1 LOCs with optical detections

Biomolecules and nanoparticles with optical properties in order to realize optical detection in a microfluidic system have been reported. Some examples are: (a) the development of a platform with 3D microfluidics used to culture metastatic prostate cancer cells. In addition a microchannel with side-chamber design to keep the spheroids in a stationary way is built. This device was designed with the purpose to evaluate the effect of drugs in the mentioned cells culture using fluorescence dyes for the detection step [67] (Figure 18.4a); (b) use of pillars geometry inside the channel, where circulating tumour cells (CTs) have been identified in peripheral blood from cancer patients. This kind of analytical system can be a potential alternative to invasive biopsies as source of cancer detection through an optical technique [68] (Figure 18.4b); (c) use of a passive mixer, without external forces [69]. The authors studied the hydrodynamics inside the channels at different geometries and did experimental tests using gold nanoparticles and $CuSO_4$ solutions to visualize the mixing tendencies. The results were coherent with the theoretical calculus exhibiting a good alternative to integrate this system into a lab-on-a-chip system (Figure 18.4c); (d) another example using optical detection is the integration of QDs into a modular microfluidic biosensor for the quantification of biomarkers such as: carcinoembryonic antigen (CEA), cancer antigen 125 (CA125), and Her-2/Neu (C-erbB-2). This detection was applied for both serum and whole saliva specimens [70] (see Figure 18.4d).

18.4.2 LOCs with electrochemical detectors

Electrochemical detectors inside the microfluidic channels for different applications are also described in the literature. Some examples include: (a) paper microfluidics, where micropaper-based microfluidic electrochemical devices (μPEDs) patterned by photolithography or wax printing, and electrodes screen-printed from conducting inks [71]. This device is applied in the heavy metal detection or glucose in aqueous solutions showing a low-cost alternative for analytical applications in environmental monitoring and public health respectively (Figure 18.5a); (b) a microfluidic system with an integrated control and microelectrochemical modules for detection of urinary proteins was reported. This system has shown a high sensitivity in the urinary proteins detection, including lysozyme and albumin, demonstrating a good alternative for (bio) chemical detections [72] (Figure 18.5b); (c) an amperometric glucose sensor based on glucose oxidase enzyme encapsulated into a sol-gel derived zirconia/Nafion matrix to demonstrate its biosensing capability is also reported. The amperometric measurements were carried out using a three-electrode system. The microchannel was fabricated using SU8 epoxy over the electrode layer and the enzyme was immobilized by sealing with a PDMS membrane [73] (Figure 18.5c); (d) a label-free immunosensor for detection of *E. coli* O157:H7 cells using faradic impedance spectroscopy with an interdigitated microelectrodes array was developed [74] (Figure 18.5d).

The developed electrochemical and LOC based systems show advantages in terms of a better system integration due to the easy miniaturization of the detectors and their compatibility with the detection platform.

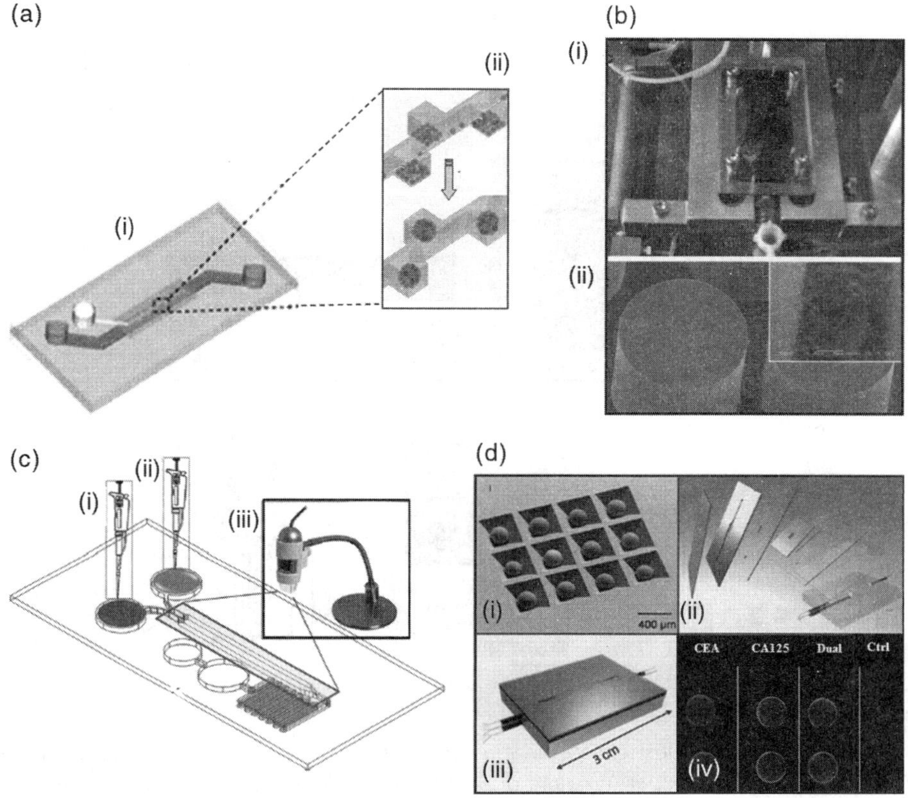

Figure 18.4 Some examples of LOCs with optical detection. *(a) 3D PDMS microfluidic for cell culture. Detection using fluorescence dyes (i), co-culture spheroids within 1 day of culture (ii). Adapted from [67] with permission from Elsevier © 2009; (b) whole blood flowing through the microfluidic device (i), scanning electron microscope image of a captured NCI-H1650 lung cancer cell spiked into blood. The inset shows a high magnification view of the cell (ii). Adapted from [68] with permission from Macmillan Publishers Ltd © 2007; (c) sample introduction (i), reagents introduction (ii), optical visualization (iii). Adapted from [69] with permission from Elsevier © 2009; (d) TEM image of biosensing device (i), layer of the whole microdevice (ii), integration of the microfluidic system (iii), and fluorescence image of each channel using quantum dots as biomarkers. Adapted from [70] with permission from Elsevier © 2009.*

18.4.3 LOCs with other detections

Hybrid or multiple detection based LOC systems have also been developed. Figure 18.6 shows some typical examples using: (a) an array of ion sensitive field effect transistors, where the gate of the transistor is open and in contact with the solution to be detected. As a result of this work a microfluidic purification chip for capturing of multiple biomarkers from blood samples is presented. The selective and quantitive detection of two model cancer antigens in less than 20 min was demonstrated [75] (Figure 18.6a); (b) CMOS-compatible approach to fabricate silicon nanowires (SiNW) array biosensor is represented, capable of sensing human cardiac troponin-TCTnT in a buffer solution. Measurements taken were conductance changes, thus providing a method of label-free detection of biomolecular species [76]

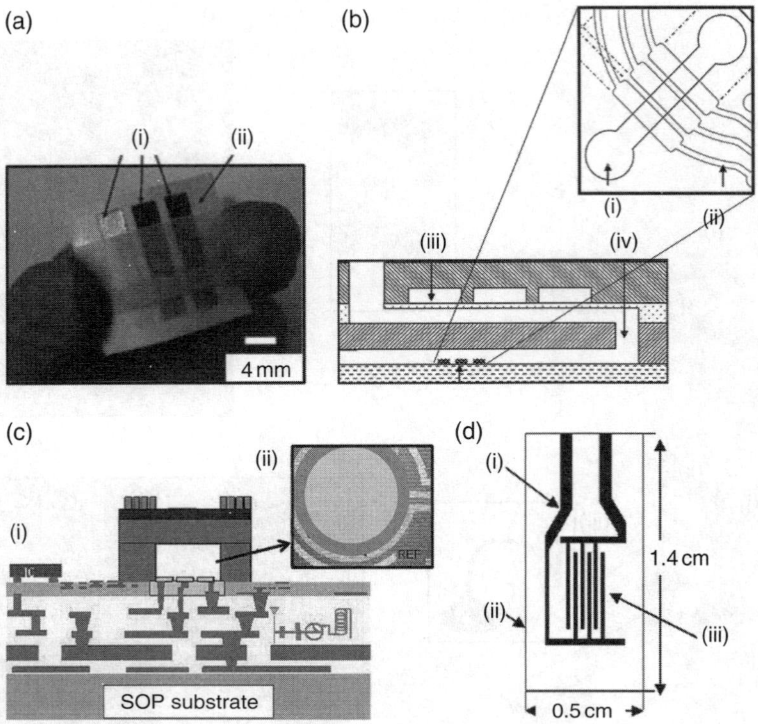

Figure 18.5 *Some examples of LOCs with electrochemical detection. (a) Paper-based microfluidic system; electrodes (i), paper channel (ii), paper (iii), tape (iv), hydrophobic barrier (v) (detection using fluoresce dyes) [71]. Adapted by permission of The Royal Society of Chemistry © 2010. (b) Microfluidic channel (i), electrochemical electrodes (ii), air chamber (iii), microchannel (iv). Adapted from [72] © 2003 American Chemical Society. (c) Amperometric glucose sensor (i), TEM lithography electrode. Adapted from [73] © 2007 IEEE. (d) Impedance interdigitated electrode; ITO coating (i), glass substrate (ii), 25 pairs of finger electrodes with 15 μm of digit and interdigit width (iii). Adapted from [74] © 2004 American Chemical Society.*

(Figure 18.6b); contactless detection using an electrode to measure the change of impedance when the influenza virus concentration change is also reported [77] (See Figure 18.6c).

18.5 Conclusions and future perspectives

Micro- and nanomaterials based detection systems applied in lab-on-a-chip technology are showing to be interesting alternative devices for Green Analytical Chemistry. Their benefits in relation with the system integration are being combined with the greatly reducing of the waste amount of solvents and samples provided by the LOC systems. The developed systems provide both environmental and cost/efficiency benefits making the analytical procedures more sustainable.

The micro- and nanomaterial based detection systems integrated into LOC show several advantages in comparison to the classical analytical methods. Nanomaterials are showing to be of particular interest in developing high sensitive (bio)detection devices with multidetection capability and compatibility with LOCs. Various detection techniques ranging from optical, electrochemical and mass changes can be applied. They offer different advantages in terms of the (bio)assay formats, sensitivity, selectivity, linear range of response

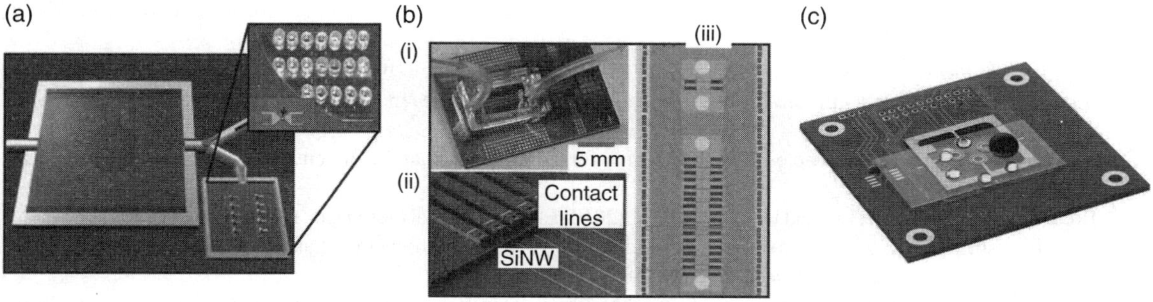

Figure 18.6 (a) Whole blood is injected into the chip with the valve set to the waste compartment (black arrow shows the direction of fluid flow) and, if present in the sample, biomarkers bind their cognate antibodies. Adapted from [75] with permission from Macmillan Publishers Ltd © 2010; (b) clockwise from top: image of SiNW device array chip, integrated with microfluidic system for fluid exchange, used in experiments (i), SEM image of a cluster of nanowires (ii), each nanowire is individually addressable by oxide-passivated metal contact lines running out to the external edge of the chip (iii). Adapted from [76] © 2009 American Chemical Society; and (c) example of a contactless microsystem for influenza virus detection including sample preparation and RT-PCR on a disc developed by Pipper and colleagues. Adapted from [77] with permission from Wiley-VCH Verlag GMbH & Co. KGaA © 2009.

beside the reproducibility of measurements. Some of the techniques (i.e. electrical/electrochemical) are inherently prone to a better miniaturization in comparison to optical/spectroscopic techniques which make the first one the 'highest' of integrated systems.

The efforts in the development of this novel generation of analytical techniques are related with the breakthroughs in microelectronics and micro/nanotechnology. This synergy is allowing the development of powerful analytical devices that are contributing to the Green (bio) Analytical Chemistry due to the minimization of hazardous substances consumption and the amount of waste generated during assays.

Problems such as the application of some rather complicated fabrication and measuring technologies probably not in agreement with mass production and user-friendly technologies need to be resolved so as to bring these devices to end-users interested in green analytical processes. Overcoming this and other drawbacks (i.e. detection limits, sensibility etc.) of the systems would bring new benefits for our society: better point-of-care diagnostics for healthcare, safety and security beside other applications.

Acknowledgements

Ministerio de Ciencia e Innovación. **MICIN** (Madrid) for the projects 2008–03079/NAN and PIB2010JP-00278 is acknowledged.

References

1. Koel, M. and Kaljurand, M. (2006) Application of the principles of Green Chemistry in analytical chemistry, Pure Appl. Chem., 78, 1993–2002.
2. Valcárcel, M.; Simonet, B. and Cárdenas, S. (2008) Analytical nanoscience and nanotechnology today and tomorrow, Anal. Bioanal. Chem., 391, 1881–1887.
3. Merkoçi, A.; Aldavert, M.; Marín, S. and Alegret, S. (2005) New materials for electrochemical sensing. V. Nanoparticles for DNA labelling, Trends in Anal. Chem., 24, 341–349.

4. de la Escosura-Muñiz, A.; Parolo, C. and Merkoçi, A. (2010) Immunosensing using nanoparticles, *Materials Today*, **13**, 24–34.
5. Aragay, G.; Pons, J.; Ros, J. and Merkoçi, A. (2010) Aminopyrazole based ligand induces gold nanoparticle formation and remains available for heavy metals sensing. A simple 'mix and detect' approach, *Langmuir*, **26**, 10165–10170.
6. Merkoçi, A. (2010) Nanoparticles-based strategies for DNA, protein and cell sensors, *Biosens. Bioelectron.*, **26**, 1164–1177.
7. Ronkainen, N.J.; Halsall, H.B. and Heineman, W.R. (2010) Electrochemical biosensors, *Chem. Soc. Rev.*, **39**, 1747–1763.
8. de la Escosura-Muñiz, A.; Ambrosi, A. and Merkoçi, A. (2008) Electrochemical analysis with nanoparticle based biosystems, *Trac-Trend. Anal. Chem.*, **27**, 568–584.
9. Suni, I.I. (2008) Impedance methods for electrochemical sensors using nanomaterials, *TrAC-Trend. in Anal. Chem.*, **27**, 604–6011.
10. Masadome, T.; Yada, K. and Wakida, S.I. (2006) Microfluidic Polymer Chip Integrated with an ISFET Detector for Cationic Surfactant Assay in Dental Rinses, *Anal. Sci.*, **22**, 1065–1069.
11. Tsai, H.H.; Lin, C.F.; Juang, Y.Z.; Wang, I.L.; Lin, Y.C.; Wang, R.L. and Lin, H.Y. (2010) Multiple type biosensors fabricated using the CMOS BioMEMS platform, *Sensor. Actuator. B*, **144**, 407–412.
12. Waleed Shinwari, M. and Jamal Deen, M. (2008) Optimization of DNA Detection Using FETs, *IEEE*, 185–188.
13. Strambini, E.; Piazza, V.; Pingue, P.; Biasiol, G.; Sorba, L. and Beltram, F. (2010) Cantilever deflection measurement and actuation by an interdigitated transducer, *Appl. Phys. Lett.*, **96**, 173505–173508.
14. Son, A.; Dosev, D.; Nichkov, M.; Ma, Z.; Kennedy, I.M.; Scow, K.M. and Hristov, K.R. (2007) Quantitative DNA hybridization in solution using magnetic/luminescent core–shell nanoparticles, *Anal. Biochem.*, **370**, 186–194.
15. Averitt, R.D.; Westcott, S.L. and Halas, N.J. (1999) Linear optical properties of gold nanoshells, *J. Opt. Soc. Am. B*, **16**, 1824–1832.
16. Loo, C.; Lin, A.; Hirsch, L.; Lee, M.-H.; Barton, J.; Halas, N.; West, J. and Drezek, R. (2004) Nanoshell-enabled photonics-based imaging and therapy of cancer, *Technol. Cancer Res. Treat.*, **3**, 33–40.
17. I. El-Sayed, H.; Huang, X. and El-Sayed, M.A. (2005) Surface plasmon resonance scattering and absorption of anti-EGFR antibody conjugated gold nanoparticles in cancer diagnostics: Applications in oral cancer, *Nano Letters*, **5**(5), 829–834.
18. Mirkin, C.A. (2000) Programming the assembly of two- and three-dimensional architectures with DNA and nanoscale inorganic building blocks, *Inorg. Chem*, **39**, 2258–2272.
19. Chuang, Y.C.; Li, J.C.; Chen, S.H.; Liu, T.Y.; Kuo, C.H.; Huang, W.T. and Lin, C.S. (2010) An optical biosensing platform for proteinase activity using gold nanoparticles, *Biomaterials*, **31**, 6087–6095.
20. Alizadeh, A.; Khodaei, M.M.; Karami, Ch.; Workentin, M.S.; Shamsipur, M. and Sadeghi, M. (2010) Rapid and selective lead (II) colorimetric sensor based on azacrown ether-functionalized gold nanoparticles, *Nanotechnology*, **21**, 315503–315511.
21. Fan, M.; Thompson, M.; Andrade, M.L. and Brolo, A.G. (2010) Silver nanoparticles on a plastic platform for localized surface plasmon resonance biosensing, *Anal. Chem.*, **82**, 6350–6352.
22. Michalet, X.; Pinaud, F.F.; Bentolila, L.A.; Tsay, J.M.; Doose, S.; Li, J.J.; Sundaresan, G.; Wu, A.M.; Gambhir, S.S. and Weiss, S. (2005) Quantum dots for live cells, in vivo imaging, and diagnostics, *Science*, **28**, 538–544.
23. Algar, W.R. and Krull, U.J. (2010) New opportunities in multiplexed optical bioanalyses using quantum dots and donor–acceptor interactions, *Anal. Bioanal. Chem.*, **398**, 2439–2449.
24. Medintz, I.L.; Stewart, M.H.; Trammell, S.A.; Susumu, K.; Delehanty, J.B.; Mei, B.C.; Melinger, J.S.; Blanco-Canosa, J.B.; Dawson, P.E. and Mattoussi, H. (2010) Quantum-dot/dopamine bioconjugates function as redox coupled assemblies for in vitro and intracellular pH sensing, *Nature Materials*, **9**, 676–684.
25. Bijamov, A.; Shubitidze, F.; Oliver, P.M. and Vezenov, D.V. (2010) Optical response of magnetic fluorescent microspheres used for force spectroscopy in the evanescent field, *Langmuir*, **26**, 12003–12011.
26. Wang, K.; Liu, Q.; Wu, X.Y.; Guan, Q.M. and Li, H.N. (2010) Graphene enhanced electrochemiluminescence of CdS nanocrystal for H2O2 sensing, *Talanta*, **82**, 372–376.
27. Aili, D. and Stevens, M.M. (2010) Bioresponsive peptide–inorganic hybrid nanomaterials, *Chem. Soc. Rev.*, **39**, 3358–3370.
28. Knopp, D.; Tang D. and Niessner, R. (2009) Review: bioanalytical applications of biomolecule-functionalized nanometer-sized doped silica particles, *Anal. Chim. Acta*, **647**, 14–30.

29. Pumera, M.; Sánchez, S.; Ichinose, I. and Tang, J. (2007) Electrochemical nanobiosensors, *Sensor. Actuator. B: Chem.*, **123**, 1195–1205.
30. de la Escosura-Muñiz, A.; Ambrosi, A. and Merkoçi, A. (2008) Electrochemical analysis with nanoparticle-based biosystems, *TrAC-Trend. Anal. Chem.*, **27**, 568–584.
31. Wang, J.; Xu, D.; Kawde, A.N. and Polsky, R. (2001) Metal nanoparticle-based electrochemical stripping potentiometric detection of DNA hybridization, *Anal. Chem.*, **73**, 5576–5581.
32. Authier, L.; Grossiord, C. and Brossier, P. (2001) Gold nanoparticle-based quantitative electrochemical detection of amplified human cytomegalovirus DNA using disposable microband electrodes, *Anal. Chem.*, **73**, 4450–4456.
33. Wang, J.; Liu, G. and Merkoçi, A. (2003) Electrochemical coding technology for simultaneous detection of multiple DNA targets, *J. Am. Chem. Soc.*, **125**, 3214–3215.
34. Wanekaya, A.K.; Chen, W.; Myung, N.V. and Mulchandani, A. (2006) Nanowire-based electrochemical biosensors. *Electroanalysis*, **18**, 533–550.
35. Ansari, A.A.; Solanki, P.R.; Kaushik, A. and Malhotra, B.D. (2009) Recent advances in nanostructured metal oxides based electrochemical biosensors for clinical diagnostics, in *Nanostructured Materials for Electrochemical Biosensors*. Nova Science Publishers.
36. Shchipunov, Y.A.; Karpenko, T.Y. and Krekoten, A.V. (2005) Hybrid organic–inorganic nanocomposites fabricated with a novel biocompatible precursor using sol–gel processing, *Compos. Interface.*, **11**, 587–607.
37. Tang, J.; Su, B.; Tang, D. and Chen, G. (2010) Conductive carbon nanoparticles-based electrochemical immunosensor with enhanced sensitivity for α-fetoprotein using irregular-shaped gold nanoparticles-labeled enzyme-linked antibodies as signal improvement, *Biosens. Bioelectron.*, **25**, 2657–2662.
38. Munge, B.; Liu, G.; Collins, G. and Wang, J. (2006) Multiple enzyme layers on carbon nanotubes for electrochemical detection down to 80 DNA copies, *Anal. Chem.*, **77**, 4662–4666.
39. Hu, C. and Hu, S. (2009) Carbon nanotube-based electrochemical sensors: Principles and applications in biomedical systems, *Journal of Sensors* (on-line publication).
40. Pérez, B. (2007) *Carbon Nanotubes for Electrochemical (Bio)Sensing,* Universidad Autónoma de Barcelona. Ph.D Thesis. Chemistry Department. Science Faculty, Barcelona-Spain,.
41. Huang, Y.; Dong, X.; Shi, Y.; Li, C.M.; Li, L.J. and Chen, P. (2010) Nanoelectronic biosensors based on CVD grown grapheme, *Nanoscale*, **2**, 1485–1488.
42. Cheng, Z.; Li, Q.; Li, Z.; Zhou, Q. and Fang, Y. (2010) Suspended graphene sensors with improved signal and reduced noise, *Nano Lett.*, **10**, 1864–1868.
43. Wang, X.F.; Zhou, Y.; Xu, J.J. and Chen, H.Y. (2009) Signal-on electrochemiluminescence biosensors based on CdS–carbon nanotube nanocomposite for the sensitive detection of choline and acetylcholine, *Adv. Funct. Mater.*, **19**, 1444–1450.
44. Wang, F.; Banerjee, D.; Liu, Y.; Chen, X. and Liu, X. (2010) Upconversion nanoparticles in biological labeling, imaging, and therapy, *Analyst*, **135**, 1839–1854.
45. Arruda, D.L.; Wilson, W.C.; Nguyen, C.; Yao, Q.W.; Caiazzo Jr, R.J.; Talpasanu, I.; Dow, D.E. and Liu, B.C.S. (2009) Microelectrical sensors as emerging platforms for protein biomarker detection in point-of-care diagnostics, *Expert Rev. Mol. Diagn.*, **9**, 749–755.
46. Cheng, M.M.C.; Cuda, G.; Bunimovich, Y.L.; Gaspari, M.; Heath, J.R.; Hill, H.D.; Mirkin, C.A.; Nijdam, A.J.; Terracciano, R.; Thundat, T. and Ferrari, M. (2006) Nanotechnologies for biomolecular detection and medical diagnostics, *Curr. Opin. Chem. Biol.*, **10**, 11–19.
47. Dittrich, P.S.; Tachikawa, K. and Manz, A. (2006) Micro Total Analysis Systems. Latest advancements and trends, *Anal. Chem.*, **78**, 3887–3908.
48. Terry, S.C. (1975) A gas chromatographic air analyzer fabricated on silicon wafer using integrated circuit technology. PhD. Dissertation Stanford University.
49. Manz, A.; Graber, N. and Widmer, H.M. (1990) Miniaturized total chemical analysis systems: A novel concept for chemical sensing, *Sensor. Actuator. B1*, 244–248.
50. Sparreboom, W.; van den Berg, A. and Eijkel, J.C.T. (2009) Principles and applications of nanofluidic transport, *Nature nanotechnology*, **4**, 713–720.
51. Tamaki, E.; Hibara, A.; Kim, H.B.; Tokeshi, M. and Kitamori, T. (2006) Pressure-driven flow control system for nanofluidic chemical process, *J. Chromatogr. A*, **1137**, 256–262.

52. Tas, N.R.; Mela, P.; Kramer, T.; Berenschot, J.W. and van den Berg, A. (2003) Capillarity induced negative pressure of water plugs in nanochannels, *Nano Letters*, **3**, 1537–1540.
53. Hibara, A.; Saito, T.; Kim, H.B.; Tokeshi, M.; Ooi, T.; Nakao, M. and Kitamori, T. (2002) Nanochannels on a fused-silica microchip and liquid properties investigation by time-resolved fluorescence measurements, *Anal. Chem.*, **74**, 6170–6176.
54. Phan, V.N.; Nguyen, N.T.; Yang, C.; Joseph, P.; Djeghla, L.; Bourrier, D. and Gue, A.M. (2010) Capillary filling in closed end nanochannels, *Langmuir*, **26**, 13251–13255.
55. Zhang, Z.Q.; Zhang, H.W. and Ye, H.F. (2010) Pressure-driven flow in parallel-plate nanochannels, *Appl. Phys. Lett.*, **95**, DOI: 154101_1-154101_3.
56. Richards Grayson, A.C.; Shawgo, R.S.; Johnson, A.M.; Flynn, N.T.; Li, Y.; Cima, M.J. and Langer, R. (2004) A BioMEMS Review: MEMS technology for physiologically integrated devices, *Proceedings of the IEEE*, **92**, 6–21.
57. Campos, L.M.; Truong, T.T.; Shim, D.E.; Dimitriou, M.D.; Shir, D.; Meinel, I.; Gerbec, J.A.; Hahn, H.T.; Rogers, J.A. and Hawker, C.J. (2009) Applications of photocurable PMMS thiol-ene stamps in soft lithography, *Chem. Mater.*, **21**, 5319–5326.
58. Lomas, T.; Mongpraneet, S.; Wisitsoraat, A.; Jaruwongrungsee, K.; Sappat, A.; Maturos, T.; Chevasuvit, F. and Tuantranont, A. (2009) Low cost hot Embossing process for plastics microfluidic chips fabrication, *6th International Conference on Electrical Engineering/Electronics, Computer, Telecommunications and Information Technology*, 462–464.
59. Shiu, P.P.; Knopf, G.K. and Ostojic, M. (2010) Fabrication of metallic micromolds by laser and electro-discharge micromachining, *Microsyst. Technol.*, **16**, 477–485.
60. Roy, S. (2007) Fabrication of micro- and nano-structured materials using mask-less processes, *J. Phys. D: Appl. Phys.*, **40**, 413–426.
61. Nath, P.; Fung, D.; Kunde, Y.A.; Zeytun, A.; Branch, B. and Goddard, Rapid prototyping of robust and versatile microfluidic components using adhesive transfer G. (2010) tapes, *Lab. Chip*, **10**, 2286–2291.
62. Mongpraneet, S.; Wisitsora, A.; Phatthanakun, R.; Chomnawang, N. and Tuantranont, A. (2010) Low-cost X-ray mask based on micropattern sputtered lead film for X-ray lithography, *American Vacuum Society*, **27**, 1299–1303.
63. Greener, J.; Li, W.; Ren, J.; Voicu, D.; Pakharenko, V.; Tang, T. and Kumacheva, E. (2009) Rapid, cost-efficient fabrication of microfluidic reactors in thermoplastic polymers by combining photolithography and hot embossing, *Lab. Chip*, **10**, 522–524.
64. Choi, S.H.; Kim, D.S. and Kwon, T.H. (2009) Microinjection molded disposable microfluidic lab-on-a-chip for efficient detection of agglutination, *Microsyst. Technol.*, **15**, 309–316.
65. Yeo, L.P.; Huan, S.; Wang, Z.; Wang, Z. and Frans, N. (2009) Micro-fabrication of polymeric devices using hot roller embossing, *Microelectron. Eng.*, **86**, 933–936.
66. Fanjul-Bolado, P.; Hernández-Santos, D. and Lamas-Ardisana, P.J. (2008) Electrochemical characterization of screen-printed and conventional carbon paste electrodes, *Electrochim. Acta*, **53**, 3635–3642.
67. Hsiao, A.Y.; Torisawa, Y.S.; Tung, Y.C.; Sud, S.; Taichman, R.S.; Pienta, K.J. and Takayama, S. (2009) Microfluidic system for formation of PC-3 prostate cancer co-culture spheroids, *Biomaterials*, **30**, 3020–3027.
68. Nagrath, S.; Sequist, L.V.; Maheswaran, S. *et al.* (2007) Isolation of rare circulating tumour cells in cancer patients by microchip technology, *Nature*, **450**, 1235–1241.
69. Jeon, W. and Shin, C.B. (2009) Design and simulation of passive mixing in microfluidic systems with geometric variations, *Chem. Eng. J.*, **152**, 575–582.
70. Jokerst, J.V.; Raamanathan, A.; Christodoulides, N. *et al.* (2009) Nano-bio-chips for high performance multiplexed protein detection: Determinations of cancer biomarkers in serum and saliva using quantum dot bioconjugate labels, *Biosens. Bioelectron.*, **24**, 3622–3629.
71. Nie, Z.; Nijhuis, C.A.; Gong, J.; Chen, X.; Kumachev, A.; Martinez, A.W.; Narovlyanskya, M. and Whitesides, G.M. (2010) Electrochemical sensing in paper-based microfluidic devices, *Lab. Chip*, **10**, 477–483.
72. Liu, G.; Wang, J.; Kim, J. and Jan, M.R. (2003) Electrochemical coding for multiplexed immunoassay of proteins, *Anal. Chem.*, **76**, 7126–7130.
73. Huang, C.J.; Chen, Y.H.; Wang, C.H.; Chou, T.C. and Lee, G.B. (2007) Integrated microfluidic systems for automatic glucose sensing and insulin injection, *Sensors Actuator B-Chem.*, **122**, 461–468.

74. Yang, L.; Li, Y. and Erf, G.F. (2004) Interdigitated array microelectrode-based electrochemical impedance immunosensor for detection of escherichia coli O157:H7, *Anal. Chem.*, **76**, 1107–1113.
75. Stern, E.; Vacic, A.; Rajan, N.K.; Criscione, J.M.; Park, J.; Ilic, B.R.; Mooney, D.J.; Reed, M.A. and Fahmy, T.M. (2010) Label-free biomarker detection from whole blood, *Nature Nanotechnology*, **5**, 138–142.
76. Chua, J.H.; Chee, R.E.; Agarwal, A.; Wong, S.M. and Zhang, G.J. (2009) Label-Free Electrical Detection of Cardiac Biomarker with Complementary Metal-Oxide Semiconductor-Compatible Silicon Nanowire Sensor Arrays, *Anal. Chem.*, **81**, 6266–6271.
77. Schulze, H.; Giraud, G.; Crain, J. and Bachmann, T.T. (2009) Multiplexed optical pathogen detection with lab-on-a-chip devices, *J. Biophoton.*, **2**, 199–211.

19

Photocatalytic Treatment of Laboratory Wastes Containing Hazardous Organic Compounds

Edmondo Pramauro, Alessandra Bianco Prevot and Debora Fabbri

Department of Analytical Chemistry, V. Pietro Giuria 5, Torino, Italy

19.1 Photocatalysis

The problems related to the elimination of hazardous substances present in liquid wastes are of great interest, not only for people directly involved in industrial and agricultural activities, but also for the chemists operating in control laboratories, where more or less abundant stocks of toxic wastes are usually accumulated and/or produced.

The two main sources of wastewaters coming from the analytical laboratory are the residual samples stored after the analysis and the wastes arising from the application of the analytical protocols. In the first case the volume of samples taken from contaminated waters, from effluent generated by the industrial production and from a variety of other human activities could be relevant, whereas in the second case the volume of solutions containing dangerous products directly produced during the application of the analytical method is generally lower.

Taking into account that the storage and disposal of all these effluents is usually difficult and costly, the direct treatment of such wastes in the laboratory could be of practical interest in order to reduce the dangerous products content below the limits admitted by the law regulations.

In this direction the so called advanced oxidation process (AOP) could represent a promising approach and an interesting alternative to other traditional wastewater treatments, being particularly attractive for degradation of organic pollutants and their transformation into non toxic or less toxic products. Among these techniques, heterogeneous photocatalysis is one of the most interesting treatments since it operates at mild conditions of temperature and pressure and it is able to perform the complete mineralization of a great variety of organic harmful species present in different aqueous matrices [1–5].

Handbook of Green Analytical Chemistry, First Edition. Edited by Miguel de la Guardia and Salvador Garrigues.

Many semiconductor oxides can be in principle used for photocatalysis (e.g. TiO_2, ZnO, MgO, WO_3, ZrO_2, CeO_2 and Fe_2O_3); TiO_2 has generally been demonstrated to be the most active. Only ZnO is as active as TiO_2 but TiO_2 shows a better resistance to chemical and photocorrosion, moreover it is safe and cheap. TiO_2, in particular in its anatase form, has demonstrated very good performances in various remediation treatments, together with a remarkable chemical stability and non toxic properties; for these reasons it is actually one of the more studied and applied photocatalytic material. TiO_2 P25 from Degussa (composed of ca. 80% anatase and 20% rutile), having a surface area of around 55 $m^2\ g^{-1}$, is the most largely used photocatalysts and a great number of experiments concerning the treatment of wastes is based on the use of suspensions of this material. Furthermore, TiO_2 is of special interest since it can also exploit natural (solar) UV. This is due to its band gap energy, which can be surpassed by the energy content of a solar photon.

19.2 Fundamentals of the photocatalytic process

Photocatalysis is based on a series of light-induced redox reactions occurring when semiconductor particles are irradiated with light of energy higher than the band-gap. It is known that, upon irradiation of semiconductor particles with light of proper energy, the generation of electron/hole pairs occurs and these species have potentials capable to reduce or oxidize many organic compounds. Moreover, they can react with water and oxygen giving rise to the formation of other reactive species (often radicals), in turn capable to attack the substrate [6,7]. The crucial role of oxidizing radicals formed upon reaction of holes with water, which can attack and break the pollutant molecules, has been clearly recognized.

For example the formation of $e^-_{CB} - h^+_{VB}$ pairs, (Figure 19.1), occurs when TiO_2 is irradiated with light having wavelength <400 nm, inducing the formation of hydroxyl radicals and superoxide radicals from water and dissolved oxygen, according to the following sequence of reactions:

$$TiO_2 + h\nu \rightarrow TiO_2\ (e^-_{CB} + h^+_{VB})$$
$$e^-_{CB} + O_2 \rightarrow O_2^{\bullet -}$$
$$2\ O_2^{\bullet -} + H_2O \rightarrow H_2O_2 + O_2 + 2\ OH^-$$
$$e^-_{CB} + H_2O_2 \rightarrow {}^{\bullet}OH + OH^-$$
$$h^+_{VB} + H_2O \rightarrow {}^{\bullet}OH + H^+$$
$$h^+_{VB} + OH^-_{(ads)} \rightarrow {}^{\bullet}OH$$

19.3 Limits of the photocatalytic treatment

Photocatalysis can be applied to treat aqueous wastes containing low or medium concentrations of pollutants, starting from few mg l^{-1} up to around 1 g l^{-1}, depending on the nature of the compounds to be degraded. Concentrations higher than around 1 g l^{-1} normally are not suitable for the photocatalytic treatment unless previous dilution of the waste.

19.4 Usual photocatalytic procedure in laboratory practice

Most of the reported laboratory treatments operated in batch mode; the photocatalyst can be used as a suspension or coating the inner walls of the reaction device; it can coat the inner walls of a glass reactor [8], or as an alternative it can be immobilized on different substrates (e.g. silica gel) and then introduced into a

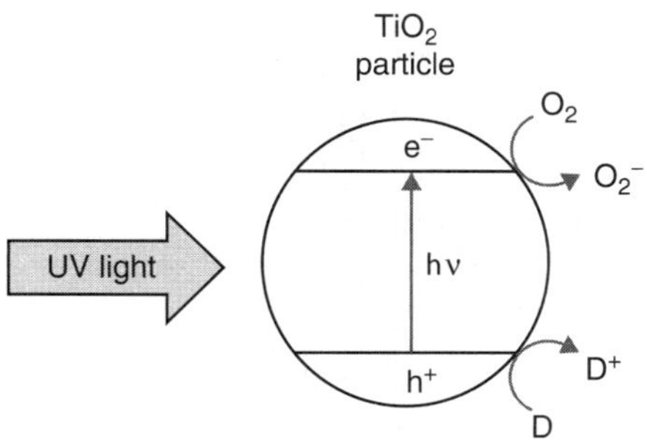

Figure 19.1 Schematic representation of the photocatalytic process: absorption of a UV photon by a TiO_2 particle and excitation of an electron from the valence to the conduction band. D is an electron donor.

borosilicate glass tube [9], polymeric membranes [10], porous alumina-silica ceramic [11], zeolite [12], activated carbon [13].

The batch approach is based on the use of suspended particles of a proper photocatalyst introduced in the aqueous waste to be treated, followed by irradiation of the stirred suspension with light of proper wavelength.

In typical degradation experiments the TiO_2 particles are previously irradiated in Solarbox or in other suitable illumination device, washed with water and then dried for several hours, in order to eliminate any organic impurity. Successively, a proper amount of the treated material is added to the waste to form a stirred suspension.

For individual preliminary tests, irradiations can be readily performed in stirred cylindrical closed cells of low volume (30–40 cm^3), made of Pyrex glass, on 5–10 ml of aqueous dispersions, containing 0.2–0.4 mM of substrates to be treated and 0.1–1 g l^{-1} of TiO_2. Within a typical laboratory Solarbox up to 4–6 cells can be irradiated, using different illumination sources (e.g. 1000–1500 W Xenon lamps) simulating AM1 solar light, equipped with a glass cut-off filter allowing the transmission of light with wavelengths ≥ 340 nm. The radiant power of the 1500 W lamp is around 60 mW cm^2 and the working temperature within the Solarbox is in the range 50–55 °C.

The treatment of higher volumes (0.5–2 l) of wastes is more adequately performed in photochemical stirred batch reactors (see Figure 19.2), equipped with a medium pressure mercury lamp (125–250 W). The Pyrex® glass jacket surrounding the lamp acts as a cut-off filter for wavelengths shorter than 300–350 nm. Within the glass jacket circulates thermostatized water in order to keep constant the reactor temperature.

19.4.1 Solar detoxification of laboratory waste

The upgrade of the process to a pilot scale in order to treat larger volumes could be done by going to solar light as radiation source, as an alternative to the lamps [14]. The design procedure for a pilot solar detoxification system requires the selection of: a reactor, catalyst operating mode (slurry or fixed matrix), reactor-field configuration (series or parallel), treatment-system mode (once-through or batch), flow rate, pressure drop, pretreatment, catalyst and oxidant loading method, pH control, and so on, so a pilot plant has to be as versatile as possible and at the same time, provide sufficient confidence in the experiments carried out in it. Solar collectors are traditionally divided in three categories: non-concentrating (or low temperature, up to 150 °C), medium concentrating (or medium temperature, from 150–400 °C) and high concentrating (or high temperature, over 400 °C). Concentrating solar systems make use of direct radiation and need solar tracking mechanisms.

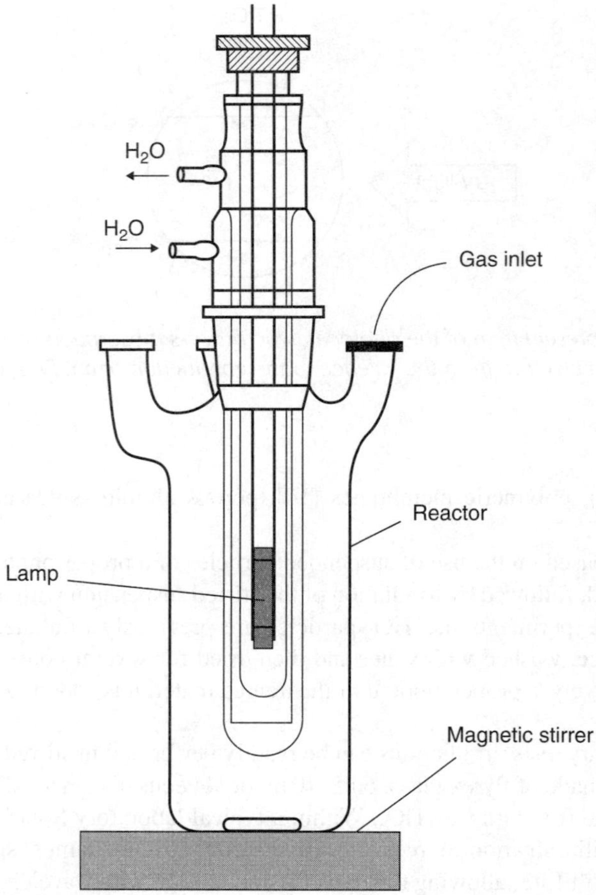

Figure 19.2 *Schematic representation of a laboratory photochemical reactor.*

Non-concentrating systems are much simpler as they do not need solar tracking and can collect direct and diffuse solar radiation with slightly lower yearly efficiencies. Non-concentrating collectors are the choice for solar photocatalytic applications.

Usually, a detoxification pilot plant is constructed with several solar collectors. All the modules are connected in series, but with valves that permit to bypass any number of them. Sampling valves are in the outlet of each of the modules. The most important sensors required for the system are temperature, pressure and dissolved oxygen. Other sensors, such as pH, selective electrodes, and so on could be useful depending on the type of experiments to be carried out. As constant pure oxygen or air is required for the oxidation of organics, an injection system at the reactor inlet allows oxygen or air to be added to the reactor either at planned intervals by opening and closing in a predefined cycle, or continuously.

Suspended TiO_2 catalyst systems give efficiencies higher than supported catalysts. After their use, TiO_2 can be agglomerated and sedimented. After sedimentation or filtration, the semiconductor can be recovered.

Figure 19.3 illustrates the photoreacting system based on compound parabolic collectors (CPC), installed at the 'Plataforma Solar of Almería' (PSA, Spain); this represents a simple piece of equipment which can exploit the solar light irradiation to treat laboratory wastes. The photoreactor configuration is a

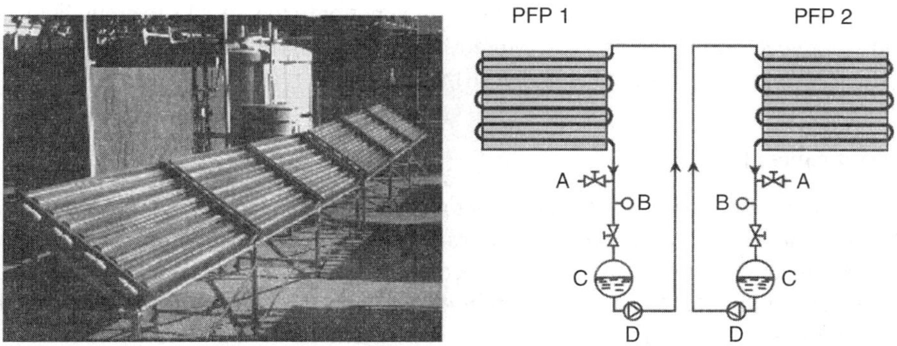

Figure 19.3 *Picture and scheme of the photoreacting system: (A) sampling valve; (B) thermocouple; (C) not-reacting tank; (D) pump. Reprinted from [41] with permission from Elsevier © 1999.*

common one in heterogeneous photocatalysis: a plug flow photoreactor (PFP) in a total recirculation loop with a not-reacting tank whose function was that of providing aeration and samples for analyses. Each photoreactor consists of three CPC modules in series placed on fixed supports inclined 37° (latitude of the PSA) with respect to the horizontal plane and facing south, in order to maximize the daily absorption of solar radiation.

The plug flow reactor consisted of UV-transparent glass tubes (i.d. 29.2 mm). All the tubes and the collectors were connected in series and the aqueous suspension was continuously fed from the not-reacting tank by means of a centrifugal pump. The total volume (V_t) of solution charged in the photoreacting system depicted in Figure 19.3 was about 40 dm^3, whereas the irradiated volume (V_i) was 22 dm^3. The catalyst used for the photoreactivity experiments was polycrystalline TiO_2.

19.5 Influence of experimental parameters

19.5.1 Dissolved oxygen

In TiO_2 photocatalysis the degradation of the pollutants is mainly an oxidative process able to mineralize the organic C to CO_2; O_2 is therefore necessary for complete mineralization and, moreover, it avoids the recombination of the e^-/h^+ couple. The concentration of oxygen also affects the reaction rate, which is faster when the partial pressure of oxygen (p_{O2}) in the atmosphere in contact with the solution increases. In any case, it seems that the difference between using air ($p_{O2} = 0.21$ atm) or pure oxygen ($p_{O2} = 1$ atm) is not drastic.

19.5.2 pH

The pH of the aqueous solution significantly affects the charge of the semiconductor particle and the size of the aggregates formed. The pH at which the surface of an oxide is uncharged is defined as the Zero Point Charge (pHzpc), which for TiO_2 (anatase form) is around 6.5. Above and below this value, the catalyst becomes negatively or positively charged, respectively.

The pH also influences the photocatalytic degradation of the ionisable organic compounds. When optimizing the pH it must to be taken into account that during the reaction, many intermediate products are produced that may modify the suspension pH and behave differently depending on the solution pH. Therefore, a detailed analysis of the best pH conditions should include not only the initial substrate, but also the rest of the compounds produced during the process. Moreover, pH also influences the concentration of the active •OH species; their formation is favoured at higher pH values, being relevant for pH > 9 [5].

19.5.3 Catalyst concentration

Whether in static, slurry or dynamic flow photoreactors, the initial reaction rates were found to be directly proportional to the catalyst mass. This indicates a truly heterogeneous catalytic regime. However, above a certain value, the reaction rate levels off and becomes independent of catalyst mass. When catalyst concentration is very high, after travelling a certain distance on an optical path, turbidity impedes further penetration of light in the reactor; moreover, under these conditions the light scattering is very intense. An optimum catalyst mass has thus to be found in order to avoid excess catalyst and ensure total absorption of efficient photons: typically not more than around $2\,g\ l^{-1}$. Moreover, consider that the minimum catalyst/substrate ratio reported in most studies is around 10, thus a preliminary dilution of the wastes to be treated could be necessary when the pollutant concentration is higher than around $200\,mg\ l^{-1}$, otherwise the degradation could be slow.

19.5.4 Degradation kinetics

It was found in most photocatalytic experiments that the primary degradation follows a pseudo-first order kinetic law, according to the equation:

$$-dC_{sub}/dt = k_{obs}C_{sub}$$

where C_{sub} is the substrate concentration and k_{obs} is the observed rate constant.

The decrease of k_{obs} by increasing the substrate concentration can be explained by assuming competition between intermediates and substrate for the semiconductor active sites. The k_{obs} values show an inverse dependence on the initial substrate concentration and can fit in the Langmuir-Hinshelwood equation:

$$k_{obs}^{-1} = (k_c K_{ads})^{-1} + k_c^{-1} C_0$$

where C_0 is the initial substrate concentration, K_{ads} is the adsorption equilibrium constant of the substrate and k_c is the product of the second order rate constant (k') by θ_{rad} (θ_{rad} represents the coverage of the sites by the reactive radicals) [15].

Previous detailed studies have demonstrated the influence of adsorption of substrates on the surface of TiO_2 particles, with a corresponding facilitated attack of such molecules from the active radical species present.

19.6 Additives reducing the e^-/h^+ recombination

In order to increase the degradation rate it is possible to add reagents capable to trap the e^-_{CB}, delaying the electron-hole recombination and to favour the formation of active oxidizing species. Among them, peroxydisulfates can offer beneficial effects since they provide both the electron scavenging and the introduction of a new active radical ($SO_4^{\bullet-}$) capable to originate •OH and to directly oxidize the organic substrates [16]. The following reactions describe the formation of the radical active species:

$$S_2O_8^{=} + e^-_{CB} \rightarrow SO_4^{=} + SO_4^{\bullet-}$$

$$SO_4^{\bullet-} + H_2O \rightarrow SO_4^{=} + {}^\bullet OH + H^+$$

The addition of $K_2S_2O_8$ has been, for example, successfully applied to increase the degradation rate and the mineralization extent of pollutants having very different molecular structures, such as organophosphorous compounds [17], chlorinated aromatics [18], anthraquinonic dyes [19] and others. Hydrogen peroxide can also be applied to accelerate the photocatalytic degradation of several compounds [20–25], however the results obtained using this cheap and common reagent are in general less satisfactory if compared with those obtained using peroxydisulfates [26,27].

19.7 Analytical control of the photocatalytic treatment

The degradation of the target compounds has to be carefully monitored in order to adjust the mass balance for the photocatalytic decomposition of the contaminants and to ensure that the organic compounds have not disappeared in some other way (evaporation, adsorption in the reactor, adsorption in the catalyst, etc.) besides photocatalysis. This monitoring is usually performed by liquid chromatography (HPLC) with UV detection since direct injection of the aqueous sample into the analytical column is allowed, avoiding sample pre-treatment except for the filtration necessary to remove TiO_2. Gas chromatography can be proposed as an alternative when pre-concentration is needed.

It is also very important to assess the extent of mineralization of the organic C, in order to establish the moment at which wastewater can be considered completely decontaminated. The basic techniques for the determination of the residual TOC in water are mainly based on the conversion of organic compounds to CO_2 and then measured using non-dispersive infrared absorption. Since many aqueous samples contain HCO_3^- and/or CO_3^{2-}, it is usually necessary to remove these species using a gas stripping technique before measuring the TOC.

In view of a safe application of the technique to treat the wastes, the identification and monitoring of transient intermediates is fundamental; they could be in principle, even more toxic and persistent than the parent compounds. Gas or liquid chromatography coupled to mass spectrometry (GC/MS and LC/MS) are suitable techniques for achieving very useful structural information.

In case of degradation products present at sub-mg l^{-1} level, the sample can be enriched by liquid-liquid extraction using an appropriate solvent; however, solid-phase extraction (SPE) is gaining in acceptance mainly because SPE generates less matrix interference and a wide range of new adsorbents are commercially available.

As additional information, in order to guarantee that the treatment is correct, toxicity tests can be performed, particularly if incomplete mineralization is observed at the time corresponding to the total disappearance of the substrate.

19.8 Examples of possible applications of photocatalysis to the treatment of laboratory wastes

The photocatalytic degradation and mineralization of a great number of organic pollutants was reported since the beginning of the environmental application studies, starting around 1980. Among them, the following classes were investigated in detail: aliphatic and aromatic hydrocarbons, haloaliphatic compounds (in particular chlorinated solvents), haloaromatic derivatives (halophenols, chlorobenzenes, PCBs, dioxins), aliphatic and aromatic amines, phenols, non chlorinated solvents, surfactants, pesticides, dyes, and so on [1–5,28,29].

Selected examples of photocatalytic treatments of aqueous laboratory wastes containing some of the mentioned pollutants are examined in the following sections.

19.8.1 Percolates containing soluble aromatic contaminants

The analysis of aqueous percolates in industrial areas is a usual task in order to verify the possible risk of contamination of subsurface water sources. Aromatic sulfonates, widely used as intermediates in dyes production, are a group of hydrophilic pollutants of great environmental concern, found in the percolates basin within certain abandoned industrial areas. Samples of these aqueous percolates were analysed and the possible photocatalytic treatment of such wastes was reported [30]. The obtained results suggest that the residual samples could be also readily treated in the laboratory before discharge.

A typical example of aromatic sulfonate found is the 2,6-naphthalenedisulfonate (2,6-NdS). Figures 19.4 and 19.5 show the evolution of the photocatalytic treatment of 2,6-NdS in closed and aerated cells respectively. The trend of the TOC decrease is faster in the second case, thus confirming the crucial role played by dissolved oxygen and by surface-adsorbed oxygen in these treatments. A favourable effect is also observed on the desulfonation, with the corresponding increase of the sulfate formation rate.

It is noteworthy that the complete abatement of 2,6-NdS occurs after 50 min irradiation, but the stoichiometric concentration of sulfate and the complete mineralization were achieved only after roughly 3 h, thus supporting the hypothesis of the presence of reaction intermediates still containing sulfonate groups, for which the degradation proceeds at a slower rate.

The substrate degradation rate increases by increasing the content of dissolved air within the irradiated cell (Figure 19.5). Under these conditions the complete mineralization of the starting substrate, with the corresponding stoichiometric formation of sulfate, can be achieved after around 2 h.

Complex percolate samples containing several different aromatic sulfonates (ca. 80 mg l^{-1} of TOC) can be readily treated in a closed laboratory reactor (capacity 500 ml, Hg-lamp 125 W). The observed TOC decrease is roughly 84% after 8 h irradiation. If bubbled air flows continuously in the reactor, the TOC decrease becomes roughly 97% after less than 5 h.

19.8.2 Photocatalytic destruction of aromatic amine residues in aqueous wastes

Aromatic amines have been recognized as high priority pollutants. The presence of such compounds, even at very low concentration levels, is often investigated in a variety of environmental and waste samples. Moreover these compounds can be also introduced as chemical reagents in analytical methods, usually as chelating ligands or coupling reagents in some spectrophotometric determinations. [31].

Photocatalysis is effective for the abatement of aniline and various aniline derivatives in aqueous wastes saturated with air [32]. The following equation describes the stoichiometric mineralization of aniline:

$$C_6H_5{-}NH_2 + 9O_2 \rightarrow 6\,CO_2 + 3\,H_2O + H^+ + NO_3^-$$

The influence of pH on the primary degradation kinetics is relevant; at low pH values (<2) the half-life of aniline is about 10 times longer than at a pH around 6. This significant effect can be attributed to the repulsion operating between the anilinium cation and the positively charged semiconductor particles On the other hand the observed large increase of degradation rate at higher pH values (ca. 10) is also due to the increase of the HO• concentration in basic solutions.

Different degradation steps can be clearly recognized: working at pH around 6 in Solarbox, the degradation of aniline (ca. 20 mg l^{-1}) in the presence of suspended TiO_2 (200 mg l^{-1}) is observed after around 40 min irradiation, the complete mineralization of its organic carbon occurs in less than 1 h, whereas the two main transient intermediates, 4-hydroxyaniline and 1,3,4-trihydroxybenzene, reach their maximum concentrations after 20 and 30 min irradiation, respectively, and are in turn completely degraded after roughly 55 min. This information is crucial to define the irradiation time since, under the reported working conditions at least 1 h irradiation becomes necessary in order to obtain a safe waste to be discharged.

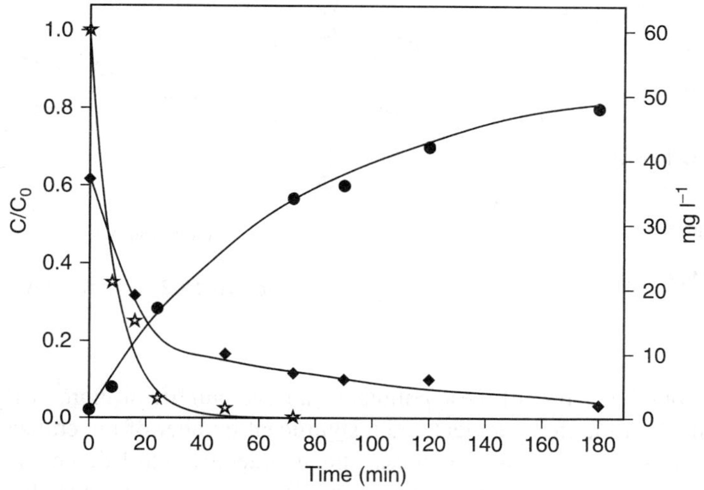

Figure 19.4 *Degradation of 2,6-NdS in closed cells. Primary process (✯), (♦) TOC (mgl^{-1}) and evolution of (•) sulfates (mgl^{-1}). Reprinted from D. Fabbri, A. Bianco Prevot and E. Pramauro, Photocatalytic degradation of aromatic sulfonates present in industrial percolates, J. Appl. Electrochem. 35, 7–8, 815–820 with kind permission from Springer Science + Business Media © 2005.*

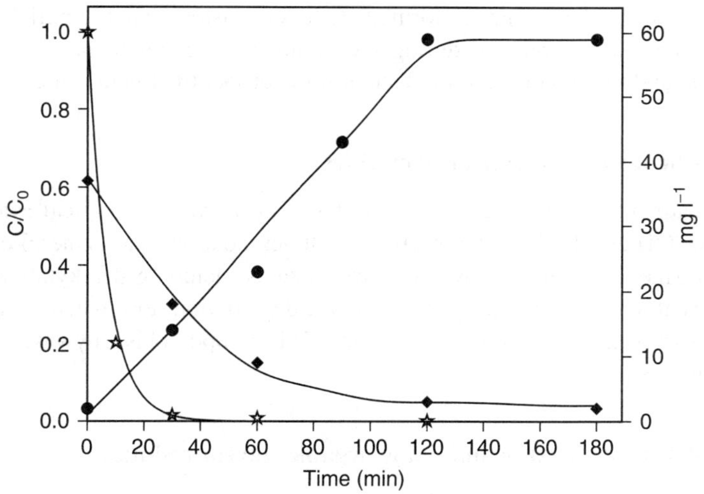

Figure 19.5 *Degradation of 2,6-NdS in aerated cells. Primary process (✯), (♦) TOC (mgl^{-1}) and evolution of (•) sulfates (mgl^{-1}). Reprinted from D. Fabbri, A. Bianco Prevot and E. Pramauro, Photocatalytic degradation of aromatic sulfonates present in industrial percolates, J. Appl. Electrochem. 35, 7–8, 815–820 with kind permission from Springer Science + Business Media © 2005.*

19.8.3 Degradation of aqueous wastes containing pesticides residue

The analysis of pesticide residues in surface and underground waters is often required by the regulations and, since the photocatalytic degradation of a great variety of pesticides present in aqueous samples was reported in past years, the photocatalytic treatment of pesticide-containing laboratory wastes can be proposed.

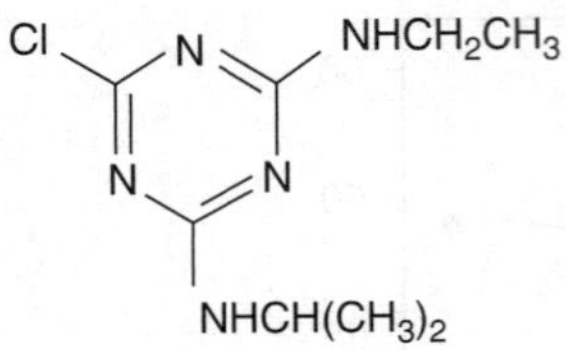

Atrazine

Structure 19.1 *Atrazine.*

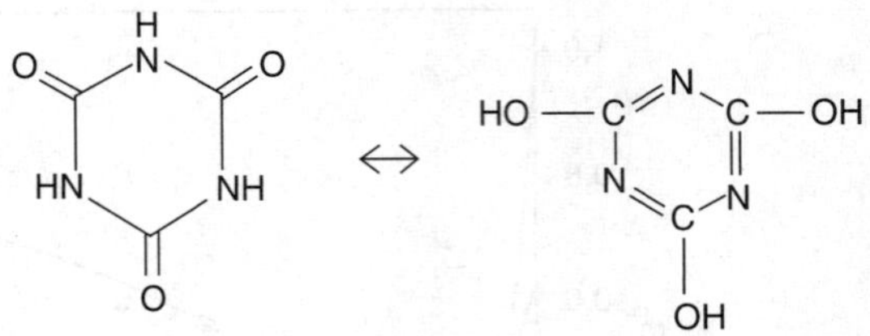

Cyanuric acid

Structure 19.2 *cynauric Acid.*

Due to the usually complex nature of these pollutants, a large number of more or less persistent reaction intermediates are formed during the treatment and a significant number of mineralization end-products are obtained. Thus, a careful analytical monitoring is essential in order to control all the transformation steps. The formation and fate of the more abundant reaction intermediates is usually followed by HPLC, leading to the observation of typical bell-shaped profiles in all the cases. An interesting point is that some intermediates can possess higher stability than the starting pollutant, thus exhibiting a higher persistency in the irradiated waste, whereas other derivatives are formed and degraded simultaneously with about the same degradation rate of the substrate.

Figure 19.6 illustrates the behaviour of the three major intermediates found during the treatment of wastewaters containing chloramben residues, and Scheme 19.1 depicts the early steps of its photocatalytic transformation. All the intermediates that are formed are less persistent than the initial substrate.

Table 19.1 reports the results obtained during the photocatalytic treatment of various pesticides. The persistence of substrate and intermediates should suggest the proper treatment time.

19.8.4 The peculiar behaviour of triazine herbicides

The photocatalytic treatment of pesticides containing the triazine ring is a typical example of the possible limits exhibited by the ·OH attack [42]. The mechanism of degradation of atrazine (6-chloro-N-ethyl-N'-(1-methylethyl)-1,3,5-triazine-2,4-diamine) involves various steps including dealkylation, dechlorination and oxidation (of aliphatic carbon and of amino groups), but the degradation stops when the very stable compound 2,4,6-trihydroxy-1,3,5-triazine (cyanuric acid) is formed. This compound is, anyway, less toxic than atrazine (see Structures 19.1 and 19.2).

19.8.5 Treatment of aqueous wastes containing organic solvent residues

Photocatalysis can also be successfully applied to treat both gaseous and aqueous streams containing organic solvents. If the treatment of laboratory wastewaters is considered, the experimental approach is essentially that previously described. Table 19.2 summarizes some of the reported applications in view of the abatement of the organic solvents residues prior to the discharge of the waste.

19.8.6 Treatment of surfactant-containing aqueous wastes

The use of aqueous surfactant solutions as alternative solvents capable of replacing the widely used but more dangerous organic liquids is of great concern in the frame of the safer green chemistry practice. In particular, the environmental applications of such solvents has become increasingly popular and various applications

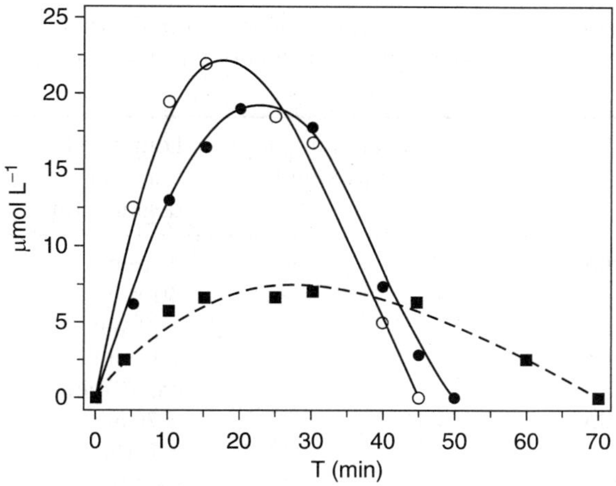

Figure 19.6 *Evolution of transient intermediates formed by photocatalytic degradation of 30 mg l^{-1} chloramben at pH 3.7. TiO2: 300 mg l^{-1} (■): 2,5-dichlorohydroquinone, (•): 3-hydroxybenzoic acid, (°): 1,2,4-trihydroxybenzene. Reprinted from [41] with permission from Elsevier © 1999.*

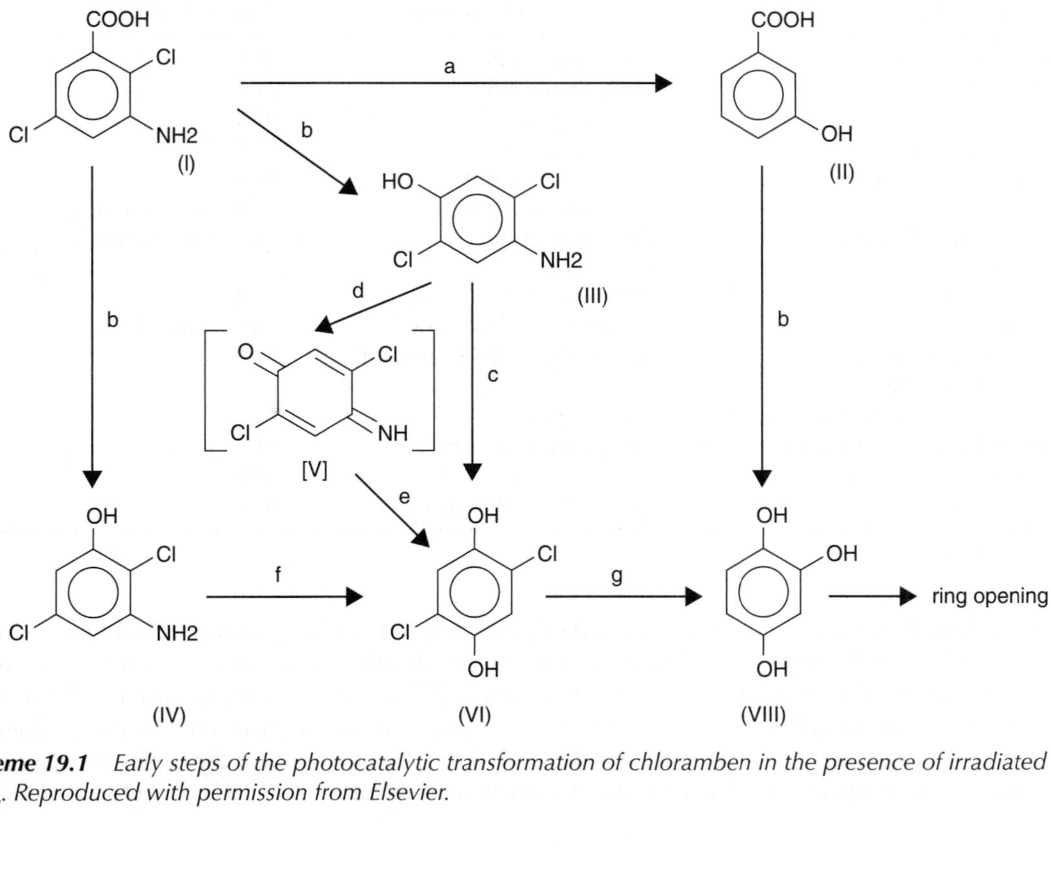

Scheme 19.1 *Early steps of the photocatalytic transformation of chloramben in the presence of irradiated TiO_2. Reproduced with permission from Elsevier.*

Table 19.1 *Persistency of some pesticides and their reaction products during the TiO_2-based photocatalytic treatment of aqueous wastes. Time necessary for completion of the degradation steps under the same working conditions.*

Substrate	Substrate degradation	Number of major transients found	Long-lived transient degradation	Mineralization	Ref.
Monuron	35 min	5	40 min	ca. 40 min	33
Carbaryl	30 min	6	40 min	ca. 50 min	34
2,3,6-TBA	60 min	5	<60 min	ca. 90 min	35
Swep	70 min	7	100 min	ca. 180 min	36
Triclosan	60 min	6	ca. 70 min	> 10 h	37
DNOC	70 min	> 10	< 70 min	> 180 min	38
Methamidophos	5 h	7	< 5 h	> 5 h	39
MCP	30 min	3	ca. 90 min	ca. 120 min	40
Chloramben	90 min	8	ca. 70 min	ca. 120 min	41

Monuron: 3-(4-chlorophenyl)-1,1-dimethylurea; Carbaryl: 1-naphthyl-N-methylcarbamate; 2,3,6- TBA: 2,3,6-trichlorobenzoic acid; Swep: methyl-N-(3,4-dichlorophenyl)carbamate; Triclosan: 5-chloro-2-(2,4-dichlorophenoxy)phenol; DNOC: 4,6-dinitro-o-cresol; Methamidophos: O,S-dimethylphosphoramidothioate; MCP: 4-chloro-2methylphenoxyacetic acid

Table 19.2 *Organic solvents treated by means of TiO_2-photocatalysis.*

Organic solvents	Photocatalytic material	Type of treatment	Ref.
CH_2Cl_2, $CHCl_3$, CCl_4	TiO_2 dispersion	BPR	43
$CHCl_3$	TiO_2 dispersion	BPR	44
$CHCl_3$	TiO_2 dispersion	BPR	45
Chlorinated ethanes	TiO_2 dispersion	BPR	46
Chlorinated methanes	TiO_2 dispersion	BPR	47
Acetonitrile	TiO_2 dispersion	solar flow photoreactor	48
Dichloroethane, dichloro-methane, trichloromethane	TiO_2 dispersion	solar flow photoreactor	49
Acetonitrile, toluene	TiO_2 dispersion	BPR	50
1,4-dioxane	immobilized TiO_2 beds	recirculating flow	51
Benzene, chlorobenzene, 1,4-dichlorobenzene	immobilized ZnO nanoparticles	BPR	52
Tetrahydrofuran, 1,4-dioxane	TiO_2 dispersion	BPR	53
Toluene and asphaltene in aqueous SDS	TiO_2 dispersion + H_2O_2	BPR	54
Crude oil in SDS emulsion	TiO_2 dispersion + H_2O_2	BPR	54
Crude oil adsorbed on Na-bentonite	TiO_2 + Na-bentonite (1:1)	BPR	55

BPR: Batch Photoreactor

can be mentioned: recovery of phenols from carbolic oils [55], washing of polychlorinated biphenyl from sandy materials using biodegradable alkyl poly(oxyethylene) glycols [56], removal of polycyclic aromatic hydrocarbons from soils using surfactants and biotreatment [57] and many other applications for soil and water remediation purposes [58]. Surfactant-containing wastes can be thus originated from these applications.

The use of surfactants is increasing also in the analytical laboratory, where they offer in many cases some unique performances, allowing the selective extraction of various organic components from complex solid matrices, such as soils, sediments, sludges, plants, foods, and so on, and enhancing the sensitivity of various analytical methods. It is well known that the extraction of hydrophobic species from many complex samples using organic solvents suffer of some drawbacks, being usually time-consuming and requiring the handling of relatively large amounts of more or less toxic, hazardous and expensive solvents. Microwave-assisted solvent extraction (MAE) [59], accelerated solvent extraction (ASE) [60] and supercritical fluid extraction (SFE) [61] have been developed in recent years and proposed to overcome most of the mentioned problems.

The combination of the MAE approach with the use of aqueous surfactant solutions as extracting phase is, in particular, a promising new extraction approach. The term microwave-assisted micellar extraction (MAME) was suggested some years ago to indicate this peculiar procedure [62], capable of offering recovery efficiencies comparable to those obtained using classical extraction methods at lower costs. For example, the cost of a MAME determination of PAHs in sediments has been estimated to be about 80–90 times lower than the Soxhlet extraction using CH_2Cl_2.

Other examples of surfactant-based procedures, successfully applied to replace the organic solvents, are the preconcentration/extraction steps exploiting the so-called cloud-point phenomenon, based on the thermal phase separation occurring in aqueous solutions of non ionic surfactants, and the formation of ionic surfactant coacervates under high ionic strength conditions [63].

In all these cases, the laboratory aqueous wastes contain relevant concentrations of surfactants which can originate some problems during the photocatalytic treatment since the surfactant itself is degraded. In particular it was found in most cases a neat reduction of the degradation rate of the pollutants, if compared with the one observed in pure water. This surfactant inhibition effect, which can be attributed to the competition of surfactant molecules for the active sites of the semiconductor particles, must be carefully considered when the treatment of surfactant-containing wastes is necessary.

19.8.7 Degradation of aqueous solutions of azo-dyes

Azo-dyes represent approximately 50–70% of the dyes available on the market today, followed by the anthraquinone group. The photocatalytic decolourization of different azo-dyes has been reported in the literature and recently reviewed [64]. Most of these papers examine in detail the primary process under different working conditions [65,66,67]. On the contrary, the reaction mechanisms involved in the photocatalytic degradation of dyes and the identification of major transient intermediates have been only more recently recognized as very important aspects of these processes [68–70].

19.8.8 Treatment of laboratory waste containing pharmaceuticals

The heterogeneous photocatalysis could be an efficient tool also in case of the treatment of laboratory wastes containing pharmaceuticals [71].

Work on this topic often offers scarce knowledge on photocatalytic intermediates of pharmaceuticals; Lambropoulou *et al.* identified 17 products of photocatalytic degradation of bezafibrate [72], Zhang *et al.* proposed a pathway for photocatalytic degradation of acetaminophen via direct hole oxidation and *ipso*-substitution [73]. As far as atenolol is concerned, Liu *et al.* [74] observed its slow direct photolysis, whereas for propranolol three photodegradation products of aromatic ring oxidation and opening were identified. In some experiments Medana *et al.* [75] identified several transformation products of atenolol, mainly mono-, di- and tri-hydroxylated derivatives.

A further study [76] describes the oxidation of atenolol at pilot-scale using Compound Parabolic Collector (CPC) under natural illumination. Two widely used AOP systems, TiO_2 and photo-Fenton solar photocatalysis, were investigated. Special attention was paid to identify the main intermediate products of atenolol formed during the treatment, by means of quadrupole time-of-flight tandem mass spectrometry coupled to ultra-high performance liquid chromatography system.

19.9 Continuous monitoring of photocatalytic treatment

In a work reported several years ago [77] the continuous flow degradation of carbaryl under UV light irradiation using a pumped TiO_2 slurry was examined. Other different manifolds were successively developed and proposed for the photo-assisted catalytic detoxification of propoxur [78], resorcinol [79] and formetanate [80] and of the reagents employed for their colourimetric determinations.

The flow injection approach can be also exploited for the monitoring of photocatalytic treatments of laboratory wastes, performed in batch photoreactors. For example, the degradation of aqueous wastes containing low concentrations of the pesticide dicamba has been followed continuously using a suitable flow-injection manifold [81] allowing a frequent sampling of low volumes of dispersion (ca. 300 µl) and the on-line filtration of TiO_2 before flowing through the UV detector.

The pesticide degradation was followed at two wavelengths (220 nm and 274 nm), corresponding to the maximum of the dicamba UV-Vis spectrum. The measurements made at the shorter wavelength evidenced that there is a neat removal (after ca. 75 min) of the herbicide and its by-products.

The described on-line procedure can be proposed in many cases as a suitable and cheap tool for monitoring the photocatalytic degradation process and for a rapid evaluation of the main experimental parameters affecting the treatment.

References

[1] Serpone, N. and Pelizzetti, E. (eds) (1989) *Photocatalysis, Fundamentals and Applications*, John Wiley & Sons, Inc., New York.

[2] Ollis, D.F. and Al-Ekabi, H. (eds) (1993) *Photocatalytic Purification of Water and Air*, Elsevier, Amsterdam.

[3] Hoffman, M.R.; Martin, S.T.; Choi, W. and Bahnemann, D.W. (1995) Environmental Applications of Semiconductor Photocatalysis, *Chem. Rev.*, **95**, 69–96.

[4] Malato, S.; Blanco, J.; Vidal, A. and Richter, H.C. (2002) Photocatalysis with solar energy at a pilot-plant scale: an overview, *Appl. Catal. B: Environ.*, **37**, 1–15.

[5] Bahnemann, D.; Cunningham, J.; Fox, M.A.; Pelizzetti, E.; Pichat, P. and Serpone, N. (1994) Photocatalytic Treatment of Water, in *Aquatic and Surface Photochemistry*. (eds G.R. Helz, R.G. Zepp and D.G. Crosby) Ch.21, Lewis, Boca Raton, FL, p.261–316.

[6] Turchi, C.S. and Ollis, D.F. (1989) Mixed reactant photocatalysis: intermediates and mutual rate inhibition, *J. Catalysis*, **119**, 483–496.

[7] Childs, L.P. and Ollis, D.F. (1980) Is photocatalysis catalytic?, *J. Catalysis*, **66**, 383–390.

[8] Vinodgopal, K.; Stafford, U.; Gray, K.A. and Kamat, P.V. (1994) Electrochemically Assisted Photocatalysis.2. The Role of Oxygen and Reaction Intermediates in the Degradation of 4-Chlorophenol on Immobilized TiO_2 Particulate Films, *J. Phys. Chem.*. **98**, 6797–6803.

[9] Ding, Z.; Hu, X.J.; Lu, G.Q.; Yue, P.L. and Greenfield, P.F. (2000) Novel Silica Gel Supported TiO_2 Photocatalyst Synthesized by CVD Method, *Langmuir*, **16,** 6216–622.

[10] Phonthammachai, N.; Gulari, E.; Jamieson, A.M. and Wongkasemjit, S. (2006) Photocatalytic membrane of a novel high surface area TiO_2 synthesized from titanium triisopropanolamine precursor, *Appl. Organometal. Chem.*, **20**, 499–504.

[11] Peng, Y. and Richardson, J.T. (2004) Properties of ceramic foam catalyst supports: one–dimensional and two-dimensional heat transfer correlations, *Appl. Catal. A*, **266**, 235–244.

[12] Atienzar, P.; Corma, A.; Garcia, H. and Scaiano, J.C. (2004) Diffuse reflectance Laser Flash Photolysis Study of Titanium-Containing Zeolites, *Chem. Mater.*, **16**, 982–987.

[13] Lam, F.L.Y. and Hu, X.J. (2003) A new system design for the preparation of copper/activated carbon catalyst by metal-organic chemical vapor deposition method, *Chem. Eng. Sci.*, **58**, 687–695.

[14] Plataforma Solar de Almeria - Centro de Investigaciones Energéticas Medioambientales y Tecnológicas (PSA-CIEMAT) (n.d.) *Solar Chemical Facilities*. Available at: http://www.psa.es (accessed 10 November, 2011).

[15] Augugliaro, V.; Palmisano, L.; Schiavello, M.; Sclafani, A.; Marchese, L.; Martra, G. and Miano, F. (1991) Photocatalytic degradation of nitrophenols in aqueous titanium dioxide dispersion, *Appl. Catal.*, **69**, 323–340.

[16] Neta, P.; Madhavan, V.; Zemel, H. and Fessenden, R.W. (1977) Rate constants and mechanism of reaction of sulfate radical anion with aromatic compounds, *J. Am. Chem. Soc.*, **99**, 163–164.

[17] Gratzel, C.K.; Jirousek, M. and Gratzel, M. (1990) Decomposition of organophosphorous compounds on photoactivated TiO_2 surfaces, *J. Mol. Catal.*, **60**, 375–387.

[18] Pelizzetti, E.; Carlin, V.; Minero, C. and Gratzel, M. (1991) Enhancement of the rate of photocatalytic degradation on TiO_2 of 2-chlorophenol, 2,7-dichlorodibenzodioxin and atrazine by inorganic oxidizing species, *New. J. Chem.*, **15**, 351–359.

[19] Bianco Prevot, A.; Baiocchi, C.; Brussino, M.C.; Pramauro, E.; Savarino, P.; Augugliaro, V.; Marcì, G.and Palmisano, L. (2001) Photocatalytic Degradation of Acid Blue 80 in Aqueous Solutions Containing TiO_2 Suspensions, *Environ. Sci. Technol.*, **35**, 971–976.

[20] Brown, G.T. and Darwent, J.R. (1984) Methyl orange as a probe for photooxidation reactions of colloidal titanium dioxide, *J. Phys. Chem.*, **88**, 4955–4959.

[21] Hisanaga, T.; Harada, K. and Tanaka, K. (1990) Photocatalytic degradation of organochlorine compounds in suspended TiO_2, *J. Photochem. Photobiol. A: Chem.*, **54**, 113–118.

[22] Jenny, B. and Pichat, P. (1991) Determination of the actual photocatalytic rate of hydrogen peroxide decomposition over suspended titania. Fitting to the Langmuir-Hinshelwood form, *Langmuir*, **7**, 947–954.

[23] O'Shea, K.E.; Garcia, I. and Aguilar, M. (1997) TiO_2 Photocatalytic Degradation of Dimethyl- and Diethyl-Methylphosphonate. Effects of Catalyst and Environmental Factors, *Res.Chem.Intermed.*, **23**, 325–330.

[24] Poulios, I.; Kositzi, M. and Kouras, A. (1998) Photocatalytic decomposition of triclopyr over aqueous semiconductor dispersions, *J. Photochem. Photobiol. A: Chem.*, **115**, 175–183.

[25] Malato, S.; Blanco, J.; Richter, C.; Milow, B. and Maldonado, M.I. (1999) Solar photocatalytic mineralization of commercial pesticides: Methamidophos., *Chemosphere*, **38**, 1145–1156.

[26] Malato, S.; Blanco, J.; Richter, C.; Braun, B. and Maldonado, M.I. (1998) Enhancement of the rate of solar photocatalytic mineralization of organic pollutants by inorganic oxidizing species, *Appl. Catal. B: Environ.*, **17**, 347–356.

[27] Malato, S.; Blanco, J.; Maldonado, M.I.; Fernandez-Ibañez, P. and Campos, A. (2000) Optimising Solar Photocatalytic Mineralization of Pesticides by Adding Inorganic Oxidizing Species: Application to the Recycling of Pesticide Containers, *Appl. Catal. B: Environ.*, **28**, 163–174.

[28] Hashimoto, K.; Irie, H. and Fujishima, A. (2005) TiO_2 Photocatalysis: A Historical Overview and Future Prospects, *Jap. J. Appl Phys.*, **44**, 8269–8285.

[29] Herrmann, J.M.; Duchamp, C.; Karkmaz, M.; Hoai, B.T.; Lachheb, H.; Puzenat, E.and Guillard, C. (2007) Environmental Green Chemistry as defined by photocatalysis, *J. Hazardous Mat.*, **146**, 624–629.

[30] Fabbri, D.; Bianco Prevot, A.; Zelano, V.; Ginepro, M. and Pramauro, E. (2008) Removal and degradation of aromatic compounds from a highly polluted site by coupling soil washing with photocatalysis, *Chemosphere*, **71**, 59–65.

[31] Clesceri, L.S.; Greenberg, A.E. and Rhodes Trussel R. (eds) (1989) *Standard Methods for the Examination of Water and Wastewater*. 17th edn. APHA-AWVA-WPCF. Washington DC.

[32] Pramauro, E.; Bianco Prevot, A.; Augugliaro, V. and Palmisano, L. (1995) Photocatalytic treatment of laboratory wastes containing aromatic amines, *Analyst*, **120**, 237–242.

[33] Pramauro, E.; Bianco Prevot, A.; Augugliaro, V. and Palmisano, L. (1993) Photocatalytic degradation of monuron in aqueous titanium dioxide dispersions,. *Environ. ci. Technol.*, **27**, 1790–1795.

[34] Pramauro, E.; Bianco Prevot, A.; Vincenti, M. and Brizzolesi, G. (1997) Photocatalytic degradation of carbaryl in aqueous solutions containing TiO_2, *Environ. Sci. Technol.*, **31**, 3126–3131.

[35] Bianco Prevot, A. and Pramauro, E. (1999) Analytical monitoring of photocatalytic treatments. Degradation of 2,3,6-trichlorobenzoic acid in aqueous TiO_2 dispersions, *Talanta*, **48**, 847–857.

[36] Fabbri, D.; Crime, A.; Davezza, M.; Medana, C.; Baiocchi, C.; Bianco Prevot, A. and Pramauro, E. (2009) Surfactant-assisted removal of swep residues from soil and photocatalytic treatment of the washing wastes, *Appl. Catal. B: Environ.*, **92**, 318–325.

[37] Rafqah, S.; Wong-Wah-Chung, P.; Nelieu, S.; Einhorn, J.and Sarakha, M. (2006) Phototransformation of triclosan in the presence of TiO_2 in aqueous suspension: Mechanistic approach, *Applied Catalysis B: Environ.*, **66**, 119–125.

[38] Fabbri, D.; Villata, L.S.; Bianco Prevot, A.; Capparelli, A.L. and Pramauro, E. (2006) Photocatalytic degradation of DNOC in aqueous TiO_2 dispersions. Investigation of the initial reaction steps. *J. Photochem. Photobiol. A: Chem.*, **180**, 157–164.

[39] Dai, K.; Peng, T.; Chen, H.; Zhang, R.and Zhang, Y. (2008) Photocatalytic degradation and mineralization of commercial methamidophos in aqueous titania suspension, *Environ. Sci. Technol.*, **42**, 1505–1510.

[40] Tanaka, K.and Reddy, K.S.N. (2002) Photodegradation of phenoxyacetic acid and carbamate pesticides on TiO_2, *Appl. Catal B: Environ.*, **39**, 305–310.

[41] Bianco Prevot, A.; Vincenti, M.; Bianciotto, A. and Pramauro, E. (1999) Photocatalytic and photolytic transformation of chloramben in aqueous solutions, *Appl. Catal. B: Environ*, **22**, 149–158.

[42] Pelizzetti, E.; Maurino, V.; Minero, C.; Carlin, V.; Pramauro, E.; Zerbinati, O.and Tosato, M.L. (1990) Photocatalytic Degradation of Atrazine and Other s-Triazine Herbicides, *Environ. Sci. Technol.*, **24**, 1559–1565.

[43] Hsiao, C.H.; Lee, C.L.and Ollis, D.F. (1983) Heterogeneous photocatalysis: Degradation of dilute solutions of dichloromethane (CH_2Cl_2), chloroform ($CHCl_3$) and carbon tetrachloride (CCl_4) with illuminated TiO_2, *J. Catal.*, **82**, 418–423.

[44] Pruden A.L. and Ollis, D.F. (1983) Degradation of chloroform by photoassisted heterogeneous catalysis in dilute aqueous suspensions of titanium dioxide, *Environ. Sci. Technol.*, **17**, 628–631.

[45] Kormann, C.; Bahnemann, D.W.and Hoffmann, M.R. (1991) Photolysis of chloroform and other organic molecules in aqueous titanium dioxide suspensions, *Environ. Sci. Technol.*, **25**, 494–500.

[46] Mao, Y.; Schoeneich, C.and Asmus, K.D. (1991) Identification of organic acids and other intermediates in oxidative degradation of chlorinated ethanes on titania surfaces in route to mineralization: a combined photocatalytic and radiation chemistry study, *J. Phys. Chem.*, **95**, 10080–10089.

[47] Calza, P.; Minero, C.and Pelizzetti, E. (1997) Photocatalytic transformations of chlorinated methanes in the presence of electron and hole scavengers, *J. Chem. Soc. Faraday Trans.*, **93**, 3765–3771.

[48] Augugliaro, V.; Bianco Prevot, A.; Caceres Vazquez, J.; Garcia-Lopez, E.; Irico, A.; Loddo, V.; Malato Rodriguez, S.; Marcì, G.; Palmisano, L. and Pramauro, E. (2004) Photocatalytic oxidation of acetonitrile in aqueous suspension of titanium dioxide irradiated by sunlight, *Adv. Environ. Res.*, **8**, 329–335.

[49] Malato Rodriguez, S.; Blanco Galvez, J.; Maldonado Rubio, M.I.; Fernandez Ibañez, P.; Gernjak, W. and Oller Alberola, I. (2005) Treatment of chlorinated solvents by TiO_2 photocatalysis and photo-Fenton: influence of operating conditions in a solar pilot plant, *Chemosphere*, **58**, 391–398.

[50] Addamo, M.; Augugliaro, V.; Coluccia, S.; Di Paola, A.; Garcia-Lopez, E.; Loddo, V.; Marcì, G.; Martra, G. and Palmisano, L. (2006) The role of water in the photocatalytic degradation of acetonitrile and toluene in gas-solid and liquid-solid regimes, *Int. J. Photoenergy*, 1–15.

[51] Yamazaki, S.; Yamabe, N.; Nagano, S. and Fukuda, A. (2007) Adsorption and photocatalytic degradation of 1,4-dioxane on TiO_2, *J. Photochem. Photobiol. A: Chemistry*, **185**, 150–155.

[52] Hariharan, C. (2006) Photocatalytic degradation of organic contaminants in water by ZnO nanoparticles: Revisited, *Appl. Catal. A: General*, **304**, 55–61.

[53] Mehrvar, M.; Anderson, W.A.and Moo-Young, M. (2000) Photocatalytic degradation of aqueous organic solvents in the presence of hydroxyl radical scavengers, *Int. J. Photoenergy*, **2**, 67–80.

[54] Pernyeszi, T. and Dekany, I. (2004) Photocatalytic degradation of hydrocarbons by bentonite and TiO_2 in aqueous suspensions containing surfactants, *Coll. Surfaces A: Physicochem. Eng. Aspects*, **230**, 191–199.

[55] Plucinski, P. (1985) Micellar extraction of phenols from carbolic oil, *Int. J. Heat Mass Transfer*, **28**, 451–458.

[56] Abdul, A.S. and Gibson, T.L. (1991) Laboratory studies of surfactant-enhanced washing of polychlorinated biphenyl from sandy material, *Environ. Sci. Technol.*, **25**, 665–671.

[57] Zheng, Z. and Obbard, J.P. (2000) Removal of polycyclic aromatic hydrocarbons from soil using surfactants and the white rot fungus *Phanerochaete chrysosporium*, *J. Chem. Technol. Biotechnol.*, **75**, 1183–1189.

[58] Paria, S. (2008) Surfactant-enhanced remediation of organic contaminated soil and water, *Adv. Coll. Interface Sci.*, **138**, 24–58.

[59] Camel, V. (2000) Microwave-assisted solvent extraction of environmental samples, *Trends Anal. Chem.*, **19**, 229–248.

[60] US Environmental Protection Agency (USEPA) (1995) *Test Methods for Evaluating Solid Waste*. Method 354. SW-846. 3rd edn. Update III. Washington DC: US GPO.

[61] Chester, T.L.; Pinkston, J.D.and Raynie, D. (1994) Supercritical Fluid Chromatography and Extraction, *Anal. Chem.*, **66**, 106R–130R.

[62] Bianco Prevot, A.; Gulmini, M.; Zelano, V. and Pramauro, E. (2001) Microwave-Assisted Extraction of Polycyclic Aromatic Hydrocarbons from Marine Sediments Using Nonionic Surfactant Solutions, *Anal. Chem.*, **73**, 3790–3795.

[63] Quina, F.H. and Hinze, W.L. (1999) Surfactant-Mediated Cloud Point Extraction: An Environmentally Benign Alternative Separation Approach, *Ind. Eng. Chem. Res.*, **38**, 4150–4168.

[64] Konstantinou, I.K. and Albanis, T.A. (2004) TiO_2-assisted photocatalytic degradation of azo dyes in aqueous solution: kinetic and mechanistic investigations. A review, *Appl. Catal. B: Environ.*, **49**, 1–14.

[65] Kuo, W.S. and Ho, P.H. (2001) Solar photocatalytic decolorization of methylene blue in water, *Chemosphere*, **45**, 77–83.

[66] Zhang, F.; Zhao, J.; Shen, T.; Hidaka, H.; Pelizzetti, E. and Serpone, N. (1998) TiO_2-assisted photodegradation of dye pollutants II. Adsorption and degradation kinetics of eosin in TiO_2 dispersions under visible light irradiation, *Appl. Catal. B: Environ.*, **15**, 147–156.

[67] Chen, F.; Xie, Y.; Zhao, J. and Lu, G. (2001) Photocatalytic degradation of dyes on a magnetically separated photocatalyst under visible and UV irradiation, *Chemosphere*, **44**, 1159–1168.

[68] Baiocchi, C.; Brussino, M.C.; Pramauro, E.; Bianco Prevot, A.; Palmisano, L.and Marcì, G. (2002) Characterization of methyl orange and its photocatalytic degradation products by HPLC/UV-VIS diode array and atmospheric pressure ionization quadruple ion trap mass spectrometry, *Int. J. of Mass Spectrometry*, **214**, 247–256.

[69] Stylidi, M.; Kondarides, D.I. and Verykios, X.E. (2003) Pathways of Solar Light-Induced Photocatalytic Degradation of Azo Dyes in Aqueous TiO_2 Suspensions, *Appl. Catal. B: Environ.*, **40**, 271–286.

[70] Bianco Prevot, A.; Fabbri, D.; Pramauro, E.; Baiocchi, C. and Medana, C. (2008) High-performance liquid chromatography coupled to ultraviolet diode array detection and electrospray ionization mass spectrometry for the analysis of intermediates produced in the initial steps of the photocatalytic degradation of sulfonated azo dyes, *J. Chromatogr. A*, **1202**, 145–154.

[71] Dalrymple, O.K.; Yeh, D.H. and Trotz, M.A. (2007) Removing pharmaceuticals and endocrine disrupting compounds from wastewater by photocatalysis, *J. Chem. Technol. Biotechnol.*, **82**, 121–134.

[72] Lambropoulou, D.A.; Hernando, M.D.; Konstantinou, I.K.; Thurman, E.M.; Ferrer, I.; Albanis, T.A. and Fernández-Alba, A.R. (2008) Identification of photocatalytic degradation products of bezafibrate in TiO_2 aqueous suspensions by liquid and gas chromatography, *J. Chromatogr. A*, **1183**, 38–48.

[73] Zhang, X.; Wu, F.; Wei Wu, X.; Chen, P. and Deng, N. (2008) Photodegradation of acetaminophen in TiO_2 suspended solution, *J. Hazard. Mater.*, **157**, 300–307.

[74] Liu, Q.T. and Williams, H.E. (2007) Kinetics and Degradation Products for Direct Photolysis of β-Blockers in Water, *Environ. Sci. Technol.*, **41**, 803–810.

[75] Medana, C.; Calza, P.; Carbone, F.; Pelizzeti, E.; Hidaka, H. and Baiocchi, C. (2008) Characterization of atenolol transformation products on light-activated TiO_2 surface by high-performance liquid chromatography/high-resolution mass spectrometry, *Rapid Commun. Mass Spectrom.*, **22**, 301–313.

[76] Radjenović, J.; Sirtori, C.; Petrović, M.; Barceló, D. and Malato, S. (2009) Solar photocatalytic degradation of persistent pharmaceuticals at pilot-scale: Kinetics and characterization of major intermediate products, *Appl. Catal. B: Environ.*, **89**, 255–264.

[77] Peris-Cardells, E.; Terol, J.; Mauri, A.R.; de la Guardia, M. and Pramauro, E. (1993) Continuous flow photocatalytic degradation of carbaryl in aqueous media, *J. Environ. Sci. Health B*, **28**, 431–445.

[78] de la Guardia, M.; Khalaf, K.D.; Carbonell, V. and Morales-Rubio, A. (1995) Clean analytical method for the determination of propoxur, *Anal. Chim. Acta*, **308**, 462–468.

[79] de la Guardia, M.; Khalaf, K.D.; Hasan, B.A.; Morales-Rubio, A. and Carbonell, V. (1995) In-line titanium dioxide-catalysed ultraviolet mineralization of toxic aromatic compounds in the waste stream from a flow injection-based resorcinol analyser, *Analyst*, **120**, 231–235.

[80] Escuriola, M.J.; Morales-Rubio, A. and de la Guardia, M. (1999) A clean analytical method for the spectrophotometric determination of formetanate incorporating on-line microwave assisted hydrolysis step, *Anal. Chim. Acta*, **390**, 147–154.

[81] Bianco Prevot, A.; Fabbri, D.; Pramauro, E.; Morales-Rubio, A. and de la Guardia, M. (2001) Continuous monitoring of photocatalytic treatments by flow injection. Degradation of Dicamba in aqueous TiO_2 dispersions, *Chemosphere*, **44**, 249–255.

Section IV
Fields of Application

20

Green Bioanalytical Chemistry

Tadashi Nishio and Hideko Kanazawa

Faculty of Pharmacy, Keio University, Tokyo, Japan

20.1 The analytical techniques in bioanalysis

High-performance liquid chromatography (HPLC) is one of the most imperative techniques in not only analytical chemistry, but in many scientific research fields; such as chemical engineering, environmental science and biological science [1]. There are many separation modes, such as adsorbent, partition, ion-exchange, gel permeation and affinity that are capable of hyphenating the mass spectrometric technique (LC-MS). Although reversed-phase liquid chromatography (RPLC) is the most widely used HPLC technique in quantitative and qualitative analysis, it requires a large amount of organic solvent as a mobile phase. The aprotic, polar and water-miscible solvent acetonitrile is a commonly used eluent in RPLC. In 2008, decreasing demand for acrylonitrile due to the financial crisis resulted in a worldwide acetonitrile shortage, since acetonitrile is a by-product of acrylonitrile production [2]. Fortunately, the acetonitrile supply resumed within several months; 'the acetonitrile crisis' seriously affected both scientific fields and industries. In addition, a dependency on fossil fuels is a worldwide problem from the view points of the depletion of fine resources and environmental problems that arise from combustion gas or the disposal of organic solvents.

It is well known in the pharmaceutical industry that the market for antibody drugs was about US$30 billion in 2007, and now gradually increasing. Though antibody drugs are effective remedies and show excellent performance, a high production cost is a major problem. It is considered that about 40% of the production process is the separation and refinement procedure. In addition, the cost for affinity chromatography accounts for a large share. Because the expensive production cost is directly reflected in pharmaceutical prices, it contributes to the escalation of medical costs. If effective and inexpensive bioseparation methods are developed and cut the cumbersome operation, production cost-cutting is expected including the purification process and waste liquid treatment.

Handbook of Green Analytical Chemistry, First Edition. Edited by Miguel de la Guardia and Salvador Garrigues.

With the advancement of medical technology and pharmacological therapy, the importance of individualized medicine has been increasing. Though molecular-targeting agents are effective remedies for particular illnesses and are well marketed in the world, they often have serious side-effects. It is important to conduct reliable diagnoses and medical treatments; not only gene analysis for gene polymorphism, but also for the analysis of individual drug metabolizing ability. For this reason, though therapeutic drug monitoring (TDM) is an indispensable analytical technique for better healthcare, it has important problems. Because several drugs are monitored by the RPLC method, waste solutions and exposure to patients or medical staff are serious problems.

A chromatography system that is independent of organic solvents is a very important and potential analytical method from the view points of the green chemistry. In this chapter, we review a green analytical method for bioseparation.Environmental-responsive chromatography is focused on in particular as a future analytical method for multidiscipline research fields. This analytical system utilizes a temperature-responsive polymer, poly(*N*-isopropylacrylamide) (PNIPAAm) and modified surfaces such as the HPLC stationary phase with an aqueous mobile phase.

20.2 Environmental-responsive polymers

'Intelligent polymers' respond to various physical and chemical stimuli, such as heat, light, pH and the addition of chemical compounds, and change their physicochemical properties rapidly and reversibly [3–8]. These functional polymers are also known as 'smart', 'stimuli responsive' or 'environmental responsive' polymers. PNIPAAm is one of the most representative polymers that exhibits thermally reversible soluble-insoluble change in the vicinity of its lower critical solution temperature (LCST) at 32°C in aqueous solution [9–10] (see Figure 20.1). PNIPAAm undergoes a sharp phase separation and a coil-globule transition of the polymer chain around its LCST. In water, PNIPAAm chains show an expanded conformation below the LCST due to strong hydration between the amido group in PNIPAAm and surrounding H_2O molecules. In contrast, PNIPAAm changes to a compact form due to strong dehydration, followed by forming macroscopic aggregation above the LCST. Additionally, it has also been confirmed that the effect of PNIPAAm precipitation by the salt concentration exists. The LCST is decreased with increasing the salt concentration, while sharp soluble-insoluble changes are maintained. The LCST remarkably shifted to 20°C in a 1 M NaCl solution. Lowering of the LCST by the addition of salt should, therefore, be due to the acceleration of dehydration, that is, salting out [11].

Based on such unique features, PNIPAAm has been utilized for drug delivery systems, bioconjugates, cell culture dishes and tissue engineering for regenerative medicine [12–16]. Furthermore, studies on the PNIPAAm-modified surface, such as evaluating the intermolecular force between the PNIPAAm surface and proteins by atomic force microscope are being actively carried out. By copolymerization with NIPAAm and different monomers, the obtained synthetic copolymer's physicochemical property can be controlled. The composition rate of monomers significantly influences the LCST. The introduction of a hydrophobic monomer to the PNIPAAm copolymer causes an increase in LCST compared to that of the PNIPAAm homopolymer. For instance, if butyl methacrylate (BMA), which is widely used as a hydrophobic molecule, is adjusted, the 1, 3, 5 mol% proportion of the PNIPAAm, LCSTs of synthetic copolymers increase to 30, 25 and 20°C, respectively [17]. On the contrary, using a hydrophilic monomer, such as acrylic acid (AAc) as a compositional unit of PNIPAAm copolymer, its transition temperature is increased compared to 32°C [18].

There have been several reports concerning the pH dependence of LCST. Hoffman *et al*. examined the LCST of a copolymer composed of NIPAAm and AAc, and showed that it increased with an increase of

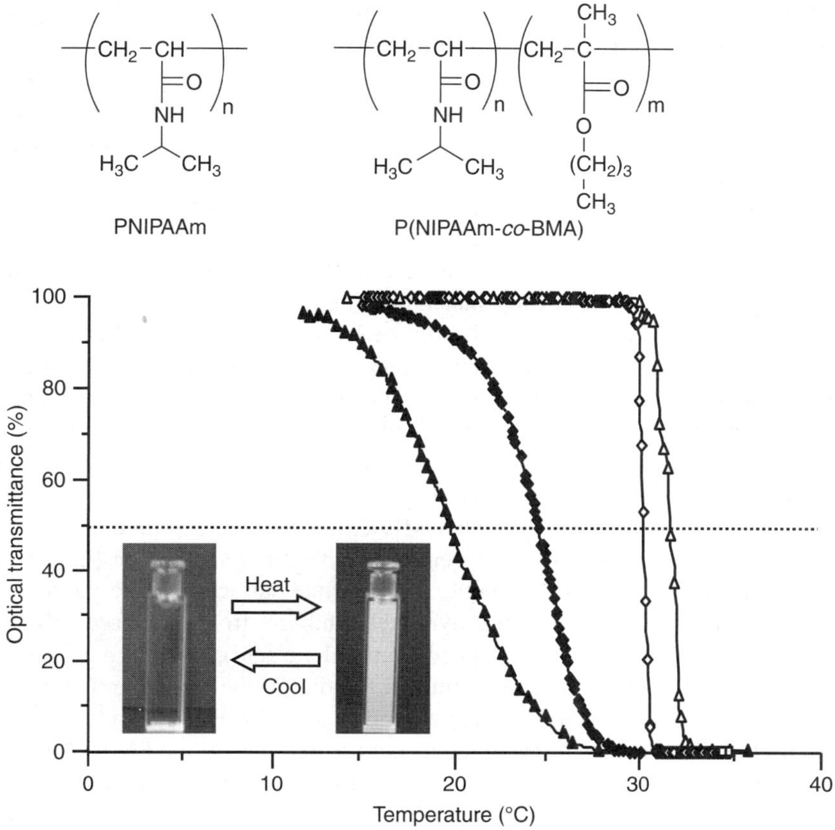

Figure 20.1 *Chemical structures and temperature-dependent optical transmittance change of PNIPAAm copolymers aqueous solution (0.5 w/v%) with various BMA compositions: PNIPAAm (LCST 32°C, Δ); BMA1% (LCST 30°C, ◊); BMA3% (LCST 25°C, ♦); BMA5% (LCST 20°C, ▲).*

the pH of the buffer solution [19]. Kim *et al.* also reported on the pH sensitivity of the LCST of a copolymer containing diethylaminoethyl methacrylate (DEAEMA) [20]. In addition, Kobayashi *et al.* made an examination of the effect of the charge density for a temperature change by measuring p*K*a and the zeta-potential. Regarding the p*K*a of the copolymer of NIPAAm, BMA and *N,N*-dimethylaminopropylacrylamide (DMAPAAm) at high temperature, dehydration of the NIPAAm isopropyl groups occurs, as well as an enhanced deprotonation of the amino group. The surface charge densities as well as the hydrophobic functional group were shown to be altered by changing the temperature. These polymers responded to both the temperature and the pH. Control of the surface charge density becomes possible with a structural change of the copolymer by changing the temperature. These facts show that control of different interactions was expected by hydrophobic and electrostatic interactions, by changing only the temperature.

We have been investigating PNIPAAm and related temperature-responsive polymers used to generate a thermally responsible stationary phase. Our major activity in this area has been to modify a temperature-responsive polymer to a packing-material surface, including a terminally, cross-linked and high density polymer brush.

20.3 Preparation of a polymer-modified surface for the stationary phase of environmental-responsive chromatography

Polymer grafting on the stationary phase appears to be an important determinant in separations. In this section, three types of polymer-grafting methods on silica surfaces to the preparation of chromatographic stationary phase are described.

First, polymer-grafted silica beads are constructed using active ester-amine coupling 'grafting to' method [11, 17, 21] (Figure 20.2a). Chain-transfer free-radical polymerization is utilized to synthesize polymers with one carboxylic acid end group, which can be derivatized to form *N*-hydroxysuccinimide ester that can be grafted to the aminopropyl silica supporting materials. The syntheses of semitelechelic PNIPAAm and its copolymer were carried out by radical polymerization using 3-mercaptopropionic acid and 2,2′-azobisisobutyronitrile as a chain transfer agent and a radical initiator, respectively. This grafting method has the advantage of a controlled molecular weight of the grafted polymer chain by adjusting the relative ratios of the monomers to the chain-transfer agent in bulk polymerization, and a dialysis treatment to a pure polymer [22]. Additionally, it also has the advantage of characterizing the properties such as LCST, molecular weight, infrared absorption and the polydispersity of synthetic polymer.

A second 'grafting from' synthetic method for PNIPAAm uses a surface-immobilized azo-initiator and cross-linker to prepare polymer layers with conventional radical polymerization [23, 24] (Figure 20.2b). Unlike the situation in the 'grafting to' technique, the substrate surface must be modified to generate an initiator functionality suitable for the polymer hydrogel synthesis from a surface. PNIPAAm hydrogel-modified matrix is prepared as follows: 4,4′-azobis(4-cyanovaleric acid), which is a polymerization initiator, was immobilized on aminopropyl silica beads. A mixed solution of the monomer and cross linker agent is added to the beads and polymerized at 70°C for several hours. The PNIPAAm gel-immobilized surface described here showed a slightly lower transition temperature than that of the PNIPAAm terminally grafted surface. We observed a drastic and reversible surface hydrophilic/hydrophobic property alternation for a PNIPAAm-terminally grafted surface due to rapid changes in the polymer hydration state around the polymer's transition temperature. Compared with a terminally grafted surface, the restricted hydrodynamic motion of a PNIPAAm segment in grafted hydrogel was considered to be due to a restricted conformational transition. The hydrogel-modified method for preparing a thermally responsive packing material was relatively simple and easy compared with the PNIPAAm terminally grafted stationary phase. Additionally, a polymer layer formed on silica beads showed resistance to an alkaline solution.

A third 'grafting from' method for PNIPAAm is high-density polymer brushes on silica-bead surfaces using controlled polymerization techniques, such as atom transfer radical polymerization (ATRP), reversible addition-fragmentation chain transfer (RAFT) polymerization and nitroxide-mediated polymerization (NMP) [25–27] (Figure 20.2c). By depositing the appropriate initiators, controlled polymerization can lead to uniform brush layers, tunable brush thickness via molecular weight control, and the ability to perform sequential polymerization steps to yield either thicker homopolymer layers or diblock copolymer layers. Among these, ATRP is an attractive polymer grafting method because it enables the preparation of surfaces with a dense polymer brush from surface immobilized ATRP initiators. The methodology allows the control of the graft chain length by varying the duration of the polymerization and regulation of the graft density by varying the concentration of the ATRP initiator on surfaces. This method incorporates a relatively large amount of polymer onto surface compare to the above two methods.

Yakushiji *et al.* reported temperature-dependent wettability changes for PNIPAAm hydrogel modified surfaces by aqueous dynamic contact-angle measurements [28]. The graft configuration of PNIPAAm produced from different grafting methods greatly influences temperature-dependent aqueous wettability changes. Okano *et al.* reported that the molecular mobility and density of a PNIPAAm chain are greatly

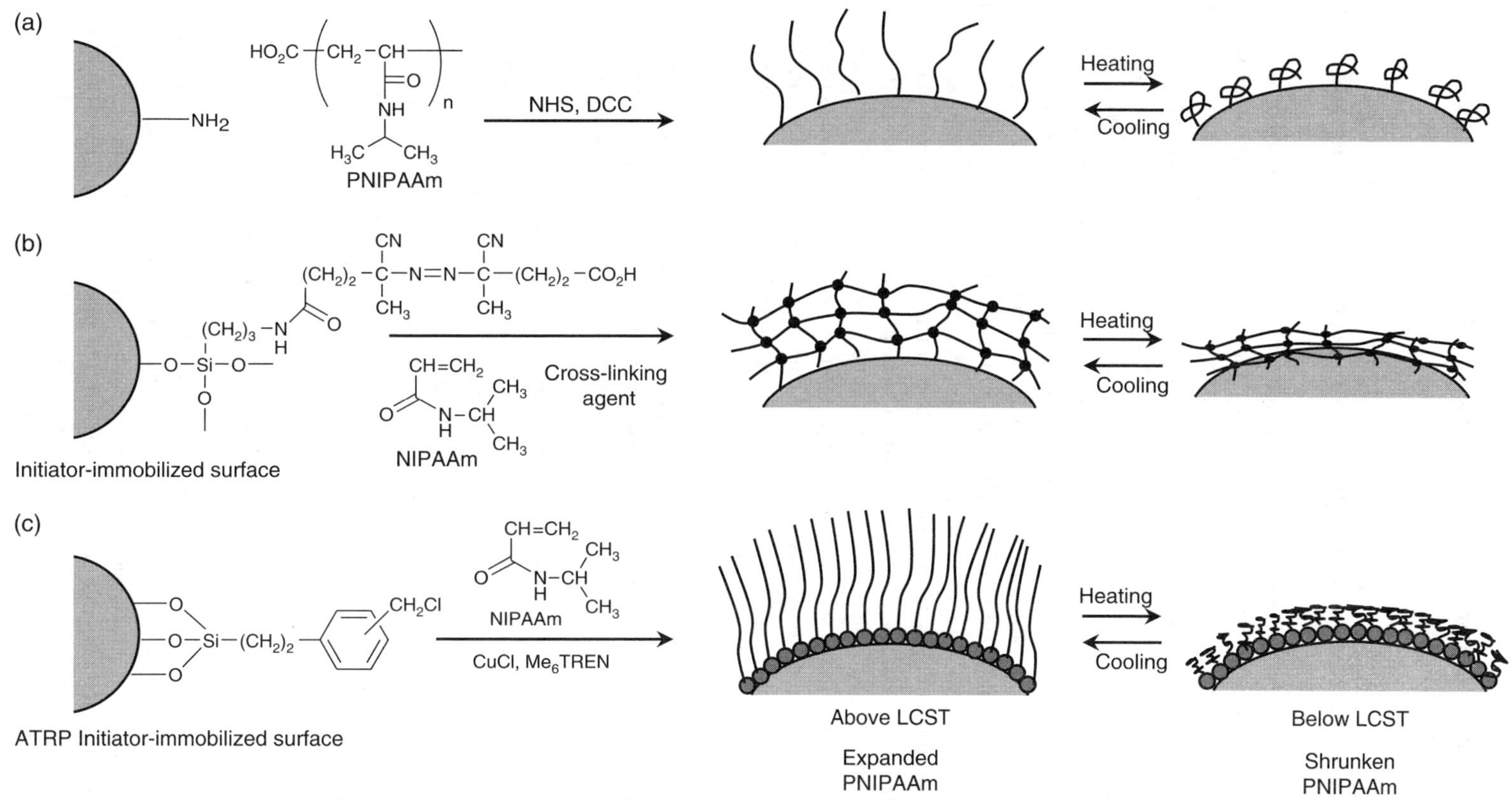

Figure 20.2 *Preparation of a PNIPAAm modified high-functionality surface using (a) terminal modification method, (b) gel modification method and (c) ATRP method. NHS; N-Hydroxysuccinimide, DCC; Dicyclohexylcarbodiimide,* Me_6*TREN; Tris(2-(dimethylamino)ethyl)amine.*

influenced by a difference in the construction of PNIPAAm modified surfaces due to the temperature [29]. It is considered that it maintains high mobility by fixing the PNIPAAm molecule at the end on the modified terminal, and quickly responds to any temperature change.

20.4 Temperature-responsive chromatography for green analytical methods

Using a column packed with PNIPAAm-modified silica, the HPLC analysis of steroids was carried out with only water as a mobile phase by changing the column temperature. Typical chromatograms of steroids analysis are shown in Figure 20.3 at a variety of temperatures (10~40°C), and extensions of retention times were observed with increasing temperature. With increasing temperature, an increased interaction between the solute components and PNIPAAm-grafted surfaces of the stationary phase was also observed [11, 17]. It was considered that a hydrophobic interaction exists between the steroids and the PNIPAAm. In contrast, in the conventional RPLC system, the opposite chromatographic behavior of the shortened retention times was observed with increasing temperature due to the solvent viscosity. A temperature-dependent resolution was achieved using only water as a mobile phase on the PNIPAAm-modified column. A drastic and reversible change in the physicochemical property of a PNIPAAm-terminally grafted silica surface from hydrophilic to hydrophobic is caused from a rapid alternation in the polymer hydration state around its transition temperature. The retention strength of steroids exhibits a linear relationship with the log$P$ values (the partition coefficients in a 1-octanol/water system). The log$P$ values for the steroids were 1.61 of hydrocortisone, 1.62 of prednisolone, 1.83 of dexamethasone, 2.30 of hydrocortisone acetate and 3.32 of testosterone. The more hydrophobic steroids show longer retention times. These results indicate that a strong hydrophobic interaction is the primary driving force for partitioning steroids into a PNIPAAm-modified surface. Since the phase transition of PNIPAAm results from the stability of the hydrophobic groups along the polymer chain in aqueous media, the LCST of the polymer should decrease with increasing polymer hydrophobicity.

A plot of the retention factor (ln$k$) versus the reciprocal temperature ($1/T$) normally shows a linear regression, with the slope representing the enthalpy change involved for the retention reaction. Figure 20.4 shows a plot for the steroids on the PNIPAAm-modified column. The slope of the plots on the PNIPAAm-modified column is negative, and it is opposite to that observed in common chromatography. This provides evidence that the interaction between steroids and the temperature-responsive surface becomes stronger at elevated temperature. Additionally, linearity in the plots is commonly observed for commercially available RPLC columns under the standard analytical condition with narrow temperature ranges. On the PNIPAAm-modified column, however, a deviation from linearity was found between ln$k$ values and $1/T$. The slope of the plots of each analyte on the PNIPAAm-modified column changed markedly at the LCST boundary. This phenomenon corresponds to a phase transition of the polymer modified on the surface, and it is indicated that the retention mechanism can be controlled by changing temperature.

20.5 Biological analysis by temperature-responsive chromatography

Therapeutic drug monitoring (TDM) has become a well-established clinical specialty [30]. Though the immunoassay auto-analyzer is well utilized for TDM, it is generally expensive, and subject to potential cross-reactivity and interference by drug metabolites and endogenous components. Furthermore, measurable drugs depended on the commercially available immunoassay kit. On the other hand, the HPLC method has the advantage of specificity and the application. In this background, HPLC, especially RPLC analysis is widely used for the simultaneous determination of a variety of drugs and accurate measurements. As previously

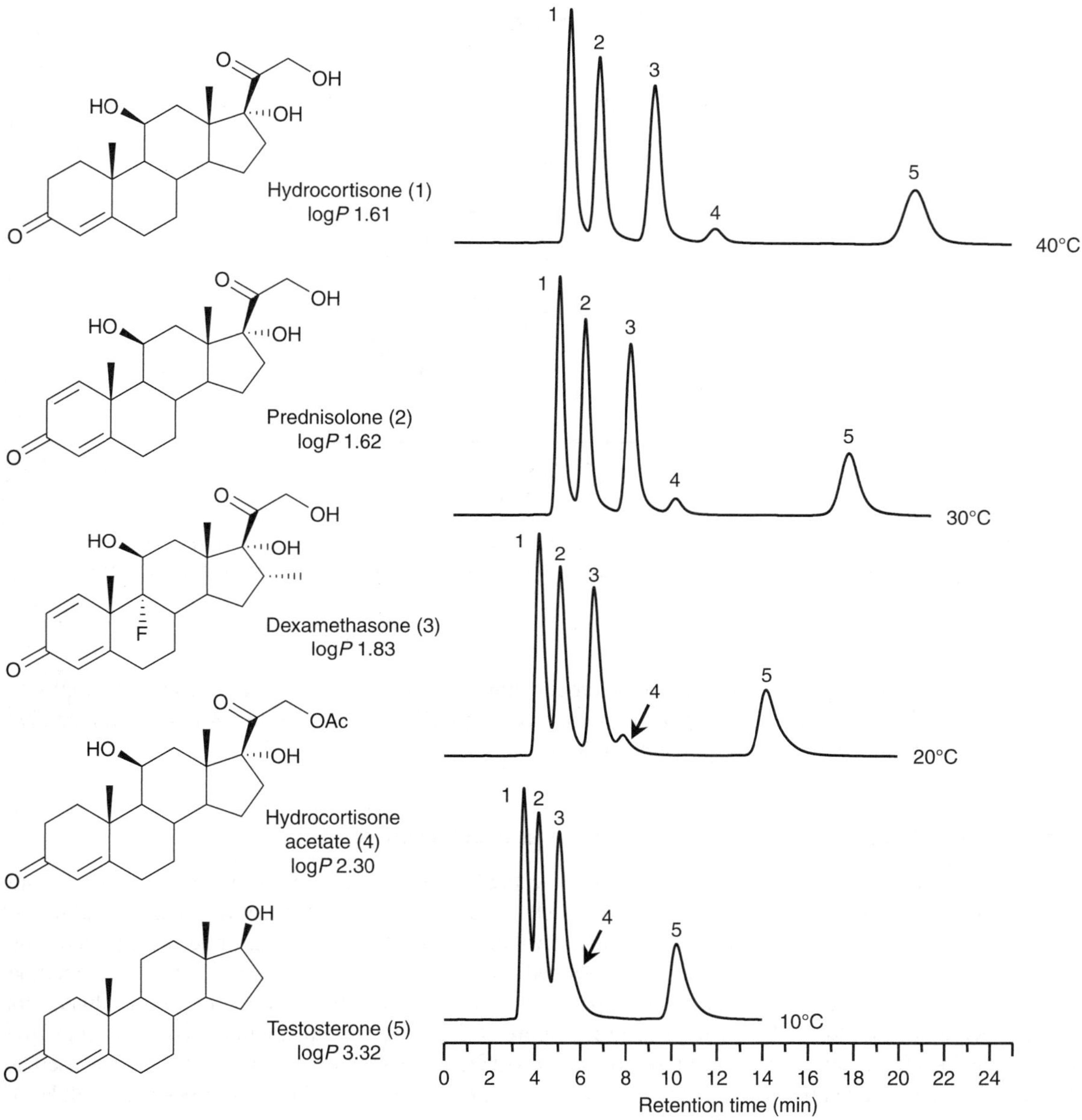

Figure 20.3 *Chemical structures and chromatograms of a mixture of five steroids with H_2O as a mobile phase at 10, 20, 30 and 40°C. HPLC condition: column, PNIPAAm modified column (4.6 mm i.d. × 150 mm); mobile phase, H_2O; flow rate, 1.0 ml min^{-1}; detection at UV 254 nm.*

described, temperature-responsive chromatography needs only water as the mobile phase, and it has the specific advantage of being applied to TDM in clinical practice and real-time drug monitoring in an operating room. Since RPLC requires a large amount of organic solvent as the mobile phase, there are some problems related to waste solutions and the exposure of patients or medical staff to harmful solvents. In this section, temperature-responsive chromatography is applied to the quantitative analysis of pharmaceutical anesthetic

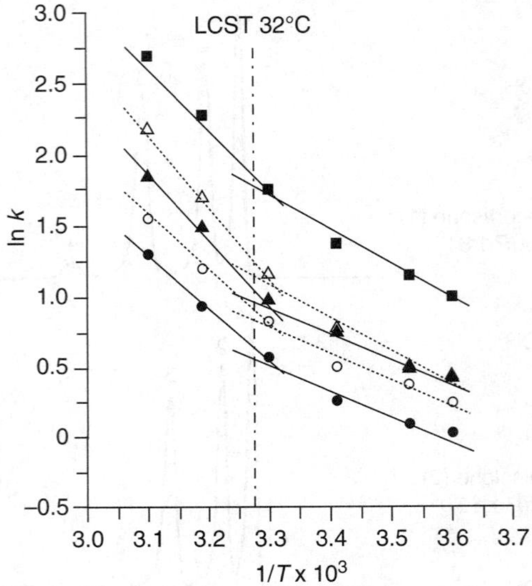

Figure 20.4 *The van't Hoff plots for five steroids on PNIPAAm modified HPLC column. Hydrocortisone (closed circle), prednisolone (open circle), dexamethasone (closed triangle), hydrocortisone acetate (open triangle) and testosterone (closed square).*

agent in a biological sample. We introduced two applications of temperature-responsive chromatography to biological analysis. First, an anesthetic agent propofol was analyzed by isocratic elution. Second, contraceptive drugs were analyzed by temperature-gradient elution.

20.5.1 Analysis of propofol in plasma using water as a mobile phase

Propofol (2,6-diisopropylphenol) is an intravenous anesthetic agent that is used for both the induction and maintenance of anesthesia [31]. Moreover, it is widely used during minor surgery of laboratory animals. Because a low propofol dosage causes intraoperative awaking, monitoring of the anesthetic agent level during anesthesia or sedation is very important. We developed a HPLC-fluorescent detection (HPLC-FL) method utilizing a temperature-responsive chromatography system for the analysis of propofol, and evaluated its performance compared to the RPLC method [32] (Figure 20.5). Drug administrated monkey serum (80 μl) was used as the biological sample. To the serum was added methanol containing thymol, which was used as an internal standard. The mixture was purified using a C18 pretreatment cartridge. After washing with H_2O, the target analyte was eluted with methanol. The recovery rate of propofol during the simple pretreatment was about 80%. P(NIPAAm-*co*-BMA) copolymer was used as a stationary phase. The obtained calibration rage was covered from 0.5 μg ml^{-1} to 10 μg ml^{-1}, and allowed a reproducible quantification of the serum propofol. The relative standard deviations of intra- and inter-assay were below than 15%. In addition, there was a good relationship of the quantification values between the temperature-responsive chromatography and the conventional RPLC method. Because the developed method used only water as the mobile phase, there were some advantages in conducting the TDM. First, the avoidance of complicated mobile phase preparation was possible, which was important for busy medical staff members to reduce their work. Second, the developed method could solve the problem related to waste solutions and the exposure of patients to harmful organic solvents, which occurs when using RPLC.

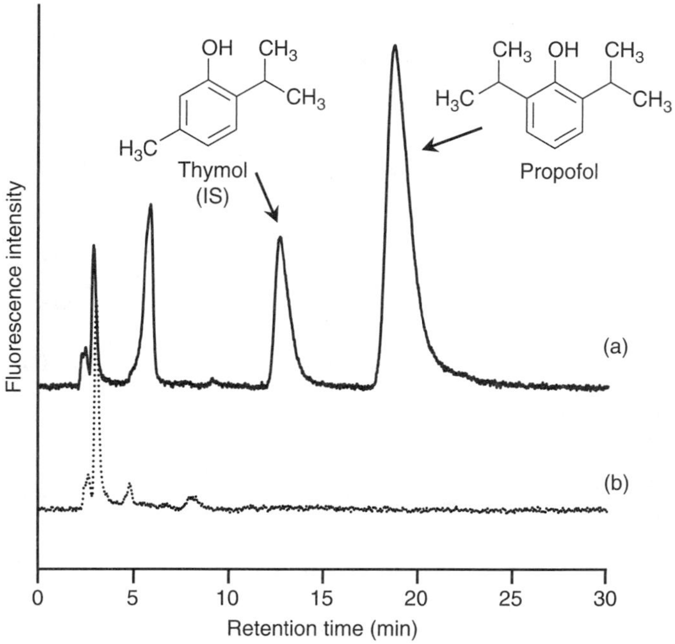

Figure 20.5 *Chemical structures and chromatograms of propofol and thymol (IS) obtained from a drug-administrated monkey serum (a), and a blank one (b), respectively. HPLC condition: column, P(NIPAAm-co-BMA5%) modified column (4.6 mm i.d. × 150 mm); column temperature, 40°C; mobile phase, H_2O; flow rate, 1.0 ml min^{-1}; fluorescent detection, excitation wavelength 267 nm, emission wavelength 310 nm.*

20.5.2 Contraceptive drugs analysis using temperature gradient chromatography

In the isocratic elution of biological fluid containing solutes with a wide range of polarity, it is sometimes difficult to achieve the desired resolution within a reasonable time. It may be necessary to use gradient elution, where volumes of an organic solvent, composition of mobile phase, or other property of solvent, such as pH or ionic strength are changed during the separation. Little attention has been paid to program the temperature changes during chromatographic analysis. Because HPLC can use the solvent strength as a functional time (solvent gradient), it has a much greater potential for changing the retention and selectivity than dose temperature change. On an HPLC column packed with temperature-responsive polymer modified silica, temperature programming can be used in lieu of a gradient solvent [33, 34]. By using a temperature-responsive polymer modified HPLC stationary phase, the gradient elution-like effect can be achieved with a single mobile phase by only controlling column temperature.

As an example, the simultaneous analysis of two kinds of oral contraceptive drugs was carried out using a temperature step gradient. Figure 20.6 shows chromatograms of levonorgestrel and ethinylestradiol, which were obtained at 10°C and 40°C, respectively. For levonorgestrel, with lower hydrophobicity, the retention time hardly changed at both temperatures. On the other hand, for ethinylestradiol, with a higher hydrophobicity, the retention time was increased too much, and influenced the hydrophobic interaction at 40°C. To move strongly retained components of the contraceptive drugs and to optimize the analysis, a temperature-programming technique was used. With a single mobile phase of water and by controlling the external temperature from 40°C to 10°C, the analytical time was dramatically reduced. The excellent resolution of the analyte was accomplished using a temperature step gradient mode and developed method could be applied to

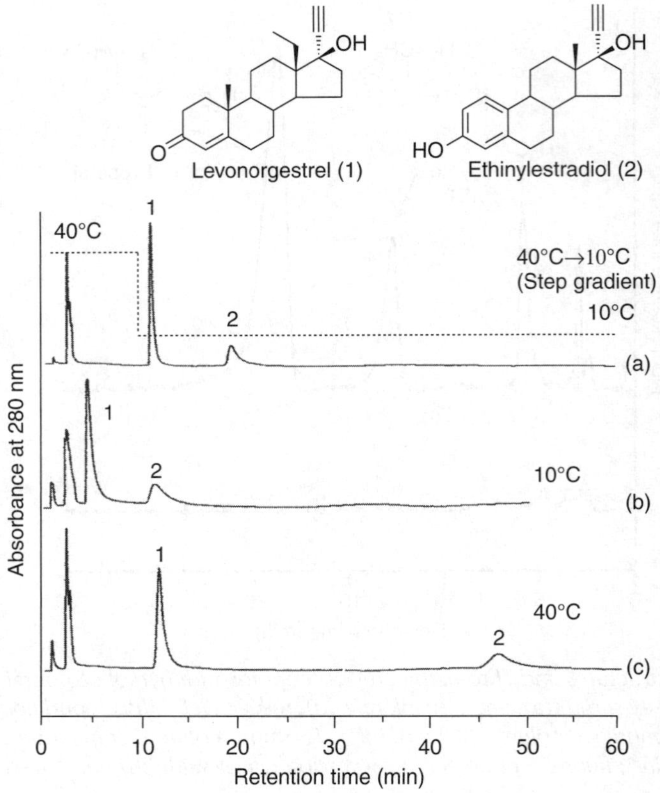

Figure 20.6 *Effect of a temperature change on the retention time of levonorgestrel (1) and ethinylestradiol (2). HPLC condition: column, P(NIPAAm-co-BMA5%) modified column (4.6 mm i.d. × 150 mm); mobile phase, H_2O; flow rate, 1.0 ml min^{-1}; detection at UV 280 nm, column temperature; (a) step gradient (40°C→10°C), (b) 10°C and (c) 40°C.*

measure urinary contraceptive drugs. This phenomenon was caused because the PNIPAAm-modified surface property of the stationary phase changed to hydrophilic at decreased temperature, and the hydrophobic interaction between the analyte component and the stationary phase was decreased. Additionally, this fact indicated the effectiveness of 'temperature-responsive' and 'thermo-reversible' properties of a PNIPAAm-modified surface.

20.6 Affinity chromatography for green bioseparation

Affinity chromatography has been widely used in biomedical research and biotechnology [35]. It is based on molecular recognition, where one recognition molecule is immobilized on a stationary phase, and the target molecule can be captured from a crude mixture. The target molecule can then be released from the base matrix and recovered in a functional form. The fundamental of elution is to decrease the affinity between the stationary phase and the analyte. A bond-breaking buffer for changing the ionic strength, solvent and pH or competitive elution is frequently used [36]. There is a serious problem that if a hard elution condition is used, the target materials undergo degeneration. For this reason, temperature-responsive chromatography, which is

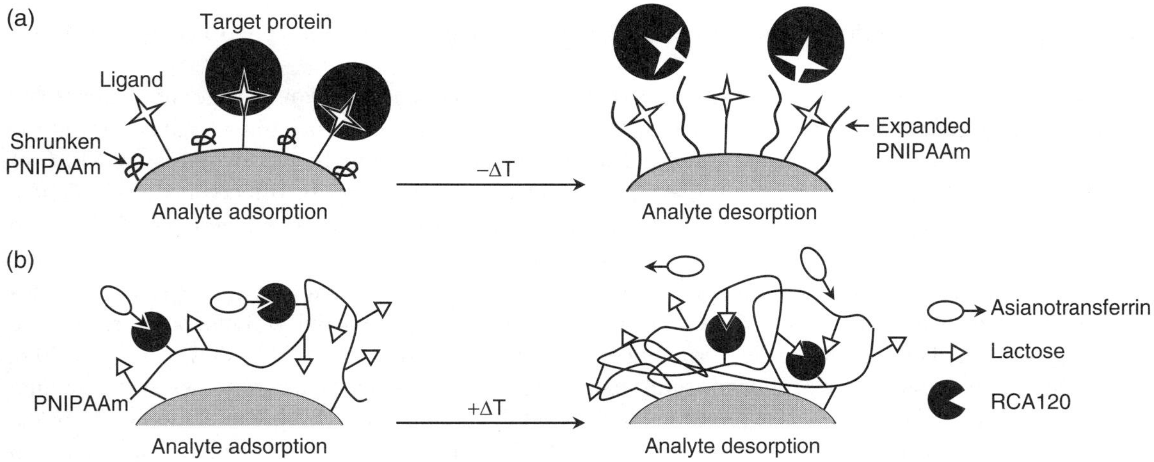

Figure 20.7 *Concepts of the temperature-responsive affinity chromatography based on PNIPAAm (a) Masking and forced-releasing effect, and (b) collapsible tether effect.*

carried out under mild conditions is expected to be suitable for affinity chromatography. There have been two reports about the temperature-responsive affinity chromatography before now.

The first method is based on the concept of 'masking and a forced-releasing effect' using a PNIPAAm [37] (Figure 20.7a). Affinity chromatographic matrixes were prepared using either poly(glycidyl methacrylate-*co*-ethyleneglycol dimethacrylate) or poly(glycidyl methacrylate-*co*-tri-ethyleneglycol dimethacrylate) beads immobilized with ligand molecules, Cibacron Blue F3G-A (CB) together with PNIPAAm. Two different length of spacer molecules (1.2 or 2.0 nm) were used for the immobilization of CB while maintaining the PNIPAAm size constant (1.7 nm), and bovine serum albumin was used as the model protein. The result showed a clear correlation between the spacer length and the binding capacity at lower than the LCST of PNIPAAm. The binding capacity under LCST was drastically diminished only when the spacer length was shorter than the extended PNIPAAm size. In addition, the adsorbed protein could be desorbed from the stationary phase by decreasing the temperature to below the LCST. Selective recovery of human albumin from human sera was accomplished by using a temperature-responsive affinity column.

Another method is based on the concept of the 'collapsible tethers effect' [38] (Figure 20.7b). A novel affinity chromatography mode in which a recognition molecule is immobilized on the media surface via collapsible tethers was developed. In normal affinity chromatography, the immobilized molecule acts passively, and its local concentration is static. In the developed method, the ligand molecule can move dynamically in response to external stimuli, a design that enables marked changes in both the local concentration of the ligand and its surrounding environment without exchanging the mobile phase. Using PNIPAAm as a scaffold for ligand and hapten attachment, controlled mobility and a microenvironment alternation of the affinity ligand *Ricinus communis* agglutinin (RCA_{120}) were accomplished. The target glycoprotein, asialotransferrin, was loaded onto a column in which PNIPAAm was partially substituted with both RCA_{120} and lactose. Though under the below LCST condition, the column retained the asialotransferrin, most of it was released above the LCST condition. This temperature-dependent elution was much greater than be explained by the temperature dependency of sugar recognition by RCA_{120}. The simplest explanation is that upon thermally induced dehydration and collapsed of the PNIPAAm chain, the co-immobilized RCA_{120} ligand and the lactose hapten are brought into closer proximity to each other, enabling immobilized lactose to displace affinity-bound asialotransferrin from the immobilized RCA_{120}.

20.7 Separation of biologically active molecules by the green chromatographic method

Biomolecules, such as peptides and proteins, are currently separated by an RPLC, ion-change chromatographic system and their combinations. RPLC is one of the most widely used chromatographic techniques in the separation, purification study of biomacromolecules. The most common mobile phase used in the RPLC of peptides and proteins involves an acetonitrile solution including 0.1% (10 mM) trifluoroacetic acid (TFA). Denaturation and a loss of bioactivity are unavoidable consequences of the elution process. Thus, these conditions should be avoided when working with most proteins.

Because temperature-responsive chromatography is performed in an aqueous environment that includes structure-stabilizing salt, this method allows for the retention of biological activity. Temperature-responsive chromatography has been applied to the separation of bioactive compounds [39–47]. As shown in Figure 20.8, the separation a mixture of three peptides, insulin chain A, chain B and β-endorphin was achieved by changing column temperature with a 0.9% NaCl aqueous solution as the mobile phase [39]. The elution order of the three peptides should reflect their hydrophobic properties. The retention times of analyte increased with increasing the temperature. The property of the surface of the stationary phase was altered from hydrophilic to hydrophobic. The increased retention times with increasing temperature were opposite to that seen in RPLC. The results provide evidence that the interaction between peptides and temperature-responsive surfaces becomes stronger at elevated temperature. Kobayashi *et al.* used cross-linked, temperature-responsive poly(*N*-isopropylacrylamide-*co*-acrylic acid-*co*-*N*-*tert*-butylacrylamide) [P(NIPAAm-*co*-AAc-*co*-*t*BAAm)] thin hydrogel layers on silica beads as column matrix modified for LC separation of basic bioactive peptides, angiotensin subtype I, II and III [40] (Figure 20.9a). Terpolymer P(NIPAAm-*co*-AAc-*co*-*t*BAAm) showed both phase transition and apparent carboxylate p*K*a shifts in water, depending on the temperature. A polymer-modified silica bead surface exhibited simultaneous thermally modulated changes in the hydrophobic-hydrophilic properties and charge densities. More effective separation of the angiotensin peptide subtype was

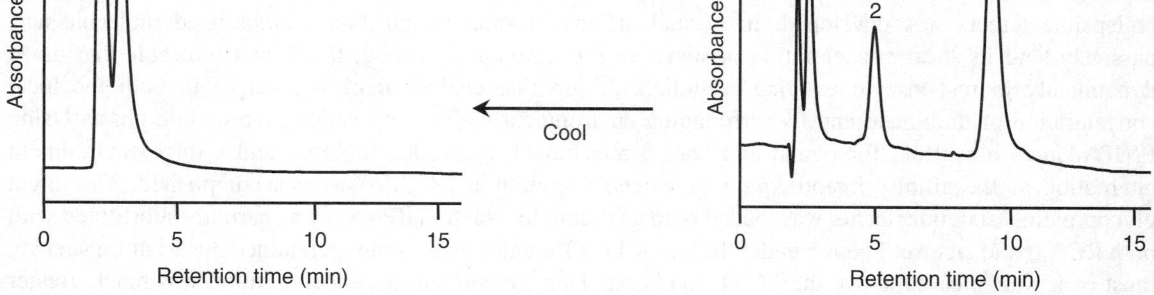

Figure 20.8 *Chromatograms of bioactive-peptides, insulin A (1), insulin chain B (2) and β-endorphin (3). HPLC condition: column, P(NIPAAm-co-BMA5%) modified column (4.6 mm i.d. × 150 mm); mobile phase, 0.9 w/v% NaCl aqueous solution; flow rate, 1.0 ml min*$^{-1}$*; detection at UV 210 nm; column temperature, (a) 10°C and (b) 40°C.*

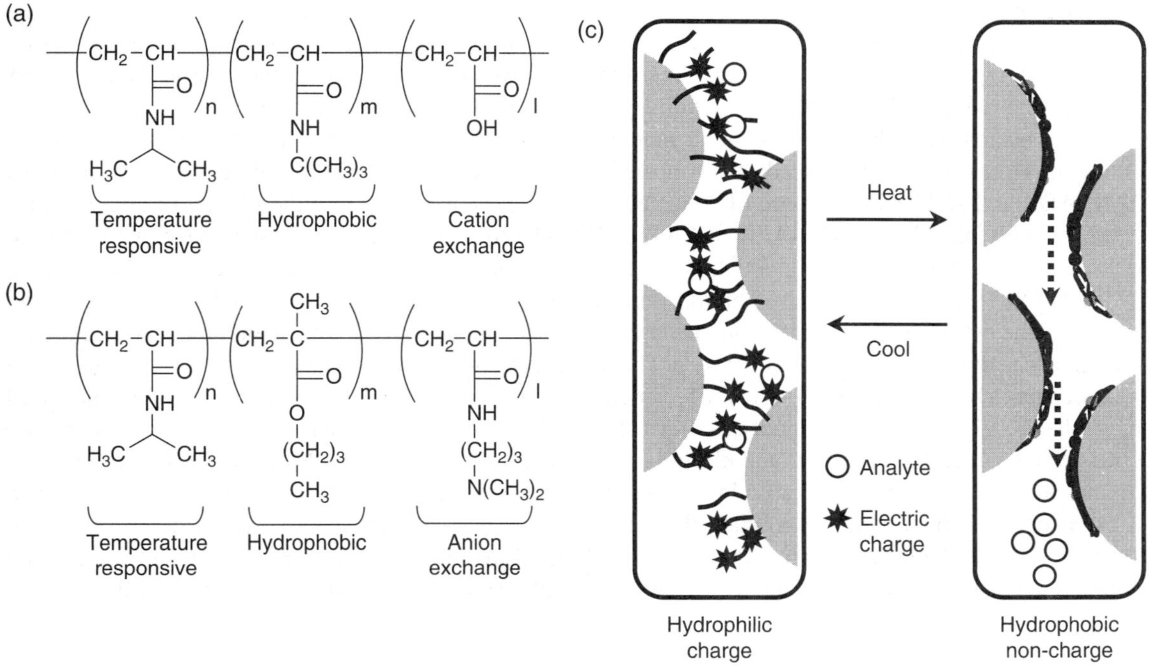

Figure 20.9 *Chemical structures of dual pH- and temperature-responsive terpolymer, P(NIPAAm-co-AAc-co-tBAAm) (a), and P(NIPAAm-co-BMA-co-DMAPAAm) (b). The concept of environmental-responsive chromatography is shown in (c).*

achieved on a column of these terpolymer thin hydrogel modified surfaces, as compared to uncharged control binary copolymers of NIPAAm and *t*BAAm. Although hydrophobic interactions effect the separation of angiotensin subtypes, combined electrostatic and hydrophobic interactions result in a more pronounced retention. At temperature below the terpolymer LCST, hydrophobic interactions predominated, and minimal changes in the electrostatic interactions were supported by a light shift in the apparent AAc carboxylate p*K*a values (Figure 20.9c). Above the LCST, electrostatic interactions were dramatically reduced as a result of the decreased charge densities of the polymer modified stationary phase. Therefore, the peptide retention times were also reduced, exhibiting a maximum at near 30–35°C. These anionic temperature-responsive polymer modified surfaces are good candidates for the improved separation of bioactive peptides under exclusively aqueous conditions. Furthermore, this developed method was also applied small molecular-weight compounds, such as intracerebral hormones [41]. Serotonin ($pK_a = 4.9, 9.8$) and its metabolite melatonine ($pK_a = 4.7$) are important chemical modulators of the biological clock of vertebrates, and are clinically used to treat jet lag and sleep diseases. Usually, in the isocratic elution of these samples containing solutes with a wide range of polarity, it is sometimes difficult to achieve the desire resolution time. It may be necessary to use gradient elution where the volumes of an organic solvent, the composition of the mobile phase, or some other property of the solvent (e.g., pH or ionic strength) are changed during the separation. In such analysis, the retention behavior changes by increasing the temperature are not observed. Figure 20.10 shows chromatograms of analytes that were obtained at 10, 30 and 50°C using a P(NIPAAm-*co*-AAc-*co*-*t*BAAm) hydrogel-modified column at pH 6.0. Because analytes are dissociated forms at pH 6.0, the electrostatic interaction between them and the polymer modified stationary phase should cause increased retention below the LCST. The polymer chain is in an extended form in aqueous media, and the AAc moiety should act as a cation-exchange group, and interact with ionized groups of the analytes. Because of the electrostatic interaction between the analytes and the polymer modified stationary phase, the retentions are bigger at low temperature. By raising the column temperature, the cation-exchange group may have been immersed in the hydrophobic polymer chain, and the reduced electrostatic interaction may have caused decreased retention times.

Another separation method using a cationic monomer, *N,N*-dimethylaminopropylacrylamide (DMAPAAm), instead of AAc, was also developed. A novel method for the qualitative analysis of adenosine nucleotides (AMP, ADP and ATP) and synthetic oligonucleotides was proposed, utilizing the pH- and a temperature-responsive polymer [42], P(NIPAAm-*co*-BMA-*co*-DMAPAAm) modified HPLC stationary phase (Figure 20.9b). On this column at pH 3.0, which is lower than the pK_a values of the adenosine nucleotides [AMP (3.80), ADP (3.9), ATP (4.1)], the retention times at low temperature were small, and decreased with an increase in the temperature. The hydrophobic interaction should contribute to the retention under the specific condition. The electrostatic interaction may not have predominated, because the adenosine nucleotides were undissociated. On the other hand, at pH 4.5, which is higher than pK_a values, the retention times increased below the LCST, while they decreased above the LCST. Because the adenosine phosphate groups are deprotonated in pH 4.5 condition, the electrostatic interaction between the adenosine nucleotides and the polymer stationary phase should cause increased retention below the LCST. Because the polymer chain is in an extended at a temperature lower than that of LCST, the DMAPAAm moiety should act as an anion-exchange group and interact with the ionized group of the adenosine nucleotides. By raising the temperature, the anion-exchange group may be immersed in the hydrophobic polymer chain, and the reduced electrostatic

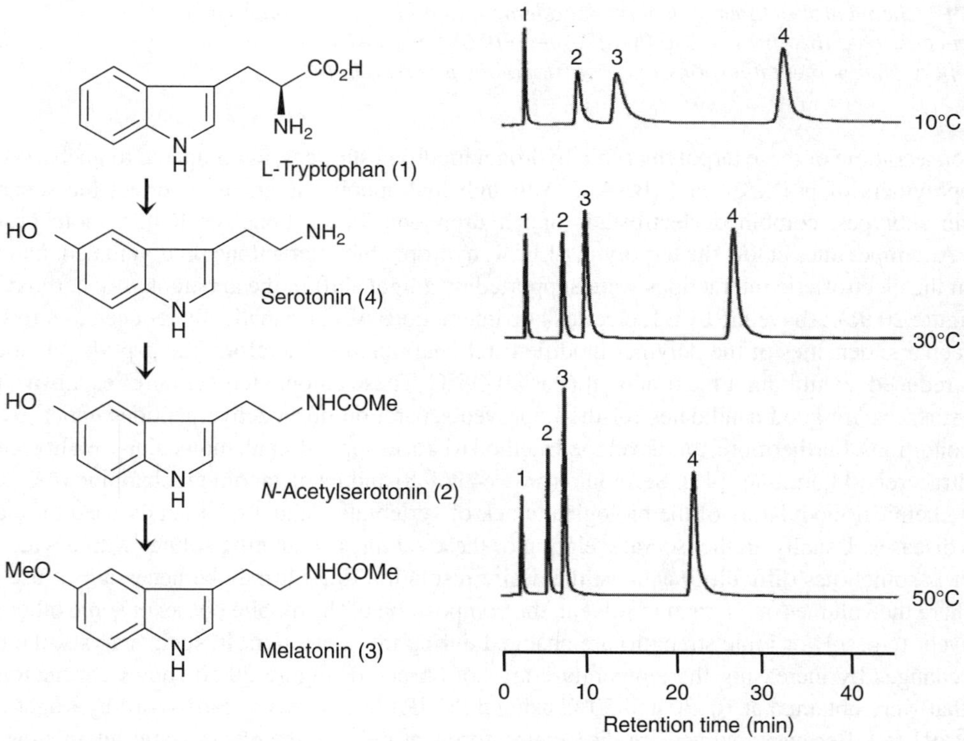

Figure 20.10 *Chemical structures and chromatograms of a mixture of tryptophan (1), N-acetylserotonin (2), melatonin (3) and serotonin (4). HPLC condition: column, P(NIPAAm-co-tBAAm-co-AAc) modified column (4.6 mm i.d. × 150 mm); mobile phase, 10 mM $CH_3CO_2NH_4$ (pH 6.0); flow rate, 1.0 ml min^{-1}; detection at UV 275 nm.*

interaction would cause decreased retention times. Since the pK_a value of the dimethylamino group in DMAPAAm is 10.4, deprotonation of the amino group of DMAPAAm moiety is enhanced under this condition. Moreover, polymer modified beads showed a decreased Zeta potential, indicating compression of the surface electrical double layer.

20.8 Protein separation by an aqueous chromatographic system

Temperature-responsive chromatography could achieve the separation of proteins under mild elution conditions, such as the physiological condition. Figure 20.11 shows chromatograms of HSA obtained

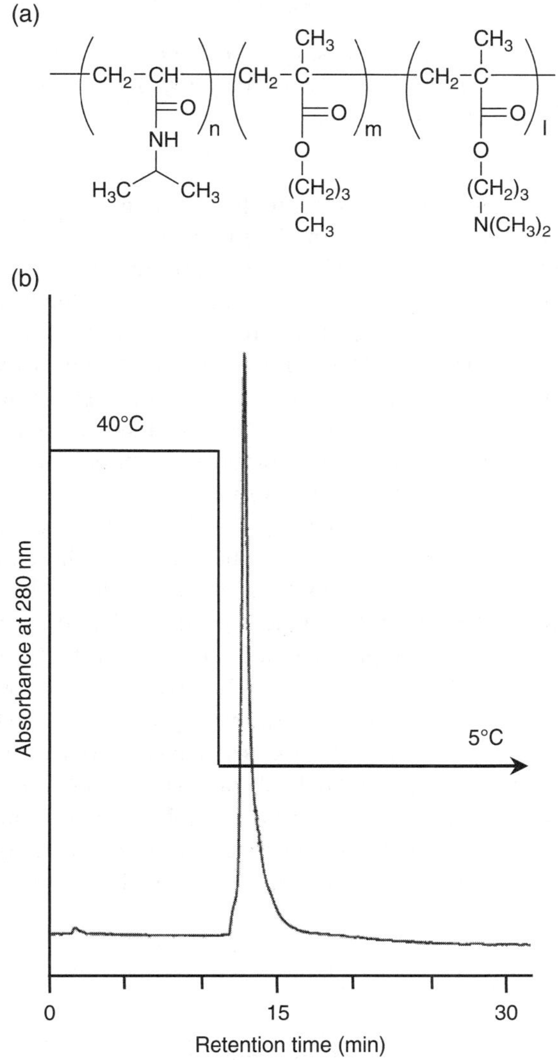

Figure 20.11 *Structure of P(NIPAAm-co-BMA-co-DMAEMA) (a) and chromatogram of human serum albumin with temperature stepwise gradient (40°C→5°C) on polymer hydrogel modified column (b). HPLC conditions: sample, HSA 20 μl (10 mg ml^{-1}); mobile phase, 16.7 mM phosphate buffer solution (pH7.0); flow rate, 1.0 ml min^{-1}; detection at UV 280 nm.*

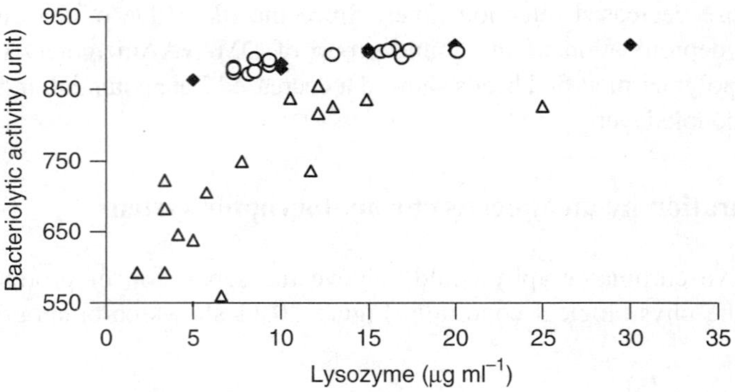

Figure 20.12 *Bacteriolytic activity of lysozyme: Lysozyme standard (♦), collected lysozyme by preparative temperature-responsive chromatography (○), collected lysozyme by preparative RPLC (Δ). Activity units are against concentration (μg ml^{-1}) of lysozyme.*

with a stepwise temperature gradient by changing the column temperature from 40°C to 5°C on P(NIPAAm-*co*-BMA-*co*-DMAEMA) hydrogel-modified column using phosphate buffer (pH 7.0) as a mobile phase. Using polymer has the dual pH and temperature responsibility, and it is composed of NIPAAm, BMA and cationic monomer, DMAEMA [2-(dimethylamino)ethylmethacrylate]. The peak of HSA was not observed at 40°C. When the temperature was changed from 40°C to 5°C, the HSA peak appeared. At the reduced temperature, the surface property of the stationary phase changed to hydrophilic and the hydrophobic interaction between HSA and the stationary phase decreased. The hydrophobic interactions are strengthened by increasing the temperature, and thus proteins are retained in the column longer. The response speed of the surface for the temperature stimulation was very rapid [34]. These features are due to rapid changes in the polymer conformation.

In a conventional hydrophobic interaction chromatography, the separations were performed using a starting mobile phase of high ionic strength to promote hydrophobic binding, followed by a gradient to a decreased the salt concentration. The problem of requiring desalting also occurs due to the use of a large amount of salt. To solve this problem, a preparative chromatographic separation of lysozyme by temperature-responsive chromatography utilizing PNIPAAm hydrogel modified column with 0.9% NaCl aqueous solution as a mobile phase was conducted [39] (Figure 20.12). The lysozyme fractions were collected and used for measuring the bacteriolytic activity of lysozyme. The recovery rate of lysozyme from the column was almost 100%, and the bioactivity of lysozyme collected by temperature-responsive chromatography was approximately the same as that of the lysozyme standard (control protein). In contrast, the activity of lysozyme collected by RPLC was reduced compared to that of the control lysozyme. Thus, the recovery of lysozyme was carried out without any loss of bioactivity by temperature-responsive chromatography. In the near future, the separation and refinement method is expected to be used for antibody drugs.

20.9 Ice chromatography

Ice chromatography, in which water-ice particles are employed as a chromatographic stationary phase, has proven to be an efficient technique for probing the solution/ice interface. The preparation of fine ice particles has allowed the analysis not only to obtain higher-resolution separation, but also to investigate the molecular process occurring on the ice surface in more detail. Okamoto *et al.* prepared water-ice particles by freezing water [48–50]. In ice chromatographic system, solutes having two or more polar groups are retained on this stationary phase with THF/hexane as the mobile phase, suggesting that multipoint interactions are required for measurable solute retention. Chromatographic separation of phenols, crown ethers and estrogens on water–ice is possible. The ice surface is expected to provide two different adsorption sites coming from the OH and O dangling bonds. Although the solute partition into the quasiliquid layers is also considered, the dependence of the retention times of the THF concentration implies that the interaction of solutes with the water-ice surface rather than the partition into the quasiliquid layers is responsible for solute retention. A retention model suggests that the number of adsorption sites for a crown ether depends on its ring size, whereas two sites are involved for the retention of phenols having two hydroxyl groups. Although hydroxyl groups can act as both a hydrogen bond donor and an acceptor, the interaction with the ice OH sites, which are exposed to the surroundings in comparison with the ice O sites, is more important. More recently, they developed chiral ice chromatography and accomplished the chiral separation of hexobarbital or 1,1′-bis-2-naphthol. Water-ice particles simultaneously doped with β-cyclodextrin and a salt enable chromatographic separation of enantiomers without synthetic processes, and enhanced chiral recognition occurring in the liquid-water phase coexistent with the solid-phase [51]. Water-ice has various interesting physical properties that other solid materials do not possess. Its low refractive index is a representative one. A liquid-core wave guide has been successfully fabricated on a water-ice slab and used for a flow-injection cell [52]. It is expected that ice chromatography on such a lab-on-ice may provide another pathway to novel analytical methodologies.

20.10 High-temperature liquid chromatography

High-temperature operation in RP-HPLC provides an opportunity to reduce the quantity of organic solvents used in a mixed organic-water mobile phase, and has promise for Green Analytical Chemistry [53–55]. High-temperature liquid chromatography (HTLC) increases the analytes mass-transfer rates, and decreases the column back pressure and total analysis time. High-temperature separation (100–200°C) has been shown to improve the analyte resolution by decreasing the mobile phase viscosity, and by increasing the diffusion rate of the sample species, thus increasing mass transfer of the analytes to the stationary phase and, thereby, decreasing the peak width. High temperature, as an optimization parameter in the separation process of the RPLC, has been widely investigated due to a recent finding of an alternative stationary phase, which has a high thermal stability at high temperature. Among the various stationary phases, zirconia-based one has received a great deal of attention, since it is extraordinary stable under a tremendous thermal and chemical condition. Sanagi *et al.* established a preliminary study on the separation of triazole antifungal agents using a polybutadiene-coated zirconia column at 100–150°C [56]. Clear separations were achieved when using 100% purified water as the organic-free eluent. Excellent limits of detection down to pg order were achieved for the separation of antifungal agents under the optimum conditions. Van't Hoff plots for the separations were linear, suggesting that no changes occurred in the retention mechanism over the temperature range studied. Boer *et al.* studied the potential of HTLC on-line combination with a screening system for bioactive compounds against the enzyme cathepsin B [57]. Samples were separated by HTLC utilizing a zirconia-based stationary phase, and subsequently analyzed by an on-line continuous-flow enzymatic assay. Detection was performed by electrospray ionization mass spectrometry, revealing both the bioactivity and the molecular mass of the bioactive compounds. Compared to conventional RPLC, the amount of methanol necessary for separation could be decreased to only 10%, which improved the compatibility of LC with a biochemical assay.

20.11 Ionic liquids

Ionic liquids (ILs), considered to be 'green' chemicals are widely used in many aspects of science and technology [58, 59]. Instead of an organic solvent, ILs have received much attention as ecologically friendly substitutes. ILs are liquid at ambient temperature, and are usually composed of relatively large organic cations and inorganic anions. They have negligible vapor pressure, good thermal stability, tunable viscosity, and primarily anion-dependent miscibility with water as well as various organic solvents. Because of these unique characteristic properties, ILs have been widely used in many research fields, and separation science is no exception.

Jiang *et al.* used 1-butyl-3-methylimidazolium tetrafluoroborate to test residual solvents in pharmaceutical preparations with static headspace gas chromatography [60]. The purpose of their work was to demonstrate the feasibility of ILs as diluents. Six solvent (acetonitrile, dichloromethane, *N*-methyl-2-pyrrolidone, toluene, *N,N*-dimethylformamide and *n*-butyl ether) were dissolved in ILs. The method of an external standard was used for quantitative analysis. Its performance was evaluated and validated, and all of the relative standard deviations were lower than 10%; the limits of detection were all of the ppm level. The developed method was both accurate and linear, and better sensitivities for the six solvents were gained with IL as diluent compared with DMSO, which is conventionally used.

Liu *et al.* used 1-butyl-3-methylimidazolium chloride/salt-based aqueous two-phase systems as a simple, rapid and effective sample pretreatment technique coupled with HPLC, for the analysis of the major opium alkaloids in *P. papaveris* [61]. To find the optimal conditions, various factors were investigated systematically; the results indicated that both the pH value and the salting-out ability of salt had a great influence on the phase separation. The recoveries of codeine and papaverine were 90.0–100.2% and 99.3–102.0% respectively. Compared with the conventional polymer-salt-water system, the extraction efficiencies obtained from the developed method were stable in a wide temperature range. In addition, this system had a relatively lower viscosity, and was convenient to be coupled with HPLC. This ILs technique was less time-consuming, and used nonvolatile solvents, which were to be distinct advantages over the conventional solid-phase extraction. The developed method was proved highly efficient and fast for the separation and enrichment of hydrophobic drug in biological samples, and would be a great potential in coupling other instruments, such as flow-injection analysis and capillary electrophoresis.

20.12 The future in green bioanalysis

The development of green bioseparation methods is indispensable to improve the problem of the exhaustion of fossil fuel resources and global environmental problems. Using an aqueous solution instead of organic solvents, in parts of these serious problems may be improved. There have been developed promising analytical methods that were introduced in this article. Environmental-responsive chromatography, which utilizes an intelligent polymer as the stationary phase, and uses only water as the mobile phase is one excellent example. This method has been applied to the separation of biologically active molecules, such as pharmaceutical agents, bioactive peptides and biomacromolecules (proteins). It is expected that various novel environmental-friendly methodologies can be accomplished by combining a number of green analytical methods, and can be applied to many research fields, such as medical and pharmaceutical sciences, biological science and environmental science.

References

1. Brown, P.R. and Grushka E. (eds) (2001) *Advances in Chromatography 41*, Marcel Dekker, New York.
2. Brettschneider, F.; Jankowski, V.; Günthner, T.; Salem, S.; Nierhaus, M.; Schulz, A.; Zidek, W. and Jankowski, J. (2010) Replacement of acetonitrile by ethanol as solvent in reversed phase chromatography of biomolecules, *J. Chromatogr. B.*, **878**, 763–768.

3. Suzuki, A. and Tanaka, T. (1990) Phase transition in polymer gels induced by visible light, *Nature*, **346**, 345–347.
4. Tanaka, T.; Nishio, I.; Sun, S.-T. and Ueno-Nishio, S. (1982) Collapse of gels in an electric field, *Science*, **218**, 467–469.
5. Known, I.C.; Bae, Y.H. and Kim, S.W. (1991) Electrically erodible polymer gel for controlled release of drugs, *Nature*, **354**, 291–293.
6. Chen, G. and Hoffman, A.S. (1993) Preparation and properties of thermoreversible, phase-separating enzyme-oligo(*N*-isopropylacrylamide) conjugates, *Bioconjugate Chem.*, **4**, 509–514.
7. Gutowska, A.; Bae, Y.H.; Jacobs, H.; Feijen, J. and Kim, S.W. (1994) Thermosensitive interpenetrating polymer networks: Synthesis, characterization, and macromolecular release, *Macromolecules*, **27**, 4167–4175.
8. Kim, Y.-H.; Bae, Y.H. and Kim, S.W. (1990) pH/Temperature-sensitive polymers for macromolecular drug loading and release, *J. Control. Release*, **28**, 143–152.
9. Fujishige, S.; Kubota, K. and Ando, I. (1989) Phase transition of aqueous solution of poly(*N*-isopropylacrylamide) and poly(*N*-isopropylmethacrylamide), *J. Phys. Chem.*, **93**, 3311–3313.
10. Schild, H.G. and Tirrell, D.A. (1990) Microcalorimetric detection of lower critical solution temperatures in aqueous polymer solutions, *J. Phys. Chem.*, **94**, 4352–4356.
11. Kanazawa, H.; Yamamoto, K.; Matsushima, Y.; Takai, N.; Kikuchi, A. and Okano, T. (1996) Temperature-responsive chromatography using poly(*N*-isopropylacrylamide)-modified silica, *Anal. Chem.*, **68**, 100–105.
12. Okano, T.; Yamada, N.; Sakai, H. and Sakurai, Y. (1993) A novel recovery system for cultured cells using plasma-treated polystyrene dishes grafted with poly(*N*-isopropylmethacrylamide), *J. Biomed. Mater. Res.*, **27**, 1243–1251.
13. Okano, T.; Yamada, N.; Okuhara, M. and Sakurai, Y. (1995) Mechanism of cell detachment from temperature-modulated, by hydrophilic-hydrophobic interaction, *Biomaterials*, **16**, 297–303.
14. Shimizu, T.; Sekine, H.; Yamato, M. and Okano, T. (2009) Cell sheet-based myocardial tissue engineering: New hope for damaged heart rescue, *Curr. Pharm. Des.*, **15**, 2807–2814.
15. Akimoto, J.; Nakayama, M.; Sakai, K. and Okano, T. (2009) Temperature-induced intracellular uptake of thermoresponsive polymeric micelles, *Biomacromolecules*, **10**, 1331–1336.
16. Kobayashi, N.; Nishikawa, M.; Sakamoto, C.; Nishio, T.; Kanazawa, H. and v Okano, T. (2009) Dual temperature- and pH-responsive fluorescence molecular probe for cellular imaging utilizing a PNIPAAm-fluorescein copolymer, *Anal. Sci.*, **25**, 1043–1047.
17. Kanazawa, H.; Kashiwase, Y.; Yamamoto, K.; Matsushima, Y.; Kikuchi, A.; Sakurai, Y. and Okano, T. (1997) Temperature-responsive liquid chromatography 2. Effect of hydrophobic groups in *N*-isopropylacrylamide copolymer-modified silica, *Anal. Chem.*, **69**, 823–830.
18. Bulmus, V.; Ding, Z.; Long, C.J.; Stayton, P.S. and Hoffman, A.S. (2000) Site-specific polymer-streptavidin bioconjugates for pH-controlled binding and triggered of biotin, *Bioconjugate. Chem.*, **11**, 78–83.
19. Feil, H.; Bae, Y.H.; Feijen, J. and Kim, S.W. (1992) Mutual influence of pH and temperature on the swelling of ionizable and thermosensitive hydrogels, *Macromolecules*, **25**, 5528–5530.
20. Kobayashi, J.; Kikuchi, A.; Sakai, K. and Okano, T. (2001) Aqueous chromatography utilizing pH-/temperature-responsive polymer stationary phase to separate ionic bioactive compounds, *Anal. Chem.*, **73**, 2027–2033.
21. Kanazawa, H.; Sunamoto, T.; Matsushima, Y.; Kikuchi, A. and Okano, T. (2000) Temperature-responsive chromatographic separation of amino acid phenylthiohydantoins using aqueous media as the mobile phase, *Anal. Chem.*, **72**, 5961–5966.
22. Takei, Y.; Aoki, T.; Sanui, K.; Ogata, N.; Okano, T. and Sakurai, Y. (1993) Temperature-responsive bioconjugates. 1. Synthesis of temperature responsive oligomers with reactive end group and their coupling to biomolecules, *Bioconjugate Chem.*, **4**, 42–48.
23. Yakushiji, T.; Sakai, K.; Kikuchi, A.; Aoyagi, T.; Sakurai, Y.; and Okano, T. (1999) Effect of cross-linked structure on temperature-responsive hydrophobic interaction of poly(*N*-isopropylacrylamide) hydrogel-modified surface with steroids, *Anal. Chem.*, **71**, 1125–1130.
24. Kanazawa, H.; Sunamoto, T.; Ayano, E.; Matsushima, Y.; Kikuchi, A. and Okano, T. (2002) Temperature-responsive chromatography using poly(*N*-isopropylacrylamide) hydrogel-modified silica, *Anal. Sci.*, **18**, 45–48.
25. Advincula, R.C.; Brittain, W.J.; Caster, K.C. and Rühe J. (eds) (2004) *Polymer Brushes*, Wiley-VCH Verlag, Weinheim.
26. Nagase, K.; Kobayashi, J.; Kikuchi, A.; Akiyama, Y.; Hideko, K. and Okano, T. (2007) Interfacial property modulation of thermoresponsive polymer brush surfaces and their interaction with biomolecules, *Langmuir*, **23**, 9409–9415.

27. Nagase, K.; Kobayashi, J.; Kikuchi, A.; Akiyama, Y.; Hideko, K. and Okano, T. (2008) Effect of graft densities and chain lengths on separation of bioactive compounds by nanolayered thermoresponsive polymer brush surface, *Langmuir*, **24**, 511–517.
28. Yakushiji, T. Sakai, K.; Kikuchi, A.; Aoyagi, T.; Sakurai, Y. and Okano, T. (1998) Graft architectural effect on thermoresponsive wettability changes of poly(*N*-isopropylacrylamide)-modified surface, *Langmuir*, **14**, 4657–4662.
29. Yoshida, R.; Uchida, K.; Kaneko, Y.; Sakai, K.; Kikuchi, A.; Sakurai, Y. and Okano, T. (2002) Comb-type grafted hydrogels with rapid deswelling response to temperature changes, *Nature*, **374**, 240–242.
30. Wong, S.Y. (1989) Advances in liquid chromatography and related methodologies for therapeutic drug monitoring, *J. Pharm. Bioamed. Anal.*, **7**, 1011–1032.
31. Shafer, A.; Doze, V.A.; Shafer, S.L. and White, P.F. (1988) Pharmacokinetics and pharmacodynamics of propofol infusions during general anesthesis, *Anesthesiology*, **69**, 348–356.
32. Nishio, T.; Suzuki, R.; Tsukada, Y.; Kanazawa, H.; Okano, T. and Miyabe-Nishiwaki, T. (2009) Aqueous chromatographic system for the quantification of propofol in biological fluids using a temperature-responsive polymer modified stationary phase, *J. Chromatogr. A*, **1216**, 7427–7432.
33. Kanazawa, H.; Matsushima, Y. and Okano, T. (2002) Temperature-responsive chromatography, in *Advanced in chromatography, 41* (eds P.R. Brown, E. Grushka), Marcel Dekker, Inc.
34. Ayano, E.; Okada, Y.; Sakamoto, C.; Kanazawa, H.; Kikuchi, A. and Okano, T. (2006) Study of temperature-responsibility on the surfaces of a thermo-responsive polymer modified stationary phase, *J. Chromatogr. A*, **1119**, 51–57.
35. Wilchek, M. and Chaiken, I. (2000) An overview of affinity chromatography, *Method Mol. Biol.*, **147**, 1–6.
36. Fier, M.A. (2001) Efficient elution of functional proteins in affinity chromatography, *J. Biochem. Biophys. Methods*, **49**, 433–442.
37. Yoshizako, K.; Akiyama, Y.; Yamanaka, H.; Shinohara, Y.; Hasegawa, Y.; Carredano, E.; Kikuchi, A. and Okano, T. (2002) Regulation of protein binding toward a ligand on chromatographic matrixes by masking and force-releasing effects using thermoresponsive polymer, *Anal. Chem.*, **74**, 4160–4166.
38. Yamanaka, H.; Yoshizako, K.; Akiyama, Y.; Sota, H.; Hasegawa, Y.; Shinohara, Y.; Kikuchi, A. and Okano, T. (2003) Affinity chromatography with collapsibly tethered ligands, *Anal. Chem.*, **75**, 1658–1663.
39. Kannazawa, H.; Nishikawa, M.; Mizutani, A.; Sakamoto, C.; Morita-Murase, Y.; Nagata, Y.; Kikuchi, A. and Okano, T. (2008) Aqueous chromatographic system for separation of biomolecules using thermoresponsive polymer modified stationary phase, *J. Chromatogr. A*, **1191**, 157–161.
40. Kobayashi, J.; Kikuchi, A. Sakai, K. and and Okano, T. (2003) Cross-linked thermoresponsive anionic polymer-grafted surfaces to separate bioactive basic peptides, *Anal. Chem.*, **75**, 3244–3249.
41. Ayano, E.; Suzuki, Y.; Kanezawa, M.; Sakamoto, C.; Morita-Murase, Y.; Nagata, Y.; Kanazawa, H.; Kikuchi, A. and Okano, T. (2007) Analysis of melatonin using pH- and temperature-responsive aqueous chromatography system, *J. Chromatogr. A*, **1156**, 213–219.
42. Ayano, E.; Sakamoto, C.; Kanazawa, H.; Kikuchi, A. and Okano, T. (2005) Separation of nucleotides with an aqueous mobile phase using pH- and temperature-responsive polymer modified packing materials, *Anal. Sci.*, **22**, 1–5.
43. Sakamoto, C.; Okada, Y.; Kanazawa, H.; Ayano, E.; Nishimura, T.; Ando, M.; Kikuchi, A. and Okano, T. (2004) Temperature- and pH-responsive aminopropyl-silica ion exchange column grafted with copolymer of N-isopropylacrylamide, *J. Chromatogr. A*, **1030**, 247–253.
44. Ayano, E.; Okada, Y.; Sakamoto, C.; Kanazawa, H.; Okano, T.; Ando, M.; and Nishimura, T. (2005) Analysis of herbicides in water using temperature-responsive chromatography and an aqueous mobile phase, *J. Chromatogr. A*, **1069**, 281–285.
45. Nagase, K.; Kobayashi, J.; Kikuchi, A.; Akiyama, Y.; Kanazawa, H. and Okano, T. (2008) Preparation of thermoresponsive cationic copolymer brush surfaces and application of the surface to separation of biomolecules, *Biomacromolecules*, **9**, 1340–1347.
46. Mizutani, A.; Nagase, K.; Kikuchi, A.; Kanazawa, H.; Akiyama, Y.; Kobayashi, J.; Annaka, M. and Okano, T. (2010) Preparation of thermo-responsive polymer brushes on hydrophilic polymeric beads by surface-initiated atom transfer radical polymerization for s highly resolutive separation of peptides, *J. Chromatogr. A*, **1217**, 5978–5985.

47. Mizutani, A.; Nagase, K.; Kikuchi, A.; Kanazawa, H.; Akiyama, Y.; Kobayashi, J.; Annaka, M. and Okano, T. (2010) Effective separation of peptides using highly dense thermo-responsive polymer brush-grafted porous polystyrene beads, *J. Chromatogr. B*, **878**, 2191–2198.
48. Tasaki, Y. and Okada, T. (2006) Ice chromatography. Characterization of water-ice as a chromatographic stationary phase, *Anal. Chem.*, **78**, 4155–4160.
49. Tasaki, Y. and Okada, T. (2009) Ice chromatography. Facilitation of applicability in ice chromatography by mechanistic considerations and by preparation of fine water-ice stationary phase, *Anal. Chem.*, **81**, 890–897.
50. Okada, T. and Tasaki, Y. (2010) Ice chromatography: current progress and future deveopment, *Anal. Bioanal. Chem.*, **396**, 221–227.
51. Shamoto, T.; Tasaki, Y. and Okada, T. (2010) Chiral ice chromatography, *J. Am. Chem. Soc.*, **132**, 13135–13137.
52. Sugiya, Y.; Harada, M. and Okada, T. (2009) Water-ice chip with liquid-core waveguide functionality. Toward lab on ice, *Lab. Chip*, **9**, 1037–1039.
53. Teutenberg, T.; Wagner, P. and Gmehling, J. (2009) High-temperature liquid chromatography. Part I. Determination of the vapour pressure of binary solvent mixtures-implications for liquid chromatographic separations, *J. Chromatogr. A*, **1216**, 6471–6480.
54. Teutenberg, T.; Wiese, S.; Wagner, P. and Gmehling, J. (2009) High-temperature liquid chromatography. Pat II: Determination of the viscosities of binary solvent mixtures-implications for liquid chromatographic separations, *J. Chromatogr. A*, **1216**, 8470–8479.
55. Teutenberg, T.; Wagner, P. and Gmehling, J. (2009) High-temperature liquid chromatography. Pat III. Determination of the static permittivities of pure solvents and binary solvent mixtures-implications for liquid chromatographic separations, *J. Chromatogr. A*, **1216**, 8480–8487.
56. Sanagi, M.M.; See, H.H.; Ibrahim, W.A.W. and Naim, A.A. (2004) High-temperature liquid chromatography of triazole fungicides on polybutadiene-coated zirconia stationary phase, *J. Chromatogr. A*, **1059**, 95–101.
57. de Boer, A.R.; Alcaide-Hidalgo, J.M.; Krabbe, J.G.; Kolkman, J.; van Emde Boas, C.N.; Nissen, W.M.A.; Lingeman, H. and Irth, H. (2005) High-temperature liquid chromatography coupled on-line to a continuous-flow biochemical screening assay with electrospray ionization mass spectrometric detection, *Anal. Chem.*, **77**, 7894–7900.
58. Marsh, K.N.; Boxall, J.A. and Lichtenthaler, R. (2004) Room temperature ionic liquid and their mixtures, *Fluid Phase Equilibr.*, **219**, 93–98.
59. Dupont, J.; Souza, R.F. and Suarez, P.A.Z. (2002) Ionic liquid (molten salt) phase organometallic catalysis, *Chem. Rev.*, **102**, 3667–3692.
60. Liu, F. and Jiang, Y. (2007) Room temperature ionic liquid as matrix medium for the determination of residual solvent in pharmaceuticals by static headspace gas chromatography, *J. Chromatogr. A*, **1167**, 116–119.
61. Li, S.; He, C.; Liu, H.; Li, K. and Liu, F. (2005) Ionic liquid-based aqueous two-phase system, a sample pretreatment procedure prior to high-performance liquid chromatography of opium alkaloids, *J. Chromatogr. B*, **826**, 58–62.

21

Infrared Spectroscopy in Biodiagnostics: A Green Analytical Approach

Mohammadreza Khanmohammadi and Amir Bagheri Garmarudi

Chemistry Department, Faculty of Science, Imam Khomeini International University, Qazvin, Iran

21.1 Infrared spectroscopy capabilities

Infrared (IR) spectroscopy is a well-known analytical method which refers to the vibrational modes of molecular bonding and results in the schematic demonstration of these vibrations for example, stretching, bending, scissoring, rocking, wagging, twisting, any combination of them and also overtones. Since the invention of first infrared spectrometer, it was believed that only pure analytes and maybe small structures are able to be analyzed by this method. Nowadays developments in accessory innovation and also trends in industrial endeavors have enabled IR spectroscopy specialists extremely well; proposing some novel capabilities of this analytical method for solution of problems in daily life. Some of the most famous developments are:

- Fourier transform
- High sensitivity detector systems
- Infrared region imaging
- Spectrometer accessories
- Fiber optic technique
- High-tech lasers
- Fast processors in computers
- Data processing and chemometrics
- Microspectroscopy instruments

Handbook of Green Analytical Chemistry, First Edition. Edited by Miguel de la Guardia and Salvador Garrigues.

- Time resolved techniques
- Apparatus parts based on nanostructures, nano materials and nano hybrids

Research and development by IR spectrometer producers is dedicated to more powerful systems, providing more capabilities while they are user friendly. The main goal of every scientific development is to facilitate life and it is desirable that efforts dedicated to life quality and health aspects are of high importance. In addition with its routine application in analysis of regular chemical and biological samples, IR spectroscopy has been gradually introduced as a common and useful approach for analysis of more complex structures such as multi component biological ones, cell lines, bio-fluids, tissues, and so on during the last two decades. This surprising and tremendous increment in IR spectroscopy favor is indebted to the professional focus on the theoretical basis for using it as a tool for structure analysis. A simple survey will provide a huge amount of publications discussing the IR spectroscopic study of biological molecules' conformations such as nucleic acids, carbohydrates, peptides and proteins. This could be referred to as the introduction to medical applications of IR spectroscopy.

Usually, most of the researchers' time in the field of analytical chemistry is spent in the sample preparation step which consists of separation, extraction, purification, condensation, concentration, and so on. This tedious step also consumes the main amount of chemicals which are utilized as the solvent, mobile phase, concentrator, extractor and phase transfer agent. Thus researchers are faced with some noteworthy subjects: lack of time and chemical wastes. Surprisingly, turning the methods which do not require sample preparation steps or need a simple one would also reduce the time consumption and chemicals' waste simultaneously. Even if the lack of time could be supposed as the common problem of all research activities and could be somehow solved by team work, the environmental problem raised by chemical pollution would remain a noticeable bug.

Global attention to the confirmation of environmental friendly activities has engaged all aspects of the human activities. Researchers are also concerned the harmful consequences of environmental pollution and try to replace the pollutant strategies with safe ones. In field of chemistry the general activities in this subject have been named 'green chemistry'. While most attention in green chemistry is paid to recycling, green analytical approaches are increasing gradually. Infrared spectroscopy as a rapid robust method which does not require a serious pre-analysis sample preparation step is capable of application to green analytical aims. IR spectroscopy, by utilization of instrumental and chemometric development could enhance the state-of-the-art in Green Analytical Chemistry.

Medical application of IR spectroscopy is an interesting example of this green analytical technique which also amplifies spectroscopists' novel ideas for invention of rapid diagnosis procedures. Medical and clinical applications of IR spectroscopy are divided in two categories: quantitative analysis and pattern recognition. Several bio-effective chemical analytes may be quantified simultaneously from the spectrum of a body fluid sample. IR spectroscopy is capable of reagent free analysis, providing valuable quantitative information for the physician about the health situation of the patient. In the other hand it is very important to diagnose the illness pattern via classification of IR spectra by an insured data base, co-analyzed by a routine diagnosis method. Pathology assisted IR spectroscopic analysis is suitable to be performed on in vitro biopsies or in vivo conditions. These analyses would reveal the molecular composition variations and correlate the IR spectral pattern with histological differences, concluding the diagnosis of disease affected or healthy normal samples. IR imaging is also capable of wider application as a green approach. In this technique microscopic functional group maps are interpretative result of chemical images providing by the apparatus. One of the main parameters influencing the diagnostic approach, allowing it to be called a green approach, is its invasive effect on the body during the procedure. According to the main aim which was to propose environmental friendly routes, it is also noticeable to apply techniques which would normally invade a patient during the clinical investigation are kept as low as possible. Taking in account the importance of a non-invasive technique to be used in the biodiagnostic method, IR spectroscopy is again a green analytical tool.

Researchers have reported the biochemical changes in the body as the main reason for any disease and IR spectroscopy has been established as a sensitive analytical route which can detect the reflex of any biochemical system to the electromagnetic ray of infrared region. It is worth mentioning that according to the complexity of IR spectral features, provision of exact information about the variation in bio-structures at molecular level is usually a very difficult task. However, several advantages of IR spectroscopy would dominate the common restrictions of this green analytical tool. IR spectroscopy is a very fast data assessing method which can collect and interpret the spectral data within few, minutes from samples in different physical states, for example; gasses, liquids, solids (pellets, powders, films, crystals, foams tissues, etc.), slurries and suspensions. The non-destructive nature of IR spectroscopy allows the analyst further experimental actions on the IR analyzed samples. The analysis procedure is free of a sample preparation step or only a very simple pre-treatment is required. On the other hand, IR spectroscopists could construct a spectral data bank containing large number of spectra which could be effective used in comparing experimental analyses, utilizing chemometric data processing strategies. The most noticeable advantage is the potential for reduction in misdiagnosis of results which is one of the main concerns in most of the diagnostic methods.

In this chapter, applications of IR spectroscopy in diagnostic analysis will be discussed. The main introduced diagnosis procedures are based on analysis of body fluids such as whole blood, serum, urine, and so on as well as tissue and cell collections which will also be reviewed. The chapter will finally report on the edge of science in this area for analytical developments.

21.2 Infrared spectroscopy of bio-active chemicals in a bio-system

Investigating the spectral characteristics of the bio-active chemicals in a bio-system is the main task of an analyst who tends to study a biological sample. Although there are large number of biochemicals in a bio-structured sample, carbohydrates, lipids, protein and nucleic acids are considered the main categories. Carbohydrates are commonly classified as mono-, di-, oligo- and polysaccharides. These biochemicals are the energy source in body which is quickly accessible. Glucose, a monosaccharide and hexose sugar, exists in bio-structure samples and is known as an intermediate energy source. It is a prevalent uncombined sugar which circulates in the blood and is essential to cell function. Its proper regulation in metabolism is of paramount importance to survival of the human. Single phase polysaccharides and their multi component structures containing proteins are also spread all around a bio-system such as the human body. Also, some pentose sugar oligomers play an important role in the backbone of nucleic acids that is, RNA and DNA. Mid-IR spectral analysis of carbohydrates such as mono- or polysaccharides demonstrateds several signal features which half of them would place in the 4000–1460 cm^{-1} spectral region with the other half below 1430 cm^{-1}. Later on we will discuss the reason by which we focus on the lower wavenumber region of mid-IR region, thus in this section, a brief list of main spectral features of the carbohydrates is shown in Table 21.1 [1–7].

Lipids are one of the other major components of a bio-structure which form cell walls and membranes. The lipid based membranes construct the scaffold stands which organize the biological active chemicals in their suitable situation. Appropriate compartmentalization of the constituent in a bio-structure is a critical matter which influences the proper functions of it. On the other hand, the biological displacement of bio-active chemicals via cell membranes is regulated by lipids. The main analyte classes of lipids which a medical analyst may be faced with are; cholesteryl esters, triglycerides and phospholipids. Infrared spectroscopy of lipids provides a huge amount of spectral features which is due to the complexity and variety of these chemicals. Most of the bio-related lipids consist of phosphorous comprising functional groups which amplify the complexity in their IR spectral pattern. Spectral-structural correlation of phospholipids has provided three divisions in mid-IR region: head functional group, interfacial structure and hydrocarbon chain. Some of the most usual mid-IR signals of lipids with their sub-structure assignment are presented in Table 21.2.

Table 21.1 *Mid-IR signal characteristics of carbohydrates.*

Spectral Region (cm^{-1})	Functional Group Vibration	Intensity
1470–1460	C–H asymmetric bending	strong
1460–1400	C–O symmetric stretching	strong
1420–1406	C–H in-plane bending	weak
1340–1280	C–O stretching	strong
1175–1125	C–H in-plane bending	weak
1150–1070	C–O–C asymmetric stretching	strong
1045–1035	C–O stretching	medium-strong
1005–990	C–H bending	very strong

Table 21.2 *Mid-IR signal assignment of phospholipid sub-structures.*

Spectral Region (cm^{-1})	Functional Group Vibration	Division
1750–1700	C=O stretching	interfacial
1500–1450	CH_2 bending	chain
1395–1360	CH_3 symmetric bending	chain
1230–1210	PO_2^- asymmetric stretching	head
1105–1070	PO_2^- symmetric stretching	head
1065–1050	C–O–P stretching	interfacial
995–970	CH_3–N asymmetric stretching	head

Proteins conduct a wide range of essential functions in body. Tissue construction and repair, water balance process, enzymatic and hormonal activities, immune service infrastructures, transport services (e.g., in hemoglobin), contractile action and antibody forming are some of the most well-known functions of polymers. The proper function of proteins is due to appropriate structural changes which would occur intramolecularly. There are several rearrangements in proteins, for example; torsion, hydrogen bonding, reorientation, nucleation, secondary structure folding, aggregation and side chain fluctuation all of which occur very fast and usually within 10^{-3}–10^{-12} seconds. IR spectroscopy has been widely applied in the study of these structural changes, especially for folding, secondary structure analysis and ligand interaction monitoring. The main reported features are related to amide functional groups of proteins which are called amide A, B and I–VII. Amides A and B's signals appear above the 3000 cm^{-1} wavenumber and the others are below 1800 cm^{-1}. It is mentionable that in sight of technical limitations only amides I and II (sometimes also amide III) could be utilized for further analytical aims as the most specific signals of protein structures.

Amide I is the most widely used band due to the protein backbone conformation and is particularly sensitive to structural conformation. This signal is in close relation with secondary protein structures such as carbonyl and C-N stretching; the major and minor contribution respectively. On the other hand nitrogen joint bonds are the main signals of amide II. The main assigned ones are C–N stretching and N–H bending. Although amide II is not as informative as amide I, it is included with amide I characteristics to improve the analysis efficiency. Table 21.3 shows more detailed information about IR spectral signals of analytically useful amide structures. Of course these bands are usually observed in coupling form and their separation is impossible.

DNA and RNA nucleic acids are vital structures which are very important for a bio-system. They carry the genetic codes which define the protein structures. The main constituents of nucleic acids are the phosphate group, saccharide sector and base structure. Base-pairing, base-stacking, base-sugar coupling and phosphate-sugar interaction are the main reasons why the IR spectrum of nucleic acid is really complex to interpret. In

Table 21.3 *Amide I and II related IR signals.*

Spectral Region (cm^{-1})	Functional Group Vibration	Amide
1690–1665	C=O stretching	I
1670–1640	N–H bending	I
1665–1630	C–N stretching	I
1590–1565	C–N stretching	II
1575–1550	N–H bending	II

Table 21.4 *Nucleic acid IR signals.*

Spectral Region (cm^{-1})	Functional Group Vibration
1705–1660	C=O stretching
1625–1600	C=C stretching
1605–1575	C=N stretching
1265–1220	PO_2^- asymmetric stretching
1160–1120	C–O stretching
1105–1080	PO_2^- asymmetric stretching
1065–1000	C–O stretching

the case of nucleic acids the asymmetric and symmetric phosphate stretching, and in-plane double-bond vibrations of bases have been reported as the most important IR spectral features. Some of the mid-IR signals are shown in Table 21.4.

Although all the above discussions about the capabilities of IR spectroscopy in analysis of bio-active chemical are really useful and interesting, there is a fact which concerns us about our goals and path. Almost all the clinical and medicine related bio-samples to be analyzed by IR spectroscopy are a combination of these chemicals and it is nearly impossible to separate and analyze them singularly. Thus we need to define some strategies which could enable the analyst to conduct a diagnostic process, taking into account all the probable variations in any of the chemical ingredients in bio-samples. When a medicine researcher decides to analyze a bio-fluid, such as blood, by IR spectroscopy, it is impossible to provide a separation system which can perform the quantitative analysis of sugar apart from other constituents which will probably cause some interference. Thus we need to discuss the capabilities of IR spectroscopy in case of complex samples. Application of IR spectroscopy for analysis of single analytes in body fluids and IR spectral pattern recognition are the introductions to this subject.

21.3 Medical analysis of body fluids by infrared spectroscopy

IR spectroscopy is very useful for clinical investigative analysis of bio-fluids according to its advantages which were discussed during the introduction of this chapter. The main procedure in a biodiagnostic route is to record the IR spectrum of the body bio-fluid sample, converting the spectral data to interpretable medical information and finally making a medical decision. The common mandatory steps are:

1) To remove the data which are not reliable due to the serious deviations from the other data which is called 'outlier detection' and to build a model of calibration, consisting of spectra from samples with known characteristics. The aim is to obtain a logical correlation between the spectral data and the object of the

analysis (desired analyte to be determined or the case situation). A gold standard method is necessary to enable comparative study and define the relation between spectra and objectives. It is very important to make a decision about the appropriate tool which can link the data to knowledge. Here is the point in which chemometric techniques will play a very important role. It is well-known that chemometrics is the supportive way by which initial data is converted to information and finally the knowledge is achieved. A simple regression, linear regressions such as classical, inverse and partial least squares or even more complex non-linear ones such as neural networks or non-linear PLS are very helpful case by case. On the other hand, supervised or unsupervised classification methods such as linear discriminant analysis, cluster analysis, principal component analysis, soft independent modeling class of analogy and even PLS or ANN are helpful for classification aims. Obtained statistical parameters must be as satisfying as possible prior to the second step. Some advanced chemometrics are also possible to accomplish for more powerful calibration models. The most famous ones are the genetic algorithm for wavelength selection and a wide range of learning algorithms in artificial neural networks. Of course the pre-treatment of recorded spectra data by some dimension reduction techniques, for example, region selection or PCA will effectively improve the quality of model, reducing the data processing time (the critical importance of chemometric data processing is discussed later).

2) Investigating the samples which are known to the analyst but are introduced as unknown ones to the constructed model. This will validate the initial model, enabling the analyst to fix the problems due to the non-conformities in the proposed model. It is regular to have a reduction in statistical reliability of the validation step but this causes some limitations in the next step.
3) Finally some samples which are unknown for both the analyst and the validated model are analyzed. The most critical statistical efficiency is reported for this step which is called 'real sample analysis'. Qualification for the generalized ability of a proposed method is based upon the obtained results in this step.

Generally, biomedical analysis methods based on IR spectroscopy are divided into two categories: quantitative analysis and pattern recognition. The first category consists of the techniques by which a bio-active chemical is quantitatively analyzed. In this kind of technique, the main aim is to define an IR spectroscopic based analysis procedure which could precisely determine the concentration of an analyte in blood, plasma, serum, urine, and so on. This could be a useful substitute for common metabolic assays which are currently used in clinical and biomedical laboratories. Of course it is noticeable that for quantitative analysis of biochemicals in bio-fluids, it is very important to provide a large population of samples for both calibration and prediction models which cover the widest range of the analyte in body. This helps the constructed model to be more effective and reliable. The second category of IR spectroscopic biomedical analyses for bio-fluids concerns the possibilities for diagnostic prediction of disease affection by interpretation of IR spectra directly. As the mid-IR spectral region consists of the fingerprint region for any analyte, it could be very informative for comparative investigations, resulting in discrimination of healthy people from those affected by a specific disease. The main proposed reason for this application is the relation between every illness and alteration in bio-fluid chemical composition. The main characteristics could be of variation in the concentration of a biochemical, structural rearrangement or chemical complexation, dissociation or degradation, presence of bio-markers and even disappearance of some bio-structures. In the other words, sometimes, the final result of IR biomedical survey is to answer the question, 'is the sample healthy or is the patient?', but in most of the situations, the final result of IR spectral analysis contains some information about the level of medically critical biochemicals. In this case, the physician is going to discover the health situation of the patient. In order to recognize the illness pattern by spectral analysis without quantitative analysis, the number of known samples which form the calibration model will dramatically affect the final reliability of the model. Thus the researchers try to provide as many samples as possible prior to this in order to define the model's structure.

21.3.1 Blood and its extracts

Blood is a very complex bio-fluid consisting huge number of chemicals. Human blood contains erythrocytes (red blood cells) and leukocytes (white blood cells) suspended in plasma. Several other bio-active structures such as hormones, anti-bodies, dissolved proteins, glucose, mineral ions and also CO_2 and CO dissolved gases exist in blood. Biochemical laboratories in which blood is analyzed use developed procedures including quantitative determination of several indices such as sugar, total protein, bilirubin, creatinine, urea, uric acid, several enzymes, lipid structures and dissolved cations. Any of the molecular indices in which concentration level in blood falls in to the detection limit of IR spectroscopy could be proposed for quantitative assay. IR spectrum of whole blood is full of interfering signals which are really hard to assign. In case of biomedical analysis, blood spectrum is usually acquired by IR sampling techniques in which the quantification process is performed. In this way, IR spectroscopic analyses are performed using fixed path-length accessories such as attenuated total reflectance (ATR) or microspectrometry.

One of the most common applications of IR spectroscopy for whole blood analysis is in monitoring of blood sugar. Diabetes is a very harmful metabolic disease. Medical concerning services report a wide range of body organ damage such as renal failure, blindness, brain damage or heart attack related to diabetes. About 5% of US citizens have diabetes and usually they have no control system which inhibits the increment in blood sugar content. Thus they need some medical treatment and also continuous monitoring of fast blood sugar as an informative parameter in disease stage recognition. One of the primary methods for measurement of glucose by IR spectroscopy was based on tissue penetration of IR ray which demonstrated some drawbacks. There were also some assays using whole blood as the sample matrix but problems due to interference of other constituents such as RBC gamma globulin and albumin lead to some limitations [8–10]. Later on and by development of data processing utilities the interference problems were resolved and IR spectroscopy could precisely monitor the whole blood glucose.

There are several reports indicating different sample physical states such as fresh blood, EDTA stabilized blood or dried samples which can be analyzed as a film. On the other hand, diffuse reflectance, attenuated total reflectance, photo acoustic and also transmission techniques have been applied for this aim [11–17]. Of course in case of liquid samples the role of water (about 90% of blood) is very important. There is a problem in mid-IR spectral region due to the intensive absorbance signals of water which restricts the extraction of useful signals and behaves as a shield which covers most of the important spectra characteristics. The common idea to avoid this problem is to use water's spectrum as the background. There is also a trend for development of more precise, accurate and sensitive procedures based on mid-IR spectroscopy. Researchers investigating the mid-IR spectroscopy for application in determining glucose in whole blood samples have reported a very different spectral region or even a single wavenumber as the best characteristic for the data extraction and analysis of signal-concentration correlation. The most indicted ones are 1200–950, 1500–750, 1119–1022, 1109, 1093, 1082 and 1040 cm^{-1} [18–22]. Thus it seems very important to select the optimal spectral region or signal by utilizing wavelength selection chemometric routes. Software supported elimination of spectral interferences and selection of more powerful instrumental hardware will also provide more reliable results. Some other constituents in whole blood samples have also been quantitatively and qualitatively analyzed by means of mid-IR spectroscopy, for example; hematocrit, albumin, glycoproteins and fibrinogen [23–25]. Recently, the suggestion was made to use the whole blood sample's mid-IR spectra for disease pattern recognition [26]. This idea will be discussed later in this chapter. It is a common pre-treatment procedure to use ethylene diamine tetra acetic acid (EDTA) as anticoagulant agent prior to analyze the whole blood samples.

Plasma and serum are two major blood components which have been commonly applied in IR spectroscopic studies as the analyzed sample. Plasma is a yellow liquid (more than 50% volume) in which blood cells are suspended. Water makes up about 90% of plasma. Glucose, hormones, proteins, clotting factors, hormones, dissolved ions and hormones are the main ingredients of plasma. The experimental procedure for preparation is to spin fresh blood in a test tube containing an anti-coagulant in a centrifuge instrument until the blood cells fall to the bottom of the tube. If the fibrinogens are removed from plasma, blood serum is prepared. Hormones, antigens and antibodies are the main constituents of blood serum.

Unlike the limited analysis subjects reported for whole blood, serum and plasma are used as the sample in different analytical aims. The same as whole blood, liquid samples or dried films of serum and plasma have been examined in IR spectroscopic analytical approaches. Several chemicals such as glucose, total protein, albumin, LDL, HDL, urea, globulin, creatinine, uric acid and triglycerides are suitable to be quantitatively analyzed. The most important advantage of these reports is to enable the analyst in the quantification of clinically important factors in serum without any sample preparation or consuming any chemical reagent. This advantage confirms the green aspect of introduced methods. Data processing chemometrical analysis plays a very important role in assessment to more precise and reliable results [27–32]. Several mid-IR spectroscopic related methods have been applied for investigational analysis of albumin in blood serum. A Japanese team has reported their studies on changes in the side chain hydrogen bonding of the secondary structure which may occur due to variations in pH of the serum samples. They have used two-dimensional ATR-FTIR spectroscopy in both synchronous and asynchronous situations. Also high pressure IR spectroscopy has been used in identification of stable misfolded states of serum albumin. A molten globule-like conformation has been proposed to be adopted in pressures of about 0.4 GPa. Analyzing the form intermediate, researchers have assumed a stable misfolded state for pressure unfolded–refolded protein [33–35]. Determination of LDL and HDL in blood serum by IR spectroscopy has provided results comparable to those of Friedewald formula for LDL and automated homogeneous assay HDL assay. On the other hand, determination of apolipoprotein B as a superior to LDL could be defined as an indirect method for quantification of this harmful cholesterol as a promoter of risk of coronary artery disease [36–38]. The same as whole blood, efforts for determination of glucose in serums is high and several spectral regions or signals have been reported as specific ones. The most commons are 1109, 1080, 1062–997 and 1035 cm^{-1}. Transmission IR, ATR-FTIR and quantum cascade laser optoelectronic absorbance have been applied for this aim [39–43].

Analytical methods based on analysis of blood plasma are more or less similar to those of serum [44–46]. Literature surveys show that serum analysis is more common than plasma. However, in both cases careful spectra manipulation and interpretation processing are mandatory to obtain useful results. Analysis for clinical goals requires being as rapid and simple as possible, because every pre-treatment step could cause a quantitative error, reducing the predictive power of the proposed method. Determination of glucose in plasma samples requires some complex data processing technique to obtain considerable results. Quantitative analysis of urea in 1520–1420 cm^{-1} and 1220–1120 cm^{-1} spectral regions by variable selection has been introduced as a feasible method to be applied in urea bio-sensors. Lipids, total protein content and protein adsorption in flow situation are the major investigated topics [47–50].

As mentioned previously, there is also a noticeable demand for prediction of health status without any quantitative result. In this type of biodiagnostics, the clinical assessor has to interpret the spectral pattern in order to define the borders between a healthy sample and a patient one. Several signal features such as intensity, shift, ratio, broadening, sharpening and other kind of tolerances will be the input to make a decision about the health situation. In case of blood and its extracts, there is a novel aspect which deals with classification of healthy people and those affected by cancer via whole blood IR spectra. This subject is discussed later in this chapter. Analysis of hamster serum samples for diagnosis of scrapie infection and also coronary related disease are some of the most common examples [51–53]. IR spectroscopic analysis of serum and synovial samples for diagnosis of diabetes and rheumatoid arthritis has also been reported [54–59]. Nowadays renal failure is also studied by IR spectroscopy via blood product analysis. It is worth mentioning that in this area the role of data processing operatives is more critical. Here, the final decision is made according to the borders defined in calibration step and more precise classification is dependent on the quality of the initial model on known samples.

21.3.2 Urine

Aside from water, urine contains a wide range of chemicals, for example; inorganic salts, proteins, hormones, and many other metabolites. The most clinically significant ingredients are urea and creatinine. Thus it could be attractive to propose that IR spectroscopic analysis is a reagent free approach which can be performed for multi component analysis. Urea, creatinine and uric acid are possible to be simultaneously determined in urine samples by IR spectroscopy. The other reported parameters are also pH, phosphate and sulfate concentration. There have been some efforts for glucose and protein determination which sometimes have been faced with some limitations. The IR detection limits are very effective in provision of satisfactory results. Of course apart from pre-concentration, extraction, aand tedious treatments, there are also some reported pre-conditioning techniques which would improve the capabilities, enabling the analyst to overcome the problems made by detection limit of IR spectroscopy. One of these new introduced techniques is micro-fluidic sample preconditioning based on laminar fluid diffusion interface. Observations have confirmed that it should improve the accuracy while extends the analytes' detection range. Clinical application of IR spectroscopy for urine samples is not limited to quantitative analysis of urine ingredients. Hyphenated analytical methods such as GC-FTIR for quantitative analysis of the drug amphetamine and its derivatives, and NMR and IR spectral analysis for distinguishing between normal renal transplants and rejected allografts are some of examples [60–65].

21.3.3 Other body fluids

Saliva as an easily accessible sample is useful for many diagnostic aims. It consists of water, enzymes, mucus, electrolytes, and antibacterial biochemicals. A wide range of analytes such as glucose, amylase, total protein, urea, immunoglobulin A, cortisol, phosphoprotein and phosphate have been precisely determined in saliva by means of ATR-FTIR spectroscopy. In case of pattern recognition studies, IR spectral analysis of saliva samples from diabetes affected patients has revealed informative differences in major metabolic components such as glucose, lipids, proteins, thiocyanate and carboxylate compared to those of healthy people. Saliva analysis also enables diagnostic investigation of enamel decalcification in orthodontic treatment and prediction of enamel decalcification risk [66–69]. IR spectroscopy of mucus as another oral body-fluid is helpful for clinical diagnosis and even environmental monitoring. Amide related signals in mid-IR region in addition with sugar side-chains could be utilized for clinical supportive decision making [70]. Kidney dysfunction (glomerulonefritis) has also been monitored via IR spectroscopy of patients' sweat. Tears from left and right eyes of human have been analyzed by IR spectroscopy to find the probable spectral variation between them [71, 72].

21.4 Diagnosis in tissue samples via IR spectroscopic analysis

The second section of this chapter confirms that IR spectroscopy is a sensitive analytical tool which can be generally used to study biomolecules. As tissues are the complex form of cells which contain the above discussed bio-structures, IR spectroscopy is proposed as a useful analytical method for tissue analysis. Of course, the variety in tissue samples is usually more complex than body fluids and the consistency in tissue structures is not as strong as bio-fluids. In a bio-fluid all parts of a homogenous sample are in the same characteristic while in a small tissue sample the biochemical structure is completely changed from one side to another one. Thus, in IR spectroscopic analysis of a tissue sample, it is absolutely critical to make an initial judgment about the biological conformation and morphology before proceeding with IR spectroscopy. In bio-fluids there is a similar distribution of cell chemicals in the liquid media and this makes the analysis easier but in tissue, the basic unit is the cell and tissue types are categorized according to cell types. The main tissue types are epithelial, muscular, connective and nerve. Most of the body's organs are compilation of these types of tissue. Thus the tissue related samples will be very heterogeneous, comprising several components and different cell types. The main variations are in cell size, shape, morphology and ingredients. In order to assume a powerful diagnosis method, it is very important to select a correct sampling method. While the analysis of body fluids is going to earn a specific position in medical diagnosis, tissue investigation is a very common route for diagnostic studies. It is well known that any disease effect will have an influence on bio-fluids but clinical analysis in some cases are completely dependent on analytical assessment of the organ in which the physician is in doubt about related to the disease. Cancer is a typical case of a disease in which most of the diagnostic decisions are made according to tissue analysis. Oncologic diagnostics relies on the efforts for differentiated observations which enable the physician to recognise of nature and determination of the situation of disease. Abnormal distortion of the cellular structures is directly caused by quantitative or qualitative changes in bio-active chemicals in tissues. The symptoms by which the clinical investigators make the illness pattern recognition are also related to these changes in the chemical composition of the cells, tissues and finally the organs. Thus IR spectroscopy could be a helpful tool which probes the changes in molecular scale. Singular structure changes, intercellular variations and interactions between the tissue matrix ingredients are all possible to monitor by IR spectroscopy. In this section we will review the possibilities of applying this method as a green diagnostic tool for the detection of cancer. As reported in all the medical reports, cancer, by about 13% of total rates, is known as the leading cause of death all around the world. During 2005, more than 7 600 000 people died because of cancer. This statistic was increased to about 8 000 000 deaths during 2007 and is dramatically increased year by year. This is continuously rising and estimated to be more in future, leading to more than 12 000 000 in 2030. It is possible to avoid many of these deaths by early detection, treatment and cure; also some cancers are possible to prevent.

Cancer is a phase disease and is spreads out over time thus detection of cancer in its early stages would dramatically increase the possibility for more successful therapy. Innovations and discoveries in cancer therapy methods are researched together with efforts in development of rapid cancer diagnosis methods. Heredity situation, environmental harm and failure patterns in cell growth are the main causes of cancer which is started by alteration in genome composition in cellular structure. Thus the initial phase is also related to variations in biochemicals.

Among several diagnostic methods for cancer detection, pathologic analysis of biopsy samples is very common and called the 'gold standard'. In this method a pathologist would tag the structural situation and cellular distribution in the tissue biopsy samples by visual observations using a light microscope. There are some serious drawbacks in this routine clinical diagnostic method. The main ones are:

- It is a time consuming process in which the pathologist must observe several parts of the biopsy and sometimes several biopsies.
- The biopsy is susceptible to environmental contamination by different agents during the sample preparation procedure.
- It is known to be a highly expensive technique.
- Obtained results are completely dependent on the pathologist's level of expertise.
- It is not usually reliable and the 'second look' confirmation by another pathologic service system is needed. This may also cost and needs further time before receiving the final result.

Thus these limitations and drawbacks have concerned the oncologic researchers in development of some new diagnosis methods which not only reduce the role of pathologist's experience on obtained results but also may be more sensitive, accurate, precise, rapid and really 'reliable'. The economic aspects of the new proposed methods would also affect the possibility of generalization for global application, even for poor nations.

According to the analytical potential in IR spectroscopy, it has also been applied as a diagnostic tool in cancer detection. In the previous sections it was discussed that IR spectroscopic analysis is aimed to be applied to several biomedical approaches which two of the main fields are quantitative analysis and pattern recognition. Application of IR spectroscopy in cancer diagnosis falls into the pattern recognition category. It may be supposed that in this field one only needs to acquire the IR spectra of healthy and patient samples, attempting to discriminate them in two classes. But the reality is a little more complex. Efforts in this subject had led to some more technical points of view. In order to define a diagnosis method as a useful one, it has to be statistically powerful. In addition with common statistical parameters which are important in analytical chemistry, clinical definitions have made two more parameters: sensitivity and specificity. Sensitivity is defined as the percentage of disease affected people who are identified as 'illness case' or proportion of actual positives which are correctly classified, while specificity is the percentage of healthy people who are identified as 'control case' or proportion of negatives which are correctly identified. In the other words, sensitivity is a criterion by which the method's ability to diagnose an illness pattern is evaluated and specificity is the parameter which explains the method's power in prevention of any conflict in diagnosis made by misdiagnosis of healthy samples. Spectral changes that must be detected in IR spectroscopic analyses are a collection of characteristics such as intensity changes, peak ratios, signal shifting, and so on. Thus the proposed strategy must be very selective to provide the final result whilst taking into account all of these changes. On the other hand, some of them must be precisely distinguished from common intrinsic variations which are very common in mid-IR spectral region, in correlation with repeatability and reproducibility of the spectrum acquisition process. Although the spectrum recording step is very important in provision of useful spectral data, the initial sample could also be impressively operative. Prepared tissue samples must cover a wide range of cellular structures in an organ which enables the investigation for the analyst of different stages in the disease. Total number of samples from patients and healthy people must be as many as possible. Schematization for acquiring the samples in a wide range of patient age from both males and females would inhibit the diagnostic model's dependency on local and geographical factors and people's living factors. All these criteria have developed into a new field of research, called 'Vibrational Pathology' [73].

21.4.1 Main spectral characteristics

Although the variation in tissue structures during the growth or spread period of a malignant or even benign tumor is really huge, researchers could have defined some effective spectral information which is commonly addressed during the IR analysis of a tissue. It is desirable that this information is also divided in to biochemical related classes. Lipids, proteins and nucleic acids have demonstrated the most serious variations in spectral data, obtained in mid-IR region. Table 21.5 provides brief information about these common signals.

As seen in Table 21.5, amide related signals are also effective in tissue sample spectra. Malignant structural disorders would also manifest in these spectral bands. The main variations have been observed in amides I and II, phosphate and amine groups, related to structural changes in protein such as amount of α-helix,

Table 21.5 *Spectral characteristics of tissue samples in mid-IR region.*

Biochemical	Functional Group	Spectral Region (cm^{-1})	Vibration Type
Lipid	CH_2	700–730	Rocking
Lipid	R–N–C	965–980	Asymmetric Stretching
Nucleic Acid	PO_2^-	1060–1080	Symmetric Stretching
Protein	CH_3, N–H, C–N	1010–1025	Stretching-Bending
Protein	C–N, N–H	1460–1470	Stretching-Bending
Several Components	O–H	1640–1645	Bending
Protein	C=O, C–N	1650–1665	Stretching
Lipid	C=O	1735–1750	Stretching

parallel and anti-parallel β-sheets, and mutations in nucleic acid. Thus it is important to assign the signals while the spectral pattern is analyzed to obtain a reliable diagnostic result. The main focused spectral differentiating information has been proposed for vibrations of methylene chains in membrane lipids, hydrogen bonding of the phosphodiester groups of nucleic acids, glycogen content, hydrogen bonding of C–OH groups in carbohydrates and proteins, symmetric phosphate stretching band of the phosphodiester groups of nucleic acids and the methylene: methyl ratio.

21.4.2 The role of data processing

Although the signal assessment process has led to useful diagnostic results, it is inevitable to propose an expandable medical method without proposing a reliable predictive model. This is the situation in which chemometrics has an effective role, making the clinical approaches powerful. In this section we will briefly review the main aspects of chemometrics as an emerging technology in all approaches to analytical chemistry in order to define the requirements in clinical biodiagnostics. Chemometrics has always been a practically oriented sub discipline of analytical chemistry for handling, interpreting and predicting chemical data. During the past three decades, raw chemical data has increasingly been processed by computers. The interdisciplinary field of chemometrics combines computing with applied mathematics, statistics and chemistry to extract useful information from raw and large chemical data through univariate or multivariate methods. In case of multivariate methods, chemometrics can be used in such areas as:

- Outlier detection
- Calibration
- Classification and pattern recognition
- Image Analysis

Of course the first three steps are the most common respective algorithms in chemometric based data processing. Even though the image analysis diagnostic methods are also based on chemometric data processing, they can be classified as pattern recognition.

21.4.2.1 Outlier detection

All results will be affected by error of input dataset. Ideal input data should be error free. However, analysts can encounter outlier data. Outliers in multivariate data can severely impact the results of statistical analyses. As referenced by Jouan-Rimbaud during the last decade, there are two types of chemical outliers:

1) The prediction samples contain the same components as the calibration samples, but in concentrations that are outside the range in the calibration set, so that their prediction would lead to an extrapolation of the model.
2) The prediction samples contain components that were not present in the calibration samples (interference causing species) [74].

Detection of data outliers and unusual data structures is one of the main tasks in the analysis of data. It is very important to detect the outliers in a very precise way. Traditionally, despite the fact that data sets are almost always multivariate, outliers are most frequently sought for each single variable in a given data set. The standard diagnostics for outlier detection in the calibration phase of a multivariate calibration experiment have been thoroughly discussed by Martens and Næs [75]. Accordingly, there are two approaches for outlier detection. The first approach is to fit the data with least squares, construct regression diagnostics and then remove the outliers. The second approach is to construct estimators that fit the majority of the data and

Table 21.6 *Literature review in development and application of outlier detecting chemometric techniques.*

Chemometrician	Year	Main idea
D. Rohlf	1975	The largest edge of the minimum spanning tree
P.J. Rousseeuw	1985	Minimum volume estimator (MVE) and minimum covariance determinant (MCD)
R.G. Garrett	1989	Chi-square plot as a tool for multivariate outlier recognition
D.L. Woodruff	1996	Detection of multiple multivariate outliers which were not always detectable in data with contamination fractions greater than 35%.
D. Jouan-Rimbaud	1999	Mahalanobis distance
A.S. Kosinski	1999	Strongly resistant to such high contamination of data with outliers
R.J. Pell	2000	Use of robust principal component regression (PCR) and iteratively reweighted partial least squares (PLS) for multiple outlier detection
K.A. Hoo	2002	Develops the approach, discusses the concept of robust statistics and winsorization, and presents the procedures for robust multivariate outlier filtering
T. Lillhonga	2005	Replicate analysis and outlier detection in multivariate NIR calibration
M. Khanmohammadi	2009	Outlier detection using the leverage method for quantitative analysis

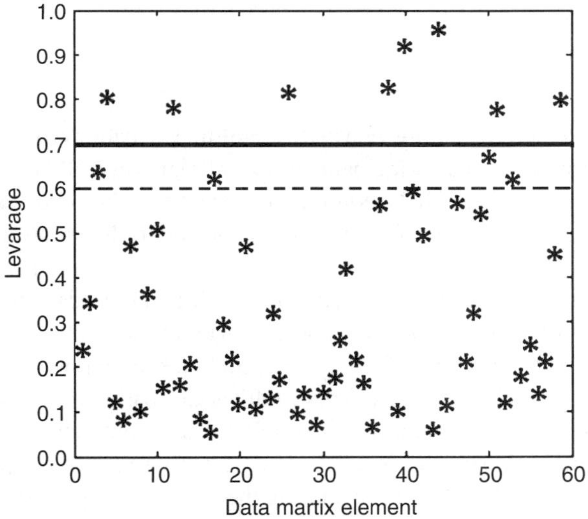

Figure 21.1 *Leverage scheme for the detection of an outlier in a data matrix.*

examine the residuals from this fit to detect outliers. Table 21.6 reviews the various outlier detection methods which have been discussed by different chemometricians [76–82].

The recent advances in outlier detection have led to development of several techniques, such as Leverage method. In this technique a data matrix (X) is formed, trying to detect outliers in the X space examining the leverage of each sample. The leverage of a sample is a measure of its spatial distance to the main body of the samples in X. The leverage of a sample in a given data matrix is determined as the elements of P matrix:

$$P = X(X^TX)^{-1}X^T$$

Large values for each element of P matrix means the sample falls into outlier data. Of course the edge of leverage discrimination graph is also effective in this step. Typically, Figure 21.1 shows the outlier detection results in Khanmohammadi *et al.*'s research. It is observed that the total number of outlier data is directly related to the fixed leverage limit.

21.4.2.2 Calibration

In analytical studies, the use of single data calibration has almost been abandoned. Obtaining additional information and selective measurements, the multivariate calibration techniques based on different types of raw data are more in use. Multivariate calibration is usually applied to complex real matrices, where the separation procedures and chemical treatments necessary to apply a simple calibration are expensive and time consuming. Multivariate calibration uses non specific predictors, generally physical information from spectra such as near infrared (NIR), UV, visible, Raman, mid-infrared (MIR), fluorescence, NMR and mass spectra. Although results may not be completely accurate, there is usually sufficient accuracy with many real problems in comparison with the results using single-data calibration. It is a multi step procedure, each step with its own importance, and having possible consequences for the steps that follow. The following steps are necessary for the accuracy of the calibration model:

- selection of a calibration set
- data pretreatment
- checking for appropriateness of calibration and validation samples
- investigating the homogeneities in data
- constructing the calibration model
- detecting the model outliers
- validating the model

Multivariate calibration does not simply consist of building a calibration model to a given data set. Multivariate calibration methods such as principal component regression (PCR), partial least squares (PLS), ridge regression (RR), generalized RR (GRR), continuum regression (CR), cyclic subspace regression (CSR), and ridge variations of PCR, PLS have been particularly subjected to the spectrophotometric determinations in many different fields of application, especially in the areas of food chemistry, pharmaceutical analysis, agriculture, environment, industrial and clinical chemistry. Multivariate calibrations are useful tools to be used in spectral analysis in order to overcome the spectral overlapping and to improve the precision and the predictive ability of the FT-IR spectrometry. PLS is one of the most common techniques, abundantly applied for quantitative aims in analytical researches. This multivariate calibration strategy has been introduced useful in spectral analysis since the overlapping signals of chemical components can be differentiated and the simultaneous inclusion of multiple spectral intensities can greatly improve the precision and the predictive ability in the analysis procedure [83–87]. Mean centering the spectral information prior to PLS data treatment will extend the capabilities [88,89]. The number and nature of the samples to be used for calibration are always critical factors in multivariate analysis. Application of some other techniques such as hierarchical cluster analysis for selection of the calibration and validation datasets has also been reported. In this sense, the selection criterion could be defined to use at least one sample from each cluster for calibration. In the case of clusters with several elements, the square root of the total number of samples included in the cluster could be assigned as the number of samples which will be used in the calibration model. In this way, the other elements of the cluster will be integrated to validation set. It is clear that the total number of samples in the validation set would be the same or more than in calibration samples [90]. More complicated chemometric approaches such as wavelength selection techniques may also introduce new advantages in calibration step. PCR, successive projection and genetic algorithm (GA) which have been reported as model improver techniques and support the final statistical parameters as being more reliable in an analytical process [82,91]. Of course, the PLS technique itself has also been reported as wavelength selection [92].

21.4.2.3 Classification

Although there are several analytical progresses resulting in quantitative data, decision making in diagnostic approaches would finally lead to positive or negative report about the presence of illness pattern. In this case one needs to define some categorical differentiable classes for the obtained data. The most common is to divide the samples in healthy or disease affected classes. Sometimes more precise separations are needed. This important area of data processing is called classification. Pattern recognition and discrimination analysis are the main aims of this chemometric prestige. Univariate analysis methods can consider only a single characteristic of the sample as the dependent variable in a selected given independent variable, thus it is usually necessary to utilize multivariate pattern recognition methodologies. Multivariate chemometric data processing allows the evaluation of several properties of the samples simultaneously. Benefits of multivariate data treatment would help to realize the correlations between thousands of analytical data points with several properties of the sample. The main multivariate classification techniques are principal component analysis (PCA), hierarchical clustering analysis (HCA), linear discriminant analysis (LDA), support vector machine (SVM) and artificial neural network (ANN).

Classification by chemometric data processing is also known as supervised pattern recognition. This field of chemometrics tries to find a precise answer for the question about the exact position of an unknown element in the data set and its main application is in assignment of the elements to a number of different classes. Thus it requires a training data set in which the situation of each member and the class to which it belongs in known. The main four steps will be:

1) Modeling the training set
2) Test sets evaluation and cross-validation
3) Improving the data
4) Applying the model to unknown data

Discrimination for the classification aims (e.g., LDA) are usually called hard modeling. LDA is utilized as a:

- predictive method for formulating a discrimination rule used to predict and allocating unknown samples in predefined classes; and an
- exploratory tool, increasing the differentiation power between classes.

Figure 21.2 shows a typical plot of LDA data processing for diagnosis of cancer via blood analysis. As shown in the figure the discriminating line would enable the researcher to see the probable misclassification.

Principal component analysis is probably the most widespread multivariate chemometric technique which allows the analyst exploration of different patterns in a data set. It may also be applied in data compression and extraction of main informative ones among the huge amount of data. Cluster analysis is a well-established approach which usually focuses on the *K nearest neighboring* (KNN) parameter as the initiator of a powerful classification technique. In this classification approach, the objects are grouped so that 'similar' objects fall into the same class. As the main criterion, objects in one cluster should be well separated from other elemental groupings and also homogenous in relation to some characteristics explaining within cluster properties. It is applied in modeling the calibration samples in groups of two or more while the most commonly selected cluster methodology is the KNN algorithm with Mahalonobis distance measuring. In case of biodiagnostics by infrared spectroscopy, cluster analysis has been rarely reported as a powerful method. Figure 21.3 demonstrates the large amount of misclassification in diagnosis of basal cell carcinoma by IR spectroscopic analysis of blood samples. There are some other chemometric classification techniques which will be discussed in Section 4.3.

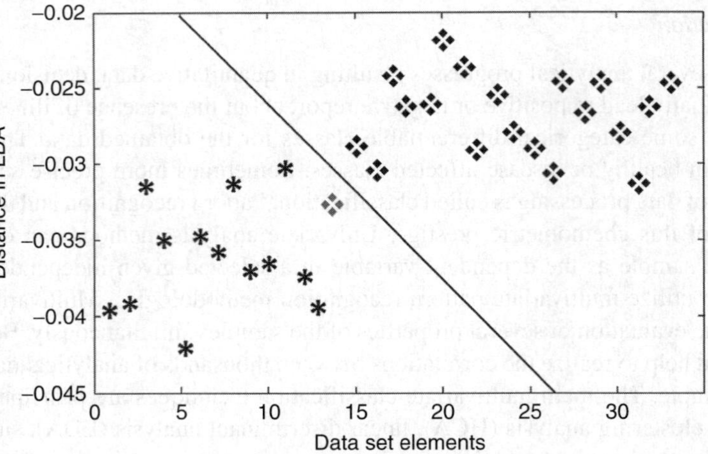

Figure 21.2 *General cancer-healthy classification of IR spectra of blood samples by LDA.*

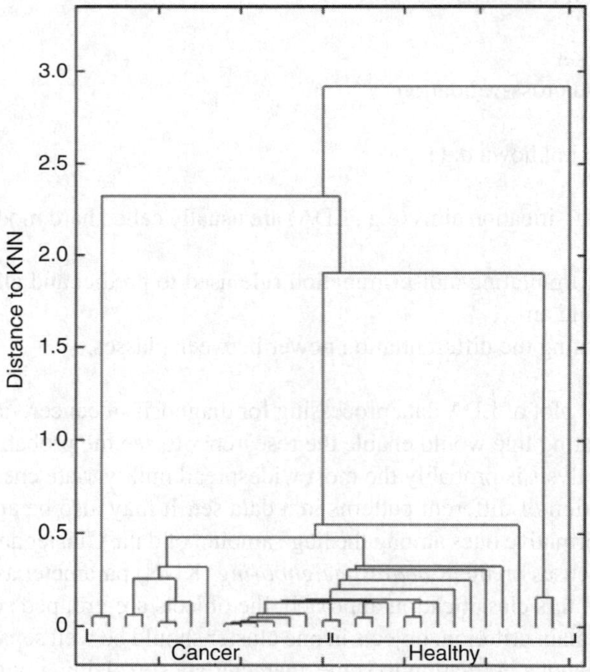

Figure 21.3 *Cluster analysis dendrogram for differentiation of cancer-healthy blood samples in case of BCCv.*

21.4.2.4 Image analysis

Biodiagnosis by IR images has been applied during the last two decades in several physician evaluations. The main fields are; cancer, dentistry, neurology, cardiovascular, drug delivery, ophthalmic, surgery, respiratory and tissue analysis. The main signals obtained in IR medical imaging are due to the thermal effect of the IR ray on the organ, bio-fluid or tissue sample. There are several criteria which must be taken into account in order to obtain more useful images. The main concerns in this area are to enable the analyst to recognise the inter-relations between the IR image signals and pathological or medical nature of them, and also to provide the standard models of known samples in order to be able to interpret the qualitative data in to quantifiable clinical information. Breast cancer is diagnosed by the IR imaging process. Of course the chemometric techniques are really effective in obtaining the most useful clinical results. Wavelet data processing and ANN have been reported as high quality techniques in disease situation classification via IR imaging [93].

21.4.3 Cancer diagnosis by FTIR spectrometry

There are several reports indicating the diagnostic capabilities of IR spectrometry for detection of malignancy in a tissue sample. Lung, ovarian, esophagus, colon, and gastric cancers among others have been successfully detected via investigating the mid-IR spectrum of tissue samples. In this section we try to discuss the capabilities of this technique in cancer diagnosis, comparing a multi step effort for diagnosis of cancer in digestive organs of human: esophagus, gastric and colon. Colon, gastric and esophagus cancers are the main malignant diseases in digestive system of human body respectively. Among these three diseases, colon cancer is the highest ranked according to the mortality rates. Comparing all malignancies, colon cancer will be the third leading cause of death (about 150 000 detected cases and 65 000 deaths in a year in the USA alone). Early diagnosis of colon malignancy may provide about five years of survival in more than 90% of cases. If colorectal cancer is detected at an early stage, then the five-year relative survival rate is 90%; however, only about 37% of colorectal cancers are diagnosed in the early stages. Abnormal cell proliferation is known as the main indicator for the initiation of colorectal malignancy which is identified by investigating the probable presence of certain molecular markers in the crypt. Gastric malignancy related diseases are also common all around the world. The main gastric diseases diagnosed by endoscopy and/or biopsy analysis are chronic superficial gastritis, gastric carcinoma and chronic atrophic gastritis. Professional characterization of biopsy samples obtained from stomach of the patient will dramatically affect the final diagnosed condition. On the other hand, esophagus cancer is still one of the concerning malignant tumors which may be called as a seriously dangerous disease. The approximated statistics of esophageal carcinoma is more than 1.5 million per year in China alone. Fiber endoscopy is the lesion analyzing device used which can also obtain the lesion sample for further pathologic inspections and is known as the best way to diagnose the disease.

Application of IR spectroscopic analyses or combination of the technology with microscopic accessories has been highly powerful in the precise detection of malignant cases in tissue biopsies. To propose a useful approach, one needs to study the cell lines in order to extract the demonstrated spectroscopic features. This strategy would be an appropriate experimental system which provides the requirements to study the origin of the spectroscopic changes observed hitherto and to apply the IR technology to other aspects of cancer biology. In case of tissue samples, the same as all other biostructures, the splitting pattern is mostly found in the 1800–1000 cm^{-1} spectral region displaying an extensive set of spectroscopic changes between malignant and healthy tissue samples. As expected, most of the clinical reports dealing with cancer detection by IR spectroscopy are based on experimental investigation of:

- Isolated signal intensities
- Signal position shift
- Intensity ratio of correlated signals

Table 21.7 *Mid-IR spectral features applied in cancer detection via tissue analysis.*

Spectral Region	Functional Group	Vibration	Related Biochemicals
960–970	C–O, C–C	Stretching	DNA
1010–1030	C–O–H	Bending	Glycogen
1040–1055	C–O	Stretching	DNA
1070–1080	PO_2-	Stretching	DNA
1075–1085	C–C	Stretching	Glycogen
1115–1120	O–H	Bending	Glycoproteins
1145–1160	C–O	Stretching	Glycogen
1165–1170	C–C	Stretching	Glycoproteins
1230–1240	C–N	Stretching	Amide III
1245–1265	N–H	Bending	Amide III
1280–1300	C–H	Stretching	Amide III
1405–1440	C–H	Bending	Amides
*1540–1550	C–N	Stretching	Amide II
1545–1560	N–H	Bending	Amide II
*1630–1640	C=O	Stretching	Amide I

*the most common signal position shifting is observed (from healthy to malignant cases the shift is from higher to lower wavenumbers).

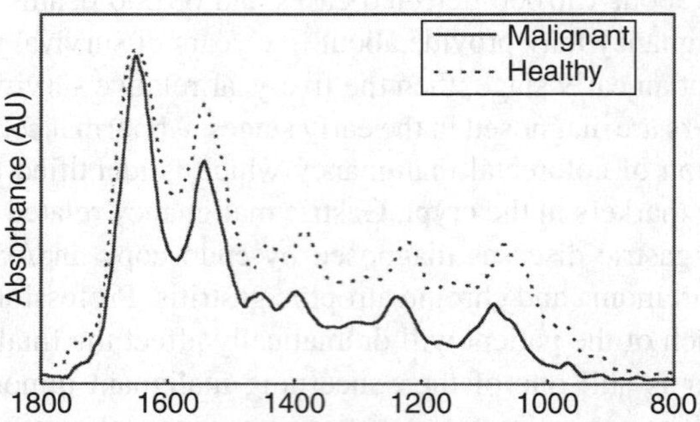

Figure 21.4 *Typical IR spectra of normal and malignant tissue samples.*

Table 21.7 details the main isolated signal variations (intensity and shifting) observed in tissue samples used to recognize the cancer or healthy cases via IR spectroscopic analysis of biopsies.

Considering all the above mentioned variations, chemometric based data processing algorithms based upon statistical-mathematical approaches would improve the reliability of diagnostic results. Of course, intensity ratios are also effective in cancer diagnosis. Reduction in amide I/amide II and RNA/DNA, and increments in glycoproteins/amide I signal intensity ratios during the progress of malignancy have been reported to be informative. Figure 21.4 shows typical IR spectra of normal and malignant tissue samples. Initial spectra data are usually transformed to matrices to be processed by chemometric classification techniques. Analysis of variance (ANOVA) is a useful starting process by which it is possible to study the

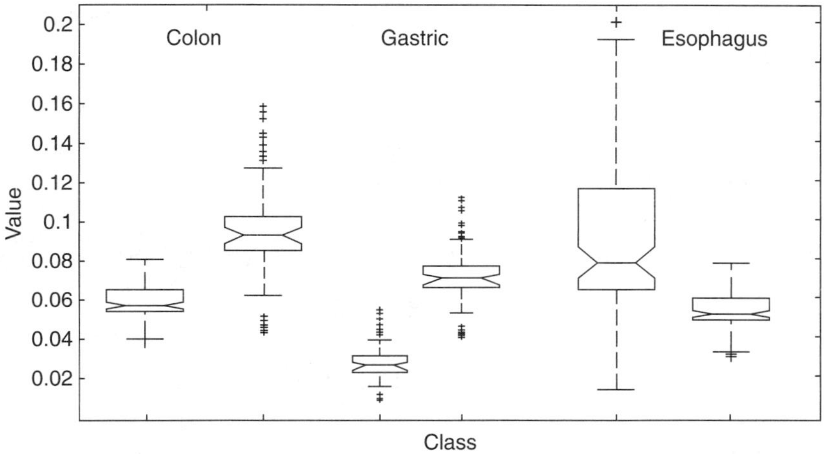

Figure 21.5 *Box plot of initial feasibility for classification of blood samples' IR spectra according to the cancer or healthy situations.*

feasibility of classification between spectral data. In this process two classes are formed consisting of a similar number in a group (e.g., 15 spectra of normal cases in class A and 15 spectra of malignant cases in class B). Then a box plot is obtained by ANOVA, and by comparing the box of each class and also the variance values the possibility for classification is concluded. Figure 21.5 shows the cumulative box plot for all three cancer cases. More distance between the mean values of the boxes of each class will ensure the researchers to better differentiate the ability of the chemometric models.

Usually a size reductive process such as PCA is also needed to provide the most informative set of data to be classified. It is mentionable that PCA itself has also been reported as a classification strategy. Of course in case of cancer diagnosis, according to the complexity of spectral features and huge amount of variations, the simple PCA would not be able to produce any noticeable result for detection of healthy or malignant cases. Score plots obtained by performing PCA on IR spectra of the organs' tissue samples which are shown in Figure 21.6 confirm the lack of differentiation. Linear discriminant analysis of spectral data obtained from samples can also be effective in reliable diagnostic classification of tissue samples. Application of LDA for this aim has been reported as a very effective strategy for diagnosis of cancer. As shown in Figure 21.7, LDA of IR spectral data from colon, gastric and esophagus tissue samples confirm this fact. Considering IR spectroscopic analysis of tissue sample as a useful biodiagnostic approach, one needs to examine the statistical parameters which are very important for clinical decision making. Two of the most famous are sensitivity and specificity which are dealing with method capabilities as:

Sensitivity: probable infection of the sample to the illness pattern while the analysis response is positive:

$$\frac{\text{Number of True Negatives}}{\text{Number of True Negatives} + \text{Number of False Positives}}$$

Specificity: probable negative response for the case which is healthy:

$$\frac{\text{Number of True Negatives}}{\text{Number of True Negatives} + \text{Number of False Positives}}$$

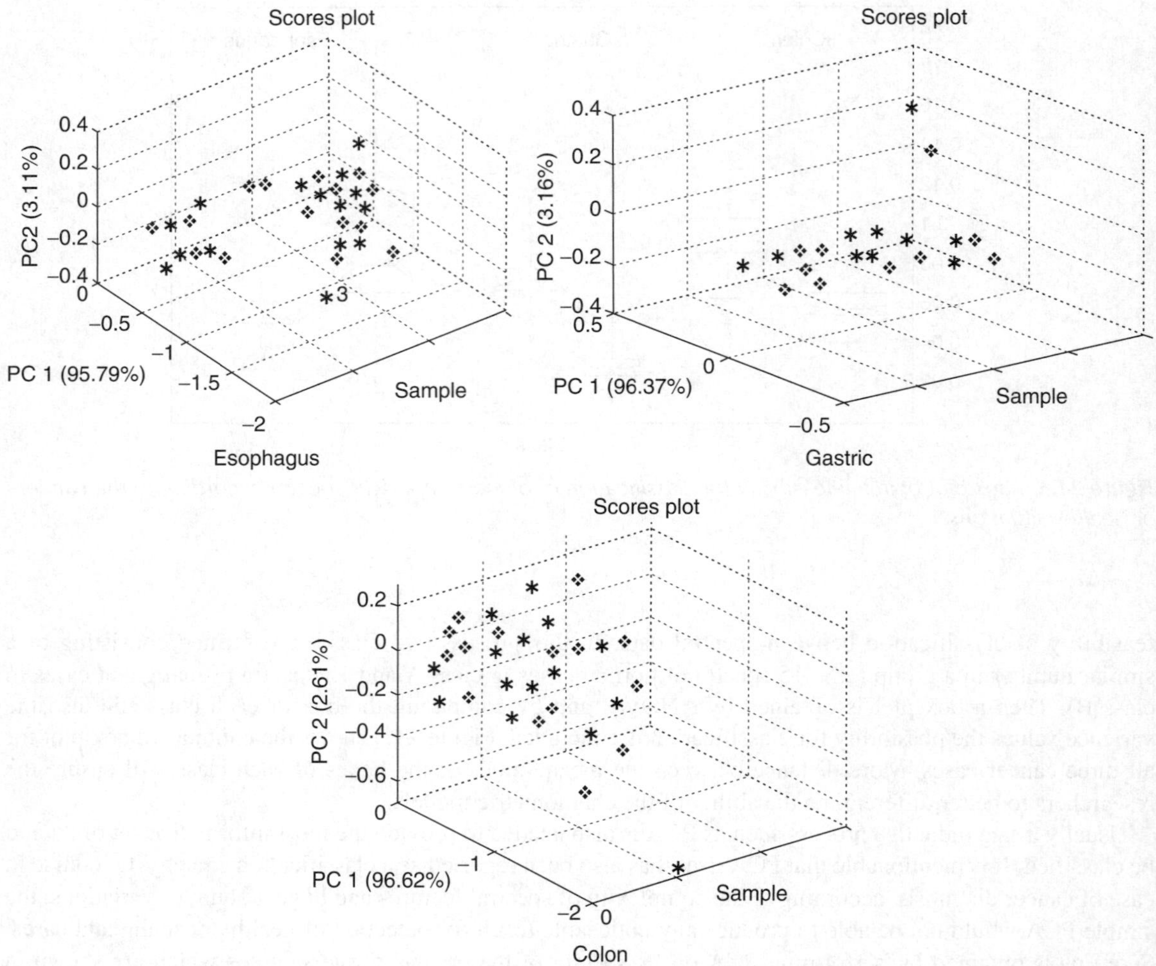

Figure 21.6 *Score plot of PCA on IR spectra of healthy and cancerous blood samples.*

21.5 New trends in infrared spectroscopy assisted biodiagnostics

Development of new hardware accessories in field of IR spectroscopy, new sampling techniques and also brilliant movements in chemometrics have all led to huge diagnostic approaches based up on IR spectroscopy. Some of these have even been transformed to technology and are utilized in pathobiologic and diagnostic laboratories. Together with technical improvements the clinical chemistry specialists are also attending to the methods by which the results are obtained:

- Statically more powerful
- Rapid
- Capable of generalization
- Not to be invasive

The last item is really important for medical science researchers because they prefer not to impose any damage on the inspected organ or any other part of body.

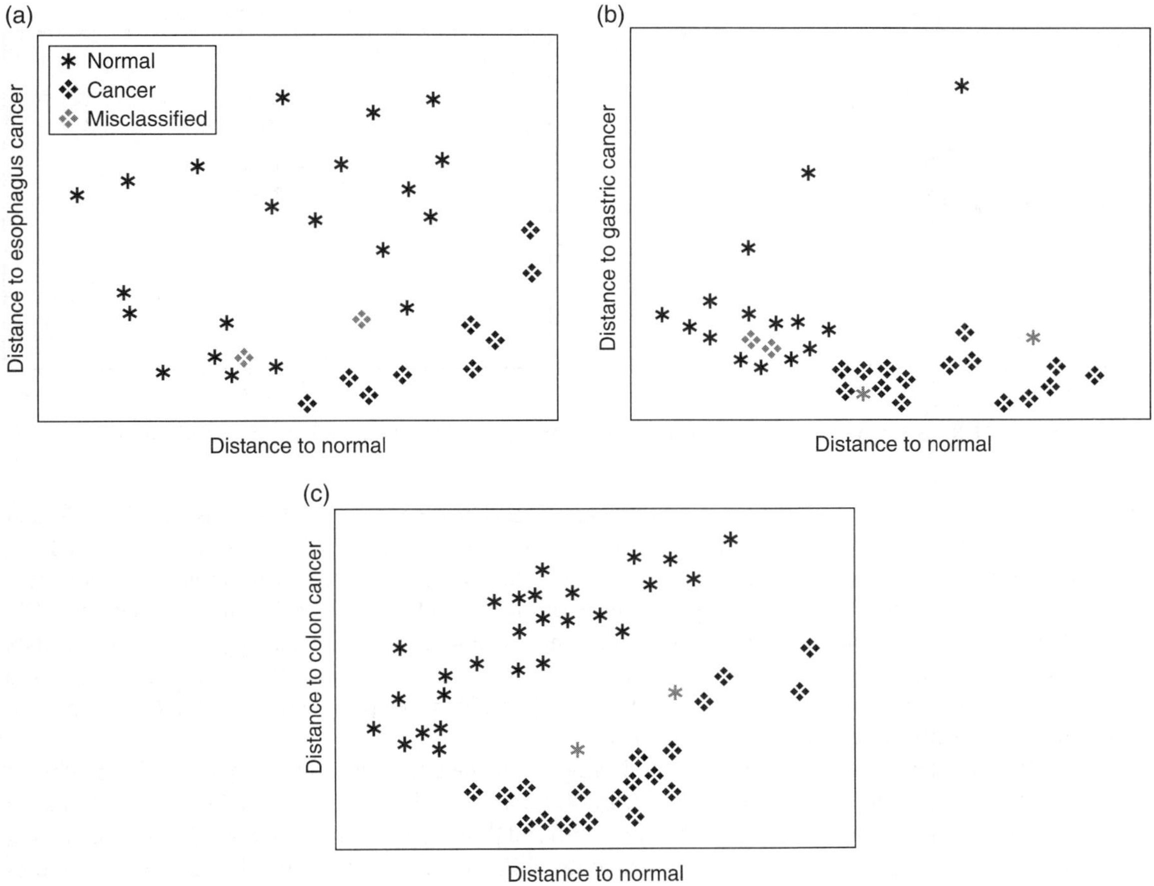

Figure 21.7 *Quality of discrimination of healthy and cancer cases by LDA.*

All the previous sections have confirmed that the main research activities for cancer diagnosis by FTIR spectrometry have been dedicated to analysis of tissue analysis. According to pathological documents there are some concerns about tissue sampling. Usually the general process is low speed for sample preparation; biopsy samples are very heterogeneous comprising several types of cells and many biochemical components; preparation procedures of biopsy samples are time consuming and difficult to operate while fixation and embedding steps in biopsy preparation procedure may cause some distorting in cell structure and even dissolve some useful biochemicals. On the other hand, prepared biopsies are very sensitive to maintaining conditions prior to analysis. In some cases accessing the body organ from which the sample is required is really difficult, causing some trouble; finally, biopsy operation is usually invasive and may accelerate the spreading process of disease.

During the last decade noticeable attention was paid to propose some new analytical approaches which could be a substitute for tissue biopsy sampling, while easier providing and analysis is possible and reliable results are achieved. The effective role of IR spectroscopy in clinical diagnostic via body fluid analysis could be named as a key for problem solving. Cancer infected cells will lose their related adhesion strength and would be separated easily, even traveling via body fluids from the malignant site to healthy ones. This process is the main argument for metastasis. Blood is a body fluid, distributed all around the body and would be called as one of the main traveling media for malignant cellular structures. Thus it is possible to propose a method

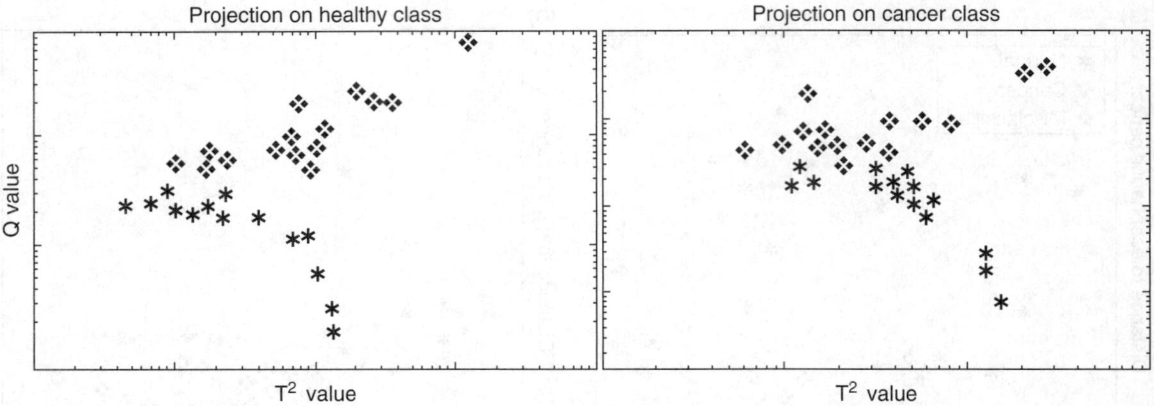

Figure 21.8 *SIMCA classification of IR spectra from whole blood samples for diagnosis of BCC.*

based upon analysis of blood sample instead of tissue in cancer diagnostic studies. It is mentionable that blood is more homogeneous and the sample preparation procedure is very simple while blood is studied.

Initial researches in this area were reported since 2007 when people affected by different types of cancer were differentiated by infrared spectroscopic analysis of their blood samples, processing the data by LDA [94]. Other chemometric techniques such as soft independent modeling of class analogy (SIMCA) and ANN have been also utilized for this aim. Basal cell carcinoma as a rare metastasis malignancy has been diagnosed by both of these chemometric techniques while ANN has provided more precise results [95,96]. Figure 21.8 demonstrates the SIMCA's capabilities schematically.

One step forward had been to analyse the blood samples obtained from healthy women and those affected by breast cancer in order to achieve the main spectral differences such as in N–H stretching, amides I and II bands. It has been also confirmed that position of α-helix and β-sheet amide I bands would shift in malignant cases. The heart of this new trend has been reported as the ability to process the spectral data of blood samples obtained from patients in different stages leading to stage classification of breast cancer via IR spectroscopic blood analysis [97].

Our main aim in this chapter was to define IR spectroscopy as a green tool in biodiagnostics. Several clinical advantages have confirmed this idea while the last trend related to blood analysis could be even greener and completely environmentally friendly due to absence of any environment pollutant chemical in all steps of the analysis. The next idea in this field could be to analyze saliva as the sample for cancer diagnosis. Finally it is interesting to say: 'Effort in surviving the human life is the Greenest aspect'!

References

1. Mantsch, H.H. (1998) Vibrational spectroscopy of lipids: a historical perspective, *Chem. Phys. Lip.*, **96**, 3–7.
2. Barth, A. (2007) Infrared spectroscopy of proteins, *Biochim. Biophys. Acta*, **1767**, 1073–1101.
3. Bandekar, J. (1992) Amide modes and protein conformation, *Biochim. Biophys. Acta*, **1120**, 123–143.
4. Fabian, H. and Mantele, W. (2002) Infrared Spectroscopy of Proteins, in *Handbook of Vibrational Spectroscopy*, Vol. 5 (eds J.M. Chalmers and P.R. Griffiths) John Wiley & Sons, Ltd, Chichester.
5. Fabian, H. (2000) Fourier Transform Infrared Spectroscopy in Peptide and Protein Analysis *in Encyclopedia of Analytical Chemistry*, Vol. 7 (ed. R.A. Meyers) John Wiley & Sons, Ltd, Chichester.
6. Ganim, Z.; Chung, H.S.; Smith, A.W.; DeFlores, L.P.; Jones, K.C. and Tokmakoff, A. (2008) Amide I Two-Dimensional Infrared Spectroscopy of Proteins, *Acc. Chem. Res.*, **41**, 432–441.

7. Surewicz, W.K.; Mantsch, H.H. and Chapman, D. (1993) Determination of protein secondary structure by Fourier transform infrared spectroscopy: A critical assessment, *Biochemistry*, **32**, 389–394.
8. Heise, M.; Marbach, R.; Janatsch, G. and Kruse-Jarres, J.D. (1989) Determination of Glucose in Whole Blood Attenuated Total Reflection Infrared Spectroscopy, *Anal. Chem.*, **61**, 2009–2015.
9. Kajiwara, K.; Uemura, T.; Kishikawa, H.; Nishida, K.; Hashiguchi, Y.; Uehara, M.; Sakakidda, M.; Ichinose, K. and Shichiri, M. (1993) Non-invasive Measurement of Blood Glucose Concentrations by Analyzing Fourier Transform Infrared Absorbance Spectra Through Oral Mucosa, *Med. & Biol. Eng. Comput.*, **31**, S17–S22.
10. Christison, G.B. and MacKenzie, H.A. (1993) Laser Photoacoustic Determination of Physiological Glucose Concentration in Human Whole Blood, *Med. & Biol. Eng. Comput.,* **31**, 284–290.
11. Ying, S.L.; Shaw, R.A.; Leroux, M. and Mantsch, H.H. (2002) Quantitation of glucose and urea in whole blood by mid infrared spectroscopy of dry films, *Vib. Spectrosc.*, **28**, 111–116.
12. Bhandare, P.; Mendelsohn, Y.; Peura, R.A.; Janatsch, G.; Kruse-Jarres, H.; Marbach, R. and Heise, H.M. (1993) Multivariate determination of glucose on whole blood using partial least-squares and artificial neural networks based on mid-infrared spectroscopy, *Appl. Spectrosc.*, **47**, 1214–1221.
13. Vonach, R.; Buschmann, J.; Falkowski, R.; Schildler, R.; Lendl, B. and Kellner, R. (1998) Application of mid-infrared transmission spectroscopy to the direct determination of glucose in whole blood, *Appl. Spectrosc.*, **52**, 820–822.
14. Kim, Y.J.; Hahn, S. and Yoon, G. (2003) Determination of glucose in whole blood samples by mid-infrared spectroscopy, *Appl. Spectrosc.*, **42**, 745–749.
15. Hosafci, G.; Klein, O.; Oremek, G. and Mantele, W. (2007) Clinical chemistry without reagents? An infrared spectroscopic technique for determination of clinically relevant constituents of body fluids, *Anal. Bioanal. Chem.*,**387**, 1815–1822.
16. Shen, Y.C.; Davies, A.G.; Linfield, E.H.; Esley, T.S.; Taday, P.F. and Arnone, D.D. (2003) The use of Fourier-transform infrared spectroscopy for the quantitative determination of glucose concentration in whole blood, *Phys. Med. Biol.*, **48**, 2023–2032.
17. Budnova, G.; Salva, J. and Volka, K. (1997) Application of molecular spectroscopy in the mid-infrared region to the determination of glucose and cholesterol in whole blood and in blood serum, *Appl. Spectrosc.*, **51**, 631–635.
18. Shen, Y.C.; Davies, A.G.; Linfield, E.H.; Taday, P.F.; Arnone, D.D. and Elsey, T.S. (2003) Determination of Glucose Concentration in Whole Blood using Fourier-Transform Infrared Spectroscopy, *J. Biol. Phys.*, **29**, 129–133.
19. Ward, K.J.; Haaland, D.M.; Robinson, M.R. and Eaton, R.P. (1992) Post-Prandial Blood Glucose Determination by Quantitative Mid-Infrared Spectroscopy, *Appl. Spectrosc.*, **46**, 959–965.
20. Zeller, H.; Novak, P. and Landgraf, R. (1989) Blood glucose measurement by infrared spectroscopy, *Int. J. Artif. Organs.*, **12**, 129–135.
21. Bhandare, P.; Mendelson, Y.; Peura, R.A.; Janatsch, G.; Kruse-Jarres, J.D.; Marbach, R. and Heise, H.M. (1993) Multivariate Determination of Glucose in Whole Blood Using Partial Least-Squares and Artificial Neural Networks Based on Mid-Infrared Spectroscopy, *Appl. Spectrosc.*, **47**, 1214–1221.
22. Kim, Y.J.; Hahn, S. and Yoon, G. (2003) Determination of Glucose in Whole Blood Samples by Mid-Infrared Spectroscopy, *Appl. Opt.*, **42**, 745–749.
23. Winters, S.; Gendreau, R.M.; Leininger, R.I. and Jakobsen, R.J. (1982) Fourier Transform Infrared Spectroscopy of Protein Adsorption from Whole Blood: II. *Ex Vivo* Sheep Studies, *Appl. Spectrosc.*, **36**, 404–409.
24. Gendreau, R.M.; S Winters, S.; Leininger, R.I.; Fink, D.; Hassler, C.R. and Jakobsen, R.J. (1981) Fourier Transform Infrared Spectroscopy of Protein Adsorption from Whole Blood: *Ex Vivo* Dog Studies, *Appl. Spectrosc.*, **35**, 353–357.
25. Kostrewa, S.; Paarmann, C.; Goemann, W. and Heise, H.M. (1998) Multivariate determination of hematocrit in whole blood by attenuated total reflection infrared spectroscopy, *AIP Conf. Proc.*, **430**, 271–273.
26. Werner, G.H.; Frueh, J.; Keller, F.; Greger, H.; Somorjai, R.L.; Dolenko, B.; Otto, M. and Boecker, D. (1998) Mid-infrared spectroscopy as a tool for disease pattern recognition from human blood, *Proc. SPIE*, **3257**, 35.
27. Martin, W.B.; Mirov, S. and Venugopalan, R. (2002) Using two discrete frequencies within the middle infrared to quantitatively determine glucose in serum, *J. Biomed. Opt.*, **7**, 613–617.
28. Shaw, R.A.; Kotowich, S.; Leroux, M. and Mantsch, H.H. (1998) Multianalyte serum analysis using mid-infrared spectroscopy, *Ann. Clin. Biochem.*, **35**, 624–632.
29. Budinova, G.; Salva, J. and Volka, K. (1997) Application of molecular spectroscopy in the mid-infrared region to the determination of glucose and cholesterol in whole blood and in blood serum, *Appl. Spectrosc.*, **51**, 631–635.

30. Hahn, S.J.; Yoon, G.W.; Kim, G.S. and Park, S.H. (2003) Reagentless Determination of Human Serum Components Using Infrared Absorption Spectroscopy. *J. Opt. Soc. Korea*, **7**, 240–244.
31. Kim, Y.J. and Yoon, G. (2002) Multicomponent Assay for Human Serum Using Mid-infrared Transmission Spectroscopy Based on Component-Optimized Spectral Region Selected by a First Loading Vector Analysis in Partial Least-Squares Regression, *Appl. Spectrosc.*, **56**, 625–632.
32. Rohleder, D.; Kocherscheidt, G.; Gerber, K.; Kiefer, W.; Köhler, W.; Möcks, J. and Petrich, W. (2005) Comparison of mid-infrared and Raman spectroscopy in the quantitative analysis of serum, *J. Biomed. Opt.*, **10**, 031108.
33. Murayama, K.; Wu, Y.; Czarnik-Matusewicz, B. and Ozaki, Y. (2001) Two Dimensional/Attenuated Total Reflection Infrared Correlation Spectroscopy Studies on Secondary Structural Changes in Human Serum Albumin in Aqueous Solutions: pH-Dependent Structural Changes in the Secondary Structures and in the Hydrogen Bondings of Side Chains, *J. Phys. Chem. B*, **105**, 4763–4769.
34. Wu, Y.; Murayama, K. and Ozaki, Y. (2001) Two-Dimensional Infrared Spectroscopy and Principle Component Analysis Studies of the Secondary Structure and Kinetics of Hydrogen–Deuterium Exchange of Human Serum Albumin, *J. Phys. Chem. B*, **105**, 6251–6259.
35. Smeller, L.; Meersman, F. Heremans, K. (2008) Stable misfolded states of human serum albumin revealed by high-pressure infrared spectroscopic studies, *Eur. Biophys. J.*, **37**, 1127–1132.
36. Liu, K.Z.; Man, A.; Dembinski, T.C. and Shaw, R.A. (2007) Quantification of serum apolipoprotein B by infrared spectroscopy, *Anal. Bioanal. Chem.*, **387**, 1809–1814.
37. Freeman, N.K. (1964) Simultaneous determination of triglycerides and cholesterol esters in serum by infrared spectrophotometry, *J. Lipid Res.*, **5**, 236–241.
38. Liu, K.Z.; Shaw, R.A. Man, A.; Dembinski, T.C.; Mantsch, H.H. (1999) Reagent-free, Simultaneous Determination of Serum Cholesterol in HDL and LDL by Infrared Spectroscopy, *Clin. Chem.*, **48**, 499–506.
39. Hahn, S. and Yoon, G. (2006) Identification of pure component spectra by independent component analysis in glucose prediction based on mid-infrared spectroscopy, *Appl. Opt.*, **45**, 8374–8380.
40. Petibois, C.; Rigalleau, V.; Melin, A.M.; Perromat, A.; Cazorla, G.; Gin, H. and Déléris, G. (1999) Determination of Glucose in Dried Serum Samples by Fourier-Transform Infrared Spectroscopy, *Clin. Chem.*, **45**, 1530–1535.
41. Martin, W.B.; Mirov, S. and Venugopalan, R. (2005) Middle Infrared, Quantum Cascade Laser Optoelectronic Absorption System for Monitoring Glucose in Serum, *Appl. Spectrosc.*, **59**, 881–884.
42. Martin, W.B.; Mirov, S. and Venugopalan, R. (2002) Using two discrete frequencies within the middle infrared to quantitatively determine glucose in serum, *J. Biomed. Opt.*, **7**, 613–617.
43. Takashi, K.; Daisuke, I.; Toshiharu, N.; Shiki, Y.; Masanori, S.; Fumio, N. and Akihide, K. (2005) Determination of the Glucose in Serum by the FT-IR-ATR Method, *Bunseki Kagaku*, **54**, 149–154.
44. Janatsch, G.; Kruse-Jarres, J.D.; Marbach, R. and Heise, H.M. (1989) Multivariate calibration for assays in clinical chemistry using attenuated total reflection infrared spectra of human blood plasma, *Anal. Chem.*, **61**, 2016–2023.
45. Heise, H.M.; Marbach, R.; Koschinsky, Th. and Gries, F.A. (1994) Multicomponent Assay for Blood Substrates in Human Plasma by Mid-Infrared Spectroscopy and its Evaluation for Clinical Analysis, *Appl. Spectrosc.*, **48**, 85–95.
46. Balakhovskiĭ, I. and Lebedev, I. (1982) Simultaneous determination of the cholesterol and triglyceride content of the blood plasma by infrared spectroscopy, *Lab. Delo.*, **5**, 281–285.
47. Kondepati, V.R.; Damm, U. and Heise, H.M. (2006) Infrared Transmission Spectrometry for the Determination of Urea in Microliter Sample Volumes of Blood Plasma Dialysates, *Appl. Spectrosc.*, **60**, 920–925.
48. Petibois, C.; Cazorla, G.; Cassaigne, A. and Déléris, G. (2001) Plasma Protein Contents Determined by Fourier-Transform Infrared Spectrometry, *Clin. Chem.*, **47**, 730–738.
49. Zhou, Y.P.; Xu, L.; Tang, L.J.; Jiang, J.H.; Shen, G.L.; Yu, R.Q. and Ozaki, Y. (2007) Determination of Glucose in Plasma by Dry Film-based Fourier Transformed-Infrared Spectroscopy Coupled with Boosting Support Vector Regression, *Anal. Sci.*, **23**, 793–798.
50. Gendreau, R.M. and Jakobsen, R.J. (1979) Blood-Surface interactions: Fourier transform infrared studies of protein surface adsorption from flowing blood plasma and serum, *J. Biomed. Mat. Res.*, **13**, 893–906.
51. Schmitt, J.; Beekes, M.; Brauer, A.; Udelhoven, T.; Lasch, P. and Naumann, D. (2002) Identification of Scrapie Infection from Blood Serum by Fourier Transform Infrared Spectroscopy, *Anal. Chem.*, **74**, 3865–3868.
52. Lasch, P.; Beekes, M.; Schmitt, J. and Naumann, D. (2007) Detection of preclinical scrapie from serum by infrared spectroscopy and chemometrics, *Anal. Bioanal. Chem.*, **387**, 1791–1800.

53. Haas, S.L.; Müller, R.; Fernandes, A. *et al.* (2010) Spectroscopic Diagnosis of Myocardial Infarction and Heart Failure by Fourier Transform Infrared Spectroscopy in Serum Samples, *Appl. Spectrosc.*, **64**, 262–267.
54. Petrich, W. Dolenko, B. Früh, *et al.* (2000) J. Disease Pattern Recognition in Infrared Spectra of Human Sera with Diabetes Mellitus as an Example, *Appl. Opt.*, **39**, 3372–3379.
55. Eysel, H.H.; Jackson, M.; Nikulin, A.; Somorjai, R.L.; Thomson, G.T.D. and Mantsch, H.H. (1997) A novel diagnostic test for arthritis: Multivariate analysis of infrared spectra of synovial fluid, *Biospectroscopy*, **3**, 161–167.
56. Petrich, W.; Staib, A.; Otto, M. and Somorjai, R.L. (2002) Correlation between the state of health of blood donors and the corresponding mid-infrared spectra of the serum, *Vib. Spectrosc.*, **28**, 117–129.
57. Ellis, D.I. and Goodacre, R. (2006) Metabolic fingerprinting in disease diagnosis: biomedical applications of infrared and Raman spectroscopy, *Analyst*, **131**, 875–885.
58. Mordehai, J.; Ramesh, J.; Huleihel, M. *et al.* (2004) Studies on acute human infections using FTIR microspectroscopy and cluster analysis, *Biopolymers*, **73**, 494–502.
59. Martin, T.C.; Moecks, J.; Belooussov, A. *et al.* (2004) Classification of signatures of Bovine Spongiform Encephalopathy in serum using infrared spectroscopy, *Analyst*, **129**, 897–901 (2004).
60. Rozelle, L.T.; Hallgren, H.J.; Bransford, J.E. and Koch, R.B. (1965) The Identification of Major Infrared Absorbing Components of Human Urine, *Appl. Spectrosc.*, **19**, 105–139.
61. Heise, H.M.; Voigt, G.; Lampen, P.; Küpper, L.; Rudloff, S. and Werner, G. (2001) Multivariate Calibration for the Determination of Analytes in Urine Using Mid-Infrared Attenuated Total Reflection Spectroscopy, *Appl. Spectrosc.*, **55**, 434–443.
62. Shaw, R.A.; Ying, S.L.; Leroux, M. and Mantsch, H.H. (2000) Toward Reagent-free Clinical Analysis: Quantitation of Urine Urea, Creatinine, and Total Protein from the Mid-Infrared Spectra of Dried Urine Films, *Clin. Chem.*, **46**, 1493–1495.
63. Mansfield, C.D.; Man, A.; Ying, S.L. and Shaw, R.A. (2005) laminar Fluid Diffusion Interface Preconditioning of Serum and Urine for Reagent-Free Infrared Clinical Analysis and Diagnostics, *Appl. Spectrosc.*, **59**, 10–15.
64. Platoff, G.E.; Hill, D.W.; Koch, T.R. and Caplan, Y.H. (1992) Serial Capillary Gas Chromatography/Fourier Transform Infrared Spectrometry/Mass Spectrometry (GC/IR/MS): Qualitative and Quantitative Analysis of Amphetamine, Methamphetamine, and Related Analogues in Human Urine, *J. Anal. Toxicol.*, **16**, 389–397.
65. Somorjai, R.L.; Dolenko, B.; Nikulin, A.; Nickerson, P.; Rush, D.; Shaw, A.; Glogowski, M.; Rendell, J. and Deslauriers, R. (2002) Distinguishing normal from rejecting renal allografts: application of a three stage classification strategy to MR and IR spectra of urine, *Vib. Spectrosc.*, **28**, 97–102.
66. Yumi, S.; Hiroshi, U. and Satoshi, Y. (2005) Non-destructive analysis of human saliva phosphoprotein by Fourier transform infrared spectroscopy, *Seikagaku*, **77**, 1074.
67. Khaustova, S.A.; Shkurnikov, M.U.; Grebenyuk, E.S.; Artyushenko, V.G. and Tonevitsky, A.G. (2009) Assessment of Biochemical Characteristics of the Saliva Using Fourier Transform Mid-Infrared Spectroscopy, *Bull. Exper. Biol. Med.*, **148**, 841–844.
68. Scott, D.A.; Renaud, D.E.; Krishnasamy, S.; Meriç, P.; Buduneli, N.; Çetinkalp, S. and Liu, K.Z. (2010) Diabetes-related molecular signatures in infrared spectra of human saliva, *Diabetol. Metab. Syndr.*, **2**, 48.
69. Qi, K.; Zou, M. and Liu, Z. (2009) Saliva analysis by Fourier transform infrared spectrum in the enamel decalcification of orthodontic treatment of the teenagers, *Shanghai Kou Qiang Yi Xue*, **18**, 136–141.
70. Skingsley, D.R.; White, A.J. and Weston, A. (2000) Analysis of Pulmonate mucus by infrared spectroscopy, *J. Moll. Stud.*, **66**, 363–372.
71. Nagase, Y.; Yoshida, S. and Kamiyama, K. (2005) Analysis of human tear fluid by Fourier transform infrared spectroscopy, *Biopolymers*, **79**, 18–27.
72. Pyrih, L.A.; Berezovskyi, V. and Dudar, I.O. (2003) Infrared sweat secretion stimulation as a means of homeostatic correction in patients with kidney dysfunction, *Fiziol. Zh.*, **49**, 25–29.
73. Lasch, P. and Kneipp, J. (2008) *Biomedical Vibrational Spectroscopy*, John Wiley & Sons, Inc., Hoboken, NJ.
74. Rimbaud, D.J.; Bouveresse, E.; Massart, D.L. and de Noord, O.E. (1999) Detection of prediction outliers and inliers in multivariate calibration, *Anal. Chim. Acta.*, **388**, 283–301.
75. Martens, H. and Næs, T. (1989) *Multivariate Calibration*, John Wiley & Sons, Inc., New York.
76. Pell, R.J. (2000) Multiple outlier detection for multivariate calibration using robust, *Chemom. Intell. Lab. Syst.*, **52**, 87–104.

77. Caroni, C. (1995) On Rolf's method for the detection of outliers in multivariate data, *J. Multivariate Anal.*, **52**, 295–307.
78. Kosinski, A.S. (1999) A procedure for the detection of multivariate outliers, *Comput. Stat. Data Anal.*, **29**, 145–161.
79. Garret, R.G. (1989) The chi-square plot: a tool for multivariate outlier Recognition, *J. Geochem. Explor.*, **32**, 319–341.
80. Hoo, K.A.; Tvarlapati, K.J.; Piovoso, M.J. and Hajare, R. (2002) A method of robust multivariate outlier replacement, *Comput. Chem. Eng.*, **26**, 17–39.
81. Lillhonga, T. and Geladi, P. (2005) Replicate analysis and outlier detection in multivariate NIR calibration, illustrated with biofuel analysis, *Anal. Chim. Acta.*, **544**, 177–183.
82. Khanmohammadi, M.; Bagheri Garmarudi, A.; Ghasemi, K.; Garrigues, S. and de la Guardia M. (2009) Artificial neural network for quantitative determination of total protein in yogurt by infrared spectrometry, *Microchem. J.*, **91**, 47–52.
83. Kargosha, K.; Khanmohammadi, M. and Ghadiri, M. (2001) Fourier transform infrared spectrometric determination of thiourea in the presence of sulphur dioxide in aqueous solution, *Anal. Chim. Acta*, **437**, 139–143.
84. Kargosha, K.; Khanmohammadi, M.; Sarokhani, M.; Ansari, F. and Ghadiri, M. (2003) Application of attenuated total reflectance Fourier transform infrared spectrometry to the determination of cephalosporin C in complex fermentation broths, *J. Pharma. Biomed. Anal.*, **31**, 571–577.
85. Khanmohammadi, M.; Nemati, H.; Rafienia, M.; Jamshidi, A. and Bagheri Garmarudi, A. (2008) Investigation of Drug Release from Biodegradable Polymeric Delivery System by Infrared Spectrometry, *Int. J. Polym. Anal. Charact.*, **13**, 353–368.
86. Khanmohammadi, M.; Hosseinnia Kojidi, M.; Bagheri Garmarudi, A.; Ashuri, A. and Soleymani, M. (2009) Quantitative Monitoring of the Amidation Reaction Between Coconut Oil and Diethanolamine by Attenuated Total Reflectance Fourier Transform Infrared Spectrometry, *J. Surfact. Deterg.*, **12**, 37–41.
87. Khanmohammadi, M.; Moeini, M.; Bagheri Garmarudi, A.; Sotudehnia, A. and Zarrabi, M. (2009) Simultaneous determination of sucrose and trehalose in olive leaves by spectrophotometry utilizing partial least squares method, *Acta. Physiol. Plant.*, **31**, 865–869.
88. Khanmohammadi, M. and Kargosha, K. (2005) Application of attenuated total reflectance Fourier transform infrared spectrometry to the determination of sodium percarbonate in washing powder detergent, *Talanta*, **65**, 824–827.
89. Khanmohammadi, M.;, Ashori, A.; Kargosha, K. and Bagheri Garmarudi, A. (2007) Simultaneous Determination of Sodium Tripolyphosphate, Sodium Sulfate and Linear Alkylbenzensulfonate in Washing Powder by Attenuated Total Reflectance: Fourier Transform Infrared Spectrometry, *J. Surfact. Deterg.*, **10**, 81–86.
90. Moros, J.; Iñón, F. A.; Khanmohammadi, M.; Garrigues, S. and de la Guardia, M. (2006) Evaluation of the application of attenuated total reflectance - Fourier transform infrared spectrometry (ATR–FTIR) and chemometrics to the determination of nutritional parameters of yogurt samples, *Anal. Bioanal. Chem.*, **385**, 708–715.
91. Khanmohammadi, M.;Karimi, M.A.; Ghasemi, K.; Jabbari, M. and Bagheri Garmarudi, A. (2007) Quantitative determination of Malathion in pesticide by modified attenuated total reflectance-Fourier transform infrared spectrometry applying genetic algorithm wavelength selection method, *Talanta*, **72**, 620–625.
92. Khanmohammadi, M.; Mobedi, H.; Mobedi, E.; Kargosha, K.; Bagheri Garmarudi, A. and Ghasemi, K. (2009) Quantitative determination of naltrexone by attenuated total reflectance – FTIR spectrometry using partial least squares (PLS) wavelength selection, *Spectroscopy*, **23**,113–121.
93. Diakides, N.A. and Bronzino, J.D. (2008) *Medical Infrared Imaging*, CRC Press, New York.
94. Khanmohammadi, M.; Ansari, M.A.; Bagheri Garmarudi, A.; Hassanzadeh, G. and Garoosi, G. (2007) Cancer Diagnosis by Discrimination between Normal and Malignant Human Blood Samples Using Attenuated Total Reflectance-Fourier Transform Infrared Spectroscopy, *Cancer Invest.*, **25**, 397–404.
95. Khanmohammadi, M.; Nasiri, R.; Ghasemi, K.; Samani, S. and Bagheri Garmarudi, A. (2007) Diagnosis of basal cell carcinoma by infrared spectroscopy of whole blood samples applying soft independent modeling class analogy, *J. Cancer Res. Clin. Oncol.*, **133**, 1001–1010.
96. Khanmohammadi, M.; Bagheri Garmarudi, A. and Ghasemi, K. (2009) Back-propagation artificial neural network and attenuated total reflectance-Fourier transform infrared spectroscopy for diagnosis of basal cell carcinoma by blood sample analysis, *J. Chemom.*, **23**, 538–544.
97. Khanmohammadi, M.; Rajabi, F.H.; Bagheri Garmarudi, A.; Mohammadzadeh, R. and Mohammadzadeh, R. (2010) Chemometrics assisted investigation of variations in infrared spectra of blood samples obtained from women with breast cancer: a new approach for cancer diagnosis, *Eur. J. Cancer Care.*, **19**, 352–359.

22

Environmental Analysis

Ricardo Erthal Santelli[1], Marcos Almeida Bezerra[2], Julio Carlos Afonso[1], Maria de Fátima Batista de Carvalho[3], Eliane Padua Oliveira[4] and Aline Soares Freire[1]

[1] *Departamento de Química Analítica, Universidade Federal do Rio de Janeiro, Rio de Janeiro, Brazil*
[2] *Departamento de Química e Exatas, Universidade Estadual do Sudoeste da Bahia, Jequié, Brazil*
[3] *Centro de Pesquisa e Desenvolvimento, Cidade Universitária, Rio de Janeiro, Brazil*
[4] *Departamento de Geoquímica, Universidade Federal Fluminense, Niterói, Brazil*

22.1 Pollution and its control

In recent decades, increasing attention has been focused on the pollution of the natural environment. The reduction of pollutants from the air, water, sediment and soil is an important and very complex task, requiring the cooperation of specialists from various areas of expertise, including analytical chemistry. Data generated from environmental analyses provide information that forms the basis from which it is possible to make decisions on the procedures to be adopted for the preservation or remediation of a region.

The purpose of environmental analysis can be separated into two topics: (1) to determine the natural concentrations of substances in the environment with the objective of performing background monitoring and (2) to determine the concentration of harmful pollutants in the environment for pollution monitoring [1]. For these purposes, analytical chemists have used classic instrumental methods developed for obtaining rapid and accurate results. The methods for environmental and monitoring are currently well established, and some are adopted as standard methods recommended for international environmental agencies.

Although these methods present analytical characteristics (such as sensitivity, accuracy, precision, detection and quantification limits) suitable for the determination of several chemical species in environmental samples of concern, a paradoxical situation emerges due to the chemical waste generated by most of these analytical methods, which can impact nature. In some circumstances, the chemicals employed in an environmental analysis can be even more toxic than the species being monitored. As a consequence, some analytical chemists have focused their efforts on the development of methodologies that are less harmful to humans and to the

Handbook of Green Analytical Chemistry, First Edition. Edited by Miguel de la Guardia and Salvador Garrigues.

environment. Currently, in the development of new analytical procedures, the quantity and toxicity of the waste is as important as any other analytical characteristic [2].

Therefore, the traditional interest in the identification and quantification of substances of environmental concern needs to be supplemented by yet another aspect, which includes the impact of its own analytical procedures on the ecosystems and lives of organisms in nature. Although environmental analysis in itself is an important step in the direction of ecological preservation, the present focus on the measurement of the influence of human activities does not address the negative aspect of current practices in routine research laboratories [3]. The development of environmentally friendly analytical methods for environmental analysis is a coherent attitude for these laboratories. The principles of green chemistry, suggested by Anastas and Warner [4], can be directly related to analytical chemistry, generating the term 'Green Analytical Chemistry'. The most important principles of Green Analytical Chemistry include the following: (1) prevention of waste generation; (2) use of safety solvents and auxiliaries; (3) design for energy efficiency; and (4) development of safer chemistry to minimize the potential of chemical accidents [5].

An order for priority can be established for the development of cleaner environmental analytical procedures: (1) chemical waste should not be produced; (2) waste that is produced should not be toxic; (3) the quantity of waste should be minimized; (4) toxic waste should be recycled and, if possible, reused and (5) the analytical procedure should include a suitable method for waste treatment and disposal.

This chapter will be dealing with some environmental green methods applied to water, wastewater, effluent, soil, sediment, gas and particulate samples, for the determination of metals, metalloids, and organic compounds of environmental concern.

22.2 Steps of an environmental analysis

An environmental analysis should be proficient in carrying out all the following different stages: (1) sample collection; (2) sample treatment and (3) analysis. Each of these steps can affect the environment, make different contributions towards pollution and have different potential for becoming greener.

22.2.1 Sample collection

Sampling is generally one of most problematic stages of an environmental analysis, and its main difficulty is in obtaining representative samples from the water, air, effluent gases, dust, soil and sediments. Considerations of site accessibility, time and expense are often more influential than purely scientific considerations. Green sampling procedures must also be considered for the possibility of eliminating some processes that require materials and energy input, such as the transport of the sample to the laboratory, conservation and storage [1, 6]. In routine analysis, the problems of sample collection can be avoided using analytical devices that generate results decreasing the delay between successive analyses (implantation of in-line and on-line systems). In-line manifold measurements can be made directly in the flow stream, for example, in pipelines or rivers, eliminating some sampling steps that can cause pollution. Off-line analysis should be avoided due the requirement of several steps, which are sources of generation of residuals. In situ sampling and analysis can also eliminate some of these problems [7].

22.2.2 Sample preparation

Although some environmental samples may be analyzed directly using an instrumental technique (e.g., metal determination in rain and river waters), most of the sample has to be prepared for analysis. A variety of sample treatment methods are used, depending on the type of sample, the analyte to be determined and the kind of analytical method that will be used. The purpose of sample treatment can be explained by the following reasons: (1) to convert the sample and analyte into a form suitable for analysis; (2) to eliminate interfering substances and (3) to concentrate the analyte, such that the analytical technique can determine it within the limits of detection [1, 6].

Sample preparation is considered to be potentially the most polluting step of analysis because it can generate dangerous residuals during and after the execution of the preparatory steps. Two traditional sample treatment methods that generate toxic residuals include the following: (1) acid digestion, in which a solid sample is decomposed and dissolved using acid digestion on a hot plate, refluxed and microwave digested, liberating the analytes for posterior elemental analysis and (2) solvent extraction, in which organic analytes are usually extracted into an organic solvent, avoiding degradation of the target substance that can occur if a drastic treatment such as acid digestion is employed.

Whenever possible, these pre-treatment steps should be avoided by choosing direct analytical methods. Unfortunately, in many cases, digestion, preconcentration or solvent extraction steps are required to perform a valid analysis. Thus, strategies, such as the use of diluted acids in pressurized microwave digestion, the use of green solvent extraction techniques or other alternatives, should be adopted for minimizing the generation of residuals and the environmental impact from such residuals [8].

Particularly regarding organic compounds, in recent years, there has been steady progress in extraction technology with the development of newer and simpler sample preparation.

22.2.2.1 Assisted solvent extractions

Microwave-assisted extraction (MAE) is one of the milestones in the development of sample preparation strategies. Microwaves are used in the solvent extraction of organic analytes from solid samples. Enhancement of extraction is based on the absorption of microwave energy by molecules of chemical compounds. The main advantages of MAE from the point of view of green chemistry are the following: significant reduction in the amount of solvent used (25–50 ml), which reduces waste generation; shortened extraction times (3 min–1 h); and reduction in the amount of the sample required, which, correspondingly, reduces energy input and cost. Ultrasonic extraction (USE) is a promising technique that can speed up and simplify sample treatment. Ultrasonic energy, when imparted to solutions, causes acoustic cavitation (i.e., bubble formation and subsequent implosion). The collapse of bubbles created by the sonication of solutions generates extremely high local temperature and pressure gradients, which may be regarded as localized 'hot spots', and can be employed for extracting analytes from solid matrices. This technique is regarded as environmentally friendly because not only is the amount of solvent decreased and the extraction time shortened but also the ultrasound research has the potential for new applications in environmental remediation [7]. Supercritical fluid extraction (SFE) has overcome the difficulties of solid-sample extraction and has gained attention as a potential replacement for conventional liquid-solvent extractions [9] because SFE can extract organic pollutants quantitatively from environmental solids in a short time [7]. SFE uses solvents that are non-toxic and available at high purity at relatively low costs. The most popular solvent for SFE is carbon dioxide because of its low critical parameters (T_c = 321°C, P_c = 72 atm), availability at high purity, low toxicity and ability to dissolve a wide spectrum of organics, including high molecular-mass and moderately polar compounds [7]. Unfortunately, many organic pollutants are polar; therefore, they cannot be extracted quantitatively from environmental solids with $SFCO_2$ [7, 10]. Pressurized liquid extraction (PLE), pressurized solvent extraction (PSE) and accelerated solvent extraction (ASE) [9] have recently been accepted by the US Environmental Protection Agency (USEPA) for extracting organic compounds. PLE is based on using solvents at a high pressure and/or high temperature without reaching the critical point [11]. PLE is similar to Soxhlet extraction, except the solvents are used near their supercritical region, in which high temperatures produce high solubility and high diffusion rates of solutes in the solvent, while the high pressure, in keeping the solvent below its boiling point, enables high penetration of the solvent in the sample. PLE provides high extraction efficiency with low solvent volumes and a short extraction time [9]. In this context, pressurized hot water extraction (PHWE)/superheated water extraction (SWE, also known as subcritical water extraction) is a technique based on the use of (pressurized) superheated water as a solvent for extraction [7, 10, 12–14]. If the pressure is raised so that the water remains in a condensed state, the superheated water can have a permittivity very similar to typical organic solvents, thus, allowing better penetration of matrix particles and resulting in enhanced solubility of intermediate to low polarity organic compounds. PHWE is performed at temperatures between 100°C and the critical point, 374.1°C [15]. The major advantage of PHWE is the reduction in the consumption of organic solvents. Moreover, water is easily available, non-toxic and can be recycled or disposed with minimal environmental problems. Hence, PHWE has steadily become an efficient and low-cost method of extraction for less polar organic compounds [14]. One disadvantage of PHWE is that the extract is a relatively dilute aqueous solution and, although free of the matrix, often needs to be concentrated/extracted before any subsequent assay steps. There are some alternatives for the concentration/extraction of the aqueous solution. Some authors have successfully proposed the coupling of PHWE with solvent trapping, which consists of trapping analytes in a small volume of organic solvent [15] or with solid=phase extraction/solid-phase microextraction (SPE/SPME). In the last case, the isolation process occurs by the partition of compounds dissolved in water between a solid sorbent and an aqueous phase, on which the analytes are selectively trapped [7]. After the trapping, the analytes are eluted with a minimal volume of adequate solvent.

Ionic liquids (IL) are gaining wide recognition as novel solvents in chemistry. They are also known as room temperature (RT) IL (not all IL are RTIL), non-aqueous IL, molten salts, liquid organic salts or fused salts. The application of IL in analytical chemistry, especially to separate analytes, is worthwhile because these solvents have some unique properties (e.g., negligible vapor pressure, good thermal stability, tunable viscosity and miscibility with water and organic solvents, and extractability for various organic compounds and metal ions). Because they are non-volatile and non-flammable, they have been proposed as environmentally friendly solvents for 'clean processes' and 'Green Chemistry', and they are good substitutes for traditional organic solvents that are volatile and flammable [7, 16].

22.2.2.2 *Virtually no solvent extraction group*

Gas extraction is used in static or dynamic headspace modes. The static headspace (gas-phase extraction) technique includes transfer of the partial headspace or gas phase above the sample in a condensed phase by means of a gas-tight syringe or transfer line into a gas chromatography machine. Dynamic headspace is the collection of the analytes in a trap from the flowing gas phase above or through the sample and then the trapped analytes by desorption into gas chromatography for their determination. Headspace techniques are completely free of solvents because inert gas is used as an extracting agent. The other advantage is that only volatiles are injected into the gas chromatograph (GC) instrument. This is very important because non-volatile compounds, if present, could decrease the column quality. A new column requires material and energy inputs; by applying gas extraction, this problem is avoided [7].

Liquid-phase micro-extraction (LPME) combines the principles of liquid-liquid extraction (LLE) and the miniaturized nature of SPME to take advantage of both techniques [17]. Here, there is the solvent-minimized sample pretreatment procedure of LLE, in which only several microliters of solvent are required to concentrate analytes from various samples rather than hundreds of milliliters needed in traditional LLE. It is compatible with capillary gas chromatography (GC), capillary electrophoresis (CE) and high performance liquid chromatography (HPLC). In LPME, the extraction normally occurs in a small amount of a water-immiscible solvent (acceptor phase) from an aqueous sample containing analytes (donor phase). LPME can be divided into three main categories:

1) single-drop micro-extraction (SDME) based on the suspension of a micro-drop of a water-immiscible organic solvent in an aqueous donor solution;
2) dispersive liquid–liquid micro-extraction (DLLME); and
3) hollow-fiber micro-extraction (HF-LPME) [18].

The main difference between SDME and LPME is the form of the analyte-acceptor phase. In SDME, the solvent is in the form of a suspended drop; whereas, in LPME, the solvent is placed in a hollow fiber. Analytes are transferred from the donor phase (usually aqueous) to a receptor phase through an organic phase immobilized in pores in the hollow fiber. In two-phase mode, the solvent in the pores of the fiber is the same as that present in the fibre. In three-phase mode, the solvent placed in the pores of the fiber differs from that inside the fiber. The three-phase mode is applied for extracting polar analytes, while the two-phase mode is applied for extracting non-polar and semi-polar analytes. The amount of solvent in the fiber is approximately 5 or 10–25 or 30 µl [7,19]. After extraction, the acceptor solution is directly subjected to a final chemical analysis by HPLC, GC, CE, or mass spectrometry (MS) [19]. The advantage of LPME over SDME is the presence of the fiber that supports organic solvents, decelerating the process of dissolution or evaporation (in HS-SDME mode) of the solvent [7].

In membrane techniques, the samples may be preconcentrated due to passing the analytes through a polymeric membrane. Some of the techniques require small amounts of organic solvents, while others are virtually solventless. Membrane extraction may be applied as an efficient tool in many modes, including microporous membrane liquid-liquid extraction (MMLLE), membrane extraction with sorbent interface (MESI), membrane-assisted solvent extraction (MASE) and supported liquid membrane extraction (SLME) [7].

22.2.3 Analysis

After the separation from the sample matrix and other interfering compounds, the detection and identification of an analyte are areas in which the chemistry can be made green. The selection of the appropriate analytical method is generally based on the following criteria: (1) expected concentration of the analyte in the sample; (2) number of samples to be analyzed; (3) time that can be devoted to the analysis; and (4) cost of analysis. In accordance with the Green Analytical Chemistry criteria, the development of instrumental methods has generally led to an efficient use of energy, especially when the method is highly automated and uses a minimal amount of sample. Instrumental methods in environmental analysis generally involve electrochemical, spectroscopic and chromatographic analyses [1, 2].

Spectroscopic methods used in the analysis of environmental samples include mainly UV-visible molecular absorption spectroscopy, atomic absorption spectrometry (AAS) and atomic emission spectrometry (AES). The most widely used techniques are flame atomic absorption spectrometry (F AAS), graphite furnace atomic absorption spectrometry (GF AAS), inductively coupled plasma optical emission spectroscopy (ICP OES) and inductively coupled plasma-mass spectrometry (ICP-MS). Atomic spectroscopic methods have been extensively used for the determination of pollutant elements, such as toxic heavy metals, because they generate low volumes of residuals and because it is possible perform fast analysis with low sample consumption [8].

F AAS is a technique that is relatively inexpensive and easy to operate with little interference. When interferences occur, they are easily identified and usually easily controlled. On the negative side, refractory elements cannot be determined with good sensitivity because the flame temperatures are not high enough to atomize a large fraction of these analytes. It is essentially a single-element technique because, in the traditional equipments, each element requires its own hollow-cathode lamp. The major advantage of GF AAS is that its detection limits are 10–100 times better than for F AAS or for ICP OES, sharing the limitations of F AAS for refractory elements. These refractory elements can be determined with good sensitivity by ICP OES because the high temperature that is reached in the plasma. Elements of interest can be detected either simultaneously or sequentially, make this technique a fast and multi-elemental. ICP-MS is an expensive technique but combines the fast multi-element features of ICP OES with limits of detection that surpass those of GF AAS [8, 20, 21].

In multi-component analysis of complex matrices, chromatography is used most often for organic pollutant determination. Since the mobile phase in HPLC may be a source of pollution, GC should be chosen whenever possible. However, some 'green' mobile phases (e.g., ethanol-water) can also be used in HPLC. Since chromatography can determine many components in a single run, the material and energy input used 'per analyte' is smaller. This makes final analysis by chromatography and foregoing pre-treatment more beneficial [7].

Additionally, the performance of several methods for water analysis can be improved according to Green Analytical Chemistry principles in their automation. This approach reduces sample size, and solvent and reagent consumption become greener than existing methods. Automation can be carried out from a simple autosampler to complex flow injection analysis (FIA), sequential flow injection analysis (SIA) and multi-commutation flow systems. These last devices are controlled by a microcomputer and allow the insertion of samples and reagents only in the instants and in the amounts necessary to implement the analytical procedure, resulting in minimal reagent consumption [2, 7, 8, 22].

22.3 Green environmental analysis for water, wastewater and effluent

In the quantification of several chemical constituents in natural water (lakes, rivers, oceans, rainwater and groundwater), waste and effluents are of concern because they can cause environmental damage if their concentration is modified, being hazardous to biota and affecting the quality of drinking water sources.

Chemical analysis in these aqueous matrices is already well established, and some standard methods from agencies have been based on classic instrumental techniques. Some methods based on classical analysis consume large quantities of reagents or energy and frequently generate large volumes of residuals. These methods often do not present sufficient limits of quantification and sensitivity for determination of some analytes in concentrations established by recent environmental legislations. Despite analytical methods based on instrumental techniques modified by electrochemical, spectroscopic and chromatographic analysis to satisfy some aspects of Green Analytical Chemistry criteria and to present sufficient analytical characteristics, the classic methods still need improvements to become environmentally friendly [7–9, 20, 23, 24].

22.3.1 Major mineral constituents

Major mineral constituents are commonly found in the milligram per liter concentration level in natural water, and the concentrations depend on the mineral deposits in the locality, the anthropogenic action and the type of natural water (river, sea, rain and groundwater). The major cations are mainly Ca^{2+}, Na^{+}, Mg^{2+} and K^{+}, and the anions are Cl^{-}, SO_4^{2-}, HCO_3^{-}, NO_3^{-}, PO_4^{3-} and NO_2^{-}. Waste and effluents, depending on the composition, can be sources of pollution by altering these major constituents [24].

Electrochemical detectors can be seen as examples of environmentally friendly analytical approaches for determining these major constituents. A variety of electrochemical techniques can be used for environmental analysis. The most common are ion selective electrodes (ISE), an example of which is the pH electrode, which can be found in all research and routine analytical laboratories. Measurements of water pH illustrate the application of a membrane electrode to carry out a simple, fast and trustworthy determination of this important parameter for the aquatic environment. Other ions, such as Cl^-, NO_3^-, Na^+, K^+, NH_4^+, F^-, and some metals may also be determined using ion selective electrodes. The classic determination of dissolved oxygen in water employs the Winkler method, which involves a titration procedure. However, the electrode method has the advantage of the possibility to use it for field measurements and in the laboratory for determination of biochemical oxygen demand (BOD). To determine the total inorganic salt content in a sample, the analyst could evaporate the sample and then weigh the resultant solid. This procedure is opposite to a green chemistry principle, which is the energy economy. It is more convenient to use a conductivity cell to make this measurement. This determination has furnished satisfactory values for total salt contents in natural water with the advantage of decreasing the throughput of analysis [8,24–26].

Different analytical separation methods have also been widely used for the simultaneous determination of these constituents in water. These include, mainly, ion chromatography (IC) and capillary electrophoresis (CE). Using less or no pollutant reagents for promoting an efficient separation between analyte peaks is the main challenge of ion chromatography. On the other hand, capillary electrophoresis is based on the electrical separation of substances and presents a low sample and solvent consumption, higher selectivity, fast analysis and is mechanically simpler than high pressure-liquid chromatography [4,5,8,27].

Table 22.1 presents some relevant environmental analytical methods for determining major constituents in natural water, wastewater and effluents, which present the green characteristics discussed above.

22.3.2 Trace metal ions

Trace metal ions are found in natural water in the microgram per liter concentration level. However, iron, zinc and manganese can sometimes approach the milligram per liter levels. Trace metals ions can originate from leaching of ore deposits or from anthropogenic sources, such as metal refining, industrial and domestic effluents, and solid waste disposal. Thus, trace metal ion concentration can be an indication of aquatic pollution [24].

Ultraviolet-visible molecular spectrometry is present in the laboratories for environmental analysis, but the majority of determination of trace elements is carried out by atomic spectrometry techniques due their characteristics, such as sensitivity, low interferences, easy of operation and versatility for a large number of environmentally interesting elements. The main individual techniques based on atomic spectrometry are flame atomic absorption spectrometry (F AAS), graphite furnace atomic absorption spectrometry (GF AAS), hydride generation atomic absorption spectrometry (HG-AAS), inductively coupled plasma optical emission spectrometry (ICP OES) and inductively coupled plasma-mass spectrometry (ICP-MS).

Despite the availability of sensitive analytical instrumental techniques, methods for separation of analytes from interferences in the complex matrices (e.g., sea water and effluents) and sometimes selective preconcentration are necessary for a valid analysis. Preconcentration uses a less expensive technique for trace analysis, but it does not have sufficiently low detection limits for trace determinations. The demand for faster, more cost-effective and environmentally friendlier analytical methods is the major incentive to improve the usual procedures used for separation and preconcentration before an environmental analysis. The green friendly separation/preconcentration methods for metal determination are solvent micro-extraction, cloud-point extraction (CPE) and solid-phase extraction (SPE). The main characteristics of these techniques are the reduction of adverse environmental impact of organic solvents and improvement of the analytical performance [7–9, 27, 42].

Table 22.1 *Some green analytical methods for determining major constituents in natural waters, wastes and effluents.*

Analytes	Samples	Analytical system	Green aspects	Reference
Sulfate	Natural waters and industrial effluents	Automated flow analysis with turbidimetric detection	Economy of chemicals using a computerized flow system and high sampling frequency	[28]
Fluoride and calcium	River, sea and drinking waters	Potentiometric – ion selective electrode	Economy of chemicals using flow injection analysis and clean potentiometric detection	[29]
Nitrate and nitrite	Drinking waters	Ion-interaction liquid chromatography with UV detector	Use of less pollutant reagents and high sampling frequency using flow system	[30]
Calcium and total hardness	Natural waters	Potenciometric sensor array with ion-selective electrodes for Ca^{2+}, Mg^{2+}, NH_4^+, K^+, Na^+, Li^+, and H^+	Clean detection step in the presence of interferents using chemometric tools	[31]
Chloride, sulfate and nitrate	Groundwaters	Ion chromatography with conductometric electrode and suppression system	Use of non-pollutant reagents and a clean detection step	[32]
Fluoride, Chlorite, Bromate, Chloride, Nitrite, Bromide, Chlorate, Nitrate, Phosphate, Sulfate and calcium and magnesium as EDTA Complexes	Mineral waters	Anion-exchange chromatography and conductometric electrode with suppression system	Use of non-pollutant reagents and a clean detection	[33]
Nitrite	Natural waters	Flow system for UV detection and residuals treatment	Replacement of toxic reagents, waste minimization and treatment and high sampling frequency	[34]
Free chlorine	Natural and tap waters	Automated flow system for UV-Vis detector	Solenoid micro-pumps to improve mixing conditions by pulsed flow and to minimize reagent consumption as well as waste generation	[35]
Calcium and fluoride	Natural and borehole waters	Automated flow system with ion selective electrode detection	Minimization of reagent consumption, clean detection technique	[36]

(*continued*)

Table 22.1 (*continued*)

Analytes	Samples	Analytical system	Green aspects	Reference
Chlorine	Natural waters	Flow system with spectrophotometric detection	The analysis become environmental friendly because $Hg(SCN)_2$ is immobilized in a epoxy resin	[37]
Ammonium	Lake waters	Flow injection with spectrophotometry	Flow analysis allows reduction of chemicals and reduction of residues	[38]
Calcium and magnesium	Underground waters	Capillary zone electrophoresis	Low consume of sample and reagents and fast analysis	[39]
Nitrate, nitrite, bromide and iodide	Seawaters	Capillary zone electrophoresis	Low consume of sample and reagents and fast analysis	[40]
Chemical Oxygen Demand (COD)	Waste waters	Copper electrode as an electrocatalytic sensor	The end products of determination do not contain toxic metals such as conventional method	[41]
Weak acid dissociable cyanide	Natural water samples	Flow system with solenoid valves using a highly fluorescent isoindole derivative.	Minimization of the drawbacks related to the reference bath procedure, i.e., use of toxic reagents, high reagent consumption and waste generation, low sampling rate, and poor sensitivity.	[42]

Table 22.2 presents some recent green analytical methods for determining trace elements in natural water, waste and effluents.

22.3.3 Organic pollutants

To encompass the main principles of the Green Analytical Chemistry in the analysis of organic pollutants, the major drawback is in sample preparation. Classic methods of sample preparation use large volumes of solvents, toxic substances and energy. Modern technologies for analyte extraction, including reduction of solvents and toxic reagents (miniaturization), maintain the analytical efficiencies (sometimes improving it) in the extraction of these analytes from environmental matrices. These new sample preparation techniques include the use of microwave and ultrasonic energies, supercritical and subcritical fluids, solid-phase and micro solid-phase extraction, stir-bar sorptive extraction, liquid-liquid micro-extraction with the small variations of single drop microextraction and liquid-liquid-liquid micro-extraction, cloud point extraction and molecularly imprinted polymers, among others. There are many reviews in the literature related to several aspects of Green Analytical Chemistry, in particular, involving theoretical applications of these new extraction techniques for environmental issues [5, 7–9, 57–62].

Table 22.2 *Some green analytical methods for determining trace elements in natural waters, wastes and effluents.*

Analytes	Samples	Analytical system	Green aspects	Reference
Fe	Waters	Flow system with spectrophotometric detection	Use of Guava extract as colorimetric reagent and generating less toxic chemical wastes	[43]
Fe	Fresh waters from lake, river or weirs	Flow system with spectrophotometric detection	Immobilized reagent in a solid-phase allow reuse it several times, low reagent consume and reduced effluent generation	[44]
Cu (II)	Wastewater from electroplating industry	Micro flow system consisted of a planar glass chip with fiber optic probe as optical sensor	System with less reagents and samples consumption and diminutive waste generation	[45]
As (III)	River and lake waters	Cloud-point extraction/ preconcentration with GFAAS detection	Use of surfactant solution to avoid organic solvent	[46]
Cd, Cr, Cu, Mn, Ni and Pb	Saline effluents	Cloud-point extraction/ preconcentration with ICP OES detection	Use of surfactant solution to avoid organic solvent and sample economy by simultaneous detection	[47]
Cd, Cu, Pb and Zn	River, sea and tap waters	Cloud-point extraction/ preconcentration with FAAS detection	Use of surfactant solution to avoid organic solvent	[48]
Cr (III) and Cr (VI)	Sea water	Cloud-point extraction/ preconcentration with FAAS detection	Use of surfactant solution to avoid organic solvent	[49]
Zn (II)	Tap water	Flow system with on line solid-phase extraction preconcentration coupled with ICP OES	Economy of chemicals using flow analysis and reduced effluent generation	[50]
Hg (II)	Tap, well and mineral waters	Flow system with on line solid-phase extraction preconcentration coupled with ICP OES	Economy of chemicals using flow analysis and reduced effluent generation	[51]

(continued)

Table 22.2 (continued)

Analytes	Samples	Analytical system	Green aspects	Reference
Cd (II)	Water	Flow system with on line solid-phase extraction preconcentration coupled with FAAS	Economy of chemicals using flow analysis and reduced effluent generation	[52]
Cu(II), Ni(II), Pb(II), Cd(II) and Mn(II)	Wastes and lake waters	Bath solid-phase extraction preconcentration coupled with FAAS	Reuse of solid-phase impregnated with the complexing reagent in several separation/ preconcentration cycles	[53]
Cd (II)	Tap, sea and rivers water	Dispersive liquid–liquid microextraction combined with GFAAS	Low consume of organic solvent and sample, low residual generation	[54]
Pb (II)	Tap, well, river and seawater samples	Dispersive liquid–liquid microextraction combined with FAAS using micro-sample introduction	Low consume of organic solvent and sample, low residual generation	[55]
Ag	Waster, tap and spring water	Ligandless dispersive liquid–liquid microextraction combined with FAAS	Low consume of organic solvent without use of complexing reagent	[56]

Table 22.3 displays some relevant green analytical methods used for the determination of organic compounds in water and wastewater that could be useful for environmental practices.

22.4 Green environmental analysis applied for solid samples

22.4.1 Soil

Nutrients, major constituents and trace elements are normally found in several chemical forms in soils. Nutrients including nitrogen, phosphorus, major constituents (calcium, magnesium, sodium and potassium), and trace metals (zinc, iron, copper, cobalt, molybdenum, and manganese) are commonly found [24]. The trace element contents of soils depend largely on the mineralogical composition of the parent material and on the processes of chemical and physical weathering. The concentrations of available trace elements and nutrients would be expected to be in the milligram per kilogram level. Organic contaminants are typically found in the microgram per kilogram concentration level or below. Anthropogenic sources of trace elements and organic contaminants are of concern due to potential harm to the ecosystem and health risks [1].

In recent years, efforts have been made to develop new instrumental methods to replace the traditional methods of sample preparation to use analytical procedures that reduce hazardous waste and energy and

Table 22.3 *Some representative green analytical methods for determining organic pollutants in waters and wastewaters.*

Analytes	Samples	Analytical system	Green aspects	Reference
Pesticides (chloropyrifos, lambda-cyhalothrin, cyfluthrin, cypermethrin, fenvalerate, delatamethrin)	Waters	Vortex-assisted liquid–liquid microextraction (VALLME) coupled with gas chromatography micro electron-capture detector (GC-µECD)	Economy of chemicals and consuming of only 30 µl of toluene	[63]
Endocrine disrupting chemicals and pharmaceuticals	Water and Wastewaters	Multi-stir bar sorptive extraction–single desorption–capillary gas chromatography/mass spectrometry	Simultaneous screen of several EDC and pharmaceuticals with different structures, method fully automated, small volumes of samples are required.	[64]
Sulfonamides (sulfacetamide, sulfadiazine, sulfathiazole, sulfamerazine, sulfadimidine, sulfamonomethoxine, sulfamethoxazole, sulfaquinoxaline)	Water	Liquid–liquid–liquid microextraction and high-performance liquid-chromatography	Low solvent volume and the possibility to use less toxic solvents	[65]
Diclofenac	Water and Wastewater	Solid-phase extraction with a molecularly imprinted polymer and HPLC photodiode array detector	Low solvent volume is required	[66]
Pesticides, polycyclic aromatic hydrocarbons, polychlorinated biphenyls	Seawater and Interstitial water	Stir bar sorptive extraction–thermal desorption–gas chromatography–mass spectrometry	Simultaneous analysis, low sample volume required (10–100 ml), MS spectrometer could permit screening for other pollutants	[67]
Polycyclic aromatic hydrocarbons	Water and wastewater	Coacervative extraction and liquid chromatography–fluorimetry	Low solvent volume is required (4 ml), low sample volume (36 ml), no clean-up is needed.	[68]
Benzimidazolic fungicides (Carbendazim, thiabendazole, fuberidazole)	River and underground water	Supramolecular solvent-based extraction and liquid chromatographic/fluorimetric	Low sample volume, avoids the use of toxic organic solvents	[69]
Triazine herbicides (simetryn, atrazine, ametryn, propazine, terbutylazine, terbutyrn)	Sludge water and watermelon	Liquid–liquid–solid microextraction based on membrane-protected molecularly imprinted polymer fiber coupling with HPLC	Sample pre-treatment steps were reduced, analytical speed was increased	[70]

Pyrethroids (allethrin, cypermethrin, prallethrin, tetramethrin, transfluthrin, imiprothrin)	River water	Ultrasound-assisted dispersive liquid–liquid microextraction combined with liquid chromatography-UV detector	Low sample volume and low volume of solvent required (20 μl)	[71]
Carbaryl and 1-naphthol	Water	Anionic surfactant micelle-mediated extraction with sodium dodecylsulfate and fluorescence spectrophotometry	Low volume of less toxic organic solvents	[72]
Estrogens (estriol, estradiol, estrone, progesterone)	Water and Wastewater	Cloud-point extraction and high performance liquid chromatography separation with ultraviolet detection	No organic solvent is necessary	[73]
24 priority substances from the European Framework Directive (alachlor, anthracene, atrazine, benzo[b]fluoranthene, benzo [g,h,i]perylene, benzo[k]fluoranthene, bezo[a]pyrene, chlorfenvinphos, chlorpyriphos, α-endosulfan, fluoranthene, hexachlorobenzene, hexachlorobutadiene, α-hexachlorocyclohexane, β-hexachlorocyclohexane, δ-hexachlorocyclohexane, indenol [1,2,3-cd]pirene, lindane, naphthalene, pentachlorobenzene, 4-[para]-nonylphenol, para-tert-octylphenol, simazine, trifluralin)	Estuarine and seawater	Stir-bar sorptive extraction followed by thermal desorption using capillary gas chromatography–mass spectrometry	No organic solvent is needed	[74]
Organophosphate and pyrethroid pesticides (λ-cyhalothrin, methyl parathion, permethrin, ethion)	Water	SDME/gas chromatography with flame ionization detection GC-FID	Requirement of only small volumes of both organic solvent (1 μl) and sample	[75]
Endocrine-disrupting phenols (bisphenol-A, 4-cumylphenol, 4-tertbutylphenol, 4-Octylphenol, 4-n-nonylphenol)	Seawater	Dispersive liquid–liquid microextraction and single-drop microextraction in combination with high-performance liquid chromatography (HPLC) with UV detection	Low sample volume (5 ml) and only 20 μl of organic solvent	[76]
Acidic pharmaceuticals (salicylic acid, ibuprofen, diclofenac)	wastewater	Hollow fiber-based liquid-phase microextraction followed by a HPLC/MS-MS	Very low (several μl) organic solvent consumption	[77]

that are safer for the environment. Significant environmentally friendly method innovations have been developed in the field of environmental analysis for determining trace elements and organic contaminants in soil [8, 9, 21, 58].

Instrumental methods that can manage solid samples for direct analysis, such as electrothermal atomic absorption spectrometry (ET AAS), glow discharge-mass spectrometry (GD-MS), laser ablation-mass spectrometry (LA-MS), electrothermal vaporization (ETV), X-ray fluorescence spectrometry (XFS), and laser-induced breakdown spectroscopy (LIBS) are good examples of the environmentally friendly analytical approaches for determining trace elements and major constituents in soil samples. The main advantage of these methods is avoiding sample pre-treatments, such as chemical decomposition and dissolution. If sample pre-treatment is mandatory for trace metal analysis, the utilization of microwave- and ultrasonic-assisted treatments could be interesting due to the use of reduced quantities of acids and solvents coupled with energy conservation.

Most methods for sample treatment and further separation of organic compounds use organic solvents, and this is the main source of organic waste. An ultimate goal in green chemistry is to eliminate or minimize the use of volatile organic solvents [78]. Thus, a variety of solvent reduction, replacement and elimination has been developed to extract and concentrate analytes [5, 9, 27, 78]. These include pressurized fluid extraction (PFE), also known as accelerated solvent extraction (ASE) [79], ultrasound extraction (USE), microwave assisted extraction (MAE), and supercritical fluid extraction (SFE) [9]. The main characteristic of these techniques is the reduction of organic solvents and speed of extraction times compared to traditional liquid-liquid extractions [8]. An example of an analytical procedure used for environmental analysis is the extraction of pesticides from soil samples using ASE. This procedure can take minutes rather than the hours required for Soxhlet extraction [80]. Ultrasound extraction uses high frequency acoustic waves to create microscopic bubbles in liquids [8]. Ultrasonic irradiation may have a strong effect on several sample treatments. Microwave-assisted extraction is an alternative method, in which digestion or extraction can be sped up by introducing energy to heat the sample in the form of microwaves. Microwaves allow reaction times to be significantly reduced [9]. SFE applied to solid environmental matrices containing trace pollutants has become a routine method. The most useful supercritical fluids to green chemists are water and carbon dioxide, which are renewable and non-flammable. SFE has been employed in a great number of soil applications covering many organic pollutants, such as polychlorinated biphenyls [80], polycyclic aromatic hydrocarbons [81], chlorinated benzenes (CB) and hexachlorocyclohexanes (HCH) [82].

Different analytical practices are developing rapidly. In this context, new measuring techniques and methods have been widely used for the identification and determination of major constituents, trace elements and organic contaminants in soil samples. The evolution of new methods for soil sample analysis include solid-phase extraction (SPE), molecularly imprinted polymers (MIP), cloud-point extraction (CPE), ion chromatography-inductively coupled plasma-sector field mass spectrometry (IC-ICP-SFMS), thermal extraction cone penetrometry and immunoassays (IA), that is, enzyme-linked immunosorbent assay (ELISA).

Table 22.4 presents some relevant green analytical methods for determining major constituents, trace elements and organic contaminants in soil samples.

22.4.2 Sediments

Sediment is any particulate matter that can be transported by fluid and that eventually is deposited as a layer of solid particles on the bed or bottom of a body of water or other liquid.

Sample pretreatment is the most critical stage of the analytical process, mainly when solid matrices are analyzed. Although there are references in the literature for several fast separation and detection methods, many of these involve time-consuming sample pre-treatment, commonly, extraction procedures. The extraction step has been the critical part in developing approaches for solid and semi-solid samples. The more usual, simple and classic extraction procedures are based on solvents and/or diluted acids assisted by

Table 22.4 *Some green analytical methods for determining major constituents, trace elements and organic contaminants in soil samples.*

Analytes	Samples	Analytical system	Green aspects	Reference
PAH	Floodplain soils	Supercritical fluid extraction	Reduction of the reagents volumes, reduction of working extraction time	[81]
Hg	Soils	Solid sampling-graphite furnace atomic absorption spectrometry	Use of direct measurement to avoid sample pre-treatment	[83]
Mo	Coal fly ash, sediment, soil and urban dust	Slurry extraction electrothermal atomic absorption spectroscopy	Use of direct measurement to avoid sample pre-treatment	[84]
Cd(II), Pb(II), Cu(II), Zn(II), Co(II), Ni(II), Cr(VI), and Mo(VI)	Soil and indoor-airborne particulate matter	Square-wave voltammetry	Use of direct measurement to avoid sample pre-treatment and reduced the time of analysis	[85]
Cd	Soil	Micro-sampling cup/slurry with atomic absorption spectrometry detection	Use of direct measurement to avoid sample pre-treatment	[86]
Zn, Cd, Tl and Pb	Soil	Ultrasonic slurry sampling electrothermal vaporization with inductively-coupled plasma mass spectrometry detection	Use of direct measurement to avoid sample pre-treatment	[87]
As, Cd and Pb	Soil	Solid sampling electrothermal vaporization with ICP OES detection	Use of direct measurement to avoid sample pre-treatment	[88]
U	Soil	Glow discharge mass spectrometry	Use of direct measurement to avoid sample pre-treatment	[89]
Pb and As	Soil	Laser ablation combined with fast electric pulse discharge plasma spectroscopy	Use of direct measurement to avoid sample pre-treatment	[90]
As, Ba, V, Mn, Pb and Hg	Soil	Laser-induced breakdown spectrometry	Use of direct measurement to avoid sample pre-treatment	[91]
Be, Si, Ca, V, Cr, Fe, Ni, Cu, Zn, As, Ag, Cd, Sn, Sb, Tl and Pb	Soils and sediments	Laser ablation inductively coupled plasma mass spectrometry	Use of direct measurement to avoid sample pre-treatment	[92]
DMA, MMA, As (III) and As (V)	Soil	Microwave assisted extraction with ion chromatography combined with ICP-MS	Use of less reagents, savings of energy	[93]

(continued)

Table 22.4 (*continued*)

Analytes	Samples	Analytical system	Green aspects	Reference
V, Cr, Fe, Ni, Cu, Zn, As, Al, Pb and U	Sediments, soil, urban particulate matter and coal fly ash	Microwave-assisted acid digestion with ICP-MS	Reduction of reagents, shorter digestion time analysis, savings of energy	[94]
As, Cd, Cu, Pb and Zn	Soil	Portable X-ray fluorescence	Fast analysis, large numbers of trace metals can be simultaneously detected with high sensitivities, low residual generation	[95]
Pb, Cu, Zn and As	Soil	Portable X-ray fluorescence	Fast analysis, large numbers of trace metals can be simultaneously detected with high sensitivities, low residual generation	[96]
PAHs, PCBs, organochlorine pesticides, di- and trinitrated toluene (DNT and TNT)	Soil	Thermal extraction cone penetrometry gas chromatography/ mass spectrometry	Reduction of organic solvent, possibility to detect subsurface contaminants in situ.	[97]
Acetaminophen, Carbamazepine, Ciprofloxacin, Codeine, Diazepam, Fenofibrate, Metoprolol, Norfloxacin, Ofloxacin, Oxytetracycline, Propanolol, Sulfamethoxazole, Tetracycline, Trimethoprim, Clofibric acid, Diclofenac and Ibuprofen	Soils and sediments	PLE using water as extractant, clean-up with SPE and determination by LC–ESI-MS/MS	Reduction of the use of organic solvents, use of a very small amount of toxic chemicals and reagents, low waste production.	[98]

Mirex, α- and γ-chlordane, p,p´-DDT, heptachlor, heptachlor epoxide isomer A, γ-hexachlorocyclohexane, dieldrin, endrin, aldrine and hexachlorobenzene	Soil	Microwave-assisted extraction using headspace solid phase microextraction and gas chromatography–tandem mass spectrometry	Reduction of the use of organic solvents, reduced sample preparation time.	[99]
Organochlorine pesticides	Soil	Miniaturized ultrasonic extraction with gas chromatography	Low consume of organic solvent and sample, low residual generation	[100]
Estrogens	Soil	Pressurized liquid extraction and gas chromatography–mass spectrometry	Reduction of the use of organic solvents, reduced sample preparation time. Savings of energy compared to Soxhlet	[101]
Pesticides	Soil	Microwave assisted extraction with LC-fluorescence detection	Not require previous treatment of the samples, reduction of working extraction time and organic solvent consumption	[102]
Prometryne	Soil	Cloud-point extraction/ preconcentration with HPLC–UV	Use of surfactant solution to avoid organic solvent	[103]
Fluoroquinolones	Soil	Molecularly imprinted polymers	High selectivity, allow complex samples to be analyzed without any previous clean-up.	[104]
Dibenzo-p-dioxins and polychlorinated dibenzofurans	Soil	Accelerated solvent extraction and purification with an enzyme-linked immunosorbent assay (ELISA) detection	Reduction of the organic solvent, high selective reactions	[105]

mechanical shaking, Soxhlet distillation, ultrasonic baths or probe devices. These procedures use large solvent volumes, are tedious, and the extracts must be handled before analyzing them, which can result in a low percent recovery for the analyte being assessed [11, 14]. Using harmful chemicals and large volumes of solvents cause environmental pollution, health hazards to laboratory personnel and extra operational costs for waste treatment. Ideally, sample preparation techniques should be fast, easy to use, inexpensive and compatible with a range of analytical instruments; therefore, the current trend is towards simplification and miniaturization of the sample preparation steps and decreasing the quantities of organic solvents used [18]. Some green solvent extraction techniques for the analysis of solid matrices have been developed to shorten the analytical procedures and minimize waste solvents, requiring much smaller volumes of solvents; in some techniques, no solvent or virtually no solvent is used, which are called solventless [7] (see Section 2.2.2).

Table 22.5 presents some relevant environmental analytical methods applied for sediment samples, which present the green characteristics mentioned above.

22.4.3 Wastes

Waste management must be faced under the principles of minimization, recovery, reuse and recycling. Taking into account that one of the main aspects of the Green Analytical Chemistry is to avoid waste generation, the best way to overcome this target is relatively simple in principle: the best waste is the one that is not generated. Therefore, in-field direct sample analysis is the ideal choice because no reagents are used [126]. However, it is usually impossible to avoid waste generation in industrial, teaching and analytical procedures. Many of these are official and certified methods adopted by environmental organisms [127]. Due to this aspect, a new green alternative must be developed, validated and certified.

Similar to analyses of sediments and soils, solid waste analysis can become greener if some strategies are adopted [126, 127]:

(1) miniaturization and automation of analytical methods;
(2) development of less-consuming energy and reagent methods;
(3) the use of less toxic (or even non-toxic) solvents and reagents; and
(4) on-line decontamination of wastes.

Strategies (1) through (3) are applicable for teaching laboratories. This is of utmost importance because students will be given an opportunity to discover Green (Analytical) Chemistry long before reaching their first job. Also, an important difference between industrial and laboratory waste is the amount generated in each case: the former is generated in much higher amounts.

Waste characterization demands many analytical tools. Supercritical fluid extraction, solid-phase extraction, microwave- and ultrasound-assisted extraction, pressurized fuel extraction, CG-MS, HPLC-MS, and MS-MS are currently in use for the analysis of biodegradable and non-degradable wastes [127–129]. Heavy metals, phenols, chlorophenols, linear alkylbenzene sulfonates, antibiotics, aromatic hydrocarbons and fluorinated compounds have been successfully analyzed by these techniques with minimum waste generation. Solid waste analysis follows the same tendencies for sediment and soil analyses, but it appears that solid waste management remains an interesting field to develop green analytical procedures because there are apparently few green procedures currently in use.

However, what should be done when dealing with waste? The availability of areas for landfills are limited and subjected to increasing environmental restraints. The options are numerous, and the choice depends on the nature of the waste. Thermal degradation is a classic final destination for organic compounds. Incineration and co-processing are common alternatives, but the latter is better from an environmental viewpoint. It accepts a wide variety of waste for energy recovery during clinker manufacturing. Also,

Table 22.5 *Some green analytical methods for determining different analytes in sediments.*

Analytes	Analytical system	Green aspects	Reference
Pesticides	Superheated water extraction with solvent trap and gas chromatography with flame ionization detector	Water as extraction solvent; low extraction cost, reduced solvent consumption and decreased extraction time	[10]
Al, As, Cd, Co, Cr, Cu, Fe, Mn, Ni, Pb, Sr, Zn (marine sediment certificate reference material)	Pressurized liquid extraction and inductively coupled plasma optical emission spectrometry	Use of diluted weak acids, as well as less corrosive, as extraction solvents	[11]
Polycyclic Aromatic Hydrocarbons	Pressurized hot water extraction with solid-phase microextraction and gas chromatography–mass spectrometry	Water as extraction solvent; minimal quantity of solvents; fully automated method	[12]
Linear alquilbenzene sulfonates	Superheated water extraction with flow injection and liquid chromatography with fluorescence detector	Water as extraction solvent; minimal quantity of solvents; drastic reduction of the extraction time	[13]
Organochlorine pesticides	Pressurized hot water extraction with solid-phase microextraction and gas chromatography–mass spectrometry	Water as extraction solvent; minimal quantity of solvents	[15]
Pharmaceuticals	Pressurized liquid extraction and solid-phase extraction followed by liquid chromatography–electrospray tandem mass spectrometry	Simple and inexpensive method; water as extraction solvent and minimal use of organic solvents in order to eliminate matrix components during the clean-up.	[98]
Polycyclic Aromatic Hydrocarbons (sediment certificate reference material)	Ionic liquid (IL)-based aggregates with solid-phase microextraction and gas chromatography–mass spectrometry after focused microwave-assisted extraction	The overall extraction method requires short extraction times and uses surfactant solution to avoid organic solvents.	[106]
Cd(II), Pb(II) and Sb(III)	Nonpathogenic green fluorescent protein-based bacterial biosensors	Water as extraction solvent; sensitive, fast, inexpensive and less laborious.	[107]
Andiandrogen vinclozolin	Gas chromatography with electron capture detector after hollow-fiber liquid-phase micro-extraction assisted by ultrasonic extraction	Faster than conventional extraction techniques; lower consumption of organic solvents; higher efficiency in cleaning the sample	[108]
Tricyclazole	Subcritical water extraction and reverse phase high performance liquid chromatography and liquid scintillation counting detector	Water as extraction solvent and minimal use of organic solvents	[109]
Persistent organic Pollutants (including organochlorine pesticides)	Microwave assisted solvent extraction and hollow-fiber liquid-phase microextraction	Water as extraction solvent; minimal use of organic solvents	[110]

(continued)

Table 22.5 *(continued)*

Analytes	Analytical system	Green aspects	Reference
Brominated flame retardants	Pressurized hot water extraction with solid-phase trap and on line liquid-chromatography-gas chromatography	Water as extraction solvent; minimal use of organic solvents	[111]
Organotin compounds	Accelerated solvent extraction and gas chromatography–mass spectrometry	Sodium acetate and acetic acid in methanol as extraction solvent	[112]
Speciation of butyl- and phenyltin compounds	Accelerated solvent extraction and liquid chromatography–inductively coupled plasma mass spectrometry	Tropolone and acetic acid in methanol as extraction solvent; reduce the extraction time from several hours to just few minutes.	[113]
Organotin compounds	Headspace single-drop micro-extration with gas chromatography–mass spectrometry	Solvent-free compared with classical liquid–liquid extractions; significantly less expensive	[114]
Industrial and anthropogenic waste-indicator compounds	Pressurized liquid extraction coupled with solid-phase extraction and gas chromatography-mass spectrometry	Reduced extraction time, and reduced solvent volume	[115]
Persistent organochlorine pesticide	Headspace solid-phase microextraction and gas chromatography with electron capture detector	Cheap, solvent free, easy to automate	[116]
Polycyclic aromatic hydrocarbons	On-line coupled pressurized hot water extraction, hollow fiber microporous membrane liquid–liquid extraction and gas chromatography with flame ionization detector	Water as extraction solvent; low consumption of organic solvents	[117]
Polycyclic aromatic hydrocarbons (sediment certificate reference material)	Pressurized hot water extraction and gas chromatograph-mass spectrometer	Water as extraction solvent; minimal use of organic solvents	[118]
Semivolatile organic compounds, polycyclic aromatic hydrocarbons, alkylated, polycyclic aromatic hydrocarbons homolog groups	Pressurized liquid extraction coupled with solid-phase extraction and gas chromatography–mass spectrometry	Sample automation, reduced extraction time, reduced solvent volume	[119]

Organophosphate triesters	Pressurized liquid extraction and solid-phase extraction followed by gas chromatography-mass spectrometry	Water as a component of extraction solvent; low consumption of organic solvents	[120]
Pesticides, pharmaceutical and personal care products	Sequential superheated liquid extraction and gas chromatography–mass spectrometry	Water as extraction solvent; shorter extraction times; lower extractant volumes and easier rupture of analyte–matrix interaction	[121]
Polycyclic Aromatic Hydrocarbons	Static pressurized liquid extractor coupled to a flow injection and high performance liquid chromatography with fluorimeter detection	Water as extraction solvent; minimal use of organic solvents	[122]
Polychlorinated biphenyls	Static subcritical water extraction coupled with solid-phase microextraction and gas chromatography with electron capture detector	Simple and inexpensive apparatus; water as extraction solvent; no sample drying steps are required and no organic solvent wastes are generated;	[123]
Polycyclic Aromatic Hydrocarbons	Static subcritical water extraction with simultaneous solid-phase extraction and gas chromatography with flame ionization detection	Very simple and rapid method; water as extraction solvent; minimal use of organic solvents	[124]
Polycyclic Aromatic Hydrocarbons	Ultrasonic micellar extraction and high performance liquid chromatography with ultraviolet detection	More rapid than other conventional extraction methods; not require the use of potentially hazardous organic solvents; extracts from sonication can be chromatographed without subsequent clean-up steps; analysis time considerably reduced.	[125]

inorganic solids may partially replace the minerals used. Ashes are incorporated to the clinker. Therefore, this procedure saves and sustains non-renewable natural resources. Unfortunately, co-processing presents many restrictions. For example, mercury, cadmium fluorine and chlorine compounds are to be avoided. Therefore, the waste generated in an industrial or laboratory site must be separated to avoid placing undesirable waste for co-processing. Mercury, cadmium and fluorine compounds must be passivated via precipitation to be disposed of in industrial dumps. However, it is much better to avoid their use in industrial and laboratory procedures, replacing them by other green alternatives.

Organic wastes containing low concentrations of heavy metals can be treated by a biological process, producing ground fertilizer (composting).

Recycling is a logical option for non-biodegradable materials, such as metals, plastic and glass. Selective collecting is essential to ensure a successful recycling program. Laboratory waste, which falls under this category, must be clean before sending to a recycling site.

However, a great tendency is to use waste as a raw material for other processes [127]. For example, food waste and compost are used for production of biofuels, mitigation of green house gases and in agriculture. Waste fibers from the tannery industry may be employed for removal of pollutants from water. Even remediation of contaminated soils or other environmental sites may be performed using appropriate waste [126]. The physico-chemical characterization of the waste determines its possible use in other industrial processes.

Despite this impressive aspect, true green chemistry means no generation of wastes. Waste remediation in a short-term measure is necessary but must not be regarded as the final solution.

22.5 Green environmental analysis applied for atmospheric samples

Historically, atmospheric compounds were measured by wet chemical methods. For instance, ozone present in an air sample was bubbled through an acidic solution containing iodide ions. The I_2 formed was determined using wet chemical techniques [130].

These classical wet methods were subjected to much potential interference. In the example given above, SO_2 presents a negative interference (it reduces I_2 to I^-), whereas NO_2 gives a positive interference (it also oxidizes I^- to I_2). As a result, these methods were abandoned, being replaced by instrumental methods of analysis. These are used in both tropospheric and stratospheric analyses [130].

22.5.1 Gases

Instrumental methods for gas analysis include the following: (1) optical spectroscopic techniques, such as chemiluminescence, fluorescence, infrared spectroscopy (multi-pass cells, FTIR, tunable diode laser spectroscopy (TDLS), non-dispersive infrared spectroscopy (NDIR), matrix isolation spectroscopy (MI), and ultraviolet-visible absorption spectroscopy (UV-vis); and (2) mass spectrometry. In these circumstances, the sample is introduced to the instrument without the aid of any reagent [130]. These methods usually offer real-time measurements and require low quantities of sample.

Some compounds, particularly those at very low concentrations in the air [130–133], can be collected from gas streams for subsequent quantification by analytical techniques (HPLC, GC, GC-MS, colorimetric methods, etc.). The media available are filters, denuders, transition flow reactors, mist chambers and scrubbers. The sample must be passed for sufficient periods of time to collect measurable amounts of the substance to be analyzed [130, 132]. Volatile organic compounds (hydrocarbons, carbonyl compounds, etc.) are typical examples of compounds collected from air samples prior to analysis [130, 133]. This analysis may be via spectroscopy or derivatization methods [130, 132, 133]. These may employ toxic reagents, such as 2,4-dinitrophenylhydrazine (DNPH), for determination of aldehydes and ketones [130].

22.5.2 Particulates

Particles are of great interest in atmospheric analysis because of their impact on human health. Most particle analyses fall into two categories: mass determination and chemical composition [130, 133].

Particles are collected based on filtration, gravitational and centrifugal sedimentation, inertial impaction and impingement, diffusion, interception or electrostatic or thermal precipitation. The choice of the method depends on parameters such as composition and size of the particles.

After collection, the mass of the collected particles can be determined by gravimetric methods, β-ray attenuation, piezoelectric microbalance and oscillating microbalance. Particle size is also of great interest. Optical methods, atomic force microscopy, scanning tunneling microscopy, electrical mobility analyzers, diffusion separators, aerodynamic particle size and condensation particle counters are the choices for analysis [130]. Currently, there is no technique capable of measuring the size distribution of the atmospheric aerosols. A combination of methods must be used for such analysis.

Chemical composition of the particles may be determined by numerous methods. The choice depends primarily on the nature of these particles (inorganic and organic components).

Colorimetric methods were extensively used in the past to determine inorganic elements [130] but have been replaced by other characterization techniques [130, 131]. The main reason is that these methods are subject to interferences, particularly in the case of complex environmental samples. Although this replacement was not based on green aspects, one of the consequences of the abandonment of colorimetric methods was to develop greener methods of analysis for particulates. These methods fall into two categories: X-ray fluorescence (XRF), particle-induced X-ray emission (PIXE), neutron activation (NA), X-ray photoelectron spectroscopy (XPS); and mass spectrometry (MS), which do not require leaching of the sample prior to analysis, making them more green procedures. Atomic absorption spectrometry (AAS), optical emission spectrometry (OES), inductively coupled plasma spectroscopy (ICP), and anodic stripping voltammetry (ASV) require the extraction of elements into a solution that is analyzed.

Inorganic ions present in particulates are analyzed by colorimetric methods, ion chromatography, and selective ion electrodes. These methods require solubilization of the sample. Infrared, Raman spectroscopy and mass spectrometry do not require previous solubilization [130].

Organic elemental carbon can be determined by thermal, optical, digestion and extraction techniques. Digestion has not shown to present clear advantages over thermal methods, but all of the above procedures may be subjected to serious errors. Speciation of organic compounds may be performed by thermal desorption or solvent extraction followed by GC-MS or HPLC. Solvent extraction may require the use of toxic solvents and inevitably generates liquid waste [130]. A green approach was to replace ordinary solvents by supercritical fluids, such as CO_2 [131, 132]. Thus, the use of large volumes of solvent is avoided, and it is not necessary to concentrate the extract (in this step losses of analytes may occur). Supercritical fluids can be used at lower temperatures, thus minimizing thermal decomposition and/or formation of artifacts that may take place using thermal desorption techniques [131, 132]. The new green techniques have potential applications in atmospheric analysis [133].

References

1. Radojevic, M. and Bashkin, V.N. (1999) *Practical Environmental Analysis*, The Royal Society of Chemistry, Cambridge.
2. Rocha, F.R.P.; Nóbrega, J.A. and Fatibello-Filho, O. (2001) Flow analysis strategies to greener Analytical Chemistry. An overview, *Green Chem.*, **3**, 216–220.
3. de la Guardia, M. and Ruzicka, J. (1995) Guest editorial. Towards environmentally conscientious Analytical Chemistry through miniaturization, containment and reagent replacement, *Analyst*, **120**, 17N.

4. Anastas, P.T. (1999) Green Chemistry and the Role of Analytical Methodology Development, *Crit. Rev. Anal. Chem.*, **29**, 167–175.
5. Koel, M. and Kaljurand, M. (2006) Application of the principles of Green Chemistry in Analytical Chemistry, *Pure Appl. Chem.*, **78**, 1993–2002.
6. Fifield. F.W. and Haines, P.J. (2000) Introduction, in *Environmental Analytical Chemistry* (eds F.W. Fifield and P.J. Haines), Blackwell Science Ltd, Cambridge.
7. Tobiszewski, M.; Mechlińska, A.; Zygmunt, B. and Namieśnik, J. (2009) Green Analytical Chemistry in sample preparation for determination of trace organic pollutants, *TrAC-Trend. Anal. Chem.*, **28**, 943–951.
8. Keith, L.H.; Gron, L.U. and Young, J.L. (2007) Green analytical methodologies. *Chem. Rev.*, **107**, 2695–2708.
9. Armenta, S.; Garrigues, S. and de la Guardia, M. (2008) Green Analytical Chemistry, *TrAC-Trend. Anal. Chem.*, **27**, 497–511.
10. Chienthavorn, O. and Su-in, P. (2006) Modified superheated water extraction of pesticides from spiked sediment and soil, *Anal. Bioanal. Chem.*, **385**, 83–89.
11. Alonso-Rodríguez, E.; Moreda-Piñeiro, J.; López-Mahía, P. *et al.* (2006) Pressurized liquid extraction of organometals and its feasibility for total metal extraction, *TrAC-Trend. Anal. Chem.*, **25**, 511–519.
12. Fernández-González, V.; Concha-Graña, E.; Muniategui-Lorenzo, S.; López-Mahía, P. and Prada-Rodríguez, D. (2008) Pressurized hot water extraction coupled to solid-phase microextraction–gas chromatography–mass spectrometry for the analysis of polycyclic aromatic hydrocarbons in sediments, *J. Chrom. A*, **1196–1197**, 65–72.
13. Priego-López, E. and Luque de Castro, M.D. (2004) Superheated water extraction of linear alquilbenzene sulfonates from sediments with on-line preconcentration/derivatization/detection, *Anal. Chim. Acta*, **511**, 249–254.
14. Teo, C.C.; Tan, S.N.; Yong, J.W.H.; Hew, C.S. and Ong, E.S. (2010) Pressurized hot water extraction (PHWE), *J. Chromatogr. A*, **1217**, 2484–2494.
15. Concha-Graña, E.; Fernández-González, V.; Grueiro-Noche, G.; Muniategui-Lorenzo, S.; López-Mahía, P.; Fernández-Fernández, E. and Prada-Rodríguez, D. (2010) Development of an environmental friendly method for the analysis of organochlorine pesticides in sediments, *Chemosphere*, **79**, 698–705.
16. Aguilera-Herrador, E.; Lucena, R.; Cárdenas, S. and Valcárcel, M. (2010) The roles of ionic liquids in sorptive microextraction techniques, *TrAC-Trend. Anal. Chem.*, **29**, 602–616.
17. Lambropoulou, D.A. and Albanis, T.A. (2007) Liquid-phase micro-extraction techniques in pesticide residue analysis, *J. Biochem. Biophys. Methods*, **70**, 195–228.
18. Sarafraz-Yazdi, A. and Amiri, A. (2010) Liquid-phase microextraction, *TrAC-Trend. Anal. Chem.*, **29**, 1–14.
19. Pedersen-Bjergaard, S. and Rasmussen, K.E. (2008) Liquid-phase microextraction with porous hollow fibers, a miniaturized and highly flexible format for liquid–liquid extraction, *J. Chromatogr. A*, **1184**, 132–142.
20. Vandecasteele, C. and Block, C.B. (1993) Modern *methods for trace element determination*, John Wiley & Sons, Ltd., Chichester.
21. Armenta, S. and de la Guardia, M (2009). Green Spectroscopy: A Scientometric Picture, *Spectros. Lett.*, **42**, 277–283.
22. Maya, F.; Estela, J.M. and Cerdà, V. (2009) Multisyringe Flow Injection Technique for Development of Green Spectroscopic Analytical Methodologies. *Spectros. Lett.*, **42**, 312–319.
23. Tobiszewski, M.; Mechlinska, A. and Namiesnik, J. (2010) Green Analytical Chemistry theory and practice. *Chem. Soc. Rev.*, **39**, 2869–2878.
24. Reeve, R. (2002) *Introduction to Environmental Analysis*, Analytical Techniques in the Sciences, John Wiley & Sons, Ltd., Chichester.
25. Wang, J. (2002) Real-time electrochemical monitoring: Toward Green Analytical Chemistry, *Accounts Chem. Res.*, **35**, 811–816.
26. Desai, R.; Villalba, M.M.; Lawrence, N.S. and Davis, J. (2009) Green Approaches to Field Nitrate Analysis: An Electroanalytical Perspective, *Electroanalysis*, **21**, 789–796.
27. Curylo, J.; Wardencki, W. and Namiesnik, J. (2007) Green aspects of sample preparation - a need for solvent reduction, *Pol. J. Environ. Stud.*, **16**, 5–16.
28. van Staden, J. F. and Taljaard, R.E. (1996) Determination of sulphate in natural waters and industrial effluents by sequential injection analysis *Anal. Chim. Acta*, **331**, 271–280.
29. Trojanowicz, M.; Alexander, P.W. and Hibbert, D.B. (1998) Flow-injection analysis with potentiometric detection for the speciation of fluoride and calcium, *Anal. Chim. Acta*, **366**, 23–33.

30. Connolly, D. and Paull, B. (2001) Rapid determination of nitrate and nitrite in drinking water samples using ion-interaction liquid chromatography, *Anal. Chim. Acta*, **441**, 53–62.
31. Saurina, J.; López-Aviles, E.; Le Moal, A. and Hernández-Cassou, S. (2002) Determination of calcium and total hardness in natural waters using a potentiometric sensor array, *Anal. Chim. Acta*, **464**, 89–98.
32. Morales, J.A.; de Graterol, L.S. and Mesa, J. (2000) Determination of chloride, sulfate and nitrate in groundwater samples by ion chromatography *J. Chromatogr. A*, **884**, 185–190.
33. García-Fernández, R.; García-Alonso, J.I. and Sanz-Medel, A. (2004) Simultaneous determination of inorganic anions, calcium and magnesium by suppressed ion chromatography, *J. Chromatogr. A*, **1033**, 127–133.
34. Melchert, W.R.; Infante, C.M.C. and Rocha, F.R.P. (2007) Development and critical comparison of greener flow procedures for nitrite determination in natural waters *Microchem. J.*, **85**, 209–213.
35. Melchert, W.R.; Oliveira, D.R. and Rocha, F.R.P. (2010) An environmentally friendly flow system for high-sensitivity spectrophotometric determination of free chlorine in natural waters, *Microchem. J.*, **96**, 77–81.
36. van Staden, J.F. and Stefan, R.I. (1999) Simultaneous flow injection determination of calcium and fluoride in natural and borehole water with conventional ion-selective electrodes in series, *Talanta*, **49**, 1017–1022.
37. Silva, C.R.; Vieira, H.J.; Canaes, L.S.; Nóbrega, J.A. and Fatibello-Filho, O. (2005) Flow injection spectrophotometric method for chloride determination in natural waters using $Hg(SCN)_2$ immobilized in epoxy resin, *Talanta*, **65**, 965–970.
38. Liu, R.; Wang, H.; Sun, A. and Liu, D. (1997) Reversed flow injection spectrophotometric determination of trace amount of ammonia in natural water by oxidation of ammonia to nitrite, *Talanta*, **45**, 405–410.
39. Motellier, S.; Petit, S. and Decambox, P. (2000) Quantitative capillary electrophoretic analysis for calcium and magnesium in sodium–matrix waters, *Anal. Chim. Acta*, **410**, 11–23.
40. Mori, M.; Hu, W.; Haddad, P.R.; Fritz, J.S.; Tanaka, K.; Tsue, H. and Tanaka, S. (2002) Capillary electrophoresis using high ionic strength background electrolytes containing zwitterionic and non-ionic surfactants and its application to direct determination of bromide and nitrate in seawater, *Anal. Bioanal. Chem.*, **372**, 181–186.
41. Silva, C.R.; Conceição, C.D.; C.Bonifácio, V.G.; Fatibello-Filho, O. and Teixeira, M.F.S. (2009) Determination of the chemical oxygen demand (COD) using a copper electrode: a clean alternative method, *J. Solid State Electrochem.*, **13**, 665–669.
42. Infante, C.M.C.; Masini, J.C. and Rocha, F.R.P. (2008) A green flow-based procedure for fluorimetric determination of acid-dissociable cyanide in natural waters exploiting multicommutation, *Anal Bioanal. Chem.*, **391**, 2931–2936.
43. Settheeworrarit, T.; Hartwell, S.K.; Lapanatnoppakhun, S.; Jakmunee, J. Christian, G.D. and Grudpan, K. (2005) Exploiting guava leaf extract as an alternative natural reagent for flow injection determination of iron, *Talanta*, **68**, 262–267.
44. Teixeira, L.S.G. and Rocha, F.R.P. (2007) A green analytical procedure for sensitive and selective determination of iron in water samples by flow-injection solid-phase spectrophotometry, *Talanta*, **71**, 1507–1511.
45. Leelasattarathkul, T.; Liawruangrath, S.; Rayanakorn, M.; Liawruangrath, B.; Oungpipat, W. and Youngvises, N. (2007) Greener analytical method for the determination of copper(II) in wastewater by micro flow system with optical sensor, *Talanta*, **72**, 126–131.
46. Tang, A.; Ding, G. and Yan, X. (2005) Cloud point extraction for the determination of As(III) in water samples by electrothermal atomic absorption spectrometry, *Talanta*, **67**, 942–946.
47. Bezerra, M.A.; Bruns, R.E. and Ferreira, S.L.C. (2006) Statistical design-principal component analysis optimization of a multiple response procedure using cloud point extraction and simultaneous determination of metals by ICP OES, *Anal. Chim. Acta*, **580**, 251–257.
48. Chen, J. and Teo, K.C. (2001) Determination of cadmium, copper, lead and zinc in water samples by flame atomic absorption spectrometry after cloud point extraction, *Anal. Chim. Acta*, **450**, 215–222.
49. Paleologos, E.K.; Stalikas, C.D.; Tzouwara-Karayanni, S.M.; Pilidis, G.A. and Karayannis, M.I. (2000) Micelle-mediated methodology for speciation of chromium by flame atomic absorption spectrometry, *J. Anal. At. Spectrom.*, **15**, 287–291.
50. Escudero, L.A.; Martinez, L.D.; Salonia, J.A. and Gasquez, J.A. (2010) Determination of Zn(II) in natural waters by ICP-OES with on-line preconcentration using a simple solid phase extraction system, *Microchem. J.*, **95**, 164–168.
51. Faraji, M.; Yamini, Y. and Rezaee, M. (2010) Extraction of trace amounts of mercury with sodium dodecyle sulphate-coated magnetite nanoparticles and its determination by flow injection inductively coupled plasma-optical emission spectrometry, *Talanta*, **81**, 831–836.

52. Zhao, S.; Liang, H.; Yan, H.; Yan, Z.; Chen, S.; Zhu, X. and Cheng, M. (2010) Online Preconcentration and Determination of Trace Levels Cadmium in Water Samples Using Flow Injection Systems Coupled with Flame AAS, *Clean – Soil, Air, Water*, **38**, 146–152.
53. Tokalıoğlu, Ç.; Yılmaz, V. and Kartal, Ç. (2009) Solid phase extraction of Cu(II), Ni(II), Pb(II), Cd(II) and Mn(II) ions with 1-(2-thiazolylazo)-2-naphthol loaded Amberlite XAD-1180, *Environ. Monit. Assess.*, **152**, 369–377.
54. Jahromi, E.Z.; Bidari, A.; Assadi, Y.; Hosseini, M.R.M. and Jamali, M.R. (2007) Dispersive liquid–liquid microextraction combined with graphite furnace atomic absorption spectrometry: Ultra trace determination of cadmium in water samples, *Anal. Chim. Acta*, **585**, 305–311.
55. Naseri, M.T.; Hemmatkhah, P.; Hosseini, M.R.M. and Assadi, Y. (2008) Combination of dispersive liquid–liquid microextraction with flame atomic absorption spectrometry using microsample introduction for determination of lead in water samples, *Anal. Chim. Acta*, **610**, 135–141.
56. Mohammadi, S.Z.; Afzali, D.; Taher, M.A. and Baghelani, Y.M. (2009) Ligandless dispersive liquid–liquid microextraction for the separation of trace amounts of silver ions in water samples and flame atomic absorption spectrometry determination, *Talanta*, **80**, 875–879.
57. Namiesnik, J. (2001) Green Analytical Chemistry - Some remarks, *J. Separ. Sci.*, **24**, 151–153.
58. David, F. and Sandra, P. (2007) Stir bar sorptive extraction for trace analysis, *J. Chromatogr. A*, **1152**, 54–69.
59. Jeannot, M.A.; Przyjazny, A. and Kokosa, J.M. (2010) Single drop microextraction - Development, applications and future trends, *J. Chromatogr. A*, **1217**, 2326–2336.
60. Augusto, F.; Carasek, E.; Silva, R.G.C.; Rivellino, S.R.; Batista, A.D. and Martendal, E. (2010) New sorbents for extraction and microextraction techniques, *J. Chromatogr. A*, **1217**, 2533–2542.
61. Pena-Pereira, F.; Lavilla, I. and Bendicho, C. (2010) Liquid-phase microextraction techniques within the framework of Green Chemistry, *TrAC-Trend. Anal. Chem.*, **29**, 617–628.
62. Farré, M.; Pérez, S.; Gonçalves, C.; Alpendurada, M.F. and Barceló, D. (2010) Green Analytical Chemistry in the determination of organic pollutants in the aquatic environment, *TrAC-Trend. Anal. Chem.*, **29**, 1347–1362.
63. Chunhong, J.; Xiaodan, Z.; Jihua, W.; Ercheng, Z.; Min, H.; Li, C. and Pingzhong, Y. (2010) Extraction of pesticides in water samples using vortex-assisted liquid–liquid microextraction, *J. Chromatog. A*, **1217**, 5868–5871.
64. Van Hoeck, E.; Canale, F.; Cordero, C.; Compernolle, S.; Bicchi, C. and Sandra, P. (2009) Multiresidue screening of endocrine-disrupting chemicals and pharmaceuticals in aqueous samples by multi-stir bar sorptive extraction–single desorption–capillary gas chromatography/mass spectrometry, *Anal. Bioanal. Chem.*, **393**, 907–919.
65. Lin, C.-Y. and Huang, S.-D. (2008) Application of liquid–liquid–liquid microextraction and high-performance liquid-chromatography for the determination of sulfonamides in water, *Anal. Chim. Acta*, **612**, 37–43.
66. Sun, Z.; Schüssler, W.; Sengl, M.; Niessner, R. and Knopp, D. (2008) Selective trace analysis of diclofenac in surface and wastewater samples using solid-phase extraction with a new molecularly imprinted polymer, *Anal. Chim. Acta*, **620**, 73–81.
67. Pérez-Carrera, E.; León, V.M.L.; Parra, A.G. and González-Mazo, E. (2007) Simultaneous determination of pesticides, polycyclic aromatic hydrocarbons and polychlorinated biphenyls in seawater and interstitial marine water samples, using stir bar sorptive extraction–thermal desorption–gas chromatography–mass spectrometry, *J. Chromatogr. A*, **1170**, 82–90.
68. Ballesteros-Gómez, A.; Rubio, S. and Pérez-Bendito, D. (2008) Determination of priority carcinogenic polycyclic aromatic hydrocarbons in wastewater and surface water by coacervative extraction and liquid chromatography–fluorimetry, *J. Chromatogr. A*, **1203**, 168–176.
69. Moral, A.; Sicilia, M.D. and Rubio, S. (2009) Supramolecular solvent-based extraction of benzimidazolic fungicides from natural waters prior to their liquid chromatographic/fluorimetric determination, *J. Chromatogr. A*, **1216**, 3740–3745.
70. Hu, Y.; Wang, Y.; Hu, Y. and Li, G. (2009) Liquid–liquid–solid microextraction based on membrane-protected molecularly imprinted polymer fiber for trace analysis of triazines in complex aqueous samples, *J. Chromatogr. A*, **1216**, 8304–8311.
71. Yan, H.; Liu, B.; Du, J.; Yang, G. and Row, K.H. (2010) Ultrasound-assisted dispersive liquid–liquid microextraction for the determination of six pyrethroids in river water, *J. Chromatogr. A*, **1217**, 5152–5157.
72. Jia, G.; Li, L.; Qiu, J.; Wang, X.; Zhu, W.; Sun, Y. and Zhou, Z. (2007) Determination of carbaryl and its metabolite 1-naphthol in water samples by fluorescence spectrophotometer after anionic surfactant micelle-mediated extraction with sodium dodecylsulfate, *Spectrochim. Acta Part A*, **67**, 460–464.

73. Wang, L.; Cai, Y.-Q.; He, B.; Yuan, C.-G.; Shen, D.-Z.; Shao, J. and Jiang, G.-B. (2006) Determination of estrogens in water by HPLC–UV using cloud point extraction, *Talanta*, **70**, 47–51.
74. Huertas, C.; Morillo, J.; Usero, J. and Gracia-Manarillo, I. (2007) Validation of stir bar sorptive extraction for the determination of 24 priority substances from the European Water Framework Directive in estuarine and sea water, *Talanta*, **72**, 1149–1156.
75. Pinheiro, A.S. and de Andrade, J.B. (2009) Development, validation and application of a SDME/GC-FID methodology for the multiresidue determination of organophosphate and pyrethroid pesticides in water, *Talanta*, **79**,1354–1359.
76. López-Darias, J.; Germán-Hernández, M.; Pino, V. and Afonso, A.M. (2010) Dispersive liquid–liquid microextraction versus single-drop microextraction for the determination of several endocrine-disrupting phenols from seawaters, *Talanta*, **80**, 1611–1618.
77. Payán, M.R.; López, M.A.B.; Fernández-Torres, R.; Mochón, M.C. and Ariza, J.L.G. (2010) Application of hollow fiber-based liquid-phase microextraction (HF-LPME) for the determination of acidic pharmaceuticals in wastewaters, *Talanta*, **82**, 854–858.
78. Varma, R.S. and Ju, Y. (2005) Solventless reactions (SLR), in *Green Separation Processes – Fundamentals and Applications* (eds C.A.M. Afonso and J.G. Crespo), Wiley-VCH Verlag GmbH & Co., Weinheim.
79. Richter, B.E.; Jones, B.A.; Ezzell, J.L. and Porter, N.L. (1996) Accelerated Solvent Extraction: A Technique for Sample Preparation, *Anal. Chem.*, **68**, 1033–1039.
80. Morselli, L.; Sabbioni, M.; Zappoli, S. and Quattroni, G. (1995) Comparison between SFE and Soxhlet extraction in the determination of PCB in soil samples. *Fresenius Environ Bull.*, **8**, 463–468.
81. Yang, Y.; Cajthaml, T. and Hofmann, T. (2008) PAH desorption from river floodplain soils using supercritical fluid extraction, *Environ. Poll.*, **156**, 745–752.
82. Maio, G.; von Holst, C.; Wenclawiak, B.W. and Darskus, R. (1997) Supercritical Fluid Extraction of Some Chlorinated Benzenes and Cyclohexanes from Soil: Optimization with Fractional Factorial Design and Simplex, *Anal. Chem.*, **69**, 601–606.
83. Resano, M.; García-Ruiz, E.; Aramendía, M. and Belarra, M.A. (2005) Solid sampling-graphite furnace atomic absorption spectrometry for Hg monitoring in soils. Performance as a quantitative and as a screening method, *J. Anal. At. Spectrom.*, **20**, 1374–1380.
84. Felipe-Sotelo, M.; Cal-Prieto, M.J.; Carlosena, A.; Andrade, J.M.; Fernández, E. and Prada, D. (2005) Multivariate optimization for molybdenum determination in environmental solid samples by slurry extraction-ETAAS, *Anal. Chim. Acta*, **553**, 208–213.
85. Farghaly, O.A. and Ghandour, M.A. (2005) Square-wave stripping voltammetry for direct determination of eight heavy metals in soil and indoor-airborne particulate matter, *Environ. Res.*, **97**, 229–235.
86. Rygh, G. and Jackson, K.W. (1987) Direct Determination of Cadmium in Soil Slurries by Micro-sampling Cup Atomic Absorption Spectrometry, *J. Anal. At. Spectrom.*, **2**, 397–400.
87. Lu, H.H. and Jiang, S.J. (2001) Organic acids as the modifier to determine Zn, Cd, Tl and Pb in soil by slurry sampling electrothermal vaporization inductively-coupled plasma mass spectrometry, *Anal. Chim. Acta*, **429**, 247–255.
88. Asfaw, A. and Wibetoe, G. (2009) Potential of solid sampling electrothermal vaporization for solving spectral interference in inductively coupled plasma optical emission spectrometry, *Spectrochim. Acta Part B*, **64**, 363–368.
89. Duckworth, D.C.; Barshick, C.M.; Bostick, D.A. and Smith, D.H. (1993) Direct measurement of uranium isotopic ratios in soils by glow discharge mass spectrometry, *Appl. Spectros.*, **47**, 243–245.
90. Kexue, L.I.; Zhou, W.; Shen, Q.; Shao, J. and Qian, H. (2010) Signal enhancement of lead and arsenic in soil using laser ablation combined with fast electric discharge, *Spectrochim. Acta Part B*, **65**, 420–424.
91. Popov, A.M.; Colao, F. and Fantoni, R. (2010) Spatial confinement of laser-induced plasma to enhance LIBS sensitivity for trace elements determination in soils, *J. Anal. At. Spectrom.*, **25**, 837–848.
92. Arroyo, L.; Trejos, T.; Gardinali, P.R. and Almirall, J.R. (2009) Optimization and validation of a laser ablation inductively coupled plasma mass spectrometry method for the routine analysis of soils and sediments, *Spectrochim. Acta Part B*, **64**, 16–25.
93. Rahman, M.M.; Chen, Z.L. and Naidu, R. (2009) Extraction of arsenic species in soils using microwave-assisted extraction detected by ion chromatography coupled to inductively coupled plasma mass spectrometry, Environ. Geochem. Health, 31, 93–102.

94. Hassan, N.M.; Rasmussen, P.E.; Dabek-Zlotorzynska, E.; Celo, V. and Chen, H. (2007) Analysis of Environmental Samples Using Microwave-Assisted Acid Digestion and Inductively Coupled Plasma Mass Spectrometry: Maximizing Total Element Recoveries, *Water Air Soil Pollut.*, **178**, 323–334.
95. Jang, M. (2010) Application of portable X-ray fluorescence (pXRF) for heavy metal analysis of soils in crop fields near abandoned mine sites, *Environ. Geochem. Health*, **32**, 207–216.
96. Carr, R.; Zhang, C.; Moles, N. and Harder, M. (2008) Identification and mapping of heavy metal pollution in soils of a sports ground in Galway City, Ireland, using a portable XRF analyser and GIS, *Environ. Geochem. Health*, **30**, 45–52.
97. Gorshteyn, A.; Smarason, S. and Robbat Jr, A. (1999) Speciation of subsurface contaminants by cone penetrometry gas chromatography/mass spectrometry, *Environ. Sci. Technol.*, **33**, 2474–2480.
98. Vazquez-Roig, P.; Segarra, R.; Blasco, C.; Andreu, V. and Picó, Y. (2010) Determination of pharmaceuticals in soils and sediments by pressurized liquid extraction and liquid chromatography tandem mass spectrometry, *J. Chromatogr. A*, **1217**, 2471–2483.
99. Herbert, P.; Morais, S.; Paíga, P.; Alves, A. and Santos, L. (2006) Development and validation of a novel method for the analysis of chlorinated pesticides in soils using microwave assisted extraction–headspace solid phase microextraction and gas chromatography–tandem mass spectrometry, Anal. Bioanal. Chem., 384, 810–816.
100. Ozcan, S.; Tor, A. and Aydin, M.E. (2009) Application of miniaturised ultrasonic extraction to the analysis of organochlorine pesticides in soil, *Analytica Chimica Acta*, **640**, 52–57.
101. Beck, J.; Totsche, K.U. and Kögel-Knabner, I. (2008) A rapid and efficient determination of natural estrogens in soils by pressurised liquid extraction and gas chromatography–mass spectrometry, *Chemosphere*, **71**, 954–960.
102. Fernandez-Alvarez, M.; Llompart, M.; Lamas, J.P.; Lores, M.; Garcia-Jares, C.; Cela, R. and Dagnac, T. (2008) Simultaneous determination of traces of pyrethroids, organochlorines and other main plant protection agents in agricultural soils by headspace solid phase microextraction–gas chromatography, *J. Chromatogr. A*, **1188**, 154–163.
103. Zhou, J.; Chen, J.; Cheng, Y.; Li, D.; Hu, F. and Li, H. (2009) Determination of prometryne in water and soil by HPLC–UV using cloud-point extraction, *Talanta*, **79**, 189–193.
104. Turiel, E.; Martín-Esteban, A. and Tadeo, J.L. (2007) Molecular imprinting-based separation methods for selective analysis of fluoroquinolones in soils, *J. Chromatogr. A*, **1172**, 97–104.
105. Nording, M.; Nichkova, M.; Spinnel, E.; Persson, Y.; Gee, S.J.; Hammock, B.D. and Haglund, P. (2006) Rapid screening of dioxin-contaminated soil by accelerated solvent extraction/purification followed by immunochemical detection, *Anal. Bioanal. Chem.*, **385**, 357–366.
106. Guerra-Abreu, L.; Pino, V.; Anderson, J.L. and Afonso, A.M. (2008) Coupling the extraction efficiency of imidazolium-based ionic liquid aggregates with solid-phase microextraction-gas chromatography–mass spectrometry: Application to polycyclic aromatic hydrocarbons in a certified reference sediment, *J. Chromatogr. A*, **1214**, 23–29.
107. Liao, V.H.-C.; Chien, M.-T.; Tseng, Y.-Y. and Ou, K.-L. (2006) Assessment of heavy metal bioavailability in contaminated sediments and soils using green fluorescent protein-based bacterial biosensors, *Environ. Poll.*, **142**, 17–23.
108. Lambropoulou, D.A. and Albanis, T.A. (2004) Sensitive trace enrichment of environmental andiandrogen vinclozolin from natural waters and sediment samples using hollow-fiber liquid-phase microextraction, *J. Chromatogr. A*, **1061**, 11–18.
109. Krieger, M.S.; Cook, W.L. and Kennard, L.M. (2000) Extraction of Tricyclazole from Soil and Sediment with Subcritical Water, *J. Agric. Food Chem.*, **48**, 2178–2183.
110. Basheer, C.; Obbard, J.P. and Lee, H.K. (2005) Analysis of persistent organic pollutants in marine sediments using a novel microwave assisted solvent extraction and liquid-phase microextraction technique, *J. Chromatogr. A*, **1068**, 221–228.
111. Kuosmanen, K.; Hyötyläinen, T.; Hartonen, K. and Riekkola, M.-L. (2002) Pressurised hot water extraction coupled on-line with liquid chromatography–gas chromatography for the determination of brominated flame retardants in sediment samples, *J. Chromatogr. A*, **943**, 113–122.
112. Arnold, C.G.; Berg, M.; Müller, S.R.; Dommann, U. and Schwarzenbach, R.P. (1998) Determination of organotin compounds in water, sediments, and sewage sludge using perdeuterated internal standards, accelerated solvent extraction, and large-volume-injection GC/MS, *Anal. Chem.*, **70**, 3094–3101.
113. Chiron, S.; Roy, S.; Cottier, R. and Jeannot, R. (2000) Speciation of butyl- and phenyltin compounds in sediments using pressurized liquid extraction and liquid chromatography–inductively coupled plasma mass spectrometry, *J. Chromatogr. A*, **879**, 137–145.

114. Colombini, V.; Bancon-Montigny, C.; Yang, L.; Maxwell, P.; Sturgeon, R.E. and Mester, Z. (2004) Headspace single-drop microextration for the detection of organotin compounds, *Talanta*, **63**, 555–560.
115. Burkhardt, M.R.; ReVello, R.C.; Smith, S.G. and Zaugg, S.D. (2005) Pressurized liquid extraction using water/isopropanol coupled with solid-phase extraction cleanup for industrial and anthropogenic waste-indicator compounds in sediment, *Anal. Chim. Acta*, **534**, 89–100.
116. Chang, S.M. and Doong, R.A. (2006) Concentration and fate of persistent organochlorine pesticides in estuarine sediments using headspace solid-phase microextraction, *Chemosphere*, **62**, 1869–1878.
117. Kuosmanen, K.; Hyötyläinen, T.; Hartonen, K. and Riekkola, M.L. (2003) Analysis of polycyclic aromatic hydrocarbons in soil and sediment with on-line coupled pressurised hot water extraction, hollow fibre microporous membrane liquid–liquid extraction and gas chromatography, *Analyst*, **128**, 434–439.
118. Andersson, T.; Pihtsalmi, T.; Hartonen, K.; Hyötyläinen, T. and Riekkola, M.-L. (2003) Effect of extraction vessel geometry and flow homogeneity on recoveries of polycyclic aromatic hydrocarbons in pressurised hot water extraction, *Anal. Bioanal. Chem.*, **376**, 1081–1088.
119. Burkhardt, M.R.; Zaugg, S.D.; Burbank, T.L.; Olson, M.C. and Iverson, J.L. (2005) Pressurized liquid extraction using water/isopropanol coupled with solid-phase extraction cleanup for semivolatile organic compounds, polycyclic aromatic hydrocarbons (PAH), and alkylated PAH homolog groups in sediment, *Anal. Chim. Acta*, **549**, 104–116.
120. García-López, M. and Cela, R.R. (2009) Pressurized liquid extraction of organophosphate triesters from sediment samples using aqueous solutions, *J. Chromatogr. A*, **1216**, 6986–6993.
121. Morales-Muñoz, S.; Luque-García, J.L.; Ramos, M.J.; Fernández-Alba, A. and Luque de Castro, M.D. (2005) Sequential superheated liquid extraction of pesticides, pharmaceutical and personal care products with different polarity from marine sediments followed by gas chromatography mass spectrometry detection, *Anal. Chim. Acta*, **552**, 50–59.
122. Morales-Muñoz, S.; Luque-García, J.L. and Luque de Castro, M.D. (2002) Static extraction with modified pressurized liquid and on-line fluorescence monitoring – Independent matrix approach for the removal of polycyclic aromatic hydrocarbons from environmental solid samples, *J. Chromatogr. A*, **978**, 49–57.
123. Hawthorne, S.B.; Grabanski, C.B.; Hageman, K.J. and Miller, D.J. (2000) Simple method for estimating polychlorinated biphenyl concentrations on soils and sediments using subcritical water extraction coupled with solid-phase microextraction, *J. Chromatogr. A*, **814**, 151–160.
124. Hawthorne, S.B.; Trembley, S.; Moniot, C.L.; Grabanski, C.B.; and Miller, D.J. (2000) Static subcritical water extraction with simultaneous solid-phase extraction for determining polycyclic aromatic hydrocarbons on environmental solids, *J. Chromatogr. A*, **886**, 237–244.
125. Pino, V.; Ayala, J.H.; Afonso, A.M. and González, V. (2001) Ultrasonic micellar extraction of polycyclic aromatic hydrocarbons from marine sediments, *Talanta*, **54**, 15–23.
126. Dutta, S. and Das, A.K. (2010) Analytical Perspective on waste management for environmental remediation, *TrAC-Trend. Anal. Chem.*, **29**, 636–644.
127. Garrigues, S.; Armenta, S. and de la Guardia, M. (2010) Green strategies for decontamination of analytical wastes, *TrAC-Trend. Anal. Chem.*, **29**, 592–601.
128. Albaseer, S.S.; Rao, R.N.; Swamy, Y.V. and Mukkanti, K. (2010) An Overview of sample preparation and extraction of synthetic pyrethroids from water, sediment and soil, *J. Chromatogr. A*, **1217**, 5537–5554.
129. Leonelli, C. and Mason, T.J. (2010) Microwave and ultrasonic processing: now a realistic option for industry, *Chem. Eng. Process.*, **49**, 885–900.
130. Finlayson-Pitts, B.J. and Pitts Jr., J.N. (2000) *Chemistry of the Upper and Lower Atmosphere*, Academic Press, San Diego.
131. Oukebdane, K.; Portet-Koltalo, F.; Machour, N.; Dionnet, F. and Desbène, P.L. (2010) Comparison of hot Soxhlet and accelerated solvent extractions with microwave and supercritical fluid extractions for the determination of polycyclic aromatic hydrocarbons and nitrated derivatives strongly adsorbed on soot collected inside a diesel particulate filter, *Talanta*, **82**, 227–236.
132. Sunarso, J. and Ismadji, S. (2009) Decontamination of hazardous substances from solid matrices and liquids using supercritical fluids extractions: a review, *J. Hazard. Mater.*, **161**, 1–20.
133. Walgraeve, C.; Demeestere, K.; Dewulf, J.; Zimmermann, R. and Van Langenhove, H. (2010) Oxygenated polycyclic aromatic hydrocarbons in atmospheric particulate matter: Molecular characterization and occurrence, *Atmos. Environ.*, **4**, 1831–1846.

23

Green Industrial Analysis

Sergio Armenta and Miguel de la Guardia

Department of Analytical Chemistry, University of Valencia, Valencia, Spain

23.1 Greening industrial practices for safety and cost reasons

Is green chemistry or, in this case, Green Analytical Chemistry something necessary or it is hype? From its definition in 1991 [1], green chemistry managed to reduce the impact of chemicals on human health and the environment. Are the mentality changes of modern society the responsibility of the present importance of green chemistry? Probably: however, the environmental reasons and benefits of green chemistry are not enough for skeptical scientists who consider the green concept as an uninteresting research niche. Nonetheless, this mentality change combined with past environmental disasters has moved the authorities to implement severe and strict environmental policies which force chemical and related industries to reduce side effects. For instance in the US, the Environmental Protection Agency (US EPA) ensures the protection of the population from significant risks to human and environment health; focusing their efforts to reduce environmental risks, enforcing US policies concerning natural resources, human health, economic growth, energy, transportation, agriculture, industry and international trade [2]. US EPA promotes green chemistry through the Presidential Green Chemistry Challenge Awards and supports green chemistry through projects and programs, including educational activities and research and development. In a similar way, the European Union created the European Environment Agency in 1990 (EEA) [3] which is responsible for producing environmental data and indicator sets, assessments and thematic analyses in order to provide a sound decision basis for environmental policies in the EU and Member countries.

Environmental laws and regulations around the world target hazardous chemicals and, as has been the case of the Register Evaluation Authorization and Restriction of Chemicals (REACH) norm in Europe [4], there is a tremendous social and economical effort to prevent environmental pollution and that could create the impression that the new scenarios for the industrial activities are more complicated and expensive than the previous ones. In addition, many companies have found that it could be cheaper and even profitable to meet

Handbook of Green Analytical Chemistry, First Edition. Edited by Miguel de la Guardia and Salvador Garrigues.

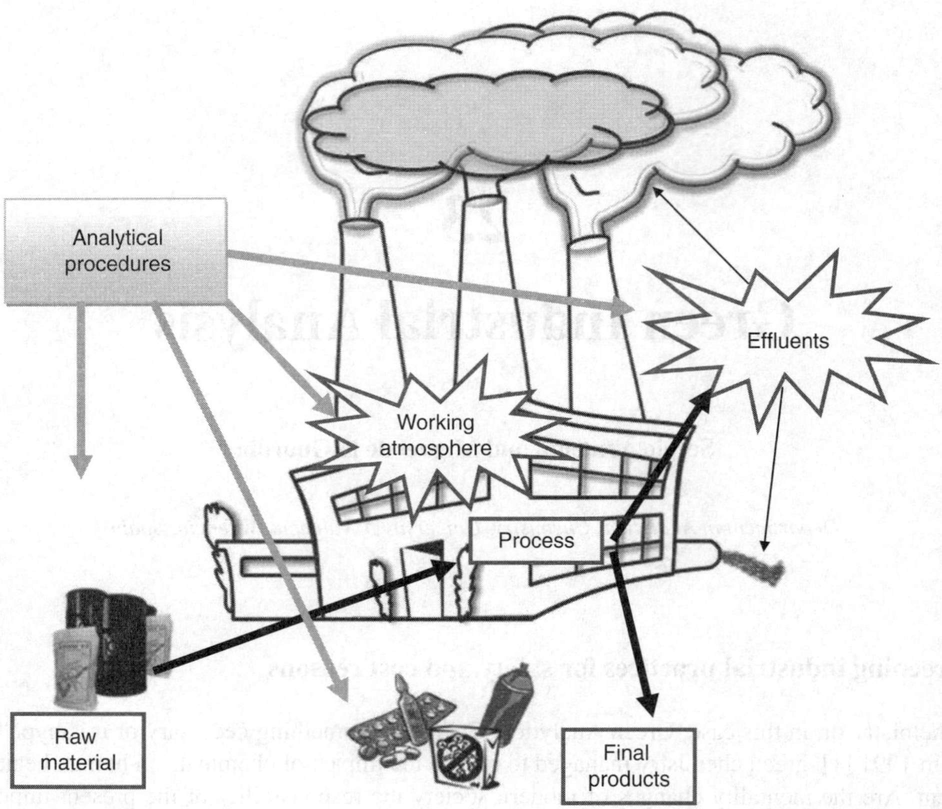

Figure 23.1 The synergistic relationships between the industrial plant and the laboratory indicating hot points from an environmental perspective.

environmental goals by using green chemistry rather than correcting pollution effects. Profits derive from an increased efficiency, a reduction of wastes, an enhanced product quality and reduced liability of processes that follow the green chemistry rules and it allows companies to comply with the law in much simpler and cheaper ways than those used in the past which were based on remediation activities more than on prevention ones.

So, in this chapter we will consider the industrial lab in its close relationship with the industrial plant process and final products also considering the hot points from an environmental perspective (see Figure 23.1). As we will try to explain in the following sections, we are convinced that on greening the industrial laboratory practices, the advantages of meeting the analysis of both raw and finished products and also making process analysis could be improved regarding environmental safety and operators' health as well as adoption of present and future norms regarding environmental protection. In this sense, pay attention to the fact that an ethical compromise with the environment provides image and economic benefits to an industry and it can assure the persistence of a good beneficial level in a strongly competitive market without creating trouble for the future.

23.2 The quality control of raw materials and end products

The quality control of raw material and finished products is an important issue in the food, drug, and pharmaceutical areas but also in any industrial field you can imagine. The purpose of determining the quality of finished products is based on the need to provide a high level of satisfaction of our customers, which is the

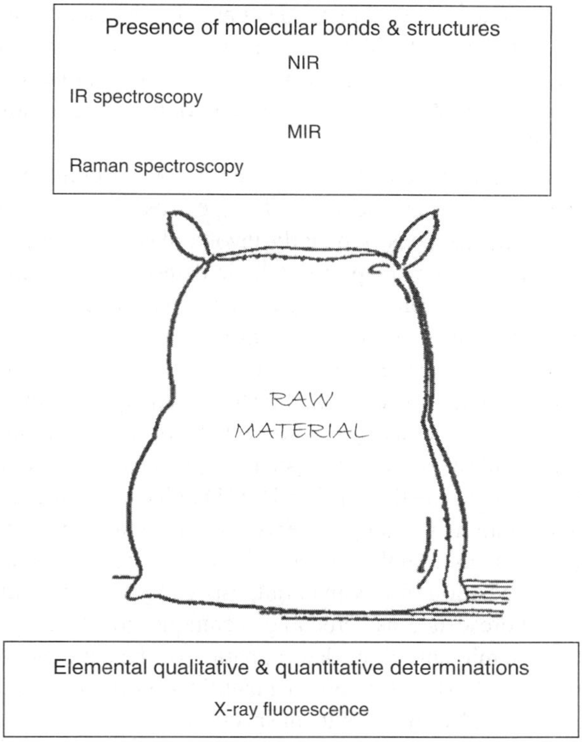

Figure 23.2 *Main green analytical tools employed for the quality control of raw materials.*

definition of the term quality from ISO 9001 standard [5]. Moreover, end products quality is important for ensuring safety and efficacy to consumers, and different organisms, such as the US Food and Drug Administration (US FDA) [6], the European Medicines Agency (EMA) [7] and the European Food Safety Authority (EFSA) [8] among others, ensures that the basic standards of safety, efficacy, and quality are kept in the corresponding end products. So to do it, several analytical techniques have been developed for a fast, economic and sustainable evaluation of the quality of finished products (see Figure 23.2).

However, in addition to the quality control analysis, assuring good production quality is of a mandatory importance to certify the quality of the raw materials in order to prevent problems during the production process.

In fact, the analytical control of raw materials does not need to be an exhaustive control but just a verification of the nature of the new products arriving at the plant in order to evaluate the parameters of relevance for their transformation and to do fast multiparametric analytical methodologies based on fingerprint signals of the raw material composition; these are enough for a screening purpose.

A really useful tool for the characterization of raw materials is IR spectroscopy, a technique which has some unique benefits: non-contact and non-destructive, highly reproducible, rapid and does not require any sample preparation. IR spectroscopy enables a non-destructive analysis of the starting material several times inside its original packaging, thus reducing the risk of cross-contamination and abolishing the need to conduct the analysis within a designated sampling area. This leads to a significant decrease in the time involved for each analysis. The IR based technique most widely employed in the industry sector is the Near-IR spectroscopy which has found an increased use in the confirmation of raw material identity and quality in the pharmaceutical, polymer and chemical industries in recent years [9]. Near-IR can be used to collect the spectra of raw materials being sensitive to their chemical and physical properties [10]. A number of recent sets of guidelines advise the use of Near-IR spectroscopy as an universal method suitable for the identification of raw materials in the pharmaceutical industry [11, 12, 13]. Faced with the problems of maintaining regulatory compliance with an increasing amount of raw material analysis, many pharmaceutical companies focused on Near-IR spectroscopy to save time and cost, and increase productivity in preference to of the most commonly employed methods in the past based on chromatography.

Early versions of Pharmacopoeia methods specified a number of different analytical methods for testing individual raw materials including UV-Visible (UV-Vis) spectroscopy and Gas Chromatography (GC). However, the aforementioned methodologies frequently involve the use of organic solvents to dissolve the sample or to leachate the analytes before to perform their determination, reducing the sample analysis frequency, increasing the costs of analysis and creating toxic wastes and environmental side effects.

Moreover, Near-IR spectroscopy is used in the food industry for the prediction of a wide range of specific chemical and physical properties such as water, protein, starch and grain hardness among others [14]. Due to its intrinsic properties, Near-IR is really useful to the brewery industry [15], oil mills [16], feed production industry [17] and dairy products [18] as a quality assurance tool for raw materials.

The most important advantages offered by Near-IR spectroscopy in the quality control of raw material, which can be extrapolated to the mid-IR range, are those related to: (1) reduction of the analysis time, (2) elimination of the consumption and concomitant disposal of solvents used in sample preparation and (3) elimination of worker exposure to noxious solvents used in the wet chemistry and toxic components of raw materials.

Concerning the elemental composition of raw material, especially in metallurgical, ceramic and concrete production, the use of X-ray fluorescence as screening technique, together with that of IR spectroscopy permits a fast identification and semiquantitative determination of the elements present in the raw products by their secondary X-ray emission [19] and a correct identification of structures from their specific links clearly identified by the IR [20] and also by the Raman spectra [21].

In short, the advantage of the aforementioned green analytical tools for raw material fast characterization is the possibility to obtain a complete picture of material composition without any previous chemical treatment directly from the samples. So, from both the environmental and the industrial points of view it is clear that a fast characterization of raw materials provides an important tool for preventing manufacturing problems and contributes to assure the quality of the final products.

On the other hand, quality control is traditionally the most important task assigned to the industrial laboratories and to do it, the laboratories must be equipped with tools for both the testing of the physical and practical properties of finished products and as well as their chemical composition and here, once again, greening the laboratory practices has the double interest of avoiding environmentally deleterious side effects of the use of toxic reagents, reducing the cost of the quality control analysis and increasing laboratory productivity.

The use of liquid chromatography (LC) as analytical method for the quality control of finished products is widely extended, being the reduction of the volume and toxicity of the employed reagents and generated wastes one of the key points of Green Analytical Chemistry.

Pfizer is one of the pioneering companies who are worried about climate change and in 1996, started a research program named *Pfizer's Energy and Climate Change Program* [22] to focus on greener synthesis processes and analytical methods for the quality control of raw materials and finished products. The main goal of the Pfizer's project was to green its processes to help the environment, saving production costs at the same time. For instance, the company is promoting ethanol based HPLC methods to replace acetonitrile as a mobile phase. Acetonitrile is considered as an EPA Hazard Air Pollutant and diverse sources affirm that NO_2, linked to acid rain, is a byproduct of its incineration process. In this sense, pharmaceutical companies using acetonitrile in their compliance with good manufacture practices (CGMP) validate HPLC methods which aim to modify their procedures to remove or reduce the acetonitrile use. Unfortunately, acetonitrile has been traditionally considered the best solvent in reverse-phase (RP) HPLC due to its physical and chemical properties and its UV absorbance characteristics. However, recently, more and more specialized voices claim

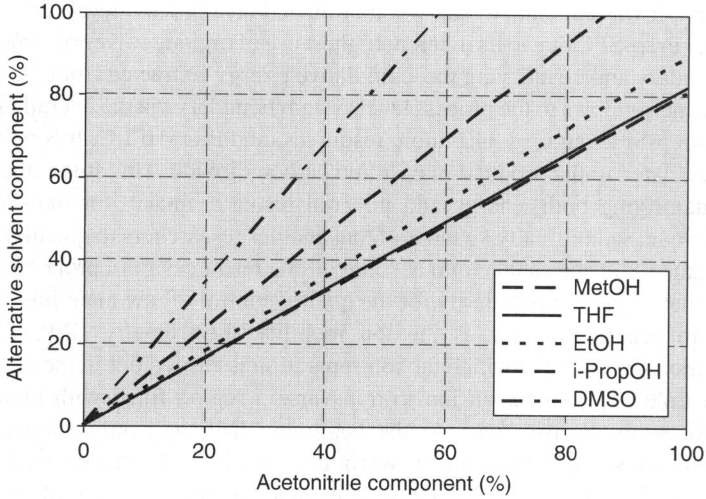

Figure 23.3 *Water-miscible solvents proposed to replace acetonitrile in chromatographic mobile phase composition.*

the benefits of replacing acetonitrile with methanol or longer chain alcohols [23, 24, 25]. Figure 23.3 shows a plot which correlates the composition of several water-miscible solvents which could be used to replace the acetonitrile as a component of the mobile phase as recommended according their eluotropic strength [26].

Sometimes, solvent replacement is not technically feasible and a reduced solvent consumption could be the greenest solution. Reduced consumption patterns can be achieved following different strategies such as (1) reduction of the pre-equilibration time and optimization of the analysis time after final peak elution, (2) solvent recycling technologies and (3) reduction of the column's internal diameter (ID).

Replacing traditional LC methods by ultra performance liquid chromatography (UPLC) ones provided an increased resolution in shorter run times which involves obtaining the same information faster than with LC. UPLC reduces at the same time the volume of generated wastes and can be considered as a greener alternative. In a press article of the February 2, 2010 [27], it was announced that Waters Corporation had installed thousands of UPLC systems, replacing the workload of tens of thousands of traditional LC systems, supported more than 500 peer-reviewed papers, demonstrated reduced solvent consumption by up to 95% for greener laboratories, and had served the needs of regulatory agencies around the globe. One of the areas in which the high resolution obtained by UPLC in extremely short analysis times is a really attractive characteristic is the pharmaceutical development industry [28]. Novartis Pharmaceuticals Corporation highlights the benefits of faster analyses by UPLC; which allow for a greater number of analyses to be performed in a shorter amount of time, thus, increasing sample throughput and lab productivity. In addition, as test experiments are performed more quickly, the overall method development time is decreased.

Supercritical Fluid Chromatography (SFC) considered as an alternative to normal phase HPLC, provides the following advantages: (1) greener, typically only uses methanol as an organic modifier, (2) faster, (3) useful for analysis of water labile analytes and (4) retain polar compounds. SFC uses CO_2 as supercritical fluid which is non-toxic, non-flammable and can be recovered from waste streams. Johnsson and Johnsson PRD, Janssen Pharmaceutica carried out a comparative study of the greenness of preparative supercritical fluid chromatography (Prep-SFC) and preparative HPLC (Prep-HPLC) in the fine chemical and pharmaceutical industry [29]. The comparison was made in terms of mass and energy requirements at different system boundary levels, and their integral resource consumptions were evaluated on the example of a specific enantiomeric separation. It was concluded that, when evaluated on a process level, Prep-HPLC requires 26% more resources than the Prep-SFC due to its inherent highest use of organic solvents. However, when evaluated on a larger system boundary and quantifying the Cumulative Exergy Extracted from the Natural Environment to deliver all mass and energy flows to the process level system boundary via the overall industrial metabolism, it becomes clear that Prep-SFC requires 34% more resources than Prep-HPLC. It is partially due to the high resource requirements related to the production of liquid carbon dioxide. This study illustrates the possibility and advantages of quantifying both, energy and material resource intake, for industrial processes, being exergy analysis: the unique, scientifically sound tool that enables researchers to quantify all kind of resources and products on the same scale and to take into account all the resources for a proper evaluation.

Another technique that is gaining popularity for the quality control of raw materials and finished products, especially in the pharmaceutical sector, is the ion mobility spectrometry (IMS). IMS is a gas-phase electrophoretic separation technique in which the ion separation is carried out in the millisecond scale on the basis of the different time required for an ion to transverse a region filled with inert drift gas under the influence of an homogenous electric field. At the beginning IMS was only applied to the detection of explosives [30], illegal drugs [31] and chemical warfare agents [32]. Pharmaceutical applications of IMS have been reviewed by O'Donnell *et al.* [33] and they include the determination of pharmaceutically active compounds [34, 35] and cleaning verification of the manufacturing equipment [36, 37].

So, it can be concluded that Green Analytical Chemistry enhances laboratory productivity, reduces the environmental risks and provides economic opportunities in the analysis of raw materials and finished products.

23.3 Process control

Recently, a growing number of researchers have focused their efforts in developing methodologies for the monitoring of chemical production processes. The main reasons supporting those developments are (1) the better quality of products required by the consumer, (2) more stringent regulations by the authorities in relation to production and production processes; and (3) greater economic competition, which requires a minimization of energy consumption, a reduction in the amount of raw materials used in reactions and process operations and a decrease in the generated wastes and emissions. The term which defines this research area is process analytical chemistry (PAC). PAC plays a very important role in the management of industrial processes, in refining of raw materials and minerals, in agriculture, in food and animal feedstuff manufacturing, in fertilizer processing and in the clinical and pharmaceutical field [38]. The principles of PAC are focused on the advantages of generating immediate analytical data to correct process deviations and eliminating the wait for the analytical response necessary to estimate the process status. In the traditional off-line approaches, samples are taken from the reactor, transported to the analytical laboratory, analyzed and evaluated and corrective actions are taken if required (time-delayed monitoring). Modern PAC promotes at-line, on-line or in-line process control, where samples are measured close to or inside the reactor or process chain and the corrective action can be taken in near or real time (real time monitoring).

In 2003, about 85% of chemical production processes were analyzed off-line [39], using analytical methods such as HPLC. The main disadvantages of those methodologies are the long time of analysis, the time delayed response and the associated costs of solvent consumption and waste generation. The off-line/at-line control is often used in cases in which the number of analysis is reduced and the sample preparation for their analysis is difficult to automate. However, the economic environment of the past few years has provoked a tendency amongst enterprises to move towards a high as possible degree of control over production processes in order to reduce costs and increase product quality. Moreover, governmental organizations, such as the FDA, are forcing several industries (the pharmaceutical industry is a good example of this) to implement process analytical tools. The process analytical technology (PAT) initiative, promotes the use of techniques that enable the monitoring of critical process parameters during pharmaceutical manufacturing [40]. Currently, optical techniques, mainly vibrational based ones, have proven to be very versatile in dealing with on- and in-line inorganic and organic constituents monitoring and are gaining a high share, due to their capability to fulfill the requested needs [41]. Those techniques do not require sampling and in conjunction with chemometric modeling, often replace classical analytical methods like chromatography [42].

The use of vibrational techniques for process analysis and in situ measurements is widely accepted. For instance, Near-IR spectroscopy has been used in situ to process control since the late 1970s. It has been used widely for many industrial applications such as polymer processing monitoring [43], fermentation and bottling processes in the wine industry [44] and production of oil in the mill industry [45] among others. The widespread use of NIR allows for a remarkable flexibility in selecting the optimum analyzer and sampling configuration for the target application.

The main advantages that mid-IR spectroscopy offers over the near-IR range is that direct information on the monitored process can be obtained without the need to treat the data with chemometric approaches. Moreover, mid-IR spectroscopy provides information which can be specifically related to each component rather than the reaction mixture as a whole which is an advantage over calorimetry. Mid-IR spectroscopy has been successfully used for monitoring of hydrogenation reactions in organic synthesis and the pharmaceutical industries [46]. Regarding commercial equipment, the React IR spectrometers, developed by Metler Toledo are reaction analysis systems based on the Attenuated total reflection (ATR) sampling mode, which has been designed for the real time analysis and monitoring of chemical reactions and processes in extreme reaction conditions. The main limitation of mid-IR spectrometry for process monitoring is the distance between the reactor and the instrument which cannot be more than several meters. The ReactIR™ 45 m developed by Metler Toledo provides a distance of 2.0 m between the reactor and the instrument using a flexible, silver halide (AgX) fiber optic technology [47].

Raman spectroscopy also offers a non-destructive, non-invasive and fast methodology of analysis. Moreover, the recent developments in spectrograph design, gratings, lasers and sampling probes have successfully moved Raman out of the research and analytical laboratory and into industrial and process monitoring applications. It is suited to real time reaction monitoring and to the characterization of industrial compounds and it has been established in semiconductor research, polymer characterization and pharmaceutical development. Diverse worldwide companies market Raman spectrometers devoted to process control, reaction analysis and monitoring [48, 49]. Applications of Raman spectroscopy include monitoring of polymerization processes [50] and drug formulations in the pharmaceutical industry [51]. However, the main limitation of Raman spectrometry ahead of IR is the acquisition cost of the spectrometers. That is probably one of the main reasons of the wide utilization of the Near-IR spectrometry in the industry.

Attenuated total reflection ultraviolet (ATR-UV) spectroscopy is another potential technique for monitoring industrial processes. The main applications of ATR-UV spectroscopy are those regarding quantification of hydroxide, sulfide, and carbonate in craft liquors [52], dissolution curve determination [53], detection of nucleation point [54], real time concentration monitoring during crystallization, the control of polymorphism and particle size uniformity [55] and detection of polymorphic transformation [56].

In short, as depicted in Figure 23.4, the reasons for making a fast–as-possible process control in the industrial field are highly compatible with the advantages offered by the green analytical methods which in fact are highly compatible with in-line and non-invasive strategies and, once again, demonstrate the compatibility between the economical and environmental objectives.

23.4 Effluent control

Industrial process involve, in addition to the production of final products, some solid and liquid wastes and also some gaseous emissions and, in all these cases some remaining unmodified raw materials and side reaction products could be present in those effluents: as indicated in Figure 23.5 the typical green tools such

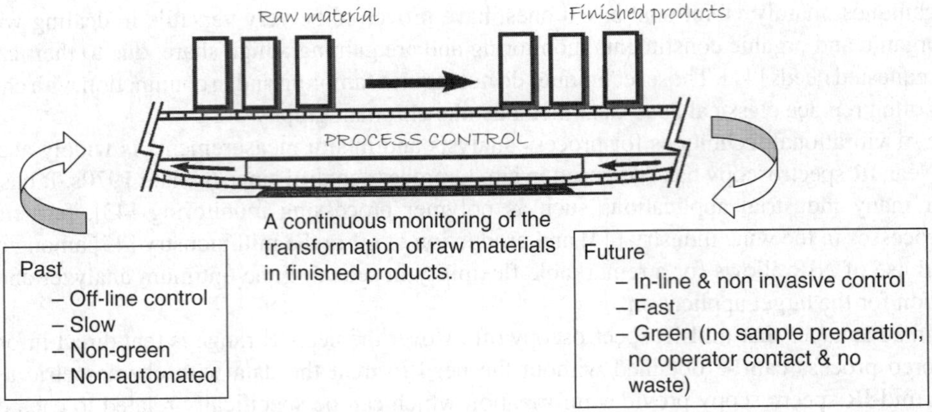

Figure 23.4 *Reasons for process control and advantages offered by green analytical tools in this field.*

as remote sensing and portable systems provide a good alternative for the on-line control of gaseous and liquid effluents, making possible a fast and environmentally friendly analysis of solid wastes by using direct analytical methods which only require a physical treatment of samples.

A given industry should assess the quality of its effluents discharged into the environment to accomplish the specified requirements and depending on the specific industry the evaluated parameters can vary. In this section, the main contaminants evaluated on the effluents discharges will be covered, focusing on the green alternatives to the available methodologies.

The European Union has laid down harmonized rules to protect the aquatic environment against industrial discharge of dangerous substances. The new regulations (2009) comprise two lists of dangerous substances to combat pollution: (1) substances in List I must be eliminated from industrial effluents and (2) substances in List II must be reduced. List I contains among others; organohalogens, organophosphorus, organotins, carcinogens, oils and hydrocarbons, and heavy metals such as mercury and cadmium. List II contains some metalloids and metals, biocides, persistent organic compounds of silicon, inorganic compounds of phosphorus, cyanides, fluorides, ammonia, and nitrites among others [57].

Oil and grease is one of the five conventional pollutants covered by the 1974 Clean Water Act and it is the second most-enforced-against parameter after pH. In 1989, the EPA was required to move from a freon extraction infrared-based determination method (included in the classification of substances that deplete the ozone layer and phase-out of production and use) to a n-hexane extraction mass-based determination method (a toxic and flammable organic solvent) [58]. To reduce the estimated use of 1.1 million liters each year of n-Hexane for EPA Method 1664A analyses a new method was developed by Orono Spectral Solutions for the solvent-free recoverable oil and grease infrared determination. The new method uses a novel extraction membrane that is infrared-amenable so that oil and grease does not need to be eluted by a solvent for measurement [59].

The use of on-line analytical methodologies to manage effluent discharges allow a more timely and efficient response to abnormal occurrences resulting from the industry operations. The chemical industry effluents, including those from pesticide, pharmaceutical, paint and dye, petrochemical, detergent, plastic and paper industries can be contaminated by feed-stock materials, by-products, product material in soluble or particulate form, washing and cleaning agents, solvents and added value products. Syngenta (called ZENECA in 1992) developed a project to install on-line analytical instrumentation, based on the determination of total carbon content and UV—visible spectrophotometry, designed to both detect and initiate a response to abnormal discharges to drain effluent [60]. The on-line analyzers were considered the most appropriate for a manufacturing site handling 4000 different chemicals.

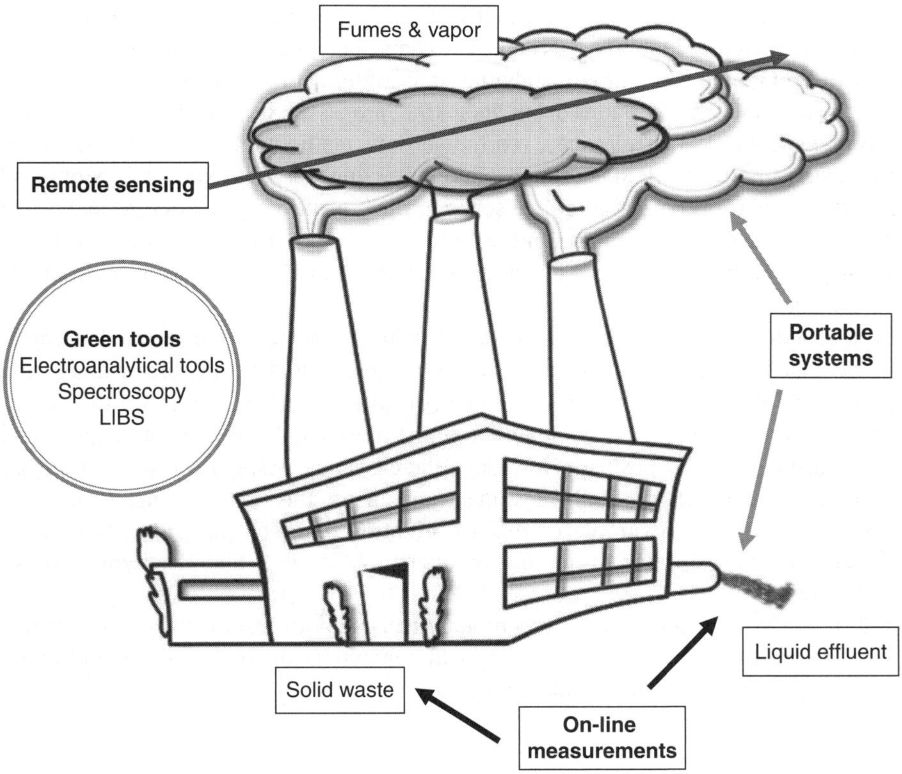

Figure 23.5 *The use of green analytical tools for effluent control in the industrial field.*

Determination of trace metals in industrial effluents is a major issue in the residue management plans of worldwide industries. 'Real-time' on-line measurements of trace metals without the need of sampling, sample transport to the laboratory, sample treatment and analysis are a major concern of those industries.

The paint industry is one of the major contributors for polluting the soil and water resources with poisonous substances such as Pb, Cr and Cd [61] and there is a great need for the development of an analytical tool which could determine the heavy metal content in paint and wastewater from paint industry accurately and rapidly. Laser induced breakdown spectroscopy (LIBS) has been successfully used for this purpose by reducing the sample treatment to the retention of undissolved particles of effluent streams into a filter paper and focusing the laser onto it. The main advantages of the methodology are those related to the possibility to perform on-site analysis, the measurement complexity is greatly reduced and there is no chance for sample loss or cross-contamination during transport or complicated preparations for laboratory analysis. Additional LIBS advantages include the ability to conduct standoff distance measurements and a small sampled size.

Moreover, the hyphenation of flow injection techniques with atomic spectrometric techniques, such as flame emission spectroscopy, inductively coupled plasma mass spectrometry, hydride-generation or cold vapor atomic absorption spectrometry, flame atomic absorption spectrometry or electrothermal atomic absorption spectrometry has proven to be a powerful analytical tool for the determination of ultratrace levels of metals in the analytical laboratories [62] but their implementation for on-line and on-site industrial effluent monitoring is still not widely and routinely adopted.

On-line sensors capable to measure parameters to assess the wastewater quality; such as the chemical and biochemical oxygen demand (COD and BOD) and total organic content (TOC) are commercially available [63].

For instance, Endress+Hauser sales COD, BOD and TOC analyzers specially designed for the quality control of effluent discharges in the food industry [64] and nowadays there are many portable systems available able to determine dissolved O_2, pH, conductivity and other interesting parameters [65].

On the other hand, gaseous emissions from industrial processes include fume vapors and suspended particles; to correctly control such matrices, remote sensing methodologies are the best choice and it is possible to find systems based on IR spectrometry in the literature for these kinds of operations [66].

On summary, the green tools based on remote sensing, flow injection spectroscopy analysis or LIBS offer fantastic media to evaluate in a fast, accurate and sensitive way the possible impact of industrial effluents thus offering unique opportunities for policy makers and responsible industries to prevent environmental pollution and to evaluate the correct behavior of systems.

A special type of industrial by-product is solid waste, which come from the incomplete transformation of products from raw materials and could contain toxic components but also valuable residues. In this case, the analytical control of industrial solid residues could be considered a starting point for the evaluation of their toxicity and must be at the core of a correct selection of the most suitable treatment method, such as reuse in another field, combustion or land deposition. Nowadays, there is an increasing interest in waste valorization [67] and thus, after a careful evaluation of solid residue components, they can be considered as raw materials for new processes. One typical example of this valorization of industrial products is the obtention of aminoacids from tanned wastes [68] after the recovery of $Cr(OH)_3$ and the hydrolysis of remaining proteinaceous residues.

For this kind of study, fast and accurate green analytical methods are mandatory to obtain the relevant information and, in this case, the appropriate use of the analytical methodologies could avoid the cost of waste treatment and provide new valuable materials.

23.5 Working atmosphere control

The working atmosphere control involves monitoring both the environment in the workplace, and the employees themselves for levels of harmful chemicals. A wide range of analytical methods can be used including; (1) those based on active sampling through pumps, filters and membranes, (2) passive sampling badges, (3) color tubes and (4) direct reading instruments for real time readout of results (see Figure 23.6).

In the first two systems, the device is set up for a given time period and the air from the workplace flow through the sampling device. At the end of the exposure period, the tube, filter or impinger is sealed and sent to the laboratory for the analysis. In the color tubes, a hand pump is used to draw a known volume of air through the tube containing a solid reagent. This reagent changes the color with the concentration level of chemical in the air which is indicated by the distance down the tube that the color changes. This method is used in the Draeger Tube for hydrogen sulfide, ammonia and chlorine exposure [69]. Finally, direct reading devices or instruments convert the concentration of chemical in the air into a reading on a meter or digital display. They are generally installed in a workplace to provide emergency warnings if gases reach a dangerous level. Uses of these devices includ monitoring oxygen, carbon monoxide/dioxide, sulfur gases, hydrocarbons, nitrogen oxide, nitric oxide, formaldehyde and ammonia levels. From the aforementioned set of tools, those involving the on-line information are the greenest ones and provide valuable data without any environmental damage. Additionally, analytical methods employed to determine volatile organic compounds and pesticides in air through the use of active and passive samplers could be considered as rearguard methodologies [70] suitable for an improved level of information which can be obtained after chromatography or spectroscopy plus chemometric analysis. Once again, on greening the back-guard methods costs could be reduced without sacrificing the main analytical figures of merit and it is our task to put the stress on the benefits of a sustainable analytical chemistry in order to spread around the industrial field good green practices.

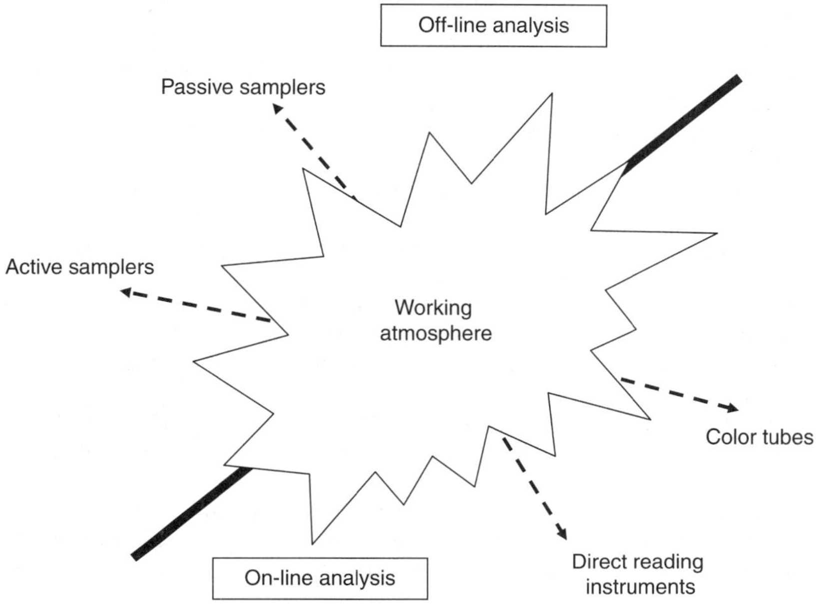

Figure 23.6 *Monitoring devices employed for working atmosphere control.*

23.6 The future starts now

The EPA's Green Chemistry Program supports fundamental research in green chemistry in order to provide to the industry with the chemically viable tools and methods necessary to develop products and processes that are more environmentally benign than those produced today. In 1992, EPA awarded six grants to fund basic research projects that consider impacts to human health and the environment in the design of chemical syntheses. This initial activity was the first of a number of research partnerships that have, to date, awarded tens of millions of dollars in the form of grants for fundamental research in green chemistry. We think that analytical chemistry is an important subdiscipline of chemistry in the industrial world that should be also greened and in this way, it would be important to convince the industry to incorporate green methods through 'Green Analytical Chemistry' grant opportunities and industry/university/government projects.

In addition, official methods of analysis of finished products and industrial effluents should be also greened by the corresponding authorities. Similar to the replacement of methods that include chlorinated hydrocarbons, official methods of analysis should be revisited taking into consideration analytical parameters such as accuracy, precision and detection limits but also green parameters such as volume and toxicity of reagents and generated waste and energy consumption. On the other hand, the correct evaluation of cost of analysis could be of great importance in demonstrating the economical opportunities offered by greening industrial laboratories.

References

[1] Anastas, P.T. and Warner, J.C. (1995) *Green Chemistry: Theory and Practice*. Oxford University Press, New York.
[2] US Environmental Protection Agency (n.d.) *Green Chemistry Home*. Available at http://www.epa.gov/greenchemistry/index.html (accessed November 4, 2011).
[3] European Environment Agency (n.d.) homepage. Available at http://www.eea.europa.eu/ (accessed November 4, 2011).
[4] Environment Directorate General, European Commission (2007) *REACH in brief*. Available at http://ec.europa.eu/environment/chemicals/reach/pdf/2007_02_reach_in_brief.pdf (accessed November 4, 2011).

[5] International Organisation for Standardization (ISO) (2008) ISO 9001:2008, ISO/TC176 Technical Committee. Available at http://www.iso.org/iso/home.html (accessed November 4, 2011).

[6] US Food and Drug Administration (n.d.) *Drugs homepage*. Available at http://www.fda.gov/Drugs/default.htm (accessed November 4, 2011).

[7] European Medicines Agency (n.d.) *homepage*. Available at http://www.ema.europa.eu/ema/index.jsp (accessed November 4, 2011).

[8] European Food Safety Authority (n.d.) *Homepage*. Available at http://www.efsa.europa.eu/ (accessed November 4, 2011).

[9] Larrechi, M.S. and Callao, M.P. (2003) Strategy for introducing NIR spectroscopy and multivariate calibration techniques in industry, *TrAC-Trend. Anal. Chem.*, **22**, 634–640.

10] Plugge, W. and van der Vliest, C. (1992) The use of near infrared spectroscopy in the quality control laboratory of the pharmaceutical industry, *J. Pharm. Biomed. Anal.*, **10**, 797–803.

[11] Council of Europe (COE) – European Directorate for the Quality of Medicines (2007) Near-infrared spectrophotometry, Section 2.2.40, 62–66.

[12] EMEA (2003) CPMP/QWP/3309/01 and EMEA/CVMP/961/01 *Note for Guidance on the use of Near infrared spectroscopy by the pharmaceutical industry and the data requirements for new submissions and variations*. EMEA, London.

[13] United States Pharmacopoeia 29 – National Formulary 24 (2006) Near-infrared spectrophotometry, 2979–2983.

[14] Miller, C.E. (2001) Chemical principles in near-infrared technology, in *Near Infrared Technology in Agricultural and Food Industries* (eds P. Williams and K. Norris). American Association of Cereal Chemists, St. Paul, Minnesota, pp. 19–38.

[15] Marte, L.; Belloni, P.; Genorini, E.; Sileoni, V.; Perretti, G.; Montanari, L. and Marconi, O. (2009) Near-infrared reflectance models for the rapid prediction of quality of brewing raw materials, *J. Agric. Food Chem.*, **57**, 326–333.

[16] Armenta, S.; Moros, J.; Garrigues, S. and de la Guardia, M. (2010) The use of near-Infrared spectrometry in the olive oil industry, *Crit. Rev. Food Sci. Nutr.*, **50**, 567–582.

[17] Cozzolino, D.; Chree, A.; Scaife, J. R. and Murray, I. (2005) Usefulness of near-infrared reflectance (NIR) spectroscopy and chemometrics to discriminate fishmeal batches made with different fish species, *J. Agric. Food Chem.*, **53**, 4459–4463.

[18] Frankuizen, R. (2001) NIR Analysis of Dairy Products, in *Handbook of Near-Infrared Analysis* (eds Burns, D.A. and Ciurczak, E.W.). Marcel Dekker, Inc., New York, pp. 499–534.

[19] Oxford Instruments (n.d.) *Hand-held portable XRF analysers*. Available at http://www.oxford-instruments.com/products/xrf-analysers/hand-held/Pages/hand-held.aspx (accessed November 4, 2011).

[20] Bosch Reig, F.; Gimeno Adelantado, J.V. and Moya Moreno, M.C.M. (2002) FTIR quantitative analysis of calcium carbonate (calcite) and silica (quartz) mixtures using the constant ratio method. Application to geological samples, *Talanta*, **58**, 811–821.

[21] Jang, M.S.; Ryu, M.K.; Yoon, M.H.; Lee, S.H.; Kim, H.K.; Onodera, A. and Kojima, S. (2009) A study on the Raman spectra of Al-doped and Ga-doped ZnO ceramics, *Curr. Appl. Phys.*, **9**, 651–657.

[22] Forte, A.I. (2006) *Climate change Program at Pfizer Inc*. Available at http://www.epa.gov/climateleaders/documents/events/oct2006/forte.pdf (accessed November 4, 2011).

[23] Welch, C.J.; Brkovic, T.; Schafer, W. and Gong, X. (2009) Performance to burn? Re-evaluating the choice of acetonitrile as the platform solvent for analytical HPLC, *Green Chem.*, **11**, 1232–1238.

[24] Brettschneider, F.; Jankowski, V.; Günthner, T.; Salem, S.; Nierhaus, M.; Schulz, A.; Zidek, W. and Jankowski, J. (2010) Replacement of acetonitrile by ethanol as solvent in reversed phase chromatography of biomolecules, *J. Chromatogr. B*, **878**, 763–768.

[25] US Food & Drug Administration (2010) *Acetonitrile Shortages: Recommendations for Reporting Changes in Analytical Procedures*. Available at http://www.fda.gov/downloads/AboutFDA/CentersOffices/CDER/UCM171776.pdf (accessed November 4, 2011).

[26] Patel, H.B. and Jefferies, T.M. (1987) Eluotropic strength of solvents - prediction and use in reversed-phase high-performance liquid-chromatography, *J. Chromatogr.*, **389**, 21–32.

[27] Water Corp. (2010) *Waters Introduces ACQUITY UPLC H-Class System; HPLC Simplicity with UPLC Performance*. Available at http://news.thomasnet.com/fullstory/UPLC-System-replaces-HPLC-systems-572073, (accessed November 4, 2011).

[28] Jerkovich, A.D.; LoBrutto, R. and Vivilecchia, R.V. (2005) *The use of Acquity UPLC in the pharmaceutical development*. LC-GC (UPLC Separation Science Redefined, suppl. May), 15–29. Available at http://chromatographyonline.findanalytichem.com/lcgc/data/articlestandard//lcgc/242005/164648/article.pdf, (accessed November 7, 2011).

[29] Van der Vorst, G. (2009) Thermodynamic Evaluation of the Greenness of Supercritical Fluid Chromatography In the Pharmaceutical Industry, in *Proceedings of the 13th Annual Green Chemistry & Engineering Conference*, MD.

[30] Ewing, R.G.; Atkinson, D.A.; Eiceman, G.A. and Ewing, G.J. (2001) A critical review of ion mobility spectrometry for the detection of explosives and explosive related compounds, *Talanta*, **54**, 515–529.

[31] Lawrence, A.H.; Nanji, A.A. and Taverner, J. (1988) Skin-sniffing/ion mobility spectrometric analysis: a potential screening method in clinical toxicology, *J. Clin. Lab. Anal.*, **2**, 101–107.

[32] Miller, R.A.; Nazarov, E.G.; Krylov, E. and Eiceman, G.A. (2006) Method and Apparatus for Control of Mobility-Based Ion Species Identification, *U.S. Patent* 7,005,632.

[33] O'Donnell, R.M.; Sun, X.B. and Harrington, P.D. (2008) Pharmaceutical applications of ion and differential mobility spectrometries, *TrAC-Trend. Anal. Chem.*, **27**, 44–53.

[34] Matz, L.M. and Hill, H.H. (2001) Evaluation of opiate separation by high-resolution electrospray ionization-ion mobility spectrometry/mass spectrometry, *Anal. Chem.*, **73**, 1664–1669.

[35] Lawrence, A.H. (1989) Characterization of benzodiazepine drugs by ion mobility spectrometry, *Anal. Chem.*, **61**, 343–349.

[36] Strege, M.A. (2009) Total residue analysis of swabs by ion mobility spectrometry, *Anal. Chem.*, **81**, 4576–4580.

[37] Payne, K.; Fawber, W.; Faria, J.; Buaron, J.; DeBono, R. and Mahmood, A. (2005) IMS for Cleaning Verification, *Spectroscopy* (The Role of Spectroscopy in Process Analytical Technologies, suppl. January), **20**, 24–27. Available at http://spectroscopyonline.findanalytichem.com/spectroscopy/data/articlestandard//spectroscopy/032005/143241/article.pdf (accessed November 7, 2011).

[38] Bakeev, K.A. (2005) *Process Analytical Technology*, Blackwell Publishing, Ltd, Oxford.

[39] Kueppers, S. and Haider, M. (2003) Process analytical chemistry – future trends in industry, *Anal. Bioanal. Chem.*, **376**, 313–315.

[40] Ewers, C.; Kueppers, S. and Weinmann, H. (2002) *Pharma Supply Chain*, Editio Cantor Verlag, Aulendorf.

[41] Bakeev, K.A. and Ciurczak, E.W. (2003) Near-infrared spectroscopy as a process analytical tool – Part 1: Laboratory applications, *Spectroscopy*, **18**, 32–35.

[42] Naes, T.; Isaksson, T.; Fearn, T. and Davies, T. (2002) *A User-Friendly Guide to Multivariate Calibration and Classification*. NIR Publications, Chichester.

[43] George, G.; Hynard, N.; Cash, G.; Rintoul, L. and O'Shea, M. (2006) Spectroscopic probes for real-time monitoring of polymer modification and degradation reactions, *C. R. Chim.*, **9**, 1433–1443.

[44] Cozzolino, D.; Parker, M.; Dambergs, R.G.; Herderich, M. and Gishen, M. (2006) Chemometrics and visible-near infrared spectroscopic monitoring of red wine fermentation in a pilot scale, *Biotechnol. Bioeng.*, **95**, 1101–1107.

[45] Armenta, S.; Moros, J.; Garrigues, S. and de la Guardia, M. (2010) Determination of Olive Oil Parameters by Near Infrared Spectrometry, in *Olives and Olive Oil in Health and Disease Prevention* (eds Preedy, V.R. and Watson, R.R.) Academic Press, Oxford, pp. 533–544.

[46] Marziano, I.; Sharp, D.C.A.; Dunn, P.J. and Hailey, P.A. (2000) On-line mid-IR spectroscopy as a real-time approach in monitoring hydrogenation reactions, *Org. Process. Res. Dev.*, **4**, 357–361.

[47] Mettler-Toledo International Inc. (n.d.) *In situ FTIR Reaction Analysis: ReactIR™*. Available at http://us.mt.com/us/en/home/products/L1_AutochemProducts/L2_in-situSpectrocopy.html (accessed November 4, 2011).

[48] Horiba,Ltd. Raman Spectroscopy, Available at http://www.horiba.com/scientific/products/raman-spectroscopy/ (accessed November 4, 2011).

[49] Kaiser optical system, Inc.(n.d.) *Raman Analyzers*. Available at http://www.kosi.com/Raman_Spectroscopy/ramananalyzers.php?ss=200 (accessed November 4, 2011).

[50] Horiba, *Real-time Monitoring of Polymerisations in Emulsions by Raman Spectroscopy – Modelling and Chemometrics*, Raman Application note, Process 02. Available at http://www.horiba.com/fileadmin/uploads/Scientific/Documents/Raman/Process02.pdf (accessed November 7, 2011).

[51] Davis, K.L.; Kemper, M.S. and Lewis, I.R. (2008) Raman Spectroscopy for Monitoring Real-Time Processes in the Pharmaceutical Industry, in *Pharmaceutical Applications of Raman Spectroscopy* (eds Šašić, S. and Ekins, S.), John Wiley & Sons, Inc., Hoboken.

[52] Chai, X.; Danielsson, L.G.; Yang, X. and Behm, M. (1998) Spectrophotometric in-line process monitoring in UV-Vis range using an ATR-probe, *Process Contr. Qual.*, **11**, 153–159.

[53] Thompson, D.; Kougoulos, E.; Jones, A. and Wood-Kaczmar, M. (2005) Solute concentration measurement of an important organic compound using ATR-UV spectroscopy, *J. Cryst. Growth*, **276**, 230–236.

[54] Anderson, J.; Moore, S.; Tarczynski, F. and Walker, D. (2001) Determination of the onset of crystallization of N1-2-(thiazolyl)sulfanilamide (sulfathiazole) by UV-Vis and calorimetry using an automated reaction platform; subsequent characterization of polymorphic forms using dispersive Raman spectroscopy, *Spectrochim. Acta A*, **57**, 1793–1808.

[55] Abu Bakar, M.R.; Nagy, Z.K. and Rielly, C.D. (2009) Seeded Batch Cooling Crystallization with Temperature Cycling for the Control of Size Uniformity and Polymorphic Purity of Sulfathiazole Crystals, *Org. Process Res. Dev.*, **13**, 1343–1356.

[56] Howard, K.S.; Nagy, Z.K.; Saha, B.; Roberston, A.; Steele, G. and Martin, D. (2009) A Process Analytical Technology Based Investigation of the Polymorphic Transformations during the Antisolvent Crystallization of Sodium Benzoate from IPA/Water Mixture, *Cryst. Growth Des.*, **9**, 3964–3975.

[57] European Parliament, Council (2006) Directive 2006/11/EC of the European Parliament and of the Council of 15 February 2006 on pollution caused by certain dangerous substances discharged into the aquatic environment of the Community, OJ L, **64**, 52–59. Available at http://eur-lex.europa.eu/LexUriServ/LexUriServ.do?uri=OJ:L:2006:064:0052:0059:EN:PDF (accessed November 4, 2011).

[58] US Environmental Protection Agency (n.d.) *Oil and Grease, Method 1664, Revision A: N-Hexane Extractable Material (HEM; Oil and Grease) and Silica Gel Treated N-Hexane Extractable Material (SGT-HEM; Non-polar Material) by Extraction and Gravimetry*. Available at http://water.epa.gov/scitech/methods/cwa/oil/1664.cfm (accessed November 4, 2011).

[59] Martin, T. (2009) *ASTM WK23240 - Validation Report: Inter-Laboratory Study to Establish Precision Statements for ASTM WK23240 – Standard Test Method for Solvent-Free Membrane Recoverable Oil and Grease by Infrared Determination*, Orono Spectral Solutions, Old Town (ME). Available at http://www.ossmaine.com/documents.php, (accessed November 4, 2011).

[60] Bramley, P.C. and Wheeler, V.A. (1994) An integrated analytical strategy for liquid effluent management, *Anal. Chim. Acta*, **291**, 243–254.

[61] Gondal, M.A. and Hussain, T. (2007) Determination of poisonous metals in wastewater collected from paint manufacturing plant using laser-induced breakdown spectroscopy, *Talanta*, **71**, 73–80.

[62] Miró, M.; Estela, J.M. and Cerdà, V. (2004) Application of flowing stream techniques to water analysis: Part III. Metal ions: alkaline and alkaline-earth metals, elemental and harmful transition metals, and multielemental analysis, *Talanta*, **63**, 201–223.

[63] Thermo Fischer Scientific Inc. (n.d.) *TOC/COD/BOD*. Available at http://www.thermofisher.com.au/show.aspx?page=/ContentAUS/Environmental-Industrial/Environmental-Monitoring-Safety/Water-Monitoring-Treatment/Water-Quality-Instruments-Test-Kits/Process/TOC-COD-BOD/TOC-COD-BOD.html (accessed November 4, 2011).

[64] Endress+Hauser Ltd. (2008) *COD, BOD, TOC and colourimetric analysers*. Available at http://www.uk.endress.com/eh/sc/europe/uk/en/resourceadditional.nsf/imgref/Download_FA050Cen.pdf/$FILE/FA050Cen.pdf (accessed November 4, 2011).

[65] RM Supply (n.d.) *Water treatment and Water Testing*. Available at http://processcontrol.rmsupply.co.uk/index.php (accessed November 4, 2011).

[66] Hammaker, R.M.; Fateley, W.G.; Chaffin, C.T.; Marshall, T.L.; Tucker, M.D.; Makepeace, V.D. and Poholarz, J.M. (1993) FT-IR Remote Sensing of Industrial Atmospheres for Spatial Characterization, *Appl. Spectrosc.*, **47**, 1471–1475.

[67] Tahiri, S. and de la Guardia, M. (2009) Treatment and valorization of leather industry solid wastes: A review, *The ALCA Journal*, **104**, 52–67.

[68] Sundar, V.J.; Gnanamani, A.; Muralidharan, C.; Chandrababu, N.K. and Mandal, A.B. (2011) Recovery and utilization of proteinous wastes of leather making: a review, *Rev. Environ. Sci. Biotechnol.*, **10**, 151–163.

[69] Cross Instrumentation (n.d.) *Draeger Short-Term Gas Detection Tubes*. Available at http://www.buydraegertubes.com/index.aspx (accessed November 4, 2011).

[70] Valcarcel, M. and Cardenas, S. (2005) Vanguard-rearguard analytical strategies, *TrAC-Trend. Anal. Chem.*, **24**, 67–74.

Index

Note: Figures are indicated by *italic page numbers*, Tables by **bold page numbers**.

Handbook of Green Analytical Chemistry, First Edition. Edited by Miguel de la Guardia and Salvador Garrigues.